APPLIED STRESS ANALYSIS

This volume consists of papers presented at the International Conference on Applied Stress Analysis, held at the University of Nottingham, Nottingham, UK, 30–31 August 1990, organised by the Department of Mechanical Engineering, University of Nottingham, to mark the retirement of Professor H. Fessler.

APPLIED STRESS ANALYSIS

Edited by

T.H. HYDE

and

E. OLLERTON

Department of Mechanical Engineering, University of Nottingham, Nottingham, UK

ELSEVIER APPLIED SCIENCE
LONDON and NEW YORK

ELSEVIER SCIENCE PUBLISHERS LTD
Crown House, Linton Road, Barking, Essex IG11 8JU, England

Sole Distributor in the USA and Canada
ELSEVIER SCIENCE PUBLISHING CO., INC.
655 Avenue of the Americas, New York, NY 10010, USA

WITH 61 TABLES AND 560 ILLUSTRATIONS

© 1990 ELSEVIER SCIENCE PUBLISHERS LTD
© 1990 BRITISH STEEL plc—pp. 14–25
© 1990 UNITED KINGDOM ATOMIC ENERGY AUTHORITY—pp. 190–200
Softcover reprint of the hardcover 1st edition 1990

British Library Cataloguing in Publication Data

Applied stress analysis.
1. Stress analysis
I. Hyde, T. H.   II. Ollerton, E.
620.1123

ISBN-13:978-94-010-6837-6   e-ISBN-13:978-94-009-0779-9
DOI:10.1007/978-94-009-0779-9

Library of Congress CIP data applied for

PREFACE

This volume records the proceedings of an international conference organised as a tribute to the contribution made by Professor H. Fessler over the whole of his professional life, in the field of applied stress analysis. The conference, held at the University of Nottingham on 30 and 31 August 1990, was timed to coincide with the date of his formal retirement from the post of Professor of Experimental Stress Analysis in the University.

The idea grew from discussions between some of Professor Fessler's academic associates from Nottingham and elsewhere. An organising committee was set up, and it was decided to invite contributions to the conference in the form of review papers and original research papers in the field of experimental, theoretical and computational stress analysis. The size of the response, both in papers submitted and in attendance at the conference, indicates that the idea proved attractive to many of his peers, former associates and research students.

A bound copy of the volume is to be presented to Professor Fessler at the conference dinner on 30 August 1990.

T.H. HYDE
E. OLLERTON

CONTENTS

Fatigue and Fracture Mechanics

Residual Stress

Yielding

Design Studies

Behaviour of Brick, Masonry and Concrete Structures

Automatic Photoelastic Fringe Reading

Photoelasticity and Photoviscoelasticity

Assessments of Techniques

Moiré Interferometry

Late Submission

Professor Henry Fessler

A SHORT BIOGRAPHY OF PROFESSOR HEINRICH FESSLER

Heinrich Fessler was born in Vienna, Austria, on the 26th of September 1924. He spent his early years in Vienna, attending the Technical Grammar School there from age 10 to 14. His parents were Jewish, and his father, a doctor, was allowed to bring his wife and two sons to Britain, where he could settle, re-qualify, and practise his profession away from the immediate attention of the Nazis. Heinrich arrived at Harwich on November 14th 1938, with his parents and his brother. On the train to Manchester, he decided that he would be called Henry, although this policy has never been a total success. In official publications such as the Calendar of the University of Nottingham he still appears as Heinrich, and a significant proportion of his many friends call him Harry.

On arrival in Lancashire, the family was taken in by the Irish Catholic manageress of a shop who felt it her Christian duty to help these Jewish victims of Nazi persecution.

They lived in the Manchester area for four years, during which time Henry studied assiduously and successfully. He passed the entrance examination of the local Junior Technical School after less than three weeks attendance at the local Secondary Modern School. In parallel with his studies there, he took a correspondence course and passed the matriculation examination of the University of London. After that, he served a year of apprenticeship with a Salford company. By this time his father had qualified to practise medicine again, and the family moved to Golcar, a village near Huddersfield in Yorkshire.

Henry worked in Jig and Tool design with Hopkinsons Ltd. from 1941 until 1947. He obtained his Ordinary National Certificate in Mechanical Engineering in 1942, and by 1947 had passed the Intermediate and Part I of the London University external degree examinations, by part time study. He attended Huddersfield Technical College full time in the 1947/48 academic year, to study for the Part II, and was awarded the external BSc of London University in Mechanical Engineering, with First Class Honours in 1948. During this period he was also active in other directions, serving as a member of the Home Guard, and taking up fell-walking, which has remained an enjoyable leisure pursuit throughout his life.

After graduation, Henry became one of the first two Science Research Council funded research students to work in the Mechanical Engineering Department of Huddersfield Technical College. He regards the two years spent there as unsatisfactory in many ways, but the research project served to introduce him to frozen stress photoelasticity, an activity which he has pursued with distinction from that day to this.

In 1950, he began his long association with the University of Nottingham, being appointed Assistant Lecturer a mere two years after the former University College had received its Royal Charter.

1950 was an important year for Henry in another respect. It was then that he married his wife, Audrey. They have three children; Martin, Paul and Catherine.

Henry was awarded the London external MSc in 1951 for a thesis on The Frozen Stress Phenomenon, and the Nottingham PhD in 1954 for a thesis on Stress Distributions in Turbo-alternator Rotors. After a period as Lecturer and then Reader, he was elected to the personal Professorship of Experimental Stress Analysis, in the Department of Mechanical Engineering "in recognition of his special eminence in this field" in 1974.

His professional life has been concerned with stress analysis, design studies, failure investigations and fracture mechanics. Much of his work employed the techniques of three-dimensional frozen stress photoelasticity, although he has also used strain gauges, Moiré interferometry, photoelastic coatings and finite element analysis. His teaching activities have been mainly in the area of solid mechanics and mechanical engineering design. Henry is an active member of the Institution of Mechanical Engineers, and of the American Society of Mechanical Engineers. He has served on many committees of the IMechE and the Institute of Physics, and on British Standard committees on Flanges and Dimensional Standards for Pressure Vessels, a Working Party of the Underwater Engineering Group, and the Applied Mechanics Sub-Committee of the Science and Engineering Research Council. He was Founding Secretary of the Permanent European Committee for Stress Analysis from 1959-61, Chairman of the Permanent European Committee from 1966-70, and has been a member of the Editorial Board of the Journal of Strain Analysis from 1966 to the present day. He has given invited lectures in Denmark, Italy, Israel, Poland, USA, China, Japan and the USSR. He has recently been invited to present the 1991 William M Murray Lecture at the Spring Conference of the Society for Experimental Mechanics in Milwaukee, Wisconsin.

In 1964 Henry was awarded a James Clayton Fund Prize by the Institution of Mechanical Engineers for a paper with R W Wilson on "Plastic-elastic strains observed near loaded fillets in stepped bars". In 1966 The IMechE. again honoured him with a Ludwig Mond Prize for a paper with P Stanley on "Stresses in torispherical drumheads: a critical evaluation". A second Ludwig Mond Prize followed in 1975, for a paper with D A Perry on "Stresses in high-pressure taper-hub flanges with recesses for nut fixings". In 1987, the Institution of Civil Engineers presented him with the David Hislop Award for a paper entitled "Parametric equations for the flexibility matrices of single brace tubular joints in offshore structures". Also in 1987 he received the Ernest William Moss Prize of the Board of the Process Industries Division of the IMechE., for a paper with T H Hyde entitled "Determination of the initial gaps between flat flanges without gaskets". All Henry's work has had relevance to the problems encountered in engineering industry. His advice has been sought by many industrial organisations, and he has undertaken consultancies with Mirrlees Blackstone, Rolls Royce, ICI Plastics, The Water Tube Boilermakers Association, Stone Manganese Marine, BP Exploration Ltd, Admiralty Metals Laboratory, National Gas Turbine Establishment, London Transport Executive, Babcock & Wilcox, International Combustion, English Electric, United Kingdom Atomic Energy Authority, Central Electricity Generating Board, Yorkshire Imperial Metals, Firth Cleveland, Nomikos ltd., British Rail, John Folkes, British Gas, Glacier Metals, The Ministry of Defence, British Steel, Howmedica International Ltd, Wimpey Offshore, British Nuclear Fuels, and others.

The attached list of his publications is a testament to his massive contribution in the field of applied stress analysis.

E Ollerton.

LIST OF PUBLICATIONS OF H. FESSLER

1. Frozen stress phenomenon in photoelasticity. **Proc. I.Mech.E.**, Vol. 1B., p 613, (1953).
2. Photoelastic determination of stresses in cylindrical shells. (with R T Rose). **B.J.A.P.**, Vol. 4., p.76, (1953).
3. On the stress distribution in the walls of pressure vessels. (with R T Rose). **J.Mech.Phys.Sol.**, Vol. 1., p.127, (1954).
4. Stress distribution in a Tee junction of thick pipes. (with B H Lewin). **B.J.A.P.**, Vol. 7., p.76, (1956).
5. Stresses in heads of thick pressure vessels. (with R T Rose). **Proc. I.Mech.E.**, Vol. 171., p.633, (1957).
6. Contact stresses in toroids under radial loads. (with E Ollerton). **B.J.A.P.**, Vol. 8., p.387, (1957).
7. Load distribution in a model of a hip joint. **J. Bone and Joint Surgery.**, Vol. 39B., p.145, (1957).
8. Contribution towards rational tee-slot design. **Engineer**, Vol. 203., p.904, (1957).
9. A photoelastic technique for strain measurement of flat aluminium alloy surfaces. (with D J Haines). **B.J.A.P.**, Vol. 9., p.278, (1958).
10. Centrifugal stresses in turbo alternator rotors. **Proc. I.Mech.E.**, Vol. 173., p.717, (1959).
11. Bending stresses in a shaft with a transverse hole. (with E A Roberts). **Proc. Int. Conf. Stress Anal.**, Delft, (1959).
12. Plasto-elastic stress distributions in lugs. (with D J Haines). **Aero. Quart.**, Vol. 10., p.230, (1959).
13. A study of large strains and the effect of different values of Poisson's ratio. (with B H Lewin). **B.J.A.P.**, Vol. 11., p.273, (1960).
14. Large strains of drum heads studied with silicone rubber models. (with J J Foreman). **J.M.E.S.**, Vol. 3., p.42, (1961).
15. Improvements of photoelastic reflection technique. (with G Clyne and R W Wilson). **B.J.A.P.**, Vol. 12., p.8, (1961).
16. Stresses in hemispherical drum heads. (with C N Lakshminarayana). **Proc. I.Mech.E.**, Vol. 175., p.127, (1961).
17. Centrifugal stresses at the bores of wheels. (with E I Hay and E A Roberts). **Engineer.**, Vol. 211., p.113, (1961).
18. Photoelastic study of stresses near cracks in thick plates. (with D O Mansell). **J.M.E.S.**, Vol. 4., p.213, (1962).
19. Stresses in branched pipes under internal pressure. (with B H Lewin). **Proc. I.Mech.E.**, Vol. 176., p.771, (1962).
20. Birefringence behaviour of an epoxy resin in compression. (with C P Bettany). **B.J.A.P.**, Vol.14., p.692, (1963).
21. On the effect of layer thickness in photoelastic reflection techniques. (with E. A A Newton and R W Wilson). **B.J.A.P.**, Vol .14., p.889, (1963).
22. Discontinuity stresses in spherical pressure vessels with cylindrical supports. (with R K Penny and D A Wright). **J.M.E.S.**, Vol. 5., p.38, (1963).

23. Stresses in internal combustion engine poppet valves. (with E Ollerton). J.M.E.S., Vol. 6., p.1, (1964).

24. Plastic-elastic strains observed near loaded fillets in stepped bars. (with R W Wilson). J.M.E.S., Vol. 6., p.9, (1964).

25. Strains near cracks in aluminium plates. (with R W Wilson). Proc. 11th Int. Cong. App. Mech., p.603, (1964).

26. Stresses in torispherical drumheads: a photoelastic investigation. (with P Stanley). J.S.A., Vol. 1., p.69, (1965).

27. Radial pressure exerted by piston rings. (with Ahmad bin Chik). J.S.A., Vol. 1, p.165, (1966).

28. Stresses in torispherical drumheads ; a critical evaluation. (with P Stanley). J.S.A., Vol. 1., p.89, (1966).

29. A contribution to the stress analysis of piston pins. (with H B Padgham). J.S.A., Vol. 1., p.422, (1966).

30. Optimisation of stress concentrations at holes in rotating discs. (with T E Thorpe). J.S.A., Vol. 2., p. 152, (1967).

31. Reinforcement of non-central holes in rotating discs. (with T E Thorpe). J.S.A., Vol. 2., p.317, (1967).

32. Centrifugal stresses in rotationally symmetrical gas-turbine discs, (with T E Thorpe). J.S.A., Vol. 3., p.135, (1968).

33. Current pressure vessel problems: a compilation of problems and needs in the analysis, design and manufacture. (Editor with P Stanley). Proc. I.Mech.E. ., Vol. 182., part 3F. (1967/68).

34. The collection of photoelastic data. J.S.A., Vol. 3., p.128, (1968).

35. A material for accelerated creep testing with models (with P A T Gill and P. Stanley). J.B.C.S.A. Conf., Roy. Aero. Soc., p.5.1-5.10., (1968).

36. Axial stresses in cylindrical tube-plates. J.S.A., Vol.3., p.281, (1968).

37. A 30 ton biaxial tensile testing machine. (with J K Musson). J.S.A., Vol. 4., p.22, (1969).

38. Shouldered plates and shafts in tension and torsion. (with C C Rogers and P Stanley). J.S.A., Vol.4., p.169, (1969).

39. Stresses at end-milled keyways in plain shafts subjected to tension, bending and torsion. (with C C Rogers and P Stanley). J.S.A., Vol. 4., p.180, (1969).

40. The application of photoelasticity to the stress analysis of an arched dam. Rilem Symposium Bucharest, Vol. 4., p.16, (1969).

41. Stresses at keyway ends near shoulders (with C C Rogers and P Stanley). J.S.A., Vol. 4., p.267, (1969).

42. Plastic-elastic strain distributions in perforated plates under uniaxial tension, (with J K Musson). First International Conference on Pressure Vessel Technology., Delft, p.571, (1969).

43. Compact loading shackles for tensile of flat plates. (with J K Musson). Strain., Vol. 5., p.223, (1969).

44. Plastic-elastic tension and bending of finned tube water-walls of boilers. (with J W Rimmington). 4th Int. Conference on Stress Analysis., Cambridge, p.56, (1970).

45. Stresses in cylindrical pressure vessels with end closures freely formed by internal pressure. (with T E Thorpe). J.S.A., Vol. 5., p.75, (1970).

46. Stresses in cylindrical pressure vessels previously shaped by free explosive forming. (with T E Thorpe). J.S.A., Vol. 5., p.255, (1970).

47. Tests of some taper-hub flanges with narrow gaskets. (with J K Musson). J. Mech. Eng. Sci., Vol 14., p.98-106, (1972).

48. On the general problems of static experimental stress analysis. CNR Seminar on **Experimental Stress Analysis of Pressure Vessels**, Bologna. Sept. 15-18, (1970).

49. Some applications of three-dimensional photoelasticity. CNR Seminar on **Experimental Stress Analysis of Pressure Vessels**, Bologna, Sept 15-18, (1970).

50. Creep strains in thick rings subjected to internal pressure. (with P A T Gill and P Stanley). **Advances in Creep Design, the A.E. Johnson Memorial Vol.**, Appl. Sci. Publishers. p.353-386, (1971).

51. Moire technique with linear mismatch for measurement of creep strains at room temperature. (with P A T Gill and P Stanley). **Beitrage zur Spannungs-und-Dehnungsanalyse.**, Akademie-Verlag, Berlin, Vol 7., p.7-20, (1973)

52. Stress distributions in some diesel engine crankshafts. (with V K Sood). **A.S.M.E. Conf. Diesel and Gas Engine Power Division Conf.**, paper no. 71-DGP-1. (1971).

53. A technique to measure deformations applied to the housings for roller bearings. (with J R Crookall and W B Heginbotham). **J.B.C.S.A. Conf. The Recording and Interpretation of Engineering Measurements.**, (1972).

54. Some practical improvements in Moiré technique. (with T E Thorpe). Strain, Vol. 8., p.1-3, (1972).

55. Deformations of some medium speed diesel engine crankshafts. (with V K Sood). **C.I.M.A.C. '73.**, 10th Int. Congress Combustion Engines, Washington DC. (1973).

56. Deformation of diesel engine frames from static model tests. (with M Perla). **A.S.M.E. Diesel and Gas Engine Power Conf.**, Washington DC. (1973).

57. Precision casting of epoxy resin photoelastic models. (with M Perla). J.S.A., Vol. 8., p.30-34, (1973).

58. Stress concentrations at dimples in crankshafts. (with P Diver). J.S.A., Vol. 9., p.78-81, (1974)

59. Photoelasticity applied to complicated diesel engine models. **Proc. Inst. Phys. Stress Anal. Group Conf.**, p.111-122, (1974)

60. An engineer's approach to the prediction of failure probability of brittle components. (with P Stanley and A D Sivill). **Proc. British Ceramic Soc. Conf. Ceramics for Turbines and Other High Temperature Engineering Applications.**, Cambridge. (1973).

61. Use of models for the prediction of creep behaviour of components. (with R A Bellamy). **A.S.M.E./I.Mech.E. Conf. on Creep and Fatigue in Elevated Temp. Applications.**, Philadelphia (1973) and Sheffield (1974).

62. Prediction of the creep behaviour of a flanged joint. (with J H Swannell). **Conf. on Creep Behaviour of Piping.**, I.Mech.E. (1974).

63. The application and confirmation of a predictive technique for the fracture of brittle components. (with A D Sivill and P Stanley). **5th Int. Conf. on Experimental Stress Analysis. Int. Centre for Mech Sciences.**, Udine, Italy. (1974).

64. Behaviour of brazed pipe flanges with separate clamping rings (with D A Perry). J.S.A., Vol. 10., p.71-76, (1975).

65. Stresses in high-pressure taper-hub flanges with recesses for nut flanges. (with D A Perry). J.S.A., Vol. 10., p.119-128, (1975).

66. Repeated use of frozen stress photoelastic models. (with M Perla and A Litewka). J.S.A., Vol. 11., p.186-190, (1976).

67. Stresses in a medium speed diesel engine frame. (with M Perla). **Proc. I.Mech.E.**, Vol. 190., p.309-318, (1976).

68. The unit strength concept in the interpretation of beam test results for brittle materials. (with P Stanley and A D Sivill). Proc. I.Mech.E., Vol. 190., p.585-595, (1976).

69. Leakage characteristics of flanged pipe joints. (with D A Perry). J.S.A., Vol. 12., p 29-36, (1977).

70. Stress analysis of finned-tube water-walls of boilers. (with P Stanley). Third Int. Conf. on Pressure Vessel Technology., A.S.M.E., Tokyo., p,169-182, (1977).

71. Applications of the four function Weibull equation in the design of brittle components. (with P Stanley and A D Sivill). Fracture Mechanics of Ceramics., Vol.3., p.51-66, Plenum Publ. Corp. (1978).

72 Prediction of creep rupture of pressure vessels. (with T H Hyde and J J Webster). A.S.M.E. Energy Technology Conf., Houston, Texas. (1977).

73. Stationary creep prediction from model tests using reference stresses. (with T H Hyde and J J Webster). J.S.A., Vol. 12., p.271-285, (1977).

74. Experimentally determined reference stresses for the prediction of creep deformation and life of components using models. (with T H Hyde and J J Webster). Proc. of Conf. on Recent Developments in High Temp. Design Methods., I.Mech.E., London. (1977).

75. Creep deformation of metals. (with T H Hyde). J.S.A. Special Issue No. 3., (1977).

76. The design of ceramic turbine blade roots (with D C Fricker). Proc. British Ceramic Soc. Conf. on Mechanical Engineering Properties and Applications of Ceramics., London, p.81-96, (1977).

77. Fillet stresses in tubular joints obtained by photoelastic techniques (with W J G Little). Proc. of Int. Symp. on Integrity of Offshore Structures, Inst. of Engrs. and Shipbuilders in Scotland., Glasgow. (1978).

78. The use of precision cast models to predict diesel engine deformations. (with P S Whitehead). Proc. 6th Int. Conf. on Experimental Stress Analysis., Munich, FRG. (1978).

79. Thermal ratchetting of a hollow stepped cylinder (with V Sagar Dwivedi, T H Hyde and J J Webster). Non-linear Problems in Stress Analysis., Applied Science Publ. Ltd., p.299-316. (1978).

80. A design exercise for a ceramic turbine disc subjected to centrifugal stresses. (with A D Sivill and P Stanley). J.S.A., Vol. 13., p.103-113, (1978).

81. Uniaxial and biaxial properties of a material for modelling creep. (with T. H. Hyde). Non-linear Problems in Stress Analysis., Applied Science Publ. Ltd., p.233-257, (1978).

82. Stresses in a tubular K-joint subjected to 'out-of-plane' bending. (with W J G Little and I J Shellard). Select Seminar, European Offshore Steels Research, UK D. of E. Offshore Steels Research Project, Commission of the European Community., Vol. 2., Welding Inst., Cambridge. (1978).

83. A comparative study of the mechanical strength of reaction-bonded silicon nitride. (with D C Fricker and D J Godfrey). Conf. on Ceramics for High Performance Applications III: Reliability., p.705-736, Plenum Publ. Corp. (1983).

84. Elastic stresses due to axial loading of tubular joints with overlap. (with W J G Little and I J Shellard). 2nd Int. Conf. on Behaviour of Offshore Structures (BOSS '79)., Imperial College, London. (1979).

85. Precision casting of epoxy resin models using expendable or reusable moulds. (with W J G Little and P S Whitehead). Proc. 8th All-Union Conf. on Photoelasticity., Vol. 1., Tallinn, USSR, (1979).

86. Dimples in pressure vessel welds. (with P D Humphrey). **Conf. on Significance of Deviations from Design Shapes.**, I.Mech.E. (1979).

87. Physical model stress analysis of tubular connections. (with N M Irvine and A C Wordsworth). **Joint I. Struct. E./B.R.S. Seminar on Physical models in Design of Offshore Structures.**, Building Research Station. (1979).

88. Deformations and stresses from statically loaded diesel engine models. (with P S Whitehead). **Universities Combustion Engine Group SRC Symp.** (1980).

89. The behaviour of a flanged pipe joint under external pressure. (with D A Perry). **4th Int. Conf on Pressure Vessel Technology.**, (1980).

90. Elasto-plastic and creep behaviour of axially loaded shouldered tubes. (with R J Dawson, T H Hyde and J J Webster). J.S.A., Vol. 15., p.21-29, (1980).

91. Comparison of finite element predictions of elasto-plastic and creep behaviour with experimental data from hemispherically ended, cylindrical pressure vessel models. (with T H Hyde and J J Webster). **4th Int. Conf. on Pressure Vessel Technology.**, (1980).

92. Practical aspects of model creep work. (with T H Hyde). **3rd I.U.T.A.M. Symp. on Creep of Structures.**, Leicester. (1980).

93. Elastic stresses due to axial loading of a two-brace tubular K joint with and without overlap. (with W J G Little). J.S.A., Vol. 16., p.67-77, (1981).

94. Design and stress analysis of a light cast, 90° - 45° K joint. (with C D Edwards). **Int. Conf. Steel in Marine Structures.**, Commission of European Community, Paris. (1981).

95. Mounting deformations of hydraulically expanded taper sleeve couplings. (with R W Stewart). J.S.A., Vol. 16., p.165-170, (1981).

96. Experimental determination of stiffness of tubular joints. (with H Spooner). **2nd Int. Symp. on Integrity of Offshore Structures.**, The University, Glasgow. (1981)

97. Static and spinning tests of reaction-bonded silicon nitride turbine blade roots. (with D C Fricker). **Conf. Engineering with Ceramics.**, Brit. Ceramic Soc. London. (1981).

98. Stress concentrations on axially loaded unsymmetrical projections on flat bars. (with P J Woods). J.S.A., Vol. 17., p.23-30, (1982).

99. Elastic stresses due to torque transmitted through the prismatic part of keyed connections. Part I: Effect of different fits and friction on standard shapes. (with M Eissa). J.S.A., Vol. 17., p.103-111, (1982).

100. The effect of variable chord wall thickness on the stresses in a light, cast 90°C - 45°C K tubular joint. (with C D Edwards). **3rd Int. Conf. on Behaviour of Offshore structures.**, Massachusettes Inst. of Technology, U.S.A. (1982).

101. A rig for creep and thermal ratchet testing of lead alloy models. (with T H Hyde and J J Webster). J.S.A., Vol. 17., p.13-22, (1982).

102. Distribution of transmitted torque and elastic stresses in keyed connections. (with M Eissa). **Proc. 7th Int. Conf. on Exp Stress Anal.**, Haifa, Israel. (1982).

103. Elastic stresses due to torque transmitted through the prismatic part of keyed connection. Part II: Effect of shape with usual fits and friction. (with M Eissa). J.S.A., Vol. 17., p.215-222, (1982).

104. Ratchetting of axially loaded tubes operating in the creep range. (with T H Hyde and J J Webster). J.S.A., Vol. 17., p.243-252, (1982).

105. The photoelastic-coating technique for plastic-elastic contact. (with M Eissa). Exp. Mech., Vol. 23., p.282-288, (1983).

106. Stresses in a bottoming stud assembly with chamfers at the ends of the threads. (with P K Jobson). J.S.A., Vol. 18., p.15-22, (1983).

107. Three-dimensional, elastic stress distributions in end-milled keyed connections. (with M Eissa). J.S.A., Vol. 18., p.143-149, (1983).

108. Comparison of stress distributions in a simple tubular joint using 3-D finite element, photoelastic and strain gauge techniques. (with C D Edwards). **15th Annual OTC Houston, Texas., Proc. A.S.M.E. (1983).**

109. Stress and stiffness analysis of tubular joints for offshore structures. **XIth Convegno Nazionale Dell'Associazione Italiana Per L'Analisi Delle Sollecitazioni.**, Torino, September. (1983).

110. Reduction of elastic stress concentrations in end-milled keyed connections. (with M Eissa). **Experimental Mechanics.**, Vol. 18., p.401-408, (1983).

111. Stress analysis of some unsymmetric screwed connections. (with Wang Joing-Hua). J.S.A., Vol. 19., p.111-119, (1984).

112. Deformation and stress analysis of engine components using models. **Design and Applications in Diesel Engineering.**, Ellis Horwood. p.110-124, (1984).

113. Parametric equation for SCFs of cast tubular T and X joints.(with C D Edwards). **Offshore Mechanics and Arctic Engineering Symp.**, Dallas, Texas, February. (1985).

114. Friction at broad and narrow contacts between silicon nitride and hardened steel. (with D C Fricker). **Proc. British Ceramic Soc.**, No. 34., p.129-144, (1984).

115. Multiaxial strength for brittle materials. (with D C Fricker). J.S.A., Vol. 19., p.197-208, (1984).

116. A theoretical analysis of the ring-on-ring loading disk test. (with D C Fricker). J.Am. Ceramic Soc., Vol. 67., p.582-588, (1984).

117. Parametric equation for SCFs of cast tubular T and X joints. (with C D Edwards). O.M.A.E. Symp., Dallas, Texas, February. (1985).

118. Stress concentrations at cast joints of tubular structures. (with C D Edwards). **4th Int. Conf. on Behaviour of Offshore Structures, (BOSS '85).**, Delft. p.465-474, (1985).

119. Ratchetting tests of a stepped beam under steady tension and cycling bending. (with T H Hyde). J.S.A., Vol. 20., p.193-200, (1985).

120. Cyclic creep testing and lead alloy beams. (with T H Hyde). **Int. Conf. on Experimental Mechanics.**, Beijing, China, September. Science Press. p.647-653, (1985).

121. Stress and stiffness analysis of tubular joints for offshore structures. **Int. Conf. on Experimental Mechanics.**, Beijing, China, September.Science Press. p32-43, (1985).

122. Fillet welds under bending and shear. (with C Pappalettere). **Trans. A.S.M.E.**, Vol. 108., p.430-435, (1986).

123. Stresses at weld toes on non-overlapped tubular joints. (with K S Elliott). **Int. Conf. on Fatigue and Crack Growth in Offshore Structures. I.Mech.E., p.1-15, (1986).**

124. Stress analysis of screwed tubular joints. (with T P Broadbent). **Int. Conf. on Fatigue and Crack Growth in Offshore Structures. I.Mech.E., p171-186, (1986).**

125. An automatic polariscope used to study a cracked thread. (with T P Broadbent). **Proc. of the Int. Symposium on Photoelasticity.**, Tokyo. Springer. p.281-292, (1986).

126. Failure probability of shouldered and notched ceramic components using Neuber notch theory. (with D C Fricker). **Non-Oxide Technical and Engineering Ceramics.**, Elsevier. p.319-339, (1986).

127. Parametric equations for the flexibility matrices of single brace tubular joints in offshore structures. (with P B Mockford and J J Webster). **Proc. Inst. Civ. Eng.** Part 2, Vol. 81., p.659-673, (1986).

128. Parametric equations for the flexibility matrices of multi-brace tubular joints in offshore structures. (with P B Mockford and J J Webster). **Proc. Inst. Civ. Eng.** Vol. 81., p.675-696, (1986).

129. An experimental technique for determining the flexibility of tubular joints. (with P B Mockford and J J Webster). **J.S.A.**, Vol. 22., p.7-15, (1987).

130. A micropolariscope for automatic stress analysis. (with R E Marston and E Ollerton). **J.S.A.**, Vol. 22., p.25-35, (1987).

131. Surface strain measurements and stress concentrations at weld toes of some tubular joints. (with K S Elliott). **Proc. 6th Int. Offshore Mech & Arctic Eng. Symp.**, p.221-229, (1987).

132. Determination of the initial gaps between flat flanges without gaskets. (with L V Lewis and T H Hyde). **Proc. I.Mech.E.**, Vol. 201., p.267-277, (1987).

133. Leakage through loaded flat-flanged joints without gaskets. (with T H Hyde and L V Lewis). **Proc. I.Mech.E.**, Vol. 202., p.1-13, (1987).

134. Bending stresses in non-overlapped tubular K-joints. (with K S Elliott). **Fatigue of Offshore Struct.** Ed. W D Dover & G Glinka., The Eng. Integrity Soc., p.77-89, (1988).

135. Bolting and loss of contact between cylindrical flat-flanged joints without gaskets. (with T H Hyde and L V Lewis). **J.S.A.**, Vol. 23., p.1-8, (1988).

136. On the effect of key edge shape on keyway edge stresses in shafts in torsion. (with T Appavoo). **J.S.A.**, Vol. 24., p.121-125, (1989).

137. Distribution of thread load in screwed tubular connections. (with T P Broadbent). **Fatigue of Offshore Struct.** Ed. W D Dover & G Glinka., The Eng. Integrity Soc., p.133-147, (1988).

138. Failure probability analysis for alumina universal heads of femoral prostheses. (with D C Fricker). **Int. Conf. The Changing Role of Eng. in Orth.**, I.Mech.E., p.19-26, (1989).

139. Stress analysis of a complex, cast node for offshore structures. (with T H Hyde). **Appl. Solid Mech 3.** Univ. of Surrey, Guildford, 5-7 April. (1989).

140. Plastic-elastic strains in two-dimensional sections of partial-penetration fillet welds. (with C Pappalettere). **J.S.A.**, Vol. 24., p.15-21, (1989).

141. Friction in femoral and photoelastic model cone taper joints. (with D C Fricker). **Proc I.Mech.E.**, Vol. 203., p.1-14, (1989).

142. A study of stresses in alumina universal heads of femoral prostheses. (with D C Fricker). **Proc. I.Mech.E.**, Vol. 203., p.15-34, (1989).

143. Flanged joints of aero-engines. (with T H Hyde and L V Lewis). 9th ISABE., Athens. September 3-8 (1989).

144. Friction at ceramic and metal contacts over a range of temperatures. (with T Dickerson and J J Webster). Int. Symp. Advances in processing and Appl. of Composites., Halifax, Nova Scotia. (1989).

145. An experimental and theoretical study of laminated beams. (with R Brooks). Proc. 7 I.C.C.M., Beijing, China. (1989).

146. Internal stresses at the saddles of YT tubular joints with three different weld fillet shapes. (with R E Marston and E Ollerton). **J.S.A.**, Vol. 25., p.85-94, (1990).

PREDICTING FATIGUE FAILURES IN THICK-WALLED CYLINDERS UNDER PULSATING INTERNAL PRESSURE

D.W.A.REES
Department of Manufacturing and Engineering Systems,
Brunel University, Uxbridge, Middlesex UB8 3PH.

ABSTRACT

High pressure cylinders and gun barrels are known to fail by fatigue when subjected to pulsating internal pressures. Prior autofrettage and compounding with interference improves the fatigue life for a given pressure range. This paper presents methods for predicting the life of each cylinder accepting that failure may arise from the propagation of one or more cracks that continuously change their geometry, from straight to elliptical, as they penetrate the cylinder wall. A bounding method is employed, giving predictions for the most severe condition of fatigue failure resulting from a single, straight-fronted crack, to that of failure under the least severe stress intensity for 30 elliptical cracks in cylinders of Ni-Cr-Mo steel. It is shown that the propagation lives observed under higher repeated pressure ranges generally lie within these bounds for plain-bored and rifled-bored monobloc cylinders in both the unautofrettaged and autofrettaged conditions. Compounding analyses apply to plain-bored, two-component, shrink-fit cylinders, with components of the same and different materials. To account for the increased lives observed under lower pressure ranges, it is necessary to account for that portion of life expended initiating a crack. However, the initial condition of a cylinder bore is much less important to that portion of life consumed by initiation following an autofrettage treatment. The effect of mean stress in the fatigue cycle and the degree of autofrettage are also examined in the case of monobloc cylinders.

INTRODUCTION

The thick-walled cylinder is employed where it is necessary to contain high pressures. The design considerations depend upon whether the cylinder is subjected to steady or cyclic pressure. In the former case the cylinder dimensions must be chosen to avoid a ductile shear failure when the pressure causes gross plasticity of the wall. In the latter case fatigue failure is likely from a propagating longitudinal crack piercing the wall thickness. This embodies a number of new design requirements; (i) avoiding or knowing when to expect fatigue failure and (ii) deciding between methods that render the cylinder more resistant to fatigue. Fatigue in thick-cylinders is examined in the present paper in relation to pressure vessels, polymer process plant, rifles and cannon. Methods of predicting the fatigue life are presented for cylinders with smooth and rifled bores, including those that derive greater fatigue strength from a prior autofrettage treatment, honing and compounding. It is shown that with the appropriate definition of a stress intensity factor, it is possible to achieve a realistic assessment of the cyclic life in each case. Previously published experimental results are employed throughout this study. The reader should refer to the original papers for details of the experimental techniques used.

By far the greatest amount of fatigue data applies to a high-strength $2\frac{1}{2}\%$ Ni-Cr-Mo alloy steel referred to as Vibrac, En25 and SAE 4335. The heat treatment consists of hardening at 850°C with an oil quench followed by tempering at 610°C for 1 hour. This treatment provides the steel with a yield stress of 825 MPa and 50% reduction in area at fracture; a suitable combination of strength and ductility for high pressure vessels. Slight property variations to strength and ductility resulting from changes to the heat treatment and material composition appear as a scatter band in crack growth data [1]. Working throughout to the upper limit of this data ensures conservative design predictions. Propagation life predictions are derived from integrating the Paris fatigue crack law; $da/dN=A(\Delta K_I)^m$ where $A=13.55\times10^{-8}$ and $m=2.25$ [2]. This gives;

$$N_f = \frac{w^{1-m/2}}{A(\Delta p)^m \pi^{m/2}} \int_{(a/w)o}^{(a/w)f} \frac{d(a/w)}{(a/w)^{m/2} [\Delta K_I/\Delta K_o]^n} \qquad (1)$$

where $\Delta K_o=\Delta p/(a\pi)$, in which a is the crack depth and w is the cylinder wall thickness. The limits of integration are the initial and final crack depths. The range of stress intensity ΔK_I will depend upon the particular cylinder, including its geometry and initial condition. That is, whether the bore is smooth or rifled, and whether the cylinder has been autofrettaged is monobloc or has been compounded. These are treated separately in the following paragraphs.

MONOBLOC CYLINDERS

There are a number of forms of thick-walled monobloc cylinder; as-received, heat-treated, honed, rifled and autofrettaged. The heat, honed and autofrettaged treatments aim to improve the fatigue strength but they may be unnecessary where the range of pressure variation is not severe enough to exceed the range of pressure at the fatigue limit. If this applies to a plain-bored cylinder it may not apply where the cylinder bore has been further machined to contain, say, O-ring sealing grooves or the helical rifling groove typical of gun barrels and cannon. Here the failure from fatigue is so enhanced with the concentration of stress at the groove root, that multiple cracking is often observed. Separate analyses of life predictions follow from experimental data pertaining to each condition of monobloc cylinder.

As-Received Cylinders
This refers to cylinders employed in the supplied condition and to those given a subsequent heat treatment either to relieve residual stress or to attain the required mechanical properties. Measurements [2] made of bore roughness in cylinders of diameter ratios ($W=d_2/d_1$) between 2 and 3 showed the greatest initial defect size to be 20µm, i.e., of the order of 0.1% of the wall thickness. The final depth of an elliptical crack, prior to static rupture of the remaining ligament, was 90% of the wall thickness. These constitute the limits of integration in eq(1). The range in the stress intensity factor under repeated and fluctuating pressure cycles is defined from combining the solution, $\Delta K_I/\Delta K_o=f(W,a/w)$, for a single, straight-fronted crack (Bowie and Freese[3]) with an elliptical shape factor $\alpha=\alpha(a/w,a/2c)$ (Rice and Levy[4]). This gives;

$$\Delta K_I=\alpha\Delta K_o f(W,a/w) \qquad (2)$$

in which a is the semi-minor axis (the crack depth) and 2c is the length of the major axis. By combining eqs(1) and (2) and integrating by the repeated application of

Simpson's rule, leads to the propagation cycles N_f, under any Δp for a cylinder of given W value. The comparison with experimental results [2,5] in Fig.1 shows that the N_f predictions are more realistic for higher ranges of pressure in as-received cylinders with W=2.4 and 2.6. The reason for this is because, at higher pressures, fewer cycles are consumed in initiation.

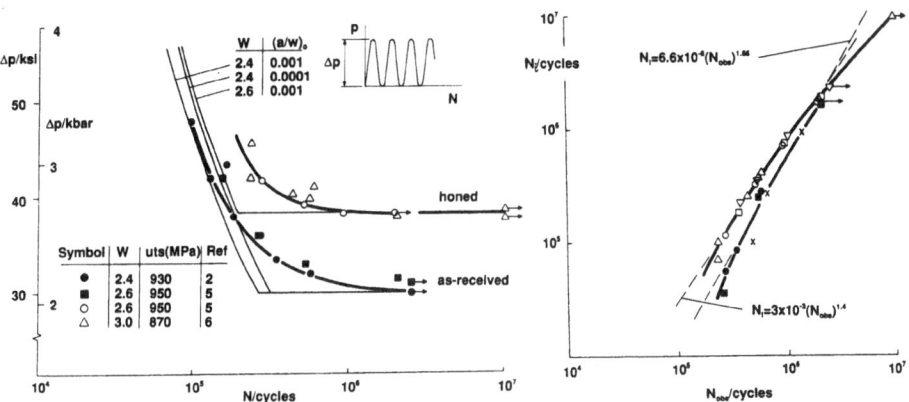

Fig.1 Repeated pressure fatigue life   Fig.2 Initiation cycles

Honed-bored cylinders

It is also evident from Fig.1 that improving the surface finish of the bores of as-received cylinders by honing leads to an impovement in the fatigue life. The test results shown are those of Rogan [5] and Morrison et al [6] for a 2½% Ni-Cr-Mo closed-ended cylinders with diameter ratio 2.4 subjected to a repeated pressure cycle. In this case eq(1) is applied with $(a/w)_o$=0.0001 for the improved surface finish. Here the added effect of the initiation cycles markedly improves fatigue life but they do not appear within a propagation life prediction. Tthe dependence of initiation cycles N_i upon the observed life N_{obs} for both as-received and honed cylinders (Fig.2) is expressed in the empirical law:

$$N_i = P(N_{obs})^q \tag{3}$$

where, within the mortal region for both repeated and fluctuating cycles, $P=3 \times 10^{-3}$ and q=1.4 for honed cylinders. In as-received cylinders the constants are $P=6.6 \times 10^{-5}$ and q=1.66 implying dependence upon surface finish. The reduction in the fatigue limit for a higher mean pressure (p_m=2.07kbar=30ksi) under the fluctuating cycle does not alter the predicted propagation lives since eq(1) depends only upon the range of applied pressure. This assumes that the Paris law constants A and m are not sensitive to mean stress which is the generally held view [7]. The different fatigue limit reflects only the dependence of the range of threshold stress intensity ΔK_{th}, a material property, upon p_m. For example, in as-received material K_{th} is approximately 5% of K_{Ic}=60 MPa√m [8].

Rifled-Bored Cylinders

Gun barrels are often manufactured with a helical rifle groove within their inner diameters. The grooves are flat bottomed connected to the two vertical sides with small corner radii. The concentration in stress around the groove corners confine initiation to these regions. Pulsating pressure experiments on grooved bored cylinders [9] show that many cracks initiate at the groove radii. These cracks subsequently propagate to penetrate the cylinder wall radially remaining aligned with the direction of the groove helix. The effect of multiple cracking on the stress intensity factor may be adapted from the load relief factor of Tweed and Rooke[10]. Thus, K_I for n elliptical cracks is derived from the single crack formula in eq(2) as follows:

$$(K_I)_n = R(K_I) \tag{4}$$

where the load relief factor R is found from applying the ratio $R=(K_I)_n/K_I$ to another multiple cracked geometry whose stress intensity is known. Tweed and Rooke chose an infinite plate with a multiply cracked central hole under uniform tension. Their derivations of R have been shown [2] to be in good agreement with a finite element determination of R for a multiply cracked thick walled ring under pressure [11]. One uncertainty in this approach is that these R values apply to straight fronted cracks where K_I and R is constant at the crack tip. F.e.m. studies [12] show that K_I varies by up to 18% around an elliptical crack contour. By confining our attention to the single point on the contour at the maximum depth of penetration, R remains valid. The effect of multiple cracking is to reduce the intensity of stress for a given depth of crack penetration. Taking R from [11] to establish K_I for the greatest number of cracks observed, n≈30 cracks, enables the application of eq(1). The corresponding life predictions for n=30 and n=1 establishes two bounds between which the experimental results may be expected to fall when 1<n<30 with constant crack geometry.

The ultrasonic crack depth measurements of Davidson et al [9] show that the crack geometry continuously changes during propagation from being straight fronted initially to a final elliptical form. Thus with the determination of a second set of bounds for single and multiple elliptical cracking, all experimental results are contained. This method is applied to Davidson's experimental results in Fig.3 for Ni-Cr-Mo (O) plain and rifled bored (▲) cylinders with W=2, $(a/w)_o=0.001$ and the same crack growth constants A and m. The figure shows that the plain bored lives lie close to the single elliptical crack life prediction confirming the analysis made previously. The rifled bored lives lie between the straight fronted bounds which are more severe. Additional honed bored results (□) [6] for the same geometry show that improved surface finish is equivalent to an R for 1<n<30 elliptical cracks. This should be associated with a change in the lower limit from $(a/w)_o=0.001$ to $(a/w)_o=0.0001$, which in itself displays only a marginal effect on the fatigue life as shown for a single elliptical crack in Fig.3.

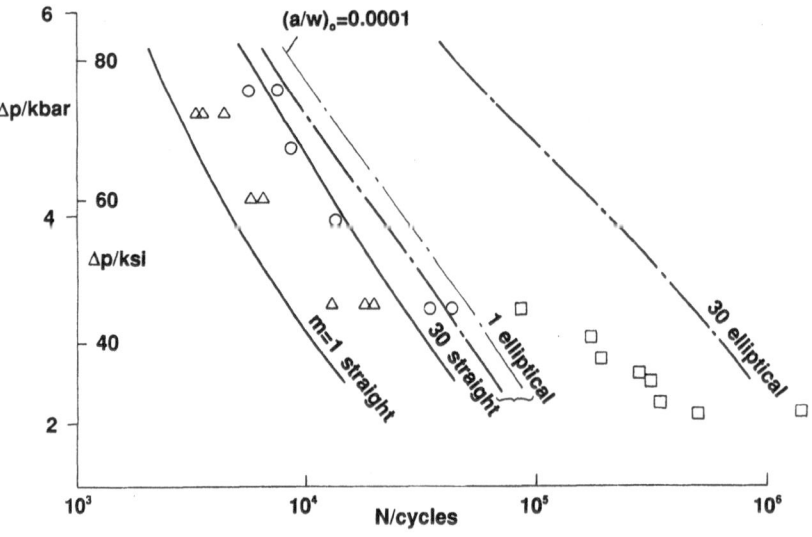

Fig.3 Fatigue life predictions

Autofrettaged Cylinders
Autofrettage refers to the method of pre-pressurising a cylinder to render the bore fibres plastic. When the autofrettage pressure is released the compressive residual stress remaining within the bore is particularlty beneficial to crack closure. Consequence the fatigue life of autofrettaged cylinders is far better than as-received cylinders irrespective of their bore condition. In the literature there is uncertainty about the choice of autofrettage pressure that will optimise fatigue strength. Some experimenters employed a fully plastic pressure [13] while others preferred a lesser pressure to spread a plastic zone to the geometric mean radius [14]. The present author believes [15] that an elastic-plasic pressure should be chosen to avoid reverse yielding of the bore fibres from occurring following pressure release. Greater pressures, when applied to cylinders with diameter ratio (W≥3), will involve a significant Bauschinger effect with a consequent loss of stability of the residual stresses.

A further factor to consider is the manner in which the ends are sealed during autofrettage. In a closed-end cylinder caps contain the pressure by stressing the cylinder wall axially. In the open-end cylinder the axial stress is absent when there is no connection between the wall and end seals around floating pistons. Since axial stress effects the distribution of residual stress it is necessary to consider each end condition separately in determining the range of stress intensity leading to fatigue failure. The choice of an appropriate stress intensity factor [16,17] showed reliable fatigue life predictions with the following modification to eq(2);

$$\Delta K_I/\Delta K_o = \alpha[f(a/w) \pm 1.12\sigma_{\theta R}/\Delta p] \qquad (5)$$

in which $\sigma_{\theta R}$ is the residual stress for a penetration (a/w) and Δp is the subsequent applied cyclic pressure. The final term in eq(3) is K_I for a straight edge crack in a semi-infinite plate. Taken with the shape factor α this form accounts for K_I due to a semi-elliptical edge crack. Normally this expression is used for shallow depths of crack penetration. It is therefore acceptable for residual stresses that change from compressive to tensile in the region of the mean wall radius under an optimum autofrettage.

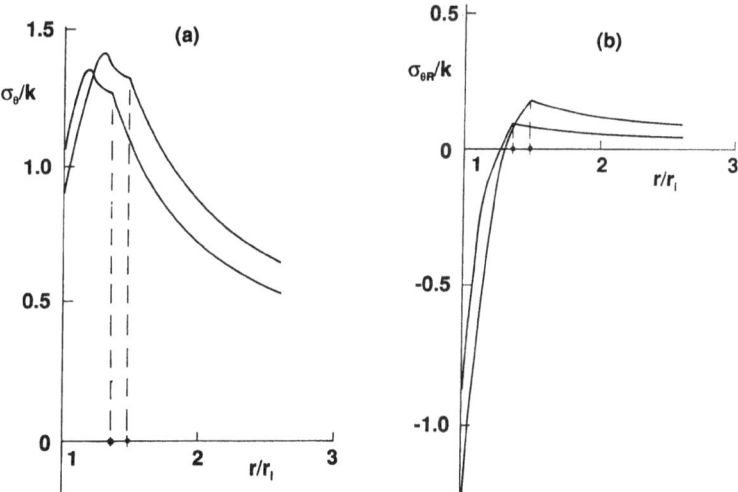

Fig.4 Hoop stress distributions during and after autofrettage

Closed Ends In Fig.4(a) the hoop stress distributions are derived from autofrettage pressures employed by Rogan [5]. Fig.4(b) gives the corresponding residual hoop stress distributions assuming elastic release from each pressure. A von Mises yield condition

6

has been used in the derivation with the account of hardening for this alloy steel given in [15]. This plot is used with eq(3) to determine the stress intensity variation with depth of elliptical crack penetration in Fig.5 which applies to each autofrettage pressure for Δp/k=0.679, where k is the shear yield stress. Plots constructed from eq(3) for all other cyclic pressures with the numerical integration of eq(1) establishes the theoretical fatigue curves for autofrettaged material in Fig.6. Modified Bowie and Freese predictions consistently lie on the safe side of Rogan's observed fatigue lives. As with unautofrettaged cylinders in Fig.1, an account of initiation is clearly necessary. The broken lines in Fig.6 reveal that portion of life expended in initiation where life at lower pressures is composed of initiation and propagation phases. With an increasing range of cyclic pressure the observed lives confirm more closely to the propagation lives since initiation is less influencial.

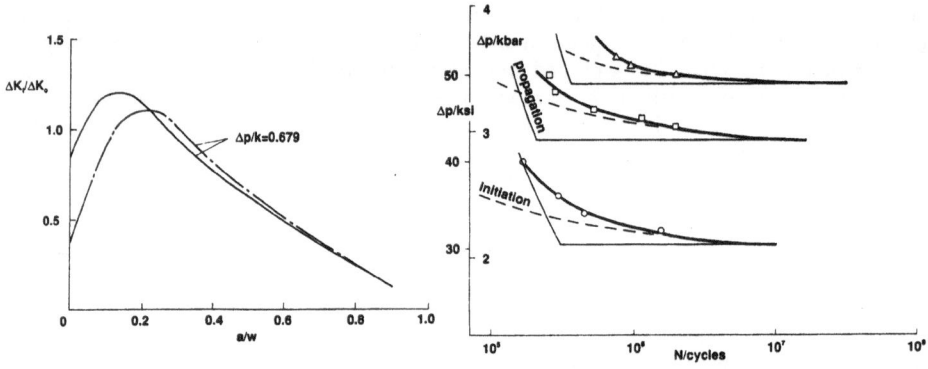

Fig.5 Stress intensity   Fig.6 Autofrettaged fatigue lives

Open Ends Results in [13] refer to a fully autofrettaged W=2 cylinder. Rees [17] showed that the residual stress distribution in thicker walled cylinders depends upon the choice of yield criterion and the nature of hardening. However, for lower diameter ratios W≤2, the influence of hardening was small enough to be ignored.

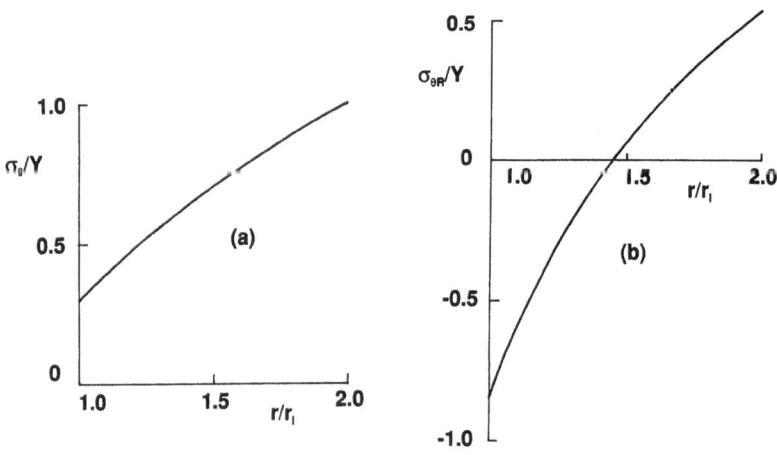

Fig.7 Hoop stress distributions during and after autofrettage

The Tresca yield criterion then leads to a closed solution to the hoop stress under a fully plastic pressure. Assuming elastic recovery of hoop stress following release of this pressure leads to the residual hoop stress (see Figs.7(a) and (b)).

$$\sigma_\theta/Y = 1 + \ln(r/r_o) \tag{6a}$$

$$\sigma_{\theta R}/Y = 1 + \ln(r/r_o) - [(r_o^2/r^2 + 1)/(r_o^2/r_i^2 - 1)]\ln(r_o/r_i) \tag{6b}$$

The fact that the residual compressive stress at the bore has not exceeded the tensile yield stress Y, confirms that reverse yielding does not occur following release of a full autofrettage pressure in this cylinder. Combining eqs(5) and (6b) leads to the ΔK_I variation in Fig.8 for the ranges of subsequently applied cyclic pressure $\Delta p = 4.49$ and 7.25 kbar. In Fig.5, ΔK_I attains a maximum value when the crack has penetrated approximately 20-30% of the wall thickness. In an un-autofrettaged cylinder the maximum ΔK_I coincides with the bore so explaining the poorer fatigue resistance. Fig.9 compares the predictions from eq(1) with experiment for as-received ($\triangle$) and autofrettaged ($\blacktriangle\blacksquare$) plain bored cylinders. The residual stresses in eq(6) when used with eq(5) are clearly adequate for the prediction of fatigue life in thinner walled autofrettaged cylinders. Putting R=0 leads to the as-received lives under a single propagating elliptical crack. Superimposed on these plots are the experimental fatigue data (O,●) for a rifled cylinder of the same composition and diameter ratio [9]. Rifling (O) reduces the fatigue strength of the as-received cylinder ($\triangle$). It is noteworthy that autofrettage renders plain and rifled bored cylinders (●) with equal fatigue strengths.

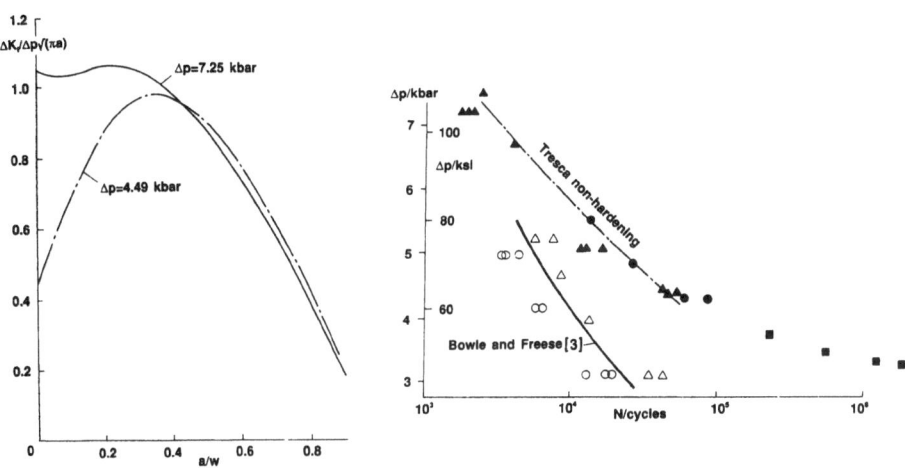

Fig.8 Stress intensities   Fig.9 Fatigue life predictions

COMPOUND CYLINDERS

Few compound cylinder fatigue results are available to enable predictions to be made from eq(1) and (5). Burns and Parry [18] manufactured two-component compounded cylinders to 1.8 in outer diameter, 1 in inner diameter with interface diameter of 1.2 in. The inner liners were either made from En 25 or a 0.15% carbon steel with their bores honed to a centre line average roughness of 8 µin before being stress relieved. They were then compounded with surrounding shells of En25 with outer diameters of 1.8 in and whose bores were ground and nitrided. Failure in these cylinders is judged to occur when a crack penetrates the full thickness of the inner liner. Therefore, in the analysis that follows, failure in a compounded cylinder with an overall K=1.8 is taken for a crack penetration depth of a/w=0.25. Comparisons are also made between the experimental fatigue lives of these compounded cylinders and the lives of En25 monobloc cylinders, taken from [5,20] with W values in the range 1.4 to 2.

Residual Stress Distributions
The interference fit was approximately constant at 0.089 mm (0.0035 in) between the En25 liner and shell. This was reduced to 0.038 mm (0.0015 in) with an 0.15% C liner. The method outlined in the Appendix was used to establish the residual stress distribution resulting from shrinking the shell on to the liner. The analysis given provides the interface pressure p_c by equating hoop strain between the two components at the common diameter d_c assuming plane strain conditions (eqs A13 and A14). A known interface pressure supplies the additional boundary condition necessary for the determination of the residual radial and hoop stress distributions in the liner (eqs A5 and A6) and in the shell (eqs A11 and A12). Fig.10(a) shows that the greatest compressive residual hoop stress lies in the bore of each liner material. It is expected that the liner residuals will serve to improve the fatigue strength in a similar manner to the residual stresses produced by autofrettage i.e. by lowering the maximum tensile hoop stress at the inner diameter under a subsequent cyclic pressure (Fig.10(b)).

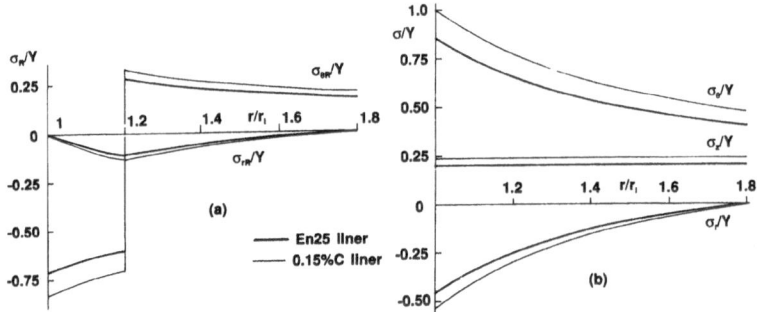

Fig.10 Residual and applied stresses for compound cylinder

Stress Intensity Factors
Equation(5) may again be employed to estimate ΔK_I when a single or elliptical crack penetrates the liner of a compound cylinder. The effect of multiple cracking on ΔK_I can further be estimated from the load reduction factor R in eq(4). Because an elliptical crack may not develop fully within a thin-walled liner, ΔK_I for a straight-fronted crack may be more appropriate in this instance; achieved simply by setting the shape factor to unity in eq(5). Crack combinations were chosen to provide bounds to the fatigue life predictions. Fig.11 illustrates the dependence of ΔK_I upon a/w under (i) a single straight crack, (ii) a single elliptical crack, (iii) thirty straight cracks and (iv) thirty elliptical cracks. The slight difference between ΔK_I for each liner material under any given combination is entirely due to the difference in liner residual stresses in Fig.10. The second set of ΔK_I estimates, for similar crack conditions, in Fig.12 are based upon a recommendation made by Underwood [19] for shallower crack depths. For a single, shallow elliptical crack, that propagates in the presence of residual stress, ΔK_I is written as:

$$\Delta K_I = \alpha[1.12\Delta\sigma_\theta\sqrt{(\pi a)} + 1.33\Delta p_r\sqrt{(\pi a)} \pm 1.12\sigma_{\theta R}\sqrt{(\pi a)}] \qquad (7a)$$

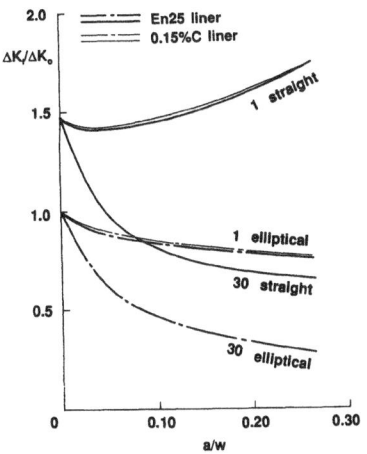

Fig.11 Stress intensities [3]

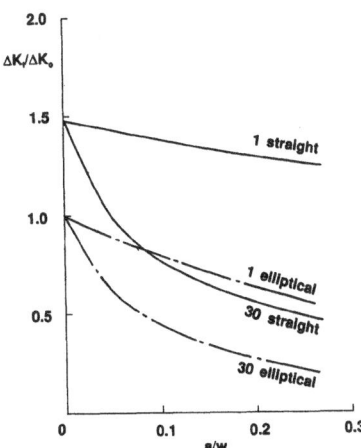

Fig.12 Stress intensities [19]

In eq(7a) the first and third terms are the standard forms of ΔK_I for a shallow edge crack in a thick semi-infinite plate. The Lamé hoop stress is responsible for opening under the applied pressure Δp, being given by:

$$\Delta\sigma_\theta = \Delta p(1+r_o^2/r^2)/(W^2-1) \tag{7b}$$

where $W=r_o/r_i=1.8$ and the relationship between a/w and r_o/r is;

$$a/w=(Wr/r_o-1)/(W-1) \tag{7c}$$

for a/w≤0.25. The second term in eq(7a) is Bueckner's solution [20] to ΔK_I with pressure applied to the crack surfaces. Under cyclic pressure conditions the range of pressure is employed in eqs(7a and b) to give, the normalised range of stress intensity:

$$\Delta K_I/\Delta K_o = 1.12\alpha[(1+r_o^2/r^2)/(W^2-1)+1.009\pm\sigma_{\theta R}/\Delta p] . \tag{7d}$$

For straight cracking again set $\alpha=1$ in eq(7d) and for multiple cracking introduce the load reduction factor coefficient R from eq(4). Comparing Figs 11 and 12 shows that the stress intensities are generally less severe by this method for crack penetrations a/w beyond 0.06. Because there is little difference between (5) and (7) with a/w<0.06, this quantifies the condition of shallow cracking. It follows from eq(7c), that eqs(5) and (7d) provide comparable fatigue lives for liners of diameter ratio 1.05 in cylinders of overall diameter ratio 1.8. Moreover, the following section shows very litle difference between fatigue lives for the present liners with a diameter ratio of 1.2.

Fatigue Life Predictions

The propagation lives were estimated from the numerical integration of eq(1) following the separate identifications of $\Delta K_I/\Delta K_o$ with the ordinates in Figs 11 and 12. For the En25 liners, the previously quoted A and m values were employed. For the 0.15% carbon steel liners the Paris law constants $A=4.71\times10^{-8}$ and $m=3.3$ were used [7]. The respective comparisons with experimental results are made in Figs 13 and 14. For a given range of pressure in each figure the fatigue life of a compound cylinder is greater than that of En25 monobloc cylinders lying in diameter ratio range 1.2 to 2.0. Monobloc test results, with W=1.8 and 2, are consistently found to fall between the fine chain-line predictions for one and thirty elliptical cracks (1e and 30e). This demonstrates the usefulness in employing bounding estimates when the precise nature of fatigue cracking is uncertain. Bounding N_f predictions (fine continuous lines) are also given for 1 and 30 straight cracks (1s and 30s) in each figure.

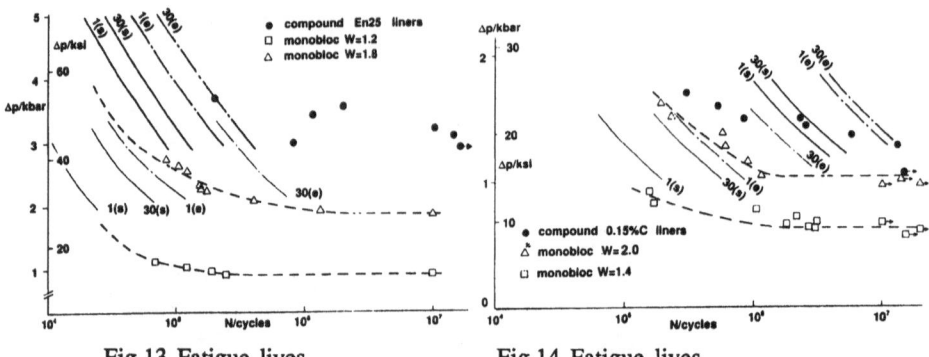

Fig.13 Fatigue lives   Fig.14 Fatigue lives

Fig 15 gives the corresponding ΔK_I plots found from eq(2) for full crack penetration through this cylinder wall. This method is, apparently, less successful when it is applied to compound cylinders. In Fig.13 the N_f bounds underestimate the observed lives for ranges of pressure in the region of the fatigue limit. The integration of eq(1) revealed no discernible difference between lives for the stress intensities of Bowie and Freese and Underwood for neither a single straight crack nor for a single elliptical crack in a compound cylinder. The predictions given in Fig.14 for a 0.15% liner material do contain the shorter lives observed.

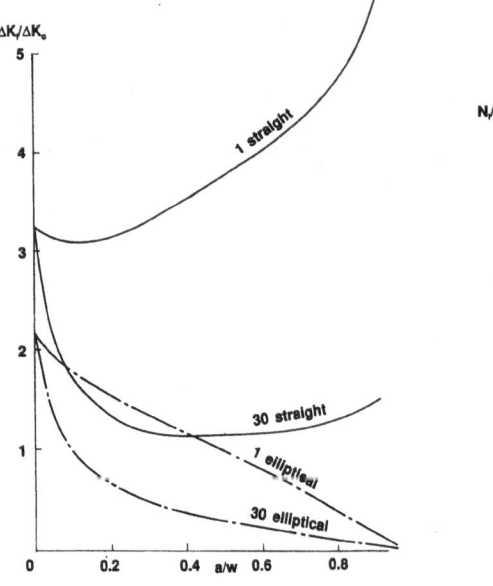

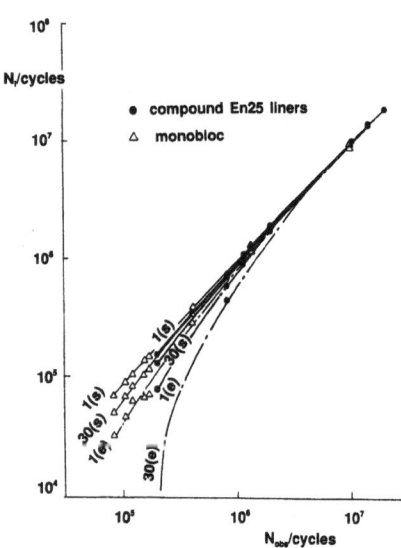

Fig.15 Stress intensities [3,4]   Fig.16 Initiation cycles

N_f predictions are sensitive to the constants in the Paris law and to the influence of initiation. The latter is examined in Fig.16 for En25 monobloc and compound cylinders and all crack conditions for W=1.8. The contribution to the observed life from initiation is not dependent upon the type of cylinder when the constants in eq(3) remain sensibly unchanged. The divergence observed at shorter lives for a large number of elliptical cracks indicates that initiation is particularly sensitive to the surface finish. Large numbers of cracks initiate from a poor surface finish with relatively few initiation cycles for which a rifled bore may be taken as a limiting condition.

CONCLUSIONS

The propagation life prediction given in eq(1) can provide realistic estimates of a wide variety of thick-walled cylinders. This includes as-received cylinders, and those with bores treated by honing, rifling and autofrettage. The life predictions of compounded cylinders appear to be less reliable by this method particularly where the liner and shell materials differ. Where the propagation lives underestimate the observed lives, at lower ranges of pressure, a suitable account of additional cycles consumed in initiating a crack may be estimated from eq(3).

REFERENCES

[1]   Birckle,A.J., Wei,R.P. and Pellissier,G.E. Analysis of plane strain fracture in a series of 0.45%C-Ni-Cr-Mo steels with different sulphur contents. *Trans ASM*, 1966, **59**, 981-989.

[2]   Rees,D.W.A. Fatigue crack growth in thick-walled cylinders under pulsating pressure, *Engng Fract Mech.*, 1989, **33**(6), 927-940.

[3]   Bowie,O.L. and Freese,C.E. Elastic analysis for a radial crack in a circular ring. *Engng Fract Mech.*, 1972, **4**, 315-321.

[4]   Rice,J.R. and Levy,N. The part through surface crack in an elastic plate, *J.Appl. Mech.*, 1972, **39**, 185-194.

[5]   Rogan,J. Fatigue strength and mode of fracture of high pressure tubing made from low-alloy high strength steels, In: Proc. 2nd Intl Conf. on *High Pressure Engineering*, I.Mech.E., 1975, 287-295.

[6]   Morrison,J.L.M., Crossland,B. and Parry,J.S.C. Strength of thick cylinders subjected to repeated internal pressure, *Proc.I.Mech.E.*, 1960, **174**(2), 95-108.

[7]   Pook,L.P. *The role of crack growth in metal fatigue*, Metals Society, London 1983.

[8]   Barsom,J.M., Imhoff,E.J. and Rolfe,S.T. Fatigue crack propagation in high strength steels, *Engng Fract Mech.*, 1971, **2**, 301-317.

[9]   Davidson,T.E., Throop,J.F., Reiner,A.N. and Austin,B.A. Analysis of the effect of autofrettage on the fatigue life characteristics of the 175 mm gun tube, WVA Tech. Rpt. 6901, 1969.

[10]   Tweed,J and Rooke,D.P. The stress intensity factor for a crack in a symmetric array originating at a circular hole in an infinite elastic solid, *Int.Jl Engng Sci.*, 1975, **13**, 653-661.

[11]   Pu,O.M. and Hussain,M.A. Stress intensity factors for a circular ring with a uniform array of radial cracks using cubic iso-parametric singular elements. STP 667, *Fract Mech.*, ASTM, 1979, 685-699.

[12]   Atluri,N. and Kathiresan,K. 3D analyses of surface flaws in thick-walled reactor vessels using displacement-hybrid finite element method, *Nuclear Eng. and Design*, 1979, **51**, 163-176.

[13] Davidson,T.E., Eisenstadt,R. and Reiner,A.N. Fatigue characteristics of open-end thick-walled cylinders under cyclic internal pressure, *Trans ASME (D)* 1963, **85**, 555-565.

[14] Parry,J.S.C. Fatigue of thick cylinders: further practical information, *Proc.I.Mech.E.*, 1965, **180**(1), 387-416.

[15] Rees,D.W.A. A theory of autofrettage with applications to creep and fatigue, *Int. J.Press.Ves and Piping*, 1987, **30**, 57-76.

[16] Rees,D.W.A. The fatigue life of thick-walled autofrettaged cylinders with closed ends, in press, *Fatigue and Fract. of Eng. Mats and Structs.*

[17] Rees,D.W.A. Autofrettage theory and fatigue life of open end cylinders, in press, *Jl Strain Analysis.*

[18] Burns,D.J. and Parry,J.S.C. Effect of mean shear stress on the fatigue behaviour of thick-walled cylinders, Proceedings: *High Pressure Engineering*, Paper **28**, 1967, I.Mech.E.

[19] Underwood,J.H. Stress intensity factors for internally pressurized thick-walled cylinders, STP **513**, *Stress Analysis and Growth of Cracks*, ASTM, 1972, 59-70.

[20] Bueckner,H.F. Boundary problems in differential equations, (ed R.E.Langer), University of Wisconsin Press, 1960, 216-230.

APPENDIX

Residual Stresses in a Two-Component, Compound Cylinder

A plane strain condition is assumed in which the axial strain is absent following shrinkage of an outer shell (O) onto an inner liner (I). This results in a common interface pressure and in a discontinuity in hoop stress between the components. The triaxial residual stress states in shell and liner may be related to the initial difference Δd between their common diameters (d_c) in the relative, elastic hoop strain expression:

$$\varepsilon_\theta = \Delta d/d_c = \{(1/E)[\sigma_\theta - \nu(\sigma_r + \sigma_z)]\}_O - \{(1/E)[\sigma_\theta - \nu(\sigma_r + \sigma_z)]\}_I \qquad (A1)$$

Now as $\varepsilon_z = 0$, the axial stress $\sigma_z = \nu(\sigma_\theta + \sigma_r)$. Substituting into eq(A1) leads to:

$$\Delta d/d_c = \{(1/E)[\sigma_\theta(1-\nu^2) + \nu p_c(1+\nu)]\}_O - \{(1/E)[\sigma_\theta(1-\nu^2) + \nu p_c(1+\nu)]\}_I \qquad (A2)$$

where Lame elastic residual stress distribution are assumed in both the shell and the liner. These are of the form:

$$\sigma_\theta = a + b/r^2, \quad \sigma_r = a - b/r^2 \qquad (A3,A4)$$

where the constants a and b are found from the known boundary conditions:

(i) For the liner; $\sigma_r = 0$ for $r = r_i$, $\sigma_r = -p_c$ for $r = r_c$. Substituting into eqs(A3) and (A4);

$$a = -p_c r_c^2/(r_c^2 - r_i^2), \quad b = -p_c r_i^2 r_c^2/(r_c^2 - r_i^2) \qquad (A5,A6)$$

Substituting eqs(A5) and (A6) into eqs(A3) and (A4) gives the liner residuals;

$$\sigma_\theta = -p_c r_c^2(1 + r_i^2/r^2)/(r_c^2 - r_i^2) \qquad (A7)$$

$$\sigma_r = -p_c r_c^2(1 - r_i^2/r^2)/(r_c^2 - r_i^2) \qquad (A8)$$

(ii) For the outer shell; $\sigma_r = 0$ for $r = r_o$, $\sigma_r = -p_c$ for $r = r_c$. Substituting into eqs(A3) and (A4) gives:

$$a = p_c r_c^2/(r_o^2 - r_c^2), \quad b = p_c r_c^2 r_o^2/(r_o^2 - r_c^2) \qquad (A9,A10)$$

Substituting eqs(A9) and (A10) into eqs(A3) and (A4) gives the shell residuals;

$$\sigma_\theta = p_c r_c^2 (1 + r_o^2/r^2)/(r_o^2 - r_c^2) \qquad (A11)$$

$$\sigma_r = p_c r_c^2 (1 - r_o^2/r^2)/(r_o^2 - r_c^2) \qquad (A12)$$

Finally, substituting eqs(A7), (A8), (A11) and (A12) into eq(A2), with $r=r_c$, leads to a relationship between the interference and the interface pressure;

$$\Delta d/d_c = p_c \{(1/E)[(1-\nu^2)(r_c^2 + r_o^2)/(r_o^2 - r_c^2) + \nu(1+\nu)]\}_o$$

$$-p_c \{(1/E)[-(1-\nu^2)(r_c^2 + r_i^2)/(r_c^2 - r_i^2) + \nu(1+\nu)]\}_I \qquad (A13)$$

from which p_c is found when the liner and shell materials are different. When the materials are the same eq(A13) the interface pressure, p_c is found from:

$$\Delta d/d_c = [(1-\nu^2)(p_c/E)][(r_c^2 + r_o^2)/(r_o^2 - r_i^2) + (r_c^2 + r_i^2)/(r_c^2 - r_i^2)] \qquad (A14)$$

For example, a common interference of 0.089 mm (0.0035 in) gives an interface pressure of p_c=82.14 MPa (5.32 tonf/in^2) for typical elastic constants E=205.4 GPa (13300 tonf/in^2) and ν=0.283 in an En25 compound cylinder with the present dimensions. Substitution p_c into eqs(A5)-(A12) will supply the constants and residual stress distributions given in Fig.10(a).

DYNAMIC STRESS ANALYSIS OF A CRACKED ROLLING MILL HOUSING

P.E.Wells, B.Sc., Research Investigator, British Steel Technical

SYNOPSIS

Knowledge of the dynamic loading and stressing of a component or assembly is essential to the design engineer to ensure satisfactory performance in a safe and economic manner. Dynamic stress/strain information can be calculated if the loading and restraining conditions are known, but these are only useful if a simple model is available to mirror these factors. In practice these simple solutions are seldom available. The alternative approach is to test an existing design, a modified design or a prototype. Such testing necessitates the use of sophisticated equipment which can monitor strains continually under dynamic conditions.

This paper describes the equipment and techniques used to monitor strain and load in the region of a crack in a rolling mill housing. The crack was discovered during routine maintenance. Production continued but urgent action was initiated to assess the effect of the crack on the continued safe operation of the mill. The crack size was measured and strain and load were recorded. Fatigue crack growth rate studies and stress analysis indicated the cause of the crack and identified areas where mill practice and design could be improved. Remedial action was recommended and undertaken.

1. INTRODUCTION

A plate rolling mill receives stock in the form of slabs which are rolled down to produce plates of various thickness. This process is continuous but downtime is required for routine maintenance work. During the maintenance of a British Steel plate mill a crack was discovered in the main mill housing. The consequence of such a crack causing failure would be the loss of mill production for many months. The immediate action was to measure the crack using a dye penetrant technique and take the calculated risk of restarting the plant. During the following maintenance period strain gauges were attached to the housing to gather more information. This paper describes the mill history, the information gathering exercise, the subsequent analysis and remedial short term and long term actions initiated.

2. MILL HISTORY

The general layout of the mill is illustrated in Figures 1 and 2. The mill is used to roll stainless slabs into plates. Slabs are typically 1.25 m x 1 m rectangles with thicknesses up to 200 mm. The mill is capable of a maximum load of 4000 ton (40 MN). The work roll is 850 mm in diameter and the typical working speed is 30 r.p.m. Plates are hot rolled to finish sizes of up to 2m wide and 4-100 mm in thickness. The mill had been in operation for about 25 years. The main mill housing is made from a Carbon, Manganese steel casting which takes the form of 4 columns connected at the top and bottom. Slab stock enters along a roller table. Close to the mill the final guide rollers (breast rollers) are attached to the main housing of the mill providing support for the stock close up to the mill stand rolls which are in two pairs (upper and lower). The work roll, comes into intimate contact with the hot slab and is designed to withstand the punishment of the contact pressures, impact forces, high temperatures and erosion. The function of the back up rolls is to carry the forces and bending moments developed.

FIGURE 1   **GENERAL LAYOUT OF PLATE MILL**   (BREAST ROLLER REMOVED)

After the first pass the direction of rolling is reversed and the stock enters the mill from the opposite direction. This process is repeated several times. At this stage the stock is rotated in the horizontal plane through 90° so that the length now becomes the width. Further passes reduce the thickness to the desired value.

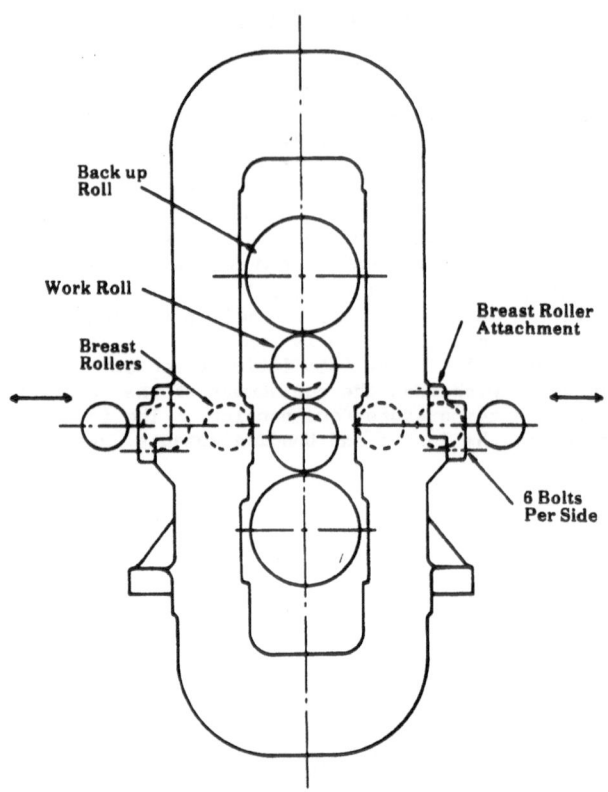

Figure 2.   **Rolling Mill - Side Elevation**

3. DISCOVERY OF CRACK

Each set of breast rollers is attached by means of 12 bolts to the mill housing (see Figures 1 and 2). The lower holes are contained within an abutment. This abutment supports the breast roller assembly during insertion and removal during maintenance. In the past these threaded holes became worn or damaged and were replaced with screw in inserts, which in turn became damaged. When the holes in the casting suffered damage the economic solution had been to use inserts having no external thread. These were joined by welding to the casting frame and allowed rapid resumption of production.

During the replacement of one of these inserts a crack was discovered by visual inspection. Immediate inspection, using dye penetrant, showed that the crack bridged two adjacent holes and extended down the hole for a depth of 100 to 150 mm (see Figures 3 and 4). The holes are about 100 mm diameter and about 200 mm deep.

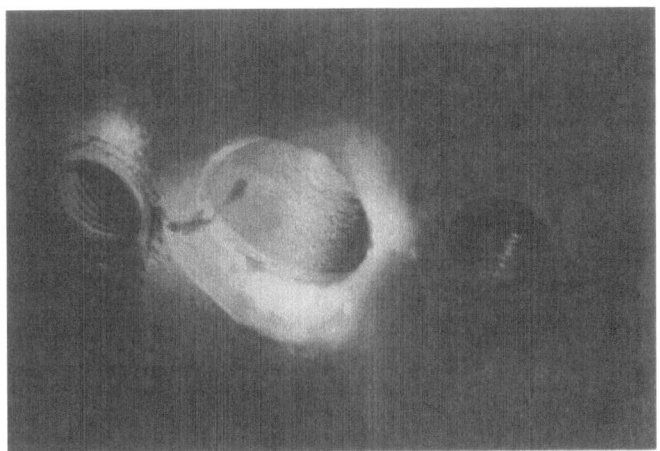

FIGURE 3   CRACK BETWEEN TWO HOLES

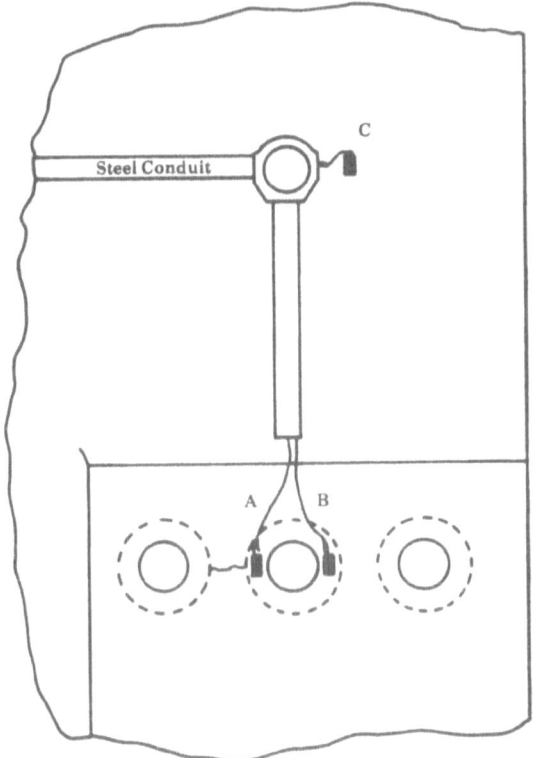

Figure 4.   Strain Gauge Positions

Inspection of such holes in the works environment is hindered by difficult access, lack of light and dirty conditions. The preliminary measurements of the crack were therefore not totally reliable. However the decision to continue production was taken on the basis that this crack had probably grown over a period of months or years and that sudden failure of the housing was unlikely within the next week. Production was restarted but urgent work was undertaken to discover why and when the crack had initiated and the rate of propagation of the crack.

4. STRAIN MEASUREMENT

4.1 Short Term Analysis

To assess the crack history, remnant life and the general and local stress levels, three strain gauges were attached to the mill housing as shown in Figure 4. Gauge C was located to assess the general stress level in the housing. Gauge A was located to determine the stress state around the cracked area near the hole. For comparison gauge B was placed adjacent to the uncracked side of the hole. Because of the limited time available it was only possible to use gauge materials designed for normal ambient temperatures. The installation was approximately one metre from the hot slabs and was exposed to significant thermal radiation. This aspect was partly alleviated by welding a metal shield adjacent to the installation to reflect the heat. A cable connected the installation to a control room 30 metres from the mill. The signal was monitored using a micro processor based digital data acquisition system.

Stress - MPa

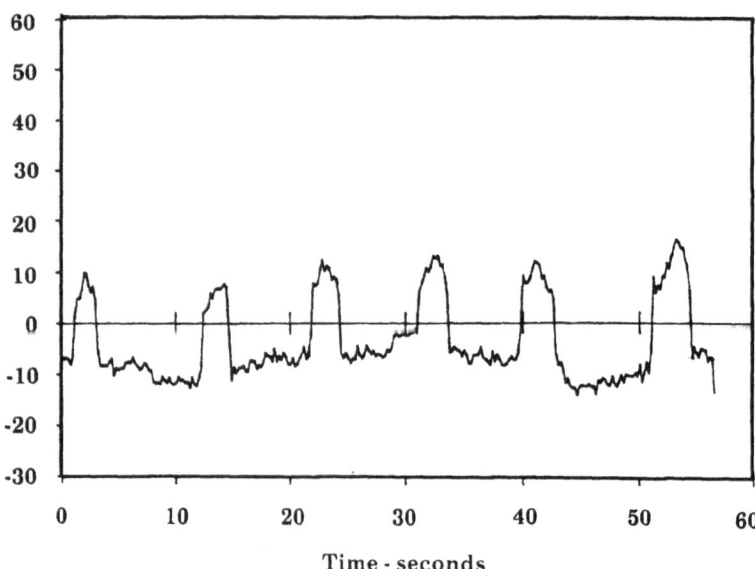

Time - seconds

Figure 5.   Plate Mill Crack - Smoothed Signal

To assess the general stress levels, the measured data were smoothed by computer. (see Figure 5). This averaging process was used to take out electrical interference present on the signal. Precise stress patterns due to rolling could not be accurately assessed because of the presence of electrical interference. Also the stress magnitudes varied with different slabs and different passes for one slab. The frequency of occurrence of the stress cycles was also an unknown factor.

For the purpose of calculation the nominal stress at the outer edge was taken to be 12 N/mm² and the inner edge was assumed to be 22 N/mm². The presence of the hole produces a stress concentration of X3. This applies to the stress from the surface to 200 mm at the bottom of the hole.

Stresses were also derived from a finite element model. This showed that the housing column was subject to bending as well as tension as shown in Figure 6.

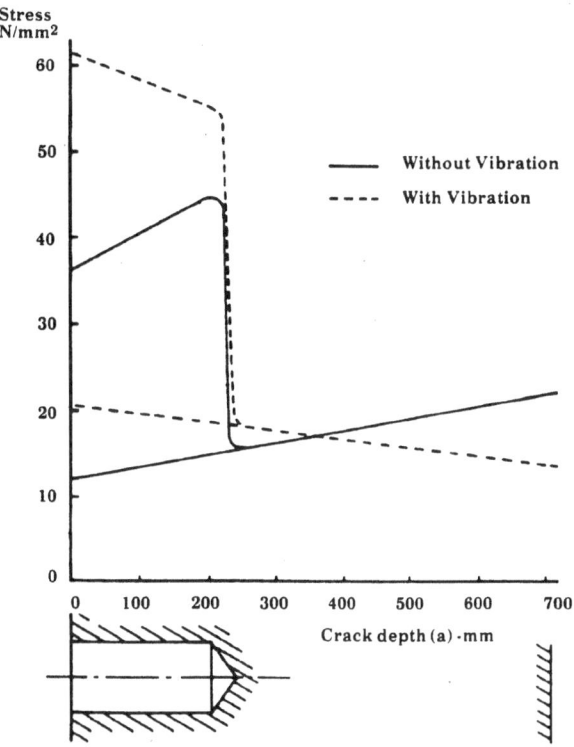

Figure 6.   Stress Variation Across Column

4.2 Medium Term Analysis

Having established the short term risk (see section 5.2) further strain gauging was undertaken to improve the quality of the stress data to refine crack growth predictions. A second installation of gauges was fitted but making use of gauges, adhesives, coatings and cables able to withstand temperatures up to 250°C. The gauges were located as previously described. The cables were run a short distance (about 3m) to the monitoring equipment which was now placed adjacent to the mill. A long cable from the monitoring equipment to the control room facilitated the power supply and computer interface connections. Also a signal from the rolling mill load measuring system was available for input.

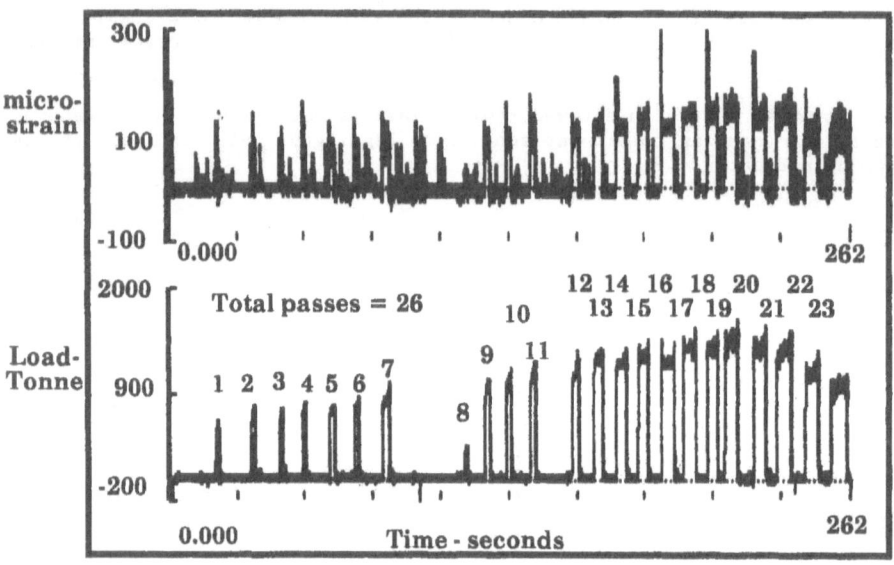

Figure 7.   Typical Strain and Load Patterns For One Plate

The signals recorded using the new system gave much improved definition. Figure 7 shows the strain and load pattern exhibited for 23 out of a total of 26 rolling passes for one slab. Previous information was collected at a sample speed of 50 Hz but features now observed, with the new system and the reduction of electrical interference, demanded an increased sampling frequency of 500 Hz. It can be seen that many of the stress cycles have spikes at the beginning of the cycle and to a lesser extent at the end of the cycle. The monitoring system allows further examination of the data. A section of the cycle can be extracted and expanded. Figure 8 shows pass No. 16. The data have been presented in terms of percentages, both load and strain values being normalised using the steady state load/strain values equal to 100%. It is now apparent that the initial 'spikes' are real events and not electrical spikes. Vibration is a clear feature, having a period of just less than 0.1 seconds, which equates to a frequency of between 11-12 Hz. It is also noticeable that the vibrations tend to be largely in the tensile direction. The vibrations present on the strain signal are not present on the load signal and are therefore not a direct result of the load. The magnitude of the strain vibrations can be as high as 280% of the steady state values.

%

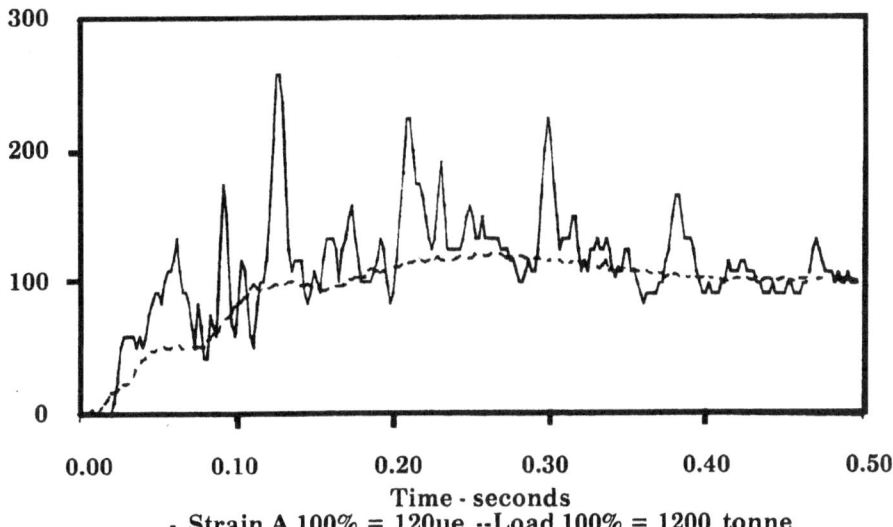

Time - seconds
- Strain A 100% = 120ue --Load 100% = 1200 tonne

Figure 8.   Plate Mill Housing - (pass 16)

This additional new information was highly significant in terms of crack initiation, crack growth, plant maintenance and plant design. All of which were re-assessed in view of these observations (see Section 5.2).

5. CRACK GROWTH

5.1. Short Term Assessment

Linear Elastic Fracture Mechanics was used to both assess the crack propagation to the discovery of the crack and the future predicted rate. Two equations were used to model the crack behaviour. The first defines the crack tip stress intensity factor.

$$\Delta K = Y . \Delta \sigma \sqrt{\pi a} \qquad \text{Equation (1)}$$

where ΔK = stress intensity at the crack tip, MPa$\sqrt{m}$
   Y = compliance function
   $\Delta \sigma$ = applied stress range, MPa
   a = crack length, m

The second equation defines the crack growth rate.

$$\frac{da}{dN} = C . (\Delta K)^m \qquad \text{Equation (2)}$$

   N = number of cycles
   C, m are constants

A distance of 150 mm was taken for the crack depth. The number of damaging cycles was not known so a range of from 1 per sec to 1 per 1000 secs was taken. The crack initiation time was not known but was assumed to be $1^1/_2$ years ago. This coincided with the welding in of the insert and an uprating of the plant. Values of C = 3 x 10^{-13} and m = 3 were used as recommended in PD 6493.(1) The calculated values of da/dN and ΔK are listed in Table 1.

TABLE 1
STRESS INTENSITY CALCULATIONS

Crack a	Freq.	Time	Cycles N	Crack Growth da/dN	m	C	Crack Intensity ΔK	
mm	Hz	Years		mm/cycle			Nmm -3/2	MPa$\sqrt{m}$
150	1.0000	1.50	4.73E+07	3.17E-06	3	3 E-13	219	6.94
150	0.3000	1.50	1.42E+07	1.06E-05	3	3 E-13	328	10.37
150	0.1000	1.50	4.73E+06	3.17E-05	3	3 E-13	473	14.95
150	0.0300	1.50	1.42E+06	1.06E-04	3	3 E-13	706	22.33
150	0.0100	1.50	4.73E+05	3.17E-04	3	3 E-13	1019	32.21
150	0.0090	1.50	4.26E+05	3.52E-04	3	3 E-13	1055	33.36
150	0.0030	1.50	1.42E+05	1.06E-03	3	3 E-13	1522	48.12
150	0.0010	1.50	4.73E+04	3.17E-03	3	3 E-13	2195	69.40

TABLE 2
LIFE PREDICTIONS

Crack a	Width b	Y	Stress σ	Intensity K	m	C	Crack Growth da/dN	Cycles	Time
mm	mm		MPa	MPa$\sqrt{m}$			mm/cycle		Years
0	711	1.12	36.00	0.00	3	3E-13	0.00E+00		
75	711	1.16	39.16	22.07	3	3E-13	1.02E-04		
150	711	1.21	42.33	35.29	3	3E-13	4.17E-04		
175	711	1.24	43.38	39.74	3	3E-13	5.95E-04	4.20E+04	0.53
200	711	1.26	44.44	44.32	3	3E-13	8.26E-04	3.03E+04	0.38
250	711	1.31	15.52	18.01	3	3E-13	5.55E-05	9.02E+05	11.44
300	711	1.37	16.22	21.61	3	3E-13	9.57E-05	5.22E+05	6.62
350	711	1.45	16.92	25.71	3	3E-13	1.61E-04	3.10E+05	3.93
400	711	1.54	17.62	30.52	3	3E-13	2.70E-04	1.85E+05	2.35
450	711	1.67	18.33	36.38	3	3E-13	4.57E-04	1.10E+05	1.39
500	711	1.84	19.03	43.84	3	3E-13	7.99E-04	6.26E+04	0.79
550	711	2.08	19.73	54.04	3	3E-13	1.50E-03	3.34E+04	0.42
600	711	2.48	20.44	69.71	3	3E-13	3.21E-03	1.56E+04	0.20
650	711	3.32	21.14	100.22	3	3E-13	9.55E-03	5.24E+03	0.07
							Total	2.22E+06	28.13

Equation (1) can be used to predict the stress intensity factor (ΔK) from the geometrical considerations of the crack and surrounding body. The Y function was modelled using an edge crack in a flat sheet (Rooke & Cartwright[2], case 1.1.20.). In practice the crack would be subject to greater restraint and therefore the Y values would be smaller. Table 2 shows the calculated K values which in turn are used to predict crack growth rates. Stress levels used are based on measured surface stresses with a stress concentration factor applied to the hole and with stress variation through the casting as indicated by Finite Element analysis. Table 2 includes a mean value taken at a crack length equal to 75 mm. The corresponding stress intensity is 22.07 MPa√m. Table 1 indicates this value equates to a damaging frequency of 0.03 or 1 cycle every 33 seconds. The time predicted for the crack to grow from 150 to 175 mm was 0.53 years. The total time predicted for the crack to propagate to the plastic collapse point was about 28 years.

This initial work indicated that the rate of crack propagation was not critical and that further more detailed work could be undertaken to establish the continued safe operation of the plant.

5.2 Medium Term Assessment

Following the discovery of higher stress levels resulting from impact loading, the calculations were repeated as detailed in Table 3. A typical impact factor, due to vibration of 1.7 was selected. Table 3 shows the average stress intensity factor at a crack depth of 75 mm to be 33.27 MPa√m. This corresponds to a damaging frequency of 0.009 or 1 cycle every 111 seconds. The predicted time for the crack to grow from 150 to 175 mm was 0.25 years. The situation was thus more critical than previously estimated. If crack propagation continued to the bottom of the hole, it would start to fan out over the full width of the casting and subsequent removal and repair would be practically impossible.

TABLE 3
LIFE PREDICTIONS INCLUDING VIBRATIONS

Crack a mm	Width b mm	Y	Stress σ MPa	Intensity K MPa√m	m	C	Crack Growth da/dN mm/cycle	Cycles	Time Years
0	711	1.12	61.20	0.00	3	3E-13	0.00E+00		
75	711	1.16	59.05	33.27	3	3E-13	3.49E-04		
150	711	1.21	56.90	47.44	3	3E-13	1.01E-03		
175	711	1.24	56.18	51.47	3	3E-13	1.29E-03	1.93E+04	0.25
200	711	1.26	55.46	55.32	3	3E-13	1.61E-03	1.56E+04	0.20
250	711	1.31	18.01	20.91	3	3E-13	8.67E-05	5.76E+05	7.31
300	711	1.37	17.53	23.36	3	3E-13	1.21E-04	4.13E+05	5.24
350	711	1.45	17.05	25.91	3	3E-13	1.65E-04	3.03E+05	3.84
400	711	1.54	16.58	28.71	3	3E-13	2.24E-04	2.23E+05	2.83
450	711	1.67	16.10	31.95	3	3E-13	3.09E-04	1.62E+05	2.05
500	711	1.84	15.62	35.98	3	3E-13	4.42E-04	1.13E+05	1.44
550	711	2.08	15.14	41.46	3	3E-13	6.76E-04	7.39E+04	0.94
600	711	2.48	14.66	50.01	3	3E-13	1.19E-03	4.21E+04	0.53
650	711	3.32	14.19	67.25	3	3E-13	2.88E-03	1.73E+04	0.22
							Total	1.96E+06	24.84

5.3 Vibration Stresses

The presence of the vibration stresses is a critical factor in the crack growth calculations and this aspect therefore merits further consideration. The vibrations are only present at the start of a rolling pass and are therefore induced by the impact of the stock entering the rolls. The stock entry speed is controlled by computer but imperfect results will cause impact. The dominating frequency (11-12 Hz) is inconsistent for the body of a casting of such high stiffness. However the top and bottom rolls are driven by independent drive shafts and motors. The vibration frequency of such a motor/shaft/roll/stock assembly is about 10-15 Hertz. It is thought that this spring/mass system imparts horizontal vibrations onto the roll end housings which in turn causes outward bending vibrations of the main mill housing.

6. COURSE OF ACTION

6.1 Immediate

It was estimated that the crack was propagating at a rate such that the crack tip would reach the end of the hole in a short period of time. Further propagation of the crack into the bulk of the cross section would render the crack practically irremovable, resulting in a predicted life of about 25 years before plastic collapse. Therefore it was deemed necessary to remove the crack. This course of action necessitated an alternative design for breast roller fixing. New housings for the breast rollers were designed and made. The cracked hole was drilled out and new holes were drilled and tapped at alternative positions. The crack had in fact reached the bottom of the hole at the time of removal.

6.2 Medium/Long Term

The possibility of further initiation and propagation of cracks may be reduced by avoiding the welding of attachments to the casting. Alternatively the driving stress behind the cracks can be controlled by seeking to improve the stock/roll speed synchronisation and also to improve and maintain the condition of the relevant moving parts.

Most industrial plants operate with a minimum of monitoring equipment. Continuous condition monitoring of load/stress etc. can be done to record cumulative damage. Such information would have been invaluable in the above situation and would be useful to assess any change in process or uprating of plant.

7. SUMMARY

Mechanical problems are a regular and ongoing feature of industrial plant. The majority of problems will be resolved through regular maintenance and will involve minimal losses to production. Occasionally serious problems occur where interruptions or risk of interruption threatens to halt production for long periods of time. The consequences of safety of operating personnel and loss of production income are very serious. The discovery of the crack in the plate mill was such an example.

The discovery of the crack in the rolling mill housing required urgent attention to assess the immediate risk. Further work was undertaken using strain gauge techniques together with crack growth studies to assess crack initiation and propagation.

The final outcome was to remove the crack to prevent further propagation. Recommendations were given for future plant operation.

8. REFERENCES

1. PD 6493: 1980 Guidance on some methods for the derivation of acceptance levels for defects in fusion welded joints, British Standards Institution.

2. Rooke D.P., Cartwright D.J., Stress Intensity Factors, 1976, Hillingdon Press, Uxbridge.

ACKNOWLEDGEMENTS

The author is indebted to Dr.R.Baker, Director of Research and Development, British Steel and Mr. J. Newborn, General Manager, British Steel Stainless for permission to publish this paper. Also to Mr. D.J. Connolly, Chief Engineer, British Steel Stainless for technical advice.

THREE-DIMENSIONAL MEASUREMENT OF OPENING DISPLACEMENT OF RAPIDLY PROPAGATING CRACKS IN PMMA

SHINICHI SUZUKI
Research Associate / Department of Energy Engineering,
Toyohashi University of Technology
Tempaku-cho, Toyohashi, 441 Japan

ABSTRACT

An optical method is developed to measure crack opening displacement (COD) in the inside of a PMMA plate specimen. The crack is a through-crack and propagating at a speed of several hundred m/sec. The crack is illuminated with a pulsed laser beam, which is weakly scattered by the two crack surfaces. The scattered light waves make interference fringes on the crack surfaces. Determination of the orders of the interference fringes are based on a result of COD measurement on a specimen surface by means of pulsed holographic microscopy. The COD in a specimen is obtained from the fringes. The COD measurement in a specimen is carried out in the vicinity of the crack front edge where the crack has three dimensional shape.

INTRODUCTION

A rapidly propagating crack appears when a plate specimen of brittle material is broken by external tensile force. Around the crack tip, there is a singular stress field which satisfies the plane stress condition [1]. A three dimensional stress field, however, appears in the region closer to the tip, because of the termination of the crack front edge at the specimen surfaces [2]-[7]. Making clear the structure of the 3-D stress field is important not only for fine understanding of the structure of a fast propagating crack tip but also for dynamic fracture toughness measurement by means of some optical methods, for example, the caustic method [6] and so on.

Recently, pulsed holographic microscopy [7] was introduced into dynamic fracture research. Pulsed holographic microscopy is able to take an instantaneous microscopic photograph of a rapidly propagating crack. From the

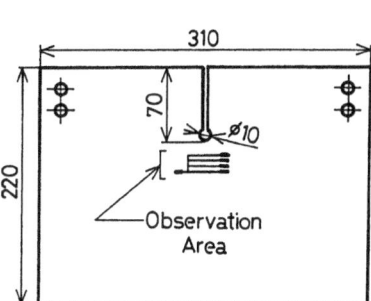

Figure 1. PMMA plate specimen.

photograph, crack opening displacement (COD) can be measured as a function of distance r from the tip. The COD which was measured with r up to 2.5mm was of the order of ten to one hundred microns in case of rapidly propagating cracks in PMMA. The dynamic stress intensity factor $K_I(v)$ is obtained from the COD data. But the COD measurement has been restricted to a specimen surface because of limits of the ability of the method. The COD in a specimen has not thus been measured though it is important and interesting.

In the present study described is an optical method for COD measurement in the inside of a PMMA specimen. The method can be applied only to transparent specimens, however, it may be useful to know the three dimensional shape of a rapidly propagating crack near the tip.

COD MEASUREMENT BY MEANS OF PULSED HOLOGRAPHIC MICROSCOPY

In this section described is the COD measurement on a specimen surface by means of pulsed holographic microscopy.

Specimen

Figure 1 shows the PMMA plate specimen used in the present study. It is 220mm long, 310mm wide and 20mm thick. There is a notch 70mm in length. A crack arises at the notch tip and propagates toward the observation area. Propagating in the observation area, the crack is recorded as a hologram by the holographic optical system described below. Four conducting strips are painted in the observation area. Crack speed is measured from the signals generated at the cutting of the conducting strips.

Pulsed holographic microscopy

Figure 2(a) shows an optical system for recording in the pulsed holographic

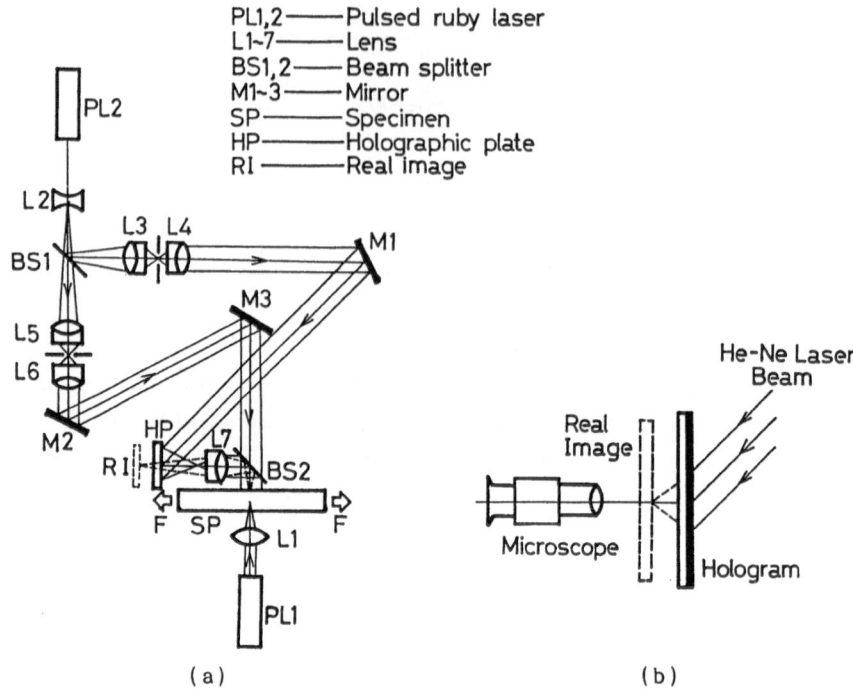

Figure 2. Optical systems for pulsed holographic microscopy. (a) Recording. (b) Reconstruction and photographing.

microscopy [7]. The tensile force F is applied to the PMMA specimen SP. The Q-switched ruby laser PL1 oscillates, and the pulsed laser light is focused in the specimen at the notch tip. A small defect arises at the notch tip, from which a rapidly propagating crack starts propagation perpendicularly to the paper plane [8]. When the crack is propagating in the observation area, the Q-switched ruby laser PL2 oscillates and emits a pulsed light beam for holographic recording of the crack. The beam is diverged by the concave lens L2 and divided into two beams by the beam splitter BS1. The reflected light from BS1 becomes a parallel light beam and impinges on the holographic plate HP; this is the reference beam in holography. The light transmitted through BS1 is collimated, half transmitted through the beam splitter BS2, and incident upon the specimen surface perpendicularly. Part of the light is reflected by the specimen surface. The reflected light is half reflected by BS2, passes through the lens L7, and falls onto the holographic plate HP. This is the object beam in holography. Behind the holographic plate, the object beam makes the real image RI of the crack magnified three times.

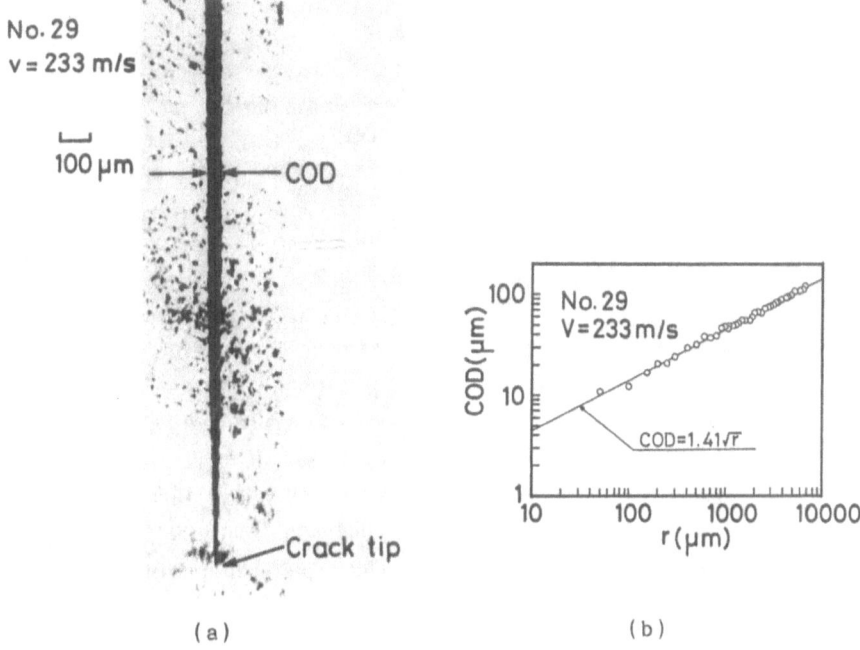

(a)

(b)

Figure 3. (a) An instantaneous microscopic photograph of a rapidly propagating crack. (b) Crack opening displacement of the crack in (a).

The real image is recorded as a hologram.

After development, illuminated with a He-Ne laser beam, the hologram reconstructs the real image RI of the crack (Fig.2(b)). The reconstructed real image is magnified and photographed through a conventional microscope which is focused on the specimen surface illuminated at the recording.

Microscopic photograph and COD measurement

An example of the microscopic photographs of rapidly propagating cracks is shown in Fig.3(a). The dark region is the opened crack. Crack speed was at 233m/sec. From the photograph we can measure crack opening displacement (COD) on the specimen surface as a function of distance r from the tip. The measured COD is shown in Fig.3(b). The COD was measured along the crack as far as 8mm from the tip. It is thought that the three dimensional stress field which appears in the vicinity of a crack tip spreads as far as the distance half the specimen thickness from the tip [5],[6]. We can therefore say that the COD measurement in the study was carried out in the three dimensional stress field. Assuming that COD is proportional to r^{λ}, we

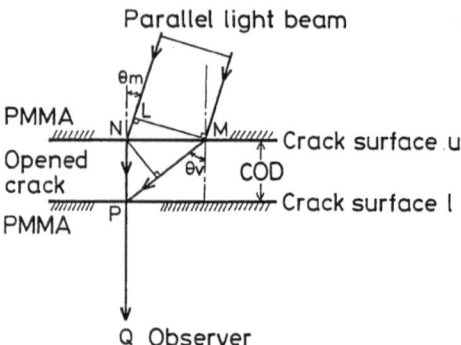

Figure 4. The principle for the COD measurement in a transparent specimen.

calculated the exponent λ by means of the least square method. As a result the λ had the value, $\lambda=0.50\pm0.04$, on the average of eight cracks. It can thus be said that, on a specimen surface, the COD of a rapidly propagating crack is proportional to square root of the distance from the tip even in the three dimensional stress field near the tip. The experimental result is consistent with the theoretical result given by Bažant and Estenssoro [4].

COD MEASUREMENT IN THE INSIDE OF SPECIMENS

In this section described is the method for measurement of COD in the inside of PMMA specimens. The specimens used here are same as that shown in Fig.1 in the previous section.

Principle of measurement

Figure 4 shows the principle for the COD measurement in a PMMA specimen. The crack propagates perpendicularly to the paper plane. The parallel beam of laser light is incident upon the crack surface u from the inside of PMMA. The incident angle of the light is θ_m. The crack is observed from the direction normal to the crack surfaces. Most of the incident light is refracted according to the Snell's law of geometrical optics, however, a little light is scattered by the crack surfaces because the crack surfaces are slightly rough. The light incident at the point N on the crack surface u is weakly scattered by the surface. The scattered light which propagates perpendicularly to the crack surface u passes through the point P on the crack surface l and goes toward the observer Q. On the other hand, most of the light incident at the point M on the crack surface u is refracted at the angle θ_v of refraction. The

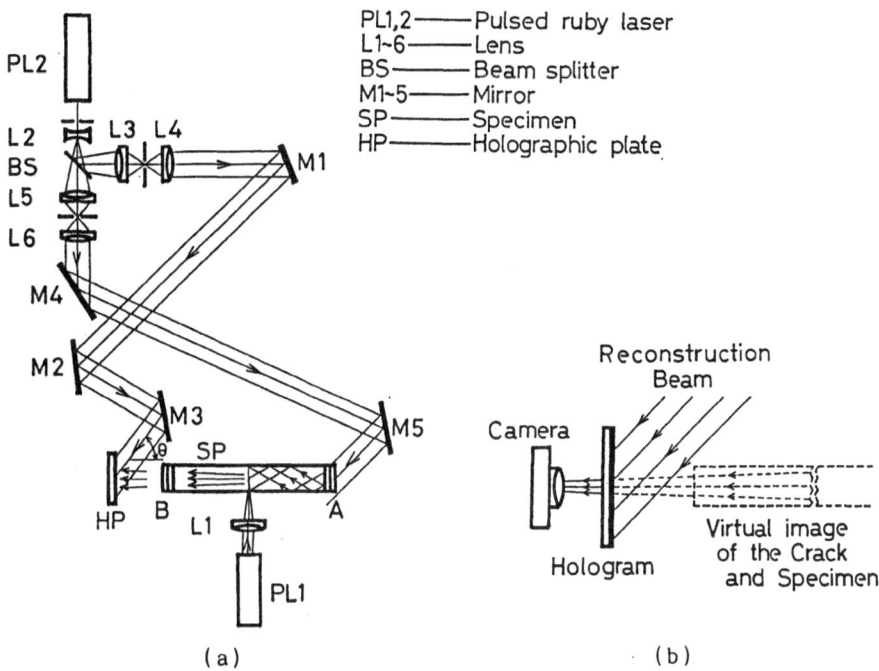

Figure 5. Optical systems for the COD measurement in a transparent specimen.
(a) Holographic recording system. (b) Reconstruction.

refracted light arrives at the point P and is weakly scattered there. The
scattered light which propagates perpendicularly to the crack surfaces goes
toward the observer Q. The observer hence observes two light waves, one
of which takes the optical path LNPQ, and the other, MPQ. The difference
ΔS of the two optical path lengths is described as follows,

$$\Delta S = COD \ (1 - \cos \theta_v). \tag{1}$$

The difference ΔS of the optical path lengths is proportional to the COD.
The two light waves interfere each other, consequently, the brightness at
the point P changes alternatively as the COD varies. Hence, the observer
Q finds interference fringes, which indicate equi-COD lines, on the crack
surface. Equation (1) doesn't contain the refractive index of PMMA, therefore,
the equation is also applicable to various transparent specimens other than
PMMA. Varying the angle θ_v of refraction of the illumination light, we can
choose the ratio of ΔS to COD, namely, the sensitivity of the interference
fringes to the change of COD.

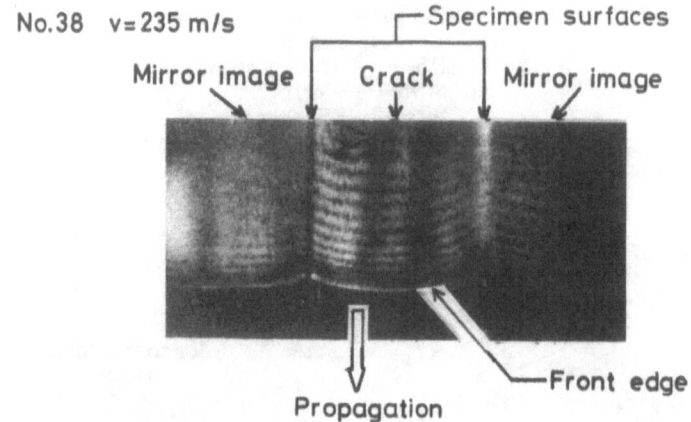

Figure 6. Interference fringes appearing on a rapidly propagating crack.

Optical method for measurement

Shown in Fig.5 is the optical system for holographic recording of the interference fringes on the crack surfaces. The method for the breaking of the specimen is the same as that described in the previous section. A crack rapidly propagates in the PMMA specimen SP perpendicularly to the paper plane. When the crack is propagating in the observation area, the Q-switched ruby laser PL2 oscillates and emits a pulsed laser beam, which is diverged by the lens L2 and divided into two parts by the beam splitter BS. The reflected beams from BS is collimated and incident upon the holographic plate HP. This is the reference beam in holography. The transmitted light beam through BS is collimated too, and impinges on the side boundary A of the specimen SP with the incident angle θ_v. Repeating total reflection a few times on the specimen surfaces, the beam propagates in the specimen and falls on a crack surface with the incident angle θ_m. The light beam is weakly scattered by the crack surfaces. The scattered light rays propagate in the specimen, pass through the side boundary B of the specimen and fall onto the holographic plate. This is the object beam in holography. The interference fringes on the crack surfaces, which are observed through the side boundary B, are recorded as a hologram.

After development, illuminated with the cw He-Ne laser beam, the hologram reconstructs the virtual image of the crack, interference fringes and specimen (Fig.5(b)). The crack and fringes are photographed through a side boundary of the specimen.

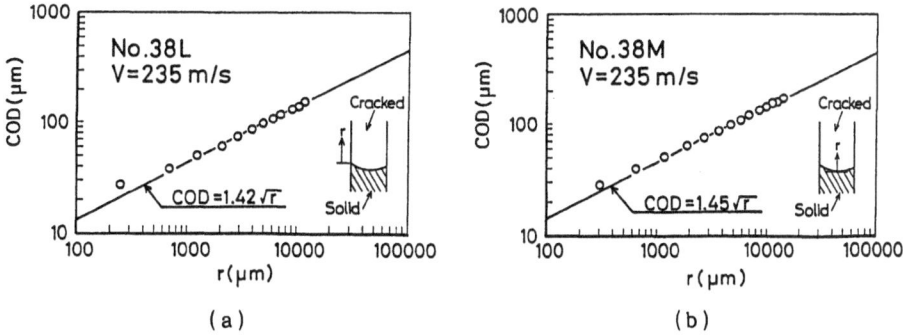

Figure 7. COD of the crack in Fig.6. (a) On the left surface of the specimen. (b) On the mid-plane between the specimen surfaces.

Interference fringes and COD measurement

Figure 6 shows an example of interference fringes of rapidly propagating cracks photographed through the above method. The crack is in the center of the photograph, on both side of which there are mirror images of the crack reflected on the specimen surfaces. The crack speed was at 235m/sec. The crack front edge is not straight but curved. The front edge at the mid in the specimen precedes that on a specimen surface [4]. Interference fringes are appearing on the crack surface. Crack opening displacement is constant along a interference fringe. Crack opening displacement can be measured from the interference fringes through the following equation,

$$COD = \frac{m\,\lambda}{1 - \cos\theta_v} \,, \qquad (2)$$

where m is the order of interference and λ is the wave length of the ruby laser (694nm), respectively. The order m is at zero on the crack front edge and is an positive integer on a bright fringe, namely, the bright fringes correspond to the integral orders. The photograph in Fig.6, however, shows that the intervals of fringes becomes shorter in the region closer to the crack front edge. As a result, it becomes difficult to distinguish each fringe in the vicinity of the crack front edge. If we can't distinguish each fringe near the crack front edge, we can't know the order of interference of every fringes even in the region fringes are distinguishable.

In the present study, the result of the COD measurement by means of pulsed holographic microscopy is utilized to determine the order of interference.

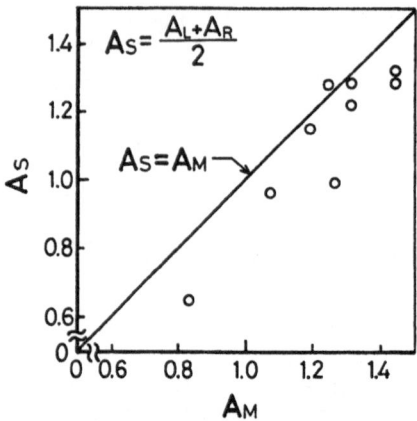

Figure 8. The ratio of the proportional constant A_s on specimen surfaces to that A_m on the mid-plane in the specimen.

The result of the COD measurement in the previous section said that the COD is proportional to $\sqrt{r}$ on a specimen surface, where r is the distance from the crack tip on the specimen surface. Thus, the order of interference m in Eq.(2) can be determined so that the COD data which is obtained from distinguishable fringes on a specimen surface becomes to be closest to $\sqrt{r}$.

Figure 7 shows the results of the COD measurement of the crack shown in Fig.6. The order of interference m is obtained so that the COD data on the left surface of the specimen becomes to be closest to $\sqrt{r}$. The COD data on the left specimen surface is shown in Fig.7(a), and is nearly proportional to $\sqrt{r}$. Figure 7(b) shows the COD data in the specimen, on the mid-plane between the two specimen surfaces. The COD is shown as a function of distance r from the crack front edge. It is found that the COD on the mid-plane in the specimen is also proportional to square root of the distance from the crack front edge.

If we assume that,

$$COD = A \sqrt{r} ,$$ \hfill (3)

then we can obtain the proportional constant, A, both on specimen surfaces and on the mid-plane in a specimen. Figure 8 shows the result, where the horizontal axis indicates the proportional constant A_m on the mid-plane, and the vertical axis, A_s on specimen surfaces. The ratio A_s/A_m was at 0.92 ± 0.08 on the average of nine cracks. The physical interpretation of the value of A_s/A_m is beyond the scope of the present study.

CONCLUSIONS

The optical method which utilizes the interference between scattered laser waves from crack surfaces makes it possible to measure, in a transparent specimen, opening displacement of a rapidly propagating crack. The crack opening displacement is proportional to square root of the distance from the crack front edge both on the specimen surfaces and on the mid-plane in a specimen, even in the near tip region where a crack has three-dimensional shape.

ACKNOWLEDGMENTS

The study was performed during the tenure of Grant-in-Aid for General Scientific Research (01550071) of The Ministry of Education, Science and Culture. The author expresses his gratitude for the grant. The author also wishes to acknowledge I. Aoki, S. Sakano and I. Yoshiyama, who are (and was) students in our university and worked together through the study.

REFERENCES

1. Yoffe, E.H., The moving Griffith crack, Phil. Mag., 1951, **42**, 739-750.

2. Folias, E.S., On the three-dimensional theory of cracked plates, Trans. ASME: J. Appl. Mech., 1975, **42**, 663-674.

3. Benthem, J.P., State of stress at the vertex of a quarter-infinite crack in a half-space, Int. J. Solids Structures, 1977, **13**, 479-492.

4. Bažant, Z.P. and Estenssoro, L.F., Surface singularity and crack propagation, Int. J. Solids Structures, 1979, **15**, 405-426, and Addendum to the paper, Int. J. Solids Structures, 1980, **16**, 479-481.

5. Yang, W. and Freund, L.B., Transverse shear effects for through-cracks in an elastic plate, Int. J. Solids Structures, 1985, **21**,977-994.

6. Rosakis, A.J. and Ravi-Chandar, K., On crack-tip stress state: An experimental evaluation of three-dimensional effects, Int. J. Solids Structures, 1986, **22**, 121-134.

7. Suzuki, S., Homma, H. and Kusaka, R., Pulsed holographic microscopy as a measurement method of dynamic fracture toughness for fast propagating cracks, J. Mech. Phys. Solids, 1988, **36**, 631-653.

8. Suzuki, S., and Nakajima, T., Development of laser inducing technique for fast propagating cracks in PMMA, In PVP-Vol.160 Dynamic Fracture Mechanics, ed. H. Homma, D.A. Shockey and G. Yagawa, ASME, 1989, 79-84.

KINKED CRACKS: FINDING STRESS INTENSITY FACTORS

D.A. HILLS AND D. NOWELL
Department of Engineering Science
University of Oxford
Parks Road, Oxford, OX1 3PJ

ABSTRACT

In the paper we describe a general method for the solution of crack-tip stress intensity factors, for cracks which are open, subject to an arbitary far field. First, the interaction of two separated cracks is treated. The case when the extremities of the two segments are brought together to form a single kinked crack is then examined, and it is shown that reliable values for the crack-tip stress intensities may be found by only a slight variation on the standard procedure, which relies on the distribution of displacement discontinuities or dislocations. The method is very economical on computing time, and it is recommended that the technique be used afresh for each problem; for this reason only sample results are given.

INTRODUCTION

A powerful, general technique for the solution of crack problems (including stress intensity factors, and, if required, stress and displacement fields), is to solve the elasticity problem for the case of an un-cracked component, and subsequently to distribute dislocations along the line of the crack so as to render the crack faces traction-free. It should be stated at the outset that these "dislocations" are simply displacement discontinuities which when distributed represent the opening displacement of the crack; they are in no way indicative of the lattice flaws which may be present, nor do they constitute a plasticity correction to an essentially elastic solution, as in the celebrated Bilby, Cottrell and Swinden solution [1]. The power of the solution lies in the closed form solutions available for the stress field created by a dislocation in a wide range of geometries. Most of these solutions originate from Dundurs' work on the solution for a dislocation in an infinite elastic medium containing an elastically dissimilar circular inclusion [2]. By specialising the choice of elastic constants or geometry we can readily find the solution for a dislocation in a half-plane, or near a circular hole [3]. Hence, any number of dislocations may be distributed within the body without influencing the surface traction conditions.

The technique of distributing dislocations has been widely used to study problems involving partially or fully closed cracks (see, e.g. [4,5]) where it has very great advantages over most other methods, but the authors have advocated its use in cases where the cracks are open [6]. In that article we considered the interaction of the crack with a free surface, using dislocation solutions appropriate for that case. Similarly, in a sequel [7] we studied kinked cracks near free edges. The solutions found were reliable and efficient, but there was a residual difficulty with the solution remote from the crack-tip: in the case of surface breaking cracks the assumed form of the dislocation distribution meant that at the crack mouth its faces were parallel. This discrepancy was subsequently removed in a small modification of the method [8]. Similarly, in our treatment of kinked cracks the complexity of the dislocation solution meant that it was difficult to incorporate behaviour of the kink with a great deal of rigour. In the present paper we hope to remove this difficulty and discuss more rigorously the nature of the solution, whilst retaining a simple engineering formulation. To this end we shall here restrict ourselves to a study of cracks within the interior of a notionally infinite region, which will permit the use of a simpler dislocation solution, whilst retaining the essential physics of the problem.

FORMULATION

The problem we shall consider first is that of two straight cracks, one of semi-length a, the other of semi-length b, at an arbitary separation r and orientated with respect to the global x-y coordinate set as indicated in Fig. 1. Dislocations will be distributed along both cracks, and we therefore need to know the state of stress induced by a

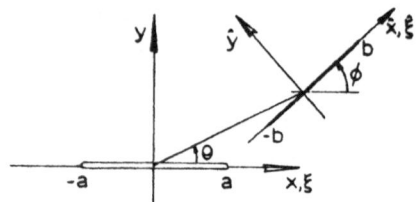

1. Global and local coordinate sets. Two arbitary straight cracks in general stress field.

single dislocation in an infinite medium, to use as a Green's function. This is given by Dundurs [9], for a dislocation having Burger's vector b_x, b_y, located at the origin, Fig. 2, as

$$
\begin{Bmatrix} \sigma_{xx} \\ \sigma_{yy} \\ \sigma_{xy} \end{Bmatrix} = -\frac{2\mu}{\pi(\kappa + 1)} \left(b_x \begin{bmatrix} y/r^2 + \frac{2x^2y}{r^4} \\ y/r^2 - \frac{2x^2y}{r^4} \\ x/r^2 - \frac{2x^3}{r^4} \end{bmatrix} + b_y \begin{bmatrix} -x/r^2 + \frac{2y^2x}{r^4} \\ -x/r^2 + \frac{2y^2x}{r^4} \\ -y/r^2 + \frac{2y^3}{r^4} \end{bmatrix} \right) \quad (1)
$$

where $r^2 = x^2 + y^2$

μ is the modulus of rigidity, $\kappa = 3 - 4\nu$ in plane strain and ν is Poisson's ratio.

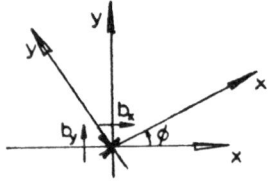

2. A single dislocation having Burger's components b_x, b_y with global and local coordinate sets.

It will subsequently be useful to determine the stress components in a new coordinate set rotated an arbitary angle ϕ with respect to the given set. Applying a routine stress transformation gives

$$
\begin{Bmatrix} \hat{\sigma}_{yy} \\ \hat{\sigma}_{xy} \end{Bmatrix}(x,y) = -\frac{2\mu}{\pi(\kappa + 1)} \left(b_x \begin{bmatrix} K_x^y(x,y;\phi) \\ K_x^x(x,y;\phi) \end{bmatrix} + b_y \begin{bmatrix} K_y^y(x,y;\phi) \\ K_y^x(x,y;\phi) \end{bmatrix} \right) \quad (2)
$$

where $K_x^{\hat{y}} = \dfrac{y}{r^2} - \dfrac{2x^2 y}{r^4} \cos 2\phi - \left(\dfrac{x}{r^2} - \dfrac{2x^3}{r^4}\right) \sin 2\phi$

$K_x^{\hat{x}} = \qquad - \dfrac{2x^2 y}{r^4} \sin 2\phi + \left(\dfrac{x}{r^2} - \dfrac{2x^3}{r^4}\right) \cos 2\phi$

$K_y^{\hat{y}} = - \dfrac{x}{r^2} - \dfrac{2y^2 x}{r^4} \cos 2\phi - \left(\dfrac{-y}{r^4} + \dfrac{2y^3}{r^4}\right) \sin 2\phi$

$K_y^{\hat{x}} = \qquad - \dfrac{2y^2 x}{r^4} \sin 2\phi + \left(\dfrac{-y}{r^4} + \dfrac{2y^3}{r^4}\right) \cos 2\phi$

The $\sigma_{\hat{x}\hat{x}}$ stress component will not be required. We are now in a position to write down the standard integral equations [6] for the nett tractions present on each crack face. They are

$$\sigma_{iy}^1(x) = \tilde{\sigma}_{iy}^1(x) - \dfrac{2\mu}{\pi(\kappa + 1)} \left\{ \int_{-a}^{a} K_y^i(x - \xi, 0; 0) B_{y1}(\xi) d\xi \right.$$

$$+ \int_{-a}^{a} K_x^i(x - \xi, 0; 0) B_{x1}(\xi) d\xi + \int_{-b}^{b} K_y^i(x_1, y_1; 0) B_{y2}(\hat{\xi}) d\hat{\xi}$$

$$\left. + \int_{-b}^{b} K_x^i(x_1, y_1; 0) B_{x2}(\hat{\xi}) d\hat{\xi} \right\} \tag{3}$$

$$i = x \text{ or } y$$

$$\sigma_{jy}^{2\hat{}}(\hat{x}) = \tilde{\sigma}_{jy}^2(\hat{x}) - \dfrac{2\mu}{\pi(\kappa + 1)} \left\{ \int_{-b}^{b} K_y^j(x_2, y_2; \phi) B_{y2}(\hat{\xi}) d\hat{\xi} \right.$$

$$+ \int_{-b}^{b} K_x^j(x_2, y_2; \phi) B_{x2}(\hat{\xi}) d\hat{\xi} + \int_{-a}^{a} K_y^j(x_3, y_3; \phi) B_{y1}(\xi) d\xi$$

$$\left. + \int_{-a}^{a} K_x^j(x_3, y_3; \phi) B_{x1}(\xi) d\xi \right\} \tag{4}$$

$$j = \hat{x} \text{ or } \hat{y}$$

where
$x_1 = x - (r \cos \theta + \hat{\xi} \cos \phi)$
$y_1 = \quad - (r \sin \theta + \hat{\xi} \sin \phi)$
$x_2 = (\hat{x} - \hat{\xi}) \cos \phi$
$y_2 = (\hat{x} - \hat{\xi}) \sin \phi$
$x_3 = \hat{x} \cos \phi + r \cos \theta - \xi$
$y_3 = \hat{x} \sin \phi + r \sin \theta$

the functions K_k^j are defined by equations (2), B_{ij} represents the dislocation density for crack segment j with Burger's vector in direction i, $\tilde{\sigma}_{iy}^j$ represents the traction in direction i on crack segment j due to the imposed stress, and in the crack's absence, and finally σ_{iy}^j represents the corresponding nett traction due to a combination of applied stresses and distributed dislocations. Note that in this formulation equations (3) and (4), and corresponding kernels (2) relate the stress components in local coordinates to Burger's components retained always in global coordinates. This short cut is acceptable providing that the cracks are always open [5]. The requirement that the crack faces be free of tractions is expressed by

$$\sigma_{iy}^1(x) = 0 \quad -a < x < a \quad i = x \text{ or } y \tag{5}$$

$$\sigma_{jy}^2(\hat{x}) = 0 \quad -b < \hat{x} < b \quad \hat{j} = \hat{x} \text{ or } \hat{y} \tag{6}$$

This gives four coupled integral equations in the unknown functions $B_{x1}(\xi)$, $B_{y1}(\xi)$, $B_{x2}(\hat{\xi})$, $B_{y2}(\hat{\xi})$. We are unable to invert these equations directly, and as a prelude to their numerical solution we normalize their ranges by setting

$$as = \xi \quad bu = \hat{\xi}$$

$$at = x \quad bv = \hat{x} \tag{7}$$

giving, in lieu of (3), (4)

$$\sigma_{iy}^1(t) = \tilde{\sigma}_{iy}^1(t) - \frac{2\mu}{\pi(\kappa + 1)} \left\{ a \int_{-1}^{+1} K_y^i (t-s,0;0)B_{y1}(s)ds + \right.$$

$$a \int_{-1}^{+1} K_x^i (t - s,0;0)B_{x1}(s)ds + b \int_{-1}^{+1} K_y^i (x_1',y_1';0)B_{y2}(u)du +$$

$$\left. b \int_{-1}^{+1} K_x^i (x_1',y_1';0)B_{x2} (u) \, du \right\} = 0 \tag{8}$$

$$i = x \text{ or } y \quad -1 \leq t \leq 1$$

$$\sigma_{iy}^{2^{\wedge}}(v) = \sigma_{iy}^{2^{\wedge}}(v) - \frac{2\mu}{\pi(\kappa + 1)} \left\{ b \int_{-1}^{+1} K_y^j (x_2', y_2'; \phi) B_{y2}(u)du + \right.$$

$$b \int_{-1}^{+1} K_x^j (x_2', y_2'; \phi) B_{x2}(u)du + a \int_{-1}^{+1} K_y^j (x_3', y_3'; 0) B_{y1}(s)ds +$$

$$a \int_{-1}^{+1} K_x^j (x_3', y_3'; 0) B_{x1}(s)ds \Bigg\} = 0 \qquad\qquad (9)$$

$$\hat{j} = \hat{x} \text{ or } \hat{y} \qquad -1 \leq v \leq 1$$

where
$$x_1' = t - ((r/a) \cos \theta + \frac{bu}{a} \cos \phi)$$

$$y_1' = \quad - ((r/a) \sin \theta + \frac{bu}{a} \sin \phi)$$

$$x_2' = \left(\frac{b}{a}\right)(v - u) \cos \phi$$

$$y_2' = \left(\frac{b}{a}\right)(v - u) \sin \phi$$

$$x_3' = \left(\frac{b}{a}\right)v \cos \phi + (r/a) \cos \theta - s$$

$$y_3' = (b/a) v \sin \phi + (r/a) \sin \theta$$

Note that the arguments of the functions K have all been normalised with respect to a.

To solve these equations we make use of the technique of Erdogan, Gupta and Cook [10]. Anticipating singularities at each end of each crack we replace the unknowns B_{ij} by the product of a fundamental function having the correct behaviour at each crack tip and unknown functions ϕ_{ij}, i.e.

$$\text{Let} \quad - \frac{2\mu}{\pi(K + 1)} B_{ij}(u) = \phi_{ij}(u) (1 - u^2)^{-1/2} \qquad \begin{array}{l} i = x,y \\ j = 1,2 \end{array} \qquad (10)$$

The discretized form of (8,9) are then [9]

$$\sum_{i=1}^{N} \frac{\pi}{N} K_y^i (t_k - s_i, 0; 0)\phi_{y1}(s_i) + \sum_{i=1}^{N} \frac{\pi}{N} K_x^i (t_k - s_i, 0; 0)\phi_{x1}(s_i) +$$

$$\left(\frac{b}{a}\right) \sum_{i=1}^{N} \frac{\pi}{N} K_y^i (x_1', y_1'; 0)\phi_{y2}(u_i) + \left(\frac{b}{a}\right) \sum_{i=1}^{N} \frac{\pi}{N} K_y^i (x_1', y_1'; 0)\phi_{x2}(u_i)$$

$$= -\sigma_{iy}' (t_k)/a \qquad\qquad i = x \text{ or } y \quad k = 1...n-1 \qquad (11)$$

$$\left(\frac{b}{a}\right)\sum_{i=1}^{N}\frac{\pi}{N}\ K_y^j\ (x_2',y_2';\phi)\phi_{y2}(u_i)\ +\ \left(\frac{b}{a}\right)\sum_{i=1}^{N}\frac{\pi}{N}\ K_x^j\ (x_2',y_2';\phi)\phi_{x2}(u_i)\ +$$

$$\sum_{i=1}^{N}\frac{\pi}{N}\ K_y^j\ (x_3',y_3';\phi)\phi_{y1}(s_i)\ +\ \left(\frac{b}{a}\right)\sum_{i=1}^{N}\frac{\pi}{N}\ K_x^j\ (x_3',y_3';\phi)\phi_{x1}(s_i)$$

$$= -\sigma_{jy}'(v_k)/a \qquad\qquad j= x\ or\ y \quad k=1\ldots n-1 \qquad (12)$$

where $\qquad s_i = u_i = \cos\left(\frac{2i-1}{2n}\ \pi\right)\ i= 1,2\ \ldots\ n$

$$t_k = v_k = \cos\left(\frac{k}{n}\ \pi\right) \qquad k= 1,2\ \ldots\ n-1$$

Equations (11, 12) represent 4n-4 equations in 4n unknowns. The four extra equations are realised by ensuring that over each crack the nett displacement between the faces in both x- and y- directions is zero, or

$$\int_{-a}^{+a} B_{i1}(\xi)d\xi = \int_{-b}^{b} B_{i2}(\hat{\xi})d\hat{\xi} = 0 \quad i= x,y \qquad\qquad (13)$$

In discretized form (13) becomes

$$\sum_{i=1}^{N}\phi_{ij}\ (u_i) = 0 \qquad \begin{array}{l} j= 1,2 \\ i= x,y \end{array} \qquad\qquad (14)$$

The crack-tip stress intensity factors are proportional to σ_{ij} (± 1), which may be found by Krenk interpolation [6,11].

Formulation - Kinked Crack

The geometry of a kinked crack, fig. 3 may readily be realised from the general two-crack case by setting

$$\tan\ \theta = \frac{(b/r)\ \sin\ \phi}{\sqrt{1-(b/r)^2\sin^2\phi}} \qquad\qquad (15)$$

and finding the more positive root of the equation

$$(r/a)^4 - 2(r/a)^2[1 + (b/a)^2] +1 +(b/a)^2[4\ \sin^2\phi\ -2] +(b/a)^4 = 0 \qquad (16)$$

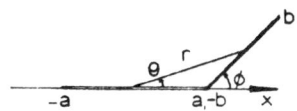

3. Kinked crack geometry.

More complex, however, is the question of the asymptotic behaviour at the kink, and the modification of side conditions (13). First, the nature of the stress state in the neighbourhood of the kink may be investigated by recourse to the classical paper of Williams [12], in which he examines the stress state near the apex of a wedge. Suppose that a wedge of included angle α is loaded by some arbitary far-field. Williams shows that in the neighbourhood of the apex, all stress components vary like $r^{\lambda-1}$ (but $\sigma_{\theta\theta}$ and $\tau_{r\theta}$ must vanish on the wedge faces), where λ is the minimum positive root of the eigen equation

$$\frac{\sin \lambda\alpha}{\lambda\alpha} = \pm \frac{\sin \alpha}{\alpha} \qquad (17)$$

Roots of this equation are displayed in fig.4. It may be seen that the value of λ is less than unity, and therefore the state of stress is singular, only if α exceeds 180°, and that the minimum value of λ (when $\alpha \rightarrow 360°$) is 1/2. This means that on the inside of the kink (region A, fig. 3) the state of stress is bounded, while outside the kink (region B, fig. 3) the stress state is singular. However, it is only very weakly so if ϕ is small and is strongest when $\phi \rightarrow -180$, so that the crack is almost folded back on itself and the kink approximates a crack tip, so that $\sigma_{ij} \sim r^{-1/2}$. Ideally, therefore, quadrature ought to be sought which incorporate an $r^{-1/2}$ singularity at one end and an $r^{\lambda-1}$ singularity at the other. There seems little hope of devising such a scheme. Suppose, however, that we write (as an example)

$$-\frac{2\mu}{\pi(\kappa + 1)} B_{ij}(u) = (1 + u)^{-1/2} (1 - u)^{1-\lambda} \phi(u) \qquad (18)$$

where the crack-tip, in this instance, lies at $u = -1$ and the kink at $u = 1$. Then, let

$$- \frac{2\mu}{\pi(\kappa + 1)}) \; B_{ij}(u) = (1 - u^2)^{-1/2} \; \psi(u) \tag{19}$$

where $\psi(u) = \phi(u)(1 - u)^{3/2-\lambda}$

From the asymptotics it is clear that $B_{ij} \to \infty$ at the kink, but $\psi(u)$ will remain finite. Therefore, one pair of conditions which may be imposed at the kink is that the "angle change" be preserved proceeding from one side of the kink to the other, i.e.

$$\psi_{11}(1) = \psi_{12}(-1) \quad i= x,y \tag{20}$$

The second side condition simply ensures that the kinked crack has no nett displacement from one end to the other, i.e.

$$\int_{-a}^{a} B_{i1}(\xi)d\xi + \int_{-b}^{b} B_{i2}(\hat{\xi})d\hat{\xi} = 0 \quad i= x,y \tag{21}$$

or in discretized form

$$\sum_{i=1}^{N} \frac{\pi}{N} (\phi_{11} + \frac{b}{a} \phi_{12}) = 0 \quad i= x,y \tag{22}$$

4. Real roots λ of eigen-equation (17)

Hence equations (14) for the crack-pair are replaced by (20,22) for the kinked crack. Note that it is far easier to implement the latter equations if the Burger's components for both crack segments are maintained in the same (global) direction.

RESULTS

Two Cracks

It is impossible to present comprehensive results with four independent parameters d/a, b/a, θ, φ, and so we illustrate the technique with two sample cases. First, we examine the case of two parallel cracks each of semi-length a in a uniform tensile· field, and whose perpendicular separation is also a. Stress intensity factors were found for a range of offsets between the cracks, and these are shown in fig. 5. The case when the two cracks are vertically above one another has been treated before [13], although only a K_I value is quoted, with which there is agreement. It is interesting to see how there is a shielding effect which reduces the mode I stress intensity until the horizontal separation is such that there is no overlap, and that the maximum mode II stress intensity is about 25% of the mode I value.

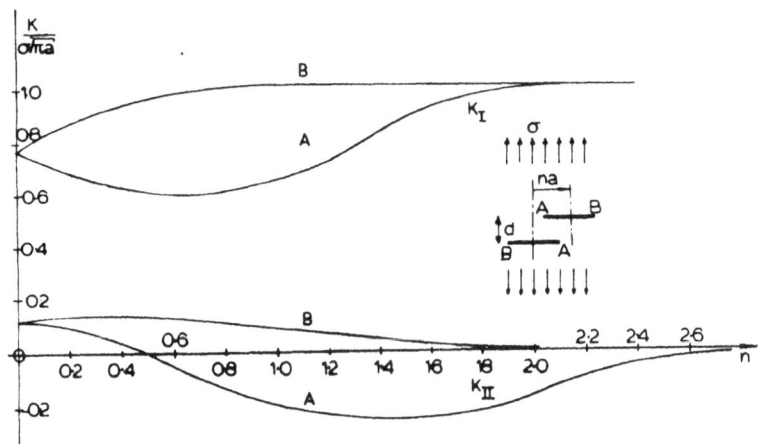

5. Stress intensity factors for a pair of equal parallel cracks with various degrees of offset under uniform tensile stress σ.

A second problem is shown in fig. 6. Here two cracks each of semi-length a have a centre-centre separation of 2a. One crack is perpendicular to the applied tension while the other is oriented at an angle ϕ. When the cracks are parallel behaviour is similar to the case cited above, although the greater separation means that shielding is less and hence K_I is higher whilst K_{II} is lower. As crack 2 (fig. 5) is rotated its crack tips suffer a monotonically decreasing mode I stress intensity whilst the magnitude of the mode II loading first increases, reaching a maximum between 30° and 50°, and then falls to zero. It is interesting to note that when crack 2 is perpendicular to the applied stress there is still a small residual tendency to open. This is, presumably, because of the disturbance to the homogeneous stress field caused by the presence of crack 1; it is not a numerical inaccuracy. It was found that in all cases convergence of the solution had occurred with

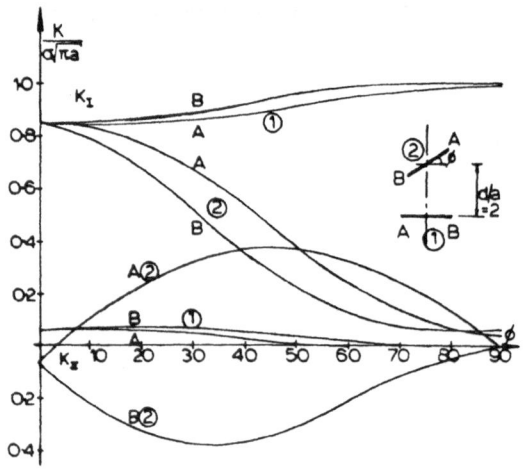

6. Stress intensity factors for two equal cracks, one perpendicular to a uniform tensile stress σ, the other oriented at some angle ϕ.

$N = 20$, and in many cases less would suffice. Inversion of the 80 x 80 matrix needed to solve the problem is therefore within the capability of many personal computers.

Kinked Cracks

The most important general result here concerns the implementation of side condition (20) to the problem. This is done by using Krenk interpolation, i.e. we find $\psi(\pm 1)$ from the following formulae

$$\psi(1) = \frac{1}{N} \sum_{i=1}^{N} \phi(u_i)\, \Lambda(i)\, g_1(i) \tag{23}$$

$$\psi(-1) = \frac{1}{N} \sum_{i=1}^{N} \phi(u_{N+1-i}) \ \Lambda(i) \ g_{-1}(i) \tag{24}$$

where $\Lambda(i) = \sin\left\{\frac{2N-1}{4N} (2i-1) \ \pi\right\}/\sin\left\{\frac{2i-1}{4N} \ \pi\right\}$

$g_1 = (1 - u_i)^{3/2-\lambda}$

$g_{-1} = (1 + u_i)^{3/2-\lambda}$

For any particular kink angle ϕ the value of λ may be found from figure 4 and substituted into equations (20,23,24). For example, when $\phi = 45°$, $\lambda = 0.674$. However, it turns out that if a wide range of values of λ are substituted into (23,24), although they have no physical relevance, the stress intensity factors are unchanged. In the present example, values of λ between 0 and 1 gave widely varying (but matched) values of $\psi_{i1}(1)$, $\psi_{i2}(-1)$, but stress intensity factors which changed only in the fifth significant figure. We conclude, therefore, that the refinement set out in the formulation is unnecessary: we assume that the crack segments show the usual square root singularities at each end and match not ψ at the kink, but ϕ. In other words we set the (fictitious) stress intensity factors at the kink to the same value.

Figures 7 and 8 depict stress intensity values for kinked cracks as a function of kink angle and kink length for representative cases. Note in fig. 7 that for the longer kink (b/a = 1.0) as $\phi \to 82°$ K_I at end B goes to zero, i.e. the crack tip closes and values of K for greater values of ϕ are invalid.

CONCLUSIONS

It has been shown that the technique of distributed dislocations for the solution of crack problems can be readily extended to the case of two adjacent cracks by formulating four coupled integral equations and solving them numerically. A single kinked crack can also be dealt with by this technique and the refinement of incorporating explicitly the correct stress singularly at the kink is shown to be unneccessary in the sample case of a crack in a uniform tensile field. The technique is simple and efficient to implement and this allows rapid computation of results for arbitrary stress fields on small computers.·

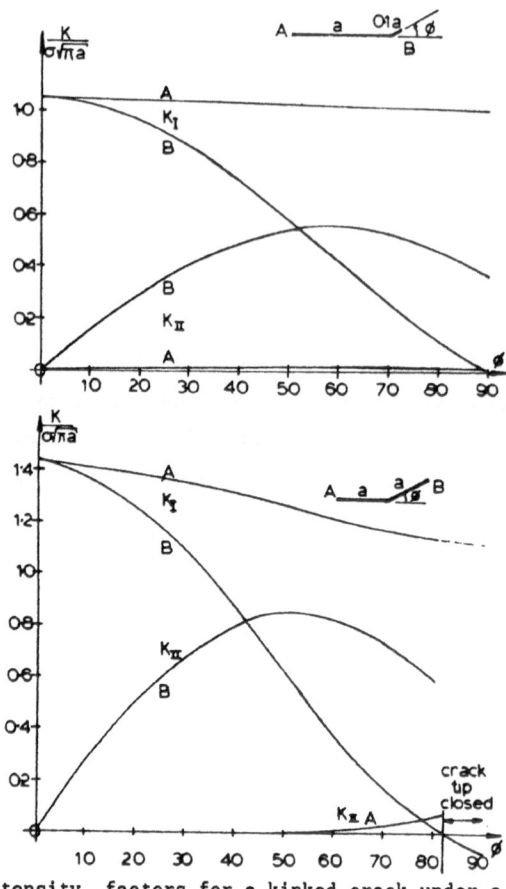

7. Stress intensity factors for a kinked crack under a uniform tensile stress σ. (a) b/a = 0.1 (b) b/a = 1.0

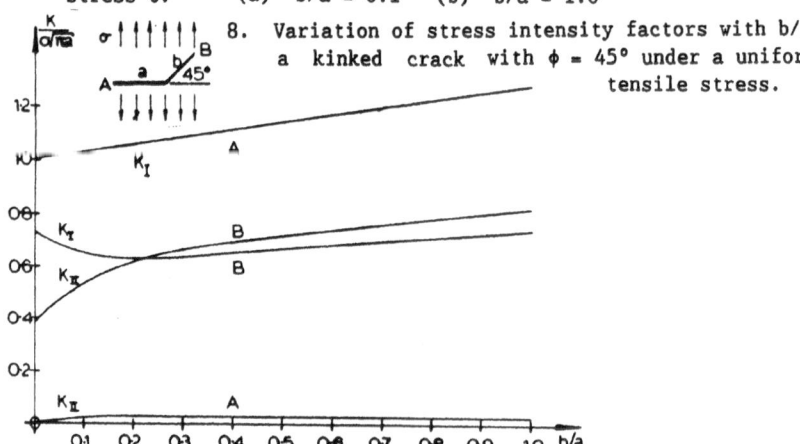

8. Variation of stress intensity factors with b/a for a kinked crack with φ = 45° under a uniform tensile stress.

References

1. Bilby, B.A., Cottrell, A.H. and Swinden, K.H. The spread of plastic yield from a notch. Proc. Roy. Soc. Ser. A., 1963, 272, 304-314.

2. Dundurs, J. and Mura, T. Interaction between an edge dislocation and a circular inclusion. J. Mech. Phys. Solids, 1964, 12, 177-189.

3. Dundurs, J. and Sendeckyi, G.P. Behaviour of an edge dislocation near a bimetallic interface. J. Appl. Phys, 1965, 36, 3353-3354.

4. Hills, D.A. and Comninou, M. An analysis of fretting fatigue cracks during the loading phase. Int. J. Solids Structures, 1985, 21, 721-730.

5. Hills, D.A. and Nowell, D. Stress intensity calibrations for closed cracks. J. Strain Anal., 1989, 24, 37-43.

6. Nowell, D. and Hills, D.A. Open cracks at or near free edges. J. Strain Anal., 1987, 22, 3, 177-185.

7. Li, Yingzhi and Hills, D.A. Stress intensity factor solutions for kinked surface cracks. J. Strain Anal. In press.

8. Dewynne, J., Hills, D.A. and Nowell, D. The opening displacement of surface-breaking plane cracks. Comp. Meth. in Appl. Mech. and Eng. (under review).

9. Dundurs, J. Elastic Interaction of Dislocations with Inhomogenities in Mathematical theory of Dislocations, Ed. T. Mura, ASME, New York, 1969.

10. Erdogan, F., Gupta, G.D. and Cook, T.S. Numerical solution of singular integral equations. In Methods of Analysis and Solutions of Crack Problems. (1973, Ed. G.C. Sih), Noordhoff, Leyden, pp. 368-425.

11. Krenk, S. On the use of interpolation polynomials for solutions of singular integral equations. Q. Appl. Maths. 1975, 32, 479-484.

12. Williams, M.L. Stress singularities resulting from various boundary conditions in angular corners of plates in extension. Jnl. Appl. Mech., 1952, 19, 526-528.

STRAIN BEHAVIOR NEAR FATIGUE CRACK TIP IN POLYMERS AND THEIR LIFE EVALUATION

AKIRA SHIMAMOTO
Saitama Institute of Technology, Okabe, Saitama 369-02, Japan

EISAKU UMEZAKI
Nippon Institute of Technology, Miyashiro, Saitama 345, Japan

FUMIO NOGATA
Himeji Institute of Technology, Syosha, Himeji 671-22, Japan

and
SUSUMU TAKAHASHI
Kanto Gakuin University, Mutuura, Yokohama 221, Japan

ABSTRACT

Polymers have been applied to various industrial fields and used as machine parts and structural members with the swift advance of industry, the usage conditions of polymers have been variously changed, and then the problems of fatigue fracture and of environmental fracture have been intensified. Therefore, the experiment for low cycle fatigue fracture of polycarbonate have been done in this research. Local strain in the vicinity of notch root and crack tip was measured in real time by using fine grid method. The relationships among local strain, crack initiation from notch and crack propagation of crack tip have been studied, and the method for the more precise life estimation was suggested in this paper.

INTRODUCTION

Recently, polymers with a high elastic modulus and strength are widely used as machine parts and structural members in cooperation with the rapid development of industry, and demands for scaling up of and making lighter machines and structures, and speeding up of working. As they are used under severer conditions, complex fracture phenomena are found in them. Therefore, understand of deformation and fracture phenomena characteristic to polymers is important for making better reliability of machines and structures, and preventing any accidents caused by their fracture. However, in comparison to metallic materials, there have been not many studies made on the fracture mechanics of polymers, especially on the relations among their deformation behavior and fracture, fracture by fatigue, and failure due to environmental causes.

Although initiation and propagation of fatigue cracks have a very important significance for the strength to fatigue fracture which accounts for most of fracture in polymers, the process of fatigue fracture is not well known yet because it occurs in the small zone of crack tip. Polymer, especially polycarbonate has a different fatigue fracture phase compared with that in metals at some levels of stress. To understand it, the relation between local strain behavior and fracture mechanism at the root of notch and the tip of crack has to be known.

In this study, local strain at the root of notch during process of crack initiation and local strain at the tip of crack during process of crack propagation were measured in real time under low cycle fatigue tests by using the fine grid method, the relations between local strain and crack initiation and crack propagation were investigated, and a life evaluation method for fatigue crack initiation was discussed.

SPECIMENS AND EXPERIMENTAL METHOD

Specimens used were polycarbonate (Lexan9030) on the market. Its mechanical properties are shown in Table 1, and dimensions of specimen, in Fig.1. Round notch with a depth of 1.5mm and a radius of 0.4mm was provided at its one side by machining. In order to measure local strain at the root of notch and the tip of crack, fine-dot grids (50.8μm pitch and 13μm diameter) were printed on the surface of specimen using photographic printing technique, as shown in Fig.2. The depth of photo-printed grids was approximately 2μm.

The specimens were naturally dried keeping them in a thermostatic room having conditions of temperature 20°C and humidity around 65% for sufficiently long period of time before use, and fatigue tests were conducted in the same room. Fatigue tests were carried out using a hydraulic

Table 1 Mechanical properties of polycarbonate.

Tensile strength	(MPa)	60.80
Elastic limit	(MPa)	40.21
Modulus of longitudinal elasticity	(MPa)	2099
Photoelastic sensitivity	(mm/N)	0.146

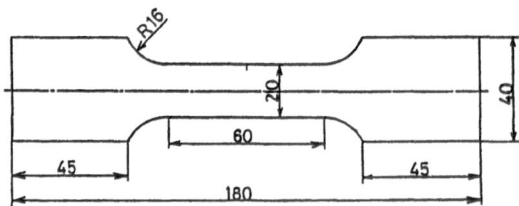

Figure 1. Shape and dimensions of specimen.

servo-controlled uniaxial-tension-compression type fatigue machine of 9807N capacity with a digital servo-controller under the condition of a constant load-amplitude at a stress ratio, $R = \sigma_{min}/\sigma_{max}$ of 0, 0.3, 0.6 and 0.8. Cyclic frequency was 0.02Hz with a sine waveform. In these fatigue tests, K-increasing method was used which implies increase in ΔK with crack propagation. The length of fatigue crack was measured on one side of specimen by using a X-Y stage with a accuracy of 0.001mm.

Photographs were taken of grids in the area of root of notch and local zone of crack tip using a camera equipped with automatic exposure and automatic film advancing functions through a relay lens and an optical microscope of 100 magnification, at regular intervals of the repetition.

Strain at the notch root and crack tip was obtained by measuring the deformation of grids on the negative film with reference to the grids in the pretested state through a enlarging profile projector with a photosensor. For strain calculation, the gauge length used was 152.4 μm (three grids long). Fig.2 shows the local region settled for measuring the notch-root and crack-tip strain.

The crack initiation cycle (Nc) was determined by two methods. One was the confirmation of initiated small crack propagation (crack length=5-10μm) from the notch root by microscopic observation, and the other used the crooked point on the curve of maximum strain (ε_{max}) at the first-line grid from the notch root vs number of cycles, as shown in Fig.4.

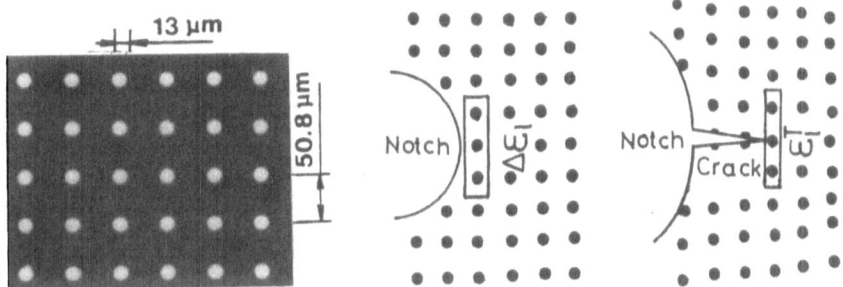

Figure 2. Fine-dot grid and schematic view of local strain measurement area.

RESULTS AND DISCUSSIONS

Fatigue crack initiation from notch root

Low cycle fatigue tests was carried out at a nominal maximum stress, σ_{max} of 32.86MPa and a nominal minimum stress, σ_{min} of 0(R=0), 9.858MPa(R=0.3), 19.716MPa(R=0.6) and 26.288MPa(R=0.8) in the parallel part of specimen. Presented in Figs.3 and 4 are the relationships between crack length (a)(i.e. notch depth was replaced by crack length) and number of load cycles (N), and local strain range ($\Delta\varepsilon_l$) at the notch root and number of load cycles (N) during the process of crack initiation, respectively. As shown in Fig.3, increasing stress ratio (R) caused an increase in fatigue crack initiation cycle (Nc), and decreased the slope of curve of crack length (a) to number of load cycles (N) during crack propagation. However, regardless of stress ratio (R), for example, as shown in Fig.4, $\Delta\varepsilon_l$ decreased at once after starting fatigue test, became about a constant after the decrease, and increased gradually with an increase in the number of load cycles. When $\Delta\varepsilon_l$ was approximately consistent with its initial value, crack initiation occurred. It was also found that only at the local zone at which crack initiated, local strain range ($\Delta\varepsilon_l$) increased, and there was not very much change in local strain at adjacent zones of the crack. From this fact, it was confirmed that fatigue crack propagation occurred at an extreme local zone[1]-[3].

For metals, fatigue crack initiation was controlled by local-strain damage accumulation was indicated, and for a quantitative expression of cumulative fatigue damage, mean local strain range ($\Delta\overline{\varepsilon}_l$) was proposed as follows[4].

$$\Delta\overline{\varepsilon}_l = \frac{1}{Nc} \int_0^{Nc} \Delta\varepsilon_l \, dN \qquad (1)$$

Then, the relationships between mean local strain range ($\Delta\overline{\varepsilon}_l$) and crack

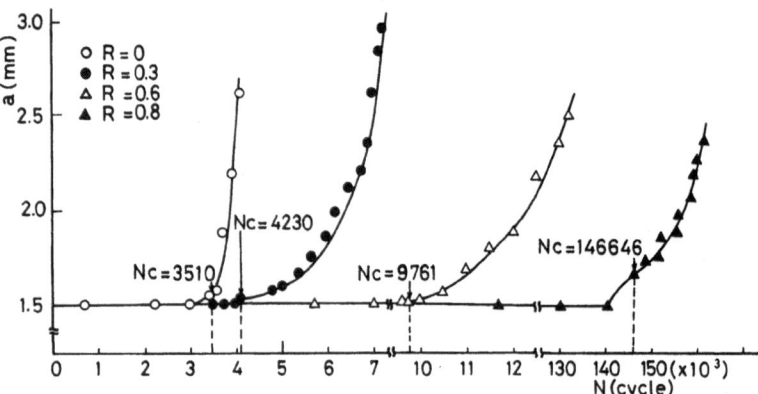

Figure 3. Relation between crack length (a) and number of load cycles (N).

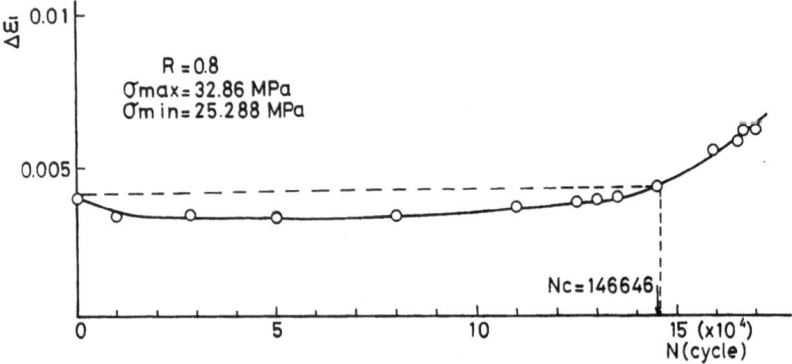

Figure 4. Relation between local strain range at the notch root ($\Delta\varepsilon_l$) and number of load cycles (N).

initiation cycle (Nc) at different stress ratios (R) are shown in Fig.5. Each data point was arranged on a certain line very well whose slope was -0.38 in a log-log coordinate graph, regardless of the values of stress ratio (R). Therefore, the linear cumulative damage law[5] based on local strain range was almost confirmed for fatigue crack initiation in polycarbonate. In Fig.6, mean local strain range ($\Delta\overline{\varepsilon}_l$) is plotted against stress ratio (R). Each data point was arranged on a certain line very well whose slope was -0.22 in a semi-logarithmic graph. This fact also showed that mean local strain range ($\Delta\overline{\varepsilon}_l$) was not affected by stress ratio (R).

Evaluation of crack initiation Life

For the investigation of the crack initiation life by parameter $\Delta\varepsilon_{N_c}$(local strain range) at crack initiation cycle (Nc), it was attempted statiscally

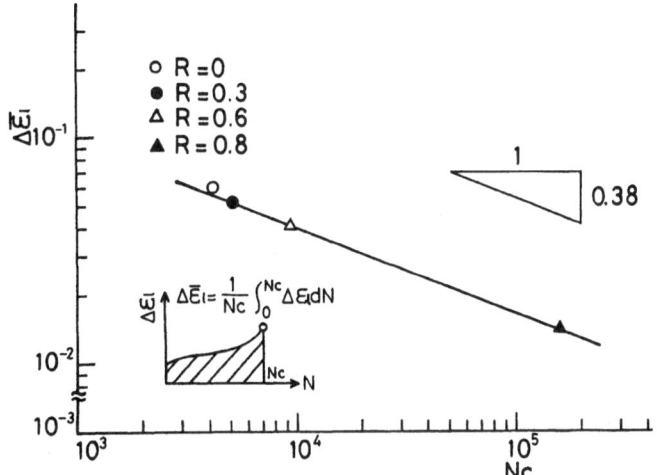

Figure 5. Relation between mean local strain range ($\Delta\overline{\varepsilon}_l$) and number of crack initiation cycles (Nc).

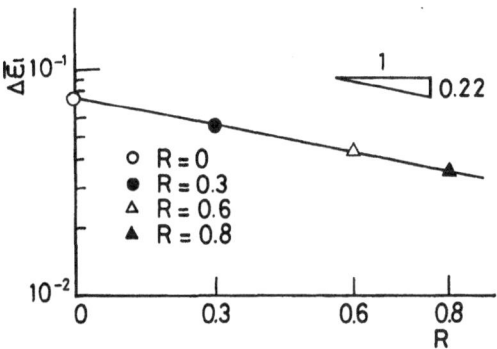

Figure 6. Relation between mean local strain range ($\Delta\overline{\varepsilon}_l$) and stress ratio (R).

to standardize the curves showing the relationship between $\Delta\varepsilon_t$ and N, for example, such as Fig.4 into one approximated curve and to present it by an equation using only two dimensionless parameters which can be expressed as

$$y(x)=(\Delta\varepsilon_N-\Delta\varepsilon_{Nc})/\Delta\varepsilon_{Nc}$$
$$x=N/Nc$$

where $\Delta\varepsilon_N$ is the value of local strain range at arbitrary cycle number (N) and $\Delta\varepsilon_{Nc}$ one at crack initiation cycle number (Nc). The standardized approximate curve shown in Fig.7 can be expressed as

$$y(x)=313.41\times10^{-4}x^3-47.75\times10^{-2}x^2+0.74x-0.03 \tag{2}$$

Using the three local strain range ($\Delta\varepsilon_{N_1}$, $\Delta\varepsilon_{N_2}$ and $\Delta\varepsilon_{N_3}$) at the notch root measured for the three cycle numbers (N1, N2 and N3) in the fatigue

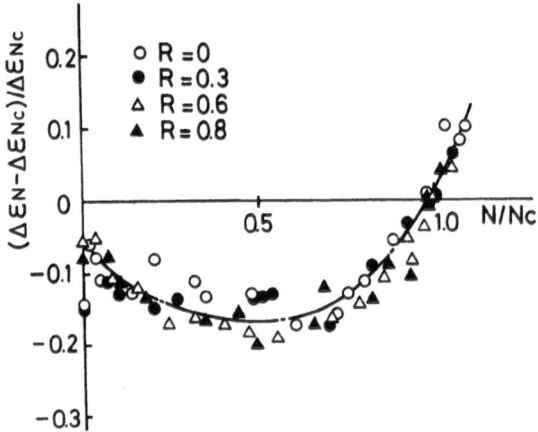

Figure 7. Approximate curve standardized with two dimensionless parameters.

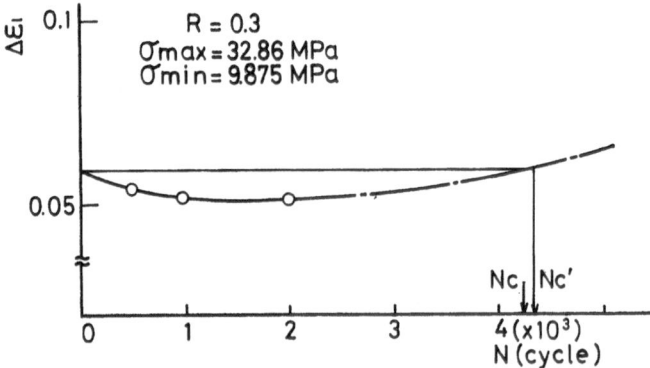

Figure 8. Relation between local strain range at the notch root ($\Delta\varepsilon_t$) and number of load cycles (N) calculated.

test, the crack initiation cycle (Nc) is determined from the equation[6],[7]

$$y(N1/Nc)-y(N2/Nc)= \beta [y(N2/Nc)-y(N3/Nc)] \tag{3}$$

where

$$\beta =(\Delta\varepsilon_{N_1}- \Delta\varepsilon_{N_2})/(\Delta\varepsilon_{N_2}- \Delta\varepsilon_{N_3})$$

The correction between local strain range ($\Delta\varepsilon_l$) and cyclic number (N) at R=0.3 estimated by this evaluation method is shown in Fig.8. In this figure, the solid line represents the approximate curve obtained experimentally, with three cycle numbers used in equation (3). The crack initiation cycle (Nc') estimated from equation (3) are in agreement with the observed crack initiation point (Nc). In consequence, the validity of this method for estimating fatigue crack initiation life was confirmed.

Relation of crack propagation rate to stress intensity range and local strain at crack tip

The relation between crack propagation rate (da/dN) and stress intensity range (ΔK) used generally for small fatigue crack initiated from the notch root is presented at different stress ratios in Fig.9. As shown in Fig.9, ΔK was about a constant, though da/dN increased in the vicinity of

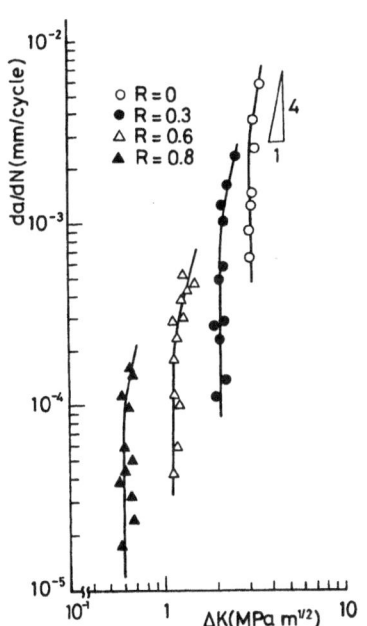
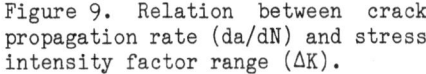
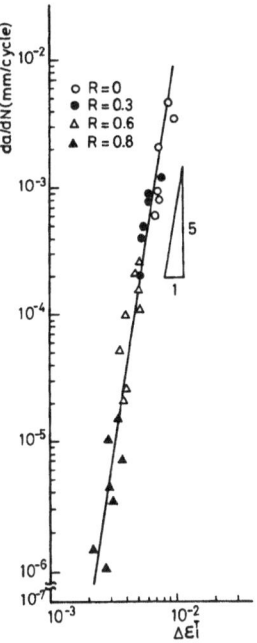

Figure 9. Relation between crack propagation rate (da/dN) and stress intensity factor range (ΔK).

Figure 10. Relation between crack propagation rate (da/dN) and local strain range at the crack tip ($\Delta\varepsilon_l^T$).

notch, regardless of the values of stress ratio (R). Therefore, initiated fatigue crack propagation behaviors were thought to be strongly affected by the states of plastic deformations near the notch root. However, the relation between da/dN and ΔK at the range of da/dN values from 1×10^{-4} to 1×10^{-2} mm/cycle was expressed by a certain line whose slope was 4 in a log-log coordinate graph, showing Paris' 4th power law in polymer. Therefore, it was confirmed that da/dN for low cycle fatigue can not be expressed linearly by ΔK, in other words, both small and large crack propagation behaviors for low cycle fatigue can not be evaluated uniformly by ΔK. This is because the small yielding condition used in linear fracture mechanics is not almost satisfied as small crack propagates with the local plastic zone due to fatigue crack containing in the plastic zone formed by a stress concentration at the notch root[8].

In Fig.10, the relation between crack propagation rate (da/dN) and local strain range at the crack tip ($\Delta \varepsilon_l^T (= \varepsilon_{max} - \varepsilon_{min})$) is presented. As shown in Fig.10, each data point was arranged on a line very well whose slope was 5 in a log-log coordinate graph. They did not show dependence on applied loads as has been experienced with plotting against ΔK, showing effectiveness thereof in the evaluation of crack propagation rate.

CONCLUSIONS

Fatigue crack initiation from the notch root and fatigue crack propagation in polycarbonate, a polymer were investigated varying the stress ratio (R) using the real-time fine grid method. Resultant findings were as follow.
(1) Local strain range ($\Delta \varepsilon_l$) at the notch root decreased at once after starting fatigue, and increased with an increase in the number of load cycles, crack initiation occurred when $\Delta \varepsilon_l$ is approximately consistent with its initial value.
(2) The relation between mean local strain range ($\Delta \overline{\varepsilon}_l$) and crack initiation cycle number (Nc) was expressed by a line whose slope was -0.38 in a logarithmic graph, and local-strain damage accumulation law in the fatigue crack initiation stage of polymers was effective, too.
(3) The relation between $\Delta \varepsilon_l$ and N was statistically standardized with two dimensionless parameters, $(\Delta \varepsilon_N - \Delta \varepsilon_{Nc}) / \Delta \varepsilon_{Nc}$ and N/Nc, and the standardized curve was used to estimate the crack initiation cycle number (Nc). In consequence, the method was effective.
(4) The relation between crack propagation rate (da/dN) and stress intensity range (ΔK) at the range of da/dN values from 1×10^{-4} to 1×10^{-2} mm/cycle was expressed by a line whose slop was 4 in a logarithmic graph, and depended on applied loads and stress ratios (R).
(5) The relation between crack propagation rate (da/dN) and local strain range at the crack tip ($\Delta \varepsilon_l^T$) was expressed by a line whose slop was 5 in a logarithmic graph without showing dependency on applied loads and stress ratios (R).

REFERENCES

[1]Shimamoto, A., Yokota, A. and Takahashi, S., Analysis on the crack initiation and strain in the vicinity of notch root under the low cycle by fine grid method (in Japanese), J. of Japan NDI, 1988, 37-9A,849-850.
[2]Shimamoto, A., Umezaki, E. and Takahashi, S., A fundamental study on rupture by low-cycle fatigue of polymers employing the fine-grid method (in Japanese), J. of Japan NDI, 1989, 38(12), 1101-1106.

[3]Sato, T., Shimada, H. and Furuya, Y., Relation between local-strain and fatigue cyclic number of crack initiation from notch-root under low cycle fatigue (in Japanese), Trans. of Japan SME, 1985, 51(466), A, 1534-1540.

[4]Shimada, H. and Furuya, Y., Local crack-tip strain concept for fatigue crack initiation and propagation, Trans. of ASME, Engng Mater. Tech., 1987, 109(4), 101-106.

[5]Kikukawa, M., Damage accumulation in fatigue process (in Japanese), Preprint of 15th Symposium on Material Strength and Fracture, 1970, 71-78.

[6]Sato, T. and Shimada, H., Evaluation of fatigue crack initiation life from a notch, Int. J. of Fracture, 1988, 10(4), 243-247.

[7]Sato, T., Kimura, K. and Iguchi, H., Local strain behavior at the notch-root and the crack-tip on corrosion fatigue (in Japanese), Preprint of 22nd Symposium on Stress and Strain Measurement, 1990, 143-148.

[8]Hudak, S. J., Small crack behavior and the prediction of fatigue life, Trans. of ASME, Engng Mater. Tech., 1981, 103(1), 26-35.

REVIEW OF SOME DEVELOPMENT OF THE HOLE DRILLING METHOD

AUGUSTO AJOVALASIT

University of Palermo, Istituto di Costruzione di Macchine,
Viale delle Scienze, 90128 Palermo, Italy

ABSTRACT

This paper contains a survey of some developments of the hole drilling method. It is mainly based on the research work carried out at the University of Palermo in the years from 1978 to 1989. The paper considers the relationship between the relaxed strain and the residual stresses for a rosette with an off-centre hole, the influence of hole eccentricity on the determination of residual stresses and of rosette calibration constants, the sensitivity of strain relaxation.

INTRODUCTION

The hole drilling method is based on: (1) the application of a strain gauge rosette on the surface of the component; (2) the drilling of a hole at the rosette centre; (3) the measurement of the relaxed strains; (4) the calculation of residual stresses using the relationships between the measured strains and the residual stresses. The hole drilling method |1-2| is the most used semidestructive mechanical method for surface stress analysis. A large number of rosettes especially developed for such a method is available.

This paper contains a survey of some developments of the hole drilling method. It is mainly based on the research work carried out at the University of Palermo in the years from 1978 to 1989.

It presents: (1) the relationship between the strain relaxed in the presence of hole eccentricity and the residual stresses; (2) the influence of hole eccentricity on the determination of residual stresses and of rosette calibration constants; (3) the sensitivity of strain relaxation.

NOTATION

A°,B°,C°	Calibration constants of rosette with off-centre hole
A,B,C,	Theoretical values of A°,B° and C° integrated along the grid length
Ao, Bo	Calibration constants of rosette with concentric hole
Ae,Be	Numerical values of Ao and Bo
As,Bs	Experimental values of Ao and Bo
A's,B's	Wrong values of As and Bs
At,Bt	Theoretical values of Ao and Bo integrated along the grid length
A"t,B"t	Theoretical values of Ao and Bo integrated over the grid area
a	Radius of hole
E	Young's modulus
E_γ	Error relative to angle γ
$E_{\sigma_1}, E_{\sigma_2}$	Errors relative to stresses σ_1 and σ_2
e	Eccentricity between hole and rosette centres
r	Distance from rosette centre to gauge centre
r_1, r_2	Distances from rosette centre to gauge ends
$\alpha_a, \alpha_b, \alpha_c$	Angles measured from σ_1 to axes of gauges a,b and c in counter-clockwise direction
β	Eccentricity angle measured from gauge a axis (x axis) in counterclockwise direction.
ϵ	Strain measured by a gauge of the rosette with concentric hole
ϵ'	Strain measured by a gauge of the rosette with off-centre hole
γ	Angle measured from gauge a axis to σ_1 in counterclokwise direction
γ'	Wrong value of γ
ν	Poisson's ratio
σ_1, σ_2	Principal stresses $(\sigma_1 \geq \sigma_2)$
σ_1', σ_2'	Wrong values of σ_1 and σ_2
Subscripts	
a,b,c,	Rosette strain gauges
i	a, b or c

SUMMARY OF THE CENTRE-HOLE METHOD

For the centre-hole method the relationship between the relaxed strain and the residual stresses is given by the well known formula:

$$\epsilon_i = (A_o/E)(\sigma_1+\sigma_2)+(B_o/E)(\sigma_1-\sigma_2)\cos2\alpha_i, \quad (i=a,b,c) \tag{1}$$

where the calibration constants Ao and Bo are determined theoretically, experimentally or numerically. The theoretical approach is strictly valid only for a through hole. Expressions for Ao and Bo taking account of the grid length (At, Bt) and also of the grid width (A"t, B"t) are available. For the blind hole the experimental constants (As, Bs) or the numerical

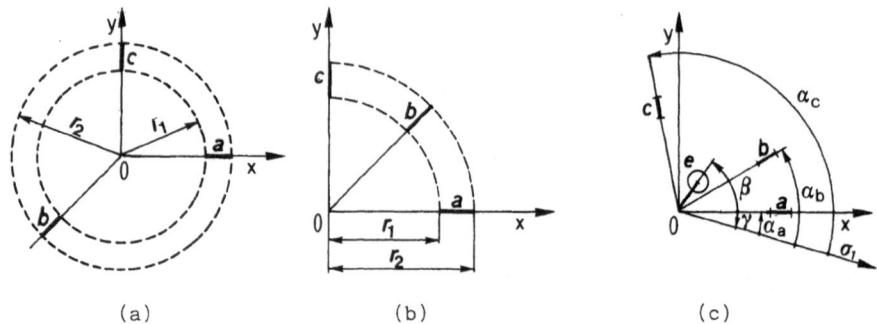

Fig.1 - Notation for the rectangular rosettes with gauges at 0°/225°/90° (a), at 0°/45°/90° (b) and for the hole eccentricity (c).

ones (Ae, Be) are preferred.

For a rectangular rosette having the gauges a, b and c at 0°,45° (or 225°) and 90° (Fig.1-a-b) the solution of equations (1) gives:

$$\sigma_{1,2} = (E/4A_o)(\varepsilon_a + \varepsilon_c) \mp (E/4B_o) \sqrt{(\varepsilon_c - \varepsilon_a)^2 + (\varepsilon_a - 2\varepsilon_b + \varepsilon_c)^2} \qquad (2)$$

$$\tan 2\gamma = (\varepsilon_a - 2\varepsilon_b + \varepsilon_c)/(\varepsilon_c - \varepsilon_a) \qquad (3)$$

where the minus and plus signs are for the maximum (σ_1) and minimum (σ_2) principal stresses.

THEORY OF THE ECCENTRIC HOLE METHOD

The misalignment between the hole and rosette centres significantly influences the strains measured by the gauges especially when small rosettes are used. Therefore the formula (1) developed for the centre-hole method becomes inaccurate. The influence of hole eccentricity has been considered by many investigators. Corrections have been proposed based on approximative methods |3|, on iterative numerical methods |4|, on formulae valid in particular cases |5|, on general formulae for rectangular |6-11| and equiangular rosettes |12|. The use of rosettes having five or six gauges has also been proposed |13|.

For a rosette with an off-centre hole the relationship between the strain measured by the gauge and the residual stresses can be expressed |6| by the formula:

$$\varepsilon_i' = A_i^o(\sigma_1 + \sigma_2)/E + (B_i^o \cos 2\alpha_i + C_i^o \sin 2\alpha_i)(\sigma_1 - \sigma_2)/E, \quad (i=a,b,c) \qquad (4)$$

where the coefficients $A^o i$, $B^o i$ and $C^o i$ depend not only on hole radius and grid geometry (as Ao and Bo) but also on gauge orientation and hole eccentricity parameters (e, β - fig. 1-c).

Theoretical expressions of A°i, B°i and C°i, which take account of gauge length, have been determined |6|. These values - named Ai, Bi and Ci - reduce to At, Bt and 0 if the hole eccentricity is zero. Theoretical values averaged over the grid area have been not available for the through hole till now. Of course theoretical values for the blind hole are not avalaible as well. In the lack of such values the following expressions have been proposed |11|:

$$A°i = Ai \ Ao/At \qquad (5)$$
$$B°i = Bi \ Bo/Bt \qquad (6)$$
$$C°i = Ci \ Bo/Bt \qquad (7)$$

where Ao and Bo are the centre hole method calibration constants.

Determination of Residual Stresses

For a rectangular rosette the solution of equations (4) gives the magnitude and direction of the principal stresses |7|:

$$\sigma_{1,2} = \frac{E}{2K}(K_7 \epsilon'_a + K_8 \epsilon'_b + K_9 \epsilon'_c) + \frac{E}{2} \sqrt{\frac{(K_1 \epsilon'_a - K_2 \epsilon'_b + K_3 \epsilon'_c)^2 + (K_4 \epsilon'_c - K_5 \epsilon'_a - K_6 \epsilon'_b)^2}{K^2}} \qquad (8)$$

$$\tan 2\gamma = (K_1 \epsilon'_a - K_2 \epsilon'_b + K_3 \epsilon'_c)/(K_4 \epsilon'_c - K_5 \epsilon'_a - K_6 \epsilon'_b) \qquad (9)$$

where the constants K, K_1 -K_9 depending on Ai, Bi, Ci are reported elsewhere |6,7|.

Instead of using the relationships (8) and (9), the solution of the off-centre rosette can be reduced to that one of a centred-hole rosette. To this end the 'standard' strains - i.e. corrected for the hole eccentricity - are calculated from the 'measured' strains by using the following relationships |6|:

$$\epsilon_a = J_{aa} \epsilon'_a + J_{ab} \epsilon'_b + J_{ac} \epsilon'_c \qquad (10)$$

$$\epsilon_b = J_{ba} \epsilon'_a + J_{bb} \epsilon'_b + J_{bc} \epsilon'_c \qquad (11)$$

$$\epsilon_c = J_{ca} \epsilon'_a + J_{cb} \epsilon'_b + J_{cc} \epsilon'_c \qquad (12)$$

where the coefficients Jij (i,j = a,b,c) depend on Ai, Bi, Ci |6|.

The residual stresses are easily calculated by substituting the standard strains in equations (2) and (3) of the concentric rosette. The previous formulae allow the correct determination of the residual stresses provided that the eccentricity parameters e and β are measured.

Influence of Hole Eccentricity on Residual Stresses

If the misalignment between the hole and rosette centres is neglected,

TABLE 1
Geometry of some strain gauge rosettes

Rosette	Type	r mm	r_1/r	r_2/r	d/r	2 a* mm min	max
MM:EA-XX-031RE-120	0°/225°/90°	1,28	0,69	1,31	0,62	0,75	1
MM:EA-XX-062RE-120	0°/225°/90°	2,57	0,69	1,31	0,62	1,5	2
MM:TEA-XX-062RK-120	0°/225°/90°	2,57	0,69	1,31	0,62	1,5	2
MM:CEA-XX-062UM-120	0°/45°/90°	2,57	0,69	1,31	-	1,5	2
MM:EA-XX-125RE-120	0°/225°/90°	5,13	0,69	1,31	0,62	3	4
HBM:RY21 3/120	0°/225°/90°	6,5	0,77	1,23	0,29	5	6
HBM:RY61 1,5/120	0°/225°/90°	2,55	0,71	1,29	0,31	1,5	1,5
BLH:FAER-03S-12-SXE	0°/45°/90°	1,77	0,77	1,23	0,57	1,5	1,75
TML:FRS-2	0°/225°/90°	2,57	0,71	1,29	0,51	1,5	2
TML:FRS-3	0°/225°/90°	5,13	0,71	1,29	0,51	3	4

*indicative values

the stresses and their directions are affected by the following errors:

$$E\sigma_1 = 100(\sigma_1'-\sigma_1)/\sigma_1 \qquad (13)$$

$$E\sigma_2 = 100(\sigma_2'-\sigma_2)/\sigma_2 \qquad (14)$$

$$E\gamma = \gamma' - \gamma \qquad (15)$$

Such errors depend on the stress field (σ_2/σ_1, γ), the hole gauge geometry (a/r_1, a/r_2), the eccentricity parameters (e,β) and the Poisson's ratio. The numerical results, which are shown in the following, refer |6,9| to the rosettes of table 1 and are calculated for ν =0.3.

Errors in the Worst Conditions of β and γ

Only the error $E\sigma$ relative to the maximum (absolute value) stress is considered.The positive error $E\sigma$ can be evaluated by using the following approximate formulae:

$$\text{for } -1 \leq \sigma_2/\sigma_1 \leq 1, \quad E\sigma = E\sigma_1 = (G_1-G_2\sigma_2/\sigma_1)e'\% \qquad (16)$$

$$\text{for } \sigma_2/\sigma_1 > 1 \text{ or } < -1, \quad E\sigma = E\sigma_2 = (G_1-G_2\sigma_1/\sigma_2)e'\% \qquad (17)$$

where e'% = 100 e/r₁ .

The coefficients G_1 and G_2 depend on r_2/r_1 , e'% and a/r (or a/r_1) which according to the ASTM standard |1| should be in the range 0.29 ÷ 0.4. Suitable expressions of coefficients G_1 and G_2 are |9|:

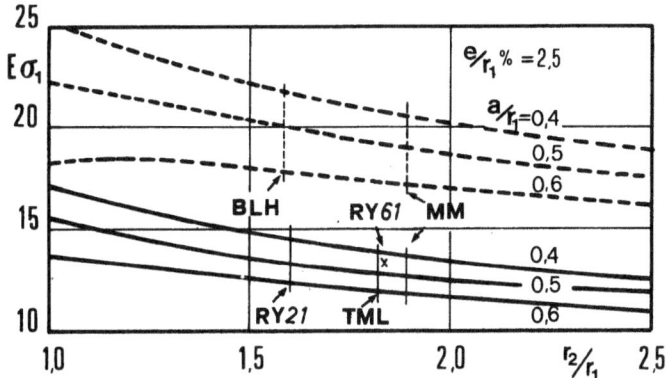

Fig.2 - Stress error E_{σ_1} in the worst conditions of σ_2/σ_1: ——— rosettes
0°/225°/90°, ------ rosettes 0°/45°/90°.

$$G_1 = p_1 + q_1 \ r_1/a \qquad (18)$$

$$G_2 = p_2 + q_2 \ r_1/a \qquad (19)$$

Table 2 shows the constants p_1, q_1, p_2 and q_2, relative to the rosettes of
table 1, valid for e'% = 0 ÷ 10 and for a/r_1 = 0.3 ÷ 0.6 which includes the
range given by the ASTM standard.
Fig.2 |9| shows the error E_{σ_1} as a function of r_2/r_1 for σ_2/σ_1 =-1, e/r_1 % =
2.5 and various values of a/r_1.

For σ_2/σ_1 = - 1 and a=a min   the stress error relative to the rosettes
of table 1 results from equations (16),(18),(19) and table 2 as follows:

- for rosettes 0°/225°/90° (fig. 1-a) $E\sigma_1 \simeq 5.6 \ (e/r_1)\%$ (20)

- for rosettes 0°/45°/90° (fig.1-b) $E\sigma_1 \simeq 8.1 \ (e/r_1)\%$ (21)

TABLE 2
Constants for the calculus of stress error due to hole eccentricity

ROSETTE	TYPE	r_2/r_1	p_1	q_1	p_2	q_2
MM: MODELS EA,TEA	0°/225°/90°	1,89	3,22	0,24	1,04	0,32
RY 61	0°/225°/90°	1,83	3,30	0,22	1,04	0,32
TML	0°/225°/90°	1,82	3,30	0,23	1,04	0,33
RY 21	0°/225°/90°	1,6	3,41	0,25	0,99	0,37
MM: MODEL CEA	0°/45°/90°	1,89	3,27	0,54	2,50	0,49
BLH	0°/45°/90°	1,58	3,35	0,62	2,53	0,57

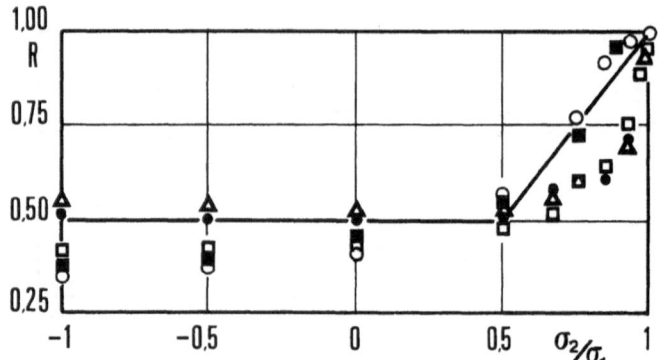

Fig. 3 - Stress error reduction R as a function of σ_2/σ_1 for rosettes having the orthogonal gauges along the principal directions (● MM:EA,TEA; ■ MM,CEA; □ HBM,RY21; ▲ HBM,RY61,TML; ○ BLH)

The previous formulae give the upper limit of stress error, i.e. the error in the worst conditions of eccentricity orientation β , stress orientation γ and stress ratio σ_2/σ_1 . They also show that: (1) 'single quadrant rosettes' (0°/45°/90°) are less favorable than 'two quadrant rosettes' (0°/225°/90°); (2) the miniature rosettes (low values of r_1) are more sensitive to eccentricity errors than the larger ones.

<u>Errors in the Worst Condition of β for γ=0°</u>
    If the directions of the principal stresses are known the stress error due to the misalignment between the hole and rosette centres can be reduced by mounting the rosette with its orthogonal gauges (a and c) along the principal stress directions.

    Fig. 3 shows, as a function of σ_2 / σ_1 , the ratio R defined by:

$$R = (E\sigma_1)_{\gamma=0}/E\sigma_1 \qquad (22)$$

where the numerator is the stress error for a rosette having the orthogonal gauges along the principal directions (γ =0° or 90°) and $E\sigma_1$ is the error in the worst condition (usually γ= 135°). Clearly the advantage of such configuration decreases as the ratio σ_2/σ_1 approaches to 1. As a practical rule Fig. 3 suggests the following values of R for all the rosettes of table 1:

$$\text{for } -1 \leq \sigma_2/\sigma_1 \leq 0.5, \qquad R = 0.5 \qquad (23)$$

$$\text{for } 0.5 \leq \sigma_2/\sigma_1 \leq 1, \qquad R = \sigma_2/\sigma_1 \qquad (24)$$

<u>Influence of Hole Eccentricity on Calibration Constants</u>
    The constants Ao and Bo, which appear in equation (1) of the concentric

TABLE 3

Experimental results of rosettes with eccentric holes

ROSETTE		1				2			
(1) Hole 2a diameter (mm)		1.52*	1.54	1.75	2.01	1.54	1.75	2.01	2a
(2) Hole e (mm) eccentr.		0.19	0.19	0.19	0.19	0.26	0.25	0.28	e
(3)	β	53°	53°	54°	54°	93°	99°	99°	β
(4) Applied	σ_1	163				163			σ_1
(5) stress	σ_2	0				0			σ_2
(6) (N/mm²)	γ	135°				90°			γ
(7) Measured	$\varepsilon'a$	−89	−82	−103	−136	55	69	83	$\varepsilon'a$
(8) strains	$\varepsilon'b$	61	61	77	96	−53	−71	−95	$\varepsilon'b$
(9) (μm/m)	$\varepsilon'c$	−87	−79	−101	−134	−181	−228	−289	$\varepsilon'c$
(10) Standard	εa	−55	−50	−62	−85	64	79	96	εa
(11) strains	εb	69	69	86	106	−35	−49	−66	εb
(12) (μ m/m)	εc	−56	−50	−65	−88	−148	−188	−237	εc
(13) Wrong	σ_1'	243	247	245	249	−0.0809	−0.1021	−0.1323	A's
(14) stresses	σ_2'	17	41	37	35	−0.1516	−0.1908	−0.2390	B's
(15) (N/mm²)	γ'	134.8°	134.7°	134.8°	134.9°	−0.0534	−0.0697	−0.0904	As
(16) Corrected	σ_1	176	175	175	180	−0.1361	−0.1718	−0.2139	Bs
(17) stresses	σ_2	−12	2.3	0.8	2.3	−0.0570	−0.0739	−0.0966	A"t
(18) (N/mm²)	γ	134.7°	135.0°	134.8°	134.8°	−0.1396	−0.1754	−0.2198	B"t

* blind hole

hole rosette are in some instances determined by experimental calibration |1| through measurement of the strains relaxed in a known stress field. In general a test bar under uniaxial loading ($\sigma_2 = 0$) is used. If the rosette is cemented to the test bar with gauge c axis directed along the applied stress σ_1, equation (1) gives:

$$A_s = (E/2\sigma_1)(\varepsilon_a + \varepsilon_c) \qquad (25)$$

$$B_s = (E/2\sigma_1)(\varepsilon_c - \varepsilon_a) \qquad (26)$$

A misalignment between the hole and rosette centres, during the calibration, influences the strains measured by the rosette and therefore gives wrong values for the constants As and Bs. The effect of such misalignment can be eliminated |7,11| by substituting in the calibration formulae (25) − (26) the standard strains, given by equations (10) − (12), instead of the measured strains.

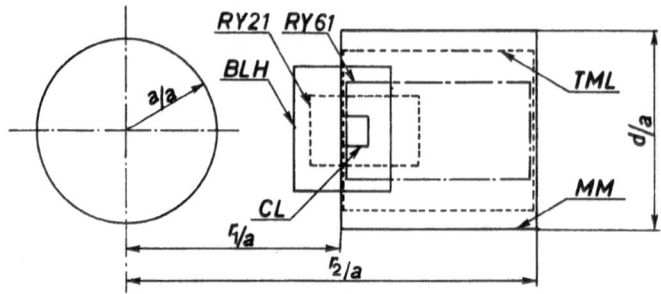

Fig.4 - Hole-gauge geometry for the rosettes of table 1 (CL indicates the MM gauge type 008CL)

Experimental Results

A steel bar under uniaxial loading was used for the experiments |11|. Precautions were taken according to the ASTM standard |1| to avoid end effects, plastic flow around the hole and bending effects. To avoid the interaction between adjacent holes the rosettes (MM: EAA-06-062RE-120 and TEA-06-062RK-120) were spaced at intervals not less then 30 a. Both blind and through holes of various radii were drilled with and without eccentricity. The effective hole diameter and eccentricity were determined with a measuring microscope. The elastic constants were determined by using the strain gauge readings before hole drilling (E= 206900 N/mm², $v = 0.28$).

Table 3 shows two of the many results |11| concerning the influence of hole eccentricity on the determination of residual stresses and of calibration constants. Hole diameter, eccentricity parameters, applied stress, measured strains and standard strains given by formulae (10) - (12) are reported in the rows from (1) to (12) for both rosettes. For the first rosette the stresses obtained without and with correction of hole eccentricity are also shown in the last rows, whereas for the second rosette the calibration constants obtained without (A's, Bs') and with correction (As, Bs) together with the theoretical values (A"t, B"t) are shown. The effect of the hole eccentricity correction is evident on both the residual stresses and the calibration constants.

STRAIN RELAXATION EFFICIENCY

The sensitivity of the hole drilling method is low. The attention of strain gauge producers and researchers |9,14-16| has been therefore

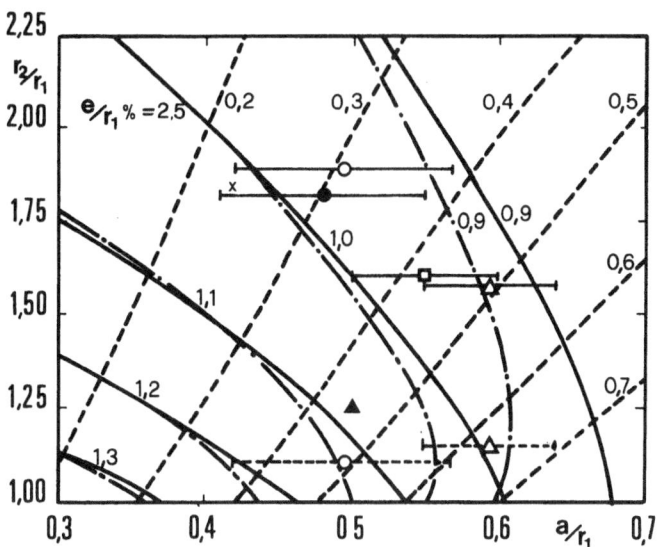

Fig.5 – Curves at constant sensitivity So(– – – –) and curves at constant stress error R' (————— ,rosettes 0°/225°/90°; ——— — ——rosettes 0°/45°/90°). Rosettes: ⊢—O—⊣ MM, ⊢—□—⊣ RY21, x RY61,⊢—△—⊣ BLH, ⊢—●—⊣TML, ⌐--O----ɪ, ⊢--△--⊣ hypothetical rosettes with 0.2 mm gauge length, Owens |16|

directed towards its improvement. Such sensitivity depends on hole-gauge geometry, which is shown in fig.4, for the rosettes of table 1. On the same figure the geometry of the smallest strain gauge (MM, type 008 CL : gauge length 0.2 mm) is shown for comparison.

To evaluate the various rosettes of table 1 the following ratio S is assumed as figure of merit of sensitivity:

$$S = \varepsilon/\varepsilon_o \qquad (27)$$

where ε is the strain, given by equation (1), due to the relaxation of residual stresses and ε_o is the full radial strain due to the same stresses (in the direction $\alpha = 0°$) i.e.

$$\varepsilon_o = (\sigma_1 - \nu\sigma_2)/E \qquad (28)$$

In particular for $\sigma_2/\sigma_1 = 0$ and $\alpha = 0$ the relationship (27) gives

$$So = Ao + Bo \qquad (29)$$

The analysis of the theoretical expressions of Ao and Bo |9| shows that the sensitivity of the hole drilling method can be increased: (1) using higher values of a/r_1 i.e. increasing the hole radius; (2) using rosettes with low values of r_2/r_1 i.e. shorter gauge length.

Fig.5 shows in the a/r_1 , r_2/r_1 coordinate system the curves at constant sensitivity (So= cost) and at constant error for $\sigma_2/\sigma_1 = -1$ ($R'=E\sigma_1/(E\sigma_1)r$ = cost) where $(E\sigma_1)r$ is a reference error calculated for a/r_1 = 0.4 and $r_2/r_1=2$. On the same fig.5 are shown the rosettes of table 1 and two hypothetical rosettes, with grid length of 0.2 mm, positioned at a distance r_1 equal to that of rosettes MM 0.62 and BLH and having the same values of the ratio a/r_1 .The gauge used in |16| is also shown. Fig. 5 allows a rapid evaluation of the influence of hole-gauge geometry (a/r_1 and r_2/r_1) on both the relaxation sensitivity and the stress error due to the hole eccentricity.

CONCLUSION

Both theory and experiments show that the hole eccentricity significantly influences the strains measured by the rosette gauges. As a general rule it is therefore recommended the accurate positioning of the hole at the rosette centre. However the effect of unwanted eccentricity can be accounted for, provided that the eccentricity parameters are measured.
Furthermore it has been shown that:
1) the miniature rosettes are very sensitive to eccentricity errors;
2) single quadrant rosettes (0°/45°/90°) are less favorable than two quadrant rosettes (0°/225°/90);
3) the stress error due to hole eccentricity can be reduced up to 50% by mounting the orthogonal gauges along the principal stresses directions.
The relationships between hole-gauge geometry, strain relaxation sensitivity and eccentricity induced stress error may help the choice of existing rosettes and the development of new ones.

REFERENCES

1. ASTM E837-85: "Standard test method for determining residual stresses by the hole-drilling strain-gage method". ASTM Philadelphia 1985 Annual Book of Standard section 3, volume 03.01, pp.810-816.
2. ROWLANDS, R.E.: "Residual stresses" in Handbook on Experimental Mechanics (A.S. Kobayashi, ed.) Prentice-Hall, Englewood Cliffs (N.J.) 1987, pp.770-790.
3. CHABENAT, A.-MARTIN,R.: "La mesure des contraintes résiduelles: methode de Mathar ed Soete, méthode de Sachs', Memoire Technique du CETIM 1975, Paris.
4. SANDIFER,J.P. e BOWIE,G.E.: "Residual stress by blind hole method with off-center hole", Experimental Mechanics 1978, 18 (5), pp.173-179.

5. HSIN-PANG WANG: "The alignment error of the hole-drilling method" Experimental Mechanics 1979, 19 (1), pp.23-27.

6. AJOVALASIT,A.: "Measurement of residual stresses by the hole-drilling method: influence of hole eccentricity "Journal of Strain Analysis 1979 vol.14, n.4 pp.171-178.

7. AJOVALASIT,A.: "The influence of hole eccentricity on the calibration constants of rosettes used in the hole drilling method" Proc. 7th Int. Conf. on Experimental Stress Analysis - Haifa 1982, pp.591-601.

8. TIEN,A.K.: "A direct method to evaluate hole-alignment error in residual - stress measurement" Exp. Mech., 1985, 25 (1), pp.43-47.

9. AJOVALASIT,A.: "Analysis of residual stresses by the hole drilling method: behaviour of strain gauge rosettes" (in italian) Proc. of 16th AIAS Conf. - Facoltà di Ingegneria - L'Aquila 1988, pp.155-168.

10. JIA-JANG WANG: "Measurement of Residual Stress by the Hole-Drilling Method: General Stress-Strain Relationship and Its Solution", Exp. Mech. 1988, 28 (4), pp.355-358.

11. AJOVALASIT,A.-PETRUCCI,G.: "Experiments on the influence of hole eccentricity on the measurement of residual stresses by the hole drilling method" (in Italian) - Proc. of the 17th AIAS Conf., Dip. di Meccanica, University of Ancona 1989, pp.429-440.

12. AJOVALASIT,A.: "Analysis of residual stresses by the hole drilling method: theory of off-centre equiangular rosette" (in Italian), Report of the Istituto di Costruzione di Macchine, Univ. of Palermo, 1980.

13. DURAN,J.-AMO,J.M.-CHAO,J.: "Determinacion de las tensiones residuales por el metodo de relajacion por orificio con rosetas de cinco o seis bandas extensometricas" CENIM, Madrid 1988 - to be published.

14. CURIONI,S.-FREDDI,A.-VESCHI,O.: "La misura delle tensioni residue come mezzo per l'ottimizzazione di cicli di fabbricazione" Proc. of the X AIAS Conf., Univ. of Calabria, Arcavacata di Rende, 1982, pp.157-167.

15. PROCTER,E.-NASH,P.R.: "A New centre-hole residual stress gauge", B.S.S.M. Twentieth Annual Conf. on Structural Integrity, Univ. of Lancaster, BSSM, Newcastle Upon Tyne 1984.

16. OWENS,A.: "Extension to the blind hole drilling technique for residual stress determination with airabrasive hole forming", Strain 1984, 20 (4), pp.159-165.

THE USE OF BARKHAUSEN NOISE TO INVESTIGATE RESIDUAL STRESSES IN MACHINED COMPONENTS

A.J. BIRKETT
University of Newcastle upon Tyne, Department
of Mechanical, Materials and Manufacturing Engineering.

ABSTRACT

The evaluation of residual stress in this work is carried out by a magnetic method based on Barkhausen noise. Some suitable systems for the detection of Barkhausen noise are outlined. Methods of analysis of Barkhausen noise are discussed.

Commercially available equipment based on Barkhausen noise was used to evaluate the stresses in gear teeth after machining and after hardening. The effects of difference cutting techniques are discussed. A comparison with an X-ray diffraction method is made for gears which had been finished by grinding. Results for gear teeth finished by shot cleaning are discussed.

INTRODUCTION

It is known that ferro-magnetic materials, which are subjected to an externally applied magnetic field, will display a magnetic hysteresis curve (Figure 1).

The shape of the curve may be explained in terms of the magnetic domain pattern in the material. Close inspection of the hysteresis loop reveals that there are distinct regions to it. There is a small region, starting from a demagnetised state, '0', on Figure 1, for which the change in magnetisation in response to an external field is completely reversible, OA. Domain walls move but return to their original position on removal of the field. Changes beyond the point A may be reversible or irreversible. There is a certain amount of rotation of domain magnetisations at points of the loop and there are changes due to the movement of domain walls from one stable position to another. The region DE on Figure 1 is the section of the loop where Barkhausen noise occurs.

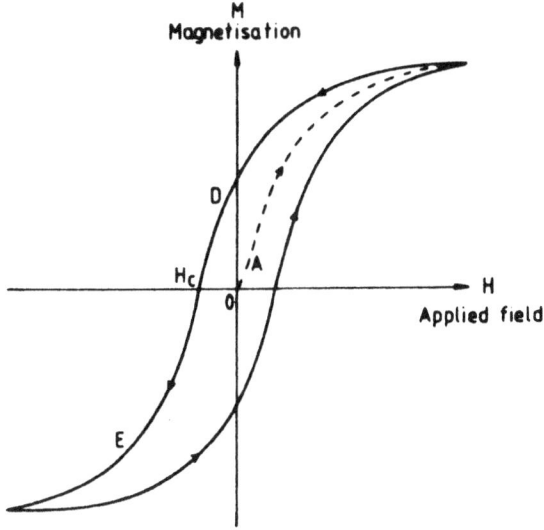

Figure 1. Typical Magnetisation Curve.

Barkhausen noise has been detected in different systems. Two examples are shown. One a surface mounted system and one a laboratory style system. The requirements to detect Barkhausen noise are a magnetising coil and a pick-up coil. In the surface mounted system, Figure 2, the pick-up coil is mounted between the legs of the C core with the magnetising coil wound onto it.

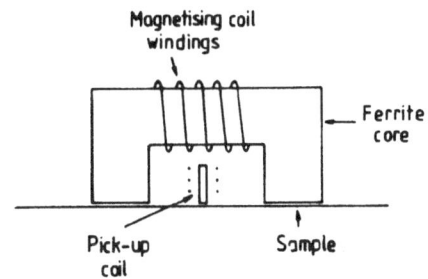

Figure 2. Surface Mounted Coil System.

The laboratory apparatus, Figure 3, [1] consists of a magnetising coil surrounding the sample with the pick up coil enclosed in the centre of the magnetising coil.

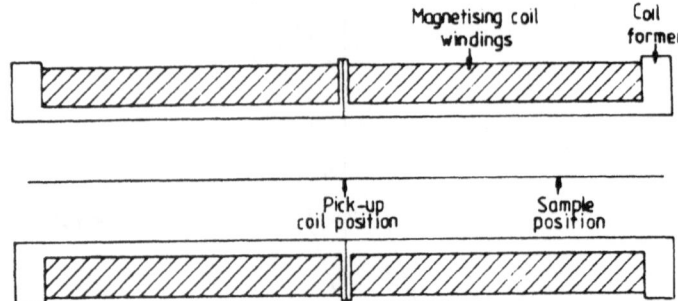

Figure 3. Laboratory Apparatus for Detecting Barkhausen Noise.

Both systems have been used successfully to detect Barkhausen noise, the surface mounted system is obviously more useful in non-destructive testing.

Analysis of Barkhausen noise has usually taken one of two forms. The first, pulse height analysis, for which the magnetic field must change very slowly and the voltage induced in the pick-up coil is proportional to the size of individual or groups of Barkhausen jumps [2]. A distribution of jump size is the output of these systems. The second, Power Spectra analysis, for which the voltage induced in the pick-up coil is continuously monitored and then Fourier Transformed to give an output in the frequency domain [1].

As Barkhausen noise is related to the crystal structure of materials it has been recognized that its study may reveal information about the material state. The most successful applications of Barkhausen noise analysis have been in the amplitude domain and are in the analysis of grain size [2] and residual stress [3]. There has been some success in frequency domain analysis of deformation [4].

The Rollscan equipment used in this study and marketed by American Stress Technologiesis based on amplitude domain measurements of Barkhausen noise. The output of the instrument is the Magnetic Parameter (MP) and with appropriate calibration the number may be related to either residual stress or microstructure depending on the application [5]. The equipment has two nominal ranges of measurement depth, 0.2 mm and 0.02 mm. The actual depth of measurement depends on the material. The equipment may be calibrated by use of a laboratory tensile/compressive test machine. The MP may be monitored for the material under test and the results related to the experimental results.

The Rollscan equipment was used on site at Eaton Ltd (Axle Division) Newton Aycliffe. The components investigated were the main driving gears of the axles which Eaton manufacture, and planetary gears which form part of a sun and planet arrangement in a transmission system.

MATERIALS AND COMPONENTS

The steel type investigated was 8822H AISI, a steel used in case carburising applications. The components analyzed were; ring gears of approximate outside diameter 16 inches, pinions which mate with the ring gears and bevel gears of approximate outside diameter 3 inches.

Both the ring gears and pinions are formed from turned, forged blanks which have teeth cut into them. Two basic methods of cutting high ring gears and pinions were investigated. The methods are named after the appropriate machine manufacturers, Gleason and Oerlikon. The Gleason method is completed on two separate machines whereas the Oerlikon is completed in one operation. The Gleason method was investigated for both ring gears and pinions and as one tooth form is very different on the two gears they are treated as separate cutting methods. It was expected that the three methods would produce different residual stress levels on the gear tooth surface. The cut ring gears and pinions are carburized in a continuous carburizing furnace and quenched in oil to produce a hard case on the gear teeth. The gears are tempered, to reduce brittleness and increase toughness, then shot cleaned. The shot clean operation has an accompanying change in the surface residual stress level.

The bevel gears are precision forged blanks which are heat treated, as above, and then hard ground to produce the required finish and the specified residual stress level.

RESULTS

(a)   Machined components

Measurements were made on the machined gear teeth at three positions on each tooth at one toe, heel and centre positions, Figure 4.

Preliminary measurements showed that an error of $\pm 10\%$ may be expected if surface contact was not achieved properly. The measurements were made on both the convex face (drive side) and the concave (coast side) of the teeth, Figure 4. It was

apparent from the measurements drive side of the tooth was
in tension and the coast side in compression. For the ring
gears the Oerlikon cutting method showed higher levels of
compression and tension. The results for the pinions were
different in that both flanks of the tooth appeared to be in
tension. Typical results of the Rollscan reading are shown
in Table 1, a figure higher than 116 indicates tension and a
figure lower than 116 indicates compression.

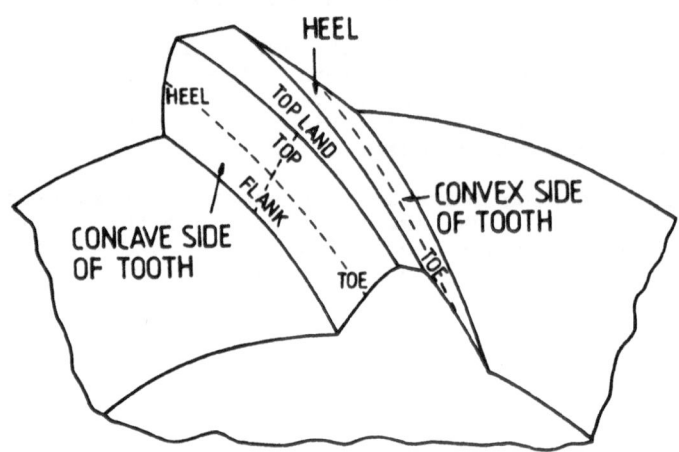

Figure 4. Typical Gear Tooth

TABLE 1
Machined Components

	Oerlikon Ring Gear		Gleason Ring Gear		Gleason Pinion	
	Drive	Coast	Drive	Coast	Drive	Coast
Heel	170	106	158	82	153	145
Centre	170	99	172	83	157	154
Toe	169	89	153	108	148	148

(b)   Hardened and tempered components

Similar positions were used for measurements on the
hardened and tempered gears. It was found that both flanks
of the tooth were in compression. Results for a hardened as
quenched and tempered ring gears are shown in Table 2. It
was noted that the magnetic parameter decreased after
temper. For this material condition a figure of less than
308 indicated compression.

TABLE 2
As quenched and tempered ring gears

| | As quenched | | Tempered | |
	Drive	Coast	Drive	Coast
Heel	99	53	32	12
Centre	86	59	30	10
Toe	102	102	30	11

(c)   Shot blasted components

An increase in magnetic parameter was observed after shot blast, the reverse had been anticipated.   Typical results are shown in Table 3. It was found that there were a large variation around the ring gears.

TABLE 3
Shot blasted ring gears (drive side of tooth)

	Heel	Centre	Toe
Tooth			
1	62	84	58
2	76	127	94
3	111	73	78

(d)   The results for three positions on the flanks of 5 teeth of a hard ground bevel gear are shown in Table 4. The same teeth were analyzed for residual stress by an X-ray ray   diffraction   technique,   the   results   are shown in Table 5 and there is an obvious correlation between the two sets of results.

TABLE 4
Magnetic parameter bevel gear

	Position 1	Position 2	Position 3
Tooth 1	54	48	43
Tooth 2	43	51	47
Tooth 3	38	39	38
Tooth 4	53	52	54
Tooth 5	47	53	52

TABLE 5
Residual Stress in kPsi by X-ray diffraction

	Position 1	Position 2	Position 3
Tooth 1	112	106	94
Tooth 2	96	110	106
Tooth 3	74	86	80
Tooth 4	112	110	114
Tooth 5	106	112	112

DISCUSSION

The results for the machined components may be explained by considering the cutting directions and the removal of material in the processes. In all the cutting processes the depth of removal should be constant along the tooth on both sides of the tooth. The Gleason cut ring gears showed more compressive stress towards the heel of the tooth on the coast side. The depth of removal is generally greater at that point so the difference in surface stress would seem sensible. It may be anticipated that a change in machine settings could change the observed stress level.

The Oerlikon method of cutting ring gears shows the opposite change in stress along the coast side of the tooth. The depth of cut in the Oerlikon method is similar along the tooth so a change in surface stress would not be anticipated. The force of the cutter in the material changes along the tooth: it is greater at the toe than the heel. The difference in stress along the tooth may be explained in this way. The drive side results for the Oerlikon method show the expected consistency.

The pinion results may be explained by considering the depth of removal and tooth shape. The depth of removal is constant on each side of the tooth so the stress level is similar. The tooth is of involute form in the pinions, not straight sided as in the ring gears and the tooth is in a spiral around the pinion rather than being a simple circle arc. These two facts may explain the consistency of the sign of the stress.

The results for the hardened and tempered components reflect the relationship between material hardness and magnetic parameter. Figure 5 [6] shows one relationship between hardness and MP and indicates the position on the curve of the components.

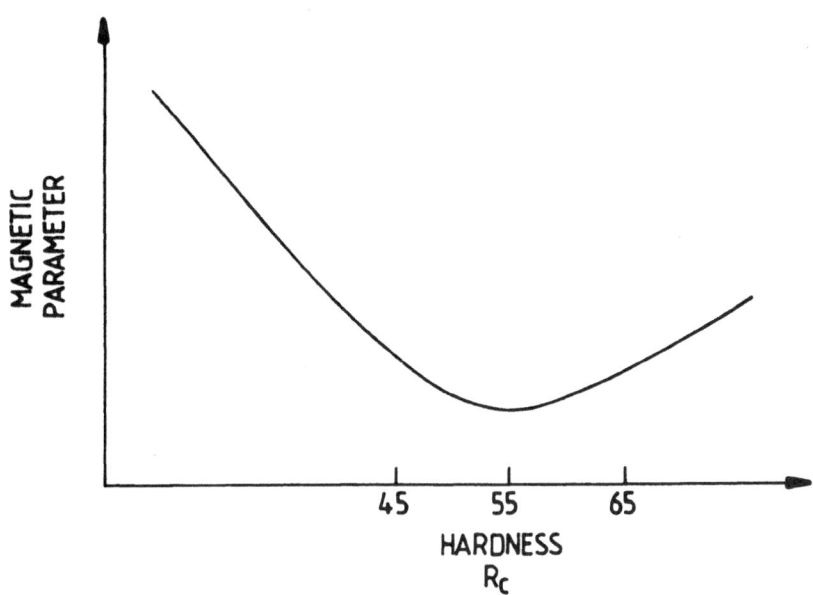

Figure 5. Relationship between Material Hardness and Magnetic Parameter.

It can be seen that the tempered components have MPs nearer to the minimum than the hardened components. This would not have been the expected result based on the magnetic properties of the material.

The increase in surface MP in the shot blasted components may be explained by the fact that there may be surface saturation of defects caused by the shot blast. The maximum compressive stress may be at some distance below the surface as shown in Figure 6.

The depth of Barkhausen noise pick up in this material condition would be nominally 20 μm [7] and the reading obtained is probably related to an average of the surface and sub-surface stress. The variation of MP around the gears is probably due to process inconsistencies in the shot blast operation.

The results for the hard ground bevel gears were interesting as they showed variations which were not expected in the grinding process. The same variations were observed in the X-ray diffraction results. The variations were attributed to the grinding wheel 'burning' the surface. The hardness in those areas was therefore lower and a similar explanation to the tempering of components may be applied. The fact that there was such good agreement between the two methods of measurement shows that the Rollscan equipment may be used to assess process consistency and is much easier to establish on the shop floor than an x-ray diffraction method. The detection of such areas is important to the running properties of the gear.

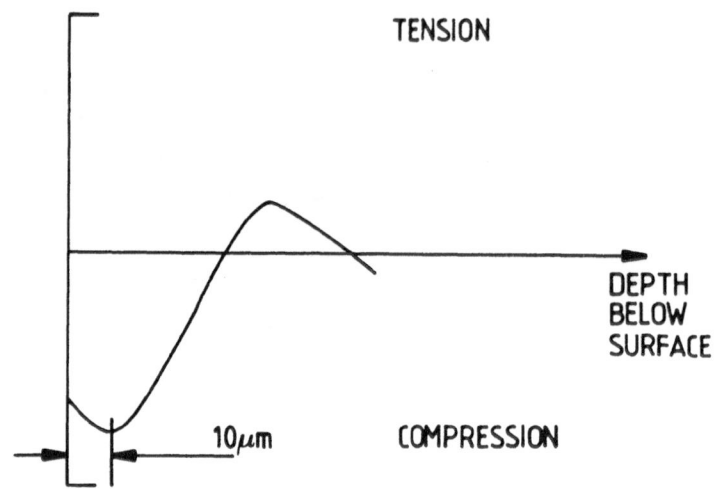

Figure 6.   Relationship between Residual Stress and Depth
Below Surface

CONCLUSION

The Rollscan equipment may be used to detect surface residual stress in gear teeth in a qualitative manner. The variations in stress found may be attributed to the deviation of the cutting processes away from the ideal conditions. The evaluation of a shot blast operation was facilitated by the equipment. There appears to be a correlation between the reading obtained on the Rollscan equipment and the residual stress measured by an x-ray diffraction method. In this case there was a variation detected which indicated the grinding process was inconsistent.

The effect of grinding burn in components is known but the effect of maximum compressive stress being at apparently different levels in the gear teeth is not. There may be a relationship with either surface wear or crack initiation and development.

ACKNOWLEDGEMENTS

The author acknowledges funding from the Science and Engineering Research Council and Eaton Ltd under the Teaching Company Scheme. The work was performed at Eaton Ltd, (Axle Division) Newton Aycliffe and their co-operation was much appreciated.

REFERENCES

[1]   Birkett, A.J., Barkhausen Noise in Steels, PhD Thesis, University of Durham, 1988.

[2]   Saynajakangas, S., A Non-destructive Electromagnetic Method for Structural Studies of Ferrous Alloys, Acta Pol. Scand. E Series, 33, 1973.

[3]   Rautioaho, R.H. and Karjalainen, L.P., Application of Barkhausen Noise Measurements to Residual Stress Analysis in Structural Steels, Acta Univ. Oul., C26, Metallurgy 4. 1983.

[4]   Birkett, A.J., Corner, W.D., Tanner, B.K. and Thompson, S.M., Influence of Plastic Deformation on Barkhausen Power Spectra in Steels, J.Phys.D., 1989, 22, 1240-1242.

[5]   Tiitto, K., Solving Residual Stress Measurement Problems by a New Non-destructive Magnetic Method. American Stress Technologies Inc., Technical Report.

[6]   Tiitto, S., Private Communication. 1988.

[7]   Tiitto, S., Depth of Measurement in Barkhausen Noise Testing, American Stress Technologies Inc., Technical Report. 1988.

RESIDUAL STRESS PATTERNS IN COLD DRAWN STEEL WIRES AND THEIR EFFECT ON FRETTING–CORROSION–FATIGUE BEHAVIOUR IN SEAWATER

R. SMALLWOOD
British Rail Technical Centre,
Brunel House, London Road, Derby DE2 8UP, UK

R.B. WATERHOUSE
Department of Materials Engineering & Materials Design,
University of Nottingham, Nottingham NG7 2RD, UK

ABSTRACT

The numerous inter-wire contacts in steel ropes are potential sites for fretting, seriously reducing the fatigue strength of the rope. The fretting behaviour is influenced by the residual stresses at points of contact which vary around the circumference of the wire, and whose magnitude depend on the reduction in the final die. Fatigue curves have been determined in air and seawater with the fretting bridges applied at different points in relation to the residual stress pattern. The effect of heat-treatment simulating the hot-dip galvanising process is also investigated.

INTRODUCTION

Steel ropes for marine applications are made up of cold-drawn wires which are usually hot-dip galvanised. Those in use for mooring off-shore oil rigs are of the single strand type so that the 5 mm dia. wires are in close-packed layers alternately wound in opposite sense. This means that there are two categories of inter-wire contact: (a) line contact between adjacent wires in the same layer and (b) trellis contact between wires in adjacent layers. When a fluctuating stress is applied to the rope, fretting, particularly at trellis contacts, leads to the initiation of fatigue cracks and thus reduces the fatigue strength of the rope [1]. In drawing the wire, the wire is drawn from the final die at an angle which causes the wire to coil. This results in a residual stress pattern in the wire which varies around its circumference. The effect of fretting is therefore very dependent on the position of the contact points in relation to the stress pattern. This has been investigated in the present work.

MATERIALS AND METHODS

The materials used were three samples of cold drawn eutectoid steel wire 5 mm dia. which had been given different percentage reductions in the final die, namely 5, 12 and 26.5%. In addition, a batch of the 12% reduction wire was given a heat-treatment of 1h at 450°C to simulate the change which occurs on hot-dip galvanising. Despite the different degrees of drawing and the heat-treatment the mechanical properties of the four steel wires were nominally the same as shown in Table I.

TABLE I

Mechanical Properties of Wire Samples

Breaking load N	UTS MPa	0.2% P.S. MPa	Hardness VHN
34000	1700	1300	520

The residual stress pattern was measured by X-ray back reflection using CrK_a radiation filtered with a V monochromator. Measurements were made using the two exposure method [2] on the reflection from (211) where $2\theta = 156.39°$. Fig. 1 represents the coiled wire as-received and indicates the points at which the stress measurements were made. Profiles of the stress distribution were made by progressively etching the surface of the specimens.

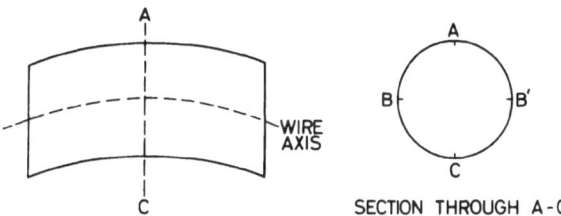

Figure 1.   Reference points on the circumference of the as-received wire.

The fretting-fatigue tests were carried out using a 20 kN servo-hydraulic fatigue machine. The ends of the wire specimen were gripped in a capstan device to minimise fretting and failure at these points. Fretting was produced by clamping a bridge composed of four pieces of the same wire as the specimen by means of a proving ring. The fretting device was surrounded by a transparent cell through which artificial seawater (made up according to BS 3900/2011) could be continuously pumped if necessary. The experimental arrangement is shown in Fig. 2. Tests were carried out at a frequency of 5 Hz with a constant maximum stress of 950 MPa and a variable minimum stress to give stress amplitudes of the range 120 to 400 MPa. Over this range of stress the slip amplitude varied between 3 and 10μm. The clamping load of 250 N at each contact was maintained constant in all tests. In most cases, the fretting points were at the B position but in certain cases the bridges were applied at the A-C positions. At the conclusion of a test the fretting site and fracture surfaces were examined in the scanning electron microscope.

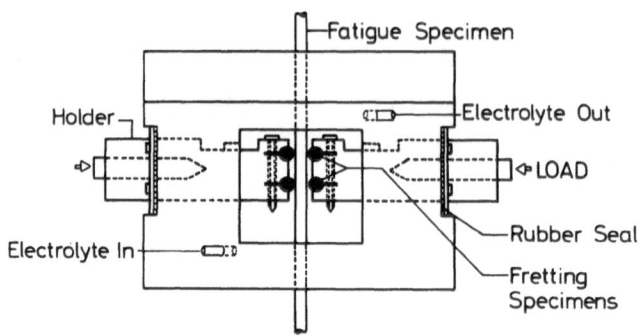

Figure 2.   Fretting-fatigue cell.

RESULTS

Fig. 3 shows the results of the residual stress measurements made at position B for the 4 wire samples. To check the stress measurements further measurements were made with an applied tensile stress. The results in Fig. 4 show that the stress measurement for the 12% wire is in good agreement with the result in Fig. 3. Fig. 5 shows the stress distribution for the 12% wire in the A and C positions in the coiled and straightened condition, and Fig. 6 summarises the surface stress level around the circumference of the wire.

Fretting fatigue curves for the 12% wire with the fretting bridges clamped at the two significant positions on the circumference of the wire, i.e. B and A-C, are shown in Fig. 7. The higher tensile residual stresses in the B position lead to a lower fatigue strength. Fracture surfaces of specimens fretted at the two locations are shown in Figs. 8a and b, where it is seen that two cracks are initiated from the points B, but only one crack from point C.

The effect of the different degrees of drawing in the final die is shown in Fig. 9 for the air environment. The higher residual stress in the 26.5% wire again leads to a lower fretting-fatigue strength. The heat-treated specimen has the lowest tensile residual stress with the expected highest fatigue strength. Results for the 12% wire in seawater are shown in Fig. 10.

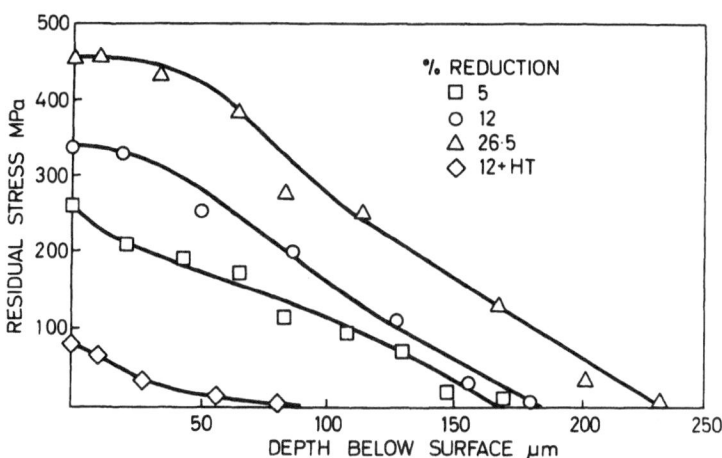

Figure 3.   Residual tensile distribution for wires drawn by different reductions in
the final die and effect of heat-treatment - position B.

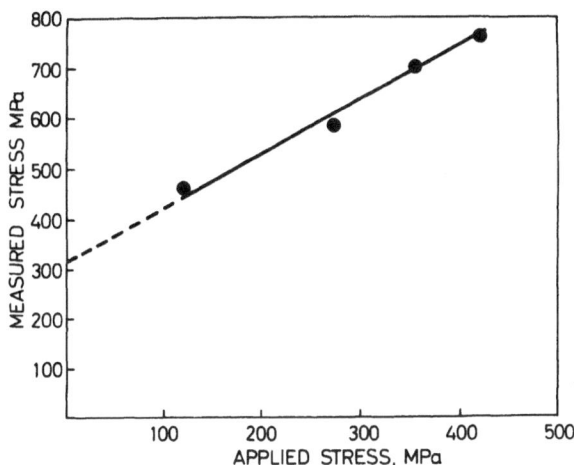

Figure 4.   Residual surface tensile stress measured by X-ray diffraction vs. applied
stress for 12% wire - position B.

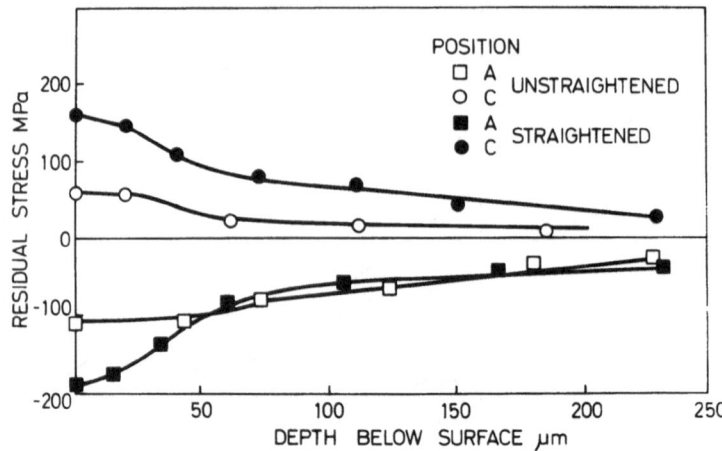

Figure 5.   Residual stress distribution for 12% wire at positions A and C in the
unstraightened and straightened conditions.

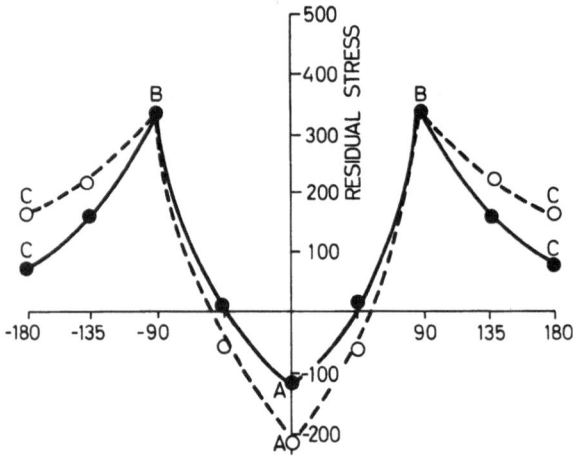

Figure 6.   Residual surface stress variation for 12% wire around the circumference
of the wire in the unstraightened ● and straightened O condition.

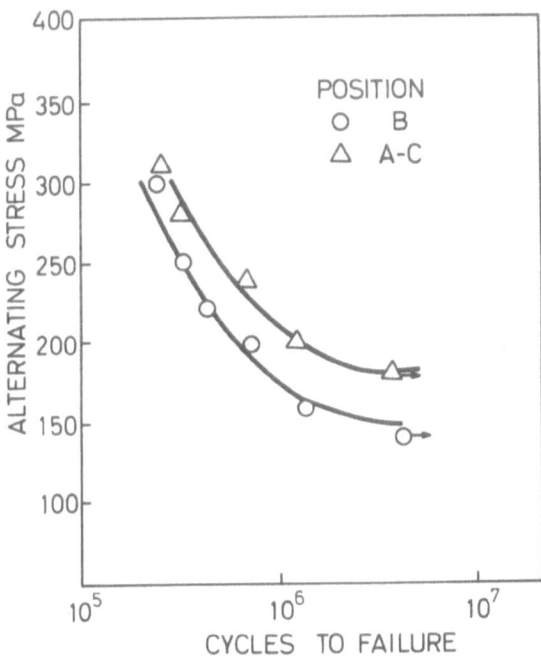

Figure 7.   Fretting-fatigue curves for 12% wire with bridges attached at B position
and A-C position in air.

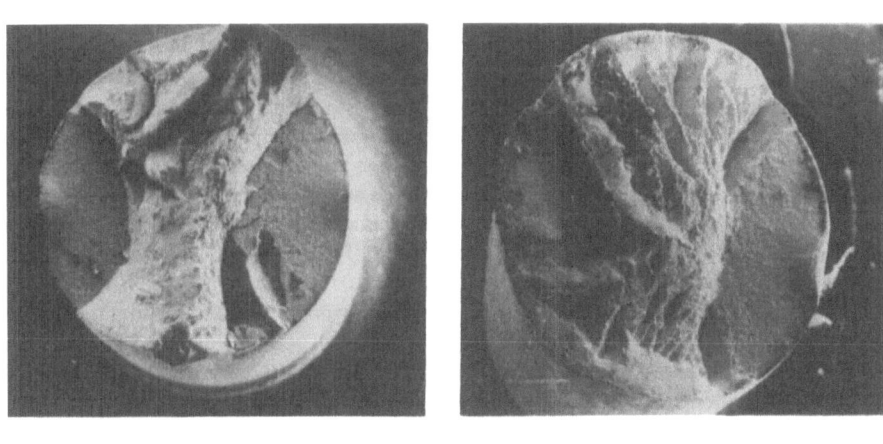

Figure 8a.   Fracture surface - bridges attached at position B.
Figure 8b.   Attached at position A-C.

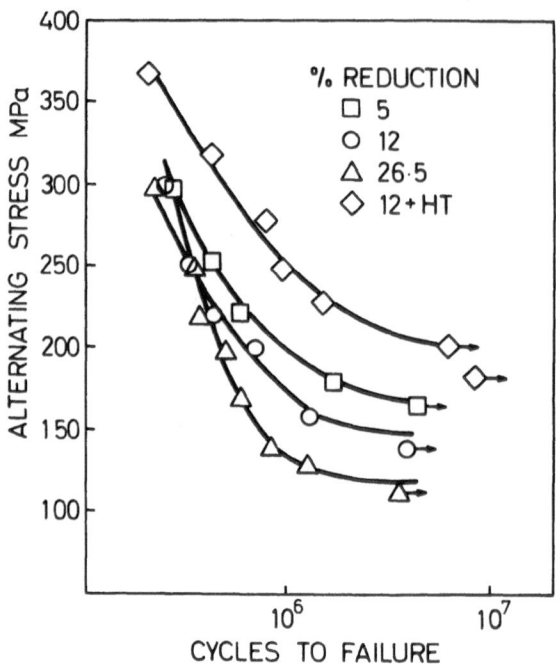

Figure 9.   Fretting-fatigue curves for wires drawn by different reductions in the
final die in air - position B.

DISCUSSION

It is perhaps surprising that the steel wires from which steel ropes are constructed
have a substantial surface residual tensile stress, particularly as these ropes are used
under tension and alternating loads in applications such as suspension bridges, lifts,
cable cars and, in the present case, moorings. The results indicate that the behaviour
under fretting-fatigue conditions, which is a common situation in the application of
these ropes, is largely governed by the magnitude of the residual tensile stresses and
the location of the fretting contact in relation to them. It is obvious that heat-
treatment to reduce the magnitude of the stresses produces a dramatic effect. It is
fortunate that it occurs adventitiously in the hot-dip galvanising process, which is
usually applied in cases where the rope is to be used in corrosive environments,
particularly marine environments.

Other methods of producing a more favourable surface stress, ideally entirely
compressive, are shot-peening [3] and diffusion treatments such as carburising and
nitriding [4]. The former would be difficult to apply to wire, although the diffusion
treatment might be possible as a batch process.

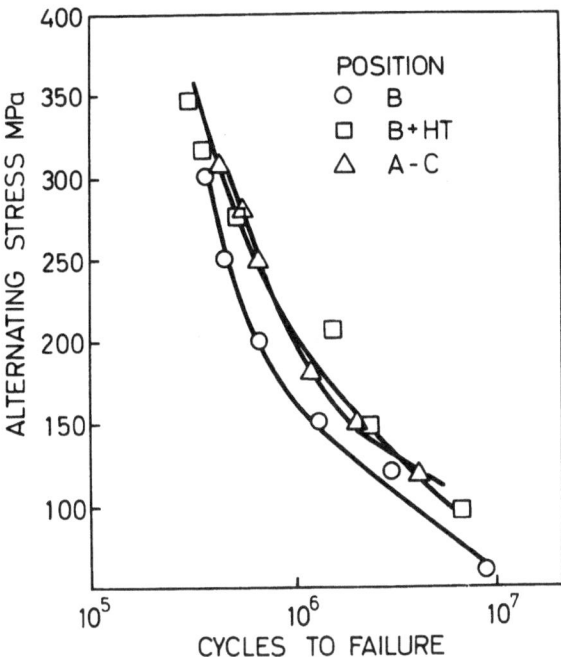

Figure 10.   Fretting fatigue curves for 12% wire in seawater with bridges applied at positions B and A-C.

Where fretting is the main sources of fatigue failures it is possible to alleviate the problem by incorporating an anti-fretting lubricant in the rope [5]. This is particularly effective in the case of lock-coil ropes where the cross-section of the outer wires is shaped so that they fit together to form a smooth outer surface.

While fretting in air produced strength reduction factors in fatigue of between 2 and 5, in seawater the effect is even further enhanced [6]. In the present case the effect of the residual stress is less marked, the corrosion factor being the major influence.

CONCLUSIONS

The following conclusions are drawn from the investigation:

(1)   the residual stress distribution around the circumference of drawn wires varies, being tensile and highest at the diametrical points at right angles to the plane of the coiled wire.

(2)   the magnitude of the residual tensile stresses depends on the degree of reduction in the final die - the greater the reduction the higher the residual stresses.

(3)   heat-treatment comparable with the effects of hot-dip galvanising produces a significant reduction in the magnitude of the stresses.

(4)   the fretting-fatigue strength of the wires depends critically on the position of the fretting contact in relation to the residual stresses. Where the fretting occurs at a point of high tensile residual stress, the fatigue strength is significantly lower.

(5)   in seawater the effects of residual stress are much reduced.

ACKNOWLEDGEMENT

The authors wish to express their appreciation to British Ropes, Doncaster, who supplied the wire samples and advice throughout. Thanks are also due to SERC for a CASE award.

REFERENCES

1.   Boleantu, L., Failure processes at the contact surfaces of wires in traction ropes, Cercetari Metalurgice, 1967, 9, 527-546.

2.   Klug, H.P. and Alexander, L.E., X-ray Diffraction Procedures for Polycrystalline and Amorphous Materials, 2 Edn., John Wiley, New York, 1974, pp. 766-8.

3.   Almen, J.O. and Black, P.H., Residual Stresses and Fatigue in Metals, McGraw-Hill, New York, 1963, pp. 129-133.

4.   Frost, N.E., Marsh, K.J. and Pook, L.P., Metal Fatigue, O.U.P., London, 1974, pp. 348-351.

5.   Waterhouse, R.B. and Taylor, D.E., Fretting fatigue in steel ropes, J. Am. Soc. Lub. Engrs., 1971, 27, 123-127.

6.   Takeuchi, M. and Waterhouse, R.B., An investigation into the fretting-corrosion-fatigue of high strength steel wire, Proc. 10th Int. Congr. on Metallic Corrosion, 7-11 Nov. 1987, Madras, India, O.U.P., 1987, 3, 1959-66.

RESIDUAL STRESSES AND WARPING IN UNSYMMETRIC LAMINATES WITH ARBITRARY LAY-UP ANGLES

W. J. JUN
Div. of Mech. Eng., KIST,
P.O. Box 131, Cheongryang, Seoul, Korea

and

C. S. HONG
Dept. of Mech. Eng., KAIST,
P.O. Box 150, Cheongryang, Seoul, Korea

ABSTRACT

Thermal residual stresses during processing incur the warping of the unsymmetrically laminated composites. While the warping of the cross-ply laminates has been characterized well, general unsymmetric laminates with arbitrary lay-up angles are still not fully understood. This paper presents the formulation which treats the curvatures and principal direction of curvature of the cured shape in unsymmetrically laminated composites. The effect of spatial dependence of in-plane strains on curvatures is significant near the bifurcation point in case of laminates having the stacking arrangement where snap-through phenomenon occurs. It is shown that the principal direction of curvature calculated from classical lamination theory agrees with this theory in the limited range of length-to-thickness ratios of laminates. Curvatures and principal direction of curvature according to the length-to-thickness ratios, the number of layer and lay-up angles of the laminates are presented.

INTRODUCTION

Thermal residual curing stresses generated during the fabrication process incur the warping of the unsymmetrically laminated composites[1]. We can observe laminated structures with various shapes due to these residual stresses induced during processing.
Most studies [2–7] for the cured shapes of unsymmetric laminates has been limited to the warping analysis of cross-ply laminates. Little work on the warping analysis of general unsymmetric laminates with twisted shapes has been done. Hyer[8] showed that classical lamination theory fails to predict the room-temperature shapes of unsymmetric laminates with arbitrary lay-up angles through the experiments. Dang and Tang[9] presented the formulation which calculates curvatures of general unsymmetric laminates with arbitrary lay-up angles. From the displacement functions of cross-ply laminate assumed on the basis of

observation, they derived strain functions with the limited spatial variation for general unsymmetric laminates. It is difficult for this formulation to obtain solutions stable in the range which two real solutions exist. It does'nt seem satisfying to investigate the behavior of the cured shapes in general unsymmetric laminate with arbitrary lay–up angles. In general, all freedom to the spatial variation of strains and displacements should be considered in the formulation. As shown in ref.[7], the effect of the shear strain is significant in the range of medium width–to–thickness ratio where the bifurcation point occurs.

From the displacement functions for the cross–ply laminate including in–plane shear strain, this paper presents a new formulation considering all spatial variation of strains and displacement functions to calculate curvatures and principal direction of curvatures at the room temperature. Therefore, the formulation in this paper holds good without the loss of generality for any unsymmetric laminates with arbitrary lay–up angles. The effect of the length–to–thickness(L/T) ratios, the number of layer(n) and lay–up angles(θ) on the curvatures is investigated for the square unsymmetric laminates. It is compared with classical lamination theory, present and existing results for the principal direction of curvature.

THEORETICAL DEVELOPMENT

Laminated composites unsymmetric to the midplane of plate are fabricated by typical curing cycle. During the cure cycle, the temperature is lowered to the room temperature and cured laminate is taken out of the autoclave. Due to the mismatch of the thermal expansions, residual stresses are built up in laminate of fig.1.

Notation and geometry used in this paper are shown in fig.2(T300/5208 Gr/Ep). Fig.3 illustrates the problem in the study. Figure 3a, 3b illustrate the cylindrical warping and anticlastic warping laminate in the cross–ply stacking arrangement while 3c, 3d shows twisted laminate of general unsymmetric laminate with arbitrary lay–up angles at the room temperature. To calculate the curvatures of laminate, the von Karman approximation to Green's strain measures is used.

Generalization of Formulation
For the cross–ply laminate, displacement functions in the midplane of the laminate can be expressed by the polynomials considering the longitudinal strains are symmetric[2,3,11]. After applying boundary conditions for the cross–ply stacking arrangement, if we select the three terms of the polynomials, displacement and strain functions can be written as the following polynomial forms[7].

$$u^{o\prime} = x' \left(a_1' - a'^2 x'^2/6 + a_3' y'^2 \right)$$
$$v^{o\prime} = y' \left(b_1' - b'^2 y'^2/6 + b_3' x'^2 \right) \tag{1}$$
$$w' = \left(a' x'^2 + b' y'^2 \right)/2$$

$$\epsilon_x' = a_1' + a_3' y'^2 - a'z$$
$$\epsilon_y' = b_1' + b_3' x'^2 - b'z \tag{2}$$
$$\epsilon_{xy}' = (a_3' + b_3' + a'b'/2)\, x'y'$$

For the square laminate, it can be assumed that the coefficient a_3' is equal to b_3'.

Through the coordinate transformation between natural body axis and principal axis, we obtain the following relations at the midplane of laminate.

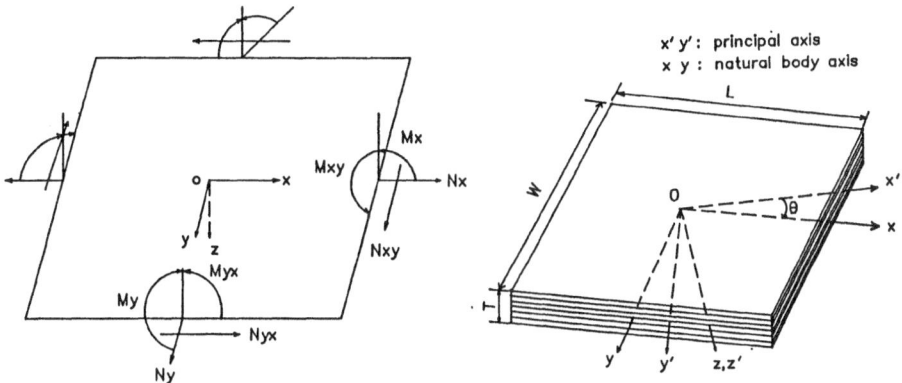

Figure 1. In–plane forces and moments loaded during the fabrication process.

Figure 2. Laminate notation and geometry.

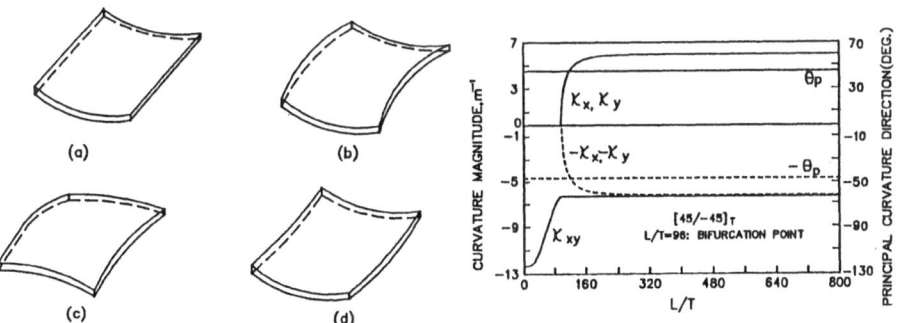

Figure 3. Laminate shapes : a) cylindrical laminate b) saddle laminate c) twisted laminate d) another twisted laminate.

Figure 4. Snap–through behavior of $[45/-45]_T$ laminate with various length to thickness ratios.

$$u^o = mu^{o\prime} + nv^{o\prime}$$
$$v^o = -nu^{o\prime} + mv^{o\prime} \tag{3}$$

and

$$x' = mx - ny$$
$$y' = nx + my \tag{4}$$

where $m = \cos\theta$, $n = \sin\theta$

While a, b, c are curvatures at the natural body axis, if a', b' are principal curvatures at the principal axis of the laminate, we have

$$\begin{Bmatrix} a \\ b \\ c \end{Bmatrix} = \begin{bmatrix} m^2 & n^2 \\ n^2 & m^2 \\ -2mn & 2mn \end{bmatrix} \begin{Bmatrix} a' \\ b' \end{Bmatrix} \tag{5}$$

After equation(4) is substituted into equation(1) and then equation(1) is substituted into equation(3), we obtain general displacement functions as follows.

$$u^\circ = a_1 x + a_2 y + a_3 x^3 + a_4 y^3 + 3a_5 x^2 y + a_6 xy^2$$
$$v^\circ = a_2 x + b_1 y + a_5 x^3 + b_3 y^3 + a_6 x^2 y + 3a_4 xy^2 \tag{6}$$
$$w = (ax^2 + by^2 + cxy)/2$$

where

$$a_1 = m^2 a_1' + n^2 b_1'$$
$$a_2 = mn\,(b_1' - a_1')$$
$$a_3 = -a'^2 m^4/6 - b'^2 n^4/6 + 2\,m^2 n^2\,a_3'$$
$$a_4 = a'^2 m\,n^3/6 - b'^2 m^3 n/6 - a_3'm\,n\,(m^2 - n^2)$$
$$a_5 = a'^2 m^3 n/6 + a_3'mn(m^2 - n^2) - b'^2 m\,n^3/6 \tag{7}$$
$$a_6 = -(a'^2 + b'^2)\,m^2 n^2/2 + a_3'\,(m^4 + n^4 - 4\,m^2 n^2)$$
$$b_1 = n^2 a_1' + m^2 b_1'$$
$$b_3 = -a'^2 n^4/6 - b'^2 m^4/6 + 2\,m^2 n^2\,a_3'$$

It is difficult to assume initial values due to too many unknowns(eleven) when we solve equation(6) by numerical approach. To minimize the number of unknowns, it is used the following trigonometric relationships.

$$2\,m\,n/(m^2 - n^2) = c/(a - b) \tag{8}$$
$$m^2 n^2 = \tfrac{1}{4} c^2/\{(a - b)^2 + c^2\} \tag{9}$$
$$\phi = a'\,b'/3 + 2a_3' \tag{10}$$

After substituting equations(8),(9),(10) into equation(7) and arranging, we get generalized displacement functions with the seven unknowns.

$$u^\circ = a_1 x + a_2 y + [\tfrac{1}{4}c^2\phi/\{(a-b)^2+c^2\} - a^2/6]x^3 - [\tfrac{1}{4}\phi c(a-b)/\{(a-b)^2+c^2\} + bc/12]y^3$$
$$+\tfrac{1}{4}[3\phi c(a-b)/\{(a-b)^2+c^2\} - ac]x^2 y + [\tfrac{1}{4}\{2(a-b)^2-c^2\}\phi/\{(a-b)^2+c^2\}$$
$$- ab/6 - c^2/12]xy^2$$
$$v^\circ = b_1 y + a_2 x + [\tfrac{1}{4}c^2\phi/\{(a-b)^2+c^2\} - b^2/6]y^3 + [\tfrac{1}{4}\phi c(a-b)/\{(a-b)^2+c^2\} - ac/3]x^3$$
$$-\tfrac{1}{4}[3\phi c(a-b)/\{(a-b)^2+c^2\} + bc]xy^2 + [\tfrac{1}{4}\{2(a-b)^2-c^2\}\phi/\{(a-b)^2+c^2\}$$
$$- ab/6 - c^2/12]x^2 y$$
$$w = (ax^2 + by^2 + cxy)/2 \tag{11}$$

where

$$a_1 = m^2 a_1' + n^2 b_1'$$
$$b_1 = n^2 a_1' + m^2 b_1'$$

$$a_2 = mn(b_1' - a_1')$$

$$a_3 = -a^2/6 + \tfrac{1}{4}c^2\phi/\{(a-b)^2+c^2\} \qquad (12)$$

$$a_4 = -bc/12 - \tfrac{1}{4}c(a-b)\phi/\{(a-b)^2+c^2\}$$

$$a_5 = -ac/12 + \tfrac{1}{4}c(a-b)\phi/\{(a-b)^2+c^2\}$$

$$a_6 = -ab/6 - c^2/12 + \tfrac{1}{4}\{2(a-b)^2-c^2\}\phi/\{(a-b)^2+c^2\}$$

Using the strain–displacement relationships with geometric nonlinearity, strain functions at the arbitrary lamina are of the form

$$\epsilon_x = a_1 + \tfrac{3}{4}c^2\phi/\{(a-b)^2+c^2\}x^2 + \tfrac{3}{2}c(a-b)\phi/\{(a-b)^2+c^2\}xy + \tfrac{1}{2}[\tfrac{1}{2}\phi\{2(a-b)^2-c^2\}$$
$$/\{(a-b)^2+c^2\}-ab/3+c^2/12]y^2-az$$

$$\epsilon_y = b_1 + \tfrac{3}{4}c^2\phi/\{(a-b)^2+c^2\}y^2 - \tfrac{3}{2}c(a-b)\phi/\{(a-b)^2+c^2\}xy + \tfrac{1}{2}[\tfrac{1}{2}\phi\{2(a-b)^2-c^2\}$$
$$/\{(a-b)^2+c^2\}-ab/3+c^2/12]x^2-bz$$

$$\epsilon_{xy} = a_2 + \tfrac{3}{4}c(a-b)\phi/\{(a-b)^2+c^2\}x^2 - \tfrac{3}{4}c(a-b)\phi/\{(a-b)^2+c^2\}y^2$$
$$+\tfrac{1}{2}[\phi\{2(a-b)^2-c^2\}/\{(a-b)^2+c^2\}+ab/3-c^2/12]xy-cz/2 \qquad (13)$$

Equation(13) is used to obtain the potential energy for the laminate. If $\phi=0$, i.e., a_3' $=-a'b'/6$, it can be simplified as strain functions with the limited spatial variation.

$$\epsilon_x = a_1 + (-ab/3 + c^2/12)y^2/2 - az$$

$$\epsilon_y = b_1 + (-ab/3 + c^2/12)x^2/2 - bz \qquad (14)$$

$$\epsilon_{xy} = a_2 + (ab/3 - c^2/12)xy/2 - cz/2$$

Minimization of Total Potential Energy
It is well known that the potential energy involving the effects of thermal expansion is given by

$$\Pi = \int_v \Omega \, dv \qquad (15)$$

We shall now consider the stress–strain law in a general thermal environment, not limited to isothermal or isentropic conditions. After the strain energy density function is expanded in a power series, when ϵ_{ij} are small and terms higher than the second order are neglected, we obtain reduced tensor form[10].

$$\Omega = \tfrac{1}{2}\overline{Q}_{ij}\,\epsilon_i\,\epsilon_j - \beta_i\,(T_r-T_c)\,\epsilon_j \qquad (16)$$

Where $\overline{Q}_{ij}$ denotes reduced transformed stiffness and the β_j are coefficients related to the elastic constants and the coefficients of thermal expansion of the material. The expression for the strain energy density function becomes

$$\Omega = \overline{Q}_{11}\epsilon_1^{\,2}/2 + \overline{Q}_{12}\epsilon_1\epsilon_2 + 2\overline{Q}_{66}\epsilon_{12}^{\,2} + \overline{Q}_{22}\epsilon_2^{\,2}/2 + 2\overline{Q}_{16}\epsilon_1\epsilon_{12} + 2\overline{Q}_{26}\epsilon_2\epsilon_{12}$$
$$(\overline{Q}_{11}\alpha_x + \overline{Q}_{12}\alpha_y + \overline{Q}_{16}\alpha_{xy})\epsilon_1\Delta T - (\overline{Q}_{12}\alpha_x + \overline{Q}_{22}\alpha_y + \overline{Q}_{26}\alpha_{xy})\epsilon_2\Delta T$$
$$-2(\overline{Q}_{16}\alpha_x + \overline{Q}_{26}\alpha_y + \overline{Q}_{66}\alpha_{xy})\epsilon_{12}\Delta T \qquad (17)$$

The expression for the total potential energy of the laminate is of the form

$$\Pi = \int_x \int_y \int_z \Omega \,(\, a_i, b_i, \phi, x, y, z, \overline{Q}_{ij}, \Delta T, \alpha_i \,) \, dxdydz \qquad (18)$$

To satisfy equilibrium of energy, the first variation of Π must be zero, i.e.

$$\delta\Pi = (\partial\Pi/\partial a_1)\delta a_1 + (\partial\Pi/\partial b_1)\delta b_1 + (\partial\Pi/\partial a_2)\delta a_2 + (\partial\Pi/\partial a)\delta a + (\partial\Pi/\partial b)\delta b$$
$$+ (\partial\Pi/\partial c)\delta c + (\partial\Pi/\partial\phi)\delta\phi = 0 \qquad (19)$$

Therefore

$$\partial\Pi/\partial a_1 = f_1 = 0, \ \partial\Pi/\partial b_1 = f_2 = 0, \ \partial\Pi/\partial a_2 = f_3 = 0, \ \partial\Pi/\partial a = f_4 = 0,$$
$$\partial\Pi/\partial b = f_5 = 0, \ \partial\Pi/\partial c = f_6 = 0 \ \text{and} \ \partial\Pi/\partial\phi = f_7 = 0 \qquad (20)$$

Equation(20) make 7 nonlinear simultaneous equations.
   To minimize the energy, the equation(21) must be positive and definite. Nonlinear Newton–Raphson's iteration is used to calculate the curvatures numerically.

$$\delta^2\Pi =
\begin{bmatrix}
\dfrac{\partial f_1}{\partial a_1} & \dfrac{\partial f_1}{\partial b_1} & \dfrac{\partial f_1}{\partial a_2} & \dfrac{\partial f_1}{\partial a} & \dfrac{\partial f_1}{\partial b} & \dfrac{\partial f_1}{\partial c} & \dfrac{\partial f_1}{\partial \phi} \\[2ex]
\dfrac{\partial f_2}{\partial a_1} & \dfrac{\partial f_2}{ab_1} & \dfrac{\partial f_2}{\partial a_2} & \dfrac{\partial f_2}{\partial a} & \dfrac{\partial f_2}{\partial b} & \dfrac{\partial f_2}{\partial c} & \dfrac{\partial f_2}{\partial \phi} \\[2ex]
\dfrac{\partial f_3}{\partial a_1} & \dfrac{\partial f_3}{\partial b_1} & \dfrac{\partial f_3}{\partial a_2} & \dfrac{\partial f_3}{\partial a} & \dfrac{\partial f_3}{\partial b} & \dfrac{\partial f_3}{\partial c} & \dfrac{\partial f_3}{\partial \phi} \\[2ex]
\dfrac{\partial f_4}{\partial a_1} & \dfrac{\partial f_4}{\partial b_1} & \dfrac{\partial f_4}{\partial a_2} & \dfrac{\partial f_4}{\partial a} & \dfrac{\partial f_4}{\partial b} & \dfrac{\partial f_4}{\partial c} & \dfrac{\partial f_4}{\partial \phi} \\[2ex]
\dfrac{\partial f_5}{\partial a_1} & \dfrac{\partial f_5}{\partial b_1} & \dfrac{\partial f_5}{\partial a_2} & \dfrac{\partial f_5}{\partial a} & \dfrac{\partial f_5}{\partial b} & \dfrac{\partial f_5}{\partial c} & \dfrac{\partial f_5}{\partial \phi} \\[2ex]
\dfrac{\partial f_6}{\partial a_1} & \dfrac{\partial f_6}{\partial b_1} & \dfrac{\partial f_6}{\partial a_2} & \dfrac{\partial f_6}{\partial a} & \dfrac{\partial f_6}{\partial b} & \dfrac{\partial f_6}{\partial c} & \dfrac{\partial f_6}{\partial \phi} \\[2ex]
\dfrac{\partial f_7}{\partial a_1} & \dfrac{\partial f_7}{\partial b_1} & \dfrac{\partial f_7}{\partial a_2} & \dfrac{\partial f_7}{\partial a} & \dfrac{\partial f_7}{\partial b} & \dfrac{\partial f_7}{\partial c} & \dfrac{\partial f_7}{\partial \phi}
\end{bmatrix} \qquad (21)$$

NUMERICAL RESULTS AND DISCUSSIONS

In this paper, to observe the behavior of the cured shapes, curvatures and principal direction of curvature are calculated for unsymmetric laminates with arbitrary lay—up angles, dimensions and the number of layer.

The effect of spatial dependence of strains is examined for the angle—ply laminate showing the snap—through phenomenon. In fig.4, the magnitude of the curvatures and principal direction of curvature are plotted with different length—to—thickness ratios. As shown in fig.4, when angle—ply $[45/-45]_T$ laminate is snapped through, xy—curvature does'nt change the magnitude and direction while x and y—curvature and principal direction of curvature represent equal magnitude and opposite sign. For these stacking arrangement where snap—through phenomenon occurs, the effect of spatial dependence of strains is significant near the bifurcation point as shown in fig.5. We can see easily that the effect of spatial dependence of strains is negligible at both ends of x—axis in fig.5.

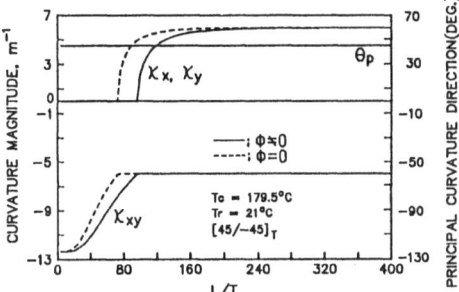

Figure 5. Effect of spatial dependence of in –plane strains on curvatures and principal direction of curvature in $|45/- 45|_T$ laminate.

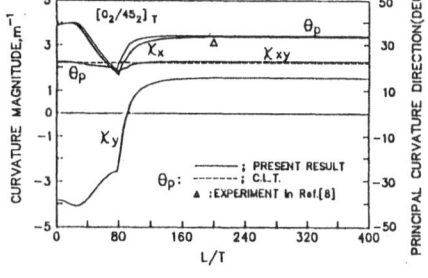

Figure 6. The behavior of principal direction of curvature in $[0_2/45_2]_T$ with various length to thickness ratios.

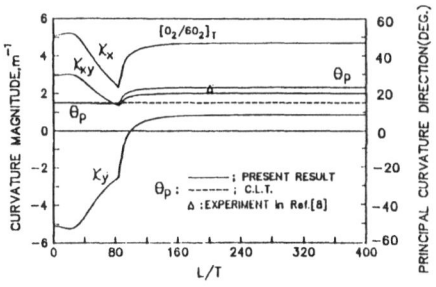

Figure 7. The behavior of principal direction of curvature in $|0_2/60_2|_T$ with various length to thickness ratios.

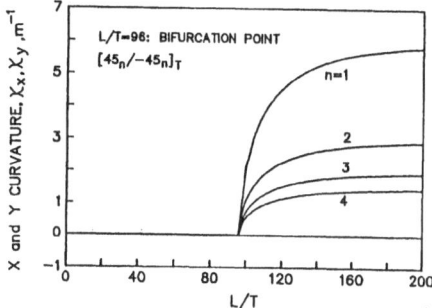

Figure 8. Number of layer effect on x and y curvature of $[45_n/-45_n]_T$ laminate.

In fig.6 and fig.7, the curvatures and principal direction of curvature are shown according to various length—to—thickness ratios. Principal direction of curvature for classical lamination theory, existing experiment and present results are compared for $[0_2/45_2]_T$ and $[0_2/60_2]_T$ laminates where snap—through phenomenon occurs. Classical lamination theory predicts that principal direction of curvature is 22, 31 degree for the experiment and 34 degree for present results in the $[0_2/45_2]_T$ 10cm x 10cm laminate of fig.6. We can here observe that principal direction of curvature (22 deg.) for classical lamination theory approaches present results before the bifurcation point beginning to show two real solutions, $L/T \leq 80$. For $[0_2/60_2]_T$ 10 cm x cm laminate in fig.7, the principal direction of curvature is 15 degree for the classical lamination theory, 21 degree for the existing experiment and 15 degree at $L/T \leq 86$, 22 degree at $L/T > 86$ of present results. We can conclude that classical lamination theory can predict the principal direction of curvature in the limited range of length to thickness ratios before snap–through phenomenon occurs.

In observing the effect of number of layer, as shown in fig.8 and 9, the snap through phenomenon exists at the $L/T = 96$ regardless of number of layer. According to the increase of the number of layer, absolute values of the curvatures decrease while principal direction of curvature is constant regardless of the number of layer. The laminate becomes flat with the change from unsymmetric(fig.8,9) laminate to antisymmetric stacking arrangement(fig.10,11,12). In this case, present results show the same results with classical lamination theory for the principal direction of curvature. It is shown that x–curvature is equal to y–curvature regardless of the number of layer for $[45/-45]_{nT}$, $[45_n/-45_n]_T$ laminates, n=1,2,3.... families of fig.8, 10,11,12.

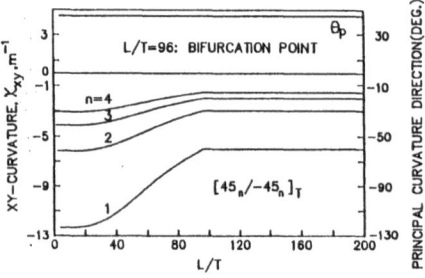

Figure 9. Number of layer effect on xy –curvature and principal direction of curvature in $[45_n/-45_n]_T$ laminate.

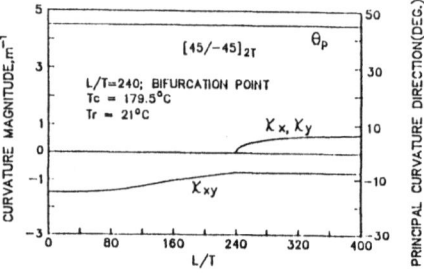

Figure 10. Room–temperature shape of antisymmetric angle ply $[45/-45]_{2T}$ with various length to thickness ratios.

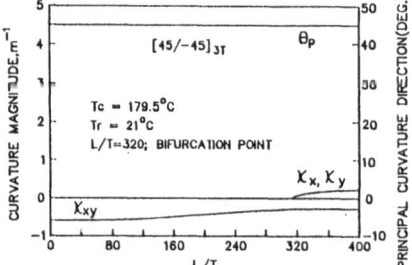

Figure 11. Room–temperature shape of antisymmetric angle ply $[45/-45]_{3T}$ with various length to thickness ratios.

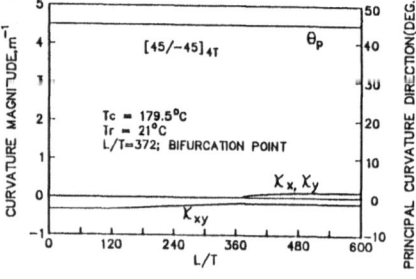

Figure 12. Room–temperature shape of antisymmetric angle ply $[45/-45]_{4T}$ with various length to thickness ratios.

Through fig. 13–16, the effect of lay–up angles on the curvatures and principal direction of curvature is shown in laminate. For $[0/90_2/\theta]_T$ laminate, the behavior of the curvatures and the principal direction of curvature are shown according to the change of lay–up angles. These laminates of families do not show snap–through phenomenon. According to the increase of lay–up angle, x–curvature increases, y and xy–curvatures increase and then decrease in the range of larger length to thickness ratios while the principal direction of curvature decreases and then becomes constant regardless of the length–to–thickness ratios.

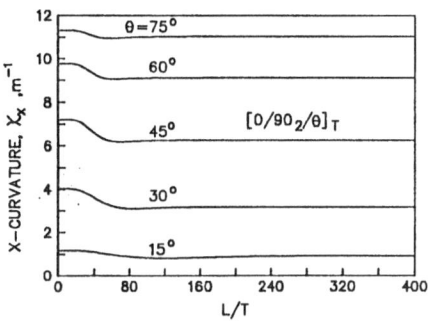

Figure 13. Effect of lay–up angle on x curvature of $[0/90_2/\theta]_T$ laminate.

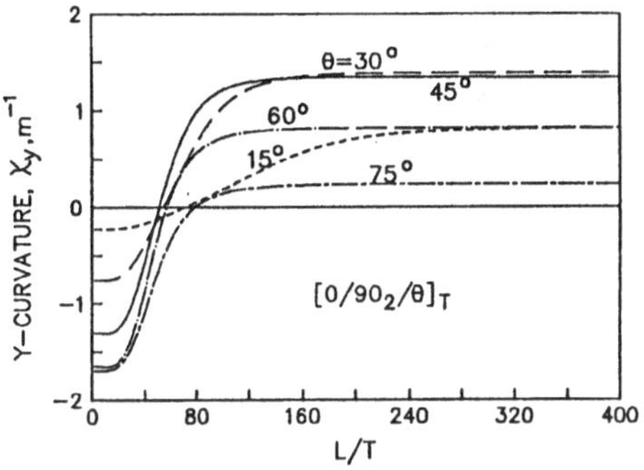

Figure 14. Effect of lay–up angle on y curvature of $[0/90_2/\theta]_T$ laminate.

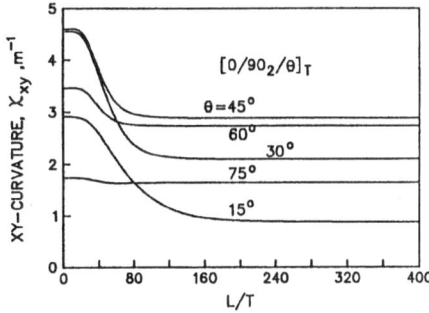

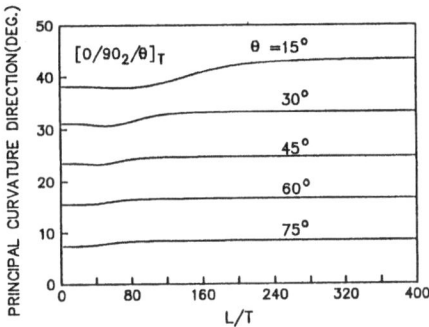

Figure 15. Effect of lay–up angle on xy curvature of $[0/90_2/\theta]_T$ laminate.

Figure 16. Effect of lay–up angle on principal direction of curvature in $[0/90_2/\theta]_T$ laminate.

CONCLUSIONS

Based upon the results of formulation involving all spatial variation of strains the following conclusion can be made. The principal direction of curvature calculated from classical lamination theory agrees with present results in the limited range of length–to–thickness ratios before the bifurcation point. The effect of spatial dependence of strains is significant near the bifurcation point especially when the snap–through phenomenon occurs. The shape of laminate became flat with the change from unsymmetric to antisymmetric stacking arrangement and the increase of number of layers as the case of cross–ply laminates. The bifurcation point where snap–through behavior begins to exist is constant regardless of the number of layers for unsymmetric angle–ply laminates with arbitrary lay–up angles.

REFERENCES

1. Tsai, S. W. and Hahn, H. T., Introduction to Composite Materials, 1980, Technomic Publishing Co.
2. Hyer, M. W., Calculations of the room–temperature shapes of unsymmetric laminates, J. of Composite Materials, 1981, 15, 296–310.
3. Hyer, M. W., The room–temperature shapes of four layer unsymmetric cross–ply laminates, J. of Composite Materials, 1982, 16, 318–40.
4. Hahn, H. T., Warping of unsymmetric cross–ply graphite/epoxy laminates, Composite Review, 1981, 3, 114–7.
5. Jun, W. J. and Hong, C. S., Warping analysis of unsymmetric laminated composites, Trans. of KSME, 1983, 7, 404–9.
6. Hyer, M. W. and Hamamoto, Temperature–curvature relations for unsymmetric laminates, Proceedings of the Third Japan–U.S. Conference on Composite Materials, 1986, 803–11.
7. Jun, W. J. and Hong, C. S., Effect of residual shear strain on the cured shape of unsymmetric cross–ply thin laminates, Int. J. of Composite Science and Technology, 1990, 38, 55–67.
8. Hyer, M. W., Some observations on the cured shape of thin unsymmetric laminates, J. of Composite Materials, 1981, 15. 175–94.
9. Jiali Dang and Yuzhang Tang, Calculation of the room–temperature shapes of unsymmetric laminates, Proceedings International Symposium on Composite Materials and Structures, China, 1986, 201–6.
10. Fung, Y. C., Foundation of Solid Mechanics, 1965, Prentice–Hall Inc.
11. Timoshenko, S. P. and Woinovsky–Krieger, Theory of Plates and Shells, 1981, McGraw–Hill, New York.

AVERAGE STRESS IN ELECTRODEPOSITS
BY THE BENDING STRIP METHOD

CHRISTO N. KOUYUMDJIEV
Department of Technical Mechanics
Technical University "Anghel Kanchev"
8, Komsomolska str., 7017 Rousse, BULGARIA

ABSTRACT

When using the strip bending method, the most frequently determined value is the average residual stress in a cross section of the deposit. There exist several relations for its calculation. In the present paper the accuracy of the different formulae is examined, using real experiment data for nickel deposits. Problems have been discussed concerning the application of these relations for greater displacements, ensuring an elastic state of the experimental sample.

INTRODUCTION

One of the oldest, however still widely used, mechanical methods for determining stress in electrodeposits is the bending strip method [7]. A thin elastic strip 0, isolated from one side, is coated on the other side with a thin layer 1 (see fig.1), deposited by electrolysis. The obtained sample bends (see fig.2). It is assumed [1, 3-7] that its deflected form is a circular arc. Then, using the measured displacement values f or δ, the radius of curvature R can be determined

$$R \approx \frac{1^2}{8f} \quad , \quad R \approx \frac{1^2}{2\delta} \quad , \tag{1}$$

as well as the stress in the deposit.

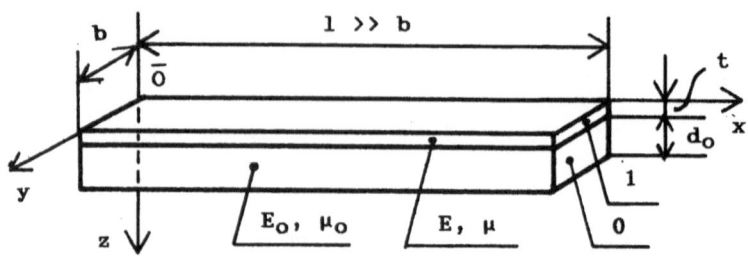

FIGURE 1.   Sample for stress determination (0 - cathode,
1 - electrodeposit)

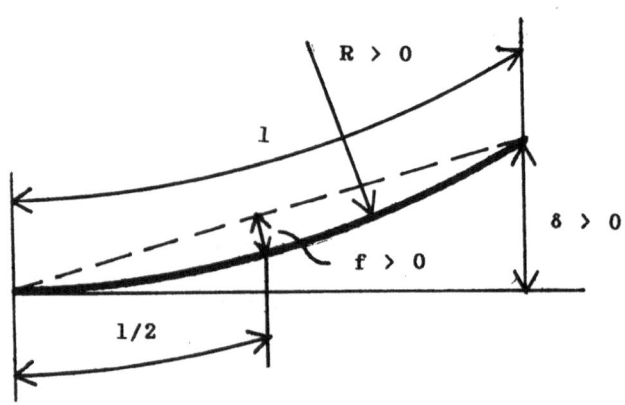

FIGURE 2.   Measured sample displacements f and δ.

There exist a number of formulae for calculating average
residual stress σ in a cross section of the deposit (Table 1).
They allow the calculation of σ by a measured final
displacement value f_{max} or δ_{max} corresponding to a final
deposit thickness value t_{max}. In [5], the shortcomings of
relations F1,...,F5 are analyzed and relation F6 is proposed.
It does not have these shortcomings, but has been derived
under the simplifying assumption that all deposit layers
emerge with identical initial strain. Therefore, its accuracy
will depend on whether real deposits conform to this

TABLE 1

Formulae for calculating average residual stress σ in the cross section of an electrodeposit

Ref.	Formula designation	Formula
7	F1	$\sigma = \sigma_{St}$
1	F2	$\sigma = \sigma_{St} (1 + \theta)^{-1}$
3	F3	$\sigma = \sigma_{St} (1 - \theta + \theta^2)$
2	F4	$\sigma = E \cdot \left(\dfrac{E\, t_{max}\, \bar{\varepsilon}}{E_0\, d_0 + E\, t_{max}} - \bar{\varepsilon} + \right.$ $\left. + \dfrac{E\, t_{max}\, \bar{\varepsilon}\, (d_0 - e + t_{max}/2)^2}{E/3\, [(d_0+t_{max}-e)^3-(d_0-e)^3] + E_0/3[(d_0-e)^3 + e^3]} \right)$ $\bar{\varepsilon} = -\dfrac{2\delta_{max}}{3l^2} \dfrac{E[(d_0+t_{max}-e)^3-(d_0-e)^3]+E_0[(d_0-e)^3+e^3]}{E\, t_{max}\, (d_0 - e + t_{max}/2)}$ $e = \dfrac{d_0}{2} + \dfrac{E\, t_{max}\, (d_0 + t_{max})}{2\, (E_0\, d_0 + E\, t_{max})}$
4	F5	$\sigma = \sigma_{St} (1 - \mu_0)^{-1}$
5	F6	$\sigma = \sigma_{St} (1 + \tilde{\gamma}\, \theta^3) (1 - \mu_0)^{-1} (1 + \theta)^{-1}$
6	F7	$\sigma = \dfrac{\tilde{E}_0 d_0^2}{l^2} \cdot \dfrac{1}{3 + 4\theta + \tilde{\gamma}\theta^2} [(1 - \tilde{\gamma}\theta^2)\delta_{max}t_{max}^{-1} +$ $+ \tilde{\gamma}\theta\, \dfrac{\tilde{\gamma}\theta_1^4 + 4\theta_1^3 + 6\theta_1^2 + 4\theta_1 + \tilde{\gamma}^{-1}}{3\,(1 + \theta_1)} \dfrac{\delta_1}{t_1}]$

Designations used:

$$\theta = \frac{t_{max}}{d_0} \quad ; \quad \theta_1 = \frac{t_1}{d_0} \quad ; \quad \gamma = \frac{E}{E_0} \quad ;$$

$$\sigma_{St} = \frac{E_0}{3} \left(\frac{d_0}{l} \right)^2 \frac{\delta_{max}}{t_{max}} = \frac{4E_0}{3} \left(\frac{d_0}{l} \right)^2 \frac{f_{max}}{t_{max}} \quad ;$$

$$\tilde{E} = \frac{E}{1 - \mu} \quad ; \quad \tilde{E}_0 = \frac{E_0}{1 - \mu_0} \quad ; \quad \tilde{\gamma} = \frac{\tilde{E}}{\tilde{E}_0}$$

assumption. In [6] a precise relation F7 is proposed which
allows a more exact calculation of average stress σ for an
arbitrary law of distribution of initial strain along the
thickness of the deposit. Calculation is carried out using two
measured displacement values δ_{max} and δ_1 corresponding to
thickness values t_{max} and t_1. If the deposit is imaginarily
divided into thin layers, parallel to the surface of the
substrate, then t_1 is the thickness of the deposit layer
adjoining the substrate.

    In the present work some points have been discussed
concerning the accuracy and the conditions of application of
formulae F1 and F5, most frequently used so far, as well as of
the new formulae F6 and F7, for real electrodeposits.

RESULTS AND DISCUSSION

The experimental data used here have been obtained by
S.Armyanov and G.Sotirova in the Institute of Physical
Chemistry at the Bulgarian Academy of Science in Sofia.
A nickel deposit (with Young's modulus E = 199900 MPa and
Poisson's ratio μ = 0.32) was deposited on an elastic platinum
strip cathode (with Young's modulus E_0 = 153800 MPa and
Poisson's ratio μ_0 = 0.44) with dimensions l = 41 mm, b = 5 mm,
d_0 = 0.22 mm. The displacement values δ_i of the free end of
the cantilever cathode were measured in relation of the
deposit thickness t_i.

TABLE 2

Measured experimental values of deposit thickness t_i
and cathode displacement δ_i; calculated residual stress
values σ_i in the i-th deposit layer

Experiment 1			Experiment 2		
t_i	δ_i	σ_i	t_i	δ_i	σ_i
μm	μm	MPa	μm	μm	MPa
1	2	3	4	5	6
1	214.3	405.0	17.3	3333.3	493.4
2	452.4	485.1	17	3285.7	655.6
3	642.9	373.0	16	3071.4	883.9
4	847.6	426.4	15	2776.2	802.2
5	1038.1	402.3	14	2500.0	765.2
6	1238.1	443.6	13	2228.6	757.3
7	1428.6	431.9	12	1952.4	651.2

TABLE 2 (cont'd)

Experiment 1			Experiment 2		
t_i	δ_i	σ_i	t_i	δ_i	σ_i
μm	μm	MPa	μm	μm	MPa
1	2	3	4	5	6
8	1619.0	446.4	11	1704.8	603.9
9	1828.6	516.9	10	1466.7	557.7
10	2023.8	491.3	9	1238.1	472.1
11	2228.6	535.2	8	1033.3	429.3
12	2404.8	466.3	7	838.1	361.4
13	2595.2	523.3	6	661.9	321.9
14	2781.0	524.8	5	495.2	244.1
15	2961.9	525.0	4	352.4	182.4
16	3133.3	510.6	3	228.6	122.7
17	3285.7	465.5	2	123.8	27.0
17.3	3333.3	493.4	1	52.4	-27.4

The results for two experiments are given in Table 2. Experiment 1 refers to deposition and experiment 2 refers to deposit dissolution. The values of residual stress σ_i in the cross section of the i-th deposit layer are given in columns 3 and 6 of Table 2. These values have been computed in the Technical University "Anghel Kanchev" in Rousse using original software based on the relations from [6].

Comparison of the Results Obtained by the Different Formulae
In [8], a numerical check has been carried out on whether average residual stress values σ, calculated by the formulae in Table 1, are close to the average value $\bar{\sigma}$ of residual stress values σ_i:

$$\bar{\sigma} = \frac{1}{t_{max}} \sum_{i=1}^{n} \sigma_i \, \Delta t_i \, , \qquad (2)$$

where Δt_i is the thickness of the i-th layer and n is the total number of layers.

The average values calculated by this formula for both experiments in Table 2 are given in the last row of Table 3. For the purpose of comparison the average stress is also calculated by formulae F1, F5, F6 and F7 from Table 1. The data in Table 3 show that the results obtained by Stoney's formula F1 are not precise. This is due to the lack of conformity to the basic assumptions under which the formula has been derived: small θ ratio, linear state of stress of the experimental sample, and equal Young's moduli for cathode and deposit. Similar are the results from the application of formulae F2,...,F4 which likewise do not take into consideration the biaxiality of the sample state of stress. Also unsatisfactory are the results obtained by the very frequently used formula F5 which in fact is a generalization of formula F1 for the case of a biaxial state of stress.

TABLE 3

Values of average residual stress σ, MPa
in the cross section of an electrodeposit,
calculated by different formulae

Formula	Experiment 1	Experiment 2
1	2	3
F1	285	285
F5	507.9	507.9
F6	471.1	471.1
F7	469.3	459.0
$\bar{\sigma}$ according to formula (2)	469.4	460.0

Formula F6 gives a very good result for experiment 1 and a good one for experiment 2. The hypothesis for identity of initial strain in all deposit layers which has been adopted for the derivation of relation F6, has been evidently better realized in experiment 1 than in 2. Despite the unfavourable circumstances in experiment 2 (where residual stress σ_i even changes its sign - see column 6 in Table 2), the error in the result obtained by formula F6 does not exceed 2.5%.

The average stress values calculated by the approximate formula F7 almost fully coincide with those obtained by relation (2). Indeed, formula F7 requires two measurements for each experiment and more calculation work, but the results practically coincide with the value of $\bar{\sigma}$, regardless of the distribution of initial strain along the thickness of the electrodeposit.

A further interest may lie in an experimental testing of the accuracy of relations F6 and F7 with other cathode-deposit material combinations.

On the Validity of the Relations for Greater Sample
Displacements

Relations F5, F6 and F7 have been derived basing on the
linear theory of thin elastic plates. One of the assumptions
of this theory is the requirement for negligibility of the
normal displacement of a plate in comparison with its
thickness, i.e. $f_{max} \ll d_o + t_{max}$. In practice this
requirement is frequently unfulfilled. For example Stoney [7]
has used a sample with dimensions $l = 102$ mm, $b = 12$ mm,
$d_o = 0.31$ mm, $t_{max} = 0.046$ mm and has measured a displacement
value $f_{max} = 3.45$ mm, i.e. several times greater than the
thickness. So the question arises whether the relations are
applicable for these practical cases as well.

The problem of the applicability of formula F5 for
greater displacements was first raised and discussed in [9]
but no final statement was reached. In [10] a study has been
made of the applicability of formula F6 using the
geometrically non-linear theory of elasticity. It has been
found that the formula is approximately accurate also for
arbitrarily great displacements (while sample deformation is
still elastic) if cathode width b satisfies the condition

$$b \leq \frac{l^2}{\delta_{max}} \arccos [1 - \lambda \frac{1-\mu-2\mu^2}{(1-\mu)^2} \frac{d_o\delta_{max}}{3\tilde{\gamma}\theta.l^2} (\frac{1+\tilde{\gamma}\theta^3}{1 + \theta} + 3\tilde{\gamma}\theta^2)] . \quad (3)$$

Here λ is the chosen admissible value for the relative
error in the value of maximal stress in the deposit due to the
linearization of the non-linear problem.

For example, if we use the numerical data from experiment
No.1 and assume that $\lambda = 0.04 = 4\%$, from relation (3) it
follows that $b \leq 5.9$ mm. So for the width $b = 5$ mm, chosen for
the experiment, the error due to linearization is less than 4%.
If this error is required to be less than 1%, the width should
be $b < 2.9$ mm.

There also stands the question whether, in case of
greater displacements, the approximate relations (1) can be
used instead of the corresponding accurate ones

$$\delta = 2R \sin^2 \frac{l}{2R} \quad , \quad f = 2R \sin^2 \frac{l}{4R} . \quad (4)$$

The conditions for the approximate validity of relations
(1) can be written as

$$\Delta_1 = \frac{(\frac{l}{2R})^2- \sin^2\frac{l}{2R}}{\sin^2\frac{l}{2R}} \leq \lambda_1 , \Delta_2 = \frac{(\frac{l}{4R})^2- \sin^2\frac{l}{4R}}{\sin^2\frac{l}{4R}} \leq \lambda_2, \quad (5)$$

where λ_1 and λ_2 are the chosen values for the admissible relative error. For example, for the experiment No.1 viewed above R $\approx$ 252 mm and from (5) it follows that the first approximate relation from (1) is valid with a relative error $\Delta_1 = 0.22$ %.

If the conditions (5) are unfulfilled, then the radius of curvature R or the curvature $\ae = R^{-1}$ should be determined by the transcendental equations (4). For this purpose the following iteration processes can be used:

$$\ae^{(j+1)} = \frac{1}{1} \arccos(1-\delta\ae^{(j)}), \quad \ae^{(j+1)} = \frac{2}{1} \arccos(1-f\ae^{(j)}),(6)$$

which are convergent for a wide range of practical cases. If in the first condition (5) it is assumed that $\lambda_1 = 1$ %, then the relative error of the first relation (1) would be $\Delta_1 \geq 1$ % and the first iteration process from (6) should be carried out if $R/l \leq 2.895$ or, according to (4), if $\delta/R \geq 0.059$. Examinations have shown that this process is convergent at $0.059 \leq \delta/R \leq 1.6$.

Criteria for Sample Elasticity

All relations viewed so far have been derived under the assumption that the materials of cathode and deposit are in an elastic state. Therefore, the calculated values of initial stress σ_{ii} and residual stress σ_i in the cross-section of the deposit [6] as well as of the maximal normal stress $\max\sigma_0$ in the cross section of the cathode have to satisfy the conditions

$$\sigma_{ii} < \sigma_E \quad , \quad \sigma_i < \sigma_E \quad , \tag{7}$$

$$\max\sigma_0 < \sigma_{Eo} \, , \tag{8}$$

where σ_E and σ_{Eo} are the maximal stress values up to which the materials of the deposit and the cathode respectively remain in an elastic state.

Taking into consideration the relation for $\max\sigma_0$ (19) from [5], from condition (8) it follows that

$$R > [R] \approx \frac{\tilde{E}_0 d_0}{6\sigma_{Eo}} \left(\frac{1+\tilde{\gamma}\theta^3}{1+\theta} + 3 \right) \approx \frac{2\tilde{E}_0 d_0}{3\sigma_{Eo}} \tag{9}$$

and from (4) we can determine approximately the maximal displacement value [δ] or [f] up to which the cathode is in an elastic state.

The [δ] value can also be determined experimentally. The free end of the cantilever cathode is loaded with a transverse force F. Statically applying greater and greater force followed by unloading makes it possible to determine the maximal displacement δ_F of the force application point, at which the cathode is still in an elastic state. Then, from the condition of equivalence of the state of stress in the most threatened cathode points for both cases (i.e. for this experiment and for an electrodeposition), we obtain

$$[\delta] = \frac{9}{8} (1 - \mu_0) \delta_F \quad . \tag{10}$$

CONCLUSIONS

1. There exist a number of formulae (Table 1) for calculating average residual stress in a cross section of an electrodeposit formed on one of the sides of a bending elastic strip cathode. A part of them (F1,...,F4) assume a linear state of stress in the sample, which is acceptable at $b \ll d_0 + t_{max}$. Samples of such dimensions are not used in experimental practice. The other relations in Table 1 take into consideration the biaxiality of the state of stress in the sample. Using real experiment data for nickel electrodeposits, it has been established that for the most unfavourable experiment (No.2), the results obtained by formulae F5, F6 and F7 differ from the average value $\bar{\sigma}$, obtained using (2), by 10 %, 2 % and 0.2 % respectively.

2. The relations F5, F6 and F7 are applicable also for greater displacements $\delta_{max} > d_0 + t_{max}$, if cathode width b satisfies the condition (3). If conditions (5) are unfulfilled and thus the approximate relations (1) are inapplicable, curvature can be determined by one of the iteration processes (6) whose convergence has been examined and confirmed.

3. Attention has been called to the necessity of checks (7) and (8) concerning the elastic state of cathode and deposit. A simple experiment and a corresponding relation (10) have been proposed for determining the maximal displacement [δ] up to which there will occur no plastic deformations in the cathode as a result of electrodeposition.

ACKNOWLEDGEMENTS

The author expresses his gratitude to S.A.Armyanov and G.S.Sotirova of the Institute of Physical Chemistry at the Bulgarian Academy of Science, Sofia, for providing their own experimental data to serve the study in the present work, as well as for discussing essential parts of it.

REFERENCES

1. Barklie, R.H.D. and Davies, H.I., The effect of surface conditions and electrodeposited metal on the resistance of materials to repeated stresses, Proc. Mech. Inst. Eng., 1930, pp. 731-750.

2. Birger, I.A., Residual Stress, Mashgiz, Moscow, 1963, p. 232 (in Russian).

3. Brenner, A. and Senderoff, S., Calculation of stress in electrodeposits from the curvature of a plated strip, J. Res. Nat. Bur. Stand., 1949, 42, pp. 105-123.

4. Finegan, J.D., A.E.C. Techn. Rept. 15, Case Institute of Technology, Cleveland, Ohio, 1961.

5. Kouyumdjiev, C.N., A calculation of average stress in electrodeposits based on the bending of a flat plate substrate, Surf. Techn., 1985, vol.26, No.1, pp. 35-44.

6. Kouyumdjiev, C.N., Residual stress distribution by the bending strip method, Surface and Coatings Technology, 1986, vol.28, No.1, pp. 39-55.

7. Stoney, G.G., The tension of metallic films deposited by electrolysis, Proc. Roy. Soc., 1909, A82, pp. 172-175.

8. Kouyumdjiev, C.N., Average stress in electrodeposits by the bending strip method, Proc. VTU "A.Kanchev", Rousse, Bulgaria, 1986, 28, No.11, pp. 143-148 (in Bulgarian).

9. Hoffmann, R.W., In Physics of Thin Films, ed. G. Hass and R.E. Thun, Academic Press, New York and London, 1966, vol.3, p. 211.

10. Kouyumdjiev, C.N., On the validity of a theoretical model for determining stress in electrodeposits for greater displacements, Theoretical and Applied Mechanics, Bulgarian Academy of Science, 1989, 20, No.3, pp. 39-54, (in Bulgarian).

THE APPLICATION OF PHOTOELASTIC TECHNIQUES IN ORTHOPAEDIC ENGINEERING

J.F.Orr, P.K.Humphreys, W.V.James, A.S.Bahrani

Department of Mechanical and Manufacturing Engineering

The Queen's University of Belfast

Ashby Building, Stranmillis Road, Belfast, BT9 5AH

ABSTRACT

The earliest applications of photoelastic methods to study the stresses in bone were recorded during the late 1930's, but studies of bone and orthopaedic implants using photoelastic models are not common. The structural characteristics of bone have interested researchers for more than one hundred years and photoelastic experiments permit ready comparison of the highly anisotropic structure of cancellous bone with principal stress directions.

The treatment of bone injuries and joint defects by implanting artificial components presents further opportunities for evaluating new devices through study of photoelastic models. Initially stresses within the implanted components were studied, but today the fixation of such devices is also of interest, in response to the recognition of causes of failure of joint replacements.

Photoelastic models offer useful predictions of the performance of orthopaedic implants and their influence on surrounding bone. More recently, other experimental methods have been developed to measure stress in components, however photoelastic techniques can still make an important contribution in the field of orthopaedic engineering.

INTRODUCTION

The phenomenon of double refraction under strain, upon which photoelastic stress analysis relies was discovered in 1815 by Brewster. The relationship between double refraction and stresses or strains were subsequently

112

derived by physicists including Neumann and Wertheim [1]. The development of photoelasticity advanced greatly after 1906 when Coker applied 'Celluloid' as a modelling material to replace glass which has low optical sensitivity and is difficult to cut [1]. The first application of the Frozen Stress Method was published by Solakian in 1935. Maxwell recorded the optical effect in a jelly material which had solidified while under elastic deformation in 1850 but did not further investigate his observation [2].

PHOTOELASTIC MODELLING OF BONE

Hip Joint

Investigation of the structure of bone was carried out during the nineteenth century, by Meyer and Wolff, but the first results of applying photoelastic stress analysis to the study of bone were described by Milch in 1940 [3]. This paper makes reference to some slightly earlier work by Alexander in 1936 but Milch was unable to find results of the work at the time of writing. He was aware of the recent development of the frozen stress method but no practical work of this type was undertaken. The experiments involved two dimensional modelling of the upper end of the femur and discussing the effects of surgical modification of bone on the principal stress directions. A further investigation of such orthopaedic procedures was reported by Dylag et al in 1964 [4].

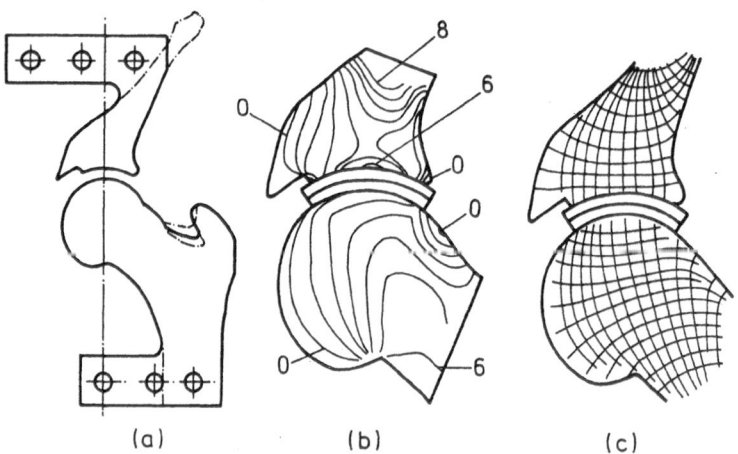

Figure 1.   Two dimensional model of the hip joint by Fessler [5].
(a) Outline of the Photoelastic model   (b) Isochromatic fringes
(c) Stress Trajectories derived from Isoclinic fringes.

In a paper published in 1957 Fessler [5] discusses an investigation of load transfer through the human hip joint, Figure 1. The important issue of the applicability of results from two dimensional models of the bones is addressed. While quantitative measurements were not found to be appropriate, conclusions were drawn regarding the distribution of joint contact stresses and the direction of principal stresses relative to bone architecture. Williams and Svensson [6] used three dimensional, frozen stress models to examine the neck of the femur. They assumed that the strain distribution in the neck of their models, as indicated by isochromatic fringes, was representative of that of the prototype bone. They then derived stresses in the cortical bone, at the surface and internal cancellous bone by applying appropriate moduli of elasticity for each material. This work, representing the non-homogeneous nature of bones, is rare in photoelastic modelling.

The representation of bone using a homogeneous material often gives rise to debate as does the consideration that bone is also non-isotropic. The latter consideration is mentioned in a paper by Holm [7], describing a two dimensional experiment to compare the trabecular orientation of cancellous bone with stress trajectories derived from isoclinic fringes. Holm concludes: "...the bones may be considered to behave as an orthotropic material, i.e. when they are loaded in a physiological way they behave as an isotropic material".

A novel use of photoelastic analysis was performed by Murikami et al [8], when they used urethane rubber to represent cartilage in a model of the hip joint and observed stresses due to joint contact and friction. This work was to investigate lubrication conditions in the joint.

A German orthopaedic surgeon, named Pauwels, used photoelastic models to demonstrate stress distributions and explain the form of bones as they are found in the body and their response to disturbance of their geometry due to disease or injury. Papers published from the 1930's to 1976 have been collated into book form and translated into English [9]. The work of Pauwels was frequently referred to by Kummer in his subsequent paper which explains the functional structure of bone with reference to photoelastic models [10].

Knee Joint

Photoelastic modelling of the knee joint has not been frequently reported. The work of Muller et al [11] involved the frozen stress method and is the earliest reference, which has been found, to three dimensional photoelastic modelling in orthopaedics. This photoelastic study of the knee involved the patella, or kneecap, a bone which is subjected to tensile forces from the patellar tendon and ligament and also to compressive forces where it contacts the femur. The stress directions which the internal structure of the bone could optimally support coincided with those predicted by the experimental method. A further study of the patella was recorded by Muller et al in 1970 [12].

The main articulation of the knee between the femur and tibia has been studied for the purposes of understanding normal and pathological stress distributions at the joint and the effects of surgical procedures [13]. Experimental models have been demonstrated to yield information relevant to the understanding of abnormalities of the knee and to the design and fixation of joint replacements [14]. Frozen stress models were required in these experiments due to the complex geometry of the knee in three dimensions.

Foot and Ankle

The authors carried out a study of the ankle joint in order to compare the internal structure of the talus bone and the stress distributions expected during walking [15]. At this time only one previous anatomical study of this region of the body included photoelastic experiments [16]. The talus is the bone which transmits all the axial limb forces from the tibia to the foot. Modelling the forces acting on the talus is simplified due to it having no attachments to muscles. Three dimensional frozen stress models were cast from an anatomical specimen and loaded between replicas of the opposing joint surfaces of the tibia and foot. The models were sectioned in three orthogonal planes and isoclinic and isochromatic fringes examined, Figure 2. The trabecular orientations within bones are often stated to indicate principal stress directions but in this case dynamic loading and joint rotations during walking cause variation of stress directions during each step. The research indicated that the bone structure is optimally aligned to support the maximum ankle joint force, during the 'push off' phase of the walking cycle.

Two recent publications by Kihara et al are of interest due to the application of scattered light photoelasticity to the study of three-dimensional models of the ankle joint [17][18]. Isochromatic fringes were recorded at the articulation of the tibia and talus for three positions of the joint. Scattered light methods would be very attractive for the study of joints and joint replacements however the specialised nature of the equipment and interpretation of fringes probably accounts for the rarity of such work.

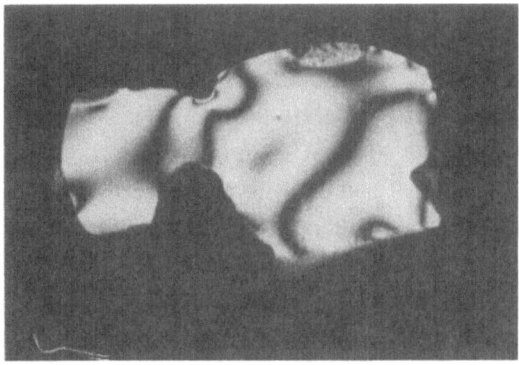

Figure 2 Isochromatic fringes from a frozen stress model of the talus [15].

FRACTURE FIXATION

It was appreciated from early work that there are considerable assumptions in modelling bone using homogeneous, isotropic casting resins but such problems do not arise in the modelling of orthopaedic implants. The modelling of both bone and implants was reported by Haboush in 1952 [19]. The object of the experiments was to examine stresses in nails and nail plates used to fix femoral neck fractures and to establish the optimum attitude of insertion. The theme of fracture fixation was further studied, using two-dimensional models, by Steen Jensen [20][21]. Initial work achieved a model of the upper femur subjected to compressive loading on the femoral head and tensile loading on the greater trochanter. Loading was verified by comparison with stress trajectories in the femur as recorded by previous anatomical researchers, notably Pauwels. This useful general model for femoral loading was used to investigate stresses in nail plates of varying design and attitudes of insertion.

JOINT REPLACEMENTS

Applications of photoelastic techniques to study joint prostheses have been reported in recent years although Kennedy et al reported studies of stresses in bone cement securing hip prostheses in 1979 [22]. This is a novel application which addresses the important problems of loosening of the hip prosthesis stem. Orr et al, investigated stresses in hip prosthesis stems in different attitudes of loading [23] and this work has been continued to examine the effects of loosening within the cement mantle on stem stresses. Cement stresses were also studied but using two-dimensional models of stem cross-sections surrounded by photoelastic material [24]. This allowed selection of stem profiles which reduced cement stress concentrations under medio-lateral loading, Figure 3. Ascough recently reported three dimensional modelling of hip prosthesis stems to complement finite element and simple bending theory approaches to evaluating fatigue performance of stems [25].

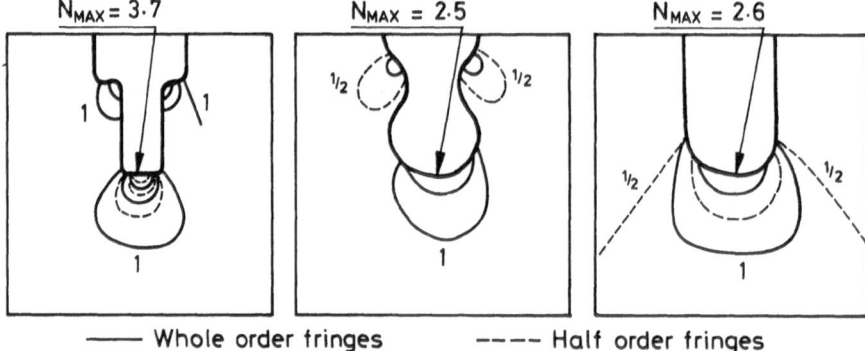

Figure 3. Isochromatic fringes around stem cross-sectional profiles [24].

All the prosthesis applications so far discussed have related to the femoral stems of hip prostheses, however the acetabular component also requires research into fixation and load transmission to the pelvis. Miles and McNamee conducted a two dimensional study to observe the effect of the joint centre position on pelvic stresses [26], Figure 4. This work demonstrates the importance of the interaction of anatomical structure, implant design and surgical technique in determining the likely success of fixation.

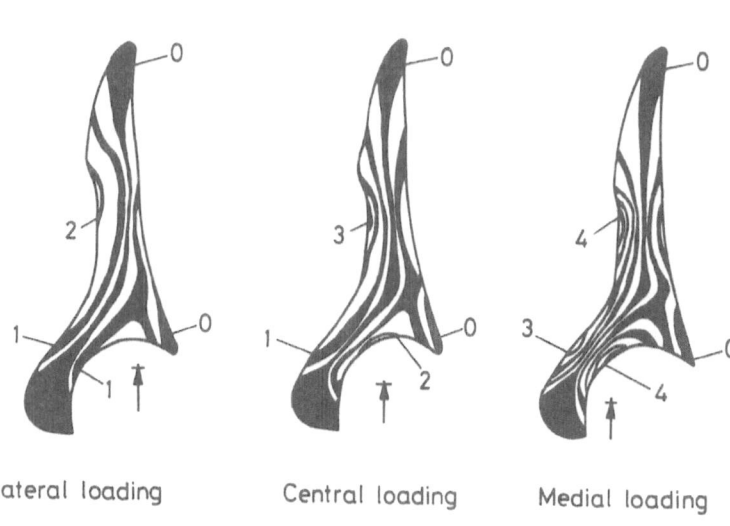

Figure 4.   Isochromatic fringe patterns in a model pelvis for three modes
of loading [26].

Photoelastic stress analysis was first applied to orthopaedic study
about the same time as the development of modern joint replacements
commenced.   Fessler was among the earliest researchers in reporting his
investigation of load transmission through the hip joint.   It is therefore
appropriate that the frozen stress method has been applied by Fessler to
optimise the design of ceramic heads for a current hip prosthesis design
[27][28].   These comprehensive publications report the methods of
determining the correct coefficient of friction at the conical head/stem
interface and the reproduction of these values in frozen stress models,
Figure 5.   The result was to identify an optimum internal profile for the
heads to reduce the probability of failure of the ceramic material due to its
exposure to tensile stresses.

118

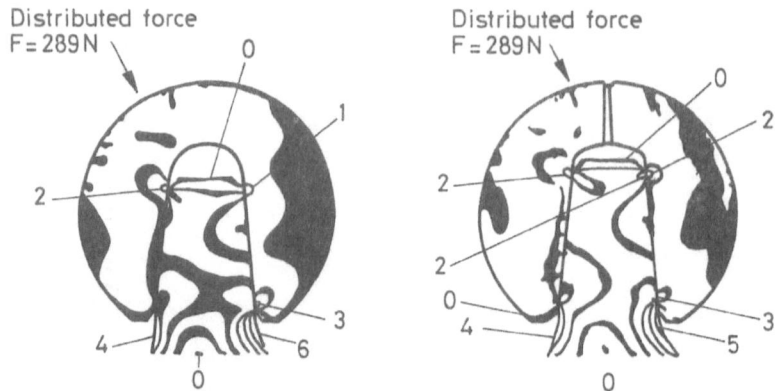

Figure 5. Isochromatic fringe patterns at the stem/head interface of the femoral component of a hip prosthesis [28].

CONCLUSIONS

At the time of this review it is fifty years since the first publication of a photoelastic investigation into bone structure. The photoelastic technique has undergone some development since that time, benefiting from the availability of new materials and in some cases the application of new technology such as coherent light from lasers. It is of interest that the numbers of publications concerning photoelastic investigation of bone and orthopaedic components have steadily increased during each decade of the period reviewed. It appears that the sustained interest in photoelastic techniques is continuing, or even expanding, with particular application to implant development.

ACKNOWLEDGEMENTS

The diagrams in Figure 1 are reproduced from the Journal of Bone and Joint Surgery by permission of the Editor.
Figures 3, 4, and 5 are reproduced from Engineering in Medicine and the Proceedings of the Institution of Mechanical Engineers by permission of the Institution.

REFERENCES

1   Frocht, M.M., Photoelasticity Vol.1, John Wiley and Sons, New York, 1965, pp 135-136, 325.

2   Drucker, D.C., Three-Dimensional Photoelasticity. In Handbook of Experimental Stress Analysis, ed. M. Hetenyi, John Wiley and Sons, New York, 1950, pp 924-976.

3   Milch, H., Photo-elastic Studies of Bone Forms, The Journal of Bone and Joint Surgery, 1940, 22, 621-626.

4   Dylag, Z., Kreczko, R., Orlos, Z., Photoelastic Study of the Effect of McMurray Osteotomy on the Mechanism of Work of the Hip Joint, Bulletin De L'Academie Polonaise Des Sciences, 1964, 12, 281-287.

5   Fessler, H., Load Distribution in a Model of a Hip Joint, The Journal of Bone and Joint Surgery, 1957, 39, 145-153.

6   Williams, J.F., Svensson, N.L., An Experimental Stress Analysis of the Neck of the Femur, Med. and Biol. Engng., 1971, 9, 479-493.

7   Holm, N.J., The Development of a Two Dimensional Stress-optical Model of the Os Coxae, Acta Orthop. Scandinavia, 1981, 41, 608-618.

8   Murakami, T., Ohtsuki, N., Chikama, H., Toyonaga, T., Nishizaki, H., Nishio, A., Tribological Study Using Human Joint Models, In Proceedings of the 8th International Congress of Biomechanics, 1981, Japan.

9   Pauwels, F., Biomechanics of the Locomotor Apparatus, Springer Verlag, Berlin 1980.

10 Kummer, B., Photoelastic Studies on the Functional Structure of Bone, Folia Biotheoretica, 1966, 6, 31-40.

11 Muller, J.M., Pupin, P., Hureau, J., Three Dimensional Photoelastic Study of the Patella. Method and Preliminary Results, Bull. Assoc. Anat., 1964, 184, 89-95.

12 Odilio, L da S, Bratt, F., Stress Trajectories in the Patella, Acta Orthop. Scandinavia, 1970, 41, 608-618.

13 Nishizaki, H., Ohtsuki, N., Murakami, T., Chikama, H., Toyonaga, T., Nishio, A., Photoelastic Study of Models Simulating Human Knee Joint, In Proceedings of the 8th International Congress of Biomechanics, 1981, Japan, pp. 144-149. (in Japanese)

14 Chand, R., Haug, E., Rim, K., Stresses in the Human Knee Joint, Journal of Biomechanics, 1976, 9, 417-422.

15 Orr, J.F., Experimental Measurement of Bone and Joint Replacements, PhD. Thesis, 1985, The Queen's University of Belfast.

16 Gierse, H., The Cancellous Structure in the Calcaneus and Its Relation to Mechanical Stressing, Anatomy and Embryology, 1976, 150, 63-83.

17 Kihara, T., Unno, M., Kitada, C., Kubo, H., Nagata, R., Three Dimensional Stress Distribution Measurement in a Model of the Human Ankle Joint by Scattered-light Polarizer Photoelasticity, Applied Optics, 1985, 24, 3363-3367.

18   Kihara, T., Unno, M., Kitada, C., Kubo, H., Nagata, R., Three Dimensional Stress Distribution Measurement in a Model of the Human Ankle Joint by Scattered-light Polarizer Photoelasticity: Part 2, Applied Optics, 1987, 26, 643-649.

19   Haboush, E.J., Photoelastic Stress and Strain Analysis in Cervical Fractures of the Femur, Bulletin Hosp Joint Diseases, 13, 1952, 252-258.

20   Steen Jensen, J., A Photoelastic Study of a Model of the Proximal Femur, Acta Orthop. Scandinavia, 1978, 49, 54-59.

21   Steen Jensen, J., A Photoelastic Study of the Hip Nail-Plate in Unstable Trochanteric Fractures, Acta Orthop. Scandinavia, 1978, 49, 60-64.

22   Kennedy, F.E., Collier, J.P., Komornik, L.A., An Experimental Study of the Stress Distribution in Bone Cement Used to Grout Standard and Porous Coated Hip Prostheses, Advances in Bioengineering, 1979 ASME Annual Meeting, 75-78.

23   Orr, J.F., James, W.V., Bahrani, A.S., A Preliminary Study of the Effects of Medio-lateral Rotation on Stresses in an Artificial Hip Joint, Engineering in Medicine, 1985, 14, 39-42.

24   Orr, J.F., James, W.V., Bahrani, A.S., The Effects of Hip Prosthesis Stem Cross-sectional Profile on the Stresses Induced in Bone Cement, Engineering in Medicine, 1986, 15, 13-18.

25   Ascough, J., Gorywoda, M. Mathematical and Experimental Modelling in Hip Joint Prosthesis Design, In The Changing Role of Engineering in Orthopaedics, The Institution of Mechanical Engineers, 1989, pp. 27-32.

26   Miles, A.W., McNamee, P.B., Strain Gauge and Photoelastic Evaluation of the Load Transfer in the Pelvis in Total Hip Replacement: The Effect of the Position of the Axis of Rotation, Proc. Instn. Mech. Engrs. Part H: Journal of Engineering in Medicine, 1989, 203, 103-107.

27   Fessler, H., Fricker, D.C., Friction in Femoral Prosthesis and Photoelastic Model Cone Taper Joints, Proc. Instn. Mech. Engrs. Part H: Journal of Engineering in Medicine, 1989, 203, 1-14.

28   Fessler, H., Fricker, D.C., A Study of Stresses in Alumina Universal Heads of Femoral Prostheses, Proc. Instn. Mech. Engrs. Part H: Journal of Engineering in Medicine, 1989, 203, 15-34.

INTERACTION OF FEMUR-HIP JOINT ENDOPROSTHESIS WITH ARTICULATED STEM

JITKA JÍROVÁ, TAMARA KAUFLEROVÁ, JIŘÍ MÁCA
Institute of Theoretical and Applied Mechanics CSAS
128 49 Prague 2, Vyšehradská 49, CZECHOSLOVAKIA

ABSTRACT

The paper deals with the interaction between the POLDI endoprosthesis with an articulated stem and femur. The stem of the endoprosthesis consists of five dia.4.5 mm bars 200 mm long. The upper ends of the bars are firmly connected with the collar, their lower ends are welded together. The endoprosthesis is made of AKV Ultra 2 austenitic stainless steel. Primary fixation of the endoprosthesis works through steady wedging of the stem into the beforehand prepared bone cavity without the application of the bone cement. Secondary fixation is ensured with bone ingrowth among the bars of the stem. The subject of the research was to gain basic knowledge about forceflow through the implant into the bone tissue and to find out stress distribution with its maximum values in the stem structure . An experimental analysis concerning the change of femur behaviour after the fitting of the endoprosthesis was carried out and a Finite-Element model was constructed.

INTRODUCTION

Replacements of the hip joint, the most important joint in the human body, have been applied successfully in Czechoslovakia since the end of the sixties even in cases of heavily pathological joints (1). The function of this joint and its replacement is being afforded considerable attention, with the purpose of ascertaining the functionality of the replacement until the end of the patient's life. From the very beginning of application of hip joint replacements, the majority of problems have been connected with anchorage of the replacement in the bone cavity; for this reason the anchorage method has been undergoing fundamental changes.

Perhaps the greatest development in the field of replacement of major human joints was the introduction of bone cement (2) as the means of connection of the individual components with the bone. It has come to light that the bone cement will enable considerable propagation of alloplasty even in complicated cases; however, it has been found that it did not have the required life.

Therefore, attention has turned once again to the methods of fixation without the application of bone cement, especially because of the ever lower age of operated patients. These methods can be classified approximately into two groups :

- close connection of the implant with the bone (so-called pressfit),
- biological fixation by bone ingrowth.

Both methods are utilized in the implantation of the cervicocapital POLDI endoprosthesis with articulated stem (5). All parts of this endoprosthesis are made of austenitic stainless steel of the AKV Ultra 2 type with the strength R_m = 950 MPa. The stem of the endoprosthesis being 200 mm long consisting of five dia.4.5 mm bars, firmly connected with the collar at the upper end and mutually welded at the lower end. The head of the endoprosthesis is a hollow sphere. To increase the corrosion resistance of the endoprosthesis the whole structure is polished. The stem of the endoprosthesis is made in two sizes with ten different head diameters. Before implantation bone cavity is prepared by special rasp and the stem inserted into this canal. The stem must be firmly pressed into the femoral diaphysis in order to prevent any movement between the stem bars and the cortical bone. On the medial side the collar must be in full contact with the resected neck. The primary fixation of the stem is based on it being firmly pressed into the diaphysis without the use of bone cement; secondarily, the stem is fixed by bone ingrowth among the individual bars of the stem.

The research carried out in this field was concerned with :

1/ Ascertainment of basic knowledge of load transfer from implant to bone;

2/ Ascertainment of the stress state of the endoprosthesis stem fitted into the femur under quasistatic load.

An experimental analysis of the change in the stress state of the femur taking place after the fitting of the endoprosthesis was made, and a Finite Element model was devised for mechanical analysis of the behaviour of the articulated stem of the endoprosthesis.

MATERIALS AND METHODS

The purpose of experimental research was the measurement of strains on the surface of an intact femur under quasistatic loads up to 2500 N, and ascertainment of the change due to the application of the endoprosthesis. The experimental method was based on the application of an extensometer to the measurements of surface strains along the medial and lateral sides. For experimental research we obtained a fresh femur from the pathological clinic. Between the individual measurement phases the femur was preserved in a towel soaked in saline solution and frozen to $-25^{\circ}C$. Before the beginning of each measurement the bone was taken out of the freezer and thawed to $+20^{\circ}C$. The freezing of the test sample in it´s moist state had no statistically significant effect on the mechanical properties of the bone tissue. The endoprosthesis was implanted into the femur at orthopaedic clinic using standard clinical practice.

For experimental investigation the femur condyles were embedded in acrylic resin. The lower part of the loading system consisted of two steel plates with a 13.5 mm dia. steel ball in between simulating a hinge. The loading force was transmitted to the femur head and to the spherical head of the endoprosthesis via a modified polyethylene acetabulum (Fig.1). From the viewpoint of static behaviour this arrangement represented a statically determinate system. The load was produced by an Instron 4301 electronic tester. Biomechanically this loading system simulates the standing position on both legs (8). The magnitude of the loading force was selected as the maximum magnitude of the resulting force applied to the hip joint during walking, given in works of reference as between 2.5 to 4-multiples of bodily weight (after the deduction ot the weight of one leg). All measurements were made with an identical crosshead rate of $1.67.10^{-2}mm.s^{-1}$.

The measuring method of strains on the femur surface is based on extensometer application. In this particular case an Instron 2620-602 extensometer was used which is compatible with the load application machine. The gauge length of the extensometer was $l_o = 12.5$ mm, the maximum extension $\Delta l_o = \pm 2.5$ mm. The extensometer was applied successively to the measurements of strains on the surface of the medial and lateral sides of the femur, in the direction of principal strains known from previous studies, using rosette strain gauges (3). The extensometer was applied because of easy and speedy relocation from one point to another. Moreover, gluing strain gauges to the greasy bone surface is laborious and takes a long preparation.

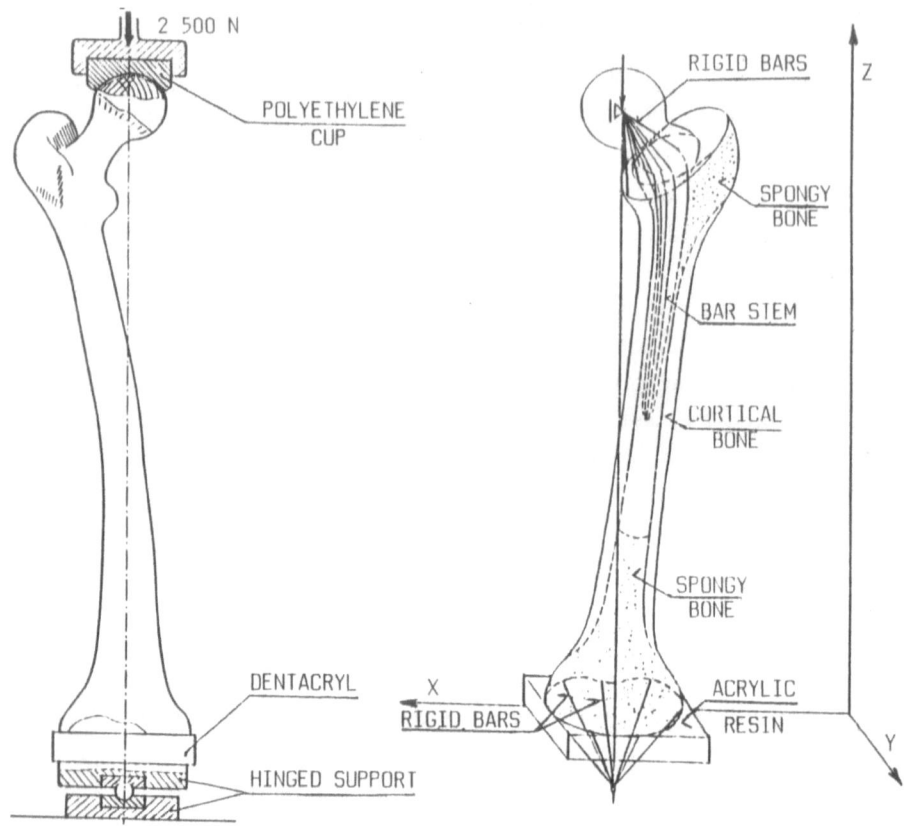

Figure 1.   Figure 2.

There is also the danger of strain gauge damage during endoprosthesis application and, consequently, the loss of previously measured values.

   After intact femur strain measurements were taken the same method was
applied to measurements on the femur with implanted endoprosthesis. The results of the measurements are shown in the Fig.3., where the dashed line
characterizes the measurement results of the intact femur and the solid line
the measurement results of the femur with endoprosthesis. There is tension
on the lateral side and compression on the medial side. The strain diagrams
have the same form on both sides; however, they differ markedly in the proximal part, especially on the medial side.

   The experiments formed the basis of the derivation of a mathematical
model, enabling the description of the behaviour of the stem structure in-

side the femur, and all stress components in the bone tissue.

Numerical analysis of femur-endoprosthesis interaction (7) was carried out by means of the SAP IV programme, currently used for solving engineering problems, on a PDP 11/44 computer. The overall number of degrees of freedom of the solved problem was 4514.

Three-dimensional Finite Element analysis was based on the application of an isoparametric tetrahedron as the basic volume element. The "anatomical" mathematical model was based on the X-ray image of the femur with implanted endoprosthesis. The individual stem bars were modelled with chains of bar elements, the head with the collar was described by five rigid bars connecting the centre of the head with the upper ends of the stem bars (Fig.2). Additional rigid bars leading from the head centre into nodes in the compressed part of the collar, described by a system of plate and membrane elements, enable them to cover the load transfer through the collar directly into the cortex. The condyles were embedded in acrylic resin and the steel plate above the ball was replaced with thirteen rigid bars converging in a fixed hinge (only seven bars are shown in Fig.2 for the sake of simplification).

The interaction of the endoprosthesis and femur is achieved by bar elements connecting the nodes of the endoprosthesis stem with the corresponding nodes of the femur. Two extreme cases were considered. The first assumed zero friction between the endoprosthesis and the bone, corresponding with the possibility of free shift of the implant in the bone ($\frac{3}{7}$H). In the second variant ($\frac{3}{7}$H), the endoprosthesis elements are firmly connected with the bone elements, and no micromovements of the replacement stem in the femur can take place. The primary fixation of the endoprosthesis, when the stem is firmly wedged in the bone cavity, lies between these two extreme cases. The secondary fixation of the implant stem by bone tissue ingrowth among the stem bars corresponds with the second variant.

Modelling of cortical and spongy bones is based on simplifying assumptions of their homogeneity and isotropy. Properties of the bone tissue and of the material of the endoprosthesis are describes by the constants E and μ, the former selected on the basis of works of reference and given in Table 1.

Boundary conditions correspond to the experiments. The sliding hinge in the centre of the endoprosthesis head describes the possibility of rotation and vertical displacement of the head. The fixed hinge in the lower part corresponds with the function of the knee joint below the condyles. The centre of the endoprosthesis head is loaded by a vertical force of 2500 N, with the line of action passing through both hinges.

TABLE 1

Material characteristics used in analysis of the bone-endoprosthesis system

Material		E /GPa/	μ
Cortical bone	proximal part	16.2	0.3
	medial part	17.6	0.3
	distal part	16.2	0.3
Spongy bone		0.39	0.3
Stem steel bars		210	0.3
Rigid bars in FE model		$2.1.10^6$	0.3
Bar elements of bone/implant interaction		210	0.3

RESULTS

Experimental research ascertained that the investigated endoprosthesis with bar stem enables a favourable transfer of forces via the endoprosthesis col- lar to the bone tissue. The stress state of the intact femur and the femur with implanted endoprosthesis is shown in Fig.3.

The stem is firmly wedged into the bone cavity, a fact testified to by the slightly higher strain values on the lateral side at the level of the stem tip (Points 21,22). The strain curve on the lateral side has the same character in the both cases; at the highest Point 14 the value measured for the femur with an implanted endoprosthesis is approximately one half of that measured for the intact femur. This value approaches very speedily in distal direction the value measured on the intact femur. The influence of the stem on this side is recognizable along its entire length.

On the medial side the strain value below the collar of the endoprost- hesis is markedly higher than the value measured on the intact bone. In the distal direction the strain values drop speedily and are lowest immidiately below the minor trochanter (Points 11,12). Towards the stem tip strain values approach those measured on the intact bone. From Point 8 the influence of the stem is no longer observable.

The strain state measured on the bone surface of a femur with an implan- ted bar stem endoprosthesis differs significantly from cemented endoprosthe- ses and non-cemented rigid endoprostheses without a collar (3),(4),(6),(8). After the implantation of the above mentioned endoprostheses in the bone ca- vity the bone tissue is un- or only slightly loaded on the medial side at the resection of the cervix. And it is in this very place that an important reaction to implantation, the so-called "calcar resorption", takes place. It is a process in which the bone tissue in the region of the calcar femoris

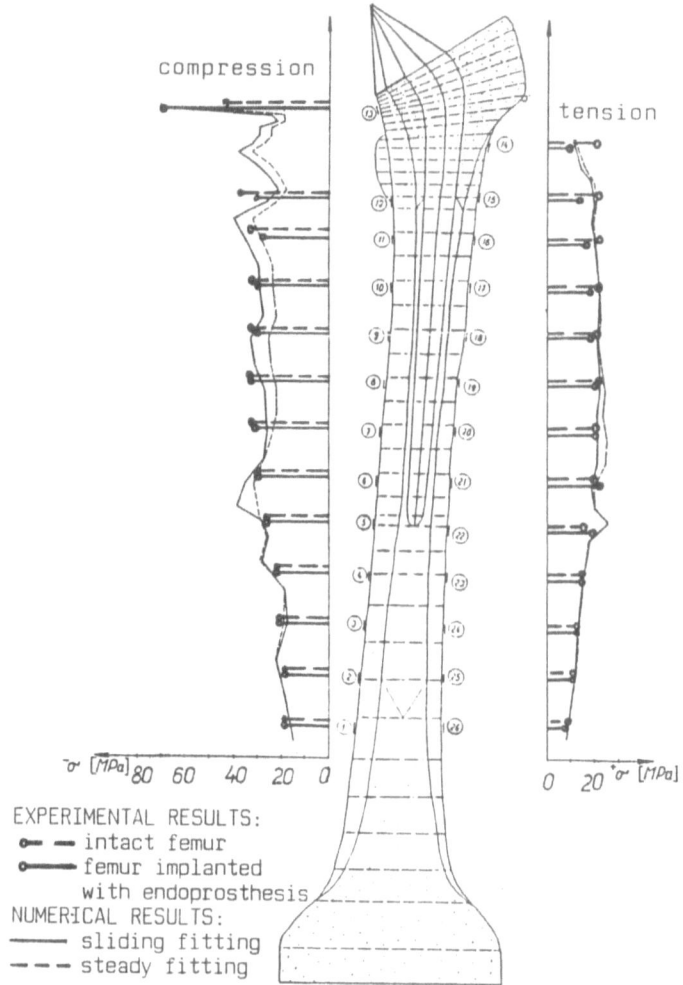

compression

tension

σ [MPa] 80 60 40 20 0

0 20 σ [MPa]

EXPERIMENTAL RESULTS:
○— — intact femur
●——— femur implanted
     with endoprosthesis
NUMERICAL RESULTS:
——— sliding fitting
— — — steady fitting

Figure 3.

resorbs to a major or minor extent (3). It is assumed that one of the causes
of resorption consists in a marked change of the stress state after implantation of the above mentioned types of endoprostheses.

With the investigated endoprosthesis with bar stem an opposite phenomenon appeared. The stress value under the collar rose after implantation from
-44.5 MPa to -70.6 MPa. This new stress value is below the value which could

result in the rupture of the bone tissue; moreover, it can be assumed that remodelling the bone tissue will result in its strengthening. Further resemblance to the state of an intact bone will be enhanced by secondary fixation by bone ingrowth which will strengthen the proximal part of the femur and integrate it. Consequently the strain value will approach the value measured on an intact bone in the same place. The influence of secondary fixation, naturally, could not be specified more clearly in this phase of experimental research.

Numerical analysis ascertained normal stresses on the surface and inside the femur and the stress state of the endoprosthesis stem. Fig.3 shows the numerical and experimental values of normal stresses on the medial and the lateral sides of the femur with implanted endoprosthesis. Numerical values refer to cases of sliding (⊰⊢) and fixed (⊰⊣) endoprosthesis positions in the bone cavity. The mathematical model with the sliding implant stem in the bone corresponded better to the experiment as had been presumed. The accordance of experimental results with the ⊰⊣ Variant in the proximity of the distal end of the endoprosthesis suggests a limited movement of the replacement, i.e. fixed fastening of the implant in the bone cavity. The theoretically ascertained stress values on the medial femur side in the region of the calcar femoris are lower than the experimental data. The maximum compressive stress on the medial side of the bone σ_{exp} = -70.7 MPa was measured below the collar, 7 mm below the femur cervix resection. This value corresponds to the theoretical stress σ_{theor} = -59 MPa. The theoretical stress values on the bone surface on the lateral side of the upper third of the diaphysis are, on the other hand, slightly higher than the experimental results.

Secondary fixation by means of bone ingrowth is described in the mathematical model by the fixed connection of the endoprosthesis with the femur. A marked change in the stress state of the bone tissue took place at the level of the distal end of the stem due to its limited movability in this region. The compression maximum below the collar dropped, in comparison with the ⊰⊣ Variant, from -59 MPa to -45 MPa. In both places the numerical values dropped to the level of experimental values measured on an intact bone. This confirmed the aforementioned statement, that secondary fixation favourably influences the stress state of the femur which, after bone ingrowth, resembles the state of an intact bone.

Generally speaking very good accordance of theoretical and experimental results can stated, so that it was possible to assess stresses of the implant structure itself by means of the FE model (Fig.4).

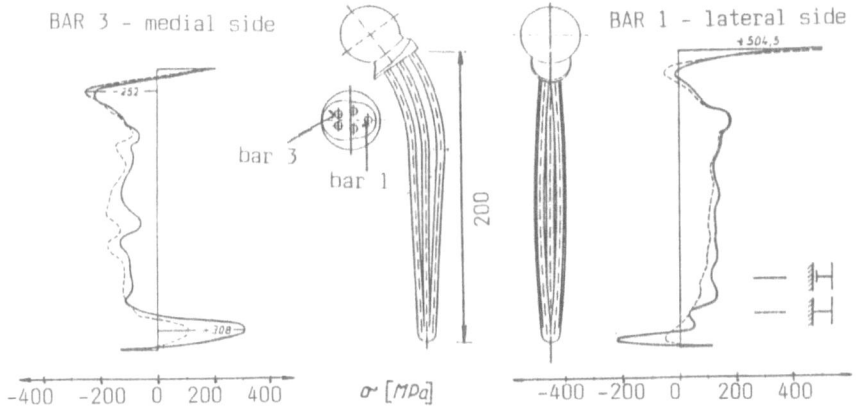

Figure 4.

Of decisive significance for the assessment of the bar stem endoprosthesis behaviour in interaction with the femur are the tensile lateral bar 1
and the compressed medial bars 3,4. Determination of the stress state of the
stem bars considers the combination of the normal force and bending moment
effects. Normal stress on lateral and medial fibres of stem bars were determined. The magnitude of stresses due to normal force along individual bars
remains almost constant. Local fluctuations of the total normal stress are
due to discrete contact with the bone tissue. Fig.4 shows the bar constrain
effect in the collar. Regarding their geometry and mode of loading, the bars
bend so that tension manifests itself mostly on their lateral sides and compression on the medial sides. The upper part of the lateral bar buckles. Local stress maxima on the lower end of the stem with opposite signs show that
the top of the endoprosthesis stem bends markedly in a lateral direction.
With the fixed connection of the stem to the bone, the effect of the bar constraint in the collar and the buckling of the lateral bar in the upper part
slightly increase and the deformation of the distal stem end is limited.

The normal stress in the endoprosthesis stem attains an absolute maximum
in the cross-section below the collar on the lateral side of lateral bar 1.
Its magnitude is +427.7 MPa in Variant 1 and +504.5 MPa in Variant 2. Both
values are far below the strength of the material R_m= 950 MPa. However, taking fatigue into account, this place may prove a weak point of the structure.

CONCLUSIONS

On the basis of the experimental method devised for measurements of strain values on the femur surface the change due to alloplasty was ascertained. Implantation of an endoprosthesis with bar stem results primarily in a significant increase of strain values on the medial side below the collar in
comparison with an intact bone. It can be assumed, however, that secondary
fixation by bone tissue ingrowth will strengthen the proximal part of the
femur with endoprosthesis, and integrate it into an integral whole with a
more favourable mode of loading the bone tissue by the collar at the calcar
femoris region. Comparison of experimental and theoretical results has revealed that the designed mathematical model describes highly satisfactorily
the behaviour of the bone/endoprosthesis system. On the basis of mathematical modelling the change of the stress state due to secondary fixation was
ascertained. This was simulated by a fixed connection of bar elements with
bone elements. Fastening of the endoprosthesis relieves the femur on the
medial side below the collar. It can be stated that the secondary fixation
of the endoprosthesis results in more favourable biomechanical conditions
in the bone, approaching the state of an intact femur. By means of the mathematical model it was possible also to describe the behaviour of the stem
structure inside the femur, and will enable the optimization of the implants
size and shape in their further development.

REFERENCES

1. Čech,O.,Pavlanský,R.,Alloarthoplastic of Hip Joint,Avicenum,Prague,1983
2. Charnley,J.,Low-friction Arthoplasty of the Hip,Springer-Verlag Berlin Heidelberg New York,1978
3. Huiskes,R.,Some Fundamental Aspects of Human Joint Replacement,Acta Orthopedica Scandinavia.No.185,1979
4. Huiskes,R.,Biomechanics of Bone-Implant Interaction. Frontiers in Biomechanics,Springer-Verlag New York Berlin Heidelberg Tokyo,1986,pp.245-262
5. Jírová,J.,Čech,O.,Beznoska,S.,Novotný,R.,Development of a New Type of Hip Joint Prosthesis.Biomechanics X-A,Human Kinetics Publishers,Champaign,Illinois,1987,pp.119-122
6. Jírová,J.,Beránek,R.,Slavík,M.,Karpíšek,M.,Force Interaction between Hip Joint and Bone,Research report,Inst.of Theoretical and Applied Mechanics CSAS,1989
7. Kauflerová,T.,Numerical Analysis of Interaction between Femur and Endoprosthesis with isoelastic Stem,Research report,Inst.of Theoretical and Applied Mechanics CSAS,1990
8. Rohlmann,A.,Mössner,U.,Bergmann,G.,Kölbel,B.,Finite-Element-Analysis and Experimental Investigation in a Femur with Hip Endoprosthesis.J.Biomechanics,Vol.16,No.9,pp.727-742

STRESS CONCENTRATION FACTORS FOR INTERSECTING ARRAYS OF NOTCHES IN BEAMS UNDER PURE BENDING

D.C. FRICKER

Howmedica International, Inc.
Raheen Industrial Estate, Limerick, Ireland

ABSTRACT

Three-dimensional frozen-stress photoelasticity was used to study two rectangular beams under pure bending loading. Regular, orthogonal arrays of different shallow notches were cut in the top and bottom surfaces, aligned at 0°/90° and also at ± 45° to the applied stress direction. The s.c.f. K_n defined as the maximum stress/nominal stress at the roots of the notches, was measured and compared with previous 2D results for multiple notches. The results for the V-notch profile Type A showed that K_n = 2.4 when the array was orientated at 90° to the applied stress, but the s.c.f. reduced to K_n = 1.6 when at ± 45° (as it was on the implanted hip stems and bone plate orthopaedic devices). Prototype profiles Types B and C at ± 45° had only slightly lower K_n = 1.4.

INTRODUCTION

In the late 1970's and early 1980's, there was a move away from the use of orthopaedic bone cement for the fixation of femoral implant stems, to cementless fixation relying on new bone ongrowth/ingrowth into a deliberately non-smooth surface. The sintered bead surface of the P.C.A. hip stem from the USA has the advantage of open pores into which bone can grow but the very high sintering temperatures (up to approximately 1300°C) reduced the fatigue strength of the substrate Vitallium® used at that time. In France, the Madreporique cast, beaded surface was developed, whilst at Limerick, a 'pyramid' textured surface was devised by machining the master pattern used in the normal investment casting process of Vitallium® hip stems.

Figure 1 shows four devices with the pyramid surface. From left to right, the Babin Chevaye prototype (not implanted) had the array of notches aligned with the stem axis for ease of machining, in the Babin Chevaye and H.P. Garches implants and the Meyrueis compression bone plate, this orientation was changed to ± 45° to reduce stress concentrations at the notch roots.

This paper describes how the stress concentration factors (s.c.f.'s)

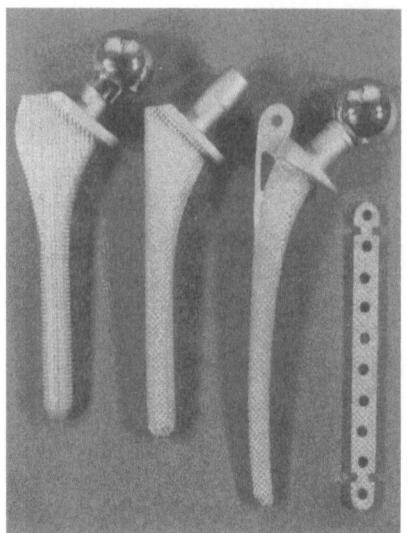

for different orientations of arrays
of different notches were measured
using 3D frozen-stress photoelasticity.
The predominant type of loading on the
proximal regions of hip stems is bending due to the downward force on the
femoral head. Because the notch depth
on implant stems was required to be
0.8 mm and stem sections were typically circular of gross dia 14 mm, the
notches could be considered 'shallow'.
Furthermore the notches were added to
the size of the counterpart smooth
stem implant (for cemented applica-
tion) and so s.c.f.'s $K_n = \sigma_{max}/\sigma_{nom}$
where σ_{nom} is the nominal, net section
bending stress, were appropriate in
estimating the fatigue performances of
these hip stems.

**FIGURE 1. Limerick orthopaedic prototype
and implants with the Type A
pyramid surface texture.**

NOTCH GEOMETRIES

Figure 2 shows the non-dimensional
shapes of the three different surface
notches studied. These shapes are sections transverse to straight axes and
so the pitch p is measured in such sections. Type A is the V-notch shape
on the surfaces of the implants shown in Figure 1. The notches were repeated in orthogonal arrays with a constant pitch p (in Sections GG and HH of
Figure 4). The orientation of an array to the direction of the nominal,
uniform, uniaxial substrate stress was either 90° (see Figure 4a) or 45°
(see Figures 4b, 4c and 4d). Four different notched surfaces were studied,
viz. Type A/90°, Type A/45°, Type B/45° and Type C/45°. The spacing p* of
notches in the x-axis direction of the applied stress is therefore p* = p
for the 90° orientation and p* = √2p for the 45° orientation.

The Type A multiple notches in Figure 2 resemble a machine screw
thread profile. Peterson [1] states that the Aero thread form of semicircular notches of pitch/radius p/r = 2.66 in the edge of a plate in tension
has K_{tn} = 1.94, whereas corresponding photoelastic tests for the Whitworth
thread shape t/r = 4.67 and p/t = 1.6 gave K_{tn} = 3.35. Therefore in order
to reduce local stress concentrations, the prototype notch Type B in Figure
2 was designed based on the Aero thread form. Peterson [1] also gives
single, 2D, shallow notch data in tension which shows that adding a flat
land of length r at the centre of a semicircular notch of radius r, reduces

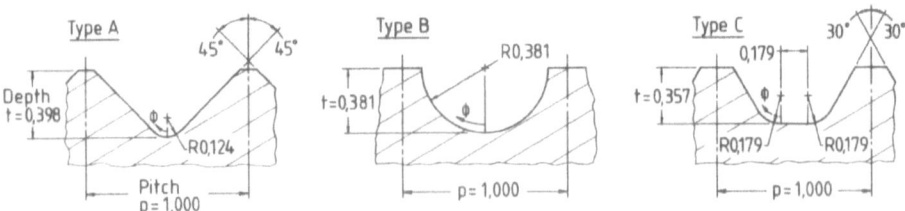

FIGURE 2. Shapes of notches studied; dimensions as multiples of the pitch p.

K_{tg} from 3.07 to 2.78. Therefore this feature was incorporated in the second prototype notch profile Type C. The Type C notch was originally given a 45° flank angle as in the Type A shape, but this was revised to 30° as shown in Figure 2 in order to maintain similar p/t values. The Type C notch root radius was chosen to be equal to that (2.5 mm) of the Type A notch in photoelastic models.

The size of photoelastic model notches was chosen to be approximately ten times prototype size, hence the Type A notch depth t = 8.0 mm. The same notch depth was chosen for Type B. However for Type C, a smaller depth t = 5.0 mm was chosen, partly because its p/t ratio was slightly larger than those of Types A and B, whilst the width of the photoelastic beam was necessarily constant (see Discussion).

When cut as a regular, orthogonal array in a gross cross-section, all three notch types produce 'pyramid' projections from the net cross-section and so could be used for orthopaedic implants requiring a textured surface.

2D THEORETICAL PREDICTIONS OF S.C.F.'S

Neuber Notch Theory
Neuber [2] considered a single, shallow notch in the edge of a thin, elastic plate under tension or bending loading. The curvilinear coordinate system he used to define the edge, resulted in a notch geometry with very rounded external corners and not sharp corners as in the notch geometries here. He gives K_{tg} values (equal for tension and bending) in terms of t/ρ where ρ is the minimum radius of curvature at the notch root. Neuber then considered a wide plate under tension with a transverse, elliptical hole; by 'splitting' the plate along its longitudinal axis, the exact solution for K_{tg} approximates that for a single, external, semi-elliptical shallow notch with square corners.

Neuber's theory for multiple, edge notches with rounded corners predicts that the proximity of the notches reduces the s.c.f. from that for a single notch, and that this reduction can be evaluated by considering an equivalent, single notch of reduced depth t' where

$$t' = t \times (p/\pi t)\tanh(\pi t/p) \qquad (1)$$

Since these notches are shallow, stress gradient effects on K_{tg} are vanishing and so for distant tension or pure bending loading,

$$K_{tg} = 1 + 2[t'/\rho]^{\frac{1}{2}} \qquad (2)$$

for multiple, semi-elliptical (or semicircular) notches with square corners, whilst

$$K_{tg} = 3[t'/2\rho]^{\frac{1}{2}} - 1 + 4/(2+[t'/2\rho]^{\frac{1}{2}}) \qquad (3)$$

for multiple notches with very rounded external corners.

For the Type A/90° notched surface, the Figure 2 geometry of longitudinal, full notch depth Sections HH at y = -p, y = 0 or y = p (see Figure 4a) is appropriate for the Neuber prediction given in Table 1. The Neuber radius of curvature ρ was taken to be equal to the root radius in Figure 2. The K_{tg} value from equation (3) was 13% less than the value from equation (2), but the Table 1 predictions were obtained using equation (2) because the semi-elliptical model is closer to the Type A and Type B shapes.

TABLE 1
Theoretical 2D stress concentration factors K_{tn}

Source	Notch type	Orientation	Parameters	S.c.f. in bending K_{tn}
Neuber	A	90°	t/ρ = 3.20, p/t = 2.51, W/w = 1.180	2.83
		45°	t/ρ = 1.60, p^*/t = 3.55, W/w = 1.180	2.34
	B	45°	t/ρ = 0.5, p^*/t = 3.71, W/w = 1.141	1.75
Peterson	B	45°	b/a = $\sqrt{2}$, p^*/b = 1.31, $2t/b$ = $1/\sqrt{2}$	1.49
	C	45°	See text	1.58

Table 1 gives K_{tn} values derived using $K_{tn} = K_{tg}/(W/w)^2$ where the gross and net beam depths were those of the photoelastic models.

For the Type A/45° and Type B/45° surfaces, longitudinal, full notch depth sections, e.g. at y = 0, have p^* = $\sqrt{2}p$ and notch root radii double the Figure 2 values. Appropriate p^* and ρ values were then used in equations (1) and (2) to calculate these Table 1 Neuber predictions.

Peterson's Data
Peterson [1] gives no general multiple, V-notch results from which K_{tn} values can be obtained for the Type A/90° or Type A/45° surfaces.

For the Type B/45° surface, the longitudinal, full notch depth section has semi-elliptical notches of major/minor axis ratio b/a = $\sqrt{2}$. Peterson gives thin plate bending data for multiple U-notches with semi-elliptical roots; his (K_{tn} - 1) values were extrapolated to the Type B/45° relative notch spacing p^*/b and depth $2t/b$, and Table 1 gives this K_{tn} prediction.

For the Type C/45° surface, Peterson gives no applicable multiple notch results. However for a single semicircular notch, adding a land reduces (K_{tg} - 1) by 14% (see above). Therefore the Type C/45° longitudinal, full notch depth section was firstly considered without the lands, in terms of multiple U-notches. Again b/a = $\sqrt{2}$ but b = $2\sqrt{2}(0.179p)$ and p^* = $\sqrt{2}(0.821p)$. Hence p^*/b = 2.29 and $2t/b$ = 1.41, and extrapolation of the Peterson data gives K_{tn} = 1.67. Reducing (K_{tn} - 1) by 14% gives the prediction in Table 1 for the Type C/45° surface.

3D PHOTOELASTIC STRESS ANALYSIS

Models
Two prismatic, rectangular cross-section beams were cast from the same mix of Araldite CT200/HT907 and were cured together. Their faces were machined to produce beams of 572 mm length, 60 mm width and gross depth W = 105 mm (see Figure 3). The arrays of notches were profile end-milled into each top 'tension' surface and bottom 'compression' surface to give notch pitches and depths p = 20.1 mm and t = 8.0 mm for Type A, p = 21.0 mm and t = 8.0 mm for Type B and p = 14.0 mm and t = 5.0 mm for Type C notches. Beam no. 1 had the Type A/90° surface on its tension face and the Type A/45° on its compression face, and beam no. 2 had corresponding surfaces Type B/45° and Type C/45° (as shown in Figure 3). The net depths for beams nos. 1 and 2 were therefore w = 89 mm and w = 92 mm respectively. Central, 120 mm long, test regions were specified to have the complete arrays of notches over the full beam width. The locations of arrays were specified so that the centres of the test regions (x = 0, y = 0 in Figure 4) coincided with a 'valley' between 'peaks' for the Type A/90° surface (see

FIGURE 3. **Loading arrangement of beams** prior
to **stress-freezing in oil.**

Figure 4a) or the intersection of two
'valleys' for the other surfaces. To
avoid the possibility of disastrous
model failures around the concentrated
loading and support rods, separate
Araldite tubular inserts 1 in OD and
$\frac{1}{2}$ in ID were made to fit into the
reamed holes through the beams.

Loading and Slicing

Four-point deadweight loading was
chosen in order to produce pure bending in the central test regions.
Figure 3 shows this arrangement. The
hogging mode (support lines within
the loading lines) and loading in
inert, transformer oil (s.g. 0.8)
were chosen to minimise Araldite (s.g.
1.25) self-weight effects. Distributed loading calculations predicted a
maximum 2% bending moment error and
so self-weight effects were neglected.
A nominal bending moment M = 15 Nm
was chosen, produced by forces of
200 N (in oil) acting at a loading
span of 520 mm and a support span of
370 mm. The nominal, uniform, uniaxial net section bending stresses
were calculated to be σ_{nom} = ±0.189
MPa and σ_{nom} = ±0.177 MPa for beams
nos. 1 and 2 respectively.

Each loaded beam in turn was taken through the stress-freezing cycle
of at least 3 h soaking at 130°C, controlled cooling at 2°C/h to 85°C and
then oven cooling to room-temperature. The deflected shapes of both beams
were accurately measured.

Figure 4 shows plan views of the different surfaces and the locations

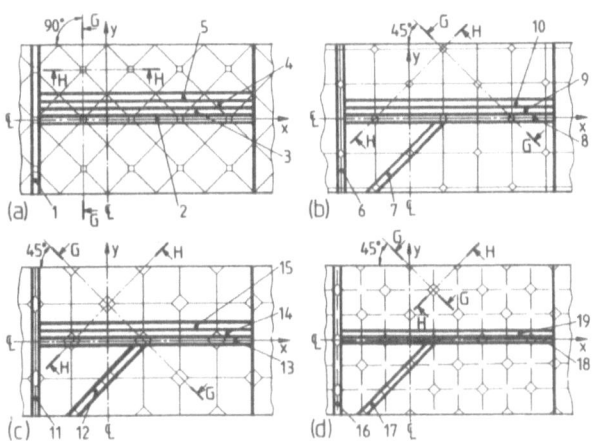

FIGURE 4. Photoelastic notch arrays and locations of slices.

and numbering of slices which were cut in planes perpendicular to the beam
faces using a 0.75 mm thickness diamond slitting wheel. Most slices were
nominally 2 mm thick, but slices nos. 17, 18 and 19 were specified to be
$1\frac{1}{2}$ mm thick.

Remnants from the centres of the beams were annealed and tensile test
strips were machined to measure the material properties of each beam above
stress-freezing temperatures.

Results

The quality of the end-milling of notches was excellent in that the surfaces had minimum roughness, smooth fillet radius/flat blends and straight
edges, which were not chipped.

The deflected shapes of both beams showed (a) they were symmetrical
about their transverse centre sections, (b) they had no measurable longitudinal twist, (c) they had negligible transverse (anticlastic) curvature
effects and (d) the radii of curvature of their neutral surfaces over their
test lengths were both estimated to equal 2820 mm, compared with the predicted lower bounds 2400 mm and 2670 mm calculated using net beam depths and
ignoring the notched surfaces. It was therefore concluded that both beams
had been successfully loaded with pure in-plane bending of the correct
magnitude.

The material properties were measured to be E = 10.2 MPa and F = 0.257
MPa/(fringe/mm) for beam no. 1 and E = 10.3 MPa and F = 0.257 MPa/(fringe/
mm) for beam no. 2.

Normal incidence, polariscope analysis of all slices showed that, as
expected, maximum fringe orders (FO's) were always located at free surfaces
and so maximum, secondary principal stress magnitudes could be obtained
without stress separations. Maximum stress magnitudes in each slice were
averaged and expressed as a stress index i. Table 2 summarises the measured
i values. Figure 5 shows typical isochromatic fringe patterns and specific
FO values. As expected, the end notch in longitudinal slices had a larger
maximum FO value (typically 15% greater) and an asymmetric fringe pattern.
Such notch roots were excluded from i averages. The locations of the maximum stresses in slices were in notch roots at $\phi = 0°$, except for transverse
slices nos. 6, 11 and 16 where they were at $\phi = \pm 45°$. At $\phi = 0°$ in these
slices, stress indices were - 0.14, 0.00 and - 0.07 respectively, i.e.
insignificant.

Analysis

The i values of the transverse slices nos. 1, 6, 11 and 16 and Section HH
slices nos. 7, 12 and 17, show no reason to doubt that the maximum principal
stress anywhere in the surfaces σ_{max} was in the longitudinal direction, with
no significant transverse stress, and was located at repeated positions.

TABLE 2
Maximum stress index values i in photoelastic slices

Type A/90°		Type A/45°		Type B/45°		Type C/45°	
Slice no.	Stress index i	Slice no.	Stress index i	Slice no.	Stress index i	Slice no.	Stress index i
1	- 0.32	6	- 0.46	11	- 0.30	16	- 0.33
2	2.42	7	0.93	12	0.83	17	0.65
3	2.31	8	1.59	13	1.35	18	1.34
4	2.08	9	1.39	14	1.27	19	1.13
5	1.47	10	1.36	15	1.23		

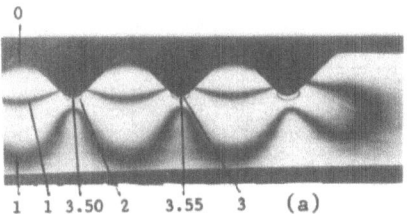 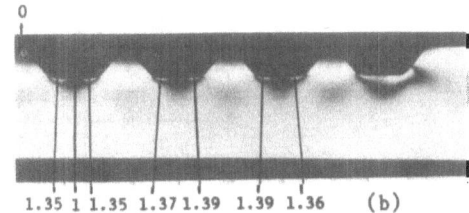

1 1 3.50 2   3.55   3   (a)   1.35 1 1.35 1.37 1.39 1.39 1.36   (b)

FIGURE 5. Fringe patterns in (a) slice no. 2 and (b) slice no. 18.

For the Type A/90° surface, these positions were $x = \pm np$, $y = \pm mp$ where m and n are integers, i.e. in transverse valleys between longitudinal peaks. For the 45° array orientations, they were at the intersection of valleys, at one point for the Type A/45° and Type B/45° surfaces, but at two points for the Type C/45° surface.

The repeating geometry of the notches, together with a uniform substrate stress, gives rise to repeating stress distributions. Figure 5 shows this repetition in two longitudinal slices. Because the thicknesses of longitudinal slices were not negligible compared to stress gradients in the y direction, a graphical curve fitting was employed on the transverse variation of i values in these slices. Figure 6 shows this, where i values have been drawn as bars, with the slice thicknesses and locations on a normalised y/p abscissa. For the Type A/90° surface, i values vary from i_{max} at y = 0 (a principal stress index due to symmetry) to i_{min} at $y = p/2$, before rising to i_{max} again at y = p. All i values are located along transverse lines $x = \pm np$. The value $i_{min} = 1$ has been assumed at y/p = 0.5 for the Type A/90° surface because this position is at the root of a longitudinal notch. For the 45° orientation of the other notch arrays, i values vary from i_{max} at y = 0 to i_{min} at $y = p/2\sqrt{2}$, before rising to i_{max} again at $y = p/\sqrt{2}$. For the Type A/45° and Type B/45° surfaces, all i values are located along lines $y = \pm x$, for example. For the Type C/45° surface, corresponding lines are $y = \pm[x \pm (0.179p/\sqrt{2})]$.

S.c.f. values $K_n = \sigma_{max}/\sigma_{nom} = i_{max}$ where i_{max} is the maximum stress index value at y = 0 (a principal plane), were estimated from Figure 6 and are given in Table 3.

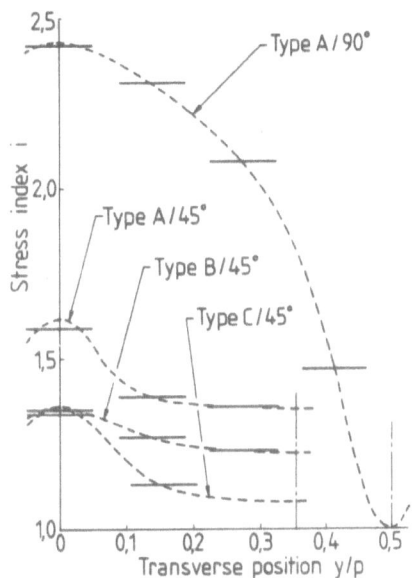

FIGURE 6. Curve fitting of photoelastic results.

DISCUSSION

Photoelastic Modelling

The sizes of model beams and notches were chosen such that
(a) the beams would be 'narrow' so that the nominal substrate stress would be uniform,
(b) the notches would be 'shallow',

TABLE 3
Photoelastically measured s.c.f.'s K_n

Notch type	Orientation	S.c.f. K_n
A	90°	2.43
A	45°	1.62
B	45°	1.36
C	45°	1.36

i.e. t/w < 0.1 so that stress disturbances near the tension and compression faces would be non-interacting, and the substrate stress would vary little near notch roots so that K_n would not depend on particular t/w values. Figure 5 shows this was achieved, since FO values are nearly uniform at one notch depth below the net surface,

(c) the pitch of notches would be small enough to achieve an array of multiple notches in both the longitudinal and transverse directions on the beam faces. Figure 4 shows this was achieved. Slices were cut in the central region bounded on both transverse sides by at least one and usually more pyramid projections. This was the reason for making t < 8 mm for the Type C/45° surface, and

(d) reasonable stress measurement accuracy would be achieved. The curve fitting, shown in Figure 6, to photoelastic data which was inherently averaged over the slice thickness, helped to achieve this.

Comparison of 2D and 3D S.c.f.'s

Peterson [1] has noted that the Neuber predictions of the stress-relieving effects of multiple, semicircular notches are underestimates. This can be seen in Table 1 for the Type B/45° surface, where the Neuber K_{tn} prediction is 17% higher than the corresponding Peterson result. For the Type A/90° surface, this is also the percentage by which the Neuber prediction in Table 1 exceeds the measured value given in Table 3, but the Neuber semi-elliptical notch shape does not model the Type A V-notch shape exactly and so part of the Neuber overestimation is probably due to this shape difference.

By re-orientating the notch from 90° to 45°, Table 3 illustrates that K_n is reduced considerably. For the Type A notch, the longitudinal section where σ_{max} occurs in the 45° orientation has an elliptical notch root profile with a central radius of curvature double the Section HH value, and a larger 55° flank angle. On the other hand, the longitudinal spacing p* is increased by a factor of $\sqrt{2}$ and so stress-relieving effects are less. Overall, the 45° orientation is probably optimum for all orthogonal arrays. The measured K_n values for the Type A/45°, Type B/45° and Type C/45° surfaces are less than the corresponding predicted values and this difference is probably due to the 3D geometry of the 45° orientation of the arrays of projecting pyramids.

Orthopaedic Implant Applications

Fatigue strength calculations for the Babin Chevaye and H.P. Garches femoral implant stems, including the s.c.f. $K_n = 1.6$, were compared with corresponding calculations for Charnley stems which are known to survive well in long-term in-vivo use. The stems were assumed proximally free and distally supported (the 'worst' case), and a normalised 1 kN femoral head load was applied, whose components had resolved angles from the femoral axis of 17° in the plane of symmetry of the stem and 12° in the anterior direction (the maximum loading situation during the walking cycle). Theoretical elastic stress analysis was employed since experimental strain-gauge results had proved its accuracy for the Charnley and other stems.

Although the notch root stresses were higher for the Babin Chevaye and H.P. Garches stems than for the comparable Charnley stem, the moderate notch sensitivity and reasonable corrosion fatigue strength of SHT cast Vitallium under repeated loading, ensured adequate safety for these stems.

Calculations of the additional surface stresses due to shear traction by typical bone ingrowth into the pyramid projections, showed the stresses to be negligible.

There have been no reported fatigue fractures of these stems, although not many were implanted because of the overwhelming popularity of bead textured surfaces which, with bone ingrowth, can support normal traction in addition to shear traction. Also the PCA stem uses the current practice of having the textured surface only over its proximal region, and having its distal region smooth; this avoids the possibility of bone support in the distal region only, which is bad for both the femur and the implant.

CONCLUSIONS

For the V-notch Type A surface geometry, the decision in the design process to re-orientate the pattern from 90° to 45° to the predominant substrate stress direction, reduced the s.c.f. K_n from 2.4 to 1.6. The alternative prototype surfaces Type B/45° and Type C/45° reduce the s.c.f. further to $K_n = 1.4$ but this small gain did not warrant their adoption.

ACKNOWLEDGEMENTS

The author thanks Howmedica International Inc. for permission to publish this paper, Professor Henry Fessler for designing the casting and loading arrangements, the late Hugh Spooner for supervising the loading and slicing, and the technicians at Nottingham University.

REFERENCES

1. Peterson, R.E., Stress Concentration Factors, John Wiley, New York, 1974, Chapter 2.

2. Neuber, H., Theory of Notch Stresses, J.W. Edwards, Ann Arbor, 1946, Chapter IV, Sections 4-5 and Chapter VII, Section 3.

THREE-DIMENSIONAL STRAIN ROSETTES FOR AN ANALYSIS OF THE GEOMEDIC
KNEE PROSTHESIS AND UNDERLYING CEMENT FIXATION

E.G. LITTLE
Department of Mechanical and Production Engineering
University of Limerick,
Republic of Ireland.

ABSTRACT

Three-dimensional strain rosettes were embedded into two models, one of
the Geomedic tibial component alone and the other including the underlying
cement. The materials used in the models were selected to provide
matching of moduli in accordance with the laws of dimensional analysis,
bone being inadequately represented. Tests permitted strain gauge heating
and reinforcement to be identified and minimised. Theoretical and
experimental strains derived from a bar containing three-dimensional
strain rosettes led to the quantification of errors in transducer
manufacture, model testing practice and model machine interaction. Data
from the investigation of the fundamental factors biasing the model test
data provided an enhanced understanding of the stresses derived from the
two models, which demonstrated failure of the cement bone interface
undoubtedly leading to tibial component loosening, an effect which has
been observed in clinical practice.

INTRODUCTION

Finite element analysis has been applied to establish trends in stress
distributions in knee prostheses and the materials underlying them, (1,2).
Unfortunately the finite element work reported in bioengineering rarely
includes experimental validation. When validity studies are carried out,
strain gauges are usually mounted onto the external surfaces of the model.
However, surface strains are of limited value when tibial component
failure is associated with contact stress, cement fracture (3) and
breakdown of the cement bone interface (4). Recognising the necessity for
an experimentally based assessment of the trend in stress distribution in

internal regions for comparison with finite element models, (5) and (6) embedded three dimensional strain rosettes (7) into large scale models of the Geomedic tibial plateau and underlying cement.

MODEL DESIGN

In order to embed three-dimensional strain rosettes and also reduce the adverse effects of strain gradients large scale models made of the . photoelastic material CT200/HT907 were used. The selection of CT200 as the material for the section to be analysed fixes the modulus of the adjacent materials by the laws of dimensional analysis. Two separate models were required, one for an analysis of the tibial plateau (model 1) and one for the cement layer (model 2). Details of the materials used in the model and methods for manufacturing the models are described by (8). The materials were isotropic and homogeneous as in the case of the finite element work cited previously. The anisotropic nature of bone was not represented. Sites for embedding the three-dimensional strain rosettes were established for model 1 from a finite element analysis of a similar product (1) and in model 2 from a two-dimensonal finite element study, which indicated that abrupt changes in section and the anterior - posterior region should be investigated.

FUNDAMENTAL INVESTIGATIONS

Strains from plastic models may be biased by strain gauge heating (9) and reinforcement (10), rosette layout, strain gradients, and errors in the model, the testing machine and the interaction between these errors (11). Each of these factors were investigated.

Strain gauge heating and reinforcement

The data in Fig. 1 shows the indicated strain from a gauge compared with the strain derived from an extensometer versus the thickness of the test specimen. Results from (12) show a dependence on strain field with strains, being more biased in bending than tension due to the shift in the

neutral axis and miniature gauges being affected to a more significant
extent than large patterns (12). Gauge backing materials (compare KFC
gauges which were embedded into model 1 and have pheonol-epoxy backings,

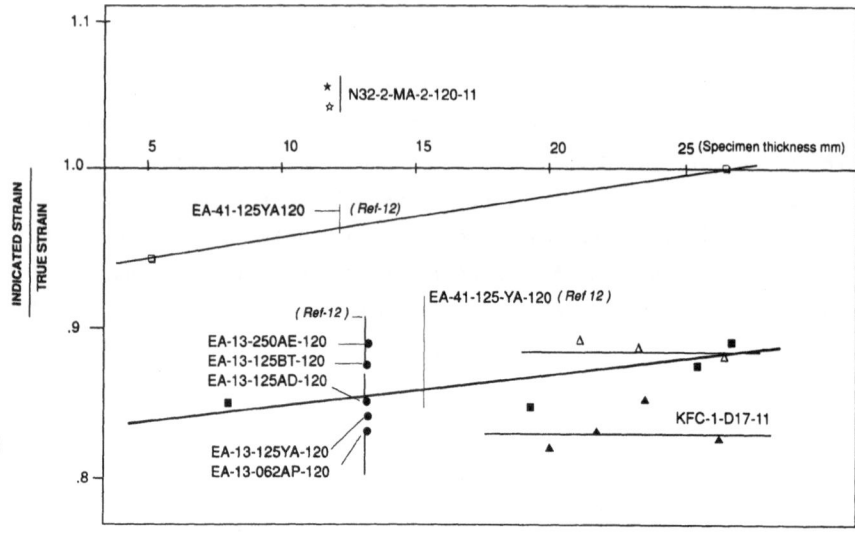

Figure 1.   Strain gauge heating and reinforcement effects from gauges
             mounted onto and embedded into CT200. Open symbols - tensile
             test and closed symbols - bend test.

with N32 which were used in model 2 and have a polyimide backing),
modulus of the model material and power density also determine the degree
of reinforcement. Mean correction factors derived from Fig 1 were applied
to the data obtained from models 1 and 2.

Rosette pattern
Results from three dimensional strain rosettes are influenced by errors in
both gauge and plane misalignment (11), with some direction cosines being
more sensitive to misalignment than others. The rosette pattern of Fig. 2
(a) was embedded in model 1 but subsequent analysis (13) indicated that
the pattern of Fig. 2 (b) gave better estimates of the tensor with
direction cosines being less adversely affected by errors in misalignment.
The pattern of Fig. 2 (b) was used in model 2. The uncertainty induced by

misalignment may be estimated by the application of Monte-Carlo methods
(11).

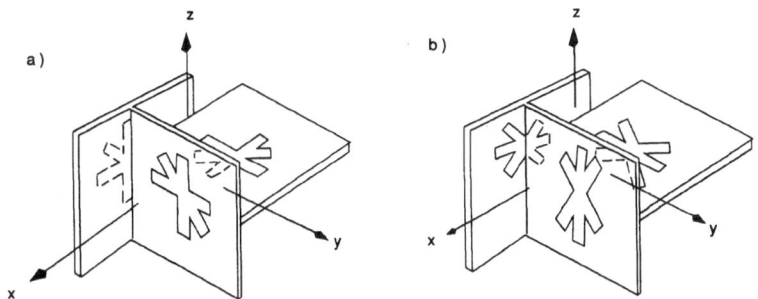

Figure 2.   Three dimensional strain rosette patterns, (a) used in
model 1 and (b) in model 2.

Strain gradients

Patterns with larger grid offsets give rise to considerable error in
tensor values when subjected to strain gradients (11).   Data is also
affected by the orientation of the gradient relative to the gauge layout
as is illustrated in Fig. 3 for the pattern of Fig 2 (b).

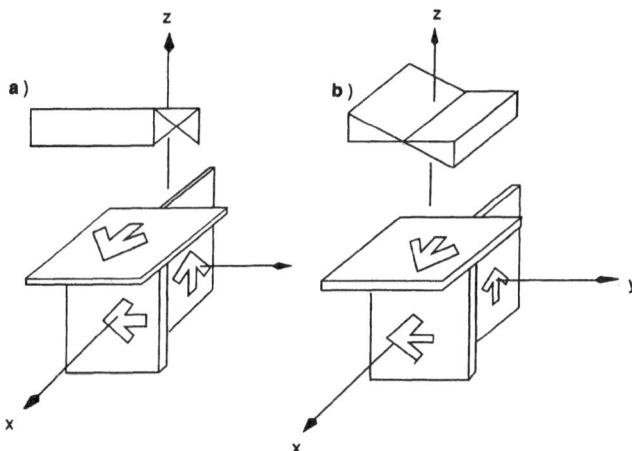

Figure 3. Typical errors in microstrain induced by the presence of 200
microstrain gradient rotated over the rosette pattern of
Fig. 2 (b).
(a) $e_x = -37$, $e_y = -37$, $e_z = -100$
(b)   no error.

The bias induced by strain gradients was minimised in models 1 and 2 by using miniature gauges with small grid offsets embedded into large scale models.

Errors in the model and testing machine and their interaction Three-dimensional strain rosettes of the type shown in Fig. 2 (b) were embedded into a prismatic bar of CT200 (see Fig. 4). The bar was mounted on a jig, the vertical axis being aligned with the centre of the upper platen of an Avery Dennison testing machine type T42U6/A and compressed. The load was applied via a ball through an end cap containing a conical

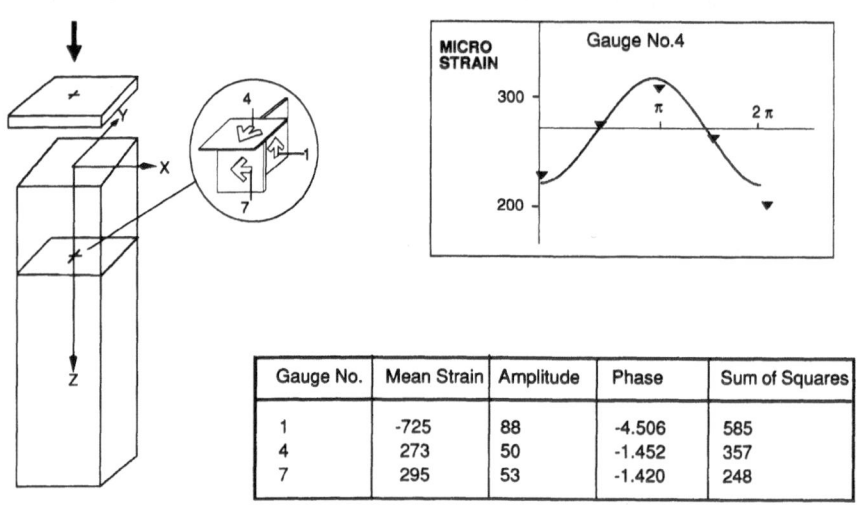

Gauge No.	Mean Strain	Amplitude	Phase	Sum of Squares
1	-725	88	-4.506	585
4	273	50	-1.452	357
7	295	53	-1.420	248

Figure 4. Sample data obtained from the compression test on the prismatic bar.

seating. Rotation of the bar through 90° increments about the vertical axis followed by reloading produced data from each gauge which was fitted to a sine wave via regression analysis, as advocated by (14). Typical data from gauges 1, 4 and 7 is shown in Fig. 4 and indicates the effect of model placement in the testing apparatus, the amplitude being influenced

by lack of parallelism in the platen and jig, minor transverse
offsets of the platen-ball and bar-end cap assembly, geometric errors in
the prismatic bar, flexibility of machine, gauge orientation and gauge
positioning. Results obtained from a similar investigation in which the
bar was offset transversely in the machine and loaded under conditions of
combined compression, bending and transverse shear showed considerable
deviation from theoretical strains (11), which was later traced to machine
frame distortion, (15). These results have implications to all strain
gauge model testing.

RESULTS FROM TESTS ON MODEL 1 AND MODEL 2

The data from model 1 is shown in Fig. 5 and results from model 2 are
shown in Fig. 6. All strains were corrected for transverse sensitivity
and reinforcement and in the case of model 1 were analysed
probabilistically using Monte-Carlo procedures (as recommended by (11))
with a mean error in gauge orientation of 0° and standard deviation of 1°.
Random errors were also induced in the analysis based on replication tests
on the prismatic bar of Fig. 4 with variance of 26 microstrain. The mean
and standard deviation of each term is given in Fig. 5 together with a
comparison of data from this test and a full field finite element analysis
of a similar product. The stresses from each of the strain gauged sites
from model 2 indicates non-linear effects in the anterior section together
with sign changes in the x direction at sites 3, 4 and 5.

DISCUSSION

The strains tabulated in Fig. 5 show small coefficients of variation for
orthogonal and principal strains. Shear strains are significantly
affected by errors in gauge orientation. Stresses in the z direction in
the anterior section are smaller in the experimental model than the finite
element model due to differences in shape between the two products and
possibly because of the model/machine interaction described above.
Experimentally derived stresses at sites 3, 4, 5, 6 and 7 are of
dissimilar form to the finite element data due to different contact
conditions (5). The non-linear data from model 2 (see Fig 6) was induced

146

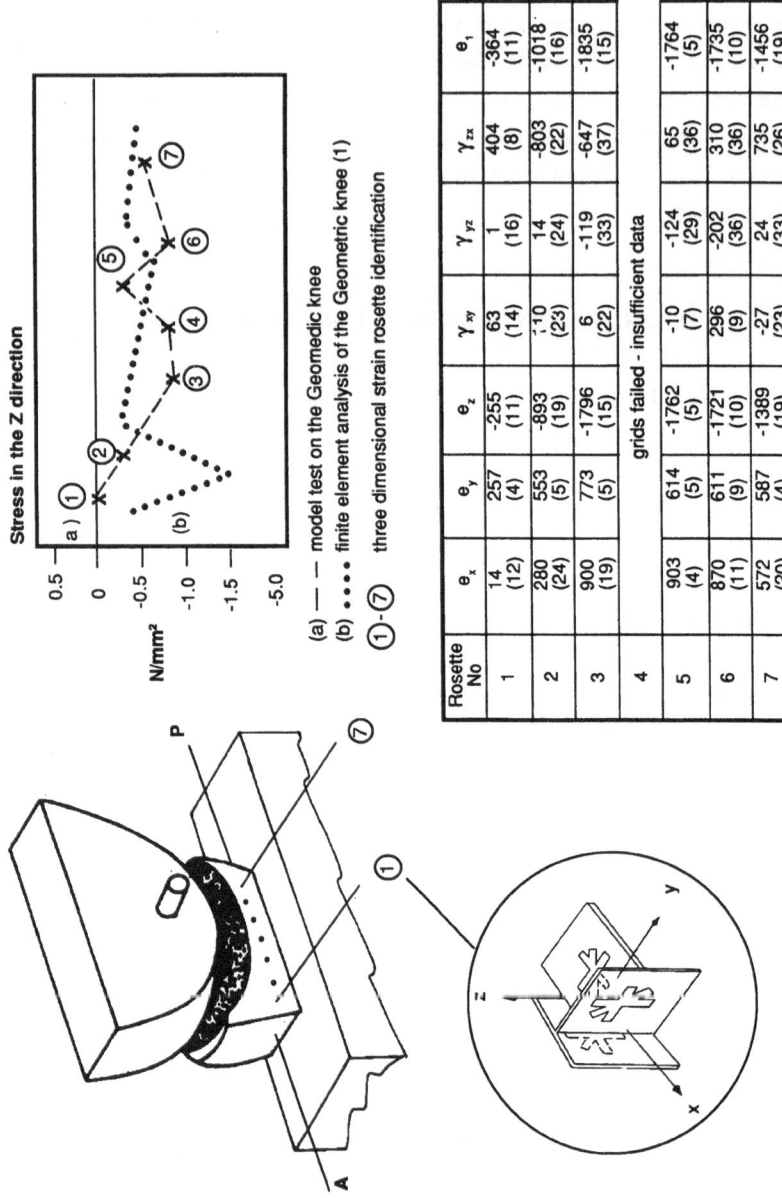

Figure 5. Mean and standard deviation ( ) of the strain tensor, maximum principal strain and comparison of stresses in the z direction, derived from rosettes (1) to (7) in the anteroposterior (A-P) axis of the Geomedic tibial component.

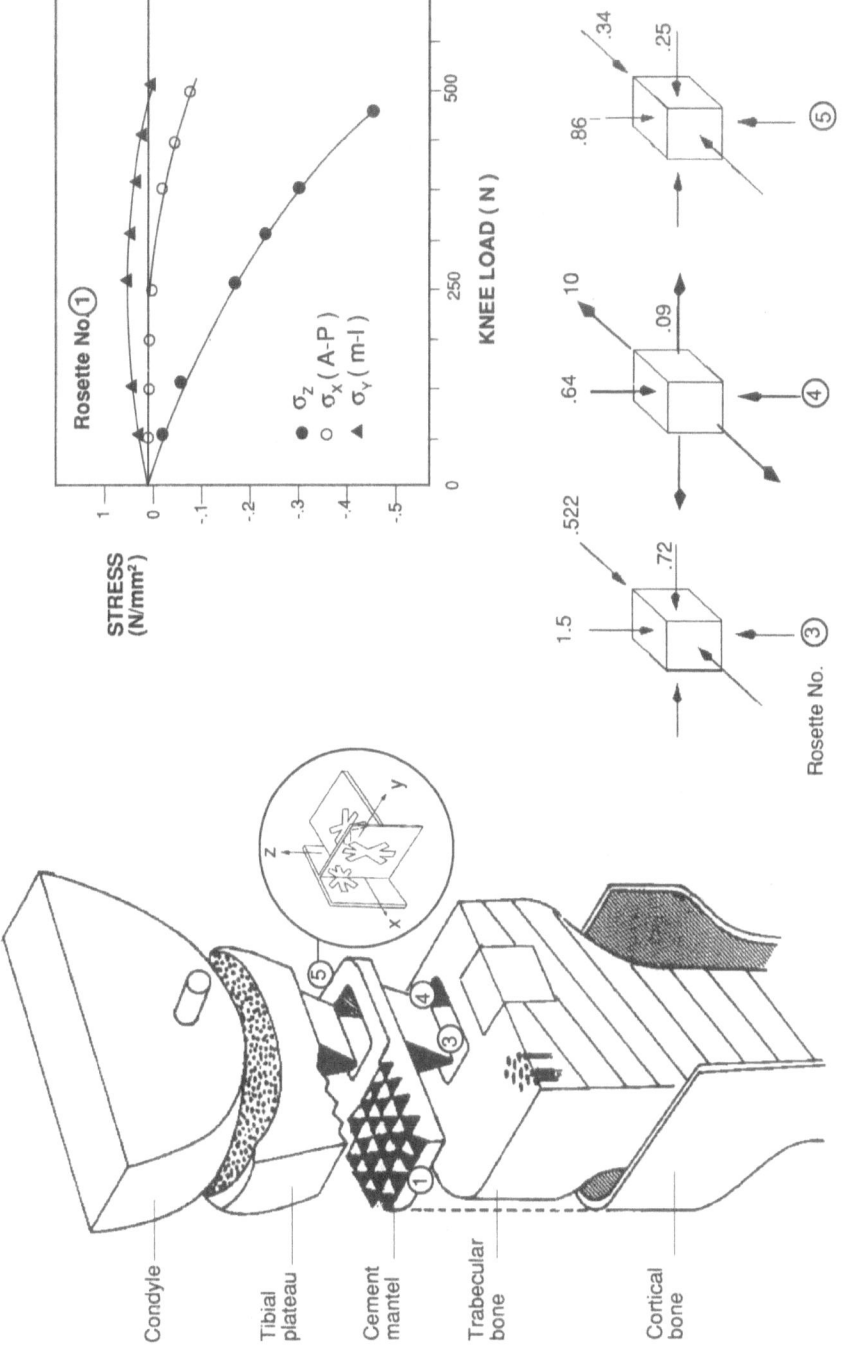

Figure 6. Stresses in the cement mantle derived from three-dimensional strain rosettes at sites (1) to (5) (Basic Figure courtesy B.S.S.M. - Strain (8)).

by a fractured interface between the modelled cement and bone, which together with sign changes in the x direction indicate tibial component tilting. Strains from model 2 were not analysed probabilistically due to the presence of strain gradients and gauge failure (6). Strain gradients were predominant due to the application of the larger grid offsets associated with the N32 rosette gauges used in the pattern of Fig. 2 (b).

CONCLUSION

Three dimensional embedded strain gauging is prone to error from various sources but can predict useful trends, providing that the data is interpreted in the light of fundamental investigations.

ACKNOWLEDGEMENT

The author would like to thank Howmedica International Incorporated Limerick for providing financial support for this work.

REFERENCES

1. Chao, E.Y., Wong, H.W., Frain, W.E. and Coventry, M.B., Stress analysis of the Geometric knee under static loading, Biomechanics symposium A.S.M.E., U.S.A., 1977, 122-141.

2. Bartel, D.L., Santavicca, E.A. and Burnstein, A.H., The effects of pegs and trays on stresses associated with loosening of knee protheses, Trans. Orthop. Res. Soc., 1980, 5, 165.

3. Steege, J.W., Polizos, T., Lewis, J.L. and Wixson, R.L., Failure mechanisms in PMMA around loaded tibial components, Trans. Orthop. Res. Soc., 1986, 355.

4   Lewis, J.L., Nicola, T., Keer. L.M., Clech, J.P., Steege, J.W. and Wixson, R.L., Failure processes at the cancellous bone PMMA interface, Trans. Orthop. Res. Soc. 1985, 144.

5. Little, E.G., Experimental stress analysis of the Geomedic knee joint using embedded strain gauges, Engineering in medicine, 1985, 14, 2, 69-74.

6. Little, E.G. and O' Keeffe, D., An experimental technique for investigation of three dimensional stress in bone cement underlying a tibial plateau, Proc. Inst. Mech. Eng., 1989, 203, 35-41.

7. Derenne, M. and Bazergui, A., Advances in the embedded strain gauge technique with an application to contact problems, Exptl. Mech., 19, 11, 105-112.

8. Little, E.G., Daly, G., Power, B., Dowson, M. and Kelly, P., The development of a model for the investigation of stresses in the cement layer underlying a tibial plateau, Strain, 1987, 23, 3, 19-25.

9. Little, E.G., Effects of self heating when using a continuous bridge voltage for strain gauging epoxy models, Strain, 1982, 18, 131-135.

10. Little, E.G., Tocher, D. and O' Donnell, P., Strain gauge reinforcement of plastics, Strain, 1990, In print.

11. Rossetto, S., Bray, A. and Levi, R., Uncertainties and errors in magnitude and direction in principal strains evaluated with three dimensional strain rosettes, Proc. 5th Int. Conf. Exptl. Stress Anal., Italy, 1974, 4, 11-4. 18.

12. Mitchell, D., Strain and temperature measurements on epoxy resin models, Transduced Technology, 1979, 1, 4, 21-25.

13. Barbato, G. and Little E.G., Performance analysis of three dimensional strain rosettes derived using computer simulation, Tech. Note: Istituto Di Metrologia,Italy, 1984.

14. Jenkins, R.F., The limitations of strain gauge load cells used as force transfer standards, Conf. Weightech 2, Institution of measurement and control, Harrogate, (U.K.), Sept. 1981, 143-158.

15. Rossetto, S., Bray, A. and Levi, R., Three dimensional strain rosettes: pattern selection and performance evaluation, Exptl. Mech., 1975, 15, 375-381.

FINITE ELEMENT ANALYSIS OF SPATE BENCHMARKS

JT BOYLE & WM CUMMINGS[§]
Department of Mechanical & Process Engineering
University of Strathclyde, Glasgow

[§] National Engineering Laboratory
East Kilbride, Glasgow

ABSTRACT

The aim of this paper is to report on some initial studies which have been made into the use of hybrid analytical and experimental techniques in the stress separation problem for thermographic stress analysis using the SPATE system. The initial part of the study, which is reported here, is based upon the use of detailed finite element analysis of a number of selected SPATE benchmark problems to provide a comparison between the accuracy of the thermographic full field contouring and smoothing.

INTRODUCTION

Thermographic stress analysis (TSA) has received considerable attention in the past fifteen years as a particularly convenient means of experimental stress analysis based on the thermoelastic effect. The thermoelastic effect is not well known amongst engineers, albeit for good reason. The simple observation of the layman that heating causes distortion, and that the amount of distortion depends on the degree of heating, has been translated into the engineering discipline known as the *theory of thermal stress* [1]. At this point the engineer would become familiar with the concept of thermal gradient and thermal strain and how these may be included in conventional stress analysis. The layman may be amused to ponder the question of whether distortion also causes a change in temperature and use his experiences to note that when a simple object is deformed, there is no appreciable change in temperature, but that if it is deformed in a cyclic manner until it breaks (a fatigue failure) the temperature of the broken surfaces does increase. The engineer would be confused by the fact that the process is not reversible (the simple object may be deformed by heating, but its temperature hardly changes during deformation) but conclude that at large deformations with plastic flow some heating can occur (as is evident in metal forming for example). In fact thermodynamic equilibrium dictates that temperature changes and straining are related, but that their effects are not equal. Indeed during recoverable elastic straining there *is* a slight change in temperature (a drop during tension), but for most engineering purposes this can be ignored. Then the assessment of the temperature field, the heat transfer analysis, can be carried out independently of the stress analysis - the mechanical and thermal behaviour are uncoupled. This uncoupling is not appropriate in an assessment of the reverse effect and the coupled *theory of thermoelasticity* [2] must be used:

In its simplest form, the theory of thermoelasticity requires that the familiar engineering heat conduction equation,

$$k\nabla^2\theta + W = C_\varepsilon\frac{\partial\theta}{\partial t}$$

where $\theta = T - T_0$ is the variation in absolute temperature T from a reference value T_0 and W is a heat source with k the thermal conductivity and C_ε the specific heat per unit volume at constant volume, be altered to include the effect of straining,

$$k\nabla^2\theta + W = C_\varepsilon\frac{\partial\theta}{\partial t} + T_0\beta\varepsilon_{ii}$$

This is known as the *generalised heat conduction equation*, where ε_{ii} is the volumetric strain and

$$\beta = \frac{E\alpha}{1 - 2\nu}$$

with α the coefficient of linear thermal expansion, and E Young's modulus and ν Poisson's ratio for an isotropic linear elastic material. This equation is complemented by the familiar engineering field equations of stress analysis which define the relation between strain and displacement, equilibrium of stress with mechanical loads and the material constitutive equation which relates stress, strain and temperature (the Duhamel - Neumann relations).

This type of coupling would cause misery to most engineers: in fact, much of the work on coupled thermoelasticity which is available in the literature is theoretical, related to the mathematical solution of simple problems [2], with little practical experience, to the extent that a comprehensive treatment and understanding of the phenomenon is absent [3]. This has not been a particular deficiency in engineering mechanics as the coupling is negligible.

It is with this background that thermographic stress analysis - an experimental technique based on thermoelasticity - should be discussed. In 1967 Belgen [4] considered the possibility of using the thermoelastic effect observed through infra-red radiometry to estimate the amplitudes of dynamic stresses. In 1974 a measurement system with the acronym *SPATE* (Stress Pattern Analysis by measurement of Thermal Emission) was patented by SIRA Ltd with funding from the UK Ministry of Defence [5] (now manufactured and marketed by Ometron Ltd). This system uses a *scanning* radiometer on a component which is rapidly loaded cyclically to produce a colour coded contour map of the temperature on a prepared surface, which may be interpreted in terms of mechanical quantities (more of which later) using the thermoelastic effect. Since 1982 numerous papers have been published investigating and using the *SPATE* system.

The advantages of the *SPATE* system are that it is *full field*, that is detailed information on a small region may be obtained ('zooming') as well as globally on a complete component, and *noncontacting* (apart from the surface preparation to increase and smooth the surface thermal emissivity) (literally 'point and shoot'). These features make this thermographic stress analysis system almost pre-eminent amongst competing full field experimental stress analysis techniques. The problem lies in the interpretation of the observed data which is related to the expectations of the engineer. It is the aim of this paper to consider these problems and to analyse some of the myths surrounding thermographic stress analysis.

THE BAD NEWS

Much criticism is levelled at thermographic stress analysis, even amongst it proponents (a good point to start is the monograph by Harwood & Cummings [6]). The two main points of contention which are usually raised are that (1) even for an isotropic homogeneous material the basic data is only in the form of bulk stress (the sum of the principal stresses) available only on the surface of the component and (2) calibration can require a fairly precise knowledge of physical data. (The other criticisms mainly relate to experimental technique, which can always be improved in any experimental system).

The first point is theoretical, but crucial, and in the writers' experience is poorly understood. The thermographic stress system (*SPATE*) measures temperature by radiation; this is related to a mechanical quantity on the basis of the theory of irreversible continuum thermodynamics, as initiated by Biot [7]. This is only a theory (and moreover one with limited experimental support). From the generalised heat conduction equation above it is assumed that straining occurs adiabatically with no conduction of heat and with no heat supply. (In the *SPATE* system heat conduction is prevented by rapid cycling). Then the amplitude of the strain cycle results in an amplitude of thermal change given by:

$$\Delta\theta = -\frac{T_0\beta}{C_\varepsilon}\Delta\varepsilon_v$$

which may be rewritten in terms of the specific heat per unit mass for constant (hydrostatic) pressure c_p and the bulk stress σ_m for an isotropic material,

$$\Delta\theta = -T_0 K \Delta\sigma_m$$

which is known as *Kelvin's formula*, where K is the *thermoelastic constant*,

$$K = \frac{\alpha}{\rho c_p}$$

and ρ is the density of the material.

Thus for an isotropic material the *SPATE* signal is related to the bulk stress (the sum of the principal stresses). It is important to appreciate this fact and the theoretical nature of the derivation. The derivation relies upon the First Law of Thermodynamics (Conservation of Energy) and the material constitutive relations (which relates entropy, strain and temperature, through the theory of continuum thermodynamics) and Fourier's Law of Heat Conduction (which relates heat flux to temperature through a phenomenological principle which defines the thermal conductivity). This theory requires that the material parameters are independent of temperature, even though it is known from experimental observation that the thermoelastic constant depends upon temperature by way of the degree of straining [8]. Theories of continuum thermodynamics which have temperature dependent material properties are scarce in the literature, often inconsistent, and without experimental support. The theory also requires that the thermal signals have an infinite speed of wave propagation, which leads to a practical paradox (in reality thermal disturbances should propagate with finite speed [9]). Finally the theory requires the concept of stress at a point (defined as load over a differential area as the area reduces to a point) which must be modified for materials with finite structure (such as composites and damaged materials associated with defects (cracks, corrosion, weldments etc)) through homogenisation [10]. The main assumption is thus that the material behaves elastically during load cycling, and that it is isotropic and homogeneous. In the absence of this assumption Kelvin's formula takes the form,

$$\Delta\theta = -\frac{T_0}{C_\epsilon}\beta_{ij}\Delta\epsilon_{ij}$$

in terms of the strain tensor ϵ_{ij} which is related to stress through the Duhamel - Neumann relations,

$$\sigma_{ij} = C_{ijkl}\epsilon_{kl} - \beta_{ij}\theta$$

where C_{ijkl} is the elasticity tensor and β_{ij} the thermoelastic tensor which are material constants. For an orthotropic material this can be rewritten as above in the form [11],

$$\Delta\theta = -\frac{T_0}{\rho c_p}\alpha_i\Delta\sigma_{ii}$$

where α_i are the orthotropic coefficients of linear thermal expansion. Thus the *SPATE* signal can only be interpreted as a linear sum of principal stresses with variable coefficients which are dependent on a detailed knowledge of the material characterisation!

At this stage the second criticism is enhanced: the thermoelastic constant for isotropic materials depends on measurements of the physical parameters of density, coefficient of thermal expansion and specific heat and on the material stress strain behaviour which may be quite complex. In the usual *SPATE* formulation this dependency on material stress strain behaviour is hidden since the thermoelastic constant does not depend on the elastic constants. For orthotropic materials even more material characterisation is required, and in general a *complete* description of the material mechanical and thermal behaviour is required to interpret the *SPATE* signal! This is true even if some calibration system is adopted (with the exception of isotropic materials).

At this point the engineer would be forgiven from retiring gracefully from further examination of the technique were it not for the fact that the *SPATE* system has proved accurate and reliable in benchmark test specimens - simply, it works. The problem in more 'advanced' applications lies in interpretation. In the absence of further information the thermographic stress analysis system could be judged inadequate and as a result is properly subject to criticism (it 'only' measures bulk stress for an isotropic elastic material). The key to a more complete understanding of the applications of thermographic stress analysis lies in a more realistic assessment of how the measurements should be assessed.

THE GOOD NEWS

It is often claimed that thermographic stress analysis techniques supply too little information, the bulk stress, and that it is difficult to accurately quantify the received information, which may be affected by variations in material properties and surface preparation [12]. While the latter (and other criticisms on the content of the data) is certainly true it is no more of a problem than in conventional methods of experimental stress analysis (the stress results derived from resistance strain gauges include similar assumptions and are only meaningful for shallow curved surfaces) and other full field techniques (photoelasticity uses a model material, moire interferometry requires smoothing of vector data to produce displacement and is also best suited to flat surfaces). The former criticism needs to be taken in context. *SPATE* does only provide bulk stress, but this is supplied full field and with high resolution if desired. Thus far from providing limited information, the method contains a considerable amount of information - it only needs to be extracted in the right way! For example, it is often stated in the literature that a failing in the *SPATE* system is that it cannot distinguish a state of pure shear from a state of zero stress in an unloaded component. While this type of statement is true, it is also absurd since the engineer using the system can clearly make the distinction - he has more information! It is thus argued here that he can also equip himself with even more information to help in interpreting and enhancing the received data.

Much is made in the literature of thermographic stress analysis on the apparent need for *stress separation* (this is discussed further by one of the authors in [13]) as a natural progression from a similar problem in traditional photoelasticity. In simple terms, for a general three dimensional component with curved surfaces, stress separation from equilibrium considerations is impossible. Further information is required, it just depends what form this should take and if stress separation is really required! No technique of experimental stress analysis should be used blindly without some independent confirmation. (This is of course a criticism often levelled at the practitioners of numerical stress analysis - how do you know the result is correct? - but which is equally directed here at experimentalists). The confirmation may be another experimental method, or a hybrid method [14] based on a numerical analysis. It is argued here that due to the information content in a system like *SPATE*, novel techniques are required.

As mentioned above some calibration procedure is usually adopted to interpret the *SPATE* signal (for example comparison with an attached strain gauge). This avoids the need to quantify the fundamental physical parameters, but reinforces the criticism that the technique may only be used for isotropic and homogeneous materials since this procedure is only transparent for such materials. Calibration is more difficult for anisotropic materials only since a detailed knowledge of the material stress strain behaviour is required (and it may be reasonably argued that this is not a problem particular to thermographic stress analysis techniques). With some lateral thinking the writers have concluded that calibration and information extraction requires a hybrid technique which is equal to the *SPATE* system in information content - namely, complementary finite element analysis.

Finite element analysis of typical *SPATE* benchmark problems is not particularly difficult with the current state of the art in personal computer and workstation based analysis systems. Such finite element analysis systems which are capable of error estimation, adaptive meshing, detail submodelling and parametric design optimisation are particularly suitable when coupled with thermographic stress analysis techniques.

The accompanying diagrams show *ANSYS* analysis of a selection of benchmarks, with contours of bulk stress. If these are compared to the *SPATE* diagrams [6] then a remarkable similarity is naturally observed.

```
ANSYS   4.2B
AUG 28 1987
14:02:57
POST1 STRESS
STEP=1
ITER=1
PSUM

ZOOM
ZV=1
* DIST=1.88
* XF=1.62
* YF=1.76
XRTO=1.09
MX=290
MN=-88.7
-49.7
-6.7

79.3
122
165

251
294
```

```
ANSYS   4.2B
AUG 28 1987
19:52:01
POST1 STRESS
STEP=1
ITER=1
PSUM

ZV=1
DIST=129
XF=69.5
YF=117
MX=1552
MN=-608
-368
-128
112

832

1312
1552
```

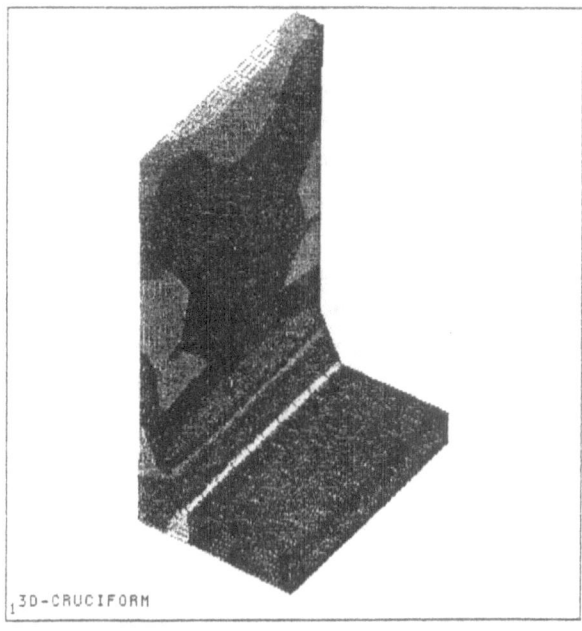

```
ANSYS  4.2B
AUG 29 1987
12:31:08
POST1 STRESS
STEP=1
ITER=1
PSUM

XV=.7
YV=1
ZV=1
DIST=106
XF=29.4
YF=66.9
ZF=41.9
HIDDEN
MX=15.3
MN=-.936
.191
2.19

6.19
8.19
10.2
12.2
14.2
15.2
```

3D-CRUCIFORM

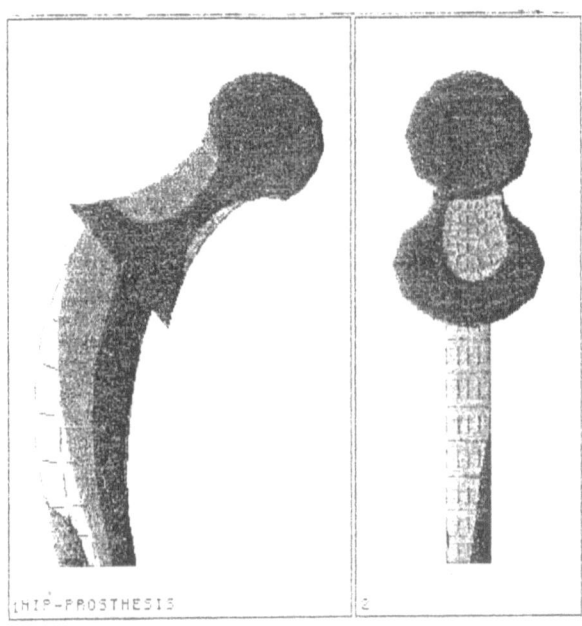

```
ANSYS  4.2B
AUG 31 1987
11:09:16
PLOT NO.   1
POST1 STRESS
STEP=1
ITER=1
PSUM

ZV=-1
DIST=45
XF=-.0993
YF=111
HIDDEN
MX=207
MN=-224
-177
-129

15.3
37.5

159
207

WIND=2
XV=-1
```

HIP-PROSTHESIS

However this simple comparison is not the key to the interpretation of *SPATE* data since much more powerful investigative procedures may be used:

There is a simple analogy between the capability of *SPATE* to examine the stress field locally at high resolution and submodelling in finite element analysis. The combination of these techniques may be used to fine tune the finite element model and simultaneously to guide *SPATE* towards regions of specific interest. Taking this concept further the basic *SPATE* data, which contains information on geometry, may be used to develop the finite element model, acting here as a guide, together with finite element error estimation techniques, to mesh density.

The most interesting procedure relates to the need for information on material properties. The interpretation of the raw *SPATE* signal for anisotropic materials requires the material behaviour to be modelled. Using the parametric design language included in *ANSYS* it is possible to assume a suitable material model and process the material constants as parameters. Subsequently the *ANSYS* design optimisation module may be used (or a simpler external program) to establish a best fit of the finite element model (suitably fine tuned with error estimation to the thermographic contour density) to the *SPATE* data (which would also depend on the material parameters). The consistency of this procedure would provide significant information on the quality of the *SPATE* data; the parallel finite element model would thus enhance this data and provide a detailed method of extracting further information. (In fact this type of procedure is a rather advanced version of the inspiring procedure used by Stanley and Chan [11] where the *SPATE* data is used to estimate the material properties). The writers are investigating these procedures further at the time of writing.

The message here is that there is much to be gained from the development of hybrid experimental numerical techniques in thermographic stress analysis, and that the existing systems should not be perceived as limited due to unrealistic or ingrained expectations of experimental stress analysis.

REFERENCES

1. Boley, B. and Weiner, J.H., Theory of Thermal Stress, Wiley, 1960

2. Nowacki, J., Thermoelasticity, 2nd Ed, Pergamon Press, 1982

3. Day, W.A, Heat Conduction within Linear Thermoelasticity, Springer Tracts in Natural Philosophy, Vol.30, 1985

4. Belgen, M.H., Structural stress measurements with an infrared radiometer, ISA Trans., 1967, 6, 49-53

5. Mountain, D.S., and Webber, J.M., Stress pattern analysis by thermal emission (SPATE), Proc. Soc. Photo-Opt. Inst. Engrs, 1978, 164, 189-196

6. Harwood, N. and Cummings, W.M., Thermoelastic Stress Analysis, Adam Hilger, 1990

7. Biot, M.A., Thermoelasticity and irreversible thermodynamics, J. Appl. Phys., 1956, 27, 241-242

8. Dunn, S.A et al The mean stress effect in metallic alloys and composites, Stress and Vibration: Recent Developments in Industrial Measurement and Analysis, SPIE Vol.1084, 1989

9. Chandrasekhharaiah, D.S., Themoelasticity with second sound, Appl. Mech. Rev., 1986, **39**, 355-376

10. Stolz, C., General relationships between micro and macro scales for nonlinear behaviour of heterogeneous materials, Chap.4, Modelling Small Deformations of Polycrystals, Gittus, J. and Zarka, J. Ed, 1985

11. Stanley, P. and Chan, W.K., The application of thermoelastic stress analysis techniques to composite materials, J. Strain Anal, 1988, **23**, 137-143

12. McKelvie, J., Consideration of the surface temperature response to cyclic thermoelastic heat generation, Stress Analysis by Thermoelastic Techniques, SPIE Vol.731, 1987

13. Boyle, J.T., Post processing SPATE data, Chap.9, Ref.[6] ibid

14. Kobayashi, A.K., Hybrid experimental stress analysis, Handbook on Experimental Mechanics, Kobayashi, A.K. Ed., Prentice Hall, 1987

FINITE ELEMENT SIMULATION OF TYRE DYNAMICS

KAZEM KORMI and MOHAMMED N ISLAM
Centre for Advanced Research in Engineering
Leeds Polytechnic, Calverley Street, Leeds LS1 3HE, UK

ABSTRACT

An idealised steady state tyre rolling process is modelled using ABAQUS. This simplified approach produces detailed response parameter values of practical importance but does not need special purpose elements or procedures. The results obtained highlight the route to successful detailed analysis.

INTRODUCTION

The pneumatic tyre is a challenging structure for the analyst or designer. The tyre in service is subjected to severe stresses and deformation whose nature and magnitude must be determined in order to accurately predict the interactions or resultant stress field. A dynamic response analysis requires that the physical properties of the component materials and their configuration are known, that the applied loads under the specified working conditions are completely characterised and that a suitable analytical technique is available to calculate the response parameters. One major problem is the difficulty of developing a unified constitutive relation that can be applied to all of the tyre material. An even bigger one is accurately defining the magnitude and distribution of loads acting under working conditions.

Inflation pressure accounts for most of the total load. Other significant loads include vehicle weight and dynamic forces applied to the tyre caused by rotation,

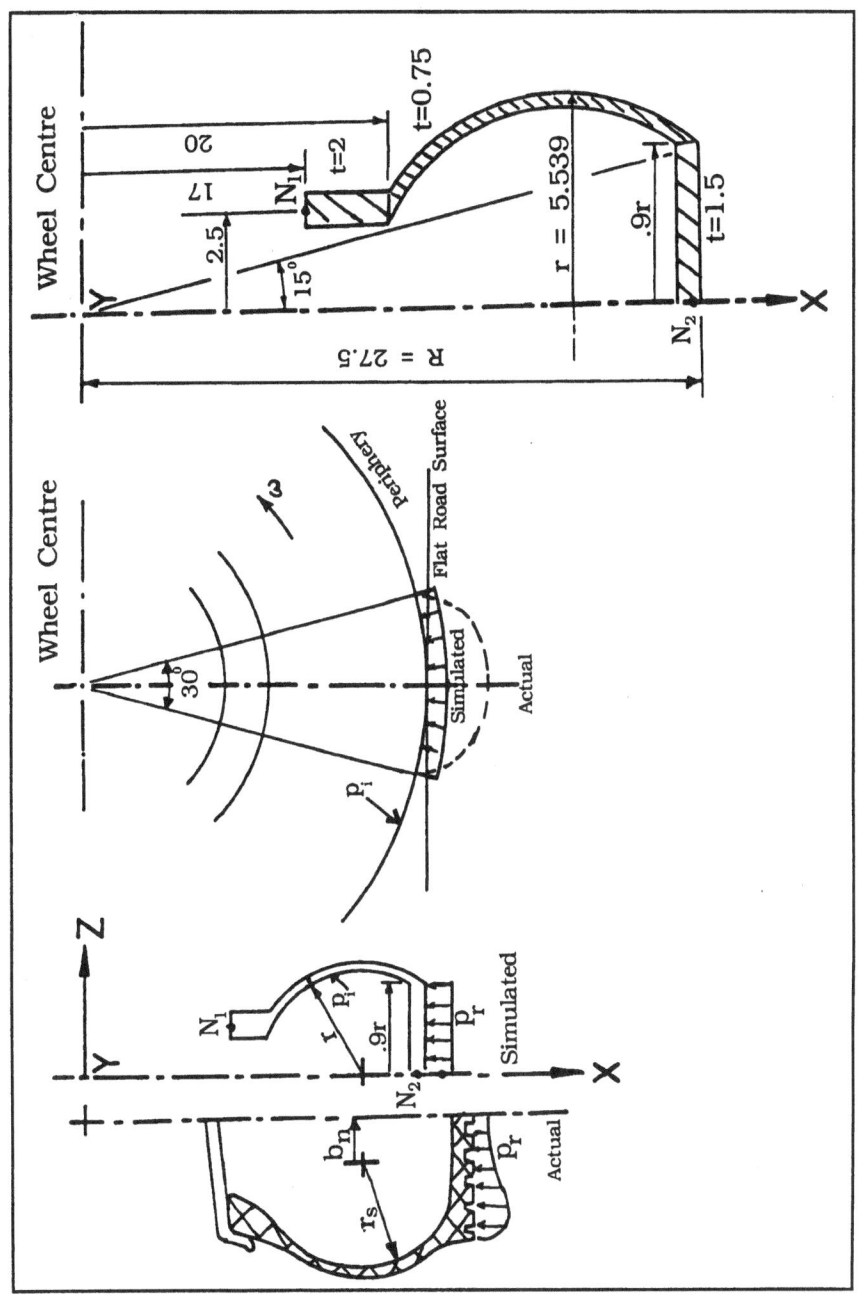

Figure 1. Geometric parameters for the tyre model.

impact and vibration. Thus the tyre not only supports the radial axle-load and transmits longitudinal braking or driving forces but also withstands the lateral cornering and camber forces. The distribution of vertical, lateral and horizontal forces over the foot-print area through which the loads are transmitted is very complex [1,2].

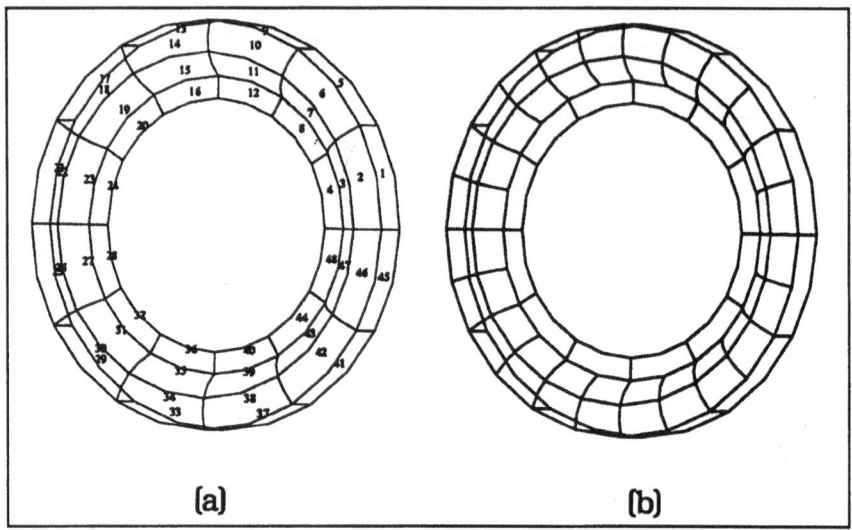

Figure 2. Discretised tyre skin (a) and reinforcing wire (b).

The shape and size of the foot-print area are affected by various factors [3]. In addition the deformation in the region of the foot-print causes abrupt changes in curvature at the foot-print boundaries which in turn lead to complicated stress fields in this region. Because of these difficulties the use of classical theory for analysing the tyre response has been confined so far to the use of simplified models and semi-empirical formulations [4,5,6]. These models cannot accommodate the transient response under non-linear dynamic load where vibration and the generation and propagation of stress waves and standing waves are important features affecting the response of the tyre. The finite element approach offers an obvious advantage here.

Recent attempts to apply the technique have been directed towards developing specialised elements and solution procedures for tackling the problem [7,8] and to studying only the load/deflection properties [9]. The theme of this paper is that the dynamic response of a rolling tyre under load can be simulated using a commercially available finite element package such as ABAQUS [10].

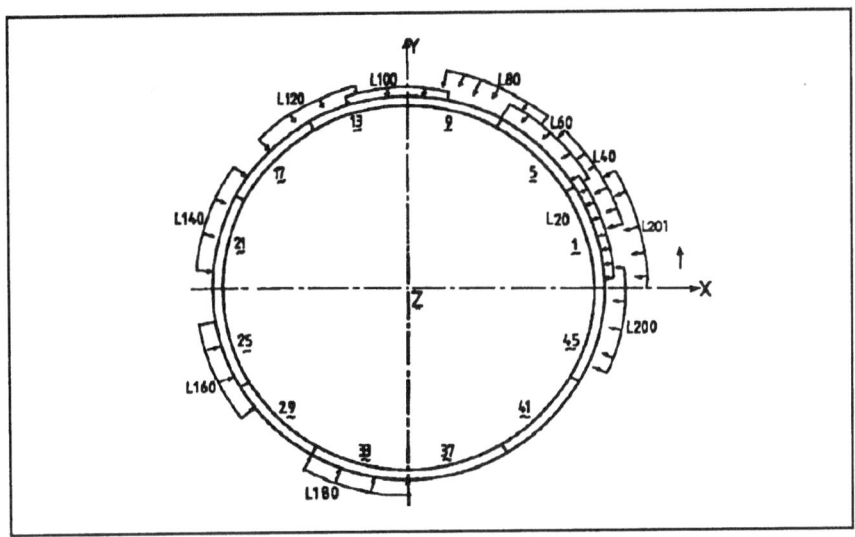

Figure 3. Positions of the foot-print load at different times.

FINITE ELEMENT MODELLING

The geometry of the model tyre is shown in Fig. 1 with the actual tyre for comparison. The tyre matrix has been discretised with S8R eight-noded shell elements. In the model the crown has thickness of 1.5 cm and is assumed to be in contact with the road surface, the sidewall has thickness .75 cm and the bead ring 2 cm. These regions are reinforced by high tensile steel wires of .5 cm, .375 cm and .75 cm diameter, respectively, discretised with three noded solid section beam

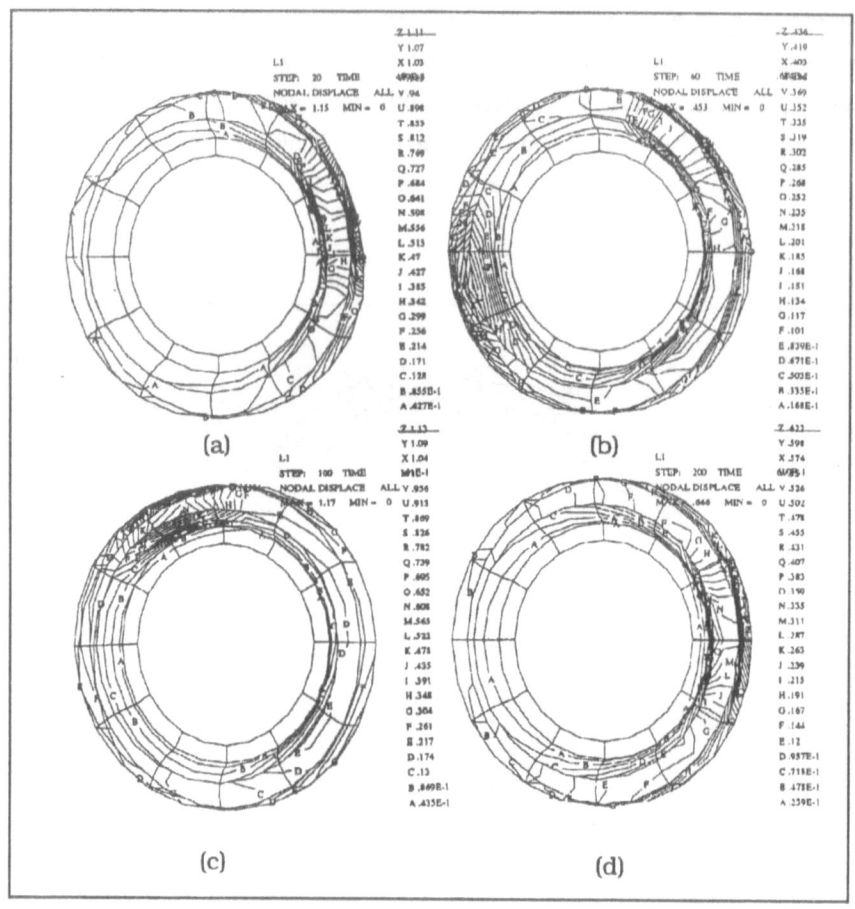

Figure 4. Distribution of total displacement at different values of total dynamic time.

elements. The tread profile has been ignored. The finite element meshes are shown in Fig. 2. The materials are assumed to be homogeneous, isotropic and elastic with Young's Modulus, Poisson's ratio and density taken respectively as 2.07 E5 N/cm², .45 and 1.9585E-5 kg/cm⁴/sec² for the rubber and 1.7E7 N/cm², .33 and 7.83E5 kg/cm²/sec² for the steel. The uniform linear speed of the wheel is taken to be 100 km/hr, the inflation pressure is 25 N/cm². The included contact angle is 30° over which the uniform road pressure of 30.58 N/cm² is assumed to act.

The rolling of the wheel is simulated by rotating the road-pressure zone around the periphery at the appropriate angular speed while keeping the wheel itself stationary. The contact angle of 30° is maintained during the rolling motion. This implies that as the motion continues, only a part of the peripheral contacting elements are in contact with the road. A special subroutine to simulate the rolling motion and to effect the non-uniformly distributed pressure loading has been developed and is used in conjunction with the parent finite element code. The algorithm in the subroutine not only rotates the pressure zone but tracks and monitors continuously the motion and hence the positions of the leading and trailing edges of the contact zone

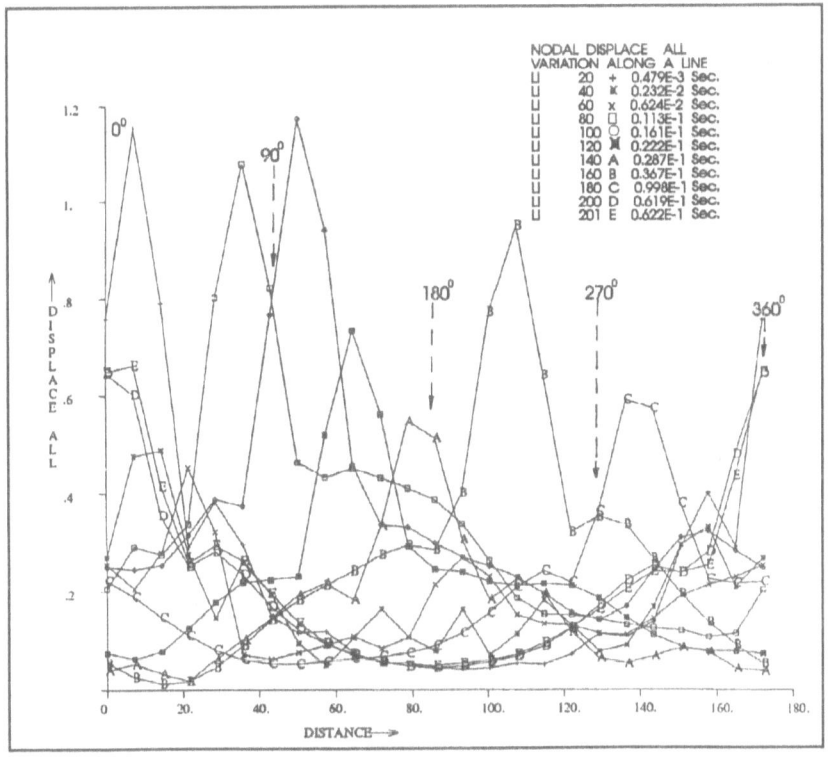

Figure 5. Variation of total displacement along the periphery of the symmetry plane at different times.

with respect to the position of the zero-time reference axes system. It then compares at any instant the position of the contact zone with those of the tyre-road contacting elements, identifies the element number and the part of it over which the road pressure has to be applied. The present version of the subroutine can be modified to accommodate rolling speed, foot-print shape and pattern of pressure distribution as functions of time and position. During the response analysis appropriate boundary conditions are applied at node sets N_1 and N_2 shown in Fig. 1.

RESULTS AND DISCUSSION

The frame of reference for the motion is shown in Fig. 3. The foot-print moves in an anti-clockwise direction from a position straddling the positive x-axis at zero time. The position of the foot-print at several specific times is also shown in Fig. 3. The deformation patterns caused by the applied load at four of these positions are shown in Fig. 4. In the early stages the deformation is confined to the region around the foot-print as shown in Fig. 4(a). As the motion proceeds the area of deformation spreads out. It initially trails the pressure zone but at the approach of one quarter turn a deformation pattern is established in the opposite half of the tyre as shown in Fig. 4(b). This secondary concentration of deformation disappears by the half-revolution position and does not recur during the second half of the revolution. The variation of total deformation with position is shown in Fig. 5 which shows the values along the periphery at various time steps. The propagation of bending waves ahead of the pressure zone can clearly be identified.

Two particular cases are shown in Fig. 6. These show how the inward deformation is more widely spread out in the later stage of loading than it is during the early stage. They also show how the peak deformation is in advance of the centre of the foot-print. The critical speed for the present case turns out to be around 137 km/hr which is below the rolling speed. Yet there is no evidence of standing waves being developed. This is contrary to the previous findings of Soedel [11].

The variations of stress components around the circumference as a function of

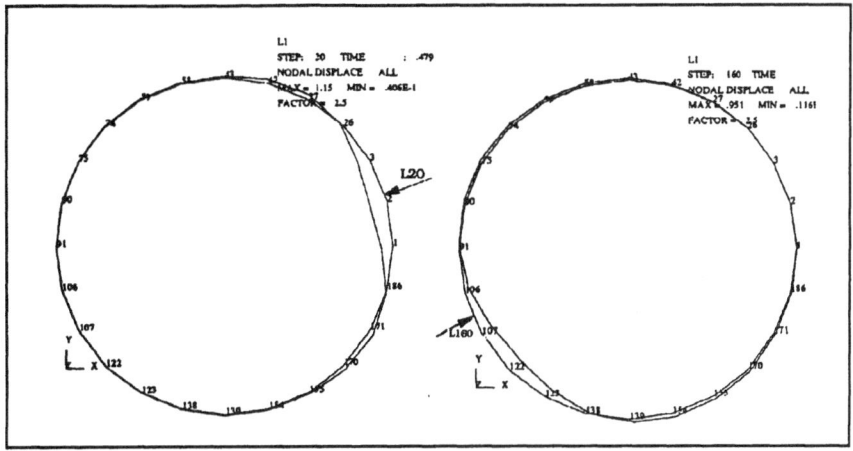

Figure 6. The deformation in the plane of radial symmetry for two time steps.

time are shown in Fig. 7. In general the transverse stress S11 is comparable to the shear stress S12 whereas the circumferential stress S22 is about 1.5 times higher. The propagation of the stress field along the periphery can be seen and in particular the characteristic that at approximately every quarter revolution the stress field is localised around the pressure zone but in between it is more evenly spaced. This indicates that the stress field induced by the load contact travels a distance equal to the rolling circumference during the time taken for one quarter turn. It can also be seen from the stress reversals in Fig. 7 that the tyre periphery vibrates in several higher modes. The distribution of S11 stress fields in the sidewall and bead rim for two illustrative time steps are shown in Fig. 8. The localised stress field in the pressure zone is accompanied by a second region of comparable strength. In Fig. 8(a) for time step L20 the second region is in advance, but in Fig. 8(b) for time step L180 it is trailing. The results show that the secondary field travels one complete revolution during the time of one half a revolution of the wheel. This difference between the sidewall behaviour and the earlier discussed tread behaviour is due amongst other things to the different geometry of the reinforcing wires in the two parts of the tyre.

168

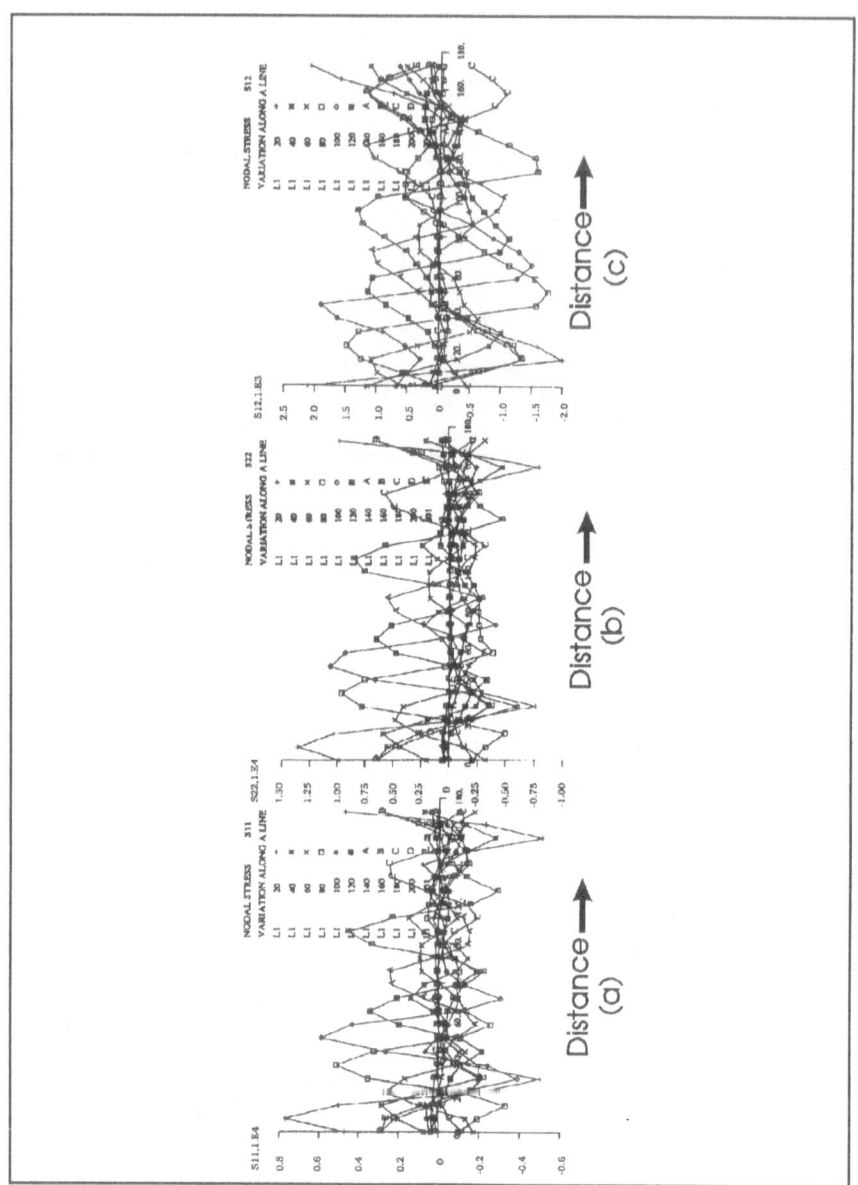

Figure 7.

Variation of (a) S11, (b) S22 and (c) S12 stress component along the circumference of the symmetry section at different times.

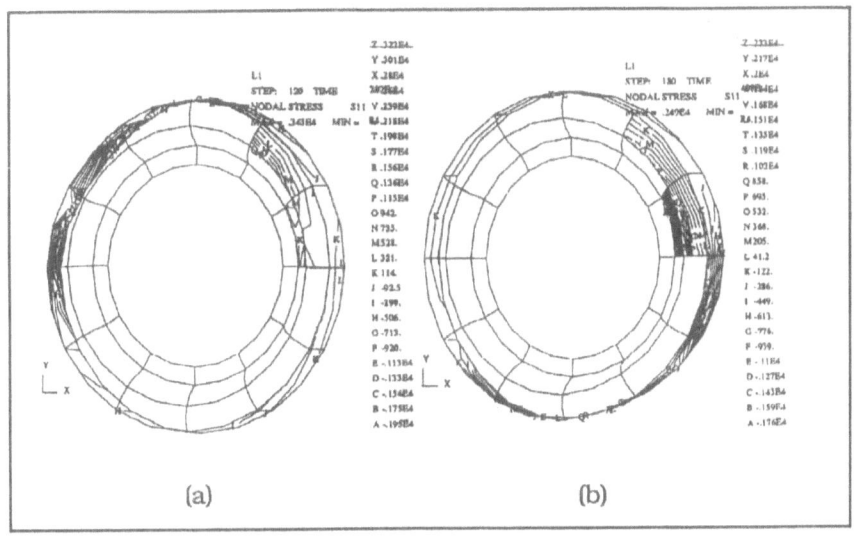

Figure 8. Distribution of the S11 stress component at two time steps.

CONCLUSIONS

The steady state idealized tyre rolling process which is modelled here is capable of producing detailed response parameters of practical importance which the enormously complicated classical and other semi-empirical methods currently available are incapable of doing. The present procedure dispenses with the need for developing special purpose elements and solution procedures for this class of problems. Extension of this model to incorporate more accurate geometry, material properties and kinematic conditions presents no serious difficulties. It is however suggested that before such an exercise is undertaken, the accuracy of the component areas mentioned above be ascertained on a unified and rational basis.

ACKNOWLEDGEMENT

The work presented here was undertaken with the active participation of the personnel of the Centre for Advanced Research in Engineering (C.A.R.E.) of Leeds Polytechnic. The authors are grateful to Phil Crowe for coding the subroutine and to Carolyn Prosser and Stephen McGill for preparing the manuscript.

REFERENCES

1. S K Clark (ed). "Mechanics of Pneumatic Tyres" U.S. Dept of Transportation, HS 805952, Washington, 1981.

2. W E Howell, et al. "Aircraft Tyre Footprint Forces" in "The Tyre pavement Interface", M G Pottinger, (Ed), ASTM STP-929, 1986.

3. W B Hornes, et al. "Recent Studies to Investigate effects of Tyre Footprint Aspect Ratio on Dynamic Hydroplaning Speed" in Ref. 2.

4. H B Pacejka "Analysis of Tire Properties" Chapter 9 of Ref. 1.

5. J E Bernard, L Segel and R E Wild, "Tire Shear Force Generation during Combined Steering and Braking Maneouvres" SAE paper 770852, 1977.

6. R S Sharp and M A El-Nashar, "A Generally Applicable Digital Computer Based Mathematical Model for the Generation of Shear Forces by Pneumatic Tyres". Vehicle System Dynamics, 15, 1986, pp 187-209.

7. A K Noor, C M Anderson & J A Tanner, "Mixed Models and Reduction Techniques for Large Rotation Analysis of Shells of Revolution with Applications to Tires" NASA TP-2343, Oct, 1984.

8. J Padovan & I Zeid, "On the Development of Travelling Load Finite Element" Compt. & Structures, 12, 1980, pp. 77-83.

9. P S Shoemaker, "Tire Engineering by Finite Element Modelling", SAE Paper No. 840065, 1984.

10. ABAQUS Finite Element Code, HKS Inc. RI. USA.

11. W Soedel "On the Dynamic Response of Rolling Tyres according to Thin Shell Approximations", J. Sound & Vibration 41(2), 1975, pp.(233-246).

DETERMINATION OF BOUNDARY VALUES IN THE INNER SURFACE OF A CYLINDER BY USING BOUNDARY ELEMENT METHOD

TOMOAKI TSUJI
Department of Mechanical Engineering
Shizuoka University
Jyoohoku 3-5-1, Hamamatsu, 432 JAPAN

NAOTAKE NODA
Department of Mechanical Engineering
Shizuoka University
Jyoohoku 3-5-1, Hamamatsu, 432 JAPAN

YOSHIO TANAKA
Department of Mechanical Engineering
Shizuoka University
Jyoohoku 3-5-1, Hamamatsu, 432 JAPAN

ABSTRACT

In order to obtain data for practical use, an inverse problem for determining boundary values in inner surface of a cylinder is investigated. The indirect boundary element method with fictitious boundary is applied to this problem. The pressure in the inner surface of the cylinder is obtained by using the tangential strain data in the outer surface, and the availability of this method is confirmed by the actual experiment.

INTRODUCTION

In recent years, a considerable effort has been devoted to inverse elastic problems. For example, authors [1]-[3] have discussed the analytical method for the inverse thermoelastic problems. Numerical methods are also used [4]-[7] by many investigators. Oda [4] obtained the contact stress distributions by using the finite element method. Tanaka and Yamagiwa [5] used the boundary element method to obtain the form of the inclusion in the elastic body. Tomishima and Yada [6] obtained the residual stress in the plain plate by the inverse method. In these papers, theoretical or numerical data are used instead of the measured data at the boundary. But measured data contain some errors naturally and these errors can not be avoided. Thus, in practical use of the inverse problem for experimental mechanics, it is important to use experimental data and to consider accuracy of the inverse problem. As a part of our study program of inverse elastic problems, we discuss an inverse problem to determine boundary conditions by using the measured data contained natural error. In this case, we consider that the boundary element method is the most benefit method for such a inverse problem.

Therefore, to obtain the data for practical use, we investigate an inverse problem to determine boundary conditions in inner surface of a cylinder. The indirect boundary method and the least-squares method are used for the formulation of this inverse problem. By the actual experiment of the cylinders subjected to concentrated load in the inner surface, efficiency and accuracy of this method in the practical case is discussed.

B.E.M. EQUATIONS

We consider the plane strain problem and introduce the fictitious boundary S^* as shown in Fig.1. Displacement u_i, traction t_i and strain ε_{ij} are shown as follows [8]:

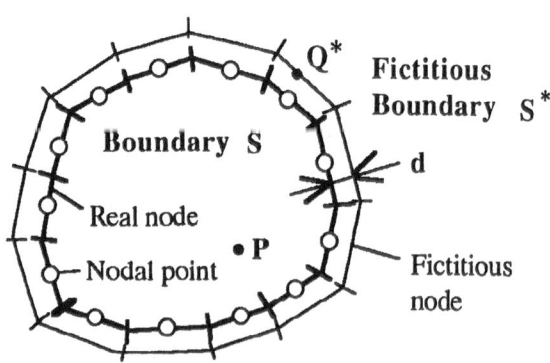

Fig.1 Fictitious Boundary.

$$u_i(P) = \int_{S^*} U_{ik}(P,Q^*)\phi_k(Q^*)dS(Q^*)$$

$$t_i(P) = \int_{S^*} T_{ik}(P,Q^*)\phi_k(Q^*)dS(Q^*) \qquad (1)$$

$$\varepsilon_{ij}(P) = \int_{S^*} E_{ijk}(P,Q^*)\phi_k(Q^*)dS(Q^*)$$

where $\phi_k(Q^*)$ is the distribution function on the fictitious boundary and the fundamental solution kernels $U_{ik}(P,Q)$, $T_{ik}(P,Q)$ and $E_{ijk}(P,Q)$ are denoted as:

$$U_{ik}(P,Q) = \frac{1}{8\pi G(1-v)}\{(3-4v)\ln(\tfrac{1}{r})\delta_{ik} + \frac{r_i r_k}{r^2}\}$$

$$T_{ik}(P,Q) = \frac{-1}{4\pi(1-v)r}[\{(1-2v)\delta_{ik} + 2\frac{r_i r_k}{r^2}\}\frac{\partial r}{\partial n} - (1-2v)(\frac{r_i}{r}n_k - \frac{r_k}{r}n_i)] \qquad (2)$$

$$E_{ijk}(P,Q) = \frac{-1}{8\pi G(1-2v)r^2}\{(1-2v)(\delta_{ik}r_j + \delta_{jk}r_i) + 2\frac{r_i r_j r_k}{r^2}\delta_{ij}r_k\}$$

$$r_i = x_i(Q) - x_i(P) \ , \ r = \sqrt{r_i r_i} \ , \ v: \text{Poisson's ratio} \ , \ G: \text{Shear module}.$$

When the boundary is separated into N linear elements and $\phi_j(Q^*)$ is assumed as constant value of $\phi_j^{(n)}$, Eq. (1) can be discretized as follows:

$$u_i(P) = \sum_{n=1}^{N} G_{ik}^{(n)}(P)\phi_k^{(n)} \ , \ t_i(P) = \sum_{n=1}^{N} F_{ik}^{(n)}(P)\phi_k^{(n)} \ , \ \varepsilon_{ij}(P) = \sum_{n=1}^{N} B_{ijk}^{(n)}(P)\phi_k^{(n)} \qquad (3)$$

where

$$G_{ik}^{(n)}(P) = \int_{S_n^*} U_{ik}(P,Q^*)dS(Q^*), \ F_{ik}^{(n)}(P) = \int_{S_n^*} T_{ik}(P,Q^*)dS(Q^*)$$

$$B_{ijk}^{(n)}(P) = \int_{S_n^*} E_{ijk}(P,Q^*)dS(Q^*) \qquad (4)$$

We consider the following conditions for the present problem.

$$u_i(P_{u,i}^m) = \overline{u}_i(P_{u,i}^m) \qquad (m=1,2,\cdots M_{u,i} \ , \ i=1,2)$$

$$t_i(P_{t,i}^m) = \overline{t}_i(P_{t,i}^m) \qquad (m=1,2,\cdots M_{t,i} \ , \ i=1,2) \qquad (5)$$

$$\varepsilon_{ij}(P_{\varepsilon,ij}^m) = \overline{\varepsilon}_{ij}(P_{\varepsilon,ij}^m) \qquad (m=1,2,\cdots M_{\varepsilon,ij} \ , \ i,j=1,2)$$

where $P_{u,i}^m$, $P_{t,i}^m$ and $P_{\varepsilon,ij}^m$ denote the points, and $M_{u,i}$, $M_{t,i}$ and $M_{\varepsilon,ij}$ denote the number of points where u_i , t_i and ε_{ij} are given respectively. By substituting Eq.(5) into Eq.(3), we can obtain $(M_{u,1} + M_{u,2} + M_{t,1} + M_{t,2} + M_{\varepsilon,11} + M_{\varepsilon,12} + M_{\varepsilon,22})$

equations. By using the least-squares method, the roots of these equations can be given as the roots of the following equation:

$$\frac{\partial S^2}{\partial \phi_\xi^{(\alpha)}} = 0 \quad (\xi=1,2\cdots N, \alpha=1,2) \tag{6}$$

where

$$S^2 = \sum_{i=1}^{2} \left[\sum_{m=1}^{M_{u,i}} \{\overline{u}_i(P_{u,i}^m) - \sum_{n=1}^{N} G_{ik}^{(n)}(P_{u,i}^m)\phi_k^{(n)}\}^2 + \sum_{m=1}^{M_{t,i}} \{\overline{t}_i(P_{t,i}^m) - \sum_{n=1}^{N} F_{ik}^{(n)}(P_{t,i}^m)\phi_k^{(n)}\}^2 \right.$$
$$\left. + \sum_{j=i}^{2} \sum_{m=1}^{M_{\varepsilon,ij}} \{\overline{\varepsilon}_{ij}(P_{\varepsilon,ij}^m) - \sum_{n=1}^{N} B_{ijk}^{(n)}(P_{\varepsilon,ij}^m)\phi_k^{(n)}\}^2 \right] \tag{7}$$

Equation (6) can be derived to 2N systems of algebraic equations as follow:

$$\sum_{k=1}^{2} \sum_{n=1}^{N} \phi_k^{(n)} \sum_{i=1}^{2} \{ \sum_{m=1}^{M_{u,i}} G_{ik}^{(n)}(P_{u,i}^m)G_{i\alpha}^{(\xi)}(P_{u,i}^m) + \sum_{m=1}^{M_{t,i}} F_{ik}^{(n)}(P_{t,i}^m)F_{i\alpha}^{(\xi)}(P_{t,i}^m) $$
$$+ \sum_{j=i}^{2} \sum_{m=1}^{M_{\varepsilon,ij}} B_{ijk}^{(n)}(P_{\varepsilon,ij}^m)B_{ij\alpha}^{(\xi)}(P_{\varepsilon,ij}^m)\} $$
$$= \sum_{i=1}^{2} \{ \sum_{m=1}^{M_{u,i}} \overline{u}_i(P_{u,i}^m)G_{i\alpha}^{(\xi)}(P_{u,i}^m) + \sum_{m=1}^{M_{t,i}} \overline{t}_i(P_{t,i}^m)F_{i\alpha}^{(\xi)}(P_{t,i}^m) $$
$$+ \sum_{j=i}^{2} \sum_{m=1}^{M_{\varepsilon,ij}} \overline{\varepsilon}_{ij}(P_{\varepsilon,ij}^m)B_{ij\alpha}^{(\xi)}(P_{\varepsilon,ij}^m)\} \qquad (\xi=1,2...N,\alpha=1,2) \tag{8}$$

By solving Eq.(8) with the known boundary conditions of u_i, t_i and ε_{ij}, unknown values of $\phi_k^{(n)}$ can be obtained and unknown boundary values can be calculated by Eq.(3).

EXPERIMENTAL ANALYSIS

We consider here the inverse problem for determining the distribution of pressure in the inner surface of the pipe by using the tangential strain data in the outer surface. Instead of this plane strain problem, we investigate the plane stress problem, because the experiment is easier for the latter problem than the former one.

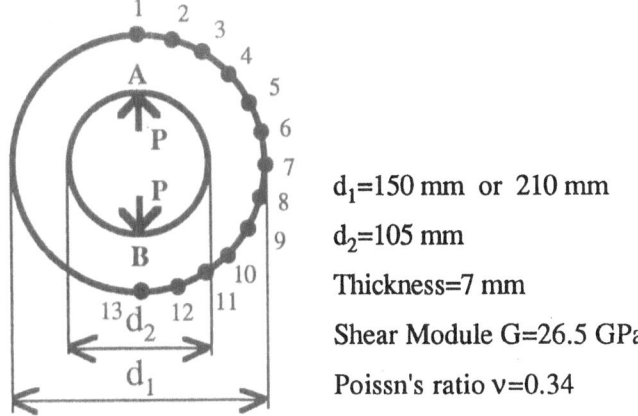

d_1=150 mm or 210 mm

d_2=105 mm

Thickness=7 mm

Shear Module G=26.5 GPa

Poissn's ratio v=0.34

Fig.2 Annular plates subjected to concentrated load in the inner surface.

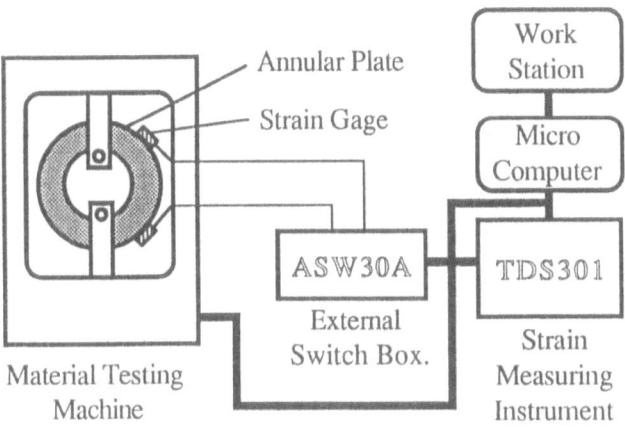

Fig.3 Measuring System

Measurement

Figure 2 show the geometry and the mechanical properties of the test material. In Fig. 2, the fill circles show the points where strain gages are attached. The circular plate is subjected to the point load P in the inner surface at points A and B by a material testing machine. The values of the strain are measured automatically by the measuring system as show in Fig.3, and are analyzed by the computer. In the practical use, it is important to obtain good result with fewer measuring points, but 13 points are too few to proceed the present inverse analysis. Thus, we intend to use the interpolation method. The measured strain

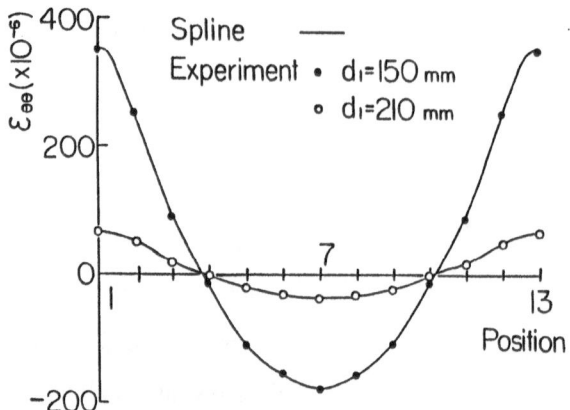

Fig.4 Distributions of hoop strain $\varepsilon_{\theta\theta}$ in outer surface.

TABLE 1

Boundary conditions for the inverse problem

Boundary	t_r	t_θ	$\varepsilon_{\theta\theta}$	Point	N. of points
Outer Surface	0	0	$\overline{\varepsilon}_{\theta\theta}$	P_O^m	M_O
Inner Surface	unknown	0	unknown	P_I^m	M_I

data and the interpolated values by using the cubic spline are shown in Fig. 4 by circles and lines respectively. The interpolated values are used as input data for the inverse analysis.

Inverse Analysis

Boundary conditions for the inverse analysis are shown in Table 1. By using these conditions, Eq.(8) can be written by the polar coordinate (r,θ) as follow:

$$\sum_{k=1}^{2}\sum_{n=1}^{N}\phi_k^{(n)}\{\sum_{i=1}^{2}\sum_{m=1}^{M_O}F_{ik}^{(n)}(P_O^m)F_{i\alpha}^{(\xi)}(P_O^m)+\sum_{m=1}^{M_O}B_{\theta\theta k}^{(n)}(P_O^m)B_{\theta\theta\alpha}^{(\xi)}(P_O^m)$$

$$+\sum_{m=1}^{M_I}B_{r\theta k}^{(n)}(P_I^m)B_{r\theta\alpha}^{(\xi)}(P_I^m)\}=\sum_{m=1}^{M_O}\overline{\varepsilon_{\theta\theta}}(P_O^m)B_{\theta\theta\alpha}^{(\xi)}(P_O^m)$$

$$(\xi=1,2,...N,\ \alpha=1,2)\qquad(9)$$

where M_I and M_O denote the number of elements for the point P_I^m in the inner surface and the point P_O^m in the outer surfaces respectively. $B_{\theta\theta k}$ and $B_{r\theta k}$ are:

⇐ subjected load

← determined pressure

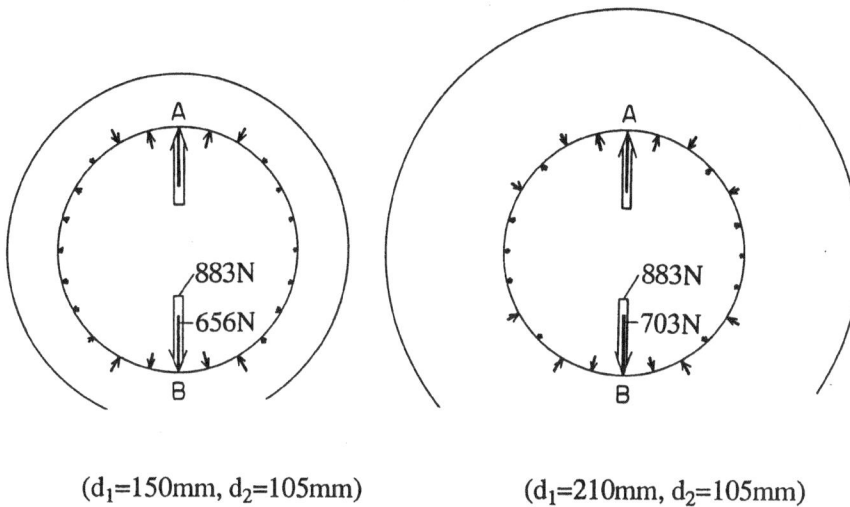

(d₁=150mm, d₂=105mm)   (d₁=210mm, d₂=105mm)

Fig.5 Determined inner traction t_r of the annular plates.
(P=883N, M_O=72, M_I=24)

$$B_{\theta\theta k}^{(n)}=\cos^{2}(\theta)\, B_{11k}^{(n)}+\sin^{2}(\theta)\, B_{11k}^{(n)}+\sin(2\theta)\, B_{12k}^{(n)}$$

$$B_{r\theta k}^{(n)}=\frac{\sin(2\theta)}{2}\,(B_{22k}^{(n)}-B_{11k}^{(n)})+\cos(2\theta)\, B_{12k}^{(n)}$$

By substituting the measured strain data $\overline{\varepsilon_{\theta\theta}}$ into Eq.(9), $\phi_k^{(n)}$ can be obtained and unknown boundary values can be calculated by Eq.(3).

Figure 5 shows the distribution of the traction t_r determined by the present inverse method with M_O=72 and M_I=24. In this figure, the point loads and the distributions of determined pressure are shown by the outlined arrows and the normal arrows respectively. It is confirmed that the present method is available for the determination of the inner pressure of the cylinder.

In the inverse calculations, it is important to know the sensitivity of M_O and M_I for the determined values. Thus, in Fig. 6, we show the relationship between the error of the determined pressure and the numbers of M_O and M_I. The mean square error is defined as follow:

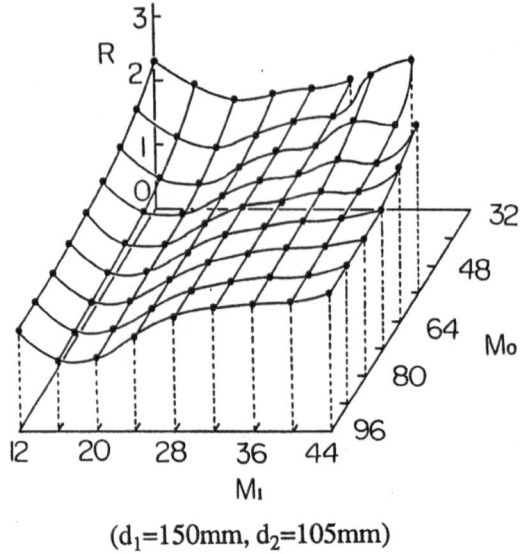

$(d_1=150mm, d_2=105mm)$

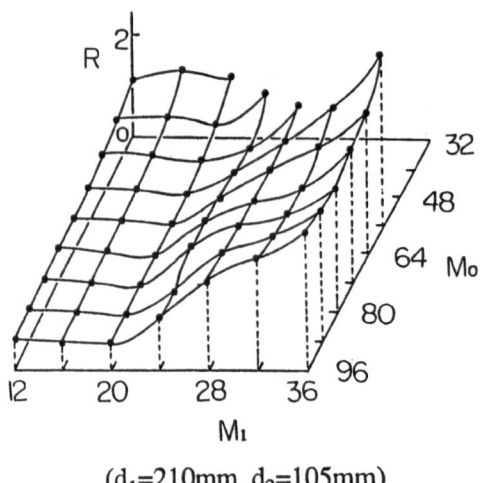

$(d_1=210mm, d_2=105mm)$

Fig.6 Relationship between the error of t_r and the number of M_O and M_I.

$$R=\sqrt{\frac{\sum_{m=1}^{M_I} \{t_r(P_I^m)-t_r(given)\}^2}{M_I}}$$

(10)

In these figures, R takes minimum values with $M_O \geq 64$ and $M_I \cong 20$. Thus, we can see that there is the optimum sets of M_O and M_I which minimize the error R. These sets of number could be related to the number of measured points and the accuracy of measured data.

CONCLUSIONS

By using the indirect boundary method, the inverse method for determining inner pressure of the cylinder is proposed. The efficiency of the present method is confirmed by the actual experiments. Moreover, the relationship between the error of the determined values and the number of nodes is shown, and it is discovered that there is the optimum number of nodes which minimize the error.

REFERENCES

1. Noda, N., Optimal heating problem for transient thermal stress in a thick plate. Thermal Stresses, **11-2**, 1988, 141-50.
2. Noda, N., On a certain inverse problem of coupled thermal stress fields in a thick plate. ZAMM, **68-9**, 1988, 411-5.
3. Noda, N., Ashida, F. and Tsuji, T., An inverse transient thermoelastic problem for a transversely isotropic body. J. Applied Mechanics, **56**, 1989, 791-7.
4. Oda, J. and Moto, S., On inverse analytical technique to obtain contact stress distributions. Trans. JSME, 1989, **55**, 872-8(in Japanese).
5. Tanaka, M. and Yamagiwa, K., Application of boundary element method to some inverse problems in elastodynamics. Trans. JSME, 1988, **54**, 1054-9(in Japanese).
6. Tomishima, T. and Yada, T., Study on an identification method of residual stresses in a plate by inverse analysis. JSME Int. J., 1989, **32-1**, 31-7.
7. Kubo, S., Ohnaka, K. and Ohji, K., Identification of heat-source and force using boundary integrals. Trans. JSME, 1988, **54**, 1329-34(in Japanese).
8. Banerjee, P.K., Boundary Element Methods in Engineering Science, McGraw-Hill, UK, 1981, pp.82-93.

MODEL-BASED CONSTITUTIVE RELATIONSHIPS FOR DESIGN AND LIFE EXTENSION OF HIGH TEMPERATURE PLANT

B Wilshire
Department of Materials Engineering, University College, Swansea SA2 8PP

ABSTRACT

Theoretical analyses of the deformation and damage processes controlling the creep behaviour of engineering alloys are shown to lead to firmly model-based constitutive relationships for design and life extension of high-temperature plant. This new approach, termed the θ Projection Concept, avoids many of the fundamental and practical problems associated with traditional methods of creep data analysis.

INTRODUCTION

Failure of high-temperature components and structures in power generation and petrochemical plant can lead to serious financial and safety problems. Indeed, plant operation can be allowed only with the reasonable assurance that catastrophic failure will not occur. In the case of the steam pipework and other high-temperature components of large-scale plant, creep is often the life-limiting factor. At the design stage, it is therefore necessary to know the stresses which the relevant engineering materials can sustain at the service temperature without creep failure occurring within the planned operational life. With service lives of up to 250,000 hours (over 30 years), the long-term stress-rupture properties must then be determined for the relevant steels. Unfortunately, with traditional parametric procedures for extrapolation of stress-rupture data, extrapolation must be limited to about three times the longest reliable test figures available. For this reason, expensive multi-laboratory test programmes must be undertaken to provide the required long-term data. Problems are then introduced because the scatter in traditional multi-laboratory data is usually around ±20% of the stress, which is the equivalent to about an order of magnitude scatter in rupture life at any specified stress level. Because failure must not occur within the planned life of plant, the minimum property values of the scatter bands are therefore used for design calculations.

Since the lower limits of the scatter bands are normally used for design purposes and a safety factor is introduced in addition to the assumption of minimum materials performance, traditional design methods may be unnecessarily conservative. In fact, operating experience in power stations and other large-scale high-temperature plant has established that service lives considerably greater than design expectations are usually achieved in practice. Because of the vast capital costs of constructing new plant, considerable attention is now being focused on refurbishment and plant life extension. Life extension generally involves 'fitness-for-purpose' evaluations which seek to demonstrate that components have an adequate margin against creep failure during planned periods of future operation. In turn, these evaluations often involve testing of samples taken from operational components defined as 'at risk' after completion of design reassessment exercises and in-situ component inspection. Reliable procedures based on testing of small samples taken from plant components are then needed to provide an indication of the remaining useful creep life of individual components and structures.

Efficient design and safe life extension of high temperature plant therefore require accurate procedures for creep data extrapolation and for remanent life estimation. In seeking to provide a comprehensive solution to these major industrial problems, a new approach to creep and creep fracture has been developed, termed the θ Projection Concept (1, 2). This new approach presents materials constitutive relationships derived from a sound theoretical understanding of the deformation and damage processes controlling high temperature creep and creep rupture. Compared with traditional theoretical and practical approaches, these model-based constitutive relationships

(i)   allow interpolation and extended extrapolation of both creep and creep rupture properties, so facilitating rapid and cost-effective acquisition of long-term design data,

(ii)   predict the behaviour patterns expected under the non-steady stress and temperature conditions encountered during plant operation and

(iii)   introduce new creep testing procedures for accurate remanent life estimation of service-exposed material.

The θ Projection Concept has the added advantage of permitting the creep and creep fracture properties of steels and other creep-resistant materials to be described in a computer-efficient form which is ideally suited to modern engineering procedures for design and life extension of high temperature components and structures.

Primary and Tertiary Creep Processes

Under high temperature creep conditions, most materials exhibit normal creep curves, as illustrated in Figure 1. Despite the complex shape of these typical curves, traditional theoretical and practical approaches to creep assume that an adequate description of a creep curve can be provided by specifying only a few standard parameters, such as the secondary creep rate ($\dot{\varepsilon}_s$), the time to fracture (t_f) and the creep ductility (ε_f). For instance, designs of power plant components are often based solely on stress-rupture measurements, which record only the variations of t_f and ε_f with stress and temperature. Similarly, deformation mechanism maps which purport to show the creep mechanisms which are dominant in different stress/temperature regimes are based exclusively on measurements or theoretical estimates of $\dot{\varepsilon}_s$. Yet, while a knowledge of t_f and ε_f precisely defines the point of failure, reliance on measurements of, say, the secondary creep rate disregards the primary and tertiary stages so that a major proportion of the information available from a creep curve is ignored.

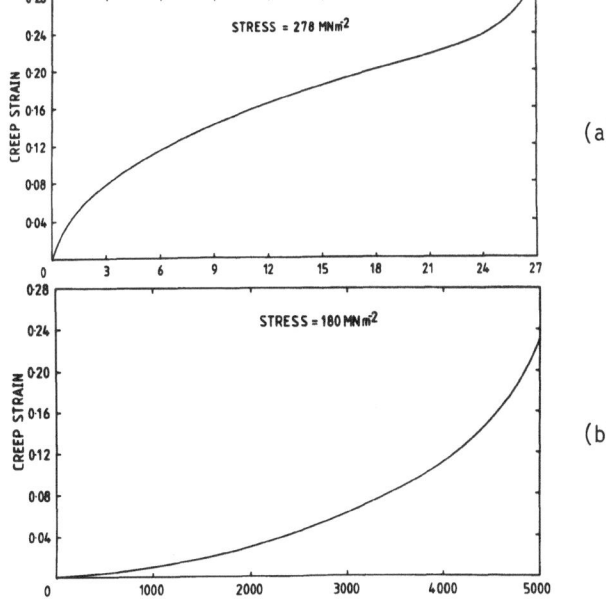

(a)

(b)

Figure 1. Creep strain against time (ks) curves recorded in high precision constant-stress tests for $\frac{1}{2}Cr\frac{1}{2}Mo\frac{1}{4}V$ ferritic steel at 838K at stresses of (a) 278MNm⁻² and (b) 180MNm⁻²

Inspection of the curves presented in Figure 1 reveals the important fact that, under high temperature creep conditions, the primary stage becomes less pronounced and the tertiary stage more dominant with decreasing applied stress. This information is totally lost by attempting to describe creep behaviour using only simple quantities such as the secondary creep rate. Clearly, a more sensible approach is to develop constitutive equations which provide a quantitative description of the shape of individual creep curves and the changes in creep curve shape with changing stress and temperature. It is therefore necessary to accurately represent the shape of individual creep curves using a function of the form

$$\varepsilon = f(t, \theta_1, \theta_2, \ldots, \theta_m) \tag{1}$$

where ε is the creep strain and the θ values are numerical parameters. Many different expressions could be chosen to describe normal creep curves but the equation selected can be used with greater confidence if it is based on a firm physical understanding of the micromechanisms of deformation and damage controlling creep and creep fracture.

When a material is loaded at high temperatures, generation and movement of dislocations normally lead to strain hardening. Simultaneously, recovery processes allow the dislocations to rearrange into low energy configurations, eg during creep of many materials, a dislocation sub-grain structure develops with regions of relatively low dislocation density within the subcells. As creep strain accumulates, the dislocation density increases and the dislocation arrangements become progressively less uniform. The creep process may then be envisaged in terms of dislocations moving in 'hard' and 'soft' regions (3). Detailed micromodelling of this process (2), leads to a kinetic description of primary creep which is essentially first order, ie the primary creep rate ($\dot{\varepsilon}_p$) is a linearly decreasing function of the primary creep strain (ε_p).

With most materials, the gradual decrease in primary creep rate with decreasing primary creep strain does not continue indefinitely. In fact, this gradually decaying creep rate can be offset in several different ways. For example, creep tests have traditionally been carried out using constant-load test procedures so that, as the specimen cross-section decreases with increasing strain, stress intensification can cause an acceleration in creep rate. However, even when more meaningful constant-stress test equipment is used, an acceleration in creep rate can be associated with (a) damage accumulation in the form of intergranular cavities and cracks, (b) microstructural instability, such as precipitate coarsening which can lead to a gradual loss of creep strength in the case of many creep-resistant alloys, and (c) mechanical instability, due to neck formation during tensile creep.

Again, detailed micromodelling of the acceleration in creep rate caused by intergranular damage accumulation and microstructural instability has shown that, for damage and degradation processes which are strain dependent (2), the tertiary creep rate ($\dot{\varepsilon}_t$) increases linearly with increasing tertiary strain ($\varepsilon_t = \varepsilon - \varepsilon_p$).

The form of normal creep curves can therefore be described in a physically meaningful way by analysis of the processes governing primary and tertiary creep. For instance, with dislocation creep processes determining the primary curve shape, the primary creep strain would be expected to decrease with decreasing applied stress. Similarly, with particle-hardened alloys, the loss of creep strength due to precipitate coarsening would become more important at an earlier fraction of the creep life in tests of longer duration, so that the tertiary stage would become more dominant with decreasing applied stress, as evident from Figure 1. Moreover, inspection of the high-precision constant-stress curves in Figure 1 indicates that the existence of a 'steady-state' period is debatable and all that can reasonably be defined is a minimum creep rate, ie the 'secondary' stage is merely the period of apparently constant creep rate which occurs when the decaying creep rate during the primary stage is offset by the acceleration in creep rate due to tertiary processes. On this basis, traditional theoretical and practical approaches based on measurements of the secondary or steady-state creep rate should be abandoned since the concept of a 'steady-state' period has no physical significance.

The θ Projection Concept

The idea that normal creep curves can be envisaged as the sum of a decaying primary and an accelerating tertiary component is an essential feature of the θ Projection Concept (1, 2). With this approach, the accumulation of creep strain with the time during a normal creep curve is described quantitatively as

$$\varepsilon = \varepsilon_t - \varepsilon_o = \theta_1\left(1 - e^{-\theta_2 t}\right) + \theta_3\left(e^{\theta_4 t} - 1\right) \tag{2}$$

where ε_t is the total strain after time, t, and ε_o is the virtually instantaneous strain which occurs on loading at the commencement of the creep test. The terms θ_1 and θ_3 then act as scaling parameters defining the extent of the primary and tertiary stages with respect to strain, while θ_2 and θ_4 are rate parameters which characterize the curvatures of the primary and tertiary components respectively. The form of equation 2 is fully compatible with the micromodelling requirements that the primary creep rate ($\dot{\varepsilon}_p$) decreases linearly with increasing primary strain, since

$$\dot{\varepsilon}_p = d\varepsilon_p/dt = \theta_2(\theta_1 - \varepsilon_p) \tag{3}$$

and that the tertiary creep rate ($\dot{\varepsilon}_t$) increases linearly with increasing tertiary strain (ε_t), since

$$\dot{\varepsilon}_t = d\varepsilon_t/dt = \theta_4(\theta_3 + \varepsilon_t) \tag{4}$$

where $\varepsilon_t = (\varepsilon - \varepsilon_p)$. Thus, equations 3 and 4 make it clear that the kinetics of the primary and tertiary processes are both first order, so that equation 2 appears to provide a suitable model-based expression for equation 1.

Computer codes have been published which allow precise estimation of the four θ parameters for any normal creep curve obtained at constant stress and temperature (1, 2). Using these procedures, equation 1 has been shown to offer an accurate description of individual creep curves. Moreover, for a wide range of metallic and ceramic materials (4-6), each θ parameter has been shown to vary systematically with stress (σ) and temperature (T). However, before concluding that the variations of the θ parameters with testing conditions represent a full description of creep properties, it is necessary to demonstrate that the stress and temperature dependences of the θ parameters are consistent with the micromechanisms invoked to explain primary and tertiary behaviour.

Consideration of equation 2 suggests that the strain-like terms θ_1 and θ_3 should vary with test conditions in a manner similar to parameters characterizing the general plastic properties of materials. Thus, for $\frac{1}{2}Cr\frac{1}{2}Mo\frac{1}{4}V$ steel, Figure 2a shows that the linear stress/logθ_1 and stress/logθ_3 plots observed at different creep temperatures can be superimposed onto single lines simply by normalization of the applied stress by the high-strain-rate yield stress. Again, inspection of equation 2 reveals that θ_2 and θ_4 take the form of first order rate constants which should depend on the micromechanisms controlling primary and tertiary creep respectively. During primary creep, dislocation generation and movement is diffusion controlled so that, at high temperatures, the temperature sensitivity of θ_2 should be determined by lattice self-diffusion. In line with this view, Figure 2b shows that the linear stress/logθ_2 relationship recorded at different creep temperatures for $\frac{1}{2}Cr\frac{1}{2}Mo\frac{1}{4}V$ steel can be rationalized by incorporating an Arrhenius term with an activation energy of 224 kJmol[-1], a value close to that expected for lattice self-diffusion in the ferrite matrix. Similarly, the linear stress/logθ_4 plots at different temperatures can also be superimposed onto a single line by incorporating an Arrhenius term, as shown in Figure 2b. However, in the case of $\frac{1}{2}Cr\frac{1}{2}Mo\frac{1}{4}V$ steel and several other particle-strengthened alloys (4, 7), superimposition of the θ_4 data is achieved only by using a stress-dependent activation energy. This is a direct consequence of the fact that several processes can all contribute to the acceleration in creep rate during the tertiary stage and each process may be characterized by a different activation energy. The observation that θ_4 for $\frac{1}{2}Cr\frac{1}{2}Mo\frac{1}{4}V$ steel is associated with a stress-dependent activation energy then merely reflects the fact that the relative importance of the various damage and degradation processes causing tertiary creep in this low-alloy steel differs under different stress-temperature conditions.

The data in Figure 2 for $\frac{1}{2}Cr\frac{1}{2}Mo\frac{1}{4}V$ steel demonstrate that the stress and temperature dependences of the four θ parameters can be expressed as

$$\left.\begin{array}{l} \theta_1 = G_1 \exp H_1(\sigma/\sigma_y) \\[6pt] \theta_2 = G_2 \exp -[(Q_2 - H_2\sigma)/RT] \\[6pt] \theta_3 = G_3 \exp H_3(\sigma/\sigma_y) \\[6pt] \theta_4 = G_4 \exp -[(Q_4 - H_4\sigma)/RT] \end{array}\right\} \tag{5}$$

where G_i and H_i (with i = 1, 2, 3, 4) are constants for the material, σ_y is the rapid yield stress at the creep temperature and Q_2 and Q_4 are the activation energies associated with the rate parameters Q_2 and Q_4 respectively. Determination of the magnitudes of parameters such as Q_2 and Q_4 therefore provides information relevant to the identification of the deformation and damage processes controlling creep behaviour. Moreover, once the constants in equation 5 are evaluated for a material, equations 1 and 5 provide a model-based quantitative description of the shape of individual creep curves and the changes in creep curve shape expected under different stress-temperature conditions, Figure 1.

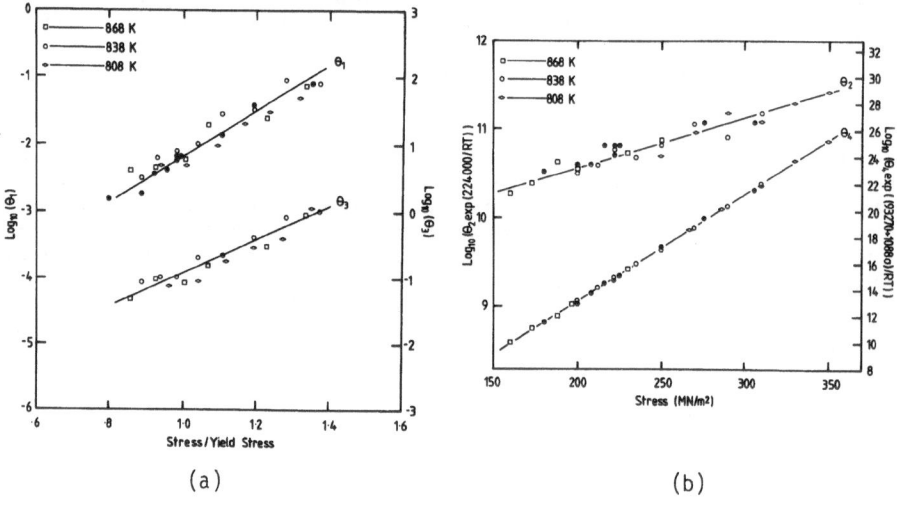

(a) (b)

Figure 2. Rationalization of (a) θ_1 and θ_3 and (b) θ_2 and θ_4 with respect to temperature for $\frac{1}{2}Cr\frac{1}{2}Mo\frac{1}{4}V$ steel

Table I
Values of the $\log_{10} \theta_i$ and fracture strain parameters according to equations 6 and 7
Units are seconds, degrees K, MNm^{-2}

Parameter	a	b	c	d
θ_1	-0.8736×10^1	0.4604×10^{-2}	-0.4489×10^{-1}	0.6814×10^{-4}
θ_2	-0.2346×10^{-2}	0.2225×10^{-1}	0.2195×10^{-1}	-0.1951×10^{-4}
θ_3	-0.1869×10^1	-0.2034×10^{-2}	-0.5497×10^{-1}	0.7990×10^{-4}
θ_4	-0.1643×10^2	0.9149×10^{-2}	-0.4723×10^{-1}	0.7139×10^{-4}
ε_f	-0.1123×10^1	0.1517×10^{-2}	0.5473×10^{-3}	-0.4721×10^{-6}

Interpolation and Extrapolation of Creep Data

From the results in Figure 2 for $\frac{1}{2}Cr\frac{1}{2}Mo\frac{1}{4}V$ ferritic steel, the values of the four θ parameters in equation 2 can be obtained for any stress and temperature, so that a full creep curve can be constructed for any test condition within the range studied experimentally, ie the θ relationships allow interpolation of data. However, the linearity of the plots in Figure 2 indicates that this new approach also allows reasonable extrapolation of creep curves (1, 2). Yet, while the data in Figure 2 suggests that the θ analysis allows both interpolation and extrapolation of creep properties, the θ expressions in equation 5 are not practically convenient for rapid computation of the parameters required for engineering design. For this reason, computer-efficient empirical relationships have been developed to describe the stress and temperature dependences of the four θ parameters (1, 2). Inspection of Figure 2 indicates that good linear relationships exist between $\log\theta_i$ and stress. Furthermore, over the relatively narrow temperature ranges of practical importance (so that $T \propto 1/T$), $\log\theta_i$ varies linearly with temperature. Consequently, $\log\theta_i$ can be written as a general linear function of stress and temperature as

$$\log\theta_i = a_i + b_i\sigma + c_iT + d_i\sigma T \qquad (6)$$

where a_i, b_i, c_i and d_i are constants for a material (with i = 1, 2, 3, 4). The values of these constants can be computed by multi-linear least squares regression analysis of the stress/$\log\theta_i$ plots at different temperatures. The values obtained for $\frac{1}{2}Cr\frac{1}{2}Mo\frac{1}{4}V$ steel are listed in Table I. Using this data, together with equations 2 and 6, a full creep curve can be constructed for any stress and temperature, but additional information is needed to define the point of failure. Fortunately, once the detailed creep curve shape is known, the rupture life is defined as the time to reach the limiting creep strain or creep ductility, ε_t. For $\frac{1}{2}Cr\frac{1}{2}Mo\frac{1}{4}V$ steel and several other creep-resistant materials (2, 4, 7), the variation of ε_t with σ and T can also be described using an equation of the form

$$\varepsilon_t = a + b\sigma + cT + d\sigma T \qquad (7)$$

where a, b, c and d are constants which are also included in Table I.

Since a full creep curve up to the point of fracture can now be specified for $\frac{1}{2}Cr\frac{1}{2}Mo\frac{1}{4}V$ steel using equations 2, 6 and 7, together with the data in Table I, any creep and creep rupture property can be computed easily. Thus, for example, the minimum creep rate ($\dot{\varepsilon}_m$) can be calculated as

$$\dot{\varepsilon}_m = \theta_1\theta_2\exp(-\theta_2 t') + \theta_3\theta_4\exp(\theta_4 t') \qquad (8)$$

where
$$t' = \frac{1}{(\theta_2+\theta_4)}\ln\frac{\theta_1\theta_2^2}{\theta_3\theta_4^2}$$

Figure 3a shows the values of $\dot{\varepsilon}_m$ predicted by analysis of short-term constant-stress creep curves compared with experimentally determined values of the minimum creep rate obtained independently at low stresses for the same batch of $\frac{1}{2}Cr\frac{1}{2}Mo\frac{1}{4}V$ steel (1). Clearly, the predicted and measured values correspond exactly. With the data in Figure 3a demonstrating the reliability of the θ Projection Concept, other creep parameters can be computed with confidence. For instance, the time to attain any given creep strain (ε^*) at a fixed stress and temperature can be obtained by numerically solving the equation

$$\theta_1(1-\exp(-\theta_2 t)) + \theta_3(\exp(\theta_4 t)-1) - \varepsilon^* = 0 \qquad (9)$$

for t when the various θ_i are calculated from equation 6. Furthermore, if ε^* is put equal to ε_t (derived using equation 7 and the data in Table I), then equation 9 can be used to determine the stress and temperature dependences of the rupture life, t_r. The calculated rupture lives, in relation to the scatter bands recorded in stress rupture programmes completed for $\frac{1}{2}Cr\frac{1}{2}Mo\frac{1}{4}V$ steel, are presented in Figure 3b. Again, the predicted behaviour patterns correspond well with the measured long-term data. The results in Figure 3 then demonstrate that data for test conditions giving creep lives of up to 100,000 hours or so can be predicted accurately by using the θ relationships to analyse high-precision creep curves with a maximum life of only around 1,000 hours. The θ Projection Concept therefore (a) offers a model-based approach which allows both interpolation and extended extrapolation of creep and creep rupture data and (b) provides the full range of creep and creep fracture properties required for engineering design in a highly computer-efficient form.

Creep under Non-Steady Conditions

Computational design methods generally require a knowledge of creep strain rate as a function of stress, strain and temperature so that analyses can be completed for components of complex shape operating under conditions resulting in thermal gradients and non-uniform loading. Moreover, procedures are required which allow design calculations to be undertaken when the stress and/or temperature varies during service. Fortunately, the θ relationships provide full materials constitutive relationships which allow straight-forward prediction of material behaviour even under non-steady conditions.

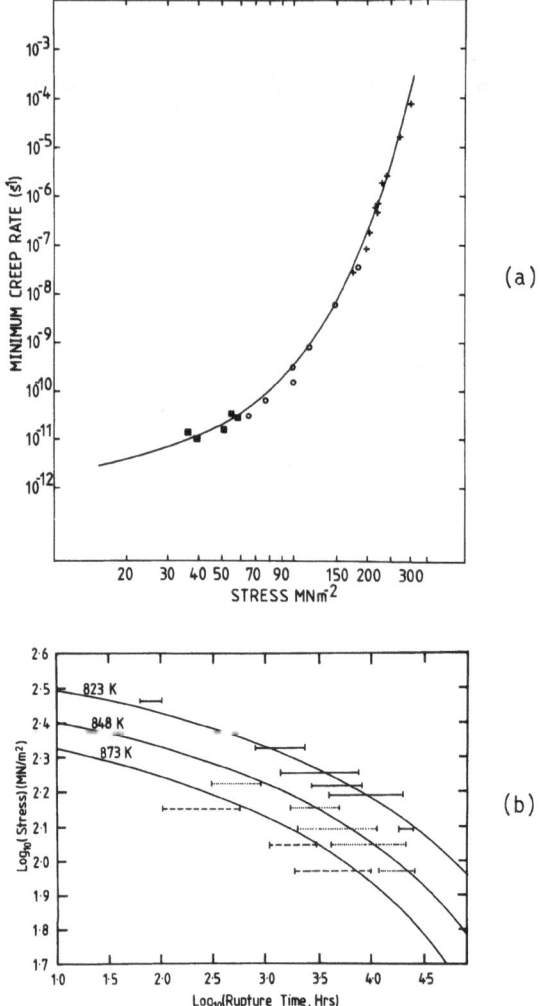

(a)

(b)

Figure 3a. The stress dependence of $\dot{\varepsilon}_m$ for $\frac{1}{2}Cr\frac{1}{2}Mo\frac{1}{4}V$ steel at 838K. The solid line is calculated from the θ relationships, showing the excellent agreement with measured long-term values. In Figure 3b, the predicted stress rupture behaviour is compared with the scatter bands associated with long-term test data.

Assume that a material is creeping under conditions $\sigma_1 T_1$. Once the coefficients in equations 6 and 7 are evaluated for the material, there is no difficulty in computing the creep curve expected at $\sigma_1 T_1$. After a time t_1, at a creep strain of ε_1, the creep rate at this point can be computed easily. Assume that the conditions are then changed from $\sigma_1 T_1$ to $\sigma_2 T_2$. There is again no problem in computing the new creep curve but it is not immediately clear how the new creep rate is to be calculated from the new curve. The problem is illustrated in Figure 4, showing two schematic creep curves corresponding to $\sigma_1 T_1$ and $\sigma_2 T_2$. It is possible to proceed from the first curve to the second along many paths, the extreme cases being referred to as time-hardening (path A) and strain hardening (path B). With steels, studies of the effects of changes in stress and temperature have shown that the strain hardening path is very nearly correct and a simple constitutive relationship can be constructed on this basis. Thus, if a strain ε_1 has been attained under conditions $\sigma_1 T_1$, the corresponding θ values are θ_i such that

$$\log \theta_i = a_i + b_i \sigma_1 + c_i T_1 + d_i \sigma_1 T_1$$

with the creep rate $\dot{\varepsilon}_1$ at the strain ε_1 given by

$$\dot{\varepsilon}_1 = \theta_1 \theta_2 \exp(-\theta_2 t^*) + \theta_3 \theta_4 \exp(\theta_4 t^*) \tag{10}$$

where t^* is the root of

$$\theta_1\left(1 - e^{-\theta_2 t}\right) + \theta_3\left(e^{\theta_4 t} - 1\right) - \varepsilon_1 = 0$$

On changing the stress/temperature conditions from $\sigma_1 T_1$ to $\sigma_2 T_2$, following a strain hardening path, the creep rate changes from $\dot{\varepsilon}_1$ to $\dot{\varepsilon}_2$, where

$$\dot{\varepsilon}_2 = \theta_1' \theta_2' \exp(-\theta_2' t') = \theta_3' \theta_4' \exp(\theta_4' t') \tag{11}$$

with the new θ values for $\sigma_2 T_2$ given by

$$\log \theta_i' = a_i + b_i \sigma_2 + c_i T_2 + d_i \sigma_2 T_2$$

and t' is the root of

$$\theta_1'\left(1 - e^{-\theta_2' t'}\right) + \theta_3'\left(e^{\theta_4' t'} - 1\right) - \varepsilon_1 = 0$$

In this way, the θ Projection Concept offers a basis for complex creep calculations under conditions of continually varying stress and temperature.

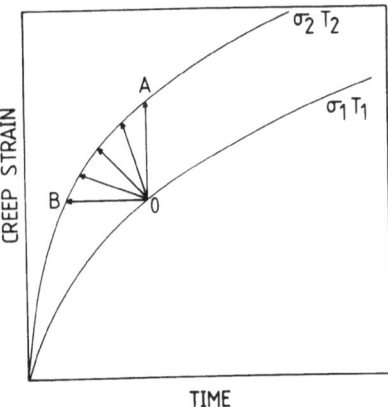

Figure 4. Schematic representation of time-hardening (path A) and strain hardening (path B) rules on changing from $\sigma_1 T_1$ and $\sigma_2 T_2$.

Remanent Life Assessment

In addition to providing a computer-efficient means of obtaining the full range of materials data required for high-temperature design, the θ relationships also introduce accurate new procedures for plant life extension (8). Currently, 'fitness-for-purpose' evaluations undertaken in order to allow planned periods of continued plant operation are normally based on a three-stage assessment procedure (9). Stage I is essentially a design reappraisal. From a knowledge of the original design calculations and plant operating records, it may be possible to estimate the approximate life fraction exhausted. Yet, even in the rare event of plant operating records being comprehensive, Stage I procedures give a pessimistic view of remaining life by not compensating for original design conservatism and by continuing the assumption of minimum materials properties. Consequently, Stage II investigates damage accumulation by in-situ component inspection methods such as determination of material composition, dimensional checks, visual and non-destructive examination, metal temperature monitoring, surface hardness testing, surface metallography and replication to assess surface cracking, microstructural changes, etc. However, the information gained from in-situ surface inspection may not necessarily be representative of the bulk of the material. As a result Stage III usually involves taking samples from components for post-exposure stress-rupture testing, using Robinson's Rule to estimate remanent life as

$$\sum_{i}^{i} \frac{t_i}{t_r} = 1 \tag{12}$$

where t_i is the time at temperature for the applied stress and t_r is the lower limit of the scatter band of the stress rupture data at that stress and temperature for the relevant material. With this approach, accelerated stress-rupture tests are usually carried out at the estimated service stress but at higher temperatures so that extrapolation to the operating temperature gives an estimate of the remaining useful creep life. The assumption inherent in this method is that the high test temperatures needed to accelerate data acquisition do not substantially modify the dislocation configurations, carbide types and dispersions and other microstructural features typically developed in steels during long-term plant exposure.

The advantage offered by adopting the θ relationships for remanent life estimation is that the test times can be minimized without using very high test temperatures simply by testing samples machined from service-exposed components at stresses and temperatures which are both higher than those experienced during plant operation, ie a post-exposure creep test is equivalent to a stress/temperature change test from $\sigma_1 T_1$ to $\sigma_2 T_2$. As illustrated schematically in Figure 5, the strain/time curve of the post-exposure creep test is essentially the portion of the stress/temperature change test under conditions, $\sigma_2 T_2$. On this basis, equations 10 and 11 apply so that, from the strain/time record at $\sigma_2 T_2$, position C under the service conditions $\sigma_1 T_1$, can be calculated precisely (Figure 5). Then, knowing the creep strain coinciding with position C, the remanent life of the material can be calculated accurately from the computed creep curve under conditions $\sigma_1 T_1$.

When a series of accelerated tests are carried out in this way for different stress/temperature combinations, the life fraction exhausted can be computed easily. The remanent life estimates obtained using the θ analysis can then be compared with those derived using Robinson's Rule, since a full post-exposure creep test also provides all the information available from standard stress-rupture methods. Studies completed for $\frac{1}{2}Cr\frac{1}{2}Mo\frac{1}{4}V$ steel (8) have then shown that the two methods give similar average results, but the scatter in the estimates from the θ analysis are far lower than with Robinsons' Rule. Moreover, once the θ analysis is completed for samples taken from an operational component, the method allows stress/temperature contours to be computed to define the conditions which must be maintained in order to ensure a specified future period of continued safe plant operation. In this way, ordering and replacement of components at the end of their service life can be scheduled safely and economically.

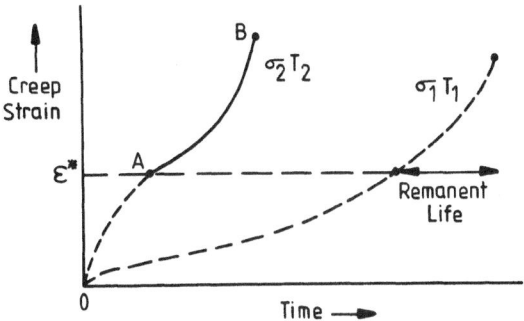

Figure 5. Schematic representation of creep curves under service conditions ($\sigma_1 T_1$) and accelerated test conditions ($\sigma_2 T_2$). The full creep curve (OB) and the post-exposure creep curve (AB) at $\sigma_2 T_2$ defines the creep strain, ε^*. Knowing ε^* allows the remanent life at $\sigma_1 T_1$ to be determined.

CONCLUSIONS

The θ Projection Concept not only offers a fundamentally sound alternative to traditional theoretical approaches to creep (2, 5) but also introduces model-based constitutive relationships of direct practical relevance for

(a)   cost-effective prediction of long-term creep and stress-rupture properties,

(b)   computer-efficient derivation of the full range of creep and creep fracture data required for design under the non-steady operating conditions experienced during service of high-temperature plant and

(c)   accurate estimation of the remanent creep life of components and structures which have operated for long periods under stress at high-temperatures in large-scale plant.

REFERENCES

1.   R.W. Evans, J.D. Parker and B. Wilshire, in 'Recent Advances in Creep and Fracture of Engineering Materials and Structures', (Ed. B. Wilshire and D.R.J. Owen), Pineridge Press, Swansea, 1982, 135.

2.   R.W. Evans and B. Wilshire, 'Creep of Metals and Alloys', The Institute of Metals, London, 1985.

3.   W.D. Nix and J.C. Gibeling, in 'Mechanisms of Time Dependent Flow and Fracture', ASM, 1983.

4.   S.G.R. Brown, R.W. Evans and B. Wilshire, Mater. Sci. Eng., 1986, 84, 147.

5.   S.G.R. Brown, R.W. Evans and B. Wilshire, Mater. Sci. Tech., 1987, 3, 23.

6.   R.W. Evans, P.J. Scharning and B. Wilshire, in 'Creep Behaviour of Crystalline Solids', (Ed. B. Wilshire and R.W. Evans), Pineridge Press, Swansea, 1985, 201.

7.   I. Beden, S.G.R. Brown, R.W. Evans and B. Wilshire, Res. Mech., 1987, 22, 45.

8.   B. Wilshire and R.W. Evans, in 'Life Assessment and Life Extension of Power Plant Components'. (Ed. T.V. Narayanam) ASME, 1989, 217.

9.   R.D. Townsend, in 'Refurbishment and Life Extension of Steam Plant', Inst. Mech. Eng., London, 1987, 223.

ANALYSIS OF CREEP TESTS ON SPRINGS

J.P. ELLINGTON, R.C.B. JUDGE & B.J. MARSDEN
Advanced Computational Mechanics
AEA Technology, Risley
Warrington, WA3 6AT

ABSTRACT

An analysis is given of the deflections of close-coiled helical springs under creep conditions, a suitable assumption allowing a very simple solution to be found without a detailed examination of the redistribution of stresses with time. The solution is intended to be of use in the analysis of data from springs used as creep specimens, and requires little more work than that involved in analysing conventional creep data. Finite element calculations have been carried out in support of this analysis, using the non-linear finite element code ABAQUS. In view of the relative simplicity of measuring spring deflections it is suggested the the proposed analysis could lead to the wider use of springs as creep specimens.

INTRODUCTION

Helically coiled springs have been used as creep specimens in the nuclear industry for a number of years [1]. This can create certain difficulties in the analysis of the results obtained. The stresses in the wire change with time owing to creep deformations, and the question arises of how to relate the spring deflection to the behaviour of the material under, say, constant tension.

Given data on the material and assumptions or knowledge of the material's behaviour under changing stresses then the stress distribution and deflection of the spring may be established by numerical methods. However, the inverse problem of calculating the material's properties from the spring deflection is only likely to be solved by a tedious process of trial and error, and even here it is difficult to see how to cope with the scatter normally encountered in creep tests.

An alternative approach [2] is to assume that there is a uniform transition from the stresses given by the elastic equations to the stresses occurring when creep strains are predominant, the arbitrary function of time involved being determined by variational methods. In this case, such an assumption allows the problem of relating deflections to material properties to be solved without a detailed knowledge of the stress distribution in the spring. An analysis based on this approach has already been proposed [3]. Developments in finite element techniques have enabled numerical analyses to be undertaken in support of the proposed method.

As is well known [4], the stresses in a close-coiled helical spring under an axial load P are approximately the same as those in a straight wire, of the same developed length as the spring, subjected to a torque $T = PR$, where R is the mean spring radius. In the torsion case it is assumed that plane cross-sections remain plane under load, and that each radius of the cross-section remains straight and rotates through the the same angle. With these assumptions the stress system consists of shear stresses acting on each cross-section normal to each radius. Now let the shear stress at any radius in a wire having only elastic properties be denoted by τ' and let the shear stress in a wire having only creep properties be τ'' (i.e. let τ' be the initial stress and τ'' be the stress occurring when creep strains are predominant). The shear stress τ, at any time t, is then assumed to be given by

$$\tau = \tau' + \lambda(t)(\tau'' - \tau') \tag{1}$$

where the function $\lambda(t)$ is to be found. Obviously $\lambda(t)$ has the value zero at time $t = 0$, and as time elapses it tends towards the value unity. Fortunately, though, for the present work $\lambda(t)$ need not be determined, and it is sufficient to note that if τ' and τ'' are equal at some radius then the stress at that radius remains constant, at $\tau = \tau'$, throughout the deformation. The strain at this radius may then be determined from the overall displacements, and thus a shear strain-time curve for a constant shear stress may be found. Having the results of several tests the equivalent stress-strain-time relationship is readily found and the problem may be considered to be solved.

BEHAVIOUR OF A WIRE UNDER TORSION

The behaviour of a wire of elastic material under a torque T is readily determined [4] and the shear stress τ' at any radius r is given by

$$\tau' = \frac{Tr}{J} \tag{2}$$

where $J = \frac{\pi r_0^4}{2}$, J being the polar second moment of area of the wire which has an outer radius of r_0 . In considering the creep behaviour of the material it is necessary to define the form of creep law to be expected, and it is assumed that for constant stresses the equivalent stress-creep strain relationship is of the form

$$\epsilon_* = k\sigma_*^n \phi(t) \tag{3}$$

where k and n are material constants and $\phi(t)$ is an observed function of time.

The equivalent stress σ_* and the equivalent creep strain ϵ_* are most readily defined in terms of principal stresses and creep strains, and are, for constant stresses

$$\left. \begin{array}{l} \sigma_* = \frac{1}{\sqrt{2}} \left[(\sigma_1 - \sigma_2)^2 + (\sigma_2 - \sigma_3)^2 + (\sigma_3 - \sigma_1)^2 \right]^{\frac{1}{2}} \\ \epsilon_* = \sqrt{\frac{2}{3}} \left[\epsilon_1^2 + \epsilon_2^2 + \epsilon_3^2 \right]^{\frac{1}{2}} \end{array} \right\} \tag{4}$$

Thus, for a constant shear stress τ, which produces an engineering creep shear strain γ, the corresponding principal stresses and strains are

$$\left. \begin{array}{l} \sigma_1 = \tau, \quad \sigma_2 = -\tau, \quad \sigma_3 = 0 \\ \epsilon_1 = \frac{\gamma}{2}, \quad \epsilon_2 = -\frac{\gamma}{2}, \quad \epsilon_3 = 0 \end{array} \right\} \tag{5}$$

Substitution in equation (4) then gives

$$\sigma_* = \sqrt{3}\tau \text{ and } \epsilon_* = \frac{\gamma}{\sqrt{3}} \tag{6}$$

and the constant shear stress-creep strain relationship is found from equation (3) to be

$$\gamma = (\sqrt{3})^{n+1} k\tau^n \phi(t) \tag{7}$$

Now, to make use of equation (1) a knowledge is required of the shear stresses τ'' which occur in a wire of material possessing creep properties only. Provided that the applied torque T does not change with time the stresses τ'' do not change either, and consequently equation (7) may be used to calculate the distribution of τ'' over the cross-section. Thus, for a constant torque, assumptions of age-hardening, strain-hardening or any other form of behaviour under changing stress conditions are not necessary. The assumptions that radii remain straight means that the shear strain γ at a radius r is directly proportional to r and so, if τ_0'' is the shear stress at the radius r_0, then equation (7) gives

$$\frac{\gamma}{\gamma_0} = \frac{r}{r_0} = \left(\frac{\tau''}{\tau_0''}\right)^n$$

$$\tau'' = \tau_0'' \left(\frac{r}{r_0}\right)^{\frac{1}{n}} \tag{8}$$

The value of τ_0'' is found by considering equilibrium over a cross-section giving

$$T = \int_0^{r_0} 2\pi r^2 \tau'' dr = \frac{2\pi \tau_0'' n r_0^{\frac{(3n+1)}{n}}}{r_0^{\frac{1}{n}} \, 3n+1}$$

$$= \frac{4J\tau_0''}{r_0} \frac{n}{3n+1} \tag{9}$$

where $J = \frac{\pi r_0^4}{2}$, and consequently

$$\tau'' = \frac{Tr_0}{J} \frac{3n+1}{4n} \left(\frac{r}{r_0}\right)^{\frac{1}{n}} \tag{10}$$

Equation (1) indicates that if at some radius r_1 the initial and steady-state stresses are equal then the stress at that radius remains constant at some value τ_1. The value of r_1 is found by equating τ' and τ'' from equations (2) and (10) giving

$$\frac{r_1}{r_0} = \frac{3n+1}{4n} \left(\frac{r_1}{r_0}\right)^{\frac{1}{n}} = \left(\frac{3n+1}{4n}\right)^{\frac{n}{(n-1)}} \tag{11}$$

The stress τ_1 at radius r_1 is then found from equation (2) to be

$$\tau_1 = \frac{Tr_1}{J} = \frac{Tr_0}{J} \left(\frac{3n+1}{4n}\right)^{\frac{n}{(n-1)}} \tag{12}$$

and from equation (7) the corresponding creep strain is

$$\gamma_1 = (\sqrt{3})^{n+1} k\tau_1^n \phi(t) = (\sqrt{3})^{n+1} k \left(\frac{Tr_0}{J}\right)^n \left(\frac{r_1}{r_0}\right)^n \phi(t) \tag{13}$$

If θ_c is the angle of twist per unit length of wire due to creep (i.e. θ_c is the time-dependent component of the overall deflection), then, since shear strain is proportional to radius

$$\theta_c = \frac{\gamma_1}{r_1} = (\sqrt{3})^{n+1} \frac{k}{r_0} \left(\frac{(Tr_0)}{J}\right)^n \left(\frac{r_1}{r_0}\right)^{n-1} \phi(t) \tag{14}$$

BEHAVIOUR OF A SPRING

Reverting to the spring problem, it has already been indicated that the stresses in a spring under an axial load are approximately the same as those in a straight wire of the same length under torsion. If R is the mean radius of the spring, which has N coils, and the axial load is P then the torque applied to the wire is PR, and the length of the wire is $2\pi RN$. Then, if θ_c is the time-dependent angle of twist per unit length, the creep component of the axial deflection at the load, δ_c, is

$$\delta_c = R(2\pi RN\theta_c) \tag{15}$$

or, on using equation (14)

$$\delta_c = \frac{2\pi R^2 N}{r_0}(\sqrt{3})^{n+1}k\left(\frac{PRr_0}{J}\right)^n\left(\frac{r_1}{r_0}\right)^{n-1}\phi(t) \tag{16}$$

Substitution from equation (11) then gives

$$\delta_c = \frac{2\pi R^2 N}{r_0}(\sqrt{3})^{n+1}\left(\frac{2PR}{\pi r_0^3}\right)^n\left(\frac{3n+1}{4n}\right)^n k\phi(t) \tag{17}$$

ANALYSIS OF CREEP TESTS ON SPRINGS

Equation (1) now allows of the complete analysis of results obtained in creep tests on springs. Firstly a plot of log δ_c against log t allows the form of $\phi(t)$ to be found. For instance, if the plot is a straight line of slope m, then $\phi(t)$ is obviously equal to t^m. Secondly, a plot of values of log δ_c at some specific time against log P gives the creep index n. (It may well turn out that the assumed form of stress dependence (σ^n) does not completely fit all the results. Nevertheless, it will be possible to find values of n to fit selected ranges of the data). Finally, the purely numerical coefficients in equation (17) may be evaluated, and the creep coefficient k can be found. Obviously curve-fitting techniques (least-squares, etc) may be used where the quality and quantity of the data available warrant the additional work.

FINITE ELEMENT ANALYSES

Existing evidence indicated that the assumption that there was some radius in the cross-section where the stress remained constant (termed a skeletal point) was not un-reasonable. Examination of the computed results of Johnson et al. [5] on bars in torsion showed that, for values of the creep index n up to say 5, a radius may be found at which the stress remained constant. However, for n greater than 5 such a radius can only be found after a certain period of stress redistribution. Similarly, Marriot and Leckie [6] have observed, through numerical studies, the existence of a skeletal point for the cases of a rectangular beam, thick cylinder and a spinning disc. However, only one value of n (=3) was used in their work, so that a range of validity of the assumption was not established.

As the success of the proposed method of calculating and analysing the creep be-haviour of springs depends on the validity of assumption of the existence of a skeletal point, finite element calculations were carried out using the commercially available code ABAQUS.

Finite Element Models

Three finite element models were developed in support of the analysis:

1. Wire in torsion, modelling the wire with 9 parabolic beam elements (Fig 1a).

2. Close-coiled helical spring under axial tensile load, modelling the spring with 18 parabolic beam elements (Fig 1b).

3. Close-coiled helical spring under axial tensile load, modelling the spring with 768 solid (20-noded brick) elements (Fig 1c).

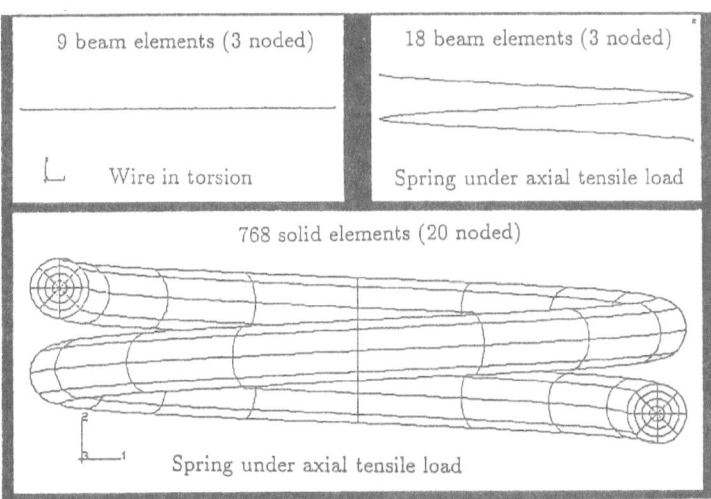

Figure 1. Finite element models.

The dimensions of the spring corresponded to springs which had previously been tested by AEA Technology under creep loading. The dimensions were as follows: mean spring radius (R) 3.81 mm, radius of the spring wire (r_o) 0.375 mm, pitch 1.07 mm. Only one and a half turns of the spring were modelled. Conditions of anti-symmetry were applied at the ends, along with constraint equations which ensured correct behaviour of the spring.

The length of the wire was equal to the unravelled length of the spring. The wire was fully fixed at one end, and a torque applied at the other. The value of the torque was chosen to be equivalent to the unit axial load applied to the spring $(T = PR)$. The first and second models were developed to correlate the behaviour of a wire in torsion with that of a spring under tension. ABAQUS, in common with other finite element codes, makes a number of assumptions in its derivation of the beam elements used in this anlysis:

Small strains (but rotations may be large).

Hoop strains are not considered.

Transverse shear deformation is considered only in the linear elastic response.

Plane cross-sections remain plane.

The third model was developed to examine the validity of results from the beam models; it was computationally more attractive to carry out most of the analyses using beam elements.

Nominal material properties were chosen. ABAQUS permits a range of creep models to be incorporated in the coding, a simple power law was chosen in this case $(\epsilon_* = k\sigma_*^n)$. The creep index (n) was in the range $1 - 5$. The creep displacement at unit time was considered to be five times the elastic displacement. Choosing a typical value for the shear stress enabled appropriate creep coefficients (k) to be chosen for the various creep indices. A plot of the creep laws is given in Fig 2.

The use of the ABAQUS non-linear finite element code enabled analyses based on both small and large displacement theory to be carried out.

Finite Element Results

The finite element models were all checked by comparing their elastic behaviour with theory. All deflections were consistent with the loads applied. The shear stresses in the two beam models were consistent with theory; a slightly higher value was noted in the solid model (Fig 3). This will be partly due to the neglecting the effects of shear force due to the spring load and to spring curvature in the beam analyses. These effects have been accounted for in an expression given by Timoshenko and Young [7], which predicts a maximum shear stress on the inner radius of the spring of 52.55 MPa (see Fig 3).

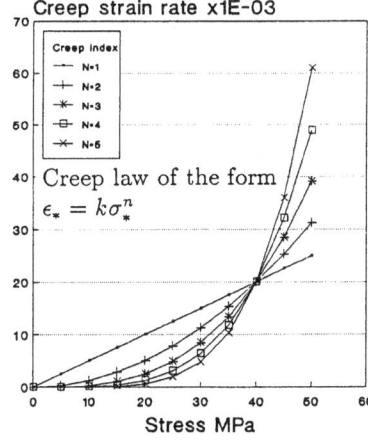

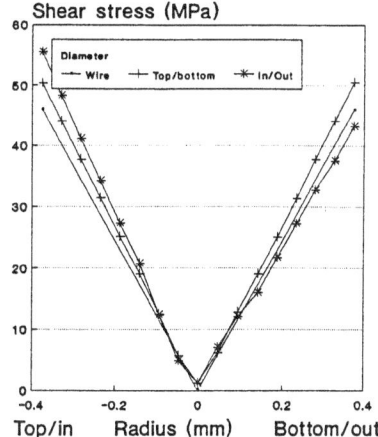

Figure 2. Assumed creep laws.   Figure 3. Shear stress in the spring.

Concentric contours of shear stress, as predicted by simple theory, were observed in the beam models. In the solid model, however, the effect of curvature was also taken into account, resulting in higher shear stresses on the inner radius of the spring (Fig 3). The shear stress distribution was approximately the same at any cross-section in both beam and solid models.

The small displacement analyses all showed the existence of a skeletal point (Figs 4a and 4b), and the radius at which this occurs is consistent with that calculated using equation 11 which gives $r_1 = 0.287$ mm for n=2.

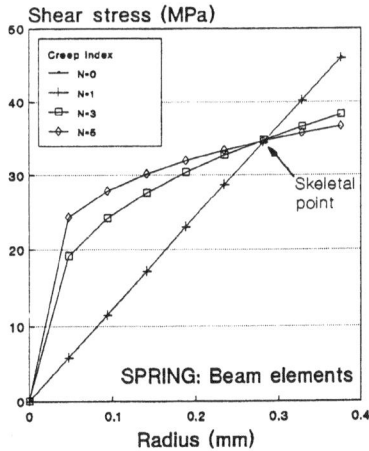

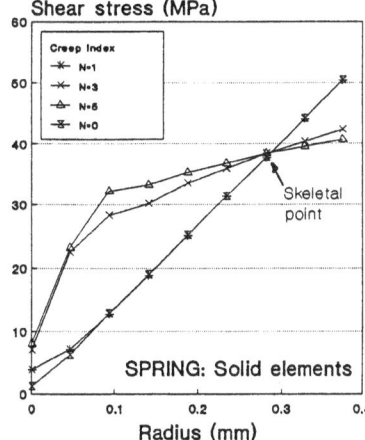

Figure 4. Spring: small displacement analyses.

Although the beam and solid elements predict slightly different values of shear stress, the skeletal point is at the same radius in both models.

The large displacement analyses for the wire under torsion and for the spring comprised of solid elements also demonstrated the existence of a skeletal point (Fig 5a and 5b).

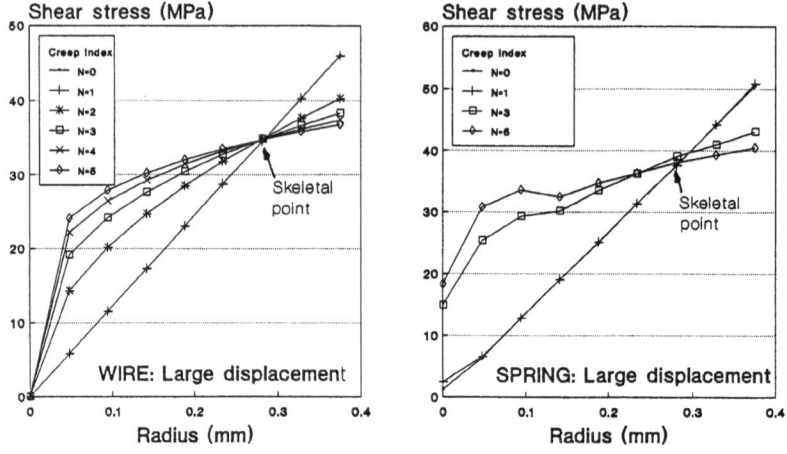

Figure 5. Wire and solid spring: large displacement analyses.

However, in the case of the spring modelled using beam elements, the use of large displacement theory resulted in markedly different values of shear stress (Fig 6a); this was attributed to the neglect of transverse shear effects in the non-linear response.

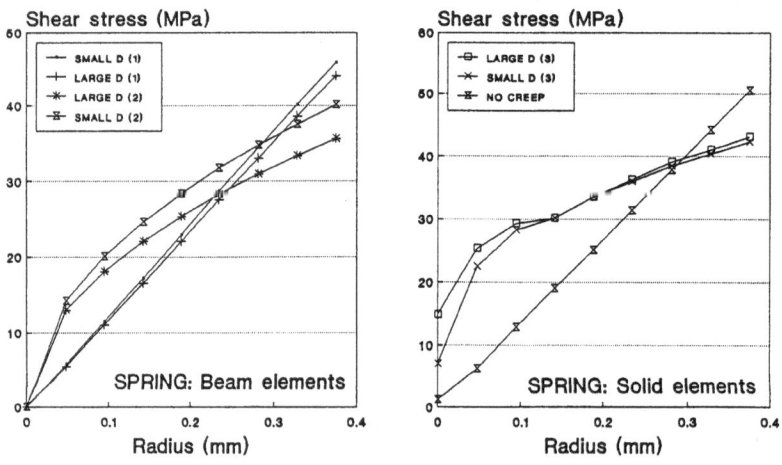

Figure 6. Spring: comparison of small and large displacement analyses

Such behaviour was not evident in the solid model, which showed consistent behaviour when using both small and large displacement theory (Fig 6b).

DISCUSSION

The analyses carried out to date have all demonstrated the validity of assuming the existence of a skeletal point in the analyses. However, it should be noted that only a limited number of creep laws have been tested; the scope remains for a more comprehensive programme of numerical analyses, which could provide a useful assessment of the applicability of the method.

For most materials in common use, the creep law index lies in the range 1 to 5. The finite element analyses carried out over this range have confirmed the existence of a skeletal point, thus confirming the validity of the proposed method for determining creep data from tests on springs. This proposed method makes the simplifying assumption that a helical spring is equivalent to a straight wire in torsion, and the finite element analyses indicate that the errors introduced by the assumption (which include neglect of shear force due to end load, concentration of shear forces at the spring inner radius, and bending stresses due to the helical shape) are not sufficiently serious to invalidate the solution (see Fig. 3).

CONCLUSIONS

1. A method of calculating the deflections of springs under creep conditions, without having to consider the redistribution of stresses with time has been presented.

2. The assumption of the existence of a point in the spring cross-section where the shear stresses remain constant (a skeletal point) was shown, using finite element calculations, to be reasonable.

3. In view of the relative simplicity of measuring spring deflections, the analysis could lead to the wider use of springs as creep specimens.

REFERENCES

1. Mosedale, D. and Lewthwaite, G.W., Irradiation Creep in Some Austenitic Stainless Steels, Nominim PE16 Alloy, and Nickel. In *Creep Strength in Steel and High-Temperature Alloys*, London: The Metals Society,1974.

2. Teregulov, I.G., On the unsteady creep of plates and shells with small displacements. *J.Appl. Maths and Mechs (PMM)*, **26**, No. 4, 1962

3. Timoshenko, S., *Strength of Materials, Part I*. Van Nostrand, 1955

4. Ellington, J.P., Deflection of Springs in Creep, *The Engineer*, **221**, 1966, 164-165.

5. Johnson, A.E., Henderson J, Khan B. Creep stress distribution in circular bars of various metallic materials under pure torque. N.E.L. Report No. 119, 1963

6. Marriot, D.L. and Leckie, F.A., Some observations on the deflections of structures during creep. *Inst. Mech. Engrs Conference on Thermal Loadings and Creep in Structures and Components*, Paper 17, 1964

7. Timoshenko, S. and Young, D.H., *Elements of materials*, 4th ed., Van Nostrand, 1962

CREEP DAMAGE AND CREEP RUPTURE OF METALS

ANDRZEJ LITEWKA and ZDZISŁAW LIS
Institute of Building Structures,
Technical University of Poznań
Piotrowo 5, 60-965 Poznań, Poland

ABSTRACT

The aim of this note is to present the theoretical model of a
creep damage and creep rupture of metals. Constitutive
equations describing these phenomena were formulated assuming
that the tertiary creep is a result of stiffness and strength
reduction of a material due to crack and void growth. A current
state of the deteriorated material microstructure is described
by the second rank damage tensor and the creep rupture
criterion proposed consists of the damage evolution equation
and the failure criterion for the material with oriented
deteriorated internal structure. The notion of the critical
amount of the damage which corresponds to the material rupture
is also discussed. To illustrate the applicability of this
stress level dependent creep rupture criterion the
complex-loading creep rupture experimental data for some metals
were used.

INTRODUCTION

Results of the stress analysis performed both for a real
structure or its model depends mainly on the actual knowledge
of the physical background of the phenomena involved. This
concerns particularly such problems like plasticity, creep and
fatigue where the applied load results in a changing material
structure [1]. The theoretical and experimental studies of
these problems performed by Hayhurst [2], Murakami and Ohno
[3], Cocks   and Ashby [4] and Chaboche [5] supplied   many
information on the nature of the material damage. However, the
problem of the constitutive equations in continuum damage
mechanics formulated within the consistent theoretical model is
still open. The aim of this paper is to discuss the
applicability of the equations which can be used to describe
the creep damage growth and creep rupture of metals. The

relevant theoretical model is based on the assumptions that the tertiary creep is the result of the stiffness and strength reduction of the material due to crack and void growth [6,7]. The constitutive equations presented here and derived in earlier paper by one of the authors [8] by employing the theory of tensor function representations can be considered as more explicit form of those proposed by Onat and Leckie [9] in their theory of materials with changing internal structure. A tensorial nature of the material damage [10] was accounted for by means of the second rank damage tensor Ω, similar to that used by Vakulenko and Kachanov [11], Murakami and Ohno [3] and Betten [12]. The specific feature of the theory presented here is the notion of a critical amount of the damage which corresponds to the material rupture. The applicability of the proposed creep rupture criterion was verified experimentally by comparing the theoretical predictions with the creep rupture test data for Nimonics 80A and 90.

CREEP RUPTURE CRITERION

Theoretical and experimental background of the creep rupture criterion employed in this paper can be found elsewhere [8,13] that is why only the final form of the relevant equations will be shown here. As was pointed out in [8] creep rupture of metals is governed by the set of two equations: failure criterion for material with deteriorated structure and damage evolution equation. The first of these equations was derived in [6] in the form of the scalar-valued function

$$c_1 \text{tr}^2\underset{\sim}{\sigma} + c_2 \text{tr}\underset{\sim}{S}^2 + c_3 \text{tr}\underset{\sim}{D}\underset{\sim}{\sigma}^2 - \sigma_s^2 = 0 \tag{1}$$

where $\underset{\sim}{\sigma}$ and $\underset{\sim}{S}$ are the stress tensor and stress deviator respectively and σ_s is the tensile strength of the material at the test temperature. The material internal state variable $\underset{\sim}{D}$ called the damage effect tensor is the second rank symmetric tensor. Its principal values D_1, D_2 and D_3 are related to those of the well known damage tensor $\underset{\sim}{\Omega}$ described by Murakami and Ohno [3] through

$$D_i = \frac{\Omega_i}{1 - \Omega_i}, \qquad i = 1,2,3 \tag{2}$$

The multipliers c_1, c_2 and c_3 are damage and temperature dependent material constants which can be calculated from the set of linear equations derived in [8,14]. The final form of these equations (11) is shown at the end of this section.

The damage evolution equation as proposed in [8,13] is the tensor-valued function

$$\dot{\underset{\sim}{\Omega}} = \left\{ \frac{1}{9}(1 - 2\nu)\text{tr}^4\underset{\sim}{\sigma} + (1 + \nu)^2\text{tr}^2\underset{\sim}{S}^2 + \frac{2}{3}(1 - 2\nu)(1 + \nu)\text{tr}^2\underset{\sim}{\sigma}\text{tr}\underset{\sim}{S}^2 + \right.$$
$$\left. + \left[\frac{2}{3}(1 - 2\nu)\text{tr}^2\underset{\sim}{\sigma} + 2(1 + \nu)\text{tr}\underset{\sim}{S}^2 \right] \frac{D_1}{1 + D_1}\text{tr}\underset{\sim}{\sigma}^2\underset{\sim}{D} \right\} k\underset{\sim}{\sigma}^* \tag{3}$$

where $\dot{\underset{\sim}{\Omega}}$ is the time derivative of the damage tensor $\underset{\sim}{\Omega}$, ν is the

Poisson ratio and k is temperature dependent material constant responsible for the damage growth. The second rank symmetric tensor σ^* is a modified stress tensor whose compressive principal stresses are replaced by zeros, whereas tensile principal stress components are left unchanged.
Further considerations presented in this paper are confined to the biaxial loading as currently all the experimental results available concern uniaxial tension, its combined effect with torsion and biaxial tension. That is why the equations (1) and (3) will be specified for a plane state of stress expressed in terms of the principal stresses σ_1 and σ_2. Then the damage evolution equation (3) takes the form of the following differential equation

$$\frac{(1 - \Omega_1)d\Omega_1}{M^2 - M^2\Omega_1 + 2M\Omega_1^2\left[1 + m^2 n\,\dfrac{1 - \Omega_1}{1 - n\Omega_1}\right]} = k\sigma_1^5 dt \tag{4}$$

where $m = \sigma_2/\sigma_1$

$$n = \Omega_2/\Omega_1 = \begin{cases} m & \text{for} \quad 0 \leq m \leq 1 \\ 0 & \text{for} \quad m < 0 \end{cases}$$

$$M = 1 - 2\nu m + m^2$$

Taking into account that the influence of the term $(1 - \Omega_1)/(1 - n\Omega_1)$ is negligible small the simplified form of the equation (4) is obtained

$$\frac{(1 - \Omega_1)d\Omega_1}{M^2 - M^2\Omega_1 + 2MN\Omega_1^2} = k\sigma_1^5 dt \tag{5}$$

where $N = 1 + m^2 n$. The solution of the differential equation (5) depends on the value of the multiplier M and eventually on the ratio m and has a form

$$t = \frac{G(m,s,\Omega_1)}{k\sigma_1^5} \tag{6}$$

where $s = \sigma_1/\sigma_s$ and

$$G = \frac{4N - M}{2MN\sqrt{M(8N - M)}}\left[\arctan\frac{4N\Omega_1 - M}{\sqrt{M(8N - M)}} + \arctan\frac{-M}{\sqrt{M(8N - M)}}\right] -$$

$$- \frac{1}{4MN}\ln\left|\frac{2N}{M}\Omega_1^2 - \Omega_1 + 1\right| \tag{7}$$

for $M < 8$

$$G = -\frac{1}{16}\ln|\Omega_1 - 2| + \frac{1}{16(\Omega_1 - 2)} + 0.074572 \tag{8}$$

for $M = 8$ and

$$G = \frac{4 - M}{4M\sqrt{M(M - 8)}} \ln\left|\frac{-M + \sqrt{M(M - 8)}}{-M - \sqrt{M(M - 8)}}\frac{4\Omega_1 - M - \sqrt{M(M - 8)}}{4\Omega_1 - M + \sqrt{M(M - 8)}}\right| -$$

$$- \frac{1}{4M} \ln\left|\frac{2}{M}\Omega_1^2 - \Omega_1 + 1\right| \tag{9}$$

for $M > 8$.

The failure criterion (1) for damaged material in the plane state of stress expressed in terms of the principal stresses has a form

$$(1 + 2m + m^2)C_1 + \frac{2}{3}(1 - m + m^2)C_2 +$$

$$+ \left(\frac{\Omega_1}{1 - \Omega_1} + \frac{n\Omega_1}{1 - n\Omega_1}m^2\right)C_3 = s^{-2} \tag{10}$$

where the constants C_1, C_2 and C_3 can be calculated from the set of three linear equations

$$(1 - \Omega_1)^2 C_1 + \frac{2}{3}(1 - \Omega_1)^2 C_2 + (1 - \Omega_1)\Omega_1 C_3 = 1$$

$$(1 - n\Omega_1)^2 C_1 + \frac{2}{3}(1 - n\Omega_1)^2 C_2 + (1 - n\Omega_1)n\Omega_1 C_3 = 1 \tag{11}$$

$$4(1 - \Omega_1)^2 C_1 + \frac{2}{3}(1 - \Omega_1)^2 C_2 + 2(1 - \Omega_1)\Omega_1 C_3 = 1$$

It is seen from Eqns (6) and (10) that the creep rupture occurs at the time t_r when the critical combination of the damage tensor components Ω_1^{cr} and $\Omega_2^{cr} = n\Omega_1^{cr}$ is achieved. The values of t_r and Ω_1^{cr} can be then easily calculated for a given plane state of stress if three material constants σ_s, k and ν are known.

EXPERIMENTAL VERIFICATION

Particularly simple form of the creep rupture criterion is obtained in the case of uniaxial tension where critical principal values of damage tensor $\underset{\sim}{\Omega}$ are as follows

$$\Omega_1^{cr} = 1 - \sigma/\sigma_s, \qquad \Omega_2^{cr} = \Omega_3^{cr} = 0 \tag{12}$$

and function G has a form

$$G = \frac{3}{2\sqrt{7}}\arctan\frac{4\Omega_1 - 1}{\sqrt{7}} - \frac{1}{4}\ln\left|\Omega_1^2 + \frac{1}{2}\Omega_1 + \frac{1}{2}\right| + 0.03159 \tag{13}$$

where σ is uniaxial tensile stress applied to the specimen.

The equations (12) and (13) were used to describe the uniaxial creep rupture experimental data [15,16] obtained for Nimonic 80A and Nimonic 90 at various temperatures. Theoretical curves presented in Figs 1 and 2 were obtained from Eqns (12) and (13) for the material constants σ_s and k shown in Table 1.

TABLE 1
Material constants

Material	Temperature (K)	σ_s (MPa)	k $(1/MPa^5 h)$
Nimonic 80A	923	700	$3.96 \cdot 10^{-17}$
	973	680	$3.39 \cdot 10^{-16}$
	1023	650	$3.03 \cdot 10^{-15}$
Nimonic 90	923	880	$2.19 \cdot 10^{-17}$
	973	785	$1.24 \cdot 10^{-16}$
	1023	700	$7.64 \cdot 10^{-16}$
	1088	590	$1.62 \cdot 10^{-14}$
	1143	530	$3.50 \cdot 10^{-13}$

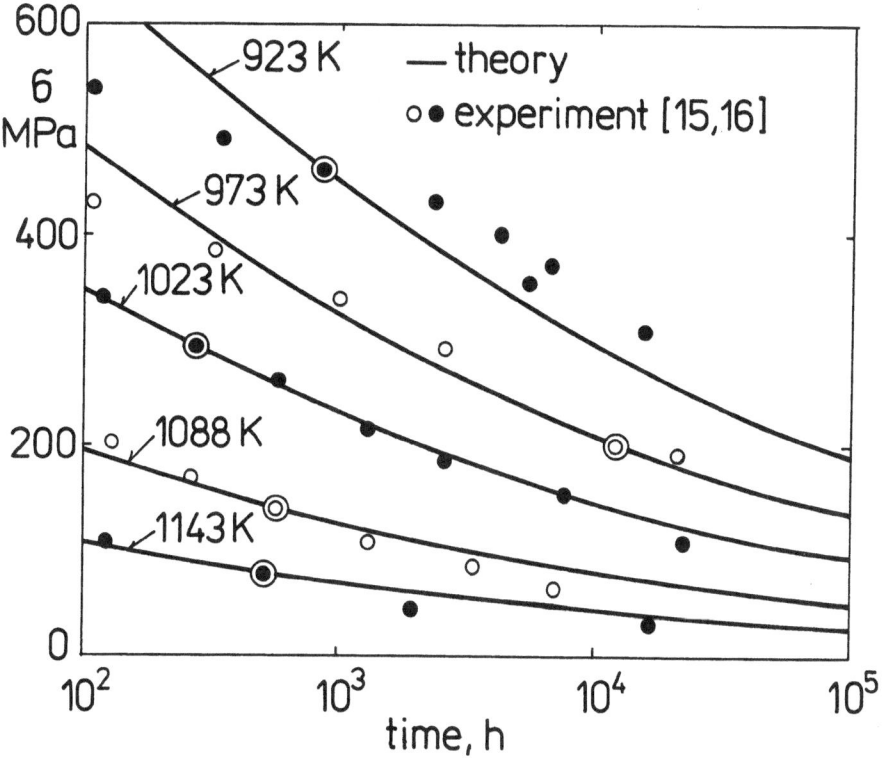

Figure 1. Creep rupture time versus tensile stress for Nimonic 90 at various temperatures.

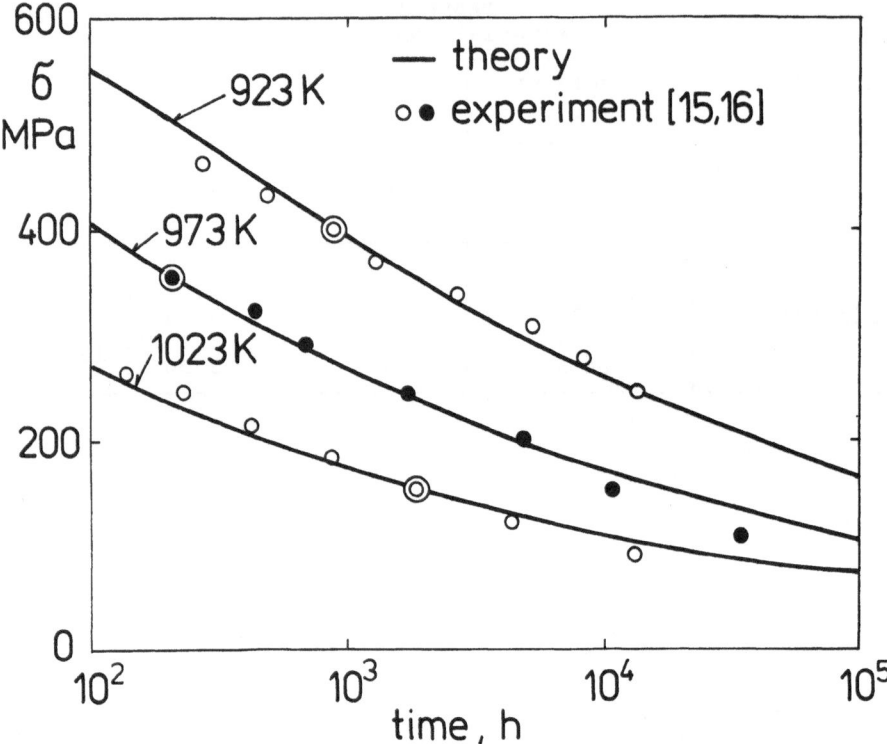

Figure 2. Creep rupture time versus tensile stress for Nimonic 80A at various temperatures.

The constant k included in the damage evolution equation (3) was calculated by employing one of the creep rupture results for each temperature marked in Figs 1 and 2 by the double circles.

The applicability of the creep rupture criterion will be now studied by comparing the biaxial experimental results with the theoretical predictions obtained from Eqns (6) and (10). To this end experimental creep rupture data for Nimonic 80A tested in torsion at 1023K by Dyson and Mc Lean [17] are used. These experimental results together with theoretical curves are shown in Fig. 3. Material constants used to construct the theoretical curve for pure shear (m = -1.0) are shown in Table 1.

STEP LOADING

The notion of the critical values of damage tensor components is particularly useful when creep rupture is caused by the loading which is a function of time. The considerations presented here will be confined to the uniaxial step loading shown in Fig. 4a where

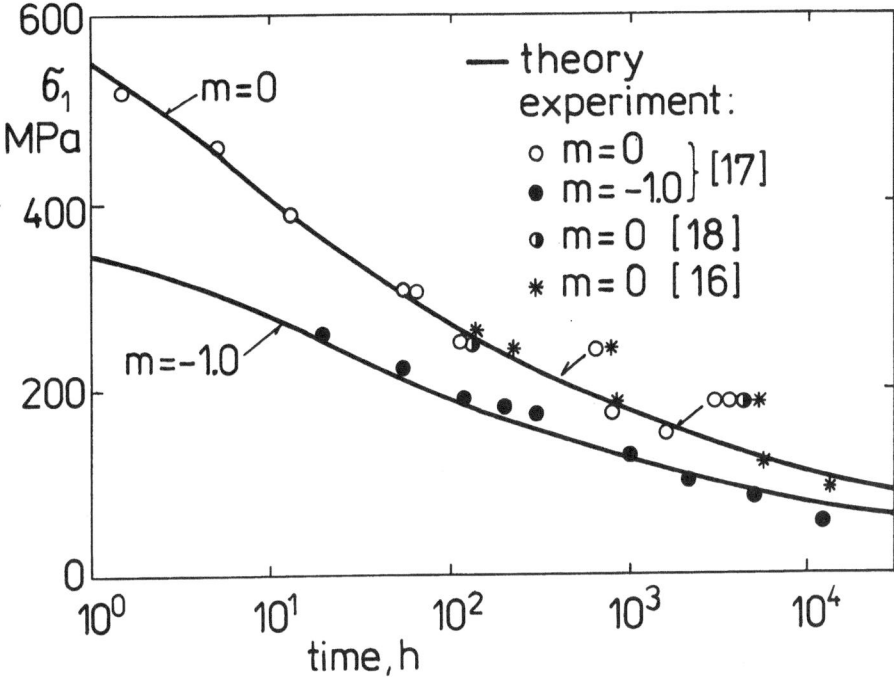

Figure 3. Creep rupture time versus maximum principal stress
for uniaxial tension and pure shear of Nimonic 80A
at 1023K.

$$\sigma(t) = \begin{cases} \sigma_A & \text{for} \quad 0 \le t \le t_A \\ \sigma_B & \text{for} \quad t > t_A \end{cases} \tag{14}$$

The damage evolution equation (5) in this case has a form of
the differential equation

$$\frac{(1 - \Omega_1) d\Omega_1}{1 - \Omega_1 + 2\Omega_1^2} = k[\sigma(t)]^5 t \tag{15}$$

The solution of this equation, taking into account (14), can be
expressed as follows

$$F(\Omega_1) + C_A = k\sigma_A^5 t \quad \text{for} \quad 0 \le t \le t_A \quad \text{and} \tag{16}$$

$$F(\Omega_1) + C_B = k\sigma_B^5 t \quad \text{for} \quad t > t_A \tag{17}$$

where $F(\Omega_1) = G - 0.03159$

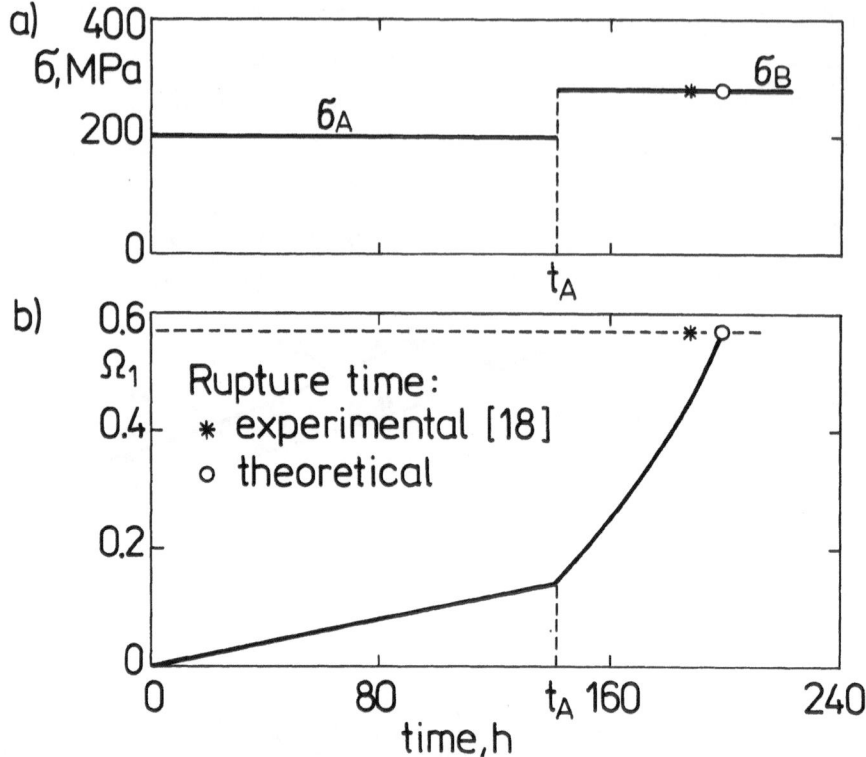

Figure 4. Step loading of Nimonic 80A at 1023K : a) uniaxial tensile stress versus time, b) maximum principal value of damage tensor versus time.

To calculate the constants c_A and c_B the following initial conditions are used

$\Omega_1 = 0$ for $t = 0$,   in equation (16) and
$\Omega_1 = \Omega_A$ for $t = t_A$, in equation (17).

Eventually one can obtain

$c_A = 0.03159$

$$c_B = k\sigma_B^5 t_A + \frac{1}{4}\ln\left|\Omega_{1A}^2 - \frac{1}{2}\Omega_{1A} + \frac{1}{2}\right| - \frac{3}{2\sqrt{7}}\arctan\frac{4\Omega_{1A} - 1}{\sqrt{7}} \quad (18)$$

where Ω_{1A} is the value calculated from equation (16) for $t = t_A$.
As an illustration of these consideration the experimental results obtained by Barlow and Ralph [18] for Nimonic 80A at the temperature of 1023K were used. Their creep rupture data for the constant uniaxial tension together with the relevant results obtained by Dyson and Mc Lean [17] and Betteridge [16]

shown in Fig. 3 indicate that the materials used in these three experiments are practically the same. That is why the numerical calculations presented here are based on the material constants gathered in Table 1 and discussed in previous section. Barlow and Ralph [18] performed two creep rupture experiments with step loading explained in Table 2. The theoretically determined damage growth for one of these experiments is shown in Fig. 4b in the form of a graph of Ω_1 versus time. The theoretically predicted creep rupture occurs when the actual amount of the material damage calculated, subsequently, from equation (16) and then from equation (17) is equal to the critical value of $\Omega_1^{cr} = 0.5692$, calculated from equation (12). Comparison of the rupture time measured experimentally and calculated from equations (17) and (18) is shown in Fig. 4 and Table 2.

Table 2
Creep rupture time for step loading

Specimen	Uniaxial stress (MPa)		t_A	Rupture time t_r, (h)	
	σ_A	σ_B	(h)	experimental [18]	theoretical
D1	200	280	141	188	199
E1	200	280	169	178	222

CONCLUSIONS

Phenomenological model of the creep rupture employed in this paper proved to be useful tool to describe the behaviour of metals at elevated temperature. Comparison of the theoretical predictions with creep rupture data obtained experimentally for two types of Nimonics corroborates the validity of the theory proposed. The most important quantity appearing in the considerations presented here seems to be the amount of damage at rupture. This amount described by means of the symmetric second rank damage tensor and refered to as the critical combination of the damage tensor components determines the creep rupture time of metals.

REFERENCES

1. Lemaitre, J., Damage modelling for prediction of plastic or creep fatigue failure in structures. Trans. 5th Int. Conf. SMiRT, Berlin 1979, North-Holland, Amsterdam, 1979,vol.L, L5/1 b, pp.1-8.

2. Hayhurst, D., On the rôle of creep continuum damage in structural mechanics. In Engineering Approaches to High Temperature Design, eds. B. Wilshire and D.R.J. Owen, Pineridge Press, Swansea, 1983, pp. 85-176.

3. Murakami, S. and Ohno, N., A continuum theory of creep and creep damage. In Creep in Structures, eds. A.R.S. Ponter and D.R. Hayhurst, Springer, Berlin, 1981, pp. 422-44.

4. Cocks, A.C.F. and Ashby, M.F., On creep fracture by void growth. In Progress in Material Science, eds.J.W. Christian, P. Haasen and T.B. Massalski, Pergamon Press, Oxford, 1983, vol. 27, pp. 189-244.

5. Chaboche, J.L., Continuum damage mechanics. Part 1: General concepts, and Part 2: Damage growth, crack initiation and crack growth. J. Appl. Mech., 1988, 55, 59-72.

6. Litewka, A., On stiffness and strength of solids due to crack development. Eng. Fract. Mech., 1986, 25, 637-43.

7. Hult, J., Stiffness and strength of damaged material. Z. Angew. Math. Mech., 1988, 68, T31-9.

8. Litewka, A., Creep rupture of metals under multi-axial state of stress. Arch. Mech., 1989, 41, 3-23.

9. Onat, E.T. and Leckie, F.A., Representation of mechanical behavior in the presence of changing internal structure. J. Appl. Mech., 1988, 55, 1-10.

10. Leckie, F.A. and Onat, E.T., Tensorial nature of damage measuring internal variables. In Physical Nonlinearities in Structures, eds. J. Hult and J. Lemaitre, Springer, Berlin, 1981, pp. 140-55.

11. Vakulenko, A.A. and Kachanov, M.L., Continuum theory of medium with cracks. Izv. Akad. Nauk S.S.S.R., M.T.T., 1971, 159-66, (in Russian).

12. Betten, J., Damage tensors in continuum mechanics. J. Méc. Théor. Appl., 1983, 2, 13-32.

13. Litewka, A., Analytical and experimental study of fracture of damaging solids. In Yielding, Damage and Failure of Anisotropic Solids, ed. J.P. Boehler, Mechanical Engineering Publ., London, 1989, pp. 653-63.

14. Litewka, A. and Hult, J., One parameter CDM model for creep rupture prediction. Eur. J. Mech. A/Solids, 1989, 8, 185-200.

15. Conway, J.B., Stress-Rupture Parameters: Origin, Calculation and Use, Gordon and Breach, New York, 1969.

16. Betteridge, W., Extrapolation of the stress-rupture properties of the Nimonic alloys. Inst. Metals, 1957-8, 86, 232-36.

17. Dyson, B.F. and Mc Lean, D., Creep of Nimonic 80A in torsion and tension. Metal Sci., 1977, 11, 37-45.

18. Barlow, C.Y. and Ralph, B., Microstructural aspects of the creep of alloys based on Nimonic 80A. In Creep and Fracture of Engineering Materials and Structures, eds. B. Wilshire and D.R.J. Owen, Pineridge Press, Swansea, 1981, pp. 447-60.

FINITE ELEMENT PREDICTION OF CREEP CRACK GROWTH FROM A SEMICIRCULAR SURFACE CRACK IN LEAD ALLOY

S. D. Smith*, J. J. Webster, T. H. Hyde
Department of Mechanical Engineering, University of Nottingham
University Park, Nottingham NG7 2RD UK

* Now at the Welding Institute, Abington, U.K.

ABSTRACT

Results are reported of a finite element simulation of the creep behaviour of a semicircular crack, on the surface of a beam, subjected to mode I loading. The results are compared with data obtained from a parallel experimental programme. At loads for which the material creep model was appropriate, the simulation gives good agreement with the experimental results. The results demonstrate that the á-C* correlation obtained from conventional fracture specimens is also applicable to this geometry. The observed crack bifurcation could be predicted from the crack tip stress field. Continuum damage calculations for a plane strain beam also provided good predictions for crack growth direction.

INTRODUCTION

Design of components operating at high temperature requires knowledge of the creep behaviour of the material. Creep may cause failure due to excessive deformations or component rupture. Also the components may contain cracks, or crack like flaws, which may grow to a critical size at which mechanical failure occurs. Techniques are required to predict crack growth in order to ensure structural integrity.

Creep crack growth has been examined in various materials using conventional fracture mechanics specimens (e.g. [1]). The tests mostly require a facility for high temperature testing involving expense and complicated test rigs. Results are expressed in terms of a parameter which correlates creep crack growth rates. It is likely that existing cracks will be of general shape possibly intersecting the free surface. This may cause a stress state variation along the crack tip. It is required to know how such a crack will grow and whether results from simpler tests can be related to its behaviour.

The use of model materials may ease the testing required to verify prediction techniques. In particular, lower loads and temperatures can be used to produce useful results. Indeed significant creep crack growth can be measured from tests performed on lead alloys at room temperature [2].

The results of these tests showed the alloy to perform in a similar manner to structural metals justifying its use in this way. The present work reports numerical analysis of lead alloy creep crack growth experiments, where the initial defect is a semicircular crack in Mode I tension situated on the surface of a beam. This identifies how a non-standard crack shape may grow and whether existing knowledge can be used to examine behaviour. The experimental results have been reported elsewhere [3]. The model material tests were run in parallel with work on 316 stainless steel at 600°C. Results for the steel are reported elsewhere [4-6].

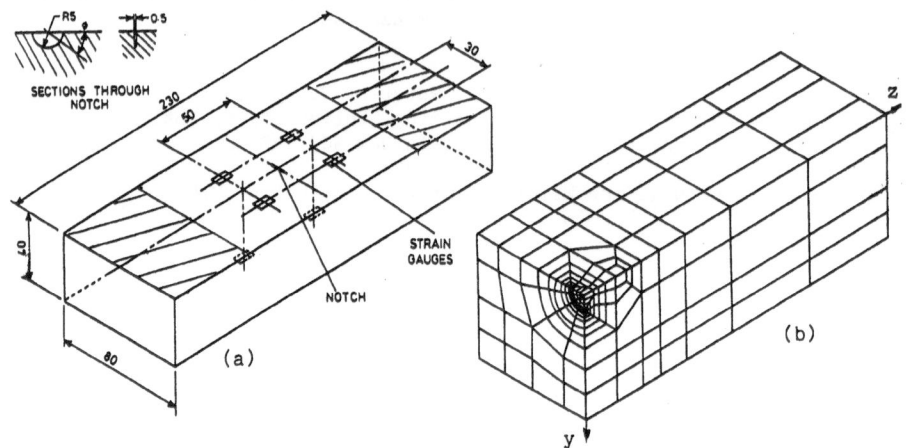

Figure 1. Cracked beam specimen and finite element mesh.

GEOMETRY AND LOADING OF SPECIMEN

A cracked lead alloy beam specimen is shown in Fig. 1(a). The specimen was clamped to loading arms over the shaded area at each end of the specimen, and the central portion was loaded in pure bending [3].

One quarter of the specimen was modelled by taking account of geometrical symmetries. The finite element mesh is shown in Fig. 1(b). This consisted of 236 twenty node brick elements. The bending load was produced by surface pressure loading in the region of the clamping.

The bending loads M on the beam are expressed in terms of a reference stress, σ_{ref}, based on the limit moment ($M_L = bd^2 \sigma_y / 4$).

i.e.
$$\sigma_{ref} = \frac{M}{M_L} \cdot \sigma_y = \frac{4M}{bd^2} \qquad (1)$$

MATERIAL BEHAVIOUR AND MODELS

The experimental work to determine the material behaviour has been reported elsewhere [3]. The initial stress-strain curve was approximated by four linear regions. In the elastic region, Young's modulus is 23.2 kN/mm^2 and Poisson's ratio is 0.44. Between the initial yield stress of 14 N/mm^2 and 20 N/mm^2 the modulus is 17.3 kN/mm^2. From 20 N/mm^2 to 22 N/mm^2 the modulus is 5.0 kN/mm^2 and above 22 N/mm^2 the modulus is 1.36 kN/mm^2. The material U.T.S. of 31.15 N/mm^2 was not included in the model.

The creep behaviour of the lead alloy was modelled by either a Norton-Bailey law:-

$$\varepsilon^c = A \sigma^n t^m \tag{2}$$

where $A = 1.471 \times 10^{-24}$, $n = 15.24$ and $m = 0.52$ or, for the continuum damage calculations, by a coupled creep/damage law:-

$$\dot{\varepsilon}^c = \frac{A_1 \sigma_e^{s(1+p)} (\varepsilon^c)^{-p}}{(1 - \omega)^r} \tag{3a}$$

$$\dot{\omega} = \frac{A_2 \sigma_e^{\chi}}{(1 + \varphi)(1 - \omega)^{\varphi}} \tag{3b}$$

where $A_1 = 7.665 \times 10^{-47}$, $A_2 = 3.311 \times 10^{-38}$, $s = r = 15.24$, $p = 0.923$, $\varphi = 15.107$ and $\chi = 25.18$. In equations (2) and (3), creep strain ε^c and damage ω are ratios and stress σ and time t have units of N/mm^2 and h, respectively. The damage hardening form of equations (3) was used.

The coupled creep damage formulation does not simulate the material failure in constrained regions, e.g. [6]. The failure can be simulated by taking the elastic modulus to be dependent on the damage. The form used here is:

$$E(\omega) = (1 - \omega^{10}) E(0) \tag{4}$$

where $E(0)$ is Young's Modulus for the undamaged material. A similar model was also used to determine plastic strain increments at points where the stresses increased above the yield stress due to stress re-distribution. In this case, equation (4) was used to include the effect of damage on the stress dependent modulus E_p.

Components of plastic and creep strain increments were obtained using the von Mises equivalent stress σ_e and the associated flow rule.

Creep crack growth can be predicted using empirical formulae. For the lead alloy material an extensive programme of experiments [2] found creep crack growth correlated with the C* parameter better than either the stress intensity factor, K_I or the reference stress, σ_{ref}. The following empirical formula was obtained:-

$$\dot{a} = 0.104 (C^*)^{0.959} \tag{5}$$

where $\dot{a}$ is in mmh^{-1} and C* in $Nmm^{-1}h^{-1}$.

FINITE ELEMENT SIMULATION

Initial elastic-plastic solutions were obtained incrementally. For each load increment an iterative procedure was used to obtain a solution for which the equivalent stresses and the strain increments fitted the uniaxial material behaviour at all points in the structure.

Creep solutions were obtained by time marching from the initial solution. Time step lengths were determined from a stability criterion [6],

$$\Delta t < \frac{4(1 + \nu)}{3 n} \quad \min \left(\frac{\sigma_e}{E \, \dot{\varepsilon}^c} \right) \tag{6}$$

For creep damage calculations n and E are replaced by $[s + \chi p/(p+1)]$ and $E(\omega)$ respectively in equation (6). Also the stiffness matrix was updated for each time step using the current local value of the damage dependent modulus. An alternative procedure in which the stiffness matrix was only updated after some specified increase in damage resulted in unstable solutions.

The plastic strain resulting from the increases in stresses due to stress redistribution were introduced as a plastic iteration when a 5% increase accumulated in the quantity $\sigma_e \, E(0)/E(\omega)$.

Distributions of the C* contour integral were calculated for the three dimensional surface crack using the relations given by Miyamoto and Kikuchi [7] for the J contour integral. The C* parameter was calculated by replacing strain and displacements by their respective rates in the J integral. The authors have published C* distributions for 3D cracks previously [eg. 5] and the calculation procedure is outlined in this paper.

ASSESSMENT OF 3-D MESH

Finite element creep, and particularly creep damage calculations are very expensive and time consuming. The 3-D mesh, Fig. 1(b), for the cracked beam was developed taking into account available computing resources. The accuracy of this mesh was assessed from finite element solutions for two plane strain finite element models of a beam with the same depth as the test beams and containing 5 mm deep through thickness crack. The mesh for one of these models was identical to that for the 3-D mesh, on the plane of symmetry, i.e. the Y-Z plane in Fig. 1(b). This coarse mesh has 58 elements; the second fine mesh had 105 elements.

A measure of mesh dependence for elastic calculations was determined by calculation of the stress intensity factor. Results were compared with the published results of Rooke and Cartwright [8] for a cracked beam in pure bending. Both the fine mesh and the coarse mesh differed from the published result by less than 0.1 per cent.

For non-linear calculations the C* contour integral was chosen to show accuracy of results. These were compared with published results, again for pure bending [9]. The coarse mesh underestimated C* by around 19% whereas the fine mesh underestimated by around 5%. This provides an indication of the accuracy of C* values for the 3-D mesh.

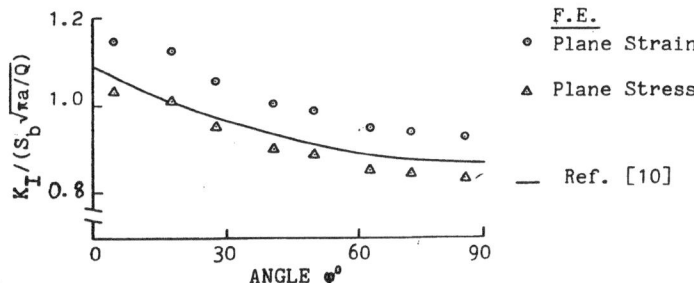

Figure 2. Variation of stress intensity factor around crack tip.

An assessment of the accuracy of the 3-D mesh was determined by comparing the distribution of the stress intensity factor around the crack tip with results from Newman and Raju [10]. Elastic distributions of the J contour integral were calculated and the stress intensity factor was obtained by assuming plane strain or plane stress conditions and equation (7):

$$K_I = \sqrt{(J\ E')}\qquad\qquad (7)$$

where E' is E for plane stress and $[E/(1-\nu^2)]$ in plane strain. A comparison of results with the published values is shown in Fig. 2. (The normalising quantities are: S_b the outer fibre stress for an uncracked beam, the crack radius a, and the shape factor, Q, which is approximately 2.464 for a semi-circular crack.) The diagram shows that the distribution of stress intensity factor compares well with the published result, and that crack tip conditions lie between plane strain and plane stress, as may be expected.

3-D ELASTIC-PLASTIC AND CREEP RESULTS

The following results are for the three-dimensional finite element mesh shown in Fig. 1(b). The material constitutive laws are those described earlier for plastic strains and by equation (2) for creep strains. Solutions for the central opening of the crack during loading are compared with experimental results in Fig. 3. At higher loads the computed results are significantly lower than the experimental results; this is probably because the material model overestimates the material hardening at stresses greater than 22 N/mm^2.

At load levels approximately equal to those of experimental tests the distribution of the J contour integral with position around the crack is shown in Fig. 4. Results are normalised by J_{90}. This is the plane strain J contour integral calculated from equation (7) with the stress intensity factor obtained from the results of Newman and Raju [10]. Fig. 4 shows that J values well exceed their elastic values at the load levels appropriate to the experimental tests.

Estimates for the redistribution times for the cracked beam, based on the transition times [11], and the transient creep finite element solutions indicate that stationary state conditions are practically achieved in times

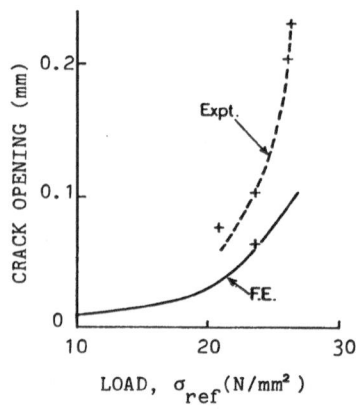

Figure 3. Crack opening.

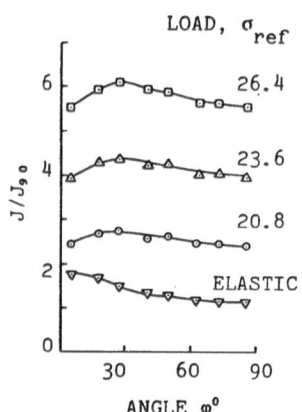

Figure 4. J Distributions

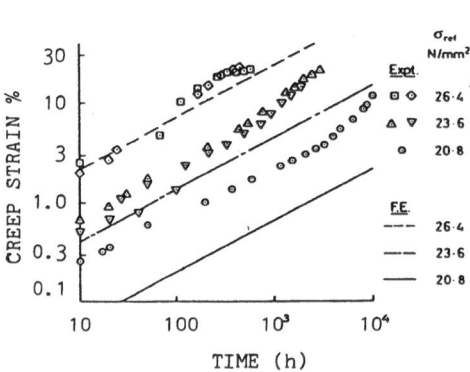

Figure 5. Surface creep strain

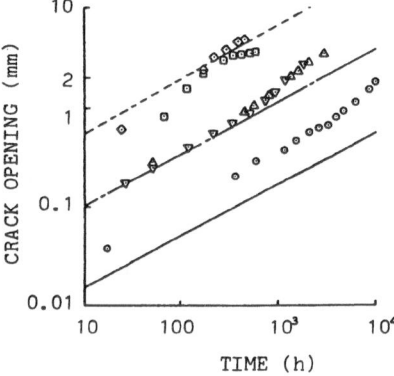

Figure 6. Crack opening

which are very much less than the test durations. Hence the predictions for the creep behaviour are based on stationary state solutions. This allowed solutions for one load level to be extrapolated to the other load levels.

The time dependent strain on the beam tensile surface from the experimental results is compared with finite element results in Fig. 5. Experimental results are from strain gauge readings and curvature measurements. There is reasonable agreement between experiment and computation, particularly at the higher loads. At short times the time exponent of equation (2) agrees well with experimental results. The low load test has the worst correlation. This may be due to a load dependence of the stress exponent in equation (2) which was found in uniaxial testing at stresses below about 18 Nmm^{-2} [3].

A similar comparison of time dependent crack opening at the free surface is shown in Fig. 6. Again the time exponent of crack opening agrees well with the model of equation (2). There appears to be a better

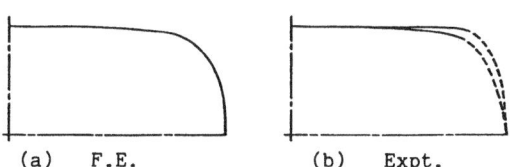

Figure 7. Normalised crack profiles.

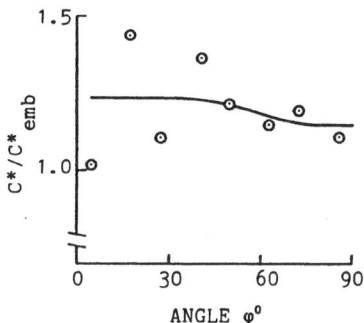

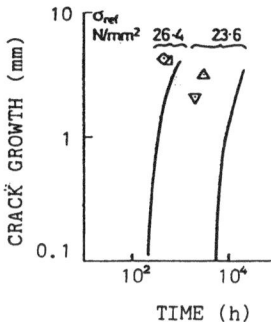

Figure 8. C* Distribution.   Figure 9. Crack growth.

correlation between experiment and computation at the low load for this measurement. If the surface strain discrepancy is due to stress exponent reduction at low stress, then the crack opening behaviour would be better predicted due to the higher stresses in the crack tip region. From the experimental results [3], it was concluded that creep crack growth did not initiate until a crack opening of around 2 mm was achieved.

Experimental and computed crack opening profiles on the top surface are compared in Fig. 7. The experimental profiles in Fig. 7(b) are normalised distributions prior to crack initiation so that results are comparable with the computations where crack growth was not simulated. Both show a large region over which there is only small variation in crack opening.

Fig. 8 shows a normalised distribution of the three dimensional C* contour integral. The normalising parameter is the C* value calculated for a circular, embedded crack under Mode I tension. This is given by He and Hutchinson [12]:

$$C^*_{emb} = a\, \sigma^\infty\, \dot{\varepsilon}^\infty\, (6/\pi)(1 + 3/n)^{-\frac{1}{2}} \qquad (8)$$

For normalising the remote tensile stress (σ^∞) and associated strain rate ($\dot{\varepsilon}^\infty$) were chosen to be the relevant beam reference stress, σ_{ref}, and associated strain rate. (Other quantities in (8) are defined elsewhere in the text.) Near the free surface the integral becomes discontinuous. This is due to the coarseness of the mesh near the free surface and the method of calculating C* [5].

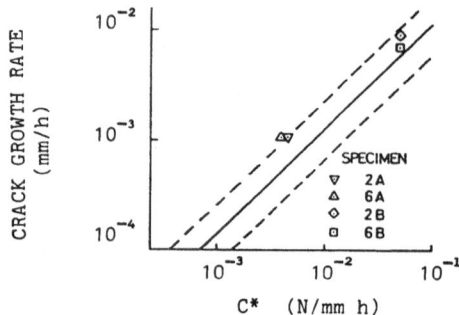

Figure 10. Crack growth rate - C* correlation.

The C* integral values at φ = 90 , together with the å - C*
correlation, equation (5), is used to obtain the crack growth predictions
shown in Fig. 9. At each load level crack growth was assumed to initiate
when the crack opening reached 2 mm. Results are not given for the low
load test since no experimental creep crack growth was found at the end of
this test. The results of Fig. 9 are dependent upon the accuracy of the
predicted crack initiation time. Fig. 6 shows that the time to produce 2
mm crack opening is progressively further under predicted at lower loads.

A time can be determined from the average of the experimentally
evaluated initiation time and the test duration. This average time can
then be used to determine a value of C* appropriate to the mid time of
creep crack growth for each test. This value of C* is plotted against the
average creep crack growth rate at φ of 90⁰ in Fig. 10. Included in Fig.
10 is the experimentally determined å - C* relationship, equation (5),
obtained for the lead alloy from a wide range of conventional fracture
mechanics specimens [2]. The comparison of the previous results with values
from the beam with a semicircular surface crack is good.

The distribution of equivalent strain rate round the crack tip was
examined to see if there was any correlation between the direction of
maximum equivalent strain rate and crack growth direction. The HRR
equivalent stress field for the creep law, equation (2), is

$$\sigma_{HRR} = \left(\frac{C^*}{I_n Am\ t^{m-1} r} \right)^{1/n+1} \tilde{\sigma}(\theta, n) \qquad (9)$$

Solutions for the constant I_n and the distribution function $\tilde{\sigma}(\theta, n)$ for
a wide range of the stress exponent n are given in reference [13]. The
plane strain normalised distribution of strain, $\tilde{\sigma}^n$, is shown in Fig. 11.

Equation (9) was used to obtain values of $\tilde{\sigma}^n$ from the equivalent
stresses at Gauss points on the plane tangent to the crack tip at φ = 85⁰.
The C* value used was that obtained from the finite element solutions,
given in figure 8 and the I_n value was taken to be the plane strain value
for n = 15.24. These results are also shown in Fig. 11. Although there is
considerable scatter in the F.E. points at θ = 80⁰, the peak value is

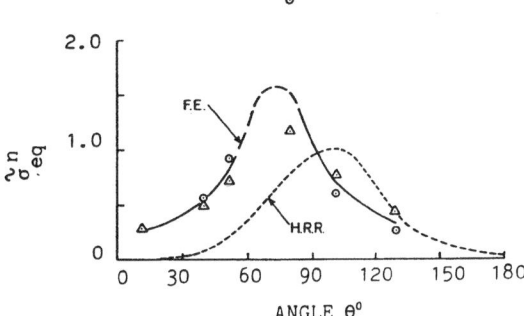

Figure 11. Equivalent strain rate distribution.

greater than the HRR plane strain solution and the maximum value occurs at a smaller value of θ, as expected, because the crack tip conditions are less constrained than the plane strain condition. Also the peak value would appear to occur in the range $60° < θ < 90°$, which compares favourably with the experiment crack bifurcation and growth direction which was in the range $50° < θ < 80°$.

TWO-DIMENSIONAL DAMAGE CALCULATIONS

The computing time required to obtain stable creep damage solutions, including elastic damage, for the 3-D finite element model was found to be prohibitive. However, solutions were obtained for the 2-D plane strain coarse mesh at the highest load level. The distribution of damage after 3500 h is shown in Fig. 12. The crack growth rate predicted by these damage calculations was approximately an order of magnitude lower than that determined from the high load beam tests. However, these results are not strictly comparable because of the different geometries. Also, previous calculations [6] have shown that damage simulations can be mesh dependent. The damage contours, Fig. 12, again indicate crack bifurcation with a crack growth direction within the range of the experimental results.

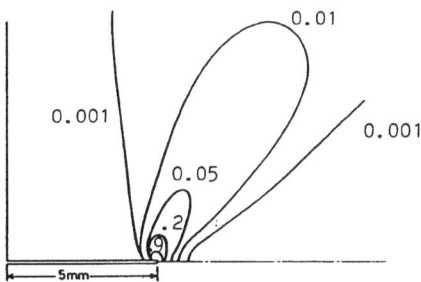

Figure 12. Plane strain damage contours, σ_{ref} = 26.4 N/mm^2, time = 3500 h.

CONCLUDING REMARKS

Experimental results for the semicircular surface crack include crack openings of 2 mm or more and up to 3.8 mm of creep crack growth [3]. The finite element model assumed small strains, small displacements and a stationary sharp crack. Another source of difference between experiment and model is the constitutive equation for creep (equation (2)). A primary creep law was used for the 3-D calculation, thus continuously reducing strain rates occur when the stresses become stationary. Uniaxial tests at similar stress levels show significant tertiary creep behaviour [3], which is not included in the present simulation. Elsewhere [3] a possible size effect between the smaller uniaxial specimens and the lead beams has been discussed. The uniaxial data indicates that a creep stress exponent of 15.24 was too large for stresses below around $18 N/mm^2$. The exponent chosen was that which was most appropriate to the load levels of the beams. However, the lower load beam test shows a significant difference from the numerical solution, which may be due to this material behaviour [3].

Despite the differences between model and test, the numerical predictions are generally good, especially for the higher loads. The C* parameter was found to correlate creep crack growth for the lead beams with previously produced results from plane stress specimens of more conventional fracture mechanics design. A similar conclusion has been made for 316 ss at 600°C [4,5]. There appears to be a useful crack growth direction prediction given by the crack tip stress field. This requires knowledge of the stress state dependence of the material creep rupture. Damage simulations require considerable computing resources, particularly for 3-D geometries. The authors' experience with the present problem is not favourable. The technique has been successfully applied by a number of workers, including the authors, to a range of cracked and uncracked structures. At present one must conclude that further work is necessary to identify the material models, geometries and solution procedures which will give reliable solutions and to develop these to reduce their computational requirements.

ACKNOWLEDGEMENT

The work reported in this paper was supported by an SERC grant and Professor H. Fessler was a joint holder of the grant.

REFERENCES

1. Sandananda, K. and Shahinian, P. Review of the fracture mechanics approach to creep crack growth in structural alloys. Eng. Frac. Mech., 1981, 15, 327-342.

2. Hyde, T. H., Kubba, B., Low, K. C. and Webster, J. J., Experimental investigation of creep crack growth in a lead alloy. Jnl. of Strain Analysis, 1985, 20, 101-110.

3. Hyde, T. H. and Smith, S. D., Creep crack growth from a semi-circular surface crack in a creep ductile material. J. Strain Analysis, 1990, 25, 1-8.

4. Hyde, T. H., Creep crack growth in 316 stainless steel at 600°C. High Temp. Techn., 1988, 6, 51-61.

5. Smith, S. D., Webster, J. J. and Hyde, T. H., Finite element investigation of creep crack growth from surface thumbnail cracks. Engng. Fract. Mech., 1989, 34, 625-635.

6. Smith, S. D., Webster, J. J. and Hyde, T. H., Three-dimensional damage calculations for creep crack growth in 316 stainless steel. Applied Solid Mechanics 3 Conference, eds. I. M. Allison and C. Ruiz, Elsevier Applied Science, 1989, 363-378.

7. Miyamoto, H. and Kikuchi, M., Evaluation of three dimensional J-integral of the C.T. specimen in elasto-plastic states. SMiRT 6th Int. Conf., 1981, paper L6/4.

8. Rooke, D. P. and Cartwright, D. J., Compendium of stress intensity factors, H.M.S.O., London, 1976.

9. Shih, C. F. and Needleman, A., Fully plastic crack problems, Part I: Solutions by a Penalty Method. Jnl. App. Mech., 1984, 51, 1984, 48-56.

10. Newman, J. C. (Jnr) and Raju, I. S., An empirical stress intensity factor equation for the surface crack. Eng. Frac. Mech., 1981, 15, 185-192.

11. Riedel, H. and Rice, J. R., Tensile cracks in creeping solids. Fracture Mechanics, Twelfth Conf., ASTM STP 700, 1980, 112-130.

12. He, M. Y. and Hutchinson, J. W., The penny shaped crack and the plane strain crack in an infinite body of power law material. Jnl. App. Mech., 1981, 48, 830-840.

13. Shih, C. F., Tables of Hutchinson-Rice-Rosengren singular field quantities.   Brown University, Report MRL E-147, 1983.

THE USE OF THE REFERENCE STRESS CONCEPT IN CREEP CRACK GROWTH STUDIES

T H Hyde
Department of Mechanical Engineering
University Park
Nottingham
NG7 2RD
England

ABSTRACT

The reference stress method used in the prediction of creep deformations and failure times of components is briefly outlined. The extension of the reference stress approach for estimating incubation times and C* values, and hence creep crack growth rates, is then described. It is concluded that estimates of incubation time and C* (and therefore creep crack growth rates), using the reference stress approach, result in reasonably accurate predictions for practical purposes.

INTRODUCTION

Many engineering components contain cracks and/or crack-like defects. If these components operate at elevated temperatures, creep crack growth may occur. Under steady loading conditions, the initial elastic or elastic-plastic stesses in the vicinity of a crack will begin to redistribute as creep strains are accumulated. Depending on the creep ductility of the material, one of a number of different failure mechanisms could occur. For example, a single crack may propagate due to a succession of localised crack tip events. Alternatively, general damage may occur and the presence of the initial crack becomes unimportant or failure may be the result of a large deformation mechanism.

A variety of fracture parameters have been used in an attempt to correlate and predict creep crack growth rates. Those most commonly used are the linear elastic stress intensity, K, the reference stress, σ_R, and the C* parameter. Although reasonably good correlations of some data have been obtained with each of these parameters, e.g. K(1), σ_R(2), C*(3), the C*-parameter seems to be the most generally applicable (e.g. 4-12). However, under some circumstances, the reference stress, σ_R, can be used to determine the failure time of a component (e.g. 13) and hence crack growth rate information may be unnecessary. Also, the reference stress approach can sometimes be used to estimate the C* parameter, e.g. (13-18).

In this paper, a brief description of the reference stress method, for predicting creep deformations and failure times is given. The uses of the reference stress approach, in estimating crack initiation times (on the basis of a critical crack tip opening displacement) and C* are described.

Reference stress and creep crack growth studies have been the subjects of an enormous number of papers. The author has therefore found it necessary to be selective when choosing the work to describe and the books and papers to cite. Also, only the behaviour of steadily loaded, isothermal components is considered. However, many of the methods described can be extended to non-isothermal and variable loading conditions. The list of references should therefore form a useful starting point for the reader who is interested in more complicated loading conditions, as well as steady loading situations.

Notation

A	constant in creep law (equation 4)
a	crack length
B	component thickness
C*	creep contour integral (equation 8)
D	reference multiplier (equation 1)
E	Youngs' modulus
I	constant in HRR stress field (equation 7)
J	Rice's path independent contour integral
K	stress intensity factor (linear elastic)
m	ratio of the collapse load of a cracked component to that of the same uncracked component
n	constant in creep law (equation 4)
P	load
P_L	limit load
R	characteristic distance
r	radial distance from crack tip
s	distance along contour for J- or C*- integrals
T_i	traction on contour for J- or C*- integrals
t	time
t_i	incubation time
$\dot{u}_i$	displacement rates on contour used for C*- integral
W*	work rate term in equation 8
x,y	Cartesian co-ordinates
β	constant relating J to δ or C* to $\dot{δ}$ in equations 14 and 15
$Δ^c$	creep displacement
$Δ^e$	elastic displacement
δ	crack tip opening displacement

$\dot{\delta}$ crack tip opening displacement-rate
δ critical crack tip opening displacement

ε^c creep strain

ε^i initial strain

ε_r strain at onset of tertiary creep

ε_u failure strain

$\dot{\varepsilon}^c$ creep strain-rate
θ crack tip angular position co-ordinate
$\mu(n)$ constant (≈ 1.5) in equation 21
σ stress
σ_R reference stress

σ_Y yield stress

χ ratio of collapse load to load at first yield
Γ contour used in J- or C*- integrals

USE OF REFERENCE STRESSES FOR PREDICTING CREEP DEFORMATIONS

The reference stress method was initially developed in order to predict the creep deformations of components, with the minimum of material data and structural analysis. Many excellent reviews of the reference stress method have appeared, e.g. (19) - (24). Knowing the reference stress, σ_R, appropriate to a particular creep deformation (displacement, twist, strain, etc.), Δ^c, at a point in a component, the creep deformation after time t, neglecting redistribution effects, is simply given by

$$\Delta^c(t) = D \; \varepsilon^c(\sigma_R, t) \tag{1}$$

where $\varepsilon^c(\sigma_R, t)$ is the creep strain obtained from a uniaxial creep test at time t, due to stress σ_R, and D is the reference multiplier. The reference multiplier is approximately constant (independent of material properties) for a given geometry and loading, but may be a function of dimensions for some types of deformation.

For some simple components, it may be possible to obtain analytical solutions for reference stresses and reference multipliers, e.g. (25) - (27). For more complex components and loadings numerical solutions can be used to determine reference stresses and reference multipliers, e.g. (28) - (31). It is also possible to obtain approximate reference stresses and reference multipliers (32) - (34) based on limit-load and linear elastic solutions, i.e.

$$\sigma_R \simeq \frac{P}{P_L} \sigma_Y \tag{2}$$

and $$D \simeq \frac{\Delta^e}{(\sigma_R/E)} \tag{3}$$

For simple components, the approximate reference stress is simply the skeletal point stress (35). In such components (e.g. beams in pure bending and thick cylinders) it is possible to accurately predict all deformations using the same reference stress; only the reference multiplier changes for different positions and types of deformation. For more complicated components and loadings the approximate reference stresses are likely to be most accurate for the maximum strains in the components. By testing model components (36-38), accurate reference stresses and reference multipliers, for any deformation, at any point in a component can be obtained, thus avoiding the necessity of using a single, approximate reference stress.

USE OF REFERENCE STRESSES FOR PREDICTING CREEP RUPTURE TIMES

Following the success of the reference stress method in predicting creep deformations, attempts have been made to obtain reference stresses for predicting creep rupture times. A rupture reference stress for a component is that stress which would cause a uniaxial specimen to fail in the same time as the component. By using a material model in which creep and damage rates are governed by the existing stress state and damage, failure times were obtained for a thick cylinder (39). It was found that the failure time based on the skeletal point stress was in close agreement with the accurately determined value (39). Thus, the approximate reference stress, equation (2), gives a reasonable prediction of failure time. Other theoretical and experimental work (e.g. (33), (34), (38), (40) - (45)) indicate that the limit load reference stress provides an upper bound to the failure times for creep ductile materials. A ductile material is defined, (43, 46, 47) as a material which is able to accommodate tertiary deformation without forming a macrocrack which propagates rapidly through the structure. Adequate ductility is obtained (46) when $\epsilon_u / \epsilon_r > 10$, where ϵ_u is the ultimate failure strain and ϵ_r is the strain at the onset of tertiary creep. For creep brittle materials (i.e. $\epsilon_u / \epsilon_r \simeq 1$ (43)), the peak stationary state stress may be used to give an accurate prediction of failure time (46). The creep strain rate for many materials can be represented reasonably accurately, by a power stress dependance (Norton creep), i.e.

$$\dot{\epsilon}^c = A\sigma^n \tag{4}$$

For materials which creep according to equation 4, the peak, equivalent, stationary state stress can be interpolated (48) from the elastic solution (n = 1) and the perfectly plastic solution (n = ∞). Therefore, the representative rupture reference stress for a component made from a creep brittle material may be obtained (41) from

$$\sigma_R \simeq \frac{P}{P_L} \sigma_y \left(1 + \frac{1}{n} (X - 1)\right) \tag{5}$$

In equation (5), X is the ratio of the collapse load to the load to first yield, which is a measure of the stress concentration.

For ductile materials, equation (5) overestimates the reference stress and equation (2) underestimates the reference stress. The difference in the predicted failure times using these two extreme reference stresses can be very large, e.g. (44). However, a theoretical and experimental investigation of typical structures (46, 49) has resulted in the following rupture reference stress being proposed for ductile materials, i.e.

$$\sigma_R = \frac{P}{P_L} \sigma_y \ (1 + 0.13 \ (X - 1))$$ (6)

In practical situations (43, 47), for which X < 2.5, it has been suggested (43, 47) that equation (6) may be conservatively simplified by using $\sigma_R = 1.2 \frac{P}{P_L} \sigma_y$.

USE OF REFERENCE STRESSES FOR DESCRIBING CRACK TIP BEHAVIOUR

Background

So far it has been assumed that the components do not contain cracks or crack-like defects. It has also been assumed that either the material is creep ductile or creep brittle. For creep ductile materials failure occurs due to a large deformation mechanism. Hence the failure time can be bounded by using the reference stresses given by equations (2) and (6). For a creep brittle material it is assumed that failure is governed by the peak stress (equation (5)) and that the crack propagation time is negligible compared to the initiation time. For materials which are neither creep ductile or creep brittle (as defined above), it is necessary to consider the possibility of creep crack growth due to a succession of crack tip events.

The initial stress field in the vicinity of a crack, for elastic or small-scale yielding conditions, is described by the elastic stress intensity factor, K. For stationary cracks, stress redistribution will occur and provided continuum damage effects are negligle, the stress field in the vicinity of the crack will approach that considered by Hutchinson (50) and Rice and Rosengren (51). In the stationary state the stress field, in the vicinity of the crack tip, for a Norton power law material, equation 4, is given by (e.g. (50)-(53))

$$\sigma_{ij}(r,\theta) = \left[\frac{C^*}{I_n Ar} \right]^{1/(n+1)} \tilde{\sigma}_{ij}(\theta)$$ (7)

where $\tilde{\sigma}_{ij}(\theta)$ are functions of angular position, θ. The path independent C*-integral is analogous to Rices' J-integral (54), i.e.

$$C^* = \int_\Gamma (W^* dy - T_i \frac{\partial \dot{u}_i}{\partial x} ds)$$ (8)

where $W^* = \int \sigma_{ij} \ d \dot{\epsilon}^c_{ij}$.

For a growing crack it is known (13, 55) that the type of stress field in the vicinity of the crack changes abruptly at n = 3. However, it has also been shown (13) that K-controlled crack growth is only likely for low n-values and high crack growth rates. Also, the C* parameter, which characterises the stationary state stress and strain-rate fields, has wide applicability (13); particularly for low crack growth rates and high n-values. There is no theoretical basis on which crack growth rates might be expected to be directly related to reference stress. However, reference stresses can be used in estimating C* (e.g. 13-18). Hence crack growth rates can be indirectly related to reference stresses.

Estimating C* using reference stresses

For a material which creeps according to equation 4, it can be shown (e.g. 14) that

$$C^* = \frac{1}{n+1} \frac{P}{B} \frac{d\dot{\Delta}^c}{da} \qquad (9)$$

for steady load situations, where $\dot{\Delta}^c$ is the load-point displacement rate.

Combining equations 1 and 9 and noting that, in general, σ_R and D are functions of crack length, a, leads (17) to the following expression for C*, i.e.,

$$C^* = \frac{n}{n+1} \frac{PD}{B} \dot{\varepsilon}^c(\sigma_R) \left[\frac{1}{n} \cdot \frac{dD}{da} \cdot \frac{1}{D} + \frac{d\sigma_R}{da} \cdot \frac{1}{\sigma_R} \right] \qquad (10)$$

It should be noted that the reference stress, σ_R, and reference multiplier, D, in equation 10 are those appropriate to the load point displacement.

For situations in which the variations of σ_R and D with crack length, a, have been obtained (e.g. 18) then equation 10 can be used to estimate C* values. In general $\frac{1}{n} \cdot \frac{dD}{da} \cdot \frac{1}{D} \ll \frac{d\sigma_R}{da} \cdot \frac{1}{\sigma_R}$ and hence equation 10 reduces to

$$C^* \simeq \frac{n}{n+1} \frac{PD}{B} \dot{\varepsilon}^c(\sigma_R) \cdot \frac{d\sigma_R}{da} \cdot \frac{1}{\sigma_R} \qquad (11)$$

Equation 11 is similar to an approximate expression for C*, derived by Harper and Ellison (14) using a limit load approach, i.e.

$$C^* \simeq \frac{-n}{n+1} \frac{P\dot{\Delta}^c}{B} \left[\frac{1}{m} \frac{dm}{da} \right] \qquad (12)$$

where m is the ratio of the collapse load of the cracked component to that of the same uncracked component.

Equations 10 and 11 will give accurate C* estimates, but for practical situations, the variations of σ_R and D (appropriate to the load-point displacement) may not be known. However, they can be estimated by using equations 2 and 3. Also, when estimating J for structures of strain hardening material, Ainsworth (56) has shown that limit load reference stresses are reasonably accurate for both tension and bend type specimens. Therefore, since experimental results indicate an approximately linear dependence of crack growth rate on C* (rather than a strong power dependence) such estimates for σ_R and D may lead to reasonably accurate predictions.

An alternative approach to estimating C* on the basis of reference stresses may be obtained by analogy with approximate relationships between J and crack tip opening displacement, δ, used in yielding fracture mechanics. By using the Dugdale model (57), for a perfectly plastic material, it can be shown (e.g. 54) that for tensile cracks under plane stress conditions,

$$J \simeq \sigma_y \delta \tag{13}$$

Rice (54) shows how the effect of strain hardening can be included in the relationship between J and δ. However, Turner (58) points out that the effects of varying constraint, work hardening, the increase of J and lack of uniqueness in the definition of δ can be approximately included by modifying equation 13, i.e.,

$$J \simeq \beta \sigma_y \delta \tag{14}$$

where $\beta \simeq 1$ is in the range $1 \leq \beta \leq 3$. The lower values of β apply to tensile loading and the higher values to bending (58).

The analogous relationship between C* and $\dot{\delta}$ is

$$C^* \simeq \beta \sigma_R \dot{\delta} \tag{15}$$

where $\beta \simeq 1$ for purely tensile situations and $\beta \simeq 3$ for bending situations.

By using equation 1, equation 15 becomes

$$C^* = \beta D \sigma_R \dot{\epsilon}^c (\sigma_R) \tag{16}$$

where D and σ_R are appropriate to the crack tip opening displacement; they may be estimated using equations 2 and 3. Equations 15 and 16 have been used (59) to process data from tests of 316 stainless steel specimens with surface, thumbnail cracks, subjected to tensile loading at $600°C$. Because the specimens were subjected to tensile loading, $\beta = 1$ was used. Also, the experimental results indicated that the crack faces opened up practically uniformly, except very close to the crack tips. Therefore β was taken to be the surface opening at the centre of the thumbnail cracks. The reference stress, σ_R, was simply taken as the net section stress. When using equation 15, average experimental $\dot{\delta}$ values were used and when using equation 16, a D-value was determined experimentally using an equation similar to equation 3 (i.e. $D = \dot{\delta} / \dot{\epsilon} (\sigma_R)$). Using both methods, the thumbnail crack data was found to fall within the scatter-band of data obtained from compact tension tests. Therefore, equations 15 and 16 offer a very simple means of estimating C* values in practical situations.

Equation 16 is similar to an expression derived by Ainsworth (13), i.e.

$$C^* \simeq R \, \sigma_R \, \dot{\varepsilon}^c \, (\sigma_R) \tag{17}$$

where R is described (13) as a 'characteristic distance', which is estimated (13) from

$$R \simeq (K/\sigma_R)^2 \tag{18}$$

The characteristic distance, R, in equation 17 is equivalent to the quantity D in equation 16. Hence, the characteristic distance, R, is related to the reference multiplier, D, appropriate to the crack tip opening displacement.

Initiation

It is known (i.e. 60) that crack growth may not start until after a significant incubation period. A critical value of crack tip opening displacement, $\hat{\delta}$, is often used (i.e. 61-63) to estimate the incubation period. Approximate solutions for crack tip opening displacements have been obtained (e.g. 61, 64).

From equatioon 1, the incubation time, t_i , can be estimated from

$$\varepsilon^c \, (\sigma_R, t_i) = \hat{\delta} / D \tag{19}$$

i.e. it is the time required to accumulate a strain of $\hat{\delta}$ /D at the reference stress, σ_R. The reference stress can be approximated by equation 3 and the reference multiplier, D, is appropriate to the crack tip opening displacement. In the absence of accurately determined D values, $D \simeq R/\beta$ can be used, where R is given by equation 18 and $\beta \simeq$ 1 in tensile cases and $\beta \simeq$ 3 in bending situations. Therefore equation 19 reduces to

$$\varepsilon^c \, (\sigma_R, t_i) \simeq \beta \, (\hat{\delta} / R) \tag{20}$$

By assuming a simple notch tip shape (i.e. semi-circular) and estimates for strain rates on the notch surface, Ainsworth (61, 64) has obtained expressions similar to equations 19 and 20. Ainsworths analyses result in initiation times predicted by

$$\varepsilon^c \, (\sigma_R, t_i) = \mu(n) \left[\frac{\hat{\delta}}{R} \right]^{\frac{n}{n+1}} \tag{21}$$

Ainsworth (64) concludes that $\mu(n)$ is sensibly independent of n and it is suggested that $\mu \simeq 1.5$ should be taken. For high n values, equation 21 reduces to equation 20. Also, taking $\mu = 1.5$ is consistent with the likely range for β (i.e. $1 \leq \beta \leq 3$).

CONCLUSIONS

By using reference stresses and appropriate reference multipliers, accurate predictions can be obtained for the creep deformations of components.

For components made of creep ductile materials, rupture reference stresses can be bounded and hence failure times can be bounded. Calladines (48) interpolation method can be used to estimate peak stationary state stresses, using elastic and perfectly plastic solutions. The peak stationary state stress can then be used to predict failure times for components made from creep brittle materials.

When component failure is controlled by creep crack growth, reference stresses cannot be used directly to predict crack growth rates. The so-called C* parameter appears to be most widely applicable for predicting creep crack growth rates. However, the reference stress concept can be used to estimate C* values. Hence, reference stresses can be used indirectly to predict creep crack growth rates. Creep crack growth rates have an approximately linear dependence on C*, rather than a strong power dependence. Hence C* values estimated on the basis of reference stresses result in reasonably accurate predictions for practical purposes.

For some materials, a critical crack tip opening displacement must be achieved before creep crack growth occurs. The reference stress approach can again be used to estimate the resulting incubation time.

REFERENCES

1. SIVERNS, M.J. and PRICE, A.T., 'Crack propagation under creep conditions in quenched 2¼ Cr-Mo steel', Int. J. Fracture, 1973.
2. WILLIAMS, J.A. and PRICE, A.T., 'A description of crack growth from defects under creep conditions', J. Eng. Mat. Tech., 1975.
3. LANDES, J.D. and BEGLEY, J.A., 'A fracture mechanics approach to creep crack growth', Mechanics of Crack Growth, ASTM STP 590, 1976.
4. ELLISON, E.G. and HARPER, M.P., 'Creep behaviour of components containing cracks - a critical review', J. Strain Analysis, 1978.
5. FU, L.S., 'Creep crack growth in technical alloys - a review', Eng. Fracture Mech., 1980.
6. SADANANDA, K. and SHAHINIAN, P., 'Creep crack growth behaviour and modelling - critical assessment', Metal Science, 1981.
7. SADANANDA, K. and SHAHINIAN, P., 'Review of the fracture mechanics approach to creep crack growth in structural alloys', Eng. Fracture Mech., 1981.
8. MALIK, S.N., 'Elevated temperature creep crack growth: state-of-the-art review and recommendations', Nucl. Eng. Des., 72, 1982.
9. WEBSTER, G.A., 'Crack growth at high temperature', Engineering Approaches to High Temperature Design (Eds. B. Wilshire and D.R.J. Owen), Pineridge Press, 1983.
10. NIKBIN, K.M., SMITH, D.J. and WEBSTER, G.A., 'Prediction of creep crack growth from uniaxial creep data', Proc. R. Soc. Lond., A396, 1984.
11. NIKBIN, K.M., SMITH, D.J. and WEBSTER, G.A., 'An engineering approach to the prediction of creep crack growth, J. Eng. Mat. Tech., 1986.

12. HYDE, T.H., KUBBA, B., LOW, K.C. and WEBSTER, J.J., 'Experimental investigation of creep crack growth in a lead alloy', J. Strain Analysis, 1985.

13. AINSWORTH, R.A., 'Some observations on creep crack growth', Int. J. Fracture', 1982.

14. HARPER, M.P. and ELLISON, E.G., 'The use of the C* parameter in predicting creep crack propagation rates', J. Strain Analysis, 1977.

15. AINSWORTH, R.A., 'The initiation of creep crack growth', Int. J. Solid Structures, 1982.

16. AINSWORTH, R.A., 'The assessment of defects in structures of strain hardening materials', Eng. Fracture Mech., 1984.

17. HYDE, T.H., 'Estimates of the creep paramerter C* in terms of reference stress', J. Strain Analysis, 1986.

18. HYDE, T.H. and LOW, K.C., 'Use of the reference stress concept to assess the effect of crack growth increment size on the accuracy of creep crack growth predictions', J. Strain Analysis, 1986.

19. PENNY, R.K. and MARRIOTT, D.L., 'Design for creep', McGraw-Hill, New York, 1971.

20. MARRIOTT, D.L., 'A review of reference stress methods for estimating creep deformations', IUTAM Symposium Creep of Structures, 1970, Gothenburg.

21. BOYLE, J.T., 'The reference stress method and its role in high temperature design'.

22. KRAUS, H., 'Reference stress concepts for creep analysis', WRC Bulletin 227, June 1977.

23. KRAUS, H., 'Creep analysis', John Wiley and Sons, New York, 1980.

24. BOYLE, J.T. and SPENCE, J., 'Stress analysis for creep', Butterworths, London, 1983.

25. ANDERSON, R.G., GARDNER, L.R.T. and HODGKINS, W.R., 'Deformation of uniformly loaded beams obeying complex creep laws', J. Mech. Eng. Sci, 1963.

26. MACKENZIE, A.C., 'On the use of a single uniaxial test to estimate deformation rates in some structures undergoing creep', Int. J. Mech. Sci., 1968.

27. JOHNSSON, A., 'An alternative definition of reference stress for creep', I.Mech.E. Conf. Pub. 13, 1973.

28. SIM, R.G., 'Reference stress concepts in the analysis of structures during creep', Int. J. Mech. Sci., 1970.

29. SIM, R.G., 'Evaluation of reference parameters for structures subjected to creep', J. Mech. Eng. Sci., 1971.

30. SIM, R.G. and PENNY, R.K., 'Plane strain creep behaviour of thick-walled cylinders, Int. J. Mech. Sci., 1971.

31. SIM, R.G., 'Reference results for plane stress creep behaviour', J. Mech. Eng. Sci., 1972.

32. SIM, R.G., 'Creep of structures', Ph.D. thesis, University of Cambridge, 1968.

33. PENNY, R.K. and MARRIOTT, D.L., 'Creep of pressure vessels', I.Mech.E. Conf. Pub. 13, 1973.

34. HAYHURST, D.R., KELLY, D.A., LECKIE, F.A., MORRISON, C.J., PONTER, A.R.S. and WILLIAMS, J.J., 'Approximate design methods for creeping structures, I.Mech.E. Conf. Pub. 13, 1973.

35. MARRIOTT, D.L. and LECKIE, F.A., 'Some observations on the deflections of structures during creep', Proc. I.Mech.E., 1963-64, 178 (Pt. 3L).

36. FESSLER, H., HYDE, T.H. and WEBSTER, J.J., 'Stationary creep prediction from model tests using reference stresses', J. Strain Analysis, 1977.

37. FESSLER, H., HYDE, T.H. and WEBSTER, J.J., 'Experimentally determined reference stresses for the prediction of creep deformations and life of components', I.Mech.E. Conf. on Recent Developments in High Temperature Design Methods, November, 1977.

38. HYDE, T.H., 'Use of lead alloy model components in plasticity and creep studies', internal report.

39. MARTIN, J.B. and LECKIE, F.A., 'On the creep rupture of structures', J. Mech. Phys. Solids, 1972.

40. GOODALL, I.W. and COCKROFT, R.D.H., 'On bounding the life of structures subjected to steady load and operating within the creep range', Int. J. Mech. Sci., 1973.

41. GOODALL, I.W. 'The creep of branch connections', I.Mech.E. Conf. on Creep Behaviour of Piping, February, 1974.

42. LECKIE, F.A., HAYHURST, D.R. and MORRISON, C.J., 'The creep behaviour of sphere-cylinder shell intersections subjected to internal pressure', Proc. R. Soc. Lond. A. 349, 1976.

43. FINDLEY, G.E. and GOODALL, I.W., 'Some observations on the design of pressure vessels operating in the creep range', I.Mech.E. Conf. on Failure of Components Operating in the Creep Range, April, 1976.

44. FESSLER, H., HYDE, T.H. and WEBSTER, J.J., 'Prediction of creep rupture of pressure vessels', ASME Energy Tech. Conf., Sept. 1977, Houston, Texas, Paper No. 77-PVP-54.

45. HENDERSON, J. and FERGUSON, F.R., 'Approximate methods for the prediction of creep rupture in components', Int. Conf. on Structural Mechanics in Reactor Technology,

46. GOODALL, I.W., COCKROFT, R.D.H. and CHUBB, E.J., 'An approximate description of creep rupture of structures', I. J. Mech. Sci., 1975.

47. GOODALL, I.W., LECKIE, F.A., PONTER, A.R.S. and TOWNLEY, C.H.A., 'The development of high temperature design methods based on reference stress and bounding theorems', J. Eng. Mat. Tech., Vol. 101, 1979.

48. CALLADINE, C.R., 'Stress concentration in steady creep: interpolation between solutions in elasticity and plasticity', Proc. I. Mech. E., 1963-64, Vol. 178, Pt. 3A.

49. AINSWORTH, R.A. and GOODALL, I.W., I. J. Solids Structures, 1976.

50. HUTCHINSON, J.W., 'Singular behaviour at the end of a tensile crack in a hardening material', J. Mech. Phys. Solids, 1968.

51. RICE, J.R. and ROSENGREN, G.F. 'Plane-strain deformation near a crack tip in power law hardening material', J. Mech. Phys. Solids, 1968.

52. RIEDEL, H. and RICE, J.R., 'Tensile cracks in creeping solids', ASTM STP 700, Philadelphia, 1980.

53. RIEDEL, H., 'The use and the limitations of C* in creep crack growth testing', Proc. ICF Int. Symp. on Fracture Mechanics, Beijing, China, 1983.

54. RICE, J.R., 'A path independent integral and the approximate analysis of strain concentration by notches and cracks', Trans. ASME, J. App. Mech., 1968.

55. HUI, C.Y. and RIEDEL, H., 'The asymptotic stress and strain field near a growing crack under creep conditions', Int. J. Fracture, 1981.

56. AINSWORTH, R.A., 'The assessment of defects in structures of strain hardening material', Eng. Fracture Mech., 1984.

57. DUGDALE, D.S., 'Yielding of steel sheets containing slits', J. Mech. Phys. Solids, Vol. 8, 1960.

58. TURNER, C.E., 'Methods for post-yield fracture safety assessment', Post Yield Fracture Mechanics (2nd Ed), Elsevier Applied Science Publishers, 1984.

59. HYDE, T.H., 'Creep crack growth in 316 stainless steel at $600^{\circ}C$, High Temperature Technology, Vol. 6, No. 2, May 1988.

60. HAIGH, J.R., 'The mechanisms of macroscopic high temperature crack growth, Part 1: Experiments on tempered CrMoV steels', Mat. Sci. Eng., 1975.

61. AINSWORTH, R.A., 'The initation of creep crack growth', Int. J. Solids Structures, 1982.

62. HYDE, T.H., LOW, K.C. and WEBSTER, J.J., 'Comparison of finite element predictions of creep crack growth with experimental data', J. Strain Analysis, 1984.

63. HYDE, T.H., LOW, K.C. and WEBSTER, J.J., 'Finite element predictions of creep crack growth in centre cracked plates and comparison with experimental results', 2nd Int. Conf. on Creep and Fracture of Engineering Material and Structures, Swansea, 1984, Pineridge Press.

64. AINSWORTH, R.A., 'Approximate blunting solutions for tensile cracks', Applied Solid Mechanics - 1 (Eds. A.S. Tooth and J. Spence), Elsevier Applied Science Publishers, 1986.

THE EXPERIMENTAL ANALYSIS OF YIELD STRESS AT THE BOTTOMS OF NOTCH ROOTS

K. TANIUCHI
Faculty of Engineering, Meiji University
1-1-1 Higashimita, Tama-ku, Kawasaki-shi
Japan

ABSTRACT

It is well known that with carbon steel, one can observe with the naked eye that the appearance of stretcher strains corresponds to yield stress. By utilizing this phenomenon investigation was made of the mechanical properties at a circular hole and between a pair of semicircular notches on a carbon steel specimen. It was possible to observe the yield progress at a notch continuously with the naked eye. The elastic-plastic stress concentration factor is affected by the hardness of the material as well as by the geometrical shape and the dimensions of the notch. Given identical shape and dimensions, the elastic-plastic stress concentration factor is decidedly smaller than the stress concentration factor for the elastic state.

INTRODUCTION

It is well known that a yield stress applied to low carbon sheet steel produces a striped pattern on the surface. This is referred to as stretcher strains [1]. By utilizing the phenomenon of striped pattern generation, it is possible to observe how the stress increases with increased load until the point where it reaches yield stress. It is possible to recognize the location of maximum stress by the naked eye and continuously observe the changes in distribution of stress around that point. By means of the generation and propagation of the striped pattern it is possible to directly and continuously observe the behavior of material in the elastic-plastic state which cannot be easily ascertained by the ordinary methods. All one needs to do by way of preparation is to give the surface of the material a smooth finish. As of yet only a few cases of the application of this method have been reported [2,3,4]. It has been attempted to investigate the effects of yield stress on a circular hole and on a pair of semicircular notches in a carbon steel strip by utilizing stretcher strains generation and propagation. In this paper the striped pattern of stretcher strains will be referred to as yield stress pattern [5].

Specimens and Method of Experiment

Figure 1 shows the shape and dimensions of two kinds of specimen used in this experiment. One specimen is of JIS 13B type having one circular hole with a diameter of 1.0 to 3.5 mm at its center. The other specimen has a pair of semicircular notches with a radius of 1.0 to 2.5 mm. Five values were randomly chosen for d and a. Table 1 shows the chemical compositions of the specimen. Two kinds of sheet steels available on the market were chosen. It was found that the size of material A in crystal grain was 16 μm while that of material B was 14 μm. After machining, the specimens were annealed with a vacuum furnace and then given a smooth finish with emery paper. The vickers hardness of material A was determined to be 125 and that of material B 195. The changes on the surface of the specimens were observed when it was subjected to a tensile load by an Instron type material testing machine with a crosshead speed of 0.5 mm/min.
A strain gauge type extensometer with a gauge length of 50 mm was used to obtain a load-elongation diagram. A video camera was used in order to get a precise observation of the behavior of the striped patterns.

Experimental Results and Discussion

Changes in the striped pattern

Figure 2 shows the changes resulting from increased load in the striped pattern near the circular hole with a diameter of 2 mm in material A. A striped pattern first appeared near the minimum section as shown in Fig. 2(a), then changed successively into the patterns shown in Figs. 2(b) and (c). It appears that this sort of propagation of the striped pattern at the circular hole indicates a change in local yield stress distribution. The striped patterns at a pair of semicircular

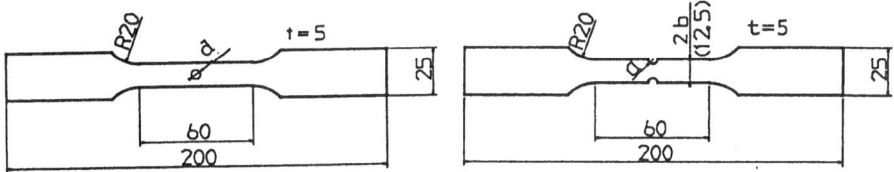

Fig. 1   Shapes and Dimensions of the specimen.

Table 1   Chemical composition of the specimens in percentage.

Composition Material	C	Si	Mn	P	S
A	0.12	0.21	0.55	0.019	0.013
B	0.53	0.17	0.61	0.29	0.16

notches generated and propagated as in Figs. 3 (a), (b) and (c); the specimen was material A with a notch radius of 3.0 mm.

The load at which the striped pattern appears will be designated as Pa, and the load at which the striped pattern grows to cover the minimum section will be designated as Pb.

The load-elongation diagram

Figure 4 shows the load-elongation diagram of a specimen of material B with a diameter of 1.5 mm and one of 3.0 mm. As shown in Fig. 4, the value of Pa, the load at the moment when the striped patterns appear, maintains a linear relation between load and elongation and comes within the scope of application of Hook's law. Pb is located when the load elongation line begins to curve slightly away from the straight line. Fig. 5 is the load-elongation diagram of a

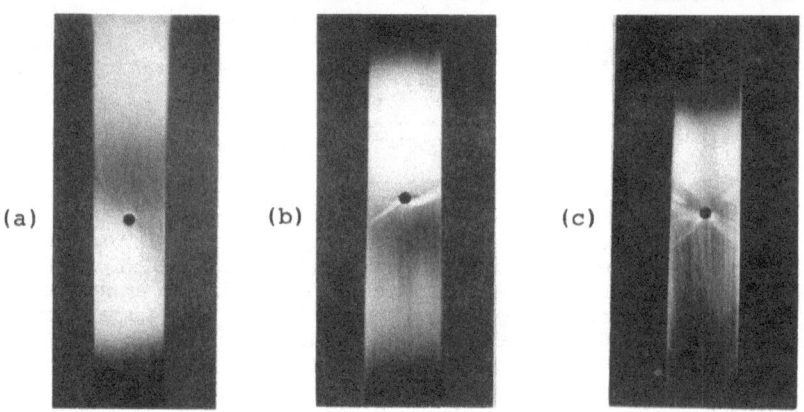

Fig. 2 **Propagation of the yield stress pattern of a specimen with a hole.**

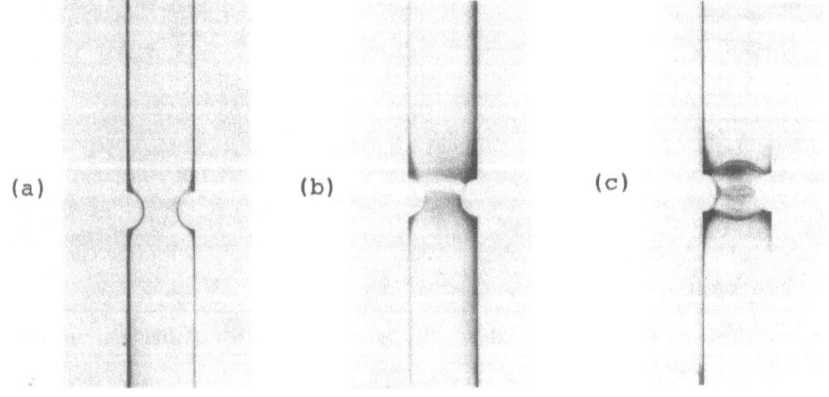

Fig. 3 **Propagation of the yield stress pattern of a specimen with a pair of semicircular notches [4].**

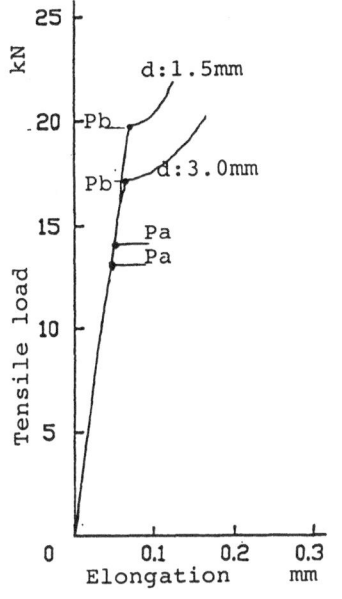

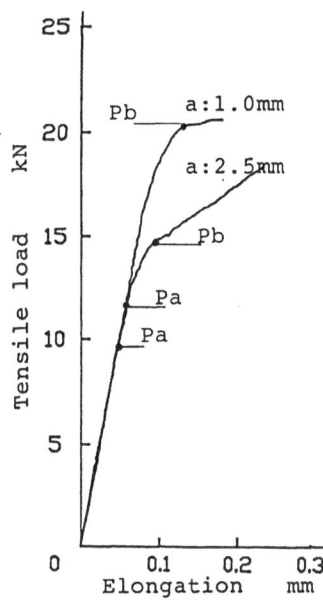

Fig. 4 Load-elongation diagram of a specimen with a hole.

Fig. 5 Load-elongation diagram of a specimen with a pair of semicircular notches.

specimen of material B in which the radius a is 1.0 mm and the radius of the semicircular notch 2.5 mm. The value of Pa in Fig. 5 is within the range of application of Hook's law, but the value of Pb is outside the range of Hook's law. As it is extremely difficult to ascertain the value of Pa from the data recorded on the load-elongation diagram either from Fig. 4 or from Fig. 5, the reading must be supplemented by the results obtained from observing the surface of the specimen with the naked eye. On the other hand, the value of Pb can easily be obtained from the data recorded in the load-elongation diagram.

Stress near the notch

Striped pattern and tensile load

Fig. 6 is the result of obtaining the experimental values of Pa and Pb while varying the size of the circular hole. The X-axis is the ratio of the diameter d of the hole and the width 2b of the parallel part of the specimen, while the Y-axis represents the value of Pa or Pb. Fig. 6(A) shows the results in the case of material A and Fig. 6(B) shows the results with material B. In Fig. 6 the measured values of Pa and Pb vary with d/2b, becoming smaller when the value of d/2b increases. Since the measured values were randomly made, they were subjected to a regression analysis. The regression equations are as follows:

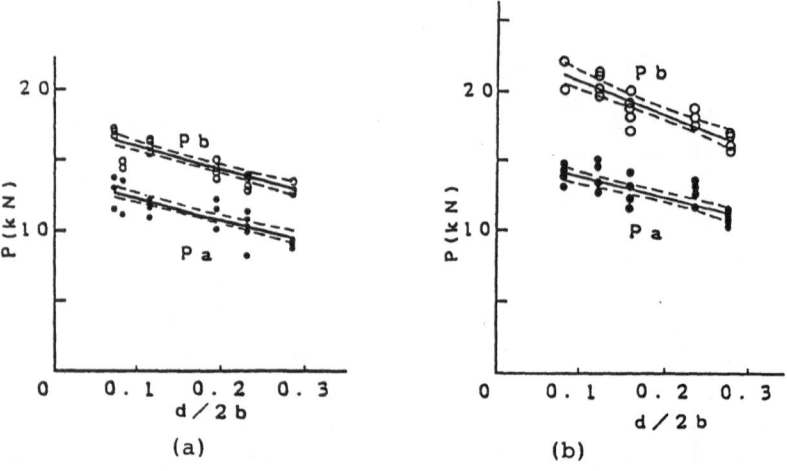

(a)   (b)

Fig. 6 Tensile load vs. d/2b values of the specimens with a
hole.

$$Pa = -14.6d/2b + 13.8 \quad (1)$$
$$Pb = -16.4d/2b + 17.7 \quad (2)$$
$$Pa = -14.5d/2b + 15.4 \quad (3)$$
$$Pb = -23.9d/2b + 23.2 \quad (4)$$

Equations (1) and (2) are the results for material A, and
Equations (3) and (4) are the results for material B. The
coefficient of correlation for Equation (1) is -0.800, that
for Equation (2) -0.906, for Equation (3), -0.797, and for
Equation (4), -0.887. The straight lines of Fig. 6 signify the
regression equations.
    Figure 7 shows the experimental values obtained from the
specimen having a pair of semicircular notches. The X-axis
denotes a/b: the ratio of the notch radius 'a' to the 1/2

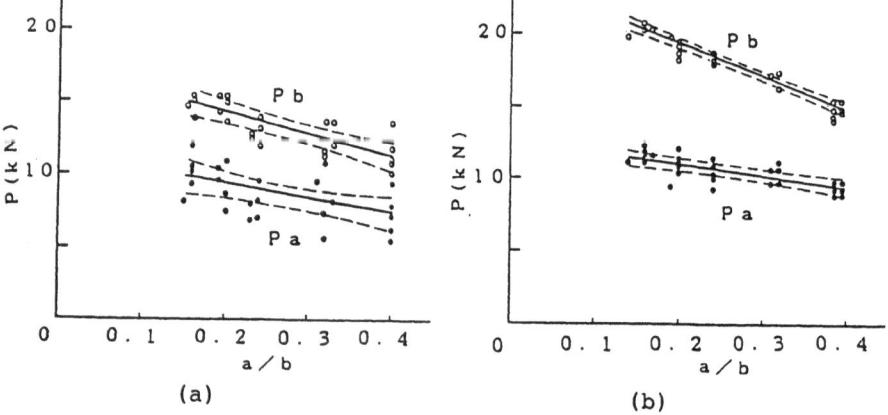

(a)   (b)

Fig. 7 Tensile load vs. a/b values of the specimens with a
pair of semicircular notches.

width 2b of the parallel part of the specimen. The measured values Pa and Pb in Fig. 7 became gradually smaller as the value of a/b increased. The following regression equations were obtained when the measured values of Pa and Pb for a specimen having a pair of semicircular notches were subjected to a regression analysis.

$$Pa = -9.8a/b + 11.5 \qquad (5)$$
$$Pb = -14.9a/b + 17.8 \qquad (6)$$
$$Pa = -8.3a/b + 12.8 \qquad (7)$$
$$Pb = -23.0a/b + 24.0 \qquad (8)$$

Equations (5) and (6) are the results for material A, and Equations (7) and (8) are results for material B. The coefficient of correlation is -0.545 for Equation (5), -0.815 for Equation (6), -0.774 for Equation (7), and -0.980 for Equation (8). The straight lines shown in Fig. 7 represent these regression equations.

Stress concentration

In order to find out the stress acting in the vicinity of the notches, it is sufficient to make an observation of the stress concentration. It was accordingly undertaken to obtain the elastic-plastic stress concentration factors for a specimen having a circular hole and one having a pair of semicircular notches.

As the load increases at the notched portion, the striped pattern appears just where the yield stress first starts to exert itself. The point where the striped pattern appears may be regarded as the place where the maximum stress is exerted.

The elastic-plastic stress concentration factor $a\sigma$ is defined by the following equation [6].

$$a\sigma = \sigma max/\sigma n \qquad (9)$$

σmax takes the value of the yield stress of the material. σn is obtained by dividing the tensile loads derived from Equations (1), (3), (5) and (7) by their respective minimum areas. Then $a\sigma$ is obtained by Equation (9). Fig. 8 is for a specimen with a circular hole. Fig. 9 is for a specimen with a pair of semicircular notches. The broken lines in Figs. 8 and 9 denote the stress concentration factor obtained from the literature [7].

The lines representing the experimental values of the elastic-plastic stress concentration factors in Figs. 8 and 9 show the same tendency as observed with the stress concentration factors. However, these lines are apparently smaller than those for the stress concentration factors.

In Figs. 8 and 9, the lines for material A are positioned lower than those for material B.

Materials with higher V_H values have higher elastic-plastic stress concentration factors even if the notch shapes are identical. Even though the stress concentration factors were determined solely by the geometrical shape of the notch, the hardness of the material

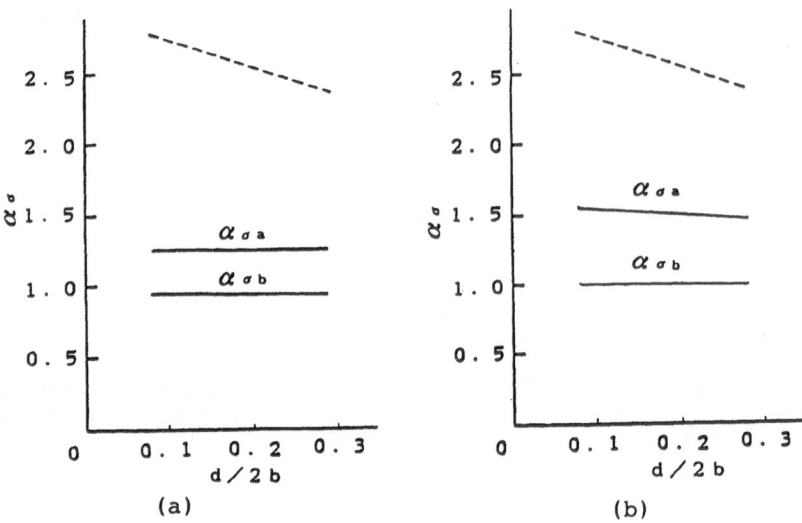

Fig. 8 Experimental values of the stress concentration factors in the elastic-plastic state of the specimens with a hole.

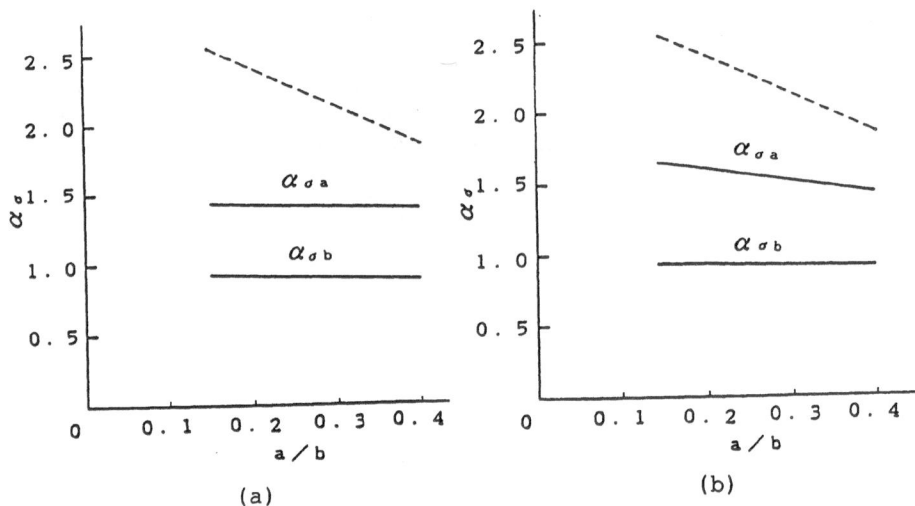

Fig. 9 Experimental values of the stress concentration factors in the elastic-plastic state of the specimens with a pair of semicircular notches.

became an additional factor for the elastic-plastic stress concentration factor.

The profile of the surface roughness of the striped pattern

The record in Fig. 10 makes up the location on the profile of surface roughness where the striped pattern occurs. Fig. 10 (a) shows the data recorded about the surface of a specimen before a load is applied. The value of Rmax is shown to be less than 1 μm. Fig. 10 (b) shows the record obtained from a specimen surface with a notch radius of 1.0 mm and whose surface is in the state denoted by Fig. 3 (c). (1) shows the results for material A and (2) the results for material B. The vicissitudes of material A are clearly larger.

(a)

(b) (1)

(b) (2)

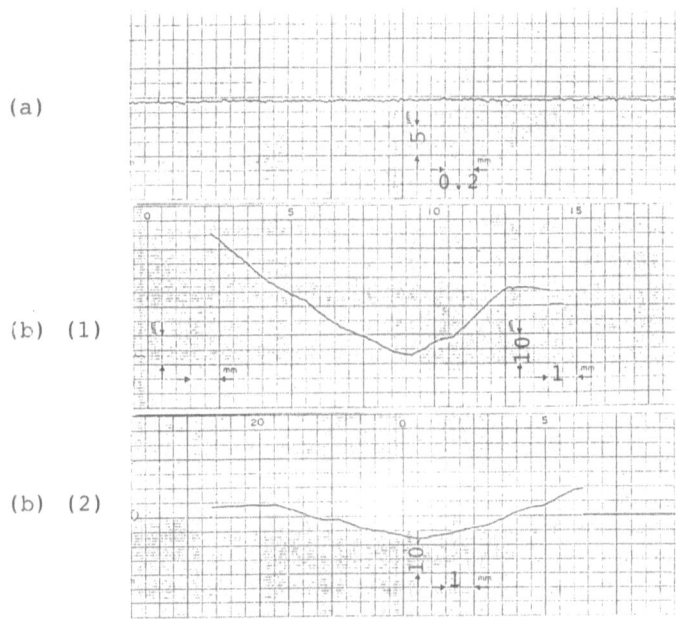

Fig. 10 The Profile of the measurements of the surface roughness of a specimen with a pair of semicircular notches.

Striped patterns are much more easily observed in material A than material B. The profile of surface roughness of the striped pattern is concave.

Conclusion

Tension testings are carried out using specimens made of a stepped carbon steel strip with a circular hole and one with a pair of semicircular notches. The following conclusions were obtained about investigating the elastic-plastic characteristics by utilizing the yield stress pattern.
It was possible to directly and continuously observe the progressive condition of yielding near a notch.

The elastic-plastic stress concentration factor is affected by the hardness of the specimen as well as by the geometrical shape and dimensions of the notch. The factor of the material with higher hardness is larger than that of the material with lower hardness. Given the identical shape and dimensions, the elastic-plastic stress concentration factor is clearly smaller than the elastic stress concentration factor

REFERENCES

1. ASM, Metals Handbook 9th Ed. vol.1 Properties and Selection: Iron and Steels, 1978, p.157.
2. Mekawa, I. and Sakurai, M., Notch Strength of Rod and Plate, Trans. Jpn. Soc. Mech. Eng., 35-271, 1969-3, pp.482-490.
3. Maekawa, I. and matsuyama, Y., Notch Strength of Rod and Plate Specimen, Part 2, Trans. Soc. Mech. Eng., 37-300, 1971-8, pp.1483-1491.
4. Taniuchi, K., The Utilization of Stretcher Strains for the Experimental Analysis of Stresses at the Bottoms of Notch Roots, Experimental Techniques, 13-8, 1989, pp.22-26.
5. Taniuchi, K., Simple Stress Sensor: Utilizing of Stretcher Stains, ASTM, STP 1025, 1989, pp.217-232.
6. JSME Handbook for Mechanical Engineers, A Fundamental, A4, Strength of materials, p.102, 1982.
7. Peterson, R.E., Stress Concentration Factors, John Wiley & Sons, New York pp.34, 150, 1974.

THE SEVERITY NEAR THE NOTCH ROOT OF NOTCHED BARS

H. HYAKUTAKE and T. HAGIO
Department of Mechanical Engineering,
Fukuoka University,
Fukuoka 814, Japan

ABSTRACT

Finite element analysis near the notch root of notched bars under tension was studied for a wide range of notch geometries. The analysis shows that the yielded zone size near the notch root is determined by both the maximum elastic stress and notch root radius. Therefore, we can express the severity near the notch root by both of the maximum elastic stress and notch root radius. In addition to the analysis, the concept of severity is subjected to further experimental scrutinization in the present paper. An experimental program is presented which examines the effect of notch root radius on the yielded zone size for the notched specimens under static tension. This is accomplished by obtaining experimental data on the notched plates and the circumferentially notched bars of polycarbonate. From the experimental results, we confirm the validity of the concept of severity for predicting the fracture strength of notched bars.

INTRODUCTION

It is surprising that so little has been published on the fracture of notched specimens of glassy polymers. After all, linear fracture mechanics concepts have been applied to the study of fracture behavior of sharply notched specimens of polymers[1]. It is important in design applications to estimate the fracture strength of notched bars over a wide range of notch geometries.

Our goal is to study the fracture behavior of engineering plastics containing various sized notches and to develop a limiting condition for predicting the load at failure. In the present paper, we show a concept of severity near the notch root by studying the elastic and elastoplastic stress analysis of notched bars subjected to tension. The limiting condition for determining the fracture strength of notched bars is based on the concept of severity. Furthermore, an experimental design is presented which proves the validity of the concept of severity for fracture of notched bars. This is accomplished by obtaining experimental data on the tension tests of the notched plates and circumferentially notched bars of polycarbonate for a wide range of notch geometries.

CONCEPT OF SEVERITY

From a practical design viewpoint it is important to have some knowledge of fracture behavior of notched bars. Some studies of the fracture of notched polymers under static load have been published previously. The fracture load will depend on many factors, such as the notch geometries; notch root radius; notch angle; notch depth; width and thickness of plate; and diameter of bar. It is well known that the fracture behavior of notched bars could not be elucidated by the classical stress concentration factor approach. In early studies[2, 3] on the fracture of notched specimens of polymers, they used a vague concept of notch sensitivity. Mills[4] investigated the yielded zone near the notch root of notched polycarbonate bars and applied to the slip line field analysis. Narisawa[5] and Kitagawa[6] modified the theory of Mills. This approach will be plagued by many difficulties due to insufficient data for the mechanical properties of the materials. An alternative approach involves using the average stress criterion. Nuismer and Whitney[7] assumed that a notched (FRP) specimen failed when the average value of stress over some fixed distance a_0 ahead of the notch root reached a constant value. However, there is no scientific interpretation the value of a_0.

In the case of a very sharply notched or cracked bar, the notch root radius is nearly zero, so that the stress at the notch root is infinity. Based on the concept of LEFM, however, we can express the severity near the crack tip by using a stress intensity factor K_I. The fracture criterion for a cracked bar, therefore, is determined by the

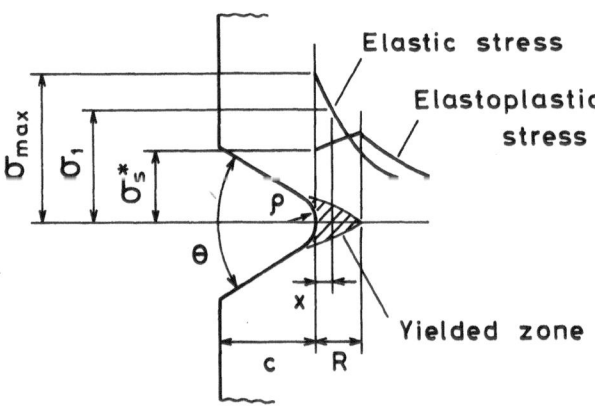

Figure 1. Stress distributions and yielded zone near the notch root.

equation: $K_I = K_{IC}$, where K_{IC} is the material constant.

On the other hand, there are many features of notch geometries in the case of notched bars having the notch root radius of larger than zero. Although fracture behavior will depend on these factors, it seems appropriate to express the severity near the notch root by using the maximum elastic stress at the notch root σ_{max} (Fig. 1) and the notch root radius ρ, as discussed in previous papers[8, 9]. The concept of severity near the notch root is similar to the one of LEFM. Both concepts are based on the similarity of relative stress distribution and the equivalence of response under a condition of small scale yielding near the notch root. It is based on the following evidence concerning the results of elastic and elastoplastic stress analysis near the notch root of notched plate[9].

(1) The relative elastic stress distribution near the notch root of a notched plate is governed predominantly by the notch root radius, and it is independent of notch depth, the width and thickness of plate.

(2) The yielded zone size at the notch root (R in Fig. 1) is determined by both the maximum elastic stress and the notch root radius, and it is independent of the notch depth.

In the present paper, the elastic and elastoplastic stress analysis of circumferentially notched bars (Fig. 2b) was performed for a wide range of notch geometries, using the finite element method. In the circumferentially notched bar as shown in Fig. 2b, the notch root radius ρ is a constant value of 1, the notch angle θ are varied from 0° to 120°, the notch depth c are varied 0.2 to 9 and the diameter of minimum section of bar $2a$ are varied from 2 to 18. In this analysis, we use an elastic-perfectly plastic material and also the Von Mises yield criterion.

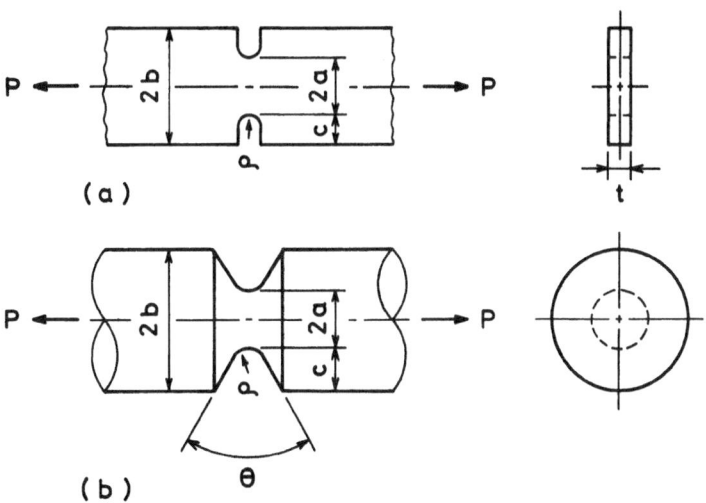

Figure 2. Geometries of notched bars: (a) notched plate, (b) circumferentially notched bar.

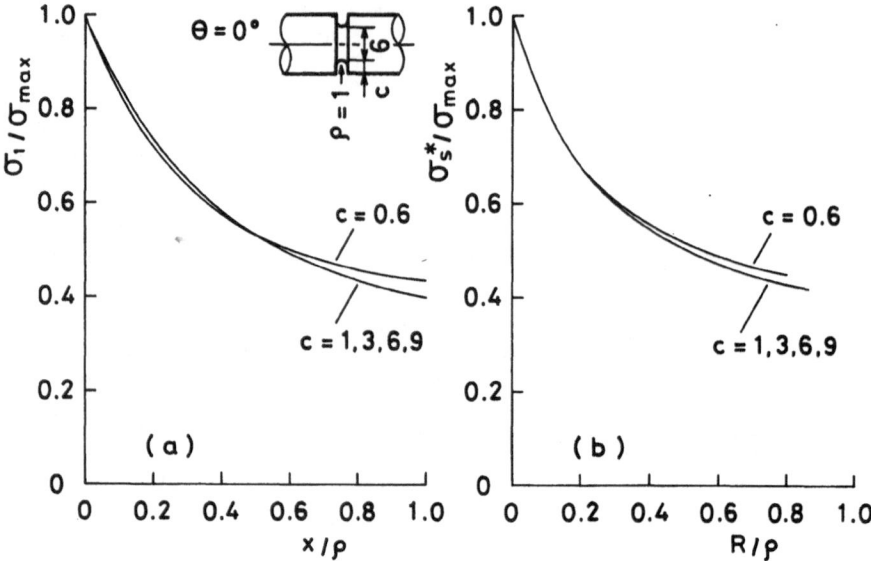

Figure 3. Elastic and elastoplastic stress analysis near the notch root
of circumferentially notched bars: (a) elastic stress distributions, (b)
the maximum elastic stress σ_{max} versus yielded zone size R.

Figure 3 shows the result of analysis of the circumferentially
notched bars for which notch angle $\theta = 0°$ and the notch depth c are the
following five different values: 0.6, 1, 3, 6 and 9. It was shown that
there was little effect of the notch angle θ on the relative elastic
stress distribution and the yielded zone size in the range of $\theta \leq 90°$.
It is evident that the elastic stress distribution near the notch root
and the yielded zone size of smaller than the value of notch root radius
are determined by both the maximum elastic stress σ_{max} and the notch
root radius ρ, and they are independent of the notch depth c of larger
than 1, as can be seen from Fig. 3a and 3b, where $\sigma_s{}^*$ is the equivalent
tensile yield stress of the material.

The evidences derived from the stress analysis of circumferentially
notched bars are similar to the evidences which are derived from the
results of analysis of notched plates. According to the results of
stress analysis mentioned above, it is presumed that the elastoplastic
stress distributions near the notch root after small scale yielding are
the same in all notched bars, for which both the maximum elastic stress
σ_{max} and notch root radius ρ are equal to each other.

Based on the concept of severity mentioned above, the fracture
criterion for a notched bar is expressed as:

$$\sigma_{max} = \sigma_{max,\,c}(\rho) \tag{1}$$

where σ_{max} is the maximum elastic stress at failure and is determined by
the product of the nominal stress and the geometrical stress

concentration factor. $\sigma_{max,c}$, on the right side of Eq. (1), is the material constant which is governed by the notch root radius ρ only and is independent of other notch geometries.

The validity of the concept of severity near the notch root and the fracture criterion of Eq. (1) are subjected to further experimental scrutinization by using the notched specimens of polycarbonate in the present paper.

EXPERIMENTAL PROCEDURE

Polycarbonate used was in the form of comercial sheets and rods. The specimens were machined to the shapes as shown in Fig. 2, and then were given a 5 h heat-treatment at 130°C to relieve the residual stresses.

Notched polycarbonate bars show two different modes of failure, that is, brittle and ductile[10, 11, 12]. Transition from ductile to brittle failure can be caused by a decrease in notch root radius and an increase in specimen thickness[13]. Details of the critical notch root radius ρ_c at the brittle-ductile transition of the failure mode of polycarbonate in static tension are given in a previous paper[9]. In the present paper, all specimens had a sharply notch root radius of smaller than ρ_c, in order to fail in a brittle manner.

In the case of notched plates, all specimens were notched in an U-shape on the both sides of plate. There were two different plates with thickness t of 2 and 5 mm. The notch radii ρ were varied from 0.2 to 0.5 mm, and the notch depths c were varied from 1 to 5 mm. Further details of experimental procedure for notched plates of polycarbonate are given in a previous paper[9]. In the case of circumferentially notched bars, all specimens had a V-shaped notch. The notch angle θ were varied from 0° to 120°. The notch root radius ρ were varied from 0.2 to 2 mm, and the notch depth c were varied from 0.5 to 2.5 mm. Tensile failure tests for notched specimens were performed using an Instron-type testing machine at a constant cross-head speed of 0.5 mm/min in a temperature controlled room at 22 ± 0.5 °C.

RESULTS AND DISCUSSION

Figure 4 shows the yielded zone near the notch root of the notched plate and circumferentially notched bar of polycarbonate just prior to brittle fracture. Figure 4a shows a transmitted light view of the inside of the notched plate specimen. On the other hands, Fig. 4b shows a reflected light view of the cut and polished surface of central section of the circumferentially notched bar. As shown in Fig. 4, the shear bands are visible inside the plastic zone and the craze extends from the tip of the plastic zone. It seems that the yielded zone size is nearly equal to the notch root radius, as can be seen from Fig. 4.

A set of notched specimens having a constant notch root radius ρ and different notch depths c and notch angle θ were tested in tension, and the size R of the yielded zone prior to fracture and the maximum elastic stress at fracture $\sigma_{max,c}$ determined. The experimental results of notched plates for which $\theta = 0°$ are shown in Fig. 5 and 6. It can be seen that R is determined by both the notch root radius ρ and the maximum elastic stress σ_{max} (Fig. 5), and $\sigma_{max,c}$ is determined by ρ only (Fig. 6). Furthermore, it is evident that both R and $\sigma_{max,c}$ are independent of notch depth c.

The experimental results of the circumferentially notched bars are shown in Fig. 7 and 8. As can be seen from Fig. 7, the yielded zone size R prior to fracture is determined by both the notch root radius ρ and the maximum elastic stress σ_{max}, and R is independent of the notch depth c and the notch angle θ of smaller than 90°. It is evident that the maximum elastic stress at fracture $\sigma_{max,c}$ is determined by the notch root radius ρ only and is independent of the notch depth c and the notch angle θ, as shown in Fig. 8. In the case of the shallow notched specimen for which the value of c/ρ is smaller than 1, however, the value of $\sigma_{max,c}$ is smaller than the one for a deep notched bar. Such experimental results can be explained by the results of stress analysis as shown in Fig. 3.

Figure 9 shows the relation between the maximum elastic stress at fracture $\sigma_{max,c}$ and the notch root radius ρ for a wide range of notch geometries. There are two kinds of specimens, notched plates and circumferentially notched bars. It can be seen that each experimental point fell in close proximity to a characteristic curve, and $\sigma_{max,c}$ is governed by the notch root radius ρ only. In other words, $\sigma_{max,c}$ has a one-to-one correspondence with a notch root radius ρ.

The experimental results mentioned above confirm the validity of the fracture criterion of Eq. (1). It is reasonable to conclude that these experimental result for notched bars of polycarbonate could be interpreted using the concept of severity near the notch root, as discussed before. The fracture criterion of Eq. (1) based on the concept of severity is the same as that expressed in the case of fracture of FRP plates containing notches[14, 15].

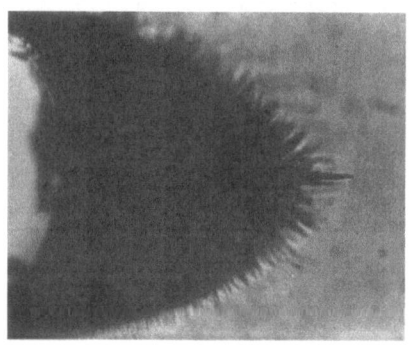 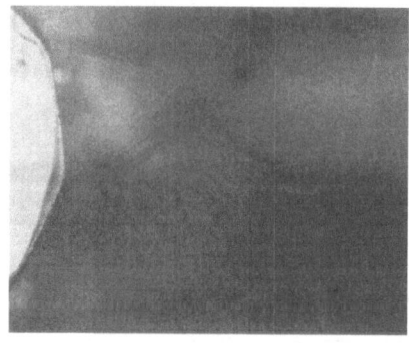

(a)   (b)

Figure 4. Yielded zone near the notch root of (a) notched plate (notch root radius ρ = 0.5 mm, notch depth c = 4 mm and thickness of plate t = 5 mm) and (b) circumferentially notched bar (ρ = 0.5 mm, c = 2 mm and notch angle θ = 60°).

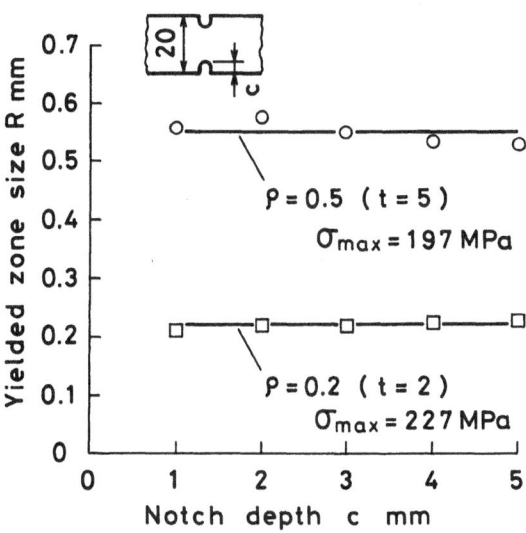

Figure 5. Yielded zone size R versus notch depth c for notched plates of polycarbonate.

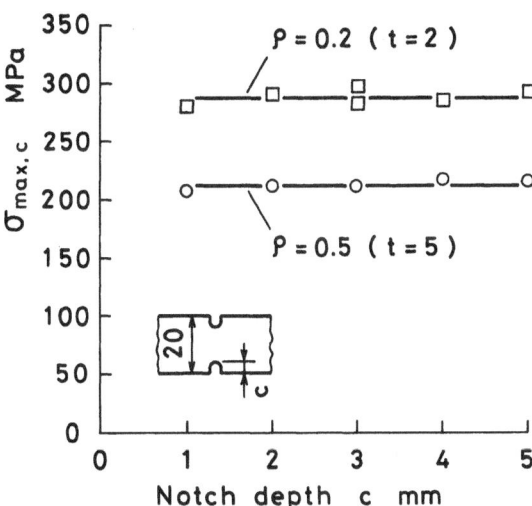

Figure 6. The maximum stress at fracture $\sigma_{max,c}$ versus notch depth c for notched plates of polycarbonate.

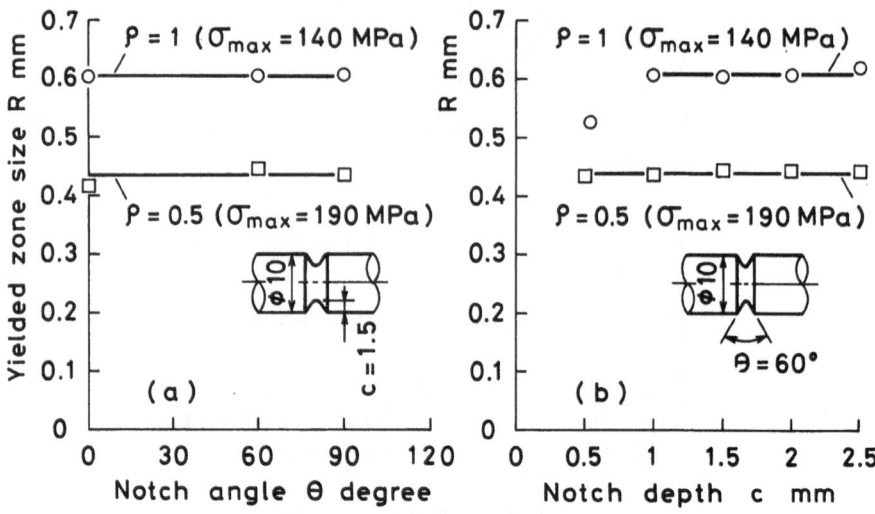

Figure 7. Yielded zone size R for circumferentially notched bars of polycarbonate: (a) R versus notch angle θ (notch depth c = 1.5 mm), (b) R versus c (θ = 60°).

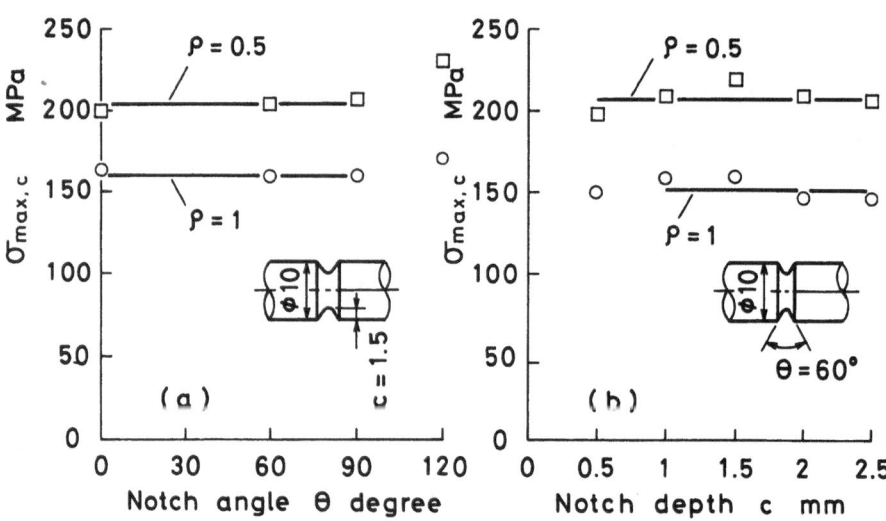

Figure 8. The maximum stress at fracture $\sigma_{max,c}$ for circumferentially notched bars of polycarbonate: (a) $\sigma_{max,c}$ versus notch angle θ (notch depth c = 1.5 mm), (b) $\sigma_{max,c}$ versus c (θ = 60°).

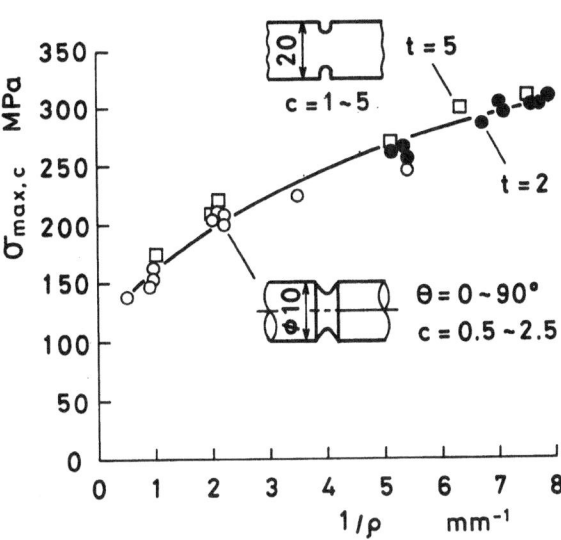

Figure 9. The maximum stress at fracture $\sigma_{max,c}$ versus notch root radius ρ for notched bars of polycarbonate.

CONCLUSIONS

Elastic and elastoplastic stress analysis near the notch root of notched bars under tension was studied for a wide range of notch geometries by using finite element method. The results of analysis are as follows: (1) The relative elastic stress distribution near the notch root is governed predominantly by the notch root radius only. (2) The yielded zone size near the notch root is determined by both the maximum elastic stress and the notch root radius.

According to the evidence mentioned above, it is concluded that we can express the severity near the notch root of notched bars by both of the maximum elastic stress and notch root radius. It is apparent the severities near the notch root are the same in all notched bars failed in a brittle manner. It is conceivable therefore that the maximum elastic stress at fracture of notched bar is governed by the notch root radius only and is independent of other notch geometries and specimen sizes. Furthermore, it is obvious that the yielded zone size near the notch root prior to fracture is governed by the notch root radius.

The concept of severity mentioned above is subjected to further experimental scrutinization in the present paper. An experimental program is presented which examines the effect of notch root radius on the fracture strength and the yielded zone size for the notched specimens under static tension. This is accomplished by obtaining experimental data on the notched plates and circumferentially notched bars of polycarbonate for a wide range of notch geometries. From the experimental results, we confirm the validity of the concept of severity for determining the fracture strength of notched bars.

REFERENCES

1. Williams, J.G., Fracture Mechanics of Polymers, Ellis Horwood, Chichester, 1984.

2. Takano, M. and Nielsen L.E., The notch sensitivity of polymeric materials. J. Appl. Polym. Sci., 1976, 20, pp. 2193-2207.

3. Prabhakaran, R., Nair, E.M.S., and P. K., Sinha, Notch sensitivity of polymers. J. Appl. Polym. Sci., 1978, 22, pp. 3011-3020.

4. Mills, N.J., The mechanism of brittle fracture in notched impact tests on polycarbonate. J. Mater. Sci., 1976, 11, pp. 363-375.

5. Narisawa, I., Ishikawa, M. and Ogawa, H., Notch brittleness of ductile glassy polymers under plane strain. J. Mater. Sci., 1980, 15, pp. 2059-2065.

6. Kitagawa, M., Plastic deformation and fracture of notched specimens due to bending in glassy polymers. J. Mater. Sci., 1982, 17, pp. 2514-2524.

7. Nuismer, R.J. and Whitney, J.M., Uniaxial failure of composite laminates containing stress concentrations. In Fracture Mechanics of Composites, ASTM STP 593, ed. S.P. Sendekyi, American Society for Testing Materials, Philadelphia, 1975, pp. 117-142.

8. Nisitani, H. and Hyakutake, H., Condition for determining the static yield and fracture of a polycarbonate plate specimen with notches. Eng. Fract. Mech., 1985, 22, pp. 359-368.

9. Hyakutake, H. and Nisitani, H., Conditions for ductile and brittle fracture in notched polycarbonate bars. JSME Inter. J., 1987, 30, pp. 29-36.

10. Fraser, R.A.W., and Ward, I.M., The impact fracture behaviour of notched specimens of polycarbonate. J. Mater. Sci., 1977, 12, pp. 459-468.

11. Pitman, G.L., Ward, I.M. and Duckett, R.A., The effects of thermal pre-treatment and molecular weight on the impact behaviour of polycarbonate. J. Mater. Sci., 1978, 13, pp. 2092-2104.

12. Parvin, M. and Williams, J.G., Ductile-brittle fracture transitions in polycarbonate. Int. J. Fract., 1975, 11, pp. 963-972.

13. Brown, H.R., A model for brittle-ductile transitions in polymers. J. Mater. Sci., 1982, 17, pp. 469-476.

14. Hyakutake, H., Nisitani, H. and Hagio, T., Fracture criterion of notched plates of FRP. JSME Inter. J., 1989, 32, pp. 300-306.

15. Hyakutake, H., Hagio, T. and Nisitani, H., Fracture of FRP plates containing notches or a circular hole under tension. The 1989 ASME Pressure Vessels and Piping Conference, Vol. 167, American Society of Mechanical Engineers, New York, 1989, pp. 141-146.

THE PREDICTION OF COLLAPSE LOADS USING FINITE DIFFERENCES

I.C. PYRAH
Department of Civil and Structural Engineering
University of Sheffield
Mappin Street
SHEFFIELD S1 3JD

ABSTRACT

The paper describes a finite difference technique which has been developed to model the behaviour of a purely cohesive elastic-perfectly plastic material. The method is used to examine the problem of a rigid punch pushed into a layer of elastic-plastic material resting on a rigid base. The effect on the collapse load of the assumed roughness of the base of the punch is examined, as is the influence of the width of the punch. The changes in the stress distributions beneath the punch as indentation takes place is also examined.

INTRODUCTION

The past twenty five years have seen substantial developments in the application of numerical techniques to engineering problems with the most significant advances being in the finite element method. A structure or continuum can be represented by elements of various shapes and sizes and due to the method's geometrical flexibility once a program, or suite of programs, has been developed a whole variety of different problems can be solved. The more general and powerful a program the more useful it is likely to be, particularly in engineering practice. Unfortunately, as programs become more versatile they also become more unwieldly and it becomes more difficult to incorporate facilities which have not been considered during their development. Whilst a systematic, modular approach simplifies the development and subsequent modification of finite element programs, the author believes that more use could be made of other methods

such as finite differences. This is particularly true in situations where
the incorporation of realistic material behaviour, accurate representation
of different boundary conditions and the reliable prediction of stresses is
more important than the ability to deal with complex geometries.

The developments in digital computing which led to the explosion of
finite element applications also encouraged renewed interest in finite
difference methods as used by Southwell and others [1]. Although the
process of relaxation is not directly suitable for computers, modern texts
deal with appropriate direct and iterative techniques for solving the
finite difference approximations to partial differential equations.
However, most publications deal with the solution of pdes in terms of a
single variable and whilst linear elastic problems can be formulated in
this way it can lead to difficulties when dealing with realistic boundary
conditions. Conventional finite difference techniques, therefore, where
the equations of compatibility and equilibrium are combined and the
resulting equation expressed in terms of displacements or a stress
function, are of limited value.

In the early days of finite elements, two finite difference methods
were proposed which overcame some of the limitations of the conventional
approach. Both techniques, the lumped parameter model [2] and dynamic
relaxation [3] use a formulation that does not combine the compatibility
and equilibrium equations and both can deal with mixed boundary conditions
and non-linear material behaviour. The latter technique forms the basis of
the method described in this paper.

DEVELOPMENT OF TECHNIQUE FOR ELASTIC-PLASTIC BEHAVIOUR

The concept of dynamic relaxation as a stress analysis technique was
described by Otter [4] with a more precise formulation given by Otter,
Casell and Hobbs [3]. The method can be divided into two parts:
a)   the specification of the variables at points throughout the
       material and
b)   a technique for solving the finite difference equations.
       Interlaced grids are used, Gilles [5] suggested this gives a better
finite difference approximation for shear stress but more
importantly this also allows displacement and/or stress boundary
conditions to be dealt with easily. Vertical and horizontal

equilibrium are satisfied at the respective displacement points and the stresses are calculated from the strains defined by the surrounding displacements (Figure 1).

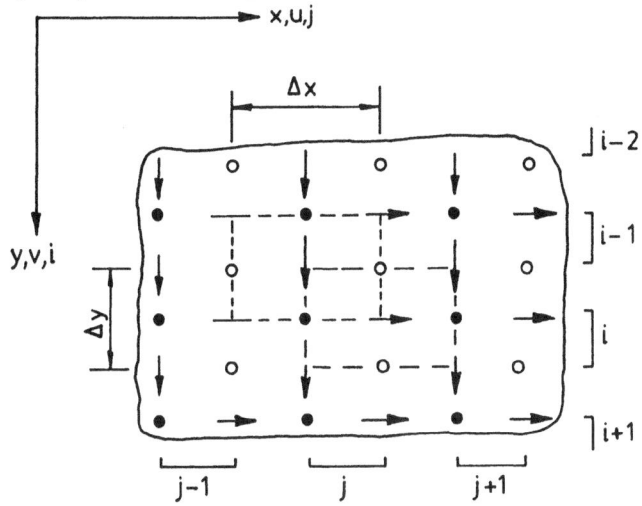

• NORMAL STRESS POINT σ_x, σ_y

o SHEAR STRESS POINT τ_{xy}

→ HORIZONTAL DISPLACEMENT u

↓ VERTICAL DISPLACEMENT v

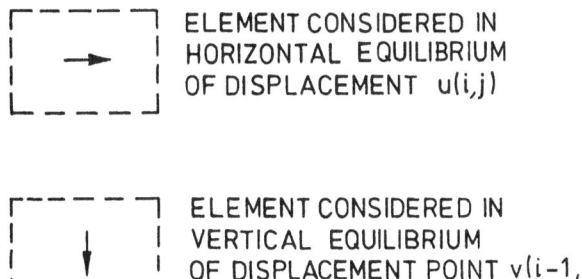

ELEMENT CONSIDERED IN HORIZONTAL EQUILIBRIUM OF DISPLACEMENT u(i,j)

ELEMENT CONSIDERED IN VERTICAL EQUILIBRIUM OF DISPLACEMENT POINT v(i-1,j)

Figure 1. Finite difference grid

Dynamic relaxation derives its name from the method used to solve the finite difference equations although it is basically an iterative rather than a relaxation technique. The problem to be solved is treated as a dynamic one, the stresses and displacements for the static problem being defined by those for the dynamic problem as the vibrations become negligible. The numerical stability and rate of convergence for this pseudo-dynamic technique are dependent upon an assumed time increment and an assumed damping factor.

The technique may be compared with the Second Order Richardson (or Frankel's) iterative procedure [6] and the modified form of dynamic relaxation used in this paper incorporates features of both methods. This modified technique uses incremental displacements as in the Richardson process but still incorporates the idea of critical damping to control the rate of convergence. The method can be extended to deal with two-phase materials e.g. soil which is composed of solid particles and void spaces filled with water [7].

The solution to a particular problem is obtained when the displacements and stresses satisfy both the stress/strain and equilibrium conditions; in the present method this is achieved by ensuring that during each iterative cycle the stress/strain equations are always satisfied and the errors in the equilibrium equations are systematically reduced until they reach an acceptable level.

At the beginning of each iteration the displacements are known and hence so too are the total strains from which the stresses can be calculated. For an elastic material there is no difficulty in calculating the stresses directly from the strains but for an elastic-plastic material it is essential to separate the elastic and the plastic components of strain. If this can be achieved the stresses can be calculated directly from the elastic strains.

For an elastic-perfectly plastic material it is usual for the total strain to be separated into elastic and plastic components

$$\varepsilon_{ij} = \varepsilon'_{ij} + \varepsilon''_{ij}$$

where ε_{ij} is the total strain

ε'_{ij} is the total elastic strain

ε''_{ij} is the total plastic strain

When the load is applied in increments the total plastic strain at the end of an increment can be split into

- ε_{ij}° plastic strain prior to application of the incremental load, and
- $d\varepsilon_{ij}^{II}$ incremental plastic strain caused by the increase in the applied load.

Thus,

$$\varepsilon_{ij} = \varepsilon_{ij}' + \varepsilon_{ij}^\circ + d\varepsilon_{ij}^{II}$$

The crux of this approach to elastic-plastic problems is the splitting up of the total strain into its three components. By assuming a yield criterion and some form of flow-rule the incremental plastic strains can be related to the stresses. As these can be expressed in terms of the elastic strain components it is possible to derive a relationship whereby the incremental plastic strains are given in terms of:

- the total strains (ε_{ij}) defined by the displacements
- the initial total plastic strains (ε_{ij}) defined as the total plastic strain at the end of the previous load increment
- the yield and flow rule parameters, and
- the elastic properties of the material.

Having defined the incremental plastic strains, the elastic strains can be calculated and the stresses evaluated. These stresses are then substituted in the equilibrium equations and the iterations continued until the error in these equations is sufficiently small. The relevant equations and their derivation for an elastic-perfectly plastic cohesive material are presented elsewhere [8].

PROBLEM DEFINITION

The program has been used to predict the behaviour of a rigid punch (width 2B) progressively pushed into an elastic-perfectly plastic material (shear strength c) of finite thickness (D).

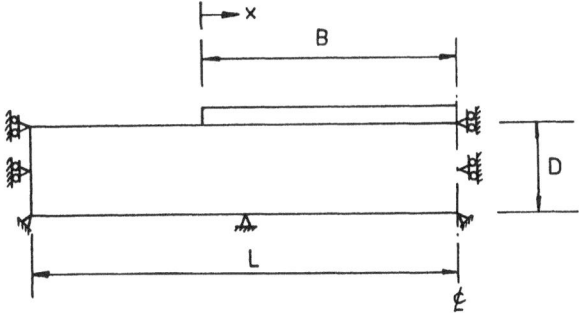

Figure 2.   Problem geometry

Due to symmetry only half the problem need be considered (Fig.2) and
different ratios of B/D have been examined (1, 3 and 10); for each geometry
a rough and a smooth punch were considered. For the smooth punch the
boundary shear stress is kept at zero. For the rough punch no horizontal
displacement is allowed and the boundary shear stress set to a value which
maintains this condition; if the required shear stress exceeds the shear
strength of the material (c), the boundary shear stress is reduced to that
of the material strength. In each analysis the width of the elastic-
perfectly plastic material extended at least 2D beyond the edge of the
punch so that the conditions of the left-hand boundary had no significant
effect on the solution (i.e. in Figure 2, L > B + 2D). The analyses
assumed that the elastic-perfectly plastic material rested on a perfectly
rough and rigid surface.

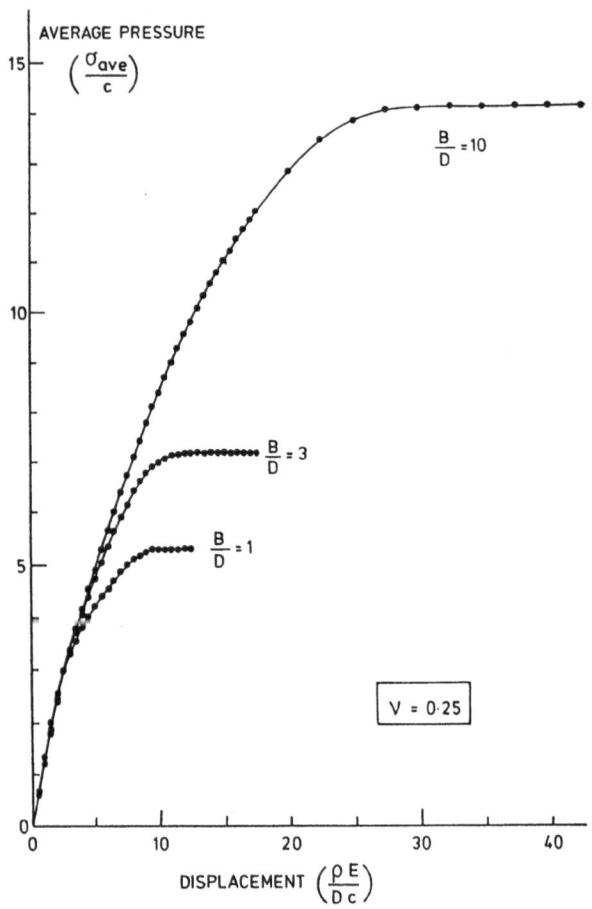

Figure 3. Load-displacement curves for a rough punch

The following material properties were adopted in the analyses: shear strength, c = 30 kPa, Young's modulus, E = 15000 kPa and Poisson's ratio, = 0.25. The thickness of the material, D, was kept constant and equal to unity; the results are presented in non-dimensional form.

RESULTS

The increase in average vertical stress (σ_{ave}) beneath a rough punch as it is pressed into the elastic-plastic material is shown in Figure 3 for B/D ratios of 1, 3 and 10. The collapse loads are well-defined and the failure stress increases as the B/D ratio increases. This is in accordance with the theoretical solutions obtained from plasticity theory (9) and Figure 4 shows the good agreement between the finite difference and plasticity solutions for both smooth and rough punchs.

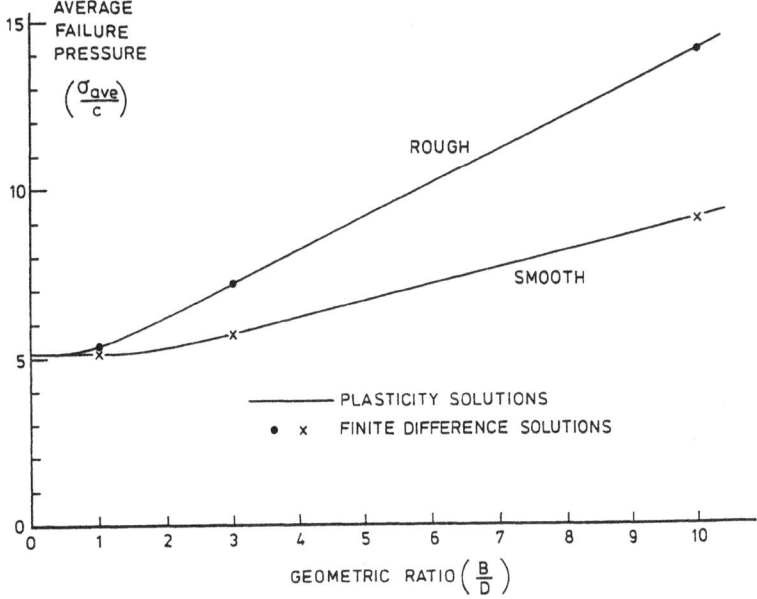

Figure 4. Failure load as a function of the geometric ratio

Plasticity theory can also be used to calculate the distribution of vertical stress on the underside of a wide punch, as shown in Figure 5 where x represents the distance in from the edge of the punch. The distribution of vertical stress at collapse produced by the finite difference program, using a (5 x 26) grid for B/D = 10, is in good agreement with the plasticity solution.

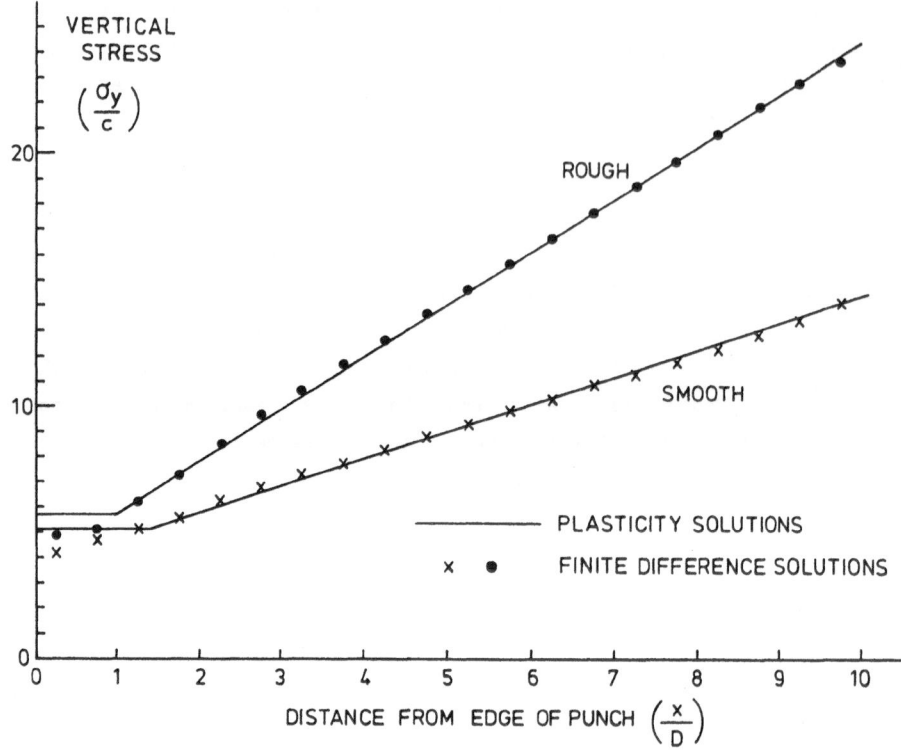

Figure 5. Variation of vertical stress beneath a wide punch

In addition to providing the stress distribution at collapse the finite difference solution allows the progressive development of the stress distribution to be examined. The results for a rough punch (B/D = 3), using a (10 x 50) grid, are shown in Figure 6. At low levels of indentation the stress distribution is basically elastic with maximum pressure at the edge of the punch; as the load increases the distribution becomes more uniform and then approximates to the plasticity solution for a wide punch, with the maximum value at the centre. The simultaneous development of shear stress beneath the punch is shown in Figure 7.

Differences between the finite difference collapse pressure distribution and the plasticity solution are presumably due to the high stress gradients at the edge of the punch and the zero shear stress (τ_{xy}) beneath its centre. This latter effect becomes less significant as B/D increases.

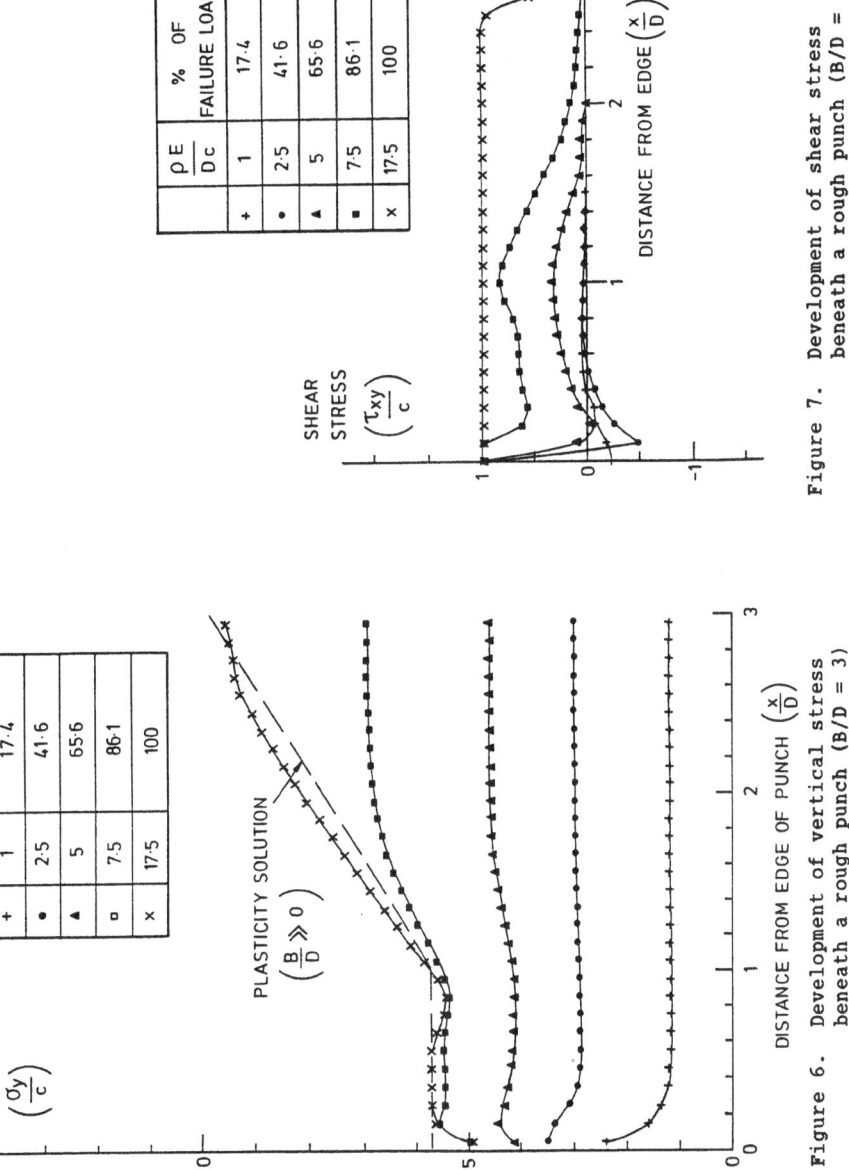

Figure 7. Development of shear stress beneath a rough punch (B/D = 3)

Figure 6. Development of vertical stress beneath a rough punch (B/D = 3)

CONCLUSIONS

A finite difference program has been used to successfully predict collapse
load and stress distributions beneath a smooth and a rough punch; the
results agree with theoretical plasticity solutions. The progressive
development of vertical and shear stress on the underside of a rough punch
has been examined.

ACKNOWLEDGEMENTS

This work has been carried out in conjunction with a study into the
behaviour of reinforced embankments on soft ground. Many useful
discussions with colleagues C C Hird and D Russell are gratefully
acknowledged.

REFERENCES

1.　Allen, D.N. De G., Relaxation Methods, McGraw-Hill, 1955.

2.　Ang, A.H.S. and Harper, G.N., Analysis of contained plastic flow in
plane solids. J.Eng. Mechs. Div. A.S.C.E., 1964, 90, 397-418.

3.　Otter, J.R.H., Cassell, A.C. and Hobbs, R.E., Dynamic relaxation.
Proc. Instn. Civ. Engrs, 1966, 35, 633-656.

4.　Otter, J.R.H., Computations for pre-stressed concrete pressure
vessels using D.R.. Nuclear Struct. Eng., 1965, 1, 61-75.

5.　Gilles, D.C. The use of interlacing nets for the application of
relaxation methods to problems involving two independent variables.
Proc. Royal Society, A. 1948, 193, 407-433.

6.　Frankel, S.P., Convergence rates of iterative treatments of partial
differential equations. Math. Tables Aids Comp., 1950, 4, 65-75.

7.　Pyrah, I.C., The solution of two-dimensional consolidation problems.
Proc. Computer Applications to Geotechnical Problems in Highway
Engineering, Cambridge, April 1980.

8.　Pyrah, I.C., Elasto-plastic analysis in geotechnical engineering
using finite differences. Int. Conf. on Computational Plasticity,
Barcelona, April 1987, 1607-1620.

9.　Mandel, J. and Salencon, J., Force portante d'un sol sur une assise
rigide (etude theoretique). Geotechnique, 1972, 22, 79-93.

A GRAPHICAL APPROACH TO SHAKEDOWN IN ROLLING CONTACT

K L JOHNSON
Cambridge University Engineering Laboratory,
Trumpington Street,
Cambridge CB2 1PZ, UK

ABSTRACT

In repeated rolling contact, even though plastic deformation may occur
during the first passage of the load, through the action of residual
stresses and strain hardening of material, the steady cyclic state may lie
within the elastic limit. Using the Tresca yield criterion it is shown in
this paper how the maximum stress for shakedown – the shakedown limit – in
line contact may be found by a simple graphical procedure.
Cases of tractive rolling, with both partial and complete slip, are
examined for (a) a perfectly-plastic and (b) a kinematically hardening
material.

INTRODUCTION

In 1957 Fessler and Ollerton published a milestone paper in Contact

Mechanics: Contact Stresses in Toroids under Radial Loads [1]. Using the

frozen stress technique of 3-dimensional photoelasticity they made the most

extensive experimental check of the internal stresses in Hertz contact

either before or since*. At that time, before finite elements, stresses

had been evaluated only at the surface and along the axis of symmetry

(z-axis) beneath the surface. Arising from a concern with rolling contact

fatigue, they obtained a closed form expression for the orthogonal shear

stresses τ_{yz} and τ_{zx} throughout the field and checked their values by

experiment.

In rolling contact a material element at a particular depth below

the surface experiences a cycle of stress. It was well known that plastic

---

* They showed that the Hertz predictions of contact area were good up
  to the maximum value of contact size to radius of curvature (0.3)
  used in their experiments, which was surprising in view of the
  idealisations of the Hertz theory.

yield is initiated at a point beneath the centre of contact by the maximum shear stress τ_{max}, which acts on planes at 45° to the surface and whose value $\approx 0.3\ p_o$, where p_o is the maximum Hertz pressure. However there was a view at the time that the orthogonal shear stress τ_{zx}, which acts parallel to the surface and changes sign from entry to exit, is more damaging since its range ($\approx + 0.25p_o$ to $-0.25\ p_o$) exceeds that of τ_{max} (range ≈ 0 to $0.3\ p_o$). For this reason the Lundberg-Plamgren theory of ball bearing life [2] was based on the orthogonal shear stress. Its significance will become apparent from the contents of this paper.

A further seminal paper was published in 1957 by A.W. Crook: Simulated gear-tooth contacts: Some experiments on their lubrication and subsurface deformation [3]. Crook showed that repeated rolling contacts at loads at which exceed the elastic limit of the material can lead to the accumulation in the near-surface layers of large plastic strains which are likely to be associated with the initiation of rolling contact fatigue. This publication stimulated theoretical work at Cambridge into plastic deformation and shakedown in repeated rolling and sliding contacts which has continued, on and off, until the present time.

Shakedown

Shakedown in repeated loading is the process whereby plastic deformation in the first few cycles of load leads to a steady cyclic state which lies within the elastic limit. The maximum load for which shakedown occurs is known as the shakedown limit. Three separate processes can contribute to shakedown:

(i)   residual stresses introduced by initial plastic flow inhibit plastic deformation in the steady state;

(ii)   strain hardening may raise the elastic limit; and

(iii) geometry changes brought about by plastic deformation may increase the conformity of the contact and reduce the contact stress.

In this paper we shall confine our attention to the plane deformation of an elastic-plastic half-space by a long cylinder in 'line contact'. (see Figure 1). In this case the surface of the half-space remains flat after deformation, so that geometry changes referred to in (iii) above are eliminated. The influence of strain hardening on shakedown will be investigated using the concept of 'kinematic hardening' as defined by Ziegler [4], in which the yield surface can be displaced in stress space, without change in size or shape. It is the simplest hardening law which models the generally observed cyclic behaviour of metals, particularly steels.

To determine shakedown limits, use is made of the shakedown theorems of the theory of plasticity:

(a)   For an elastic-perfectly plastic material Melan's theorem [5] states that: 'If <u>any</u> system of self-equilibrating residual stresses ρ_{ij} can be found which, in combination with the stresses due to the repeated load σ_{ij}, do not exceed yield at any time, then elastic shakedown will take place.'

(b)   For a kinematically hardening material Melan's theorem has been extended by Ponter [6] to state: 'If any system of fictitious residual stresses ρ_{ij}^{*} can be found which, in combination with the stresses due to the repeated load, σ_{ij} do not exceed yield at any time, then elastic shakedown will take place.'

Note (i) that the fictitious residual stresses ρ_{ij}^{*} are made up of an indeterminate combination of the real residual stresses ρ_{ij} and the displacements α_{ij} in stress space; (ii) that the stresses ρ_{ij}^{*} for each material element are independent and are not required (as in Melan's theorem) to satisfy conditions of self-equilibrium.

If at a given load, no system of residual stresses can be found which satisfies Melan's or Ponter's conditions, i.e. the load exceeds the shakedown limit, then plastic deformation will take place with every loading cycle. It can take two forms: either a closed cycle of reversing plastic strain or an open cycle in which increments of uni-directional strain accumulate with repeated cycling as demonstrated in Crook's experiment [3]. Both types of behaviour are likely to lead to failure; the first by fatigue and the second by ductile fracture. The shakedown limit, therefore, provides a rational design criterion for repeated rolling contacts.

Yield criterion

In this paper we shall follow the Tresca (maximum shear stress) yield criterion. For plane deformation in the x-z plane, it may be expressed:

$$(\sigma_1 - \sigma_2)^2 = (\sigma_{xx} - \sigma_{zz})^2 + 4\tau_{zx}^2 = 4k^2 \tag{1}$$

provided that σ_3 (= σ_{yy}) is the intermediate principal stress.

Residual Stresses

Consider the cylinder rolling on a half-space shown in Figure 1. The restriction to plane deformation ensures that any residual shear stresses are independent of y and that the shear stresses ρ_{xy} and ρ_{yz} are absent. Steady-state rolling ensures that any residual stresses are independent of the coordinate x in the rolling direction. Finally self-equilibrium with a traction-free surface requires that: $\rho_{zz} = \rho_{zx} = 0$.

We are left therefore with only two possible components of residual stress $\rho_{xx}(z)$ and $\rho_{yy}(z)$ which are functions of depth z only.

It should be noted, however, that the fictitious residual stresses ρ_{ij}^{*} associated with Ponter's theorem for a kinematically hardening material, since they include an unrestricted displacement α_{ij} of the yield locus, are not required to satisfy conditions of equilibrium. The fictitious stresses ρ_{zz}^{*} and ρ_{zx}^{*} are therefore not required to be zero.

ROLLING AND SLIDING CONTACT

The stresses in a plane-strain rolling contact, such as that shown in Figure 1, have been analysed by several authors. Expressions for the stress components under the action of both a normal load P and a tangential (tractive) force Q are given by Johnson [7], along with a brief table of values.

The normal pressure is given by Hertz:

$$p(x) = p_o\{1 - x^2/a^2\}^{\frac{1}{2}} \tag{2}$$

In sliding contact, with or without rolling, the tangential traction

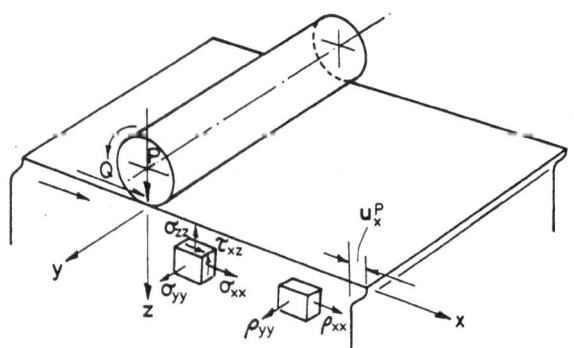

Figure 1. Rolling/Sliding contact of a cylinder with an elastic perfectly-plastic half-space.

is taken to be

$$q(x) = \mu p_o \{1 - x^2/a^2\}^{\frac{1}{2}} \tag{3}$$

where μ is the coefficient of sliding friction.

We now plot, from the above data, stress trajectories of τ_{zx}/p_o against $\frac{1}{2}(\sigma_{xx} - \sigma_{zz})/p_o$ due to the combined action of pressure p and frictional traction q for several different depths, z = constant, as shown in Figure 2 for μ = 0.2. Such trajectories trace the variations in stress experienced by material elements at different depths, as the rolling load passes over. The Tresca yield criterion, given by equation (1), maps in this figure as a circle of radius k/p_o, with its centre at the origin. The maximum Hertz stress for purely elastic behaviour p_o^y (the elastic limit) is given by the radius of the smallest circle which circumscribes the stress trajectories at all depths. For the case of μ = 0.2, shown in Figure 2, this circle has a radius k/p_o^y = 0.308, giving an elastic limit p_o^y = 3.25 k.

To apply Melan's theorem to find the shakedown limit we add freely chosen residual stresses $\rho_{xx}(z)$ and $\rho_{yy}(z)$ to the contact stresses trajectories shown in Figure 2. The stress $\rho_{yy}(z)$ is chosen to ensure that $\sigma_{a}(= \sigma_{yy} + \rho_{yy})$ is always the intermediate principal stress. Addition of a

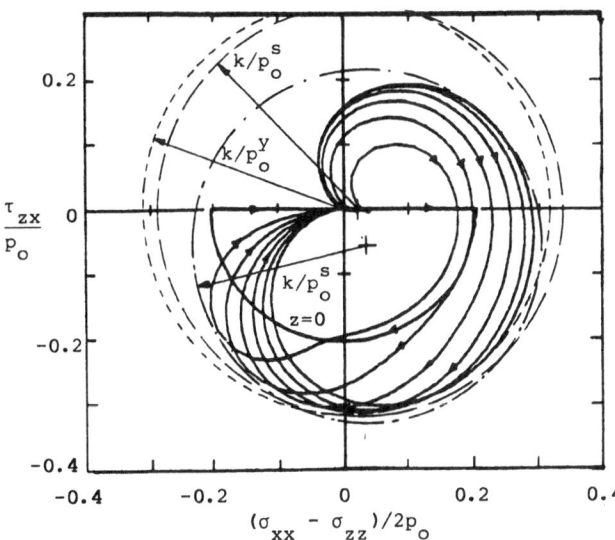

Figure 2. Stress trajectories in rolling and sliding, μ = 0.2. Yield loci: ------ First yield; ─── ─── shakedown, perf.plastic; ───·─── Shakedown, kin. hardening.

compressive $p_{xx}(z)$ has the effect of translating the trajectories in Figure 2 to the left by the magnitude of p_{xx}. The same effect would be achieved by translating the yield locus to the right by the same distance. In this way the shakedown limit p_o^s is determined by finding the smallest circle, having its centre on the horizontal axis, which circumscribes all the stress trajectories. Such a circle (drawn in Figure 2) has radius $k/p_o^s = 0.281$, giving the shakedown limit for a perfectly plastic material $p_o^s = 3.56$ k.

With a kinematically hardening material, Ponter's theorem permits the centre of the circle to be displaced by arbitrary values of p_{xx}^* and p_{zx}^*. This enables an even smaller circle of radius $k/p_o^s = 0.25$ to enclose each of the trajectories, which gives the shakedown limit $p_o^s = 4.0$ k.

It is well known that, at high friction ($\mu > 0.3$), yield first occurs at the surface ($z = 0$) rather than beneath it. The stress trajectory for $z = 0$, which takes the form of a semi-circle centred at the

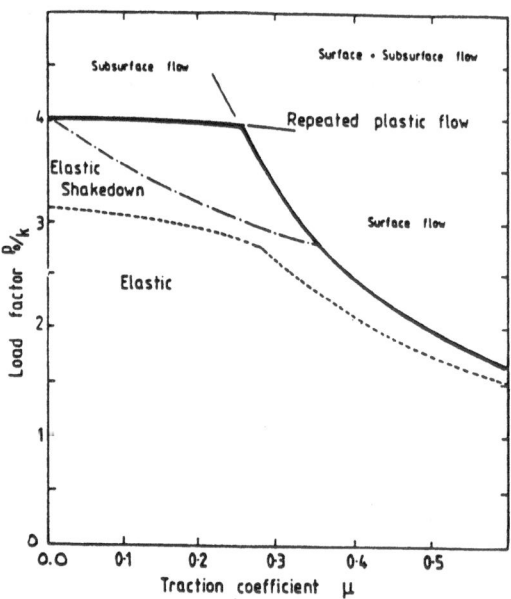

Figure 3. Shakedown map for complete slip
------ Elastic limit; ——-——Shakedown limit,el.perf.pl
——— Shakedown limit, kin. hardening.

origin and radius μ encloses all the others. The smallest circle which circumscribes this trajectory is centred at 0 and has the same radius μ. It follows, therefore, that the elastic limit and the shakedown limits are all equal whether or not the material is capable of hardening.

i.e.
$$p_o^y = p_o^s = k/\mu \qquad (4)$$

These results were obtained by Johnson and Jefferis [8]. The elastic limits and shakedown limits are plotted against traction coefficient in Figure 3, which clearly shows the reduction in cyclic load capacity with increasing traction coefficient above the critical value of 0.25. The shakedown limit is then governed by the range of σ_{xx}. The non-proportional cycle of stress experienced by a surface element: tension $(+\sigma_{xx})$ followed by orthogonal shear (τ_{zx}) followed by compression $(-\sigma_{xx})$ and represented by the semi-circular trajectory in Figure 2, is thought to be particularly damaging (see Bower and Johnson [10]).

TRACTIVE ROLLING WITH PARTIAL SLIP

When rolling takes place with a tangential force Q less than limiting friction μP, as in the driving wheel of a vehicle for example, 'microslip' occurs over part of the contact area, while there is no slip over the remainder. This problem was first analysed by Carter [11], who showed that microslip took place towards the trailing edge of the contact region, and that the stresses in the contacting bodies could be found by the superposition of a tangential traction

$$q' = \mu p_o \{1 - x^2/a\}^{\frac{1}{2}} \qquad (5a)$$

acting over the complete contact $(-a \leqslant x \leqslant +a)$ together with a traction

$$q'' = - \mu p_o \{(c^2/a^2) - (x + d)^2/a^2\}^{\frac{1}{2}} \qquad (5b)$$

acting over the strip $(- a \leqslant x \leqslant 2c - a)$. This distribution of traction is shown in Figure 4, together with the variation of surface stress σ_{xx}. Note that $c = a - d$ and

$$Q/\mu P = 1 - c^2/a^2 = (d/a)(2 - d/a) \qquad (6)$$

Tractive rolling was investigated at Nottingham by Haines and Ollerton [12], again using the technique of frozen stress photoelasticity, which gave good support for stress distribution shown in Figure 5.

Low friction

Provided that the coefficient of friction is below a critical value μ_c, first yield and shakedown are governed by subsurface stresses. Subsurface stress trajectories similar to those shown in Figure 2 have been computed and then elastic and shakedown limits found as before. The results are plotted in Figure 6 (perfectly plastic) and Figure 6 (kinematically hardening) against the traction coefficient Q/P for different values of the coefficient of friction μ.

The results are not very different from those presented in Figure 2 for complete slip. This is not surprising since the <u>subsurface</u> stresses are not likely to be much influenced by the <u>distribution</u> of surface traction.

High friction

With high friction, when plastic deformation initiates at the surface, the behaviour is influenced significantly by partial slip.

The normal pressure p(x) gives rise to equal biaxial compression at the contact surface: $\sigma_{xx} = \sigma_{zz} = - p(x)$. To examine yielding in the z-x plane, therefore, this stress state can be ignored and only stresses due to the shear traction q(x), shown in Figure 4, need to be examined. They may be expressed:

In the no-slip zone, $-a \leq x \leq a - 2d$

$$\frac{\sigma_{xx}}{\mu p_o} = \frac{2d}{a} \qquad ; \qquad \frac{\sigma_{zz}}{\mu p_o} = 0$$

$$\frac{\tau_{zx}}{\mu p_o} = - q = - \{1 - x^2/a^2\}^{\frac{1}{2}} + \{(c^2/a^2) - (x + d)^2/a^2\}^{\frac{1}{2}}$$

(7)

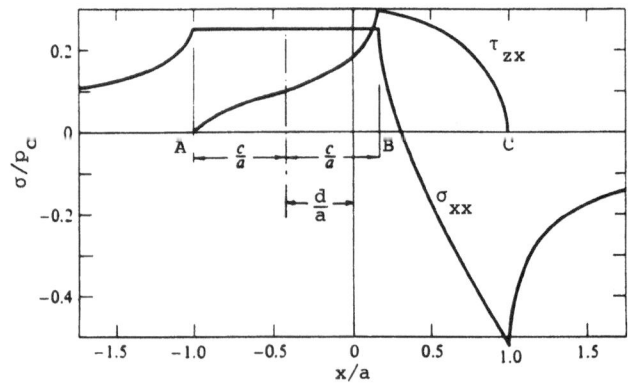

Figure 4. Partial slip: surface stresses.

In the slip zone, $a - 2d \leq x \leq a$

$$\frac{\sigma_{xx}}{\mu p_o} = \frac{2d}{a} - 2 \; \{\frac{(x + d)^2 - (a - d)^2}{a^2}\}^{\frac{1}{2}} \; ; \qquad \frac{\sigma_{zz}}{\mu p_o} = 0 \qquad (8)$$

$$\frac{\tau_{zx}}{\mu p_o} = - \; \{1 - x^2/a^2\}^{\frac{1}{2}}$$

Referring to Figure 4,

At A $(x = -a)$: $\qquad \dfrac{\sigma_{xx}}{\mu p_o} = 2 \; d/a; \qquad \dfrac{\sigma_{zz}}{\mu p_o} = \dfrac{\tau_{zx}}{\mu p_o} = 0;$

At B $(x = a - 2d)$: $\quad \dfrac{\sigma_{xx}}{\mu p_o} = 2 \; \dfrac{d}{a} \; ; \qquad \dfrac{\sigma_{zz}}{\mu p} = 0; \qquad \dfrac{\tau_{zx}}{\mu p_o} = - \; 2 \; \{\dfrac{d}{a} \; (1 - \dfrac{d}{a})\}^{\frac{1}{2}};$

At C $(x = a)$: $\qquad \dfrac{\sigma_{xx}}{\mu p_o} = 2 \; \dfrac{d}{a} - 4 \; (\dfrac{d}{a})^{\frac{1}{2}}; \qquad \dfrac{\sigma_{zz}}{\mu p_o} = \dfrac{\tau_{zx}}{\mu p_o} = 0;$

where $d/a = 1 - \{1 - Q/\mu P\}^{\frac{1}{2}}$.

A typical stress trajectory for a surface element is shown in Figure 5, where the letters A,B, and C correspond to the stress states at A, B and C in Figure 4. It is then a simple matter to draw in the circumscribing circles which represent the shakedown states.

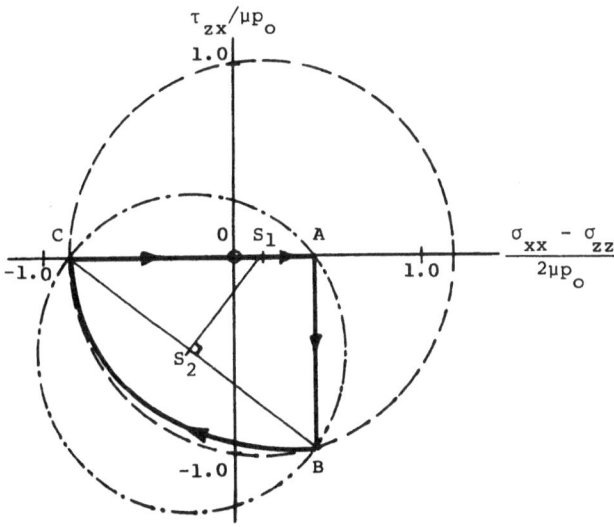

Figure 5. Surface trajectory A B C for partial slip,
— — — — Yield locus for perfect plastic solid.
—— · —— Yield locus for kin. hardening solid.

For a perfectly plastic material the centre of the circle S_1 lies on the horizontal axis, displaced from O by the residual compressive stress ρ_{xx}. The radius of the circle is thus given by

$$k/\mu p_o^s = CS_1 = (\tfrac{BC}{2})^2 / (\tfrac{AC}{2}) = (\tfrac{d}{2})^{\frac{1}{2}}(2 - \tfrac{d}{a})$$

so that the shakedown limit is given by

$$p_o^s/k = \{\mu(d/a)^{\frac{1}{2}}(2 - d/a)\}^{-1} \qquad (9)$$

Equation (9) gives p_o^s/k as a function of $Q/\mu P$. The shakedown limits are plotted against Q/P in Figure 6a for various values of μ.

Similarly, for a kinematically hardening material the centre of the circumscribing circle S_2 is free to displace by arbitrary values of ρ_{xx}^* and ρ_{zx}^*. Its radius is thus given by

$$k/\mu p_o^s = BC/2 = \{(\tfrac{d}{a})(2 - \tfrac{d}{a})\}^{\frac{1}{2}}$$

so that $\qquad\qquad p_o^s/k = \{(\tfrac{d}{a})(2 - \tfrac{d}{a})\}^{-\frac{1}{2}} \mu^{-1} \qquad (10)$

In the same way equation (10) enables the shakedown limits to be plotted as a function of Q/P and μ in Figure 6b.

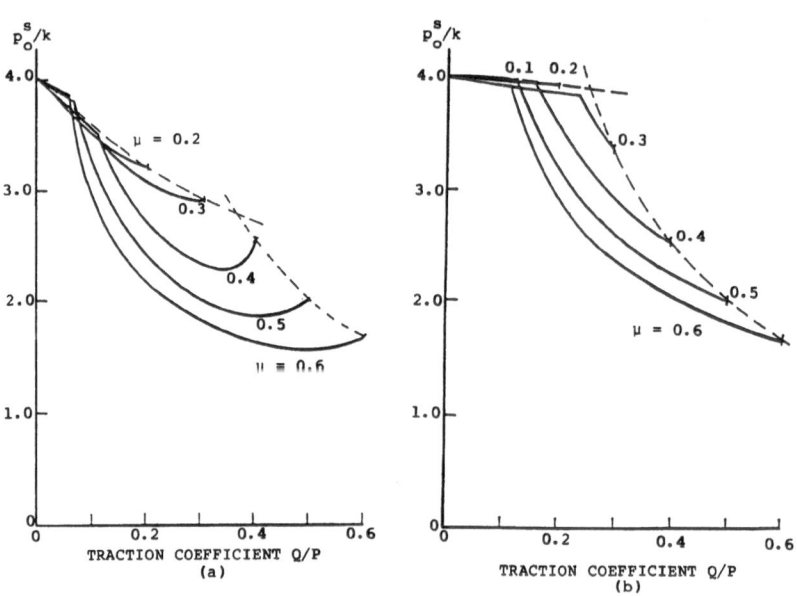

Figure 6. Shakedown maps for partial slip: (a) El.Perfect-plastic; (b) Kin. hardening.

It may be shown that, for the range of parameters displayed in Figures 3 and 6, out of plane residual stress ρ_{yy} or ρ_{yy}^* develop to ensure that $\sigma_3 = \sigma_{yy} + \rho_{yy}$ or $\sigma_{yy} + \rho_{yy}^*$ is the intermediate principal stress. Yield is then governed by equation (1) in terms of the in-plane stress plotted in Figures 2, 4 and 5, and lateral plastic flow is prevented. (The results in this section are new).

CONCLUSION

Use of the Tresca yield criterion permits a simple graphical approach to shakedown in rolling contact which has been presented in this paper. Modelling the material as an elastic-perfectly plastic solid in plane deformation shows the role of residual stresses alone in promoting shakedown. Modelling the material as a kinematically hardening solid shows how cyclic hardening properties of the material can, in some but not all circumstances, increase the shakedown limit. The diagram, in which the trajectories of stress experienced by material elements and the yield loci are represented, provides a clear visual indication of the influence of both residual stresses and hardening upon shakedown.

The method of this paper reproduces a shakedown map (Fig. 3) for rolling with complete slip, which is already in the literature. The situation of tractive rolling with partial slip, as found with a locomotive driving wheel, has been examined and new shakedown maps obtained. These results demonstrate the damaging effect (reduction in the shakdown limit) of a high coefficient of limiting friction μ, even if the traction coefficient Q/P at which the contact is operating remains rather modest.

ACKNOWLEDGEMENT

The author is grateful to Dr H R Shercliff for carrying out some of the computations.

REFERENCES

1.   Fessler, H., and Ollerton, E., Contact stresses in toroids under radial loads, Brit.J.Appl.Phys., 1957, E, 387-393.

2. Lundberg, G., and Palmgren, A., Dynamic capacity of rolling bearings, Acta Polytechnica, Stockholm, Mech.Eng.Series 1947, 1, 1-50.

3. Crook, A.W., Simulated gear-tooth contacts: some experiments on their lubriction and subsurface deformation, 1957, Proc.Inst.Mech.Engrs. 171, 187-214.

4. Ziegler, H., A modification of Prager's hardening rule, Quart.of Appl.Math. 1959, 17, 55-65.

5. Melan, E., Der spannungsgudstand eines Henky-Mises schen Kontinuums bei Verlandicher Belastung, Sitzungberichte der Ak. Wissenschaften Wien, Ser. 2A, 1938, 147, 73.

6. Ponter, A.R.S., A general shakedown theorem for elastic-plastic bodies with work hardinng, 3rd Int.Conf. on Structural Mechanics in Reactor Tech., London, 1976.

7. Johnson, K.L., Contact Mechnics, C.U.P., Cambridge 1985, pp. 103-205 and 429.

8. Johnson, K.L., A shakedown limit in rolling contact, Proc. 4th US Nat.Conference of Appl.Mech., Berkeley, 1962, ASME.

9. Johnson, K.L., and Jefferis, J.A., Plastic flow and residual stresses in rolling and sliding contact. Proc.Inst.of Mech.Engrs. Symposium on Rolling Contact Fatigue, London, p. 50.

10. Bower, A.F., and Johnson, K.L., The influence of strain hardening on cumulative plastic deformation in rolling and sliding line contacts', 1989 J.M.P.S., 37, 471-493.

11. Carter, F.W., On the action of a locomotive driving wheel, Proceedings, Royal Society, A112, 151.

12. Haines D.J. and Ollerton, E., Contact stress distributions on elliptical contact surfaces subjected to radial and tangential forces, Proceedings, Institution of Mechanical Engineers, 177, 95.

THE USE OF SHAKEDOWN CONCEPTS IN THE DEVELOPMENT OF DESIGN RULES

FOR SHELL STRUCTURES SUBJECTED TO SEVERE CYCLIC THERMAL LOADING

K.F. CARTER, A.C.F. COCKS, A.R.S. PONTER AND R.J.M. VENESS
Engineering Department, University of Leicester, Leicester, U.K.

ABSTRACT

From the shakedown and ratchetting limit analysis of a wide range of
thermal loading problems for axisymmetric thin shells it is evident that
only a limited number of possible mechanisms of incremental collapse
occur. As a result it is possible to develop simplified analysis
procedures that can be used directly in the design process. This paper
describes two such procedures. The first of these has evolved from the
examination of the full range of interaction diagrams that are available,
classifying them into groups and fitting simple equations to the shakedown
boundaries. It has emerged that a simple conservative estimate of the
elastic and plastic shakedown boundaries can be obtained by drawing a
straight line on an interaction diagram whose intercept with the thermal
loading axis is determined from the peak stress and the maximum membrane
thermo-elastic stress. The second approach involves analysing each of the
limited number of possible collapse mechanisms using the upper bound
kinematic shakedown theorem. For each mechanism it is possible to obtain a
simple expression for the shakedown boundary in terms of the major
features of the thermo-elastic stress history which can be used as a basis
of design code limits for thermally loaded thin shell structures. The
extension of these techniques to creep effects is also discussed.

INTRODUCTION

For some time it has been recognised that the current design code methods for severe cyclic thermal loading situations are, in some situations, unsafe whilst being overly conservative in others. In response to this an upper bound kinematic theory has been developed to find the boundaries between different types of structural behaviour under these loads. In particular the combinations of thermal and mechanical loading at which the onset of ratchetting (cyclic strain growth) behaviour of the structure occurs can be found, together with the mechanism of plastic deformation at the boundary.

This upper bound kinematic theory has been implemented by finite element discretization of a structure and solution for the lowest possible upper bound, which may include the exact solution, using linear programming techniques. The essential assumptions employed in the calculation are a Tresca yield surface, perfect plasticity and linear variation of plastic multipliers (and thus strains) within the finite elements. The yield surface used for these axisymmetric thin shell cases is two-dimensional in which the bending moments are disposed into plastic hinges at the nodal intersections between finite elements. The computer program takes as input the geometry of the axisymmetric thin shell, the material data, the temperature history and the thermo-elastic stress calculated by an elastic finite element program kindly supplied by the UKAEA.

Interaction diagrams show the areas corresponding to different regimes of structural behaviour for a given structure and thermal loading history, the x-axis being defined as the normalised constant mechanical load C and the y-axis as the normalised thermal load T. The constant mechanical load required to produce the onset of ratchetting as a function of the magnitude of the thermal load is calculated by means of the upper bound method of Ponter and Carter [1] by linear scaling of the thermal loading history. The other boundaries between the regimes of structural behaviour are also found as a by-product of the calculation. Four regimes of structural behaviour can be found within an interaction diagram, namely

Elastic (E): In this region any combination of mechanical load and thermo-elastic stress is always within the plastic yield surface throughout the structure and thus deformations are purely elastic.
Shakedown (S): Here the total loading initially exceeds the plastic yield

surface somewhere in the structure, but the thermo—elastic stress itself is less than twice the yield stress of the material. The effect is that for the first few loading cycles plastic strain growth occurs. However a residual stress field develops due to these strains, so that after a number of cycles the behaviour of the plastically deformed structure becomes fully elastic. Inside the shakedown region there is no low cycle fatigue.

Plastic shakedown (P): In this region the thermo—elastic stresses exceed twice yield, and plastic strains occur at two or more instants in the thermal loading cycle, but accumulate to zero strain over the complete cycle. Typically this can happen when a region of material, whose thermo-elastic stresses cannot be accommodated within the plastic yield surface, is prevented from deforming plastically by surrounding material in the elastic region. In the plastic shakedown region the lifetime of the structure is determined by fatigue.

Ratchetting (R): Here the total loading exceeds the plastic yield surface and the structure exhibits incremental cyclic plastic strain growth which settles to a constant rate after a few cycles. The usual design criteria is that the structure shall not ratchet. However if this is unavoidable, the lifetime of the structure is then determined by the number of loading cycles for which the accumulated plastic strain remains within a given tolerance.

Typical shapes of interaction diagrams for different thermal loading cases are shown in Figs 1 to 3. In Fig.1 the through thickness temperature gradient is completely dominant, thus the ratchetting boundary is close to the Bree line [2] and the lifetime of the structure at high thermal and low mechanical loads is determined by fatigue. In Fig.2 the axial temperature gradient is dominant resulting in ratchetting at zero mechanical load for moderate or high levels of thermal load where the plastic shakedown region may be small or absent. Fig.3 shows the effects of movement of a predominantly axial temperature gradient as might happen when the level of a liquid sodium pool rises in a fast reactor. Ratchetting occurs at low levels of thermal load because the maximum thermo—elastic stresses are swept along a significant length of the structure rather than being localised at a point. It should be emphasised that interaction diagrams are unique for each structure and cyclic loading history, although there is information within the diagram on the likely behaviour of the structure when the loading conditions are changed.

The use of interaction diagrams in design codes for the prevention of ratchetting and the estimation of creep deformation and rupture (lifetime) are discussed in the following sections.

278

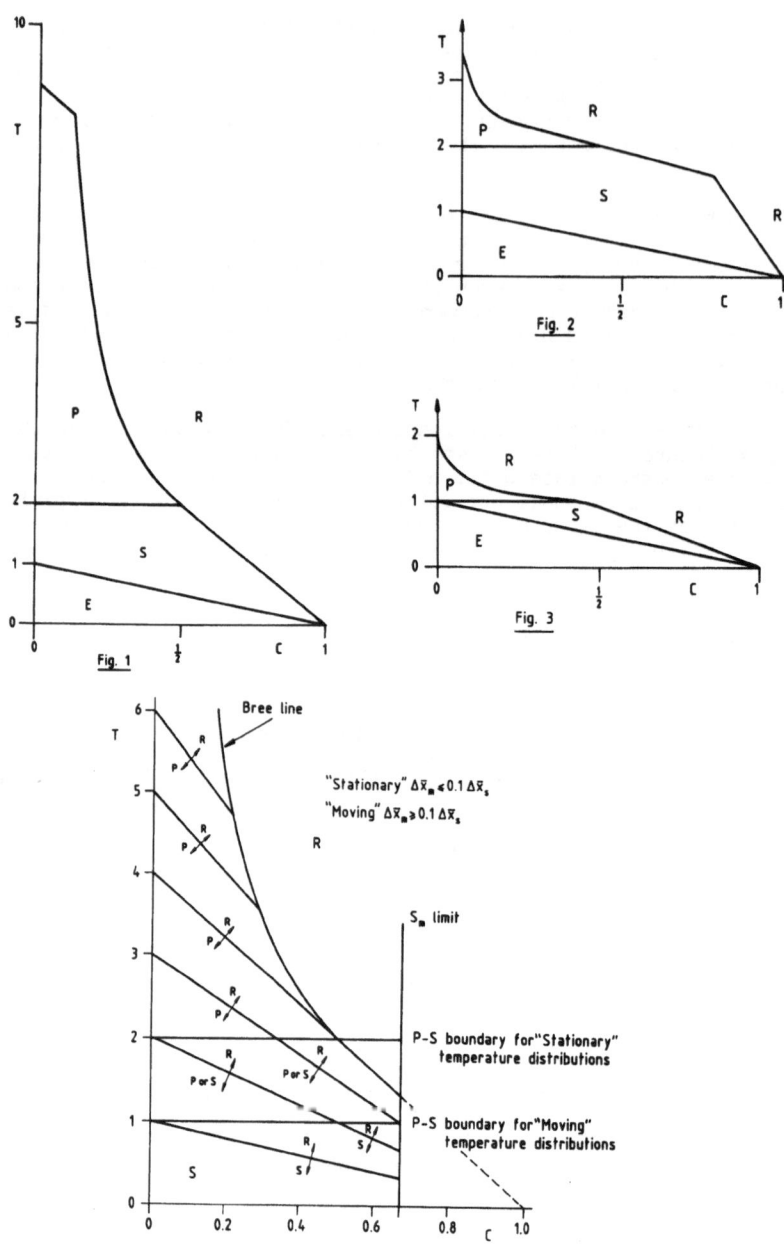

Fig. 1

Fig. 2

Fig. 3

Fig 4

"Stationary" $\Delta \bar{x}_m \ll 0.1 \Delta \bar{x}_s$
"Moving" $\Delta \bar{x}_m \geqslant 0.1 \Delta \bar{x}_s$

Bree line

S_m limit

P-S boundary for"Stationary" temperature distributions

P-S boundary for"Moving" temperature distributions

R-Freedom from ratchetting cannot be demonstrated
P-No ratchetting, but local reverse plastic straining
S-No ratchetting, shakedown to elastic behaviour.

SIMPLIFIED INTERACTION DIAGRAM APPROACH

The simplified interaction diagram design code method is based on a theoretical understanding of structural behaviour in a wide range of circumstances, which is confirmed by the available experimental evidence. From this, areas of thermo-mechanical loading have been deduced which lead to different forms of structural behaviour. For instance where through thickness temperature gradients are dominant, the behaviour will be Bree type, whilst moving severe axial temperature gradients can reduce the ratchetting boundary to the elastic line.

The temperature histories found in practice can be idealised into combinations of axial and through thickness temperature gradients, where the shape of the axial temperature distribution can be approximated as ramp like or spike like. These shapes correspond to the most commonly found types of temperature distribution within fast reactor structures. Parametric studies of mechanical and thermal loading of cylindrical thin shells, which are the most commonly found structures, have been carried out using combinations of these idealised axial and through thickness temperature distributions. The resultant interaction diagrams have been systematically categorised, and then simplified (in a conservative manner) into a single line, whose characteristics are determined by the thermal loading conditions. This approach is thought likely to be able to cover at least two-thirds of the design cases that are likely to arise in practice.

For the purpose of design codes approximate interaction diagrams can be constructed from this single line, by matching the real temperature history to the ramp like and spike like idealisations of the temperature distribution. The criteria for matching is the maximum rate of change of temperature found in the axial and through thickness directions anywhere in the structure. This is because the deformation mechanism is usually localised at the point of maximum thermo-elastic stress in the structure which normally coincides with the place where the temperature is most rapidly varying. Alternatively the maxima of the thermo-elastic stress distribution may be used directly, if it is known. This single line approximation to the ratchetting boundary is likely to be most accurate at high thermal load, low mechanical load (e.g. typical of fast reactor loading conditions) and at low thermal load, high mechanical load (e.g. typical of pressure water reactors and other types of steam plant).

In situations where the cyclic temperature gradient is in the through-wall direction ONLY (i.e. the Bree type situation where there is no thermal membrane stress at any time during the operating cycle) the Bree line is used as the boundary at which the onset of ratchetting behaviour occurs. The Bree line is defined as

$$C = 1 - T/4 \qquad\qquad T \leq 2$$

$$C = 1/T \qquad\qquad T \geq 2$$

In this case the boundary between the shakedown (S) and plastic shakedown (P) regions occurs at T = 2.

Anywhere axial cyclic temperature gradients (ramp or spike) are present, with or without through-wall temperature gradients (i.e. there are thermal membrane stresses present, however small), the ratchetting boundary is further restricted by a straight line from (1,0) to a point (0,F). The boundary at which the onset of ratchetting occurs is then the minimum value of C for a given value of T between the Bree line and the straight line as shown in Fig.4.

The parameter F is a measure of the severity of the effects of the axial temperature variation. As a guide, a value of F = 1 gives the elastic line, which remains safe in the severest possible cases, whereas a value of F greater than 6 indicates a transition back towards Bree like behaviour. Pure Bree type behaviour is equivalent to an F value of infinity, indicating that there are no thermal membrane stresses encountered in the operating period.

Should the range of oscillation $\Delta \bar{x}_m$ of an axial temperature ramp be greater than 10% of the ramp length $\Delta \bar{x}_s$ (i.e. the temperature distribution can be classed as a moving temperature front), then the boundary between the shakedown (S) and the plastic shakedown (P) regions moves from T = 2 to T = 1. The line defining the upper limit of elastic behaviour (i.e. the elastic line) within an interaction diagram is simply a straight line between (0,1) and (1,0).

The onset of ratchetting behaviour is principally determined by the relative amounts of thermal membrane and bending stresses in the axial and hoop directions. The value of F can be calculated from the results of an elastic finite element stress analysis of the shell for the operating period using

$$R = \sigma_t / \sigma_m$$

where σ_t is the elastically calculated maximum value of the thermal stress (Tresca) within the shell during the operating period and σ_m is the corresponding maximum value of the membrane component of the thermal stress. The value of R may be increased by a factor of 2, for 'stationary' temperature distributions. The F factor is then given by

$$F = 0.9 \ R \qquad\qquad\qquad R < 5/2$$

$$F = 0.9 \ \sqrt{(5R/2)} \qquad\qquad R > 5/2$$

This design code procedure [3] gives a reasonable and conservative estimation of the ratchetting boundary over the entire range of thermal loading situations, within the confines of the necessary restrictions placed on the use of this method. The disadvantage of the method is that so far it is limited to tube type structures and that because the method is conservative it will tend to overestimate creep effects.

MECHANISMS APPROACH

It is obviously not practicable to carry out parametric loading studies on every type of axisymmetric thin shell subject to every type of possible thermal loading history. Thus an alternative approach to the problem has been devised by Ponter and Veness [4] based on the observation that in all the thermal and mechanical loading histories for axisymmetric thin shells, using the Tresca yield surface, a very limited number of mechanisms of plastic deformation along the ratchetting boundary have been observed. The design rules can then be formulated in terms of the characteristics of the thermo-elastic stresses and primary load which activate each of these mechanisms. The accuracy of the interaction diagrams calculated by this method is approaching that of the full upper bound finite element calculation although care must be taken to ensure that all possible solution mechanisms have been examined.

For particular cases, an interaction diagram is constructed as a locus of individual mechanism lines, where the ratchetting boundary follows the minimum load for all possible mechanisms consistent with the primary load. The equations for each of the mechanism lines can be translated into standard design code form, which gives a system of several inequalities all of which must be satisfied for the prevention of ratchetting.

In the case of cylindrical structures, it is found that the sub-class of mechanisms of deformation is limited to only four possibilities. Two of these mechanisms involve only the extreme values of the thermo-elastic stress at a particular point in the structure for a highly localised deformation mechanism. The other two mechanisms involve deformation over an intermediate length, requiring the calculation of a mean value of the maximum hoop stress over this length.

As the ratchetting boundary calculated by this approach is more accurate than the simplified interaction diagram method, the estimation of the effects of creep on the structure, which relies in part on a ratchetting boundary as input, is considerably better.

CREEP STRAIN ESTIMATION

Techniques have been developed at Leicester University for the prediction of creep effects, with particular reference to creep deformation and creep rupture (i.e. the assessment of a component lifetime). These techniques are based solely on the use of interaction diagrams and simple uniaxial test data. A conservative estimation of accumulated creep strain can be obtained using an appropriate interaction diagram by the method of Ponter and Cocks [5]. The technique is to obtain values of the creep reference stress and reference temperature which may be applied to isochronous creep curves to give the accumulated creep strain in a given amount of time or vice versa. The assumptions are the same as those used in the calculation of the interaction diagram together with the assumption of rapid thermal cycling, which is conservative, and that the creep properties of the material are independent of temperature, above the temperature at which creep first starts to happen.

In the Ponter and Cocks method it is reasoned that any point on the ratchetting boundary must give a value of creep reference stress equal to the yield stress of the material for the corresponding value of the reference temperature. Consequently the actual creep reference stress for a particular loading point will be a linear scaling of that point from the origin to the ratchetting boundary in a straight line.

The mechanism of creep deformation will be the same as that of the corresponding point on the ratchetting boundary. This observation is then used to find a value for the creep reference temperature. This is done by integrating the maximum value of the temperature at any point within the structure $T_{max}(\underline{x})$ during the thermal loading cycle, over the region in which the plastic deformation strain ε^p occurs.

$$T_r = \frac{\displaystyle\int_V T_{max}(\underline{x}) \; d\varepsilon^p \; dV}{\displaystyle\int_V d\varepsilon^p \; dV}$$

In practise it is found that the value of the reference temperature calculated using the above formula is almost always the maximum temperature found anywhere within the structure. The main exception to this is where the thermal loading is Bree like (i.e. thermal bending stresses only) when the reference temperature takes the mean value of the through-wall temperature.

The values of the creep reference stress and reference temperature are then used together with the appropriate isochronous creep curves for the material to obtain the accumulated creep strain in a given time or the time for the structure to amass a given amount of creep strain.

CONCLUSION

Both design code methods are based a theoretical understanding of structural behaviour in a wide range of thermal loading situations, which is confirmed by the available experimental evidence. The simplified interaction diagram approach provides a very useful way of screening possible structures for potential ratchetting and creep problems in the early stages of design, although in the long term the mechanisms method may prove more accurate and complete provided that the classes of mechanisms are properly identified and solved conservatively.

ACKNOWLEDGEMENTS

The authors would like to thank the United Kingdom Atomic Energy Authority (UKAEA), the Science and Engineering Research Council (SERC) and the European Economic Community (EEC) for their interest and funding in this research over a period of years, without which the present levels of understanding of thermal loading phenomena would have been unobtainable.

REFERENCES

1. Ponter, A.R.S. and Carter, K.F. Upper Bound Methods for use in Design and Assessment of Axisymmetric Thin Shells subjected to Cyclic Thermal Loading.
   Nucl.Eng.Design 1989, 116, 239

2. Bree, J. Elasto-Plastic behaviour of Thin Tubes subjected to Internal Pressure and Intermittent High Heat Fluxes with application to Fast Nuclear Reactor Fuel Elements.
   J.Strain Analysis 1967, 2, 226

3. Ellington, J.P., Love, J.B., Porter, A.G., Carter, K.F. and Ponter, A.R.S. Rules for the Prevention of Ratchetting (Cyclic Strain Growth) - Including Creep
   Report AGT9B-90-32(UK), Issue 2,
   UKAEA, Risley, U.K.

4. Ponter, A.R.S. and Veness, R.J.M. (to appear)

5. Ponter, A.R.S. and A.C.F.Cocks, A.C.F. Computation of Shakedown Limits for Structural Components (Brussels Diagram) Part II - The Creep Range
   Final Report, Contract RAP-066-UK(AD),
   Commission for the European Communities Directorate-General for Science, Research and Development,
   Leicester University, U.K. (to appear)

ARTICULATING PIN FIXED JOINTS

L D McCONNELL
Consultant, GEC ALSTHOM Engineering Research Centre
Stafford ST17 4LN

ABSTRACT

Joints comprised of lugs connected by a single pin in shear are widely used to transmit loads. Such joints fall into two main categories.

The first provides a rigid connection between two members with the pin sitting symmetrically within the load path. The joint may be subjected to fatigue but the fluctuating loads do not deviate from the mean path.

The second type allows a measure of articulation between the two members. This produces asymmetric load distribution within the lugs even when the axial load is sensibly constant.

This paper reviews the influence of geometric variables on stresses induced by axi-symmetric loading and goes on to develop a stress amplitude analysis for the articulating joint case.

The theme is developed to examine three practical applications and shows how design may be optimised for each case.

INTRODUCTION

Pinned joints are widely used in a variety of engineering design applications. They have, in consequence, attracted a large number of technical papers.

Published data fall into three main groups dealing with theoretical or experimental stress analyses, or with fatigue assessments.

Almost without exception attention has been restricted to the case in which the load path is normal to, and disposed symmetrically about, the pin.

In many applications, however, a degree of articulation occurs between the two joined members. It is the purpose of this paper to review information on the axi-symmetric load case and to use it to optimise lugs which are subjected to transverse, ie articulating, movements.

NOTATION

a	Lug head distance from pin hole centre
C	Diametral clearance of pin = $2(R_H - R_p)$
d	Pin hole diameter = $2R_H$
D	Lug width
e	Load eccentricity
M_B	Mass of blade
M	Bending moment across section
N	Big end shell nip load
P	Big end bolt load
R_B	Spinning radius of blade centre of gravity about engine axis
R_c	Crank pin radius of action
R_H	Radius of pin hole = $d/2$
R_p	Radius of pin
S	Shear load across joint face
S_I	Inertial Shear along joint face
S_B	Bolt load shear along joint face
T	Inertial Tension across joint face
W	Inertial force due to piston, gudgeon pin and con rod
α	Position of contact point, see figure 2
β	Big end joint base angle, see figure 6
δ	Shortening of lug centre distance
η	Position at which bending moment is considered, see figure 7
θ	Articulation angle, see figure 2
μ	Coefficient of friction
Φ	Set back angle, see figure 6
Ω	Rotational angular velocity Rad s^{-1}

THE AXI-SYMMETRIC LOAD CASE - STRESSES

The classic starting point for this study is the paper by Frocht and Hill [1] published in 1940. Subsequently numerous other analytical and fatigue studies have been carried out to identify optimum lug proportions.

It is not the purpose of this paper to repeat the findings of others and these have been usefully consolidated in such sources [2] as ESDU Item No 81006. However, these studies usually consider pin clearances <0.25% and illustrate a marked effect of variations up to this value.

With any degree of clearance the stresses at the minimum section are not linear with load due to changing conformity of pin and hole periphery. In experimental investigations it is, therefore, important to use strains during testing which are representative of those in the prototype. In the following section of this paper it will be shown that pin clearances well in excess of 0.25% may be desirable when the lug articulates.

The author has analysed a large number of his own photoelastic results for the axi-symmetric load case and has produced a carpet plot given in figure 1. It covers a practical design area with d/D between 0.4 and 0.6, a/D between 0.45 and 0.6 and with pin clearances up to 5%.

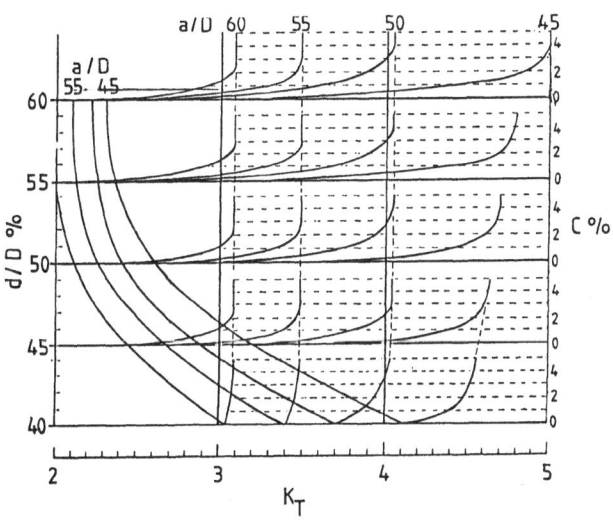

Figure 1. Stress Concentration Factors at Pin Hole.

Lug thickness also has a significant effect because of the resulting maldistribution of load along the hole axis due to pin bending. ESDU 81006 gives correction factors which range from 1.15 when the thickness equals the pin hole diameter to 1.9 when the thickness is twice the hole diameter. These factors apply to a typical mid-range of lugs in which the pin is supported on either side.

In the case of a forked joint the male member has the pin supported on one side only and the distribution factor is likely to be increased by some 20% or even more. This can be compensated by making the thickness of the male supports each equal to 75% of the female lug thickness when the two components are made from materials of equal tensile or fatigue strength.

For lightness an alternative solution is to make a multi-shear joint but this is expensive.

Throughout the published data little or no emphasis has been placed on the importance of radiusing or chamfering the ends of the pin hole. This operation reduces hard-bedding of the pin and improves the fatigue susceptibility of the high stress areas. A chamfer must be carefully controlled for quality if it is to be fully effective.

It is the author's experience that lugs with large clearance pins have optimum strength when d/D approaches 0.60.

Lugs with interference fitted bushes and used with large clearance pins can be approximated to a 1% clearance for stress purposes.

THE ARTICULATING LUG

When a pinned joint articulates each lug element rotates around the pin with a rolling action until a limiting angle, defined by friction, is reached. During this rolling action the load line becomes displaced from the geometric centreline and so disrupts the symmetry of the stress distribution.

Beyond the friction limited value slip takes place.

In figure 2 differences in hole and pin radii are exaggerated for clarity.

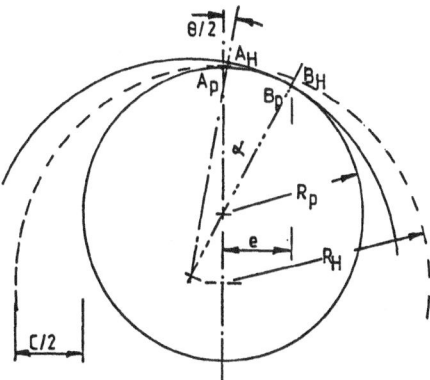

Figure 2. Rolling Action at Pin

If the pin is considered fixed in space and one element rotates through an angle $\theta/2$ clockwise then the other element will similarly move anticlockwise. From the initial central position at which points A_H and A_p are coincident the contact point moves to B. The position of the contact point is defined by

$$a = \theta R_H/2 \ (R_H - R_p) \tag{1}$$

and a load eccentricity results such that

$$e = R_p \ Sin \ [\theta R_H/2 \ (R_H - R_p)] \tag{2}$$

θ and hence e can continue to grow until

$$Tan^{-1}\mu = \theta R_p/2 \ (R_H - R_p) \tag{3}$$

and

$$\theta max = 2(R_H - R_p) \ R^{-1}_p \ Tan^{-1}\mu \tag{3a}$$

$$emax = R_p \ Sin \ (R_H \ R^{-1}_p \ Tan^{-1}\mu) \tag{2a}$$

From 3(a) we note that a close fit pin cannot articulate without slipping and that the articulation increases with increasing clearance. To be able to articulate without slipping improves the efficiency and the wear life of the joint.

Two further aspects of the geometric influence of pin clearance should be considered.

The first is that the stress on the underside of the lug (the bursting stress) is usually smaller than that at the pin hole under uniaxial loading. In the case of an articulating joint the stress here has a double frequency component as the joint moves through a $\pm\theta$ cycle. In the nature of lug manufacturing techniques this bursting stress usually lies across the grain flow whereas the pin hole stresses are coincident with the grain flow. This affects the fatigue strength and requires that the bursting stresses within the range of geometries covered by ESDU 81006 may become significant.

The second further consideration is that when a joint articulates with pin rolling the distance between the centres of the joint elements is reduced by an amount δ where

$$\delta = 2(R_H - R_p)[1 - \text{Cos } (\theta R_H/2(R_H - R_p))] \tag{4}$$

This can be used to advantage when work is required to be dissipated in the joint as in the damping of a vibrating system.

APPLICATION I - PIN FIXED COMPRESSOR BLADE

The pin fixed lug as a support member for compressor blading used extensively in the past. Its chief attribute is to eliminate the node at the base of the aerofoil under first cantilever excitation. This case has been extensively studied by Goatham and Smailes [3].

They pointed out that the pin jointed root permits rolling and the geometry can be used to control the frequency of the excitation response. The offset load produces a torque about the pin axis which is, in effect, a variable stiffness spring in which the stiffness can be defined as

$$\frac{\text{Torque about pin axis}}{\text{Angle of blade roll}} = \frac{m \, R_B \Omega^2 R_p{}^2}{2(R_H - R_p)} \tag{5}$$

At low rotational speeds or when the clearance is large the stiffness of the root is small and tends to zero. Thus the frequency of the blade in its fundamental mode also tends to zero. The frequency will be proportional to the blade speed since the blade attachment represents its only stiffness. With increasing rotational compressor speed the stiffness and hence the frequency increases but will always remain below that of a fixed root blade.

This application of a pinned joint is therefore vitally dependent on its geometry and the vibration characteristics interact with the stress analysis.

APPLICATION 2 - COMPACT JOINT

In a particular application an extremely compact articulating joint was required and this problem was solved by reducing the normally round pin to a lenticular form as shown in figure 3.

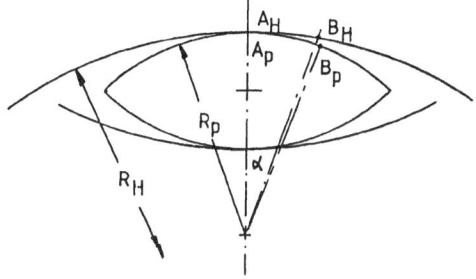

Figure 3. Lenticular Pin.

Because the bending stiffness of the pin is much reduced it was compensated by introducing a multi shear arrangement, figure 4.

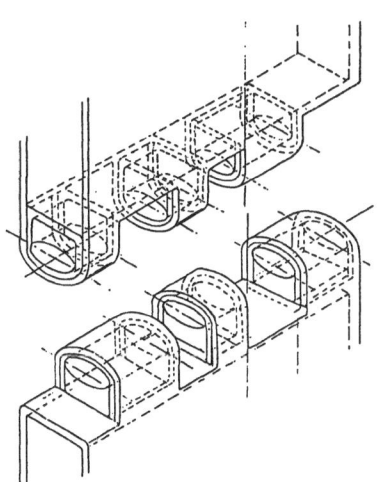

Figure 4. Compact Pinned Joint.

The unequal lug 'thicknesses' are deliberately introduced to restrict assembly to one configuration only.

All stresses in the system can be determined by the analyses given earlier in this paper.

APPLICATION 3 - CONNECTING-ROD

Each end of a conventional connecting-rod is a pin fixed lug. The load pattern is cyclic and directed generally along the axis of the rod due to inertia of the reciprocating components. In a four stroke engine the maximum tensile load occurs at the top of the scavenge and into the induction stroke. It is this part of the load cycle that usually does most damage to the integrity of the rod.

Consider particularly the big-end. Provided that the lubricant supply is sufficient to maintain an intact elasto-hydrodynamic oil film the shear load between the rod eye and the crank-pin is minimal. Thus the load offset effect is comparatively small and has a minor effect on load considerations.

Modern high speed engines using large diameter crank pins and a big-end joint face at right angles to the rod axis often do not allow the con-rod to be withdrawn through the cylinder bore. Where this is essential the joint face may be inclined at a convenient angle. To deal with force components along the joint face some feature such as serrations or dowels must be provided to avoid relative movement between the two halves of the rod.

An incorrectly designed joint face may suffer fretting and consequent fatigue break up at the face. Wear at the interface may also reduce the bolt pre-load and lead to bolt failure.

The optimization of the joint face should now concern the designer.

If the usual stress distribution in a lug is considered it will be remembered that the peak tensile stresses are at the three o'clock and nine o'clock positions. At six and twelve o'clock the radius of curvature of the pin hole is reduced and compressive tangential stresses occur. In each of the four positions there will be a maximum bending stress distribution across the section.

The most important design consideration is to position the joint face at the point of minimum bending in order to keep the interface pressure as near uniform as possible. This requires a more comprehensive knowledge of the stress distribution and photoelastic testing is probably the most effective method of analysis.

Because the connecting-rod big end usually has a large hole relative to its width it can be approximated to a thin ring and a useful design analysis can thus be carried out.

For a ring in diametral loading the bending moment across a section at any angular position is given by

$$M = WR_H \ (\pi^{-1} - \tfrac{1}{2}Sin\gamma) \tag{6}$$

and equating this to zero yields

$$\eta_o = Sin^{-1} \ (2/\pi) = 39.54° \tag{6a}$$

A shear force exists across the section which can be minimised by rotation of the split line plane.

In the practical case there are other forces to be considered. These include a bolt load P and the bearing shell nip load N, ie that load required to close the joint against the shell nip allowance. These forces are shown in figure 5.

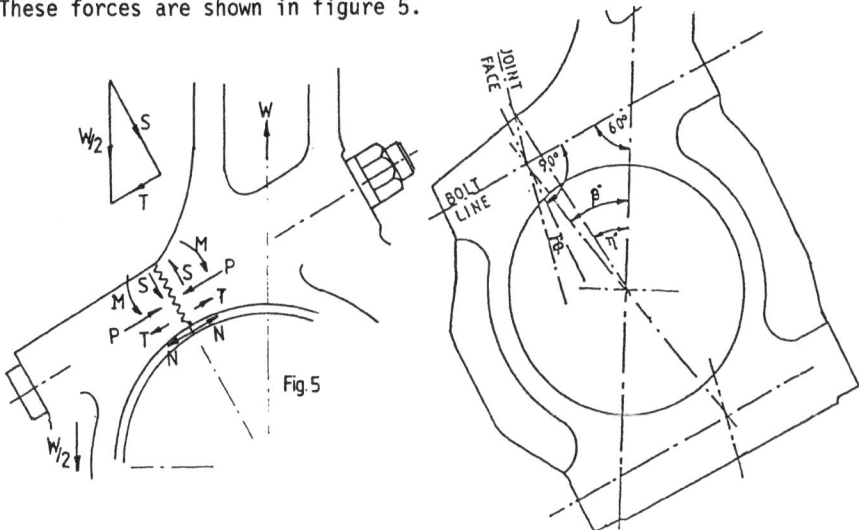

Fig.5

Figure 5. Forces Acting at Joint.   Figure 6. Joint Face Geometry.

The following cases enter into the final choice of joint geometry, the bolt line having been fixed arbitrarily at 60° to the rod axis. See figure 6.

Inertia bending moment at joint from equation 6.

Inertia shear at joint

$$S_I = \tfrac{1}{2} W \, Cos \, (30° - \phi) \qquad (7)$$

Inertia tension at joint

$$T = \tfrac{1}{2} W \, Sin \, (30° - \phi) \qquad (8)$$

Bolt compression load at joint

$$(P - N) \, Cos \, \phi \qquad (9)$$

Bolt shear load at joint

$$S_B = (P - N) \, Sin \, \phi \qquad (10)$$

β can be related to ϕ and η for which a compromise value associated with η_o gives $\beta = 42°$ approximately.

The fluctuating shear stress across the joint can be minimised by

$$0.25 \, W \, Cos \, (30° - \phi) = (P - N) \, Sin \, \phi \qquad (11)$$

The solution to this transcendental equation yields the optimum set back angle ϕ;see figure 7.

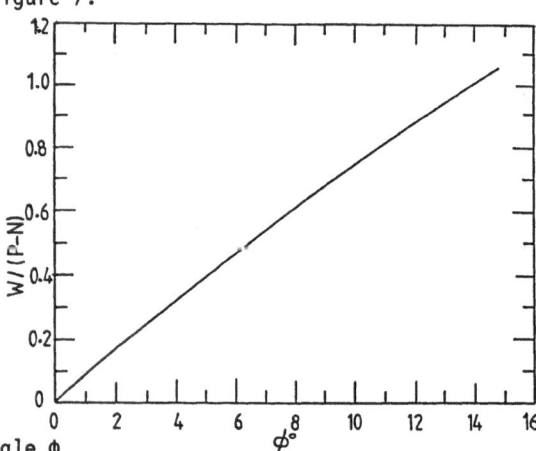

Figure 7. Set Back Angle ϕ.

It will be noted that Φ depends on the relationship W/(P − N) ie on the inertia load and the nett bolt load.

Using this analytical approach the author has improved the life of a rod which was prone to joint face serration break-up. Fatigue tests on the original, straight across, design reproduced the service failure in half a million load cycles. The modified design, capable of being produced from the same forging, ran to fifty million load cycles with no evidence of distress.

CONCLUSIONS

The articulating pin-fixed joint is subjected to dynamic forces which are additional to those usually considered in the design of pin-fixed lugs. The geometric optimization can be carried out using some basic formulae and in general require rather larger pin holes and greater clearances than is generally supposed.

Variants of the articulating joint can be designed to control the frequency characteristics of compressor blading or to provide an extremely compact filament wound joint. A force analysis of the special case of a connecting-rod big end can be used to considerably extend the service life of the rod.

ACKNOWLEDGEMENTS

The author thanks the Directors of GEC ALSTHOM Ltd for permission to publish this paper.

REFERENCES

1. Frocht M.M. and Hill N.H., Stress concentration factors around a central hole in a plate loaded through a pin in the hole. Trans ASME, J.App.Mechs, 1940, 62, A5.

2. E.S.D.U., Stress concentration factors, Axially loaded lugs with clearance fit pins, Engineering Science Data Unit Item No 81006, London 1982.

3. Goatham J.I. and Smailes G.T., Some Vibration characteristics of pin-fixed Compressor Blades, Trans ASME, J.Engineering for Power, 1967, 491-501.

AN AXI-ASYMMETRICAL ANALYSIS OF CIRCULAR FLANGE CONNECTIONS SUBJECTED TO EXTERNAL BENDING MOMENTS

TOSHIYUKI SAWA
Department of Mechanical Engineering, Yamanashi University,
4-3-11,TAKEDA,KOFU,YAMANASHI,400 Japan

TSUNESHI MOROHOSHI
ASK Co.Ltd.,Japan

AKIHIRO SHIMIZU
Epson-Seiko Co.Ltd.,Japan

KYOUICHI YAMAMOTO
Mitsubishi Electric Co.Ltd.,Japan

ABSTRACT

In designing bolted joints,it is important to examine the clamping effect (the distribution of contact stress) and to estimate the load factor(the ratio of an increment of bolt axial force to an external load). The clamping force of bolts and an external bending moment are both axi-asymmetrical loads,and no investigations regarding as axi-asymmetrical loads have been performed. In this paper,the clamping effect and the load factor in the case where clamped parts are circular flanges subjected to external bending moments are analyzed as axi-asymmetrical problems using a three-dimensional theory of elasticity. Experiments are carried out concerning the distribution of contact stress and the load factor for the external bending moment(the relationship between an increment/decrement of bolt axial force to an external bending moment). The analytical results are satisfactorily consistent with the experimental ones.

INTRODUCTION

In designing bolted joints,it is important to examine the distribution of contact stress which governs the clamping effect and to estimate the load factor(the ratio of an increment of bolt axial force to an external load). Many investigations(1)-(5) have been carried out on the characteristics mentioned above,but they have only dealt with the case where a clamped part with a nut and bolt is a hollow cylinder(1)-(4) and the case where the clamping forces by nuts and bolts are replaced with annular symmetrical loads(5). No investigations regarding axi-asymmetrical loads have,to our knowledge,been performed except for the study(6) by F.E.M. Therefore,it is necessary to clarify the effects of dispersed clamping forces by nuts and bolts on the characteristics of bolted connections. On the other hand, prior investigations have only dealt with the situation where external loads act on the connections in the axial direction(4)(5). Only some investigations(7)-(11) have been carried out on bolted connections subjected to external bending moments.

In this paper,the distribution of contact stress and the load factor for an external bending moment (the relationship between an increment/decrement of bolt axial force and the external bending moment) in the case where a circular flange connection is subjected to an external bending moment are discussed. In the analysis,replacing two circular flanges with a finite solid cylinder and regarding the clamping forces by bolts and an external bending moment as an axi-asymmetrical load, the characteristics of bolted connections,such as the distribution of contact stresses and the load factor for the external bending moment,are clarified using a three-dimensional theory of elasticity. The contact stresses are measured by photoelasticity(stress freezing method) and experiments to measure the load factor for the external bending moment are carried out. The analytical results are compared with the experimental ones.

THEORETICAL ANALYSIS

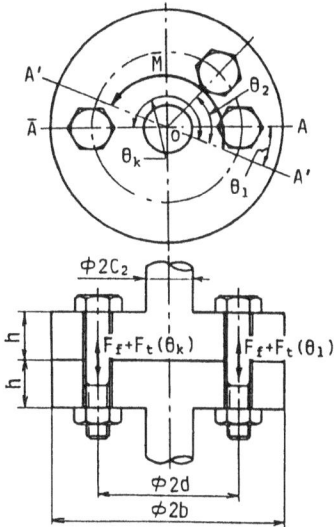

Figure 1. A circular flange connection subjected to an external bending moment

Analysis of contact stresses

Figure 1 shows a circular flange connection,which is fastened by N sets of nuts and bolts arranged in equal pitches with an initial clamping force F_f. An external bending moment M is applied to the surface A'-A' and an increment/decrement $F_t(\Theta_k)$ of bolt axial force is produced on the bolt deflected by Θ_k. Replacing two circular flanges with a finite solid cylinder,the contact stress is analyzed. In the analysis, a cylindrical coordinate (r,Θ,z) is used. The thickness of the flanges is designated by h,the outer diameter by 2d,the diameter of web by $2c_2$,Young's modulus by E and Poisson's ratio by ν. Figure 2(a) shows a model for analysis in the initial clamping state and (b) shows the case where an external bending moment is applied to the connection. The distribution of contact stress is obtained by superposing the analytical results for Figure 2(a) and (b) on the result for Figure 2(c),where the diameter of the bearing surface is denoted by $2c_1$. In the analysis,bolt holes are not considered,but the distribution of contact stress is estimated using the model for analysis shown in Figure 2. The distribution of contact stress near the bolt holes is estimated using a model(4)(11) of a hollow cylinder.

In the initial clamping, the compressive loads which act on the bearing surfaces are applied to the circle region in the diameter $2c_1$ as a uniform load P_1 as shown in Figure 2(a). Expanding the load distribution into Fourier-Dini expansions,the boundary condition of Figure 2 is expressed in the following equations,where an approximate measurement of the circle region is obtained by calculating the fan shape sections using the method shown in Figure 3.

$$r=b\;;\;\sigma_r=\tau_{zr}=\tau_{r\theta}=0$$
$$z=\pm h\;;\;\sigma_z=\sum_{L=1}^{9}\left[\bar{a}_0+\sum_{s=1}^{\infty}\bar{a}_sJ_0(\gamma_sr)\right.$$
$$\left.+\sum_{m=1}^{\infty}\sum_{i=1}^{\infty}\bar{a}_iJ_m(\beta_ir)\cos m\theta\right]$$
$$\tau_{zr}=\tau_{\theta z}=0$$

(1)

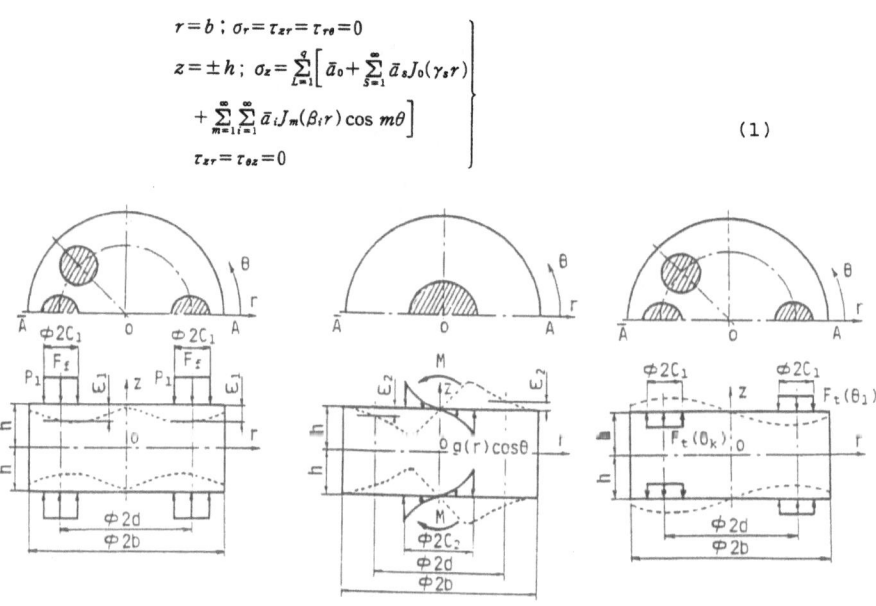

(a) initial clamping state

(b) the case where a bending moment is aplied

(c)the case where increments of bolt axial force act

Figure 2. A model for analysis

where

$$\bar{a}_0 = a_0(d_2^2 - d_1^2)/b^2$$

$$\bar{a}_s = \frac{2a_0\{d_2 J_1(\gamma_s d_2) - d_1 J_1(\gamma_s d_1)\}}{\gamma_s b^2 J_0^2(\gamma_s b)}$$

$$\bar{a}_i = \frac{2\lambda_i^2 a_m}{b^2(\lambda_i^2 - m^2)J_m^2(\lambda_i)} \int_{d_1}^{d_2} r J_m(\beta_i r) dr$$

$$a_0 = -P_1 N(a_2 - a_1)/\pi$$

$$a_m = \frac{-P_1}{m\pi}\left[\sum_{k=0}^{N-1}\{\sin m(\theta_k + a_2) - \sin m(\theta_k + a_1)\}\right.$$

$$\left. + \sum_{k=1}^{N}\{\sin m(\theta_k - a_1) - \sin m(\theta_k - a_2)\}\right]$$

$$\theta_k = 2\pi(k-1)/N + \theta_1 \quad (k = 1, 2, 3, \cdots, N)$$

$$d_1 = \frac{c_1}{\sin a}\sin\left\{-a + \sin^{-1}\left(\frac{d}{c_1}\sin a\right)\right\}$$

$$d_2 = \frac{c_1}{\sin a}\sin\left\{\pi - a - \sin^{-1}\left(\frac{d}{c_1}\sin a\right)\right\}$$

$$a = (a_1 + a_2)/2$$

$$a_1 = \frac{L-1}{q}\sin^{-1}\frac{c_1}{d}$$

$$a_2 = \frac{L}{q}\sin^{-1}\frac{c_1}{d} \quad (L = 1, 2, 3, \cdots, q)$$

and q is a number of the fan shape sections, $J_0(\gamma_s r)$ and $J_m(\beta_i r)$ are the first kind of Bessel functions of order 0 and m, respectively. γ_s and β_i are the s-th and the i-th positive roots satisfying the equations $J_1(\gamma_s b) = J'_m(\beta_i b) = 0$.

In the case of Figure 2(b), an external bending moment which acts on the webs of circular flanges is replaced with the load distribution $g(r)\cos\theta$ which distributes within the region $0 \leq r \leq c_2$.
That is,

$$M = 4\int_0^{c_1}\int_0^{\pi/2} g(r)\cos^2\theta r^2 d\theta dr$$

the boundary condition is expressed as follows;

$$\left.\begin{array}{l} r = b \quad ; \sigma_r = \tau_{zr} = \tau_{r\theta} = 0 \\[4pt] z = \pm h \,; \sigma_z = \sum_{i=1}^{\infty} a_i J_1(\beta_i r)\cos\theta \\[4pt] \tau_{zr} = \tau_{\theta z} = 0 \end{array}\right\} \tag{2}$$

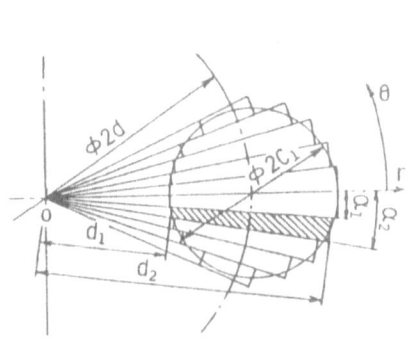

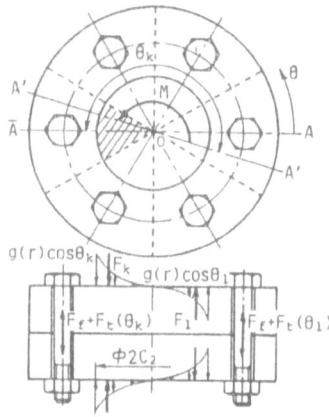

Figure 3. An approximation of load which acts on the part corresponded to bearing surface

Figure 4. An assumption of bending moment which each bolt bears

where

$$a_i = \frac{2\lambda_i^2}{b^2(\lambda_i^2-1)J_i^2(\lambda_i)} \int_0^{c_1} rg(r)J_1(\beta_i r)dr$$

The interface on the tension side of a bolted connection,which is fastened with an initial clamping force F_f, begins to separate with an increase of an external bending moment M. The bending moment M,when the interface begins to separate, is determined such that the contact stress σ_z becomes zero.

Analysis of the load factor for an external bending moment

When an external bending moment M is applied to the surface A'-A' shown in Figure 4, an increment/decrement $F_t(\theta_k)$ of bolt axial force is produced in a bolt located on the surface O-$\overline{A}$ which is deflected θ_k from the surface A'-A'. Replacing the bending moment with the sinusoidal load, it is assumed that the bolt located on the surface O-$\overline{A}$ bears the load indicated at the hatched area in the bending moment M. Namely, the integration $g(r)\cos\theta rdrd\theta$ within the region $\theta_k - \pi/N \leq \theta \leq \theta_k + \pi/N$ is denoted by F_k. F_k is expressed by Eq.(3). In reference to the calculation method(4)(5) of the load factor in the case where a bolted connection is subjected to a tensile load,the load factor Φ_b for an external bending moment(we denote the factor by L.F.B.M. hereinafter) is given by Eq.(4).

$$F_k = \int_0^{c_1} \int_{\theta_k - \pi/N}^{\theta_k + \pi/N} g(r) \cos \theta r d\theta dr$$

(3)

$$\Phi_b = \frac{F_t(\theta_k)}{F_k} = \frac{K_t}{K_t + K_c} \left(\frac{K_c'}{K_b}\right)$$

(4)

where K_t(14) is the spring constant for a bolt-nut system and K_c(4) is the compressive spring constant for the clamped part near a bolt hole. K'_c is the compressive spring constant for the clamped part and K_b is the bending spring constant(10)(11) for the clamped part. In this paper,the values K'_c and K_b are obtained from the equations $K'_c = F_f/2\varepsilon_1$ and $K_b = F_k/2\varepsilon_2$ using the mean displacements at the bearing surface ε_1 and ε_2,which are obtained from the analytical results of Figure 2(a) and (b),where ε_2 is the mean displacement at the bearing surface of a bolt located on the surface O-$\overline{A}$ in Figure 2(b).

EXPERIMENTAL METHOD

Measurement of contact stresses by photoelasticity

Photoelastic (stress freezing method) experiments(15) were carried out in order to measure the contact stresses in the case of Figure 2(a) and (b). Figure 5(a) shows an experimental model on a solid cylinder fastened with an initial clamping force. The dimensions of the cylinder are 2b=120 mm,2h=60 mm and the compressions W/N act on the bearing surface of the cylinder through N discs in the diameter of 24 mm arranged in equal pitches. Figure 5(b) shows an experimental apparatus for the case of Figure 2(b). An external bending moment M(M=45/$\overline{W}$/2=22.5$\overline{W}$) is applied to the solid cylinder using pins. The test specimens are made of epoxy resin(Young's modulus E is 2.9GPa and Poisson's ratio$\nu$is 0.36). The stresses of the specimens were frozen in a furnace.

Measurement of the load factor for an external bending moment(L.F.B.M.)
The values of L.F.B.M. were obtained in the reference (11) in which the experiments were carried out.

COMPARISONS OF ANAIYTICAL RESULTS WITH EXPERIMENTAL ONES

Analytical results of contact stress
In the numerical calculations,Young's modulus E is put as 206 GPa,Poisson's ratio ν as 0.3,and the terms i and s in the series are put as 20 and m as 10. Figure 6 shows the distribution of contact stress σ_z at $\Theta=0$ for the case of Figure 2(a) putting h/b=0.5,d/b=0.458,C_1/b=0.2 and q=100(Figure 3), where Θ_1 is put as zero. In Figure 6(a),the ordinate is the ratio of contact stress σ_z to the mean contact stress σ_{zm} and the abscissa is the ratio of the distance r to the radius b. From the results,it is seen that a slight variation in the distribution of contact stress is observed with an increase of the number of N bolts. Figure 6(b) shows the distribution of contact stress σ_z in the circumferential direction Θ. It is seen that the contact stress decreases between two bolts with a decrease of the number of N bolts.

Figure 7 shows the distribution of contact stress for the case of Figure 2(b),where the dimensions are put as h/b=0.41,0.6,1.0 and C_2/b=0.5 ,and the stress distribution at the webs is given by the equation g(r)=r/C_2. The tension side of the surface A'-A' is $\Theta=0$. Figure 7(a) shows the effects of the thickness of the cylinder on the distribution of contact stress in the radial direction.

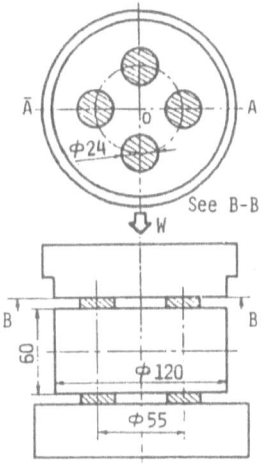

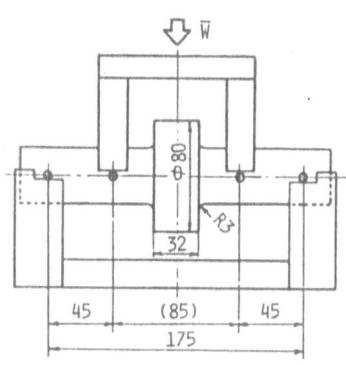

(a) initial clamping state   (b)in the case where an external bending moment is applied

Figure 5. Sketch of experimental setups in photoelastic experiments

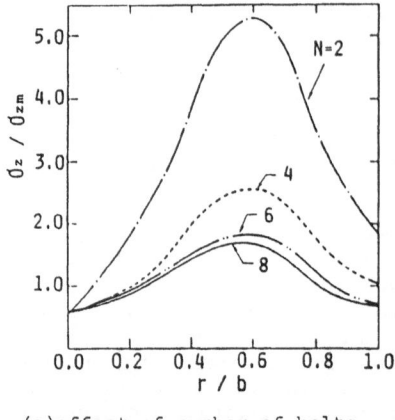

(a)effect of number of bolts
(θ=o)

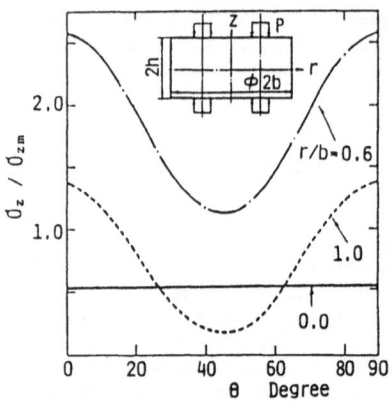

(b)distribution of contact stress
in circumferential direction

Figure 6. Analytical results of distribution of contact stress in
the initial clamping state

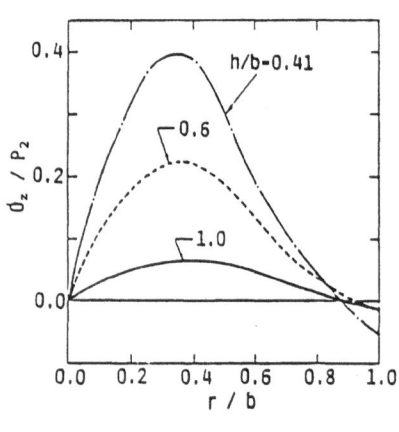

(a)in the radial direction

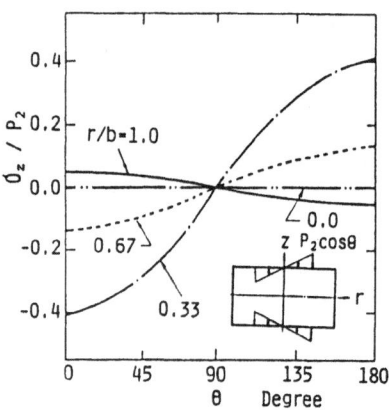

(b)in θ direction

Figure 7. Analytical results of distribution of contact stress in
the case where external bending moments are applied(Fig.2(b),z=0)

COMPARISONS WITH EXPERIMENTAL RESULTS OBTAINED BY PHOTOELASTICITY
Concerning the Principal Stress Difference

The experimental results by photoelasticity were obtained by converting
the isochromatic fringes to the value of the stress. In the numerical
calculations,Young's modulus E was put as 2.9 GPa and Poisson's ratio
as 0.36. Figure 8(a) shows the comparison in the case of Figure

2(a),where the number,N, of bolts is 2 and 4. The analytical results are satisfactorily consistent with the experimental results. Figure 8(b) shows the comparisons in the case of Figure 2(b). In the comparisons the bending moment M of 26.46 KN-mm is applied. The analytical result is shown in the case when the load distribution g(r) is put as the equation $g(r)=r^2/C_2^2$ taking into account the stress distribution at the ends of the cylinder(z=+h) which was obtained from the photoelastic experiment. The analytical result is in fairly good agreement with the experimental result.

Comparisons concerning L.F.B.M.
The numerical calculations were done using the dimensions of circular flanges used in these experiments. Young's modulus E of the circular flanges was put as 206 GPa and their Poisson's ratio as 0.3. Figure 9 shows the comparison of the analytical results with the experimental results concerning L.F.B.M. of the bolt located on Θ_1=0,in which the maximum increment of bolt axial force is produced,where the number,N, of bolts are 4 and the bolt pitch diameter 2d is 55 mm. The abscissa indicates the external bending moment M and the ordinate indicates the bolt axial force F_f+F_t. The analytical result obtained by a two-dimension model for analysis(11) is shown ,too. The analytical results are satisfactorily consistent with the experimental results. An increment F_t of bolt axial force increases with an increase of the external bending moment M. When the interface starts to separate with a further increase of M, F_t plots a curve.

Concerning the prediction when the interface starts to separate,the analytical result is fairly consistent with the experimental result. Table 1 shows the comparisons with respect to L.F.B.M. of the bolt located on Θ_1=0. It is shown that the value of Φ_b decreases with an increase of d.

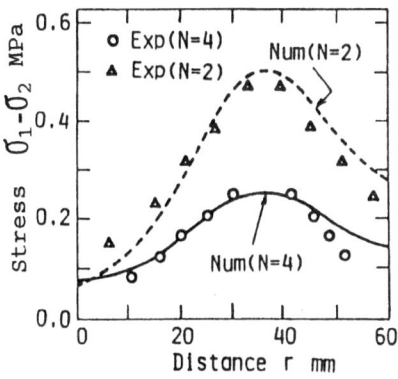
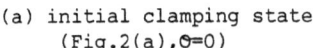

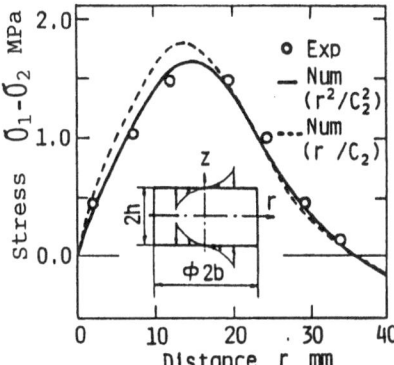

(a) initial clamping state
(Fig.2(a),Θ=0)

(b)in the case where external bending moments are applied
(Fig.2(b),θ=0)

Figure 8. Comparisons of analytical results with experimental ones concerning contact stress(z=0)

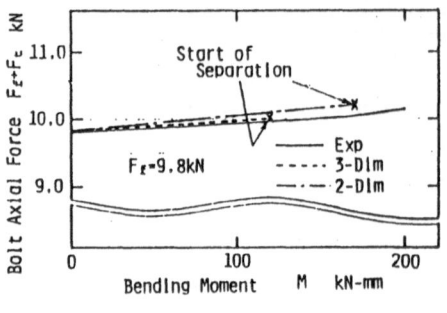

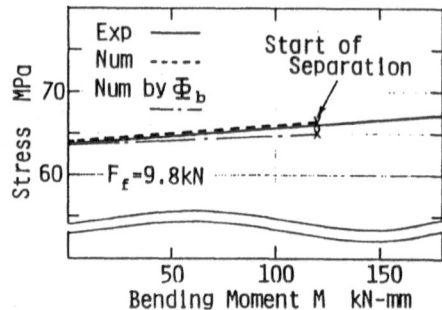

Figure 9. Comparison concerning
increment of bolt axial
force(L.F.B.M.)

Figure 10. Comparison concerning
maximum stress produced
on bolt

Table 1 Correction factor K'c/Kb and values of L.F.B.M.

	d	27.5	35.0	42.5
2-Dim	K'_c/K_b	0.1776	0.0789	——
	Φ_b	0.0312	0.0118	——
3-Dim	K'_c/K_b	0.1248	0.0556	0.0245
	Φ_b	0.0232	0.0093	0.0049
Exp	Φ_b	0.023	0.009	0.004

Figure 10 shows the comparison concerning the maximum stress
produced in the bolt. The analytical result is consistent with the
experimental result. It is seen that the maximum stress produced in the
bolt is larger than the stress obtained from L.F.B.M.

CONCLUSIONS

In this paper, the characteristics of circular flange connections
subjected to external bending moments were discussed. The following
results were obtained.
(1)Replacing two circular flanges with a finite solid cylinder and
treating initial clamping forces and an external bending moment as axi-
asymmetrical loads,the distribution of contact stress was analyzed using
a three-dimensional theory of elasticity.

305

(2)The contact stresses were measured by photoelasticity. The analytical results obtained by (1) were satisfactorily consistent with the experimental results.
(3)The load factor for the external bending moment was analyzed. Experiments were also performed. The analytical results were satisfactorily consistent with the experimental results.
(4)The maximum stress produced in a bolt was analyzed. The analytical result was consistent with the experimental result. It was found that the maximum stress was larger than the stress obtained by the load factor for the external bending moment.

REFERENCES

(1) Bradley,T.L.,Lardner,T.J. and Mikic,B.B.,Bolted Joint Interface Pressure for Thermal Contact Resistance,Trans.ASME,ser.E,1971, 38,No.2,pp542.
(2) Fernlund,I.,Druckverteilung zwiscen Dichtflachen an Verschrauben Flanschen,Konstruction,1970,22,No.6,pp218.
(3) Motosh,N., Stress Distribution in Joints of Bolted or Riveted Connections,Trans.ASME,ser.B,1975,97,No.1,pp157.
(4) Yoshimoto,I.,Maruyama,K.,Sawa,T.and Nishiguchi N.,The Force Ratio of Bolted Joints,Bulletin of JSME,1977,20,No.141,pp357.
(5) Sawa,T. and Shiraishi,H.,A Simple Method to Calculate the Force Ratio of Bolted Joints,Bulletin of JSME,1983,26,No.216,pp1088.
(6) Tanaka,M. and Yamada,H.,Behaviours of Bolted Joints under External Loads,Trans.Japan Soc. Mech. Engrs.(in Japanese),1985, 51,No.466,Ser.C,pp1362.
(7) Asai, O.,Tsukui,K. and Kawamo,K.,Study on High-Strength Bolted End Plate Connections,Trans. Japan Soc. Mech. Engrs. (in Japanese),1980,46,No.402,Ser.A,pp158.
(8) Pavlov,P.A.,Approximate Calculation for the Flexure with Tension of Flanged Joints of Schafts,Russ.Eng.J.,1976,56,No.10,pp16.
(9) Dreger,H.,Berechnung der Krafte,Biegemomente und beanspruchungen einer Exzentrich Verspanten und Exzentrich Belasteten Schraube, Draht-Welt, 1979,11,No.11,pp497.
(10)Sawa,T.,Shiraishi,H.,Minakuchi Y. andMaKINOS.,Onthe Characteristics of a Bolted joint Subjected to an External Bending Moment,Bulletin of JSME,1982,25,No.201,pp444.
(11)Sawa,T.,Taira,Z. and Shiraishi H.,On the Characteristics of a Bolted Joint Subjected to an External Bending Moment,Bulletin of JSME,1984, 27,No.224,pp348.
(12)Sawa,T.,Iwata ,A.,Kumano,H. and Maruyama,H.,On the Characteristics of Bolted Joints Subjected to Bending Moments,Bulletin of JSME,1986, 29,No.247,pp281.
(13)Yoshino,T. and Utsugi,S.,An Axi-asymmetrical Deformation Analysis of finite Solid Cylinders,1975 Joint JEME-ASME Appl.Mech.Western Conf. paper,1975,pp19.
(14)Sawa,T. and Maruyama K.,On the Deformation of Bolt Head and Nut in a Bolted Joint,Bulletin of JSME, 1976,19,No.128,pp203.
(15)James W.Dally and William F.Riley, Experimental Stress Analysis, Mcgraw-Hill,Second Edition,1978,pp490.

BEHAVIOUR OF OPEN SECTION STEEL MEMBERS SUBJECT TO COMBINED BENDING AND TORSION

M. A. EL-KHENFAS
formerly Research Student
Department of Civil & Structural Engineering, University of Sheffield, UK
D.A. NETHERCOT
Professor of Civil Engineering
University of Nottingham, University Park, Nottingham NG7 2RD, UK

ABSTRACT

A finite element analysis of the nonlinear behaviour of open section thin-walled members under compression, biaxial bending and torsional loading, including inelastic material response, has been used to obtain solutions to a number of example problems. These highlight the relative importance of the various sources of nonlinearity and may be used to assess the suitability of simplified design approaches. This leads to an adaptation of an existing combined stress approach as a suitable basis for the limit state design of steel members; the method is used in a design guide recently published by the Steel Construction Institute.

INTRODUCTION

Amongst the extensive literature devoted to the behaviour of thin-walled structural components under static loading, very little deals with problems that involve the deliberate application of torsional loading. Numerous studies of varying degrees of rigour, involving varying levels of complexity, for behaviour under bending, compression and shear, acting singly or in combination, are available. Similarly results for out of plane buckling — flexural buckling of columns, torsional buckling of columns, lateral-torsional buckling of beams and flexural torsional buckling of columns — covering various aspects of the particular problem types, are readily obtainable. However, problems in which direct torsional load is present, either singly or more particularly when it acts in combination with other load types leading to an interaction between effects, have received comparatively little attention.

Even in the elastic range considerable choice exists as to the exact basis for the analysis. This arises as a direct result of the existence of different theories of torsion, each implying different levels of coupling between the various structural actions present. Even the simplest linear theory,because it must include the effects of warping, leads to significantly more complex governing equations than does straightforward flexural behavioural. Solutions to the basic linear problem have been provided in a design oriented text by Terrington [1] for a variety of examples. Similar solutions are provided in the AISC.

Design Guide [2] as well as in a recently published SCI Guide [3] that takes the problem a stage further by including a failure criterion for combined bending and torsion.

One of the first attempts at solving the governing equations of bending and torsion in the inelastic range was that of Razzaq [4]. He used a finite difference technique, comparing the results with tests on 22 large scale beam specimens. In 1978 Kollbrunner [5] produced a theoretical ultimate strength analysis for I-section cantilevers; once again the results were shown to agree closely with specially conducted experiments. More recently several authors have applied the finite element method to this class of problem, basing their formulations on several slightly different bending/torsion theories. A review of these, including specific comparison between the differences in concept, is provided in the paper by El-Khenfas and Nethercot [6]. A more general review of literature relating to the torsional behaviour of thin-walled members is also available [7].

It is the purpose of this paper to present some of the results obtained in a recent numerical study [7] of the nonlinear behaviour of I-section steel members under combined bending and torsion that includes the effects of gradually increasing plasticity in the section. Full details of the analytical formulation and supporting verification studies have already been published [6]. Thus the present paper concentrates on the presentation and discussion of a selection of new results, concentrating on those in which deliberate torsional loading is present.

Analysis
All of the results presented herein have been obtained using the one dimensional finite element analysis developed by El-Khenfas [7] that is fully described in ref. [6]. This is so formulated that it permits the combined bending and torsion problem to be addressed at 3 different levels, depending on the type of strain-displacement relationship used as the starting point. Progressive spread of yield is allowed for in each case, with the material being assumed to follow a trilinear stress-strain curve. Because of the arbitrary way in which yield may develop, the type of cross-section assumed is completely general, the only restriction being that it is composed of a series of flat plate elements. Thus no prior assumptions regarding geometrical properties are made; these are determined at all stages of loading and thus for any degree of yielding directly from first principles.

Solution of the resulting nonlinear, incremental problem is by means of a Newton-Raphson scheme, with convergence being controlled by either an out-of-balance force check or through a norm involving both the residual forces and the displacement increments in that particular load step [6]. The method has been checked against a selection of previously available theoretical and experimental results designed to verify its accuracy and applicability across the whole of the problem range. These comparisons have been fully discussed elsewhere [6,7].

Results
The first problem is included in order to illustrate the differences that result in an elastic analysis due to the use of different strain - displacement relationships.

A standard I-section member is assumed to be simply supported at both ends against both bending and twisting [3] i.e. no warping restraint, to possess an initial lack of straightness in its weaker plane and to be subjected to a biaxially eccentric compressive load F together with a torsional moment at mid-span M_t. The load path assumed is that M_t is held constant while F is gradually increased.

308

Figs. 1 and 2 present load versus twist at the beam's centre relationships for two different values of M_t according to each of the three types of analysis. From these it is clear that neglect of the nonlinear terms in the basic strain-displacement relations leads to significant overestimates of load carrying capacity but that behaviour up to load levels of

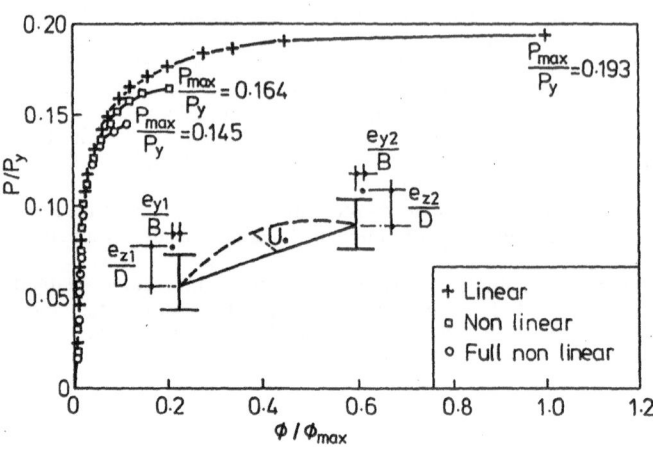

Figure 1. Comparison of elastic analyses for W12 x 43 section: $e_{y1}/B = e_{y2}/B = 0.125 e_{z1}/D = e_{z2}/D = 0.257$, $L/r_y = 25$, $u_o = L/1000$, $\sigma_y = 50$ksi, $\phi_{max} = 19°$, $M_t=0$.

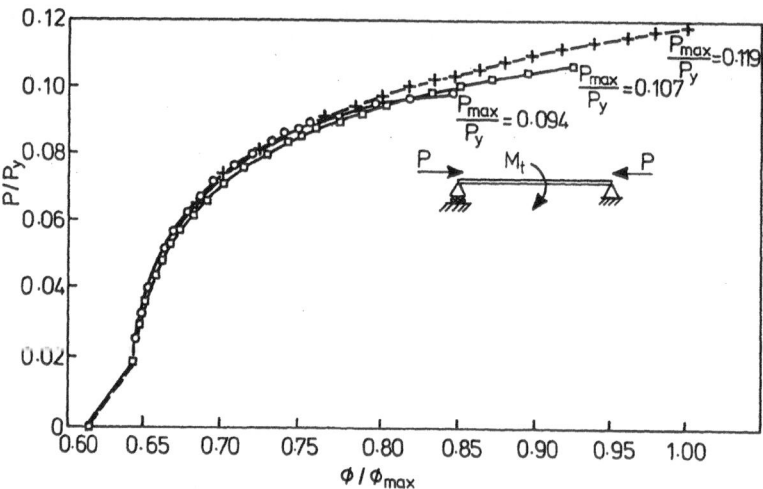

Figure 2. Comparison of elastic analyses for W12 x 43 section: $e_{y1}/B = e_{y2}/B = 0.125$, $e_{z1}/D = e_{z2}/D = 0.257$, $L/r_y = 25$, $u_o = L/1000$, $\sigma_y = 50$ksi, $\phi_{max} = 26°$, $M_t/M_p = 0.035$

(say) 90 per cent of the failure load predicted by the least rigorous analysis is quite similar for all three types of analysis. Whilst the first point is clearly a cause for concern, the second is more reassuring within the context of attempts to provide simple approximate methods of estimating deformations at working load [3] - a condition that frequently controls for practical examples.

Fig. 3 provides a complete interactive comparison of the three sets of results over the whole range of M_t values up to that at which failure would occur by M_t alone. It is clear from this that for predominantly torsional loading differences between results reduce due to the decreased opportunity for coupling between the nonlinear terms in the tangential and geometric stiffness matrices.

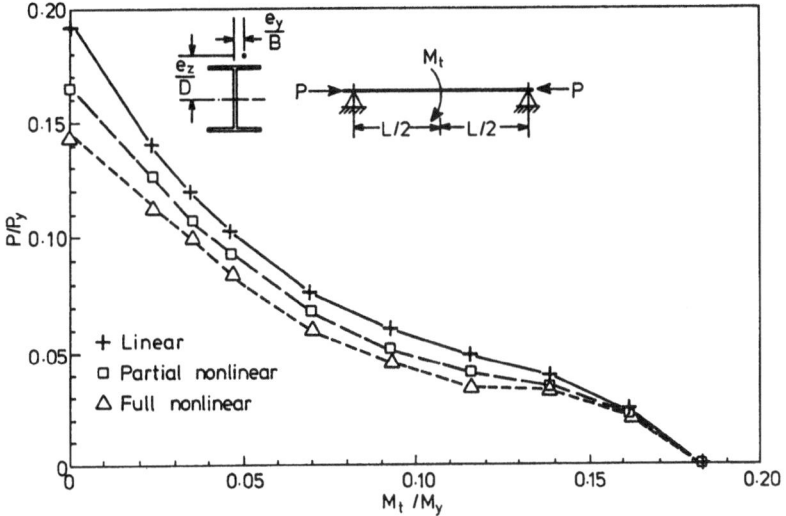

Figure 3. Interactive plot for W12 x 43 section, elastic analysis.

Referring back to the basis for the analysis [6], the full expression for longitudinal direct strain ε_x is:

$$\varepsilon_x = W_{,x} + [U_{,xx}z + V_{,xx}y + \phi_{,xx} \, \omega_n] \tag{1}$$

$$+ \quad [V_{,xx} \, \phi z + V_{,xx} \frac{\phi}{2} z + V_{,xx} \, \phi y + V_{,xx} \frac{\phi}{2} Y]$$

$$+ \quad \tfrac{1}{2} \, [U^2_{,x} + v^2_{,x} + \rho \phi^2_{,x}]$$

in which $W_{,x}$ is due to longitudinal displacement

$U_{,xx}, V_{,xx}, \phi_{,xx}$ are due to curvatures

y, z, ω_n define the point on the cross-section

$U_{,x}, V_{,x}, \phi_{,x}$ are axial nonlinear strains produced by the curvatures.

The so-called "linear analysis" uses only the first term plus the contributions in the first bracket, whilst the "nonlinear" analysis uses all the terms but omits the nonlinear geometrical and tangential parts of the stiffness matrices and the "full nonlinear analysis" includes everything. Moving through these three approaches does add considerably to the complexity of the calculations. Each approach is, of course, a nonlinear one, being capable of following an out-of-plane buckling type response for a geometrically imperfect member when subject to in-plane loading only i.e. lateral-torsional buckling under major axis bending.

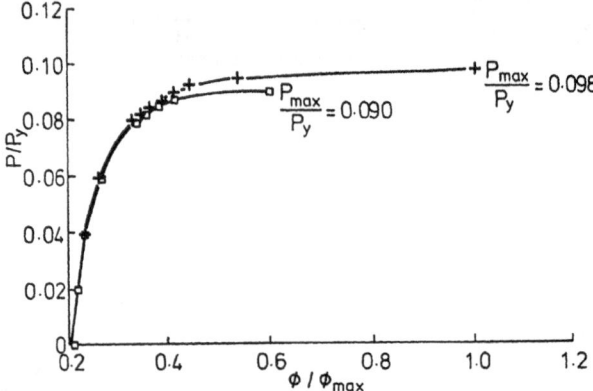

Figure 4. Comparison of inelastic analyses for W12 x 43 section: $e_{y1}/B = 0.125$, $e_{y2}/B = 0$, $e_{z1}/D = e_{z2}/D = 0.686$, $\phi_{max} = 3°$, $M_t/M_p = 0.155$.

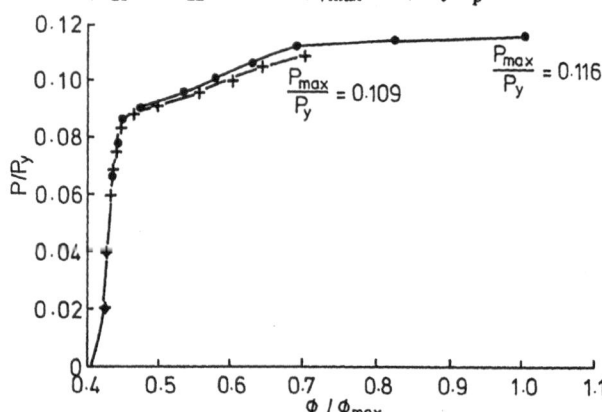

Figure 5. Comparison of inelastic analyses for W12 x 43 section: $e_{y1}/B = e_{y2}/B = 0.063$, $e_{z1}/D = e_{z2}/D = 0.172$, $\phi_{max} = 6°$, $M_t = M_p = 0.015$.

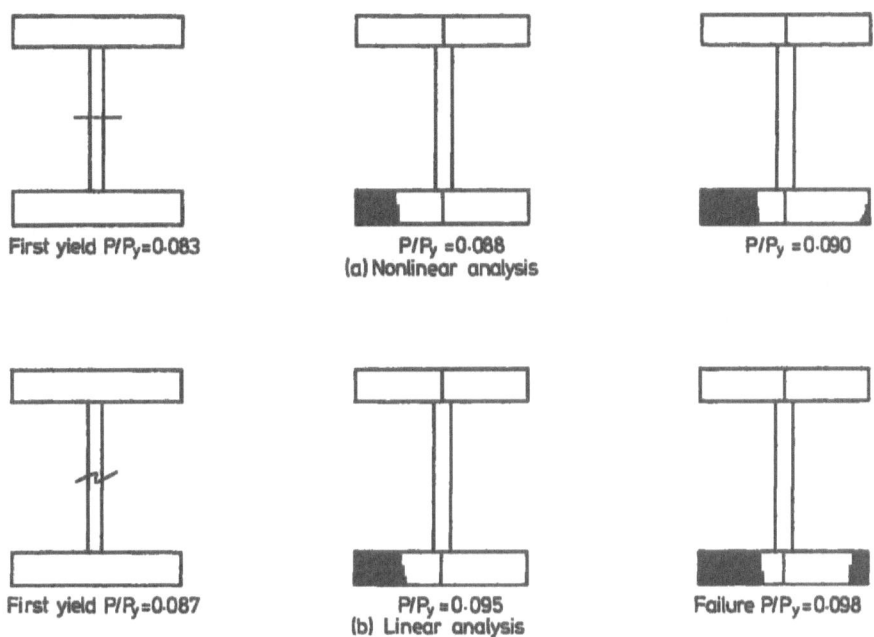

Figure 6. Spread of yield for example of Fig. 4.

When inelastic material behaviour is included differences between results from the three classes of analysis tend to reduce as illustrated by the example of Figs. 4 and 5. This gives compressive load versus twist relationships for analyses 1 and 3 for a similar problem to that considered previously. Variations in ultimate strength are 9 per cent and 6 per cent, with variations in first yield being 6 per cent and 3 per cent respectively. Comparisons between the extent of yielding at different load levels of the sort presented as Fig. 6 show, as expected, an earlier initiation of yield for the nonlinear analysis due to the larger deformations and thus larger strains and an earlier failure with less plasticity developed due to the greater loss of stiffness produced by the destabilising effect of the extra degrees of flexural-torsional coupling.

A set of results for different combinations of M_t and applied major axis moment M_y at collapse for three beams of different overall slenderness $\bar{\lambda}$ is given in Fig. 7. In the absence of M_t failure would, of course, occur by inelastic lateral-torsional buckling [8]. It is of interest to note that the results are only slightly concave when plotted on an interactive basis and thus support the notion of a linear combination as the basis for an interaction formula approach to design [3].

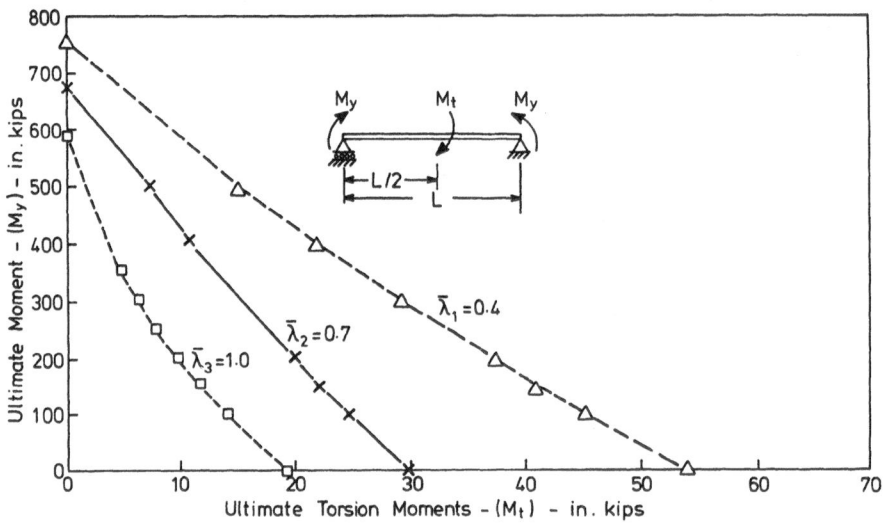

Figure 7. Interactive plot for W12 x 43 section, inelastic analysis.

Design
Pastor and DeWolf [9] have proposed an allowable stress criterion for design under combined major-axis bending and torsion based on a simplified combination of elastically calculated stresses. This may be stated as:

$$\frac{M_y}{M_{yy}} + \frac{\phi M_y}{M_{yz}} + \frac{\sigma_w}{\sigma_y} \leq 1 \qquad (2)$$

in which $\quad M_y \quad = \quad$ applied major axis moment

$\qquad M_{yy}, M_{yz} \quad = \quad$ moments at first yield

$\qquad \phi \quad = \quad$ maximum angle of twist calculated by first order theory

$\qquad \sigma_w \quad = \quad E\omega_n \phi''$ is the warping stress

$\qquad \sigma_y \quad = \quad$ material yield stress

The suitability of this condition has been investigated by comparison with three sets of results obtained from the present analysis. These all relate to beams subject to constant major axis end moments M_y and a gradually increasing mid-span torsional moment M_t.

Table 1 presents the comparison for a beam slenderness $\bar{\lambda}$ of 0.4; the cases $\bar{\lambda} = 0.7$ and 1.2 are covered in ref. [7]. The values of M_y and M_t shown correspond to the failure loads calculated by the program [7], the calculated angle of twist ϕ at failure is also given, all other quantities are simplified values obtained by following the procedures of ref. [9].

TABLE 1

Interactive Calculations for $\bar{\lambda} = 0.4$

ϕ x 10⁻³	M_y	M_t	ϕ_c x 10⁻³	ϕ'' x 10⁻⁴	$M_z = \phi_c M_y$	$\sigma_w = E\omega_n \phi''$	First Yield			
(rad)	(kip.in)	(kip.in)	(rad)		(kip.in)	(ksi)	M_y/M_{yy}	M_z/M_{yz}	σ_w/σ_y	Σ
9.017	0.0	53.785	9.676	1.481	0.0	52.759	0.0	0.0	1.055	1.055
8.264	100.0	45.117	8.117	1.243	0.812	44.256	0.136	0.013	0.885	1.034
7.461	150.0	40.889	7.356	1.126	1.103	40.109	0.204	0.018	0.802	1.024
6.908	200.0	37.332	6.715	1.028	1.343	36.620	0.272	0.022	0.732	1.026
5.564	300.0	29.697	5.342	0.818	1.603	29.131	0.409	0.026	0.583	1.018
4.162	400.0	21.841	3.929	0.601	1.572	21.424	0.545	0.025	0.429	0.999
3.121	500.0	15.14	2.724	0.417	1.362	14.851	0.681	0.022	0.297	1.00
0.409	720.8	0.0	0.0	0.0	0.0	0.0	0.981	0.0	0.0	0.981

Clearly for the case shown the design approach represents a good approximation to the true failure state. For the more slender cases (larger $\bar{\lambda}$) considered the summation tended to fall below unity, thereby implying an unconservative approach. This was largely due to a tendency for ϕ to be underestimated and also to the particular use of σ_w, since warping stresses were observed to increase rather more rapidly once yielding had started. Some modification to Eq. 2 to allow for this unconservatism at moderate and high slenderness is therefore required.

The approach taken in ref. [3] is to introduce an amplification factor of the type traditionally used in beam-column design formulae [10] to account for the effect of major axis moments acting through the torsional deformations. Thus Eq. 2 must be replaced by the pair of conditions:

i)   Buckling check:

$$\frac{M_y}{Mb} + \frac{\sigma_{byt} + \sigma_w}{p_y} \left[1 + 0.5 \frac{M_y}{M_b} \right] \le 1 \tag{3}$$

ii)   "Capacity Check"

$$\sigma_{bx} + \sigma_{byt} + \sigma_w \le p_y \tag{4}$$

The first of these allows for overall failure and therefore permits the use of an "equivalent uniform moment" $\bar{M}_y$ in the case of moment gradient loading [8]; it reduces towards the case of a normal lateral-torsional buckling as the torsional component of the applied loading is decreased. Eq. 4 is a straightforward elastic local strength check that must be satisfied at all points within the member. In both equations σ_{byt} depends upon the minor axis bending effect of the M_y moment acting through the twist ϕ and is given by:

$$\sigma_{byt} = \frac{\phi M_y}{Z_y}$$

in which Z_y is the elastic minor-axis section modulus.

CONCLUSIONS

A selection of results for the response of thin walled members subject to combinations of load that include direct torsional loading have been obtained using an ultimate strength finite element analysis based on nonlinear beam theory. These are used as the basis for an interaction formula type of design approach for combined bending and compression.

ACKNOWLEDGEMENTS

The work reported herein was conducted whilst both authors were members of the Department of Civil and Structural Engineering at the University of Sheffield.

REFERENCES

1.   Terrington, J.S., "Combined Bending and Torsion of Beams", BCSA Publication No. 31, 1970.

2.   American Institute of Steel Construction, "Torsional Analysis of Steel Members", AISC, Chicago, 1983.

3.   Nethercot, D.A., Malik, A.S., and Salter, P.R., "Design of Beams Subject to Combined Bending and Compression", SCI Publication No. 057, 1989.

4.   Razzaq, Z., "Theoretical and Experimental Study of Biaxially Loaded Thin-walled Beams of Open Section with and without Torsion", D.Sc. dissertation, Washington University, 1974.

5. Kollbruner, C.F., Hajdin, N., and Branislovc, C., "Elastic-plastic Thin-walled I-section Beam Subjected to Bending and Warping Torsion", Institute for Engineering Research, Zurich, Report 43, 1978.

6. El-Khenfas, M.A. and Nethercot, D.A., "Ultimate Strength Analysis of Steel Beam-columns Subject to Biaxial Bending and Torsion", Res Meccanica, 28, 1989, pp. 307-360.

7. El-Khenfas, M.A., "Three Dimensional Ultimate Strength Analysis of Beam-columns", Ph.D. thesis, University of Sheffield, 1987.

8. Trahair, N.S., "Inelastic Lateral Buckling of Beams", ch. 3 of "Beams and Beam Columns: Stability and Strength", ed. R. Narayanan, Applied Science Publishers, London, 1983, pp 35-70.

9. Pastor, P.T., and DeWolf, T.J., "Beams with Torsional and Flexural Loads", Journal of the Structural Division, ASCE, Vol. 105, No. 3, March 1979, pp 527-535.

10. Chen, W.F. and Atsutu, T., "Theory of Beam-Columns", Vols 1 and 2, McGraw-Hill, New York, 1976 and 1980.

EXPERIMENTAL DETERMINATION OF STRAIN CONCENTRATION FACTORS IN RHS TRUSS GAP K-CONNECTIONS

JEFFREY A. PACKER **GEORGE S. FRATER**
Professor Research Assistant
Department of Civil Engineering, University of Toronto, 35 St. George St.,
Toronto, Ontario M5S 1A4, Canada.

and

KIM S. ELLIOTT
Lecturer, Department of Civil Engineering, University of Nottingham,
University Park, Nottingham NG7 2RD, England.

ABSTRACT

Two large-scale rectangular hollow section Warren trusses with welded gap K-connections have been tested elastically under single panel point loading and extensive strain measurements have been made around three connections with different size welds. From this strain data, Strain Concentration Factors have been determined according to accepted methods and compared with values predicted by current parametric formulae recommended by researchers at Delft University of Technology and University of Karlsruhe. Agreement with both procedures is good for the hot-spot on the bracing member, provided the fillet weld size is large relative to the member thickness, which is likely to be the case in most practical applications. For the hot-spot on the chord member, however, both of the above procedures significantly overestimate the Strain Concentration Factor, for large fillet welds. The influence of fillet weld size, connection noding eccentricity and location of a connection within a truss are all shown to be important influencing parameters and are discussed in the paper.

NOTATION

b_i = External width of square or rectangular hollow section (RHS) member i (90° to plane of truss) ($i = 0, 1, 2$)

e = Noding eccentricity of connection

g = Gap between bracing members at the junction with the chord face

g' = Non-dimensional gap size = g/t_0

h_i = External depth of square or rectangular hollow section (RHS) member i (in plane of truss) ($i = 0, 1, 2$)

H_i = Leg length of fillet weld on member i

i = Subscript to denote member of connection; $i = 0$ for chord, $i = 1$ for compression bracing member, $i = 2$ for tension bracing member

$I, I' =$ Non-dimensional strain indices $= \varepsilon/\varepsilon_{nom}$ or $\varepsilon/\varepsilon'_{nom}$ respectively

SCF = Stress Concentration Factor

SNCF = Strain Concentration Factor

$t_i =$ Thickness of hollow section member i $(i = 0, 1, 2)$

$\beta =$ Width ratio between bracing member and chord $= b_i/b_0$

$\varepsilon_{ab} =$ Nominal axial strain in the bracing member

$\varepsilon_{ac} =$ Nominal (prestressing) axial strain in chord

$\varepsilon_{hs} =$ Hot spot strain at weld toe

$\varepsilon_{mb} =$ Nominal bending strain in the bracing member at reference position (see Fig. 4)

$\varepsilon_{mc} =$ Nominal (prestressing) bending strain in chord

$\varepsilon_{nom} =$ Axial strain in bracing member $= \varepsilon_{ab}$

$\varepsilon'_{nom} =$ Maximum strain (axial + bending) in bracing member at reference position (see Fig. 4)

$2\gamma =$ Width to thickness ratio of chord member $= b_0/t_0$

$\tau =$ Thickness ratio between bracing and chord members $= t_i/t_0$

$\theta =$ Angle between bracing member and chord member

$\xi =$ g/b_1

INTRODUCTION

A great deal of research has now been undertaken to determine Stress Concentration Factors (SCF) and Strain Concentration Factors (SNCF) for welded tubular (circular) connections due to the fatigue problems associated with connection design in offshore steel jacket-type structures. This has taken the form of experimental measurements of strains in the vicinity of the joint, finite element and other numerical models, and photoelastic techniques, with the results having been incorporated into many national offshore structures design codes. Unlike circular tube connections, there has been a severe lack of information available for the fatigue design of rectangular hollow section (RHS) connections due to the absence of extensive and reliable experimental data. RHS *gap* connections are much more fatigue-critical than their overlap connection counterparts, yet are much easier to fabricate and hence popular. In certain dynamically-loaded RHS structures such as bridges and crane booms this lack of design guidance for fatigue performance represents a major setback.

The typical methods available to a design engineer for estimating fatigue life of structural connections are: (i) a visual classification of such connections into stress range categories, and (ii) the "hot spot stress" method.

The former approach is very inaccurate, but is included in many national steel building codes (e.g. [1]), yet generally does not cover welded RHS connections. Until very recently the classification approach was still cited [2] as being the most appropriate for square hollow section connections (due to insufficient data), and connections were classified into groups with nearly the same fatigue resistance (K overlap, N overlap, K and N gap), subject to a particular range of validity for various geometric parameters. In a comprehensive international state-of-the-art report by CIDECT [3], this method was still cited as being the best available for *square* hollow section connections, even though a hot spot stress method was advocated for circular hollow section connections.

The hot spot stress method has been found to be a more accurate and reliable method for treating fatigue in offshore circular tube connections and this method was first tentatively recommended by the International Institute of Welding (IIW) Subcommission XV-E [4] for square and rectangular tube connections. In these recommendations SCF values were given for different connection types as a function of the thickness ratio of the tubes being joined, with a minimum SCF value of 3.0. These SCF values could then be applied to the nominal stress in the bracing member to determine the hot spot stress range (S_r), and hence the fatigue life (cycles to failure) from a set of S_r-N curves. It was recognized that more accurate SCF values were necessary, based on multiple connection parameters. This translates into obtaining more accurate SNCF values by experimental or analytical research. A European collaborative project in the Netherlands and the Federal Republic of Germany was accordingly started to rectify this dearth of SNCF data for RHS welded truss-type connections by means of undertaking finite element analyses and a limited amount of RHS *isolated* connection testing. The results of these European projects are now being published [5,6,7] and recommended SNCF parametric equations are being proposed. Revised international recommendations will soon be prepared by the IIW tubular connections committee, but these studies need to first be verified by SNCF measurements on *large-scale* RHS welded connections in *complete* trusses, which is the topic of this paper.

STRAIN CONCENTRATIONS AT WELD TOES IN RHS CONNECTIONS

Strain concentrations in RHS connections are the result of complex structural interaction between two (or more) flexible-walled, welded tubes. Strain maxima are found at weld toes because:
   (i) this is the nearest position to the intersection of the tube walls where the tube walls are unreinforced by the weld, and
   (ii) a geometric discontinuity is present.
In describing the degree of strain concentration, the weld toes on the chord and bracing member tubes must be treated separately because the loadings differ.

Axial forces and in-plane bending moments in the bracing member of a planar truss connection, (such as shown in Fig. 1), are transmitted through the body of the weld to the chord face. As the chord connecting face is relatively flexible, distortions of the chord face occur, particularly for low bracing member to chord width ratios, along with local bending of the bracing member walls. This complex three-dimensional behaviour is further complicated by the shape of the RHS members which are stiffer in their corners than in the midface. The result is a maldistribution of loading to the stiffer parts, which accounts for why the greatest strain concentrations have been found near the weld toes at the corners of the RHS bracing member, and in the gap region for the chord member [5].

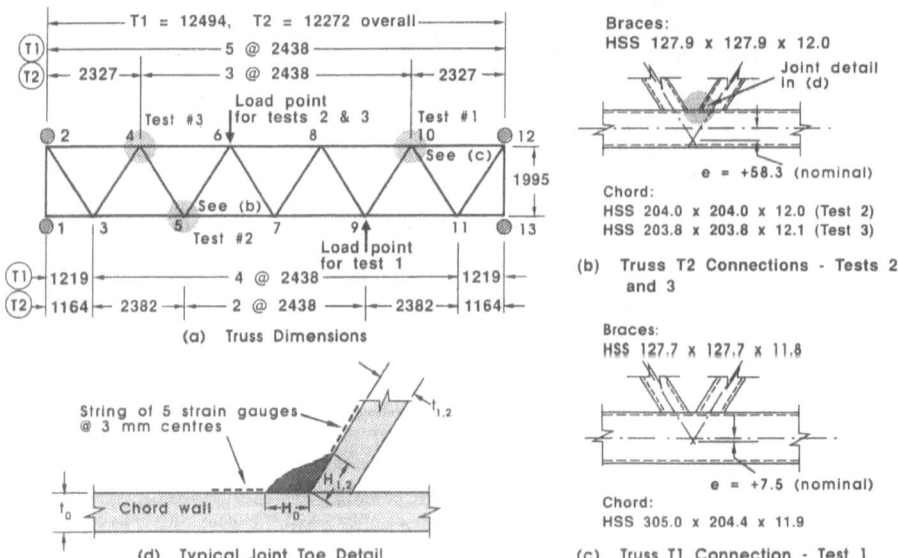

Figure 1. **Details of RHS trusses T1 and T2 tested, with measured member dimensions.**

Unfortunately it is not possible to measure these strain maxima exactly *at* the weld toes because of physical restrictions. Surface strain measurements in RHS connections are usually obtained from electrical resistance strain gauges attached to the outer surfaces of the steel tubes. The stresses computed from these strains are used to predict the maximum principal, or hot-spot, SCF. Because the hot-spot SNCF occurs right at the weld toe, it is necessary to extrapolate the strain from two or more points, as shown in Fig. 1 (d). The magnitude of the hot-spot strain then depends on:

(i) the position at which it is measured, and

(ii) the gradient and nature of the curve used in the extrapolation.

It is convenient to define the *position* by the weld leg length, (i.e. by the toe of the weld), as this can be easily measured, to an accuracy of ± 1 mm. *Gradients* must be evaluated using "best" approximations of data obtained from strings of strain gauges. These data must be collected from gauges located within prescribed distances from the weld toe. These distances, together with the most popular methods of extrapolation, (namely linear and quadratic), are now reasonably well established and are summarized in Fig. 2 [5]. In order to verify the strain concentration data generated by other authors, (for example [5,6,7,8]), these two extrapolation procedures are used herein, in the manner shown in Fig. 2.

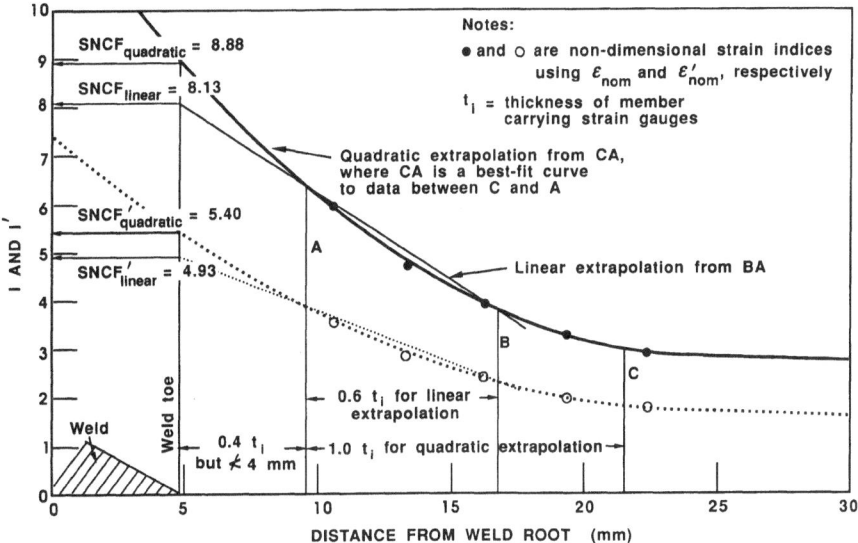

Figure 2. SNCF determination by linear and quadratic strain extrapolation methods [5] using ε_{nom} and ε'_{nom} (Fig. 4) for location B on bracing member (Fig. 3) in Test 2 (Truss T2, Connection 5).

EXPERIMENTAL WORK

Two large-scale, 12.3 m and 12.5 m span, simply-supported RHS Warren trusses as shown in Fig. 1(a) have been tested elastically under single panel point loading to produce strain data for three RHS gap K-connections. Fig. 1(b) and (c) illustrates the main difference between Truss T1 and Truss T2; i.e., the chord depth of h_0 = 304.8 mm (with e = 7.5 mm) and h_0 = 203.2 mm (with e = 58.3 mm), respectively. Horizontal dimensions for Truss T1 differ from those in Truss T2 slightly due to two 50% overlapped connections (Nos. 3 and 11) in Truss T2; otherwise all connections are gapped with a gap size (g) of 38 mm. Other pertinent SNCF non-dimensional parameters for both trusses are almost identical $(\tau = 1.0, \ \beta = 0.625, \ g' = 3.2, \ 2\gamma = 17)$.

Fig. 3 shows SNCF strain gauge "measurement lines" for the three joints tested, which co-incide with the inside wall line of the bracing member. To avoid the region of influence of notch strain the nearest SNCF strain gauge was placed $0.4t_i$ from the weld toe, or 5 mm to 6 mm. (See Fig. 2). Extensive strain gauge instrumentation elsewhere on all of the truss members enabled the determination of truss member axial loads and in-plane bending moments. At a distance of 2.5 b_i (see Fig. 4) up the bracing member from the chord face, (measured along the centreline of the bracing member), the bending strain in the tension bracing members ranged from 27% to 35% of the total strain (axial + bending). At the chord face/brace centreline position (see Fig. 4), this ratio ranges from 38% to 42% which translates into a 1.62 to 1.74 amplification factor to increase axial stresses to account for the secondary bending moment in the connection, and slightly exceeds the 1.5 factor recommended by IIW [4].

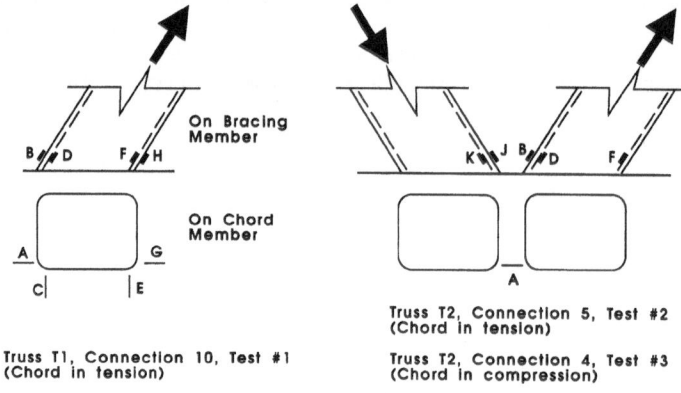

Figure 3. SNCF strain gauge string locations.

All SNCF results were established by normalizing strains in the bracing member as well as the chord with respect to a nominal longitudinal strain, ε_{nom} (axial strain only) or ε'_{nom} (axial plus bending strain), in the bracing member. In the case of SNCF strings on the chord (at A, C, E and G in Fig. 3), an adjustment was carried out according to van Wingerde [9] to make experimental truss results (with varying chord loadings) consistent with isolated K-joint test results, as follows:

$$\text{SNCF}_{chord, \ lines \ C,E} = \varepsilon_{hs} + 0.4(\varepsilon_{ac} + \varepsilon_{mc})/\varepsilon_{nom} \ (\text{or} \ \varepsilon'_{nom}), \quad \text{and} \tag{1}$$

$$\text{SNCF}_{chord, \ lines \ A,G} = \varepsilon_{hs} - 1.6(\varepsilon_{ac} + \varepsilon_{mc})/\varepsilon_{nom} \ (\text{or} \ \varepsilon'_{nom}) \tag{2}$$

By normalizing with respect to ε_{nom} and ε'_{nom}, to produce I and I', SNCF values obtained could be directly compared to those given by parametric formulae recommended by researchers at Delft [5] (using ε'_{nom}) and Karlsruhe [7] (using ε_{nom}). For comparison with the Delft formulae, the reference position for the bracing member nominal bending strain is taken as the intersection of the bracing member centreline with the chord face (see Fig. 4) [9].

SNCF values, by both linear and quadratic extrapolation, and using ε_{nom} and ε'_{nom}, for all test locations on trusses T1 and T2 are recorded in Table 1. Fig. 2 illustrates how 4 particular SNCF entries (location B, Test 2) in Table 1 were determined. Quadratic extrapolation was performed by plotting a least-squares regression curve fit to data with an algorithm by Marquardt [10]. Finally, strain gauge rosettes established the direction of the maximum principal strain at locations B, D and F on the bracing member, which were found to be generally parallel to the measurement line and having an intensity equal to the strain recorded within the SNCF strain gauge string.

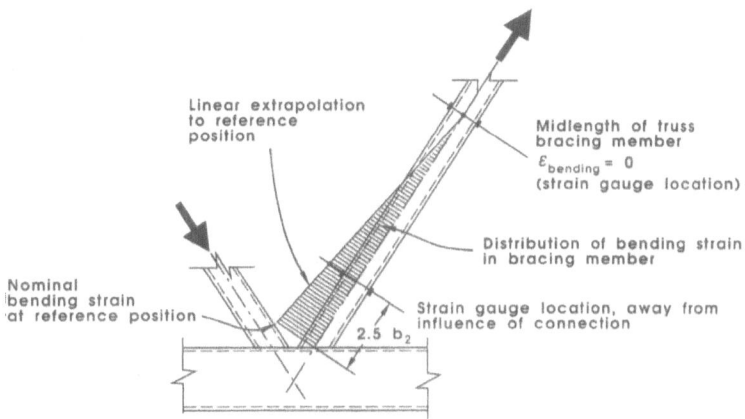

Figure 4. Determination of nominal bending strain for calculating ε'_{nom} and I'.

TABLE 1

Experimental SNCF values (linear and quadratic) using ε_{nom} and ε'_{nom}.

Test Number	Location	$H_{0,1,2}$ (mm)	SNCF' (using ε'_{nom}) Quadratic	SNCF' (using ε'_{nom}) Linear	SNCF (using ε_{nom}) Quadratic	SNCF (using ε_{nom}) Linear
1	Brace B	5.9	5.86	5.11	9.52	8.31
Truss T1	D	6.3	4.77	4.32	7.75	7.02
Connection 10	F	7.4	2.71	2.56	4.40	4.15
	H	7.6	1.45	1.30	2.35	2.11
	Chord A	6.7	2.96	2.78	4.80	4.52
	C	8.5	1.41	1.40	2.29	2.27
	E	7.6	2.76	2.46	4.48	3.99
	G	9.6	-0.07	-0.10	-0.11	-0.17
	A*	10.5	2.18	2.43	3.38	3.77
2	Brace B	4.9	5.40	4.93	8.88	8.13
Truss T2	D	5.3	3.63	3.26	5.98	5.37
Connection 5	F	5.6	2.24	2.15	3.70	3.55
	J*	16.5	3.55	3.31	5.72	5.33
	K*	15.0	2.86	2.61	4.61	4.20
	Chord A	5.6	2.89	2.84	4.76	4.69
	A*	19.5	1.68	1.71	2.72	2.76
3	Brace B	11.6	3.46	3.17	6.02	5.52
Truss T2	D	11.8	2.45	2.35	4.27	4.08
Connection 4	F	10.4	2.05	1.95	3.57	3.40
	J*	12.8	4.49	3.90	7.91	6.87
	K*	11.4	3.02	2.73	5.33	4.81
	Chord A	13.7	0.95	1.00	1.65	1.75
	A*	14.9	1.39	1.29	2.46	2.28

* Using ε_{nom} and ε'_{nom} on compression member

Note: String A in Tests 2 and 3 was offset from the measurement line by 7.5mm (63% t_o) and 11mm (92% t_o), respectively, because the gap distance at the measurement line could not accommodate the 5-element strain gauge.

EVALUATION OF RESULTS

Test results show that the maximum SNCF values occur in the gap region, in either the chord member or bracing member wall. i.e. At location B in tension bracing member, (or location J in compression bracing member), and location A in chord. This is in agreement with both Mang et al [7] and Puthli et al [5]. At these critical locations, the difference in SNCF values determined by quadratic and linear extrapolation procedures is fairly small, with the "quadratic SNCF" being upto 15% greater than the "linear SNCF". Since tests 2 and 3 or Truss T2 have connections which are nominally identical except for their weld sizes, these reveal that the SNCF at a particular location depends on relative weld size (H_i/t_i) – see Table 1 – with SNCF *decreasing* as weld leg size *increases*. This is illustrated in Fig. 5 for locations B, D and F on the tension bracing member, and locations J and K on the compression bracing member. The notable influence of relative weld size (H_i/t_i) and weld profile has also been observed by other researchers [for example 11, 12, 13] on connections between other structural members or tubular (CHS) connections, and is considered to a certain extent in the AWS D1.1 code [14].

It is interesting to note that this change in SNCF with weld size is only consistent for connections with the same noding eccentricity. (e.g. B2 to B3, or D2 to D3 in Fig. 5, which are all measured on Truss T2). If one compares B1 (location B on Truss T1 with $e \sim 0$) with B2 (location B on Truss T2 with $e \sim h_o/4$) in Fig. 5, however, it can be seen that SNCF *decreases* as weld leg size *decreases*, a reverse trend to before which must be due to the difference in noding eccentricity. These results suggest that the addition of a positive noding eccentricity in a typical gap connection has the beneficial effect of reducing the SNCF at a particular location in a tension bracing member. Parametric SNCF formulae currently proposed by both Delft [5] and Karlsruhe [7] do not include the connection noding eccentricity as a parameter, although the Delft formulae are stipulated to be valid only for $e = 0$ [5] and the Karlsruhe formulae are quoted to have the eccentricity effect "implicitly contained in the indicated SNCF values" [7].

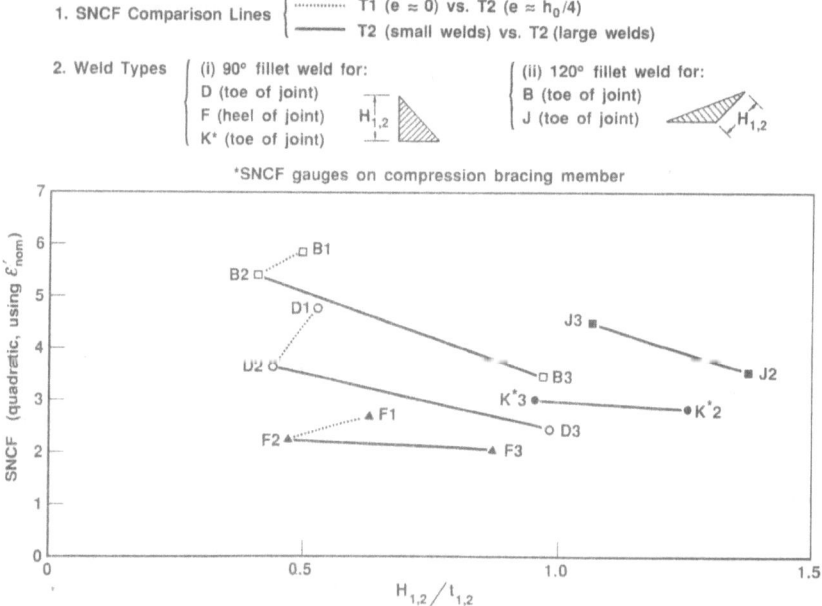

Figure 5. Experimental SNCF vs. H/t for Tests 1, 2 and 3 on Trusses T1 and T2.

The Delft parametric formulae referred to previously [5] are principally based on finite element modelling with one isolated K-connection test, ($200 \times 200 \times 8$mm chord and $120 \times 120 \times 4$mm bracing members), whereas the Karlsruhe research programme [7] involved similar computer analysis but a comprehensive series of 24 tests on gapped K-connections with varying geometrical parameters. An additional CIDECT research programme [9] by Delft University of Technology will further investigate the effect of bending moment on SNCF, and parametric formulae will probably be revised again. The following formulae were proposed by Delft and Karlsruhe:

Delft [5]:

$$\text{SNCF}^{33\%}_{chord} = (2.84 - 3.1\beta) \left(\frac{2\gamma}{12.5}\right)^{(1.02+1.1\beta)} \left(\frac{\tau}{0.5}\right)^{(0.8+0.5\beta)} \tag{3}$$

$$\text{SNCF}^{33\%}_{brace} = 1.0 + (1.49 - 0.9\beta) \left(\frac{2\gamma}{12.5}\right)^{(3.13-2.55\beta)} \left(\frac{\tau}{0.5}\right)^{(-0.25+1.5\beta)} \tag{4}$$

The validity range for these formulae is:

$$0.25 \leq \tau \leq 1.0 \qquad 0.4 \leq \beta \leq 0.6 \qquad 12.5 \leq 2\gamma \leq 25 \qquad e = 0$$

For percentages of bending strain (ε_{mb}) to total strain ($\varepsilon_{ab} + \varepsilon_{mb}$) in the bracing member other than 33%, equations (3) and (4) should be corrected as follows:

$$\text{SNCF}^{M\%} = \left[1.13 - 0.4\left(\frac{\varepsilon_{mb}}{\varepsilon_{ab} + \varepsilon_{mb}}\right)\right] \text{SNCF}^{33\%} \tag{5}$$

Karlsruhe [7]:

$$\text{SNCF}_{chord} = \tau(0.00288\gamma^3 + g') + 5.21\xi(1 - 0.178\xi^2 g') - 0.1515\beta^3 g'^2 - 1.57 \tag{6}$$

$$\text{SNCF}_{brace} = 3.3\tau(2 - \tau) + 0.305\xi\gamma^2(0.3 - 0.01\xi\gamma) + 0.04\gamma\beta(6.38 - \gamma\beta^2) - 3.8\left(\frac{\gamma g'}{100}\right)^2 - 2.0 \tag{7}$$

The validity range for these formulae is:

$$0.4 \leq \beta \leq 1.0 \qquad 0.4 \leq \tau \leq 1.0 \qquad 12.5 \leq 2\gamma \leq 25$$
$$1.6 \leq g' \leq 7.1 \qquad 0.25 \leq \xi \leq 0.75 \qquad 35° < \theta < 60°$$

The range of validity for the Delft formulae is not met by two truss parameters; β for both trusses slightly exceeds the upper limit of 0.6 and $e = 58.3$mm in Truss T2 contravenes the $e = 0$ condition. But for the sake of comparison, Table 2 shows the SNCFs predicted by all of the above formulae relative to the maximum critical SNCFs from the truss tests (as determined by quadratic extrapolation).

Table 2 shows that the Delft SNCF parametric formulae underestimate the measured value for the *bracing* member in all cases, which is an unsafe trait. This is especially noteworthy in view of the fact that a positive noding eccentricity (Test Nos. 2 and 3) has been seen to *reduce* the measured in-situ SNCF on a tension bracing member. For the chord member, however, the Delft formulae are in reasonable agreement with the test results, except for when the chord is in compression in which case the predicted SNCF is very conservative.

For the Karlsruhe SNCF parametric formulae, the predicted SNCF values for the bracing member follow the same trend as for the Delft formulae, except the average amount of underestimation in SNCF is somewhat less. For the chord member the Karlsruhe SNCF predictions are extremely inaccurate and are excessively conservative.

TABLE 2

Comparison of predicted quadratic SNCFs (by equations 3 to 7) with maximum experimental SNCFs, related to the tension bracing member.

Member	Method for Normalizing Strains	Experimental SNCF (max. value)/Predicted SNCF		
		Test No. 1	Test No. 2	Test No. 3
Bracing	Using ε'_{nom} (Delft method)	$5.86/3.35 = 1.75$	$5.40/3.30 = 1.63$	$3.46/3.22 = 1.07$
	Using ε_{nom} (Karlsruhe method)	$9.52/6.69 = 1.42$	$8.88/6.60 = 1.35$	$6.02/6.55 = 0.92$
Chord	Using ε'_{nom} (Delft Method)	$2.96/3.26 = 0.91$	$2.89/3.20 = 0.90$	$0.95/3.07 = 0.31*$
	Using ε_{nom} (Karlsruhe method)	$4.80/17.17 = 0.28$	$4.76/16.84 = 0.28$	$1.65/16.29 = 0.10*$

Notes: * = Chord in compression.

CONCLUSIONS

It has been shown that weld size relative to member thickness (H_i/t_i) has a significant influence on the gap K-connection Strain Concentration Factor (SNCF), with SNCF decreasing as weld leg size increases. It has also been observed that the addition of a positive noding eccentricity to a connection, (with positive being towards the outside of the truss), tends to reduce the maximum SNCF in the tension bracing member.

For both the Delft [5] and Karlsruhe [7] SNCF parametric formulae, the agreement with test results is not particularly good across the full range of weld sizes examined, with both methods generally underestimating the bracing member maximum SNCF and overestimating the chord member SNCF. However, it could be argued that only Test No. 3 has weld leg sizes which are representative of those which might be required in practice, for a fatigue-critical truss. (At location B which was critical for the bracing member $H_2/t_2 = 0.96$, and at location A $H_0/t_0 = 1.14$). The latest IIW fatigue design recommendations [4], for example, require fillet welds with a throat thickness of at least $1.0t_i$. Nevertheless, parametric formulae applicable over a wider range of weld sizes would be desirable. If one focuses attention solely on Test No. 3 in Table 2, which had the largest weld sizes, the predicted SNCF values for the tension bracing member agree well with the test result, for both prediction methods. However, for the *chord* member in Test No. 3 the predictions by both methods are still poor, with the Delft method [5] offering more promise. Chord SNCFs are influenced by the location of the connection within a truss and the adjustment for compression chords, as evidenced by Test No. 3, appears to warrant further refinement. The difference between SNCFs determined experimentally by linear and quadratic extrapolation techniques was found to be small for the hot-spot locations (up to 15%), and a more important issue is the range of weld sizes for which recommended parametric formulae are valid.

ACKNOWLEDGEMENTS

The experimental programme described in this paper is part of a larger research project at the University of Toronto on RHS welded connections, sponsored by the Comité International pour le Développement et l'Etude de la Construction Tubulaire (CIDECT), Ipsco Inc., University Research Incentive Fund (Ontario Provincial Government), and the Natural Sciences and Engineering Research Council of Canada (NSERC), with the RHS members supplied by Stelco Steel and Ipsco Inc. The

SNCF study reported herein has been financially supported by NSERC and a NATO International Collaborative Research Grant. The support of all these agencies, as well as the collaboration of the staff of Delft University of Technology, is gratefully acknowledged.

REFERENCES

1. Canadian Standards Association, Limit States Design of Steel Structures, CAN/CSA-S16.1-M89, CSA, Rexdale, Ontario, Canada, 1989.

2. Wardenier, J., Hollow Section Joints, Delft University Press, Delft, Netherlands, 1982.

3. CIDECT, Fatigue Behaviour of Welded Hollow Section Joints, CIDECT Monograph No. 7, Constrado, England, 1982.

4. International Institute of Welding Subcommission XV-E, Recommended Fatigue Design Procedure for Hollow Section Joints. Part I — Hot Spot Stress Method for Nodal Joints. IIW Doc. XV-582-85, IIW Annual Assembly, Strasbourg, France, 1985.

5. Puthli, R.S., Wardenier, J., de Koning, C.H.M., Van Wingerde, A.M. and Van Dooren, F.J., Numerical and Experimental Determination of Strain (Stress) Concentration Factors of Welded Joints between Square Hollow Sections. Heron, 1988, 33(2), 1–50.

6. Van Wingerde, A.M., Puthli, R.S., de Koning, C.H.M., Verheul, A., Wardenier, J. and Dutta, D., Fatigue Strength of Welded Unstiffened RHS Joints in Latticed Structures and Vierendeel Girders. CIDECT Final Report 7E+7F-89/5E, Delft University of Technology, Delft, Netherlands, June 1989.

7. Mang, F., Herion, S., Bucak, O. and Dutta, D., Fatigue Behaviour of K-Joints with Gap and with Overlap made of Rectangular Hollow Sections. In Proceedings of the International Symposium on Tubular Structures, Lappeenranta, Finland, September 1989, Elsevier Applied Science Publishers pp. P5.07-1 to P5.07-13.

8. Soh, A.K., Too, H.K. and Wong, C.F., SCF Equations for T and K Square Tubular Welded Joints. In Proceedings of the 6th International OMAE Conference, Houston, USA, March 1987.

9. Van Wingerde, A.M., Personal Correspondence between Delft University of Technology and University of Toronto, January/February 1990.

10. Marquardt, D.W., An Algorithm for Least-Squares Estimation of Nonlinear Parameters. J. Soc. Ind. Appl. Math., 1963, 11(2), 431–441.

11. Gurney, T.R., Finite Element Analyses of some Joints with the Welds Transverse to the Direction of Stress. Welding Research International, 1976, 6(4), 40–72.

12. Marshall, P.W., Connections for Welded Tubular Structures. Houdremont Lecture, in Proceedings of the International Conference on Welding of Tubular Structures, Boston, USA, July 1984, Pergamon Press pp. 1–54.

13. De Back, J., Size Effect and Weld Profile Effect on Fatigue of Tubular Joints. In Proceedings of Conference on Safety Criteria in Design of Tubular Structures, Tokyo, Japan, July 1986, pp. 331–343.

14. American Welding Society, Structural Welding Code – Steel, ANSI/AWS D1.1-90, AWS, Miami, Florida, USA, 1990.

DYNAMIC BEHAVIOUR OF SOLIDS CLARIFIED BY HIGH SPEED PHOTOELASTICITY

KOZO KAWATA
Science University of Tokyo
Noda, Chiba Prefecture 278 Japan

ABSTRACT

The main five cases of dynamic behaviour of solids clarified by high speed
photoelasticity in Kawata's laboratory with my colleagues since 1965 are
stated, including both historical review and results obtained recently. They
reveal the superior character of photoelasticity as visualization and analysis method of dynamic behaviour of solids, such as stress concentration,
plastic region propagation front, etc., in both elastic and plastic states,
differing remarkably from the corresponding static behaviour.

1. VERIFICATION OF EXISTENCE OF MACH WAVE LIKE STRESS WAVE IN SOLID BY
   SUPERSONIC DISTURBANCE SOURCE TRAVELLING IN SOLID [1]

The phenomenon of Mach wave in solid was found photoelastically first and
published in 1967[1]. A bullet as a disturbance source (Figure 2) travelling with a speed larger than the stress wave velocity of polyurethane, to
raise Mach wave in the solid (Figures 3, 4, and 5). HIMAC 16H high speed
framing camera was used (Figure 1).

2. DYNAMIC STRESS CONCENTRATION ANALYSIS IN HIGH VELOCITY TENSION OF LONG
   STRIPS WITH NOTCHES OR HOLE [2][3]

The relation of dynamic stress concentration factor f_d varying with time vs.
static stress concentration factor f_s of very long strips with notches or
hole under dynamic tension (Figures 7-9) were analyzed by high speed photoelasticity by means of a multiple spark gap camera using hard epoxy resin
(Figure 6). The phenomenon of maximum stress position deviation towards incident side at the passage of stress wave front and the behaviour that f_d
increases with time and then tends to f_s were found. The results of corresponding FEM analyses (Figure 10) well support the above mentioned behaviour
obtained photoelastically.

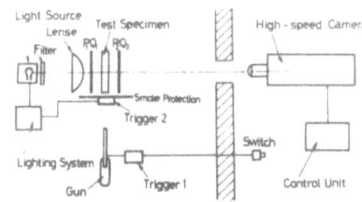

Figure 1. Experimental setup.
Light source: Ushio
Xenon lamp, UF-983.

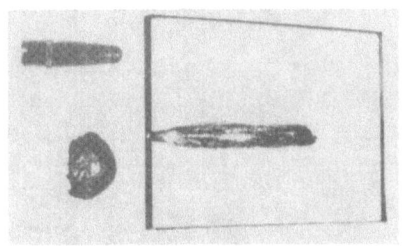

Figure 2. Disturbance source and
polyurethane plate.
Bullet: upper: original
form (diameter: 8mm,
weight: 10g), lower:
after recovery.

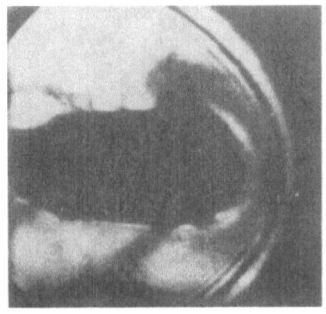

Figure 3. Mach wave like stress
wave in solid. Stress
wave velocity c=132m/s,
bullet velocity v=545
m/s, framing rate=1.18
$\times 10^5$ pps.

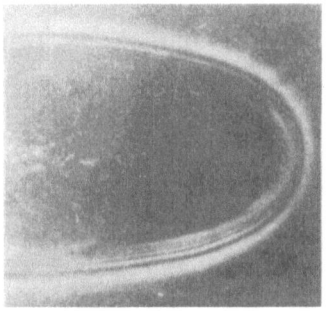

Figure 4. ditto. c=34m/s, v=540
m/s, framing rate=1.07
$\times 10^5$ pps.

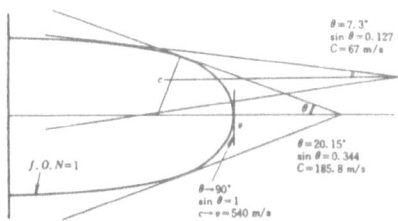

Figure 5. A sketch of isochromatic line of fringe order 1,
corresponding to Figure 4.

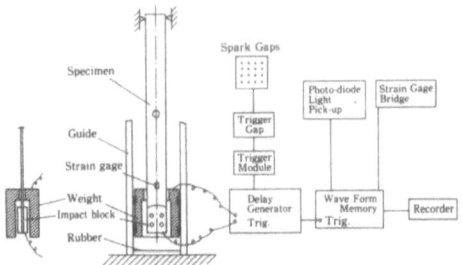

Figure 6. Experimental arrangement.

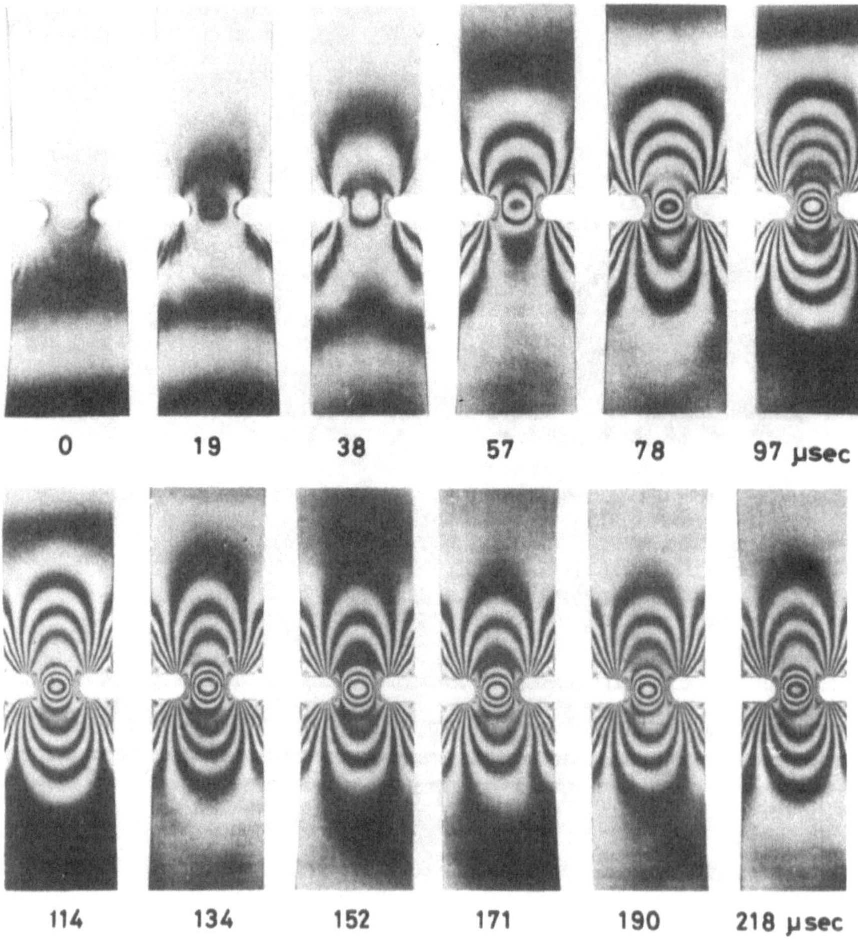

0   19   38   57   78   97 μsec

114   134   152   171   190   218 μsec

Figure 7. High speed photoelastic fringe patterns for dynamic
tension of epoxy strip specimen with U-shaped notches
at both sides. Radius of curvature R = 5mm.

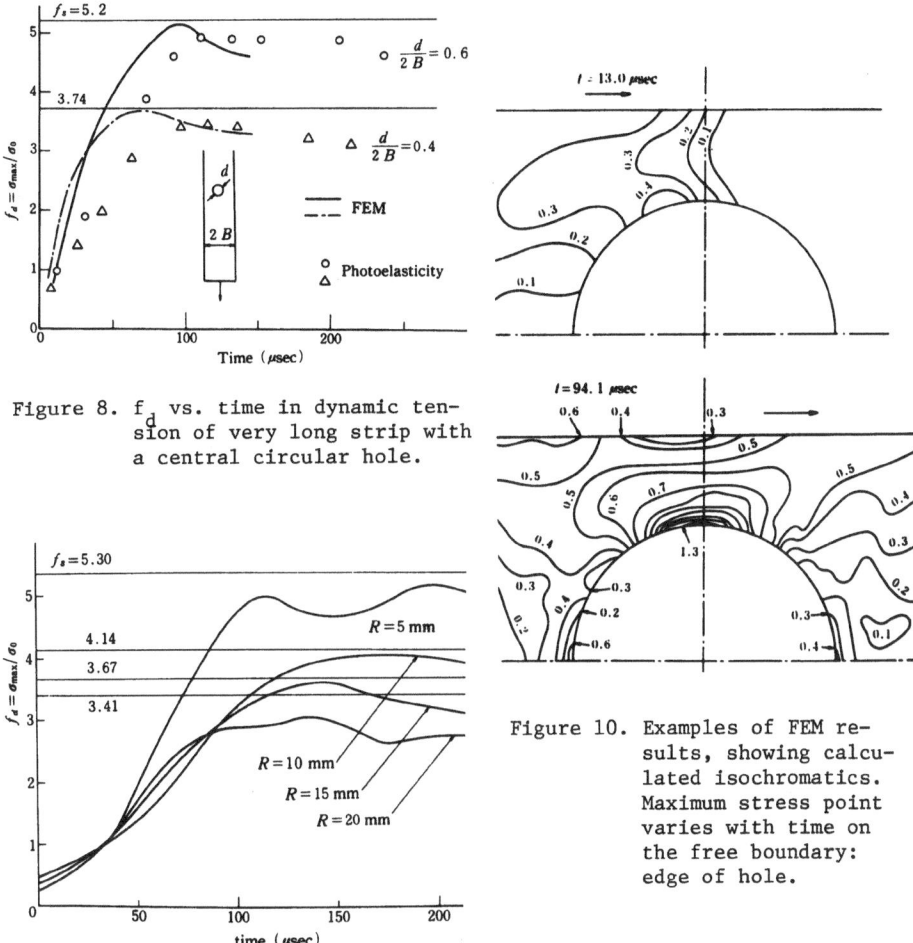

Figure 8. f_d vs. time in dynamic tension of very long strip with a central circular hole.

Figure 10. Examples of FEM results, showing calculated isochromatics. Maximum stress point varies with time on the free boundary: edge of hole.

Figure 9. f_d vs. time in dynamic tension of very long strip with U-shaped notches. R=5, 10, 15, and 20 mm. Results by photoelasticity.

3. DYNAMIC FRACTURE ANALYSIS OF COMPOSITES [4]

As the composites such as CFRP and GFRP are not transparent, photoelastic coating method (coating material: epoxy rubber) was used. Using a multiple spark gap camera (Figure 11), the relation of f_d vs. time was obtianed for composites (Figure 12). Further the initiation of dynamic fracture in composite specimen was clearly detected.

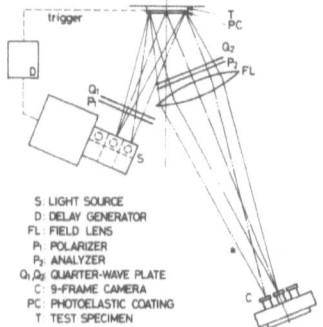

S: LIGHT SOURCE
D: DELAY GENERATOR
FL: FIELD LENS
P₁: POLARIZER
P₂: ANALYZER
Q₁,Q₂: QUARTER-WAVE PLATE
C: 9-FRAME CAMERA
PC: PHOTOELASTIC COATING
T: TEST SPECIMEN

Figure 11. Optical system of multiple spark gap reflection type camera.

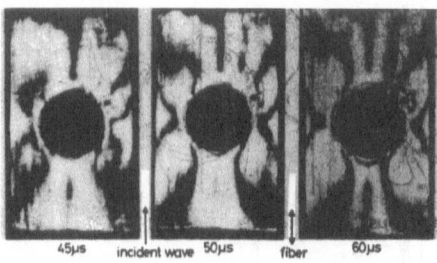

Figure 12. A sequence of high speed isochromatic fringe patterns for a unidirectional glass/polyester strip (D/W = 26/50).

4. ANALYSIS OF THE DIFFERENCES OF BCC AND FCC METALS IN DYNAMIC PLASTICITY [6]

Remarkable differences in plastic wave propagation behaviour of BCC and FCC metals were clarified and explained by Kawata [5] basing upon a micromechnical analysis, for high velocity tension of a thin bar of finite length. How about the case in which stress-concentrated area exists, is the subject of the present work. In high velocity tension of strips with notches of BCC and FCC metals (Figure 13), the difference in the propagation behaviour of plastic zone was found using photoelastic coating method (Figures 15-22). The experimental setup is shown in Figure 14. The ratio of maximum velocity in BCC metal to the one in FCC metal was about 6.4.

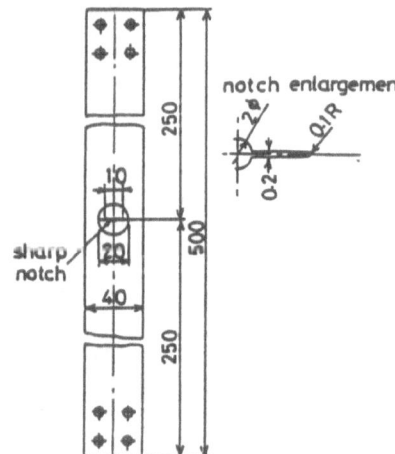

Figure 13. Specimen dimensions.

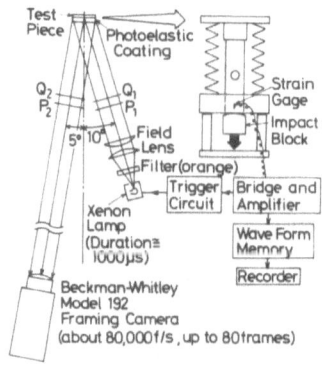

Figure 14. Experimental setup for photoelastic coating analysis of elasto-plastic wave propagation in metallic specimens.

331

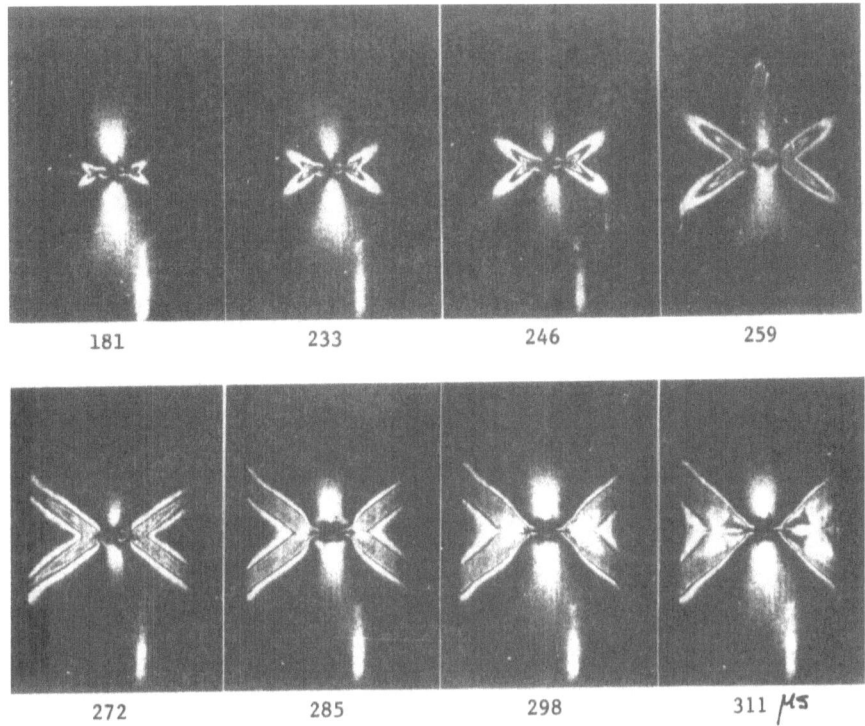

Figure 15. Photoelastic isochromatic fringe patterns in dynamic
tension of mild steel strip specimen with sharp notch.

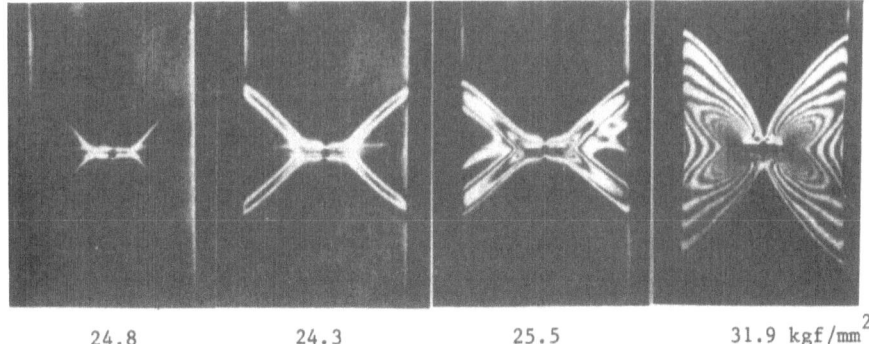

Figure 16. Photoelastic isochromatic fringe patterns in static
tension of mild steel strip specimen with sharp notch.

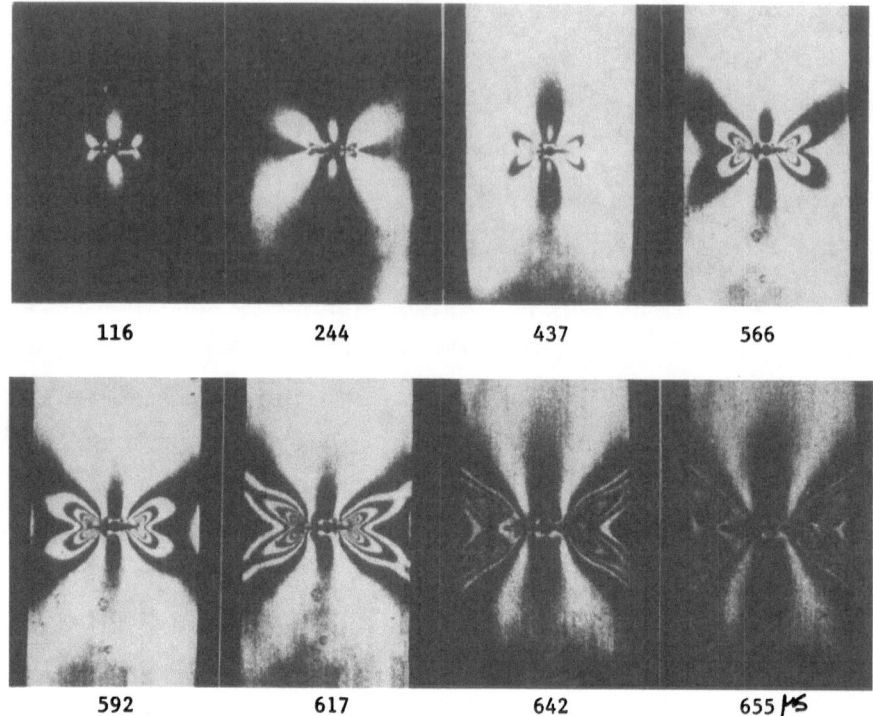

116   244   437   566

592   617   642   655 μs

Figure 17. Photoelastic isochromatic fringe patterns in dynamic tension of aluminum alloy 2024-T3 strip specimen with sharp notch.

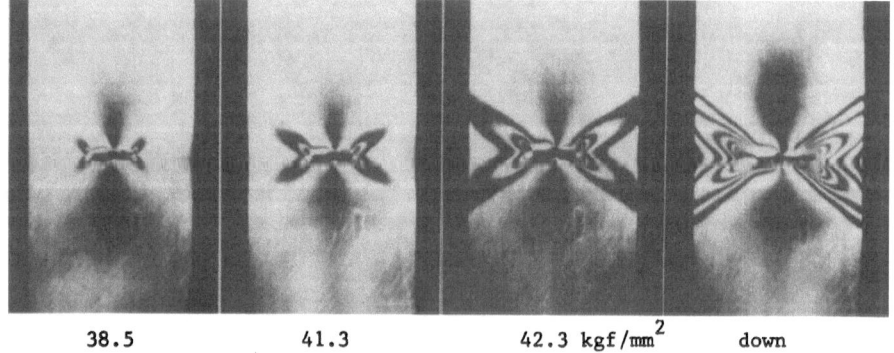

38.5   41.3   42.3 kgf/mm^2 down

Figure 18. Photoelastic isochromatic fringe patterns in dynamic tension of aluminum alloy 2024-T3 strip specimen with sharp notch.

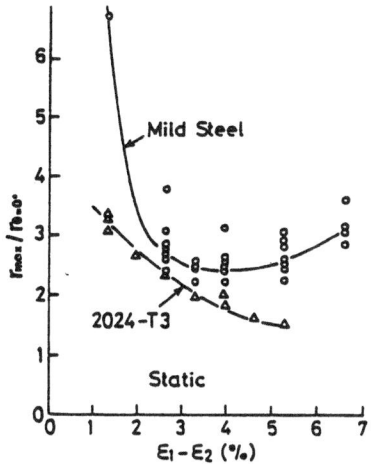

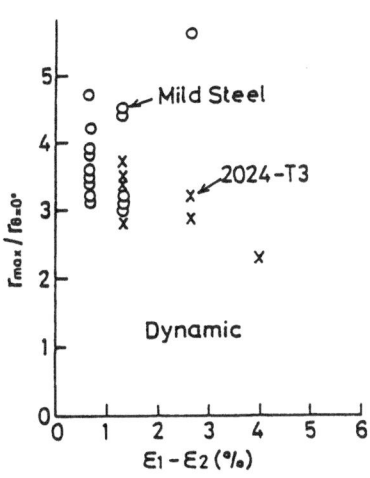

Figure 19. Plastic enclave ratio vs. principal strain difference for static tension.

Figure 20. Plastic enclave ratio vs. principal strain difference for dynamic tension.

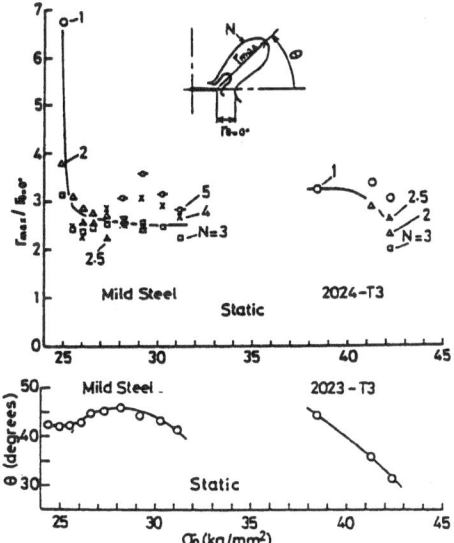

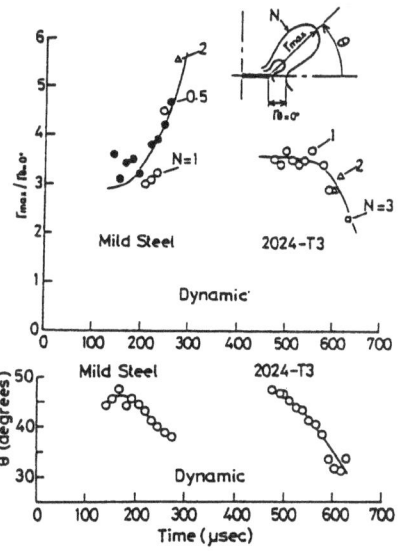

Figure 21. Plastic enclave ratio and plastic enclave angle vs. nominal tensile stress for static tension.

Figure 22. Plastic enclave ratio and plastic enclave angle vs. time measured from the time for which the fringe of order 1 appears at the sharp notch tip for dynamic tension.

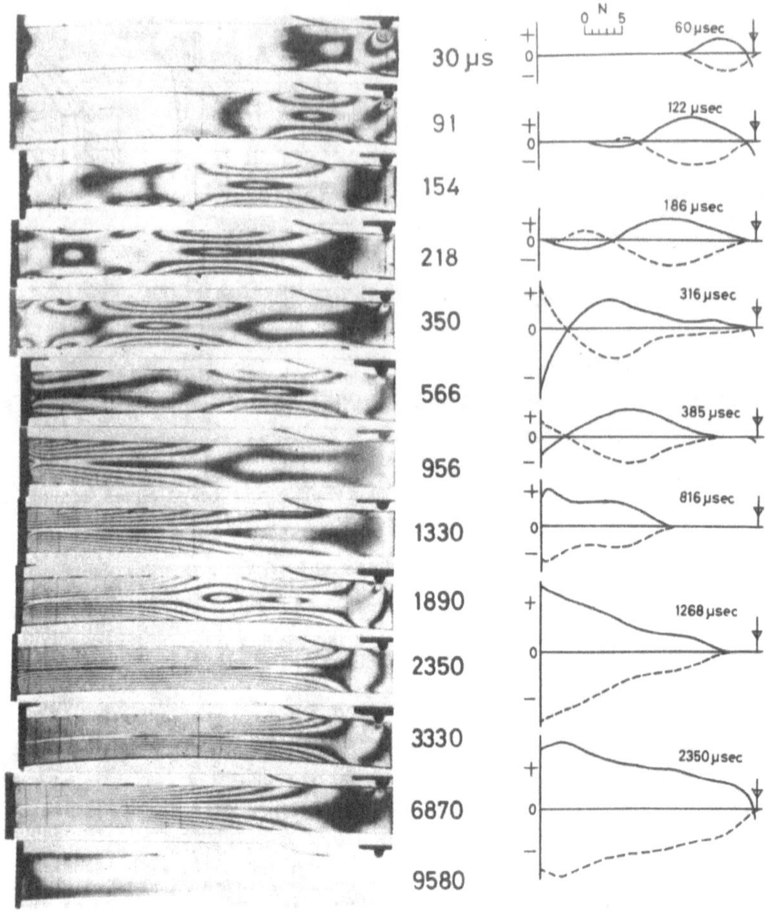

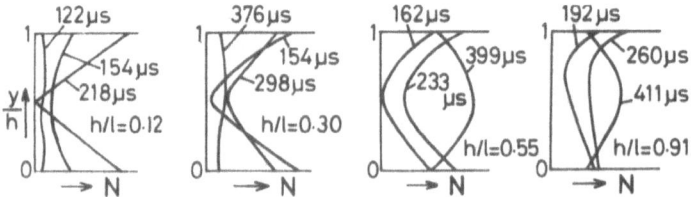

Figure 23. Photoelastic fringe patterns of beam of which depth/span ratio is 0.12 under impact bending and corresponding stress distribution along upper (solid line) and lower (dotted line) edge.

Figure 24. Fringe order distribution along the section at the middle point of span of the cantilever beam under impact bending.

5. ANALYSIS OF THE EFFECT OF DEPTH/SPAN RATIO OF CANTILEVER UNDER IMPACT BENDING [7]

Effects of depth/span ratio on bending stress wave propagation and on dynamic stress concentration factor in cantilever beam under transverse impact load were analyzed by means of a multiple spark gap camera using hard epoxy resin. The following facts were found: (1) The upper limit of depth/span ratio for generating bending wave is about 0.57 (Figure 24). (2) In the bending of thin bar, the first dynamic stress concentration occurs at the place near the impact point and next the second dynamic stress concentration occurs at the fixed end of the cantilever (Figure 23).

REFERENCES

1. Kawata, K. and Hashimoto, S., On photoelastic and theoretical analyses of some dynamic stress concentrations by elastic waves. JSME 1967 Semi-International Symposium Papers, Exp. Mech., Japan Society of Mechanical Engineers, Tokyo, 1967, Vol.1, 111–118.

2. Kawata, K., Hashimoto, S., Hondo, A. and Ide, T., On the dynamic stress concentration of very long strips with notches under high velocity tension. Bull. Institute of Space and Aeronautical Sciences, University of Tokyo, 1981, 17, 427–447.

3. Kawata, K. and Hashimoto, S., Dynamic stress concentration analysis in high velocity tension of strips with notches or hole by means of high speed photoelasticity. In Photoelasticity, ed. M. Nisida and K. Kawata, Spronger Verlag, Tokyo, Berlin, Heidelberg, New York, 1986, pp.73–79.

4. Kawata, K., Takeda, N. and Hashimoto, S., Photoelastic-coating analysis of dynamic stress concentration in composite strips. Exp. Mech., 1984, 24, 316–327.

5. Kawata, K., Micromechanical study of high velocity deformation of solids , Proc. 15th International Congress of Theoretical and Applied Mechanics , Toronto, North-Holland, Amsterdam, 1980, 307–317.

6. Unpublished.

7. Hashimoto, S. and Kawata, K., Analysis of impact bending of cantilever with various depth/span ratios by means of high-speed photoelasticity. In Photoelasticity, ed. M. Nisida and K. Kawata, Springer Verlag, Tokyo, Berlin, Heidelberg, New York, 1986, pp.81–88.

Dynamic Stress Analysis of a Three-Dimensional Solid Body

(Dynamic Stress Concentration Factor around a Cavity)

Masakatsu Sugiura*   and   Masaichiro Seika*
* Department of Mechanical Engineering,Daido Institute of Technology,2-21
Daido-cho,Minami-ku,Nagoya 457,Japan

abstract

The reflection and interference of stress-waves play important roles
in dynamic cases.It is very desirable to study the value of dynamic stress
concentration factors, in three-dimensional solid bodies.Hence,we have
analyzed the stress propagation and the dynamic stress concentration
phenomena around a spherical cavity in a cylindrical bar by utilizing the
strain gage method, the dynamic photoelastic method and also the finite
element method.Emphasis was laid on the dynamic stress concentration
analysis of bodies with inner cavities such as those often found in welded
parts or castings. The simulation was achieved by using test models of
composite specimen geometry in which a spherical cavity was introduced by a
variation of the sandwich method. We found that the three-dimensional
dynamic stress concentration factor obtained by the present three methods are
all near the static stress concentration factors already reported for
similar specimen geometries.

Introduction

In structure and machine parts subjected to dynamic load,due to the
complex reflection and interference of stress waves,the stress distribution
is remarkably different from cases subjected to static load, and sometimes
unexpected damage may occur. Accordingly, their stress analysis is
considered to be an important problem.So far, the problem of stress wave
propagation when an axial impact load is applied to a cylindrical bar having
uniform cross section has been theoretically and experimentally studied by
many researchers. As to the stress analysis of a spherical cavity inside a
cylindrical bar,especially regarding the static stress concentration problem
around a cavity,only the static theoretical solutions of Atsumi (1) and
Ling(2) and the research of Hui-Pih(3) by freezing photoelasticity have been
carried out so far, but the dynamic analysis has not yet been performed. As
to the related problem,there is only one example in which the analysis of
stress waves in a semi-infinite elastic/viscoplastic body having a spherical
cavity was carried out by Tanimura(4) by introducing the equation defining a

general strain rate tesor and using hyperbolic coordinates.Dynamic stress analysis of a cylindrical bar with finite diameter having a spherical cavity,such as those as actually used for machines and structures,have not been performed yet.

The authors paid attention to this point, and experimentally analyzed the time variation of the stress distribution around a spherical cavity which exists on the axis of a cylindrical bar. They especially analyzed the dynamic stress concentration,by the combined used of resistance wire strain gage and dynamic photoelasticity,and also separately determined the dynamic stress concentration factor by finite element method,and carried out a comparative study on the calculated values and the experimental values.

Experimental procedure and Test Model

In Figure 1, the block diagram of the experimental setup for carrying out this experimental is shown.First,in order to determine the time variation of the stress around a spherical cavity in a cylindrical bar by resistance wire strain gage method, an impact was given to the upper end of the test model.The rod I was first set at a certain height above the upper end of a test model,and then dropped freely along a vertical guide tube.

By this impact,compressive stress wave arose in the test model,and propagated downward.In the resistance wire strain gage method,in order to sense this stress wave,the authors placed two strain gages(gage length 1 mm) on the inner edge,and two strain gages on the outer edge of a spherical

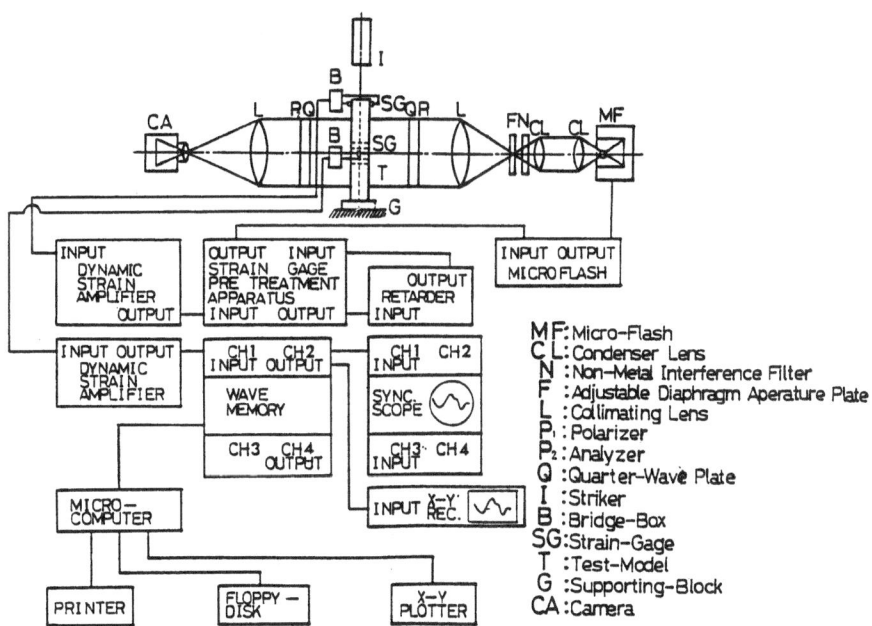

MF: Micro-Flash
CL: Condenser Lens
N : Non-Metal Interference Filter
F : Adjustable Diaphragm Aperature Plate
L : Collimating Lens
P₁ : Polarizer
P₂ : Analyzer
Q : Quarter-Wave Plate
I : Striker
B : Bridge-Box
SG: Strain-Gage
T : Test-Model
G : Supporting-Block
CA : Camera

Fig.1 Block diagram of experimental setup

cavity in the test model.The stress wave propagated was memorized in a waveform memory,and after the contents of the memory were displayed and confirmed with a synchroscope,the numerical conversion was performed with a personal computer,and the dynamic stress values were calculated.The results were printed out,and as occasion requires,the stress waveform was drawn in relation to time axis with an X-Y plotter.

In the experimental method by dynamic photoelasticity,the setup as shown in the upper part of Figure 1 was used. Compressive stress arose in a test model by impact loading,and propagated downward.One pair of polymicron strain gages of 1 mm gage length were attached to both surfaces of a test model at the positions 1 mm from the upper end,and the stress wave was detected when it reached the position of the strain gages. The signal was amplified with a dynamic strain amplifier,placed in a strain gage pretreatment unit and rectified into the uniform waveform ,and finally transmitted to a delaying equipment.When a light source was triggered after a given delay time which was required for a compressive wave to reach any intended position,a series of photographs was taken,and the state of the stress wave and the details of the concentration of fringes around a spherical cavity at each instant were observed in a light field or in a dark field. The experiment was carried out in a liquid with the same refractive index as that of a test model so that the light incident on the test model did not scatter uselessly.

The test model shown in Figure 2 is that for resistance wire strain gage method,as an example.As seen in Figure 2 ,a round bar of acrylic resin was cut to length L=150 mm,and a groove of 6 mm width and 1=30 mm length was cut one end of the bar.At one end of the groove,four half-sphere having diameter d=15, 20, 25 and 30 mm were cut by using a half-spherical end mill.The position was precisely controlled with a NC milling machine. The acrylic bars cut in this way were combined as one set. The acrylic plates and epoxy plates inserted in the round bars were finished into 1=60 mm length and W=40 mm width with a milling machine,accurately.

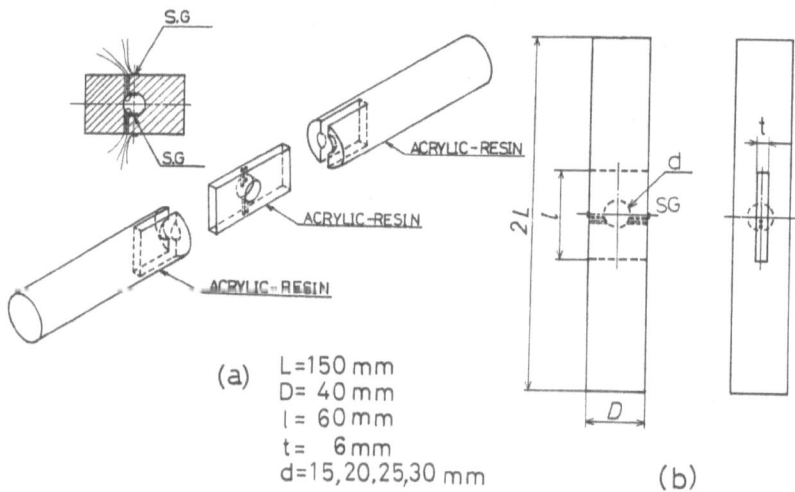

(a) L=150 mm
D= 40 mm
l = 60 mm
t = 6 mm
d=15,20,25,30 mm

(b)

Fig.2 Form of test model (for resistance wire strain gage method)

In the acrylic resin and epoxy resin plates finished to the prescribed dimensions, four circular holes having diameters d=15, 20, 25 and 30 mm were cut respectively with a lathe by one point cutting while sufficient cutting oil was always used so that thermal stress was not induced. In addition, on the acrylic resin plates for the resistance wire strain gage method , two strain gages were placed inside the circular. Besides,four guide tubes of diameter d=0.3 mm were attached to lead the four wires to outside.

The test models for resistance wire strain gage method and dynamic photoelastic method prepared in this way were assembled as shown in Figures 2(a) and 2(b). The adhesive was prepared by mixing epoxy resin adhesive (made by Cemedyne,1565) and a hardener(made by Cemendyne,1565) in a ratio of about 4:6. If the surfaces to be bonded were lightly polished with emery paper,at the time of adhesion,the separation seemed to be decreased. In order to sufficiently mix the adhesive, to promote the chemical reaction and to accelerate the hardening,the adhesive was put in a furnace kept around 60 °C,stirred sufficiently,and air bubbles were eliminated. It was uniformly applied to the acrylic resin bar,epoxy resin plate and acrylic resin plates,then those were assembled and fixed with vises and left at normal temperature for one full day to establish complete bonding without bubbles. The bonded test models were kept in a drier so that edge effect was not induced. A cylindrical rod of length L=300 mm and diameter d=57 mm was used as the falling rod for impacting. In Table 1, the physical properties of the models are shown.

Table 1 Physical properties of test model

	Epoxy resin	Acrylic resin
Young's modulus E (MPa)	3260.8	3131.4
Photoelastic sensitivity (mm/N)	0.10	0.01
Poisson's ratio ν	0.38	0.36
Specific gravity γ	1.23	1.20

Experimental results and discussions

Nominal stress at middle surfaces of cylindrical bars

For determining the nominal stress(σnom) at the middle surface of cylindrical bars the strain gage method was used. The strain gages of 1 mm gage length were placed on the circumference of the middle cross section of the cylindrical bars without cavity, and by measuring the strain detected by them, the norminal stress was determined. The effect of the waves reflected from the lower end of a test model and the upper end of the falling rod was taken into account for determining the nominal stress, and the average value of the stable stress at the hypothetical position of a spherical cavity when the delay time is not larger than T=200 μs was taken as the nominal stress. The variation of these stresses in relation to delay time were the values shown by d/D=0 in figure 3 for resistance wire strain gage method, and the values shown by d/D=0 in Figure 4 for dynamic photoelastic method.

Measurement of stress change on internal surfaces of spherical cavities in cylindrical bars in relation to time by strain gage method

Figure 3 shows the stress change on the internal surfaces of the spherical cavities in cylindrical bar test model in relation to time.When the ratio of the diameter of spherical cavities d to the diameter of cylindrical bars D was varied over d/D=0.375, 0.5, 0.625 and 0.75. It was seen that as the diameter of the spherical cavities increased, the maximum stress gradually become greater. Besides,as the model had relatively smaller ratio d/D, the time at which the maximum stress occurred seemed to decrease. Similarly,it was found that as the ratio d/D was smaller, the time during which stress became stabilized was longer.

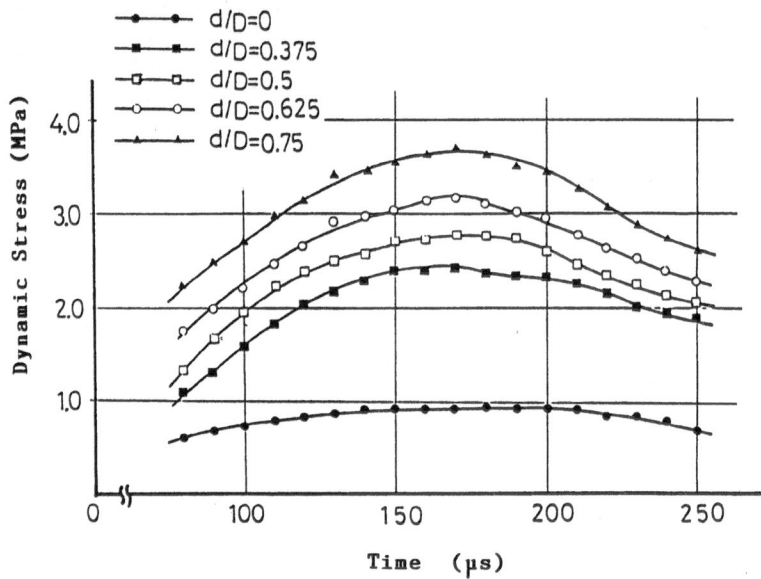

Fig.3 Relation of dynamic stress to time obtained by strain gage method

Measurement of stress change on internal surfaces of spherical cavities in cylindrical bars in relation to time by dynamic photoelastic method

Figure 4 shows the change of the stress in relation to delay time when the ratio of the diameter of spherical cavities to the diameter of cylindrical bars was varied over d/D=0.375, 0.5 and 0.625. Beside, the waveform of the stress change obtained by dynamic photoelastic method showed less variation than that obtained by strain gage method,and seemed to be rather stable.

In the strain gage method,the stress wave was sensed by 1 mm gage length while in the dynamic photoelastic method the stress in an epoxy resin plate averaged over t=6 mm thickness,therefore,the stress variation as a whole tended to appear somewhat lower.

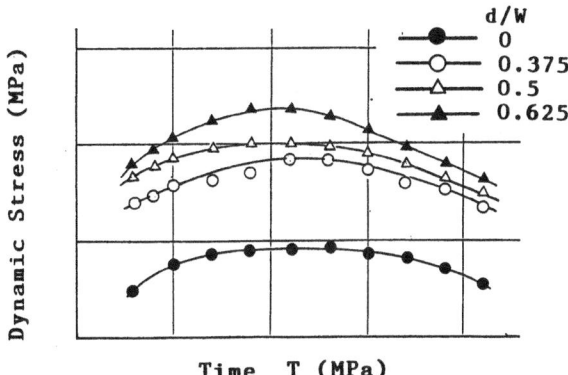

Fig.4 Relation of dynamic stress to time obtained by dynamic photoelastic
method

Analysis by finite element method and discussions

Method of analysis

The equation of motion at the time of infinitesimal deformation in an
elastic system is expressed as follows.

$$[M]\{\ddot{d}\}+[K]\{d\}=\{P\} \quad \cdots\cdots\cdots\cdots\cdots\cdots\cdots\cdots\cdots(1)$$

Where, $\{d\}$ is a nodal point displacement vector. $\{\ddot{d}\}$ is a nodal point
acceleration vector, $[M]$ is a mass matrix, $[K]$ is a stiffness matrix and $\{P\}$
is an external force vector. For solving the equation of motion(1),direct
numerical integration method(Newmark β method)is used. The nodal point
velocity vector and nodal point displacement vector at the state 1 are given
by those at the 0 as follows

$$\left.\begin{array}{l} \{\dot{d}_1\}=\{\dot{d}_0\}+\dfrac{\varDelta T}{2}[\{\ddot{d}_0\}+\{\ddot{d}_1\}] \\[2mm] \{d_1\}=\{d_0\}+\varDelta T\{\dot{d}_0\}+\left(\dfrac{1}{2}-\beta\right)\varDelta T^2 \\[2mm] \{\ddot{d}_0\}+\beta\varDelta T^2\{\ddot{d}_1\} \end{array}\right\} \quad \cdots\cdots\cdots\cdots\cdots\cdots(2)$$

When $\beta=1/4$ is assumed, Equation (2) becomes as follows.

$$\left.\begin{array}{l} \left([M]+\dfrac{\varDelta T^2}{4}[K]\right)(\{\dot{d}_1\}+\{\dot{d}_0\})=2[M] \\[2mm] \{\dot{d}_0\}-\varDelta T[K]\{d_0\}+\dfrac{\varDelta T}{2}(\{P_0\}+\{P_1\}) \\[2mm] \{d_1\}=\{d_0\}+\dfrac{\varDelta T}{2}(\{\dot{d}_1\}+\{\dot{d}_0\}) \end{array}\right\} \quad \cdots\cdots\cdots\cdots\cdots(3)$$

In calculation of Equation (3), a lumped mass matrix was used for $[M]$,
triangular ring element were used for $[K]$, and the displacement method in
which the distribution of displacement in a triangle is linear was used. The

time increment was taken as   T=2.5×10^{-6}sec.

Figure 5 is an example of the form of a model used for the calculation and the division into element.

(a) in the figure is the model without spherical cavity for obtaining nominal stress σnom and (b) is the model with a spherical cavity. As for the dimensions, L=300 mm and R=20 mm and S=150 mm were adopted. The number of elements was 271, and the number of nodal points was 170 in the model(a), Calculation was performed on four kinds of spherical cavity diameter D(2r)=15, 20, 25 and 30 mm. The material constants were assumed to be E = 3195.7 MPa, γ = 0.37 and γ = 1.22.

The lower end of the model was assumed to be supported and the upper end of the model was pulled by impact load Q. In order to obtain a stable solution, the stepwise load Q was increased according to Q=Q_0 (1-$e^{(-\alpha t)}$),where α =1.5x10^6 .

The calculation was carried out under the above condition,and the maximum stress obtained at a spherical cavity was denoted by σmax,while the stress at the center of a circular bar without spherical cavity at the hypothetical position of a spherical cavity was taken as nominal stress. From the ratio of both stresses, dynamic stress concentration factor αd was defined by the following equation.

$$\alpha_d = \sigma_{max} / \sigma_{nom} \quad \cdots\cdots\cdots\cdots\cdots(4)$$

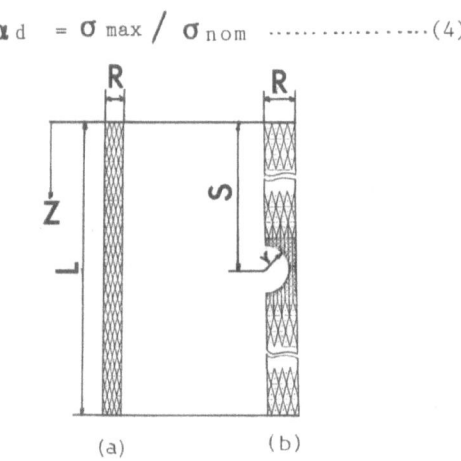

(a)   (b)

Fig.5 Division of test model into elements

Results of determining stress change on internal surfaces of spherical cavities in cylindrical bars in relation to time by finite element method

Figure 6 shows the calculated variation of the stress in relation to time in a cylindrical bar with out spherical cavity (d=0) and in the cylindrical bars with cavities,of which the values of the ratio d/D, the diameter of spherical cavities d to the diameter of cylindrical bars D, were changed variously.

As clearly seen in Figure 6, relatively stable stress variation was shown in a solid bar(d=0), but the stress in the cylindrical bars having spherical cavities varied greatly regardless of the values of d/D . Besides as the ratio d/D increased, the variation of stress in relation to time became larger, though the overall tendency was quite similar. In addition,as the ratio d/D increased, the time at which maximum stress occurred was at first delayed though slightly. The maximum stress occurred before the delay time T=250 μs when the reflected wave should reach the spot was regarded as the maximum stress(σ max) of that material.

As clearly seen in Figure 6, the stress after T=250 μs rapidly increased in all test models due to the effect of the reflected waves from the lower end of the test model.

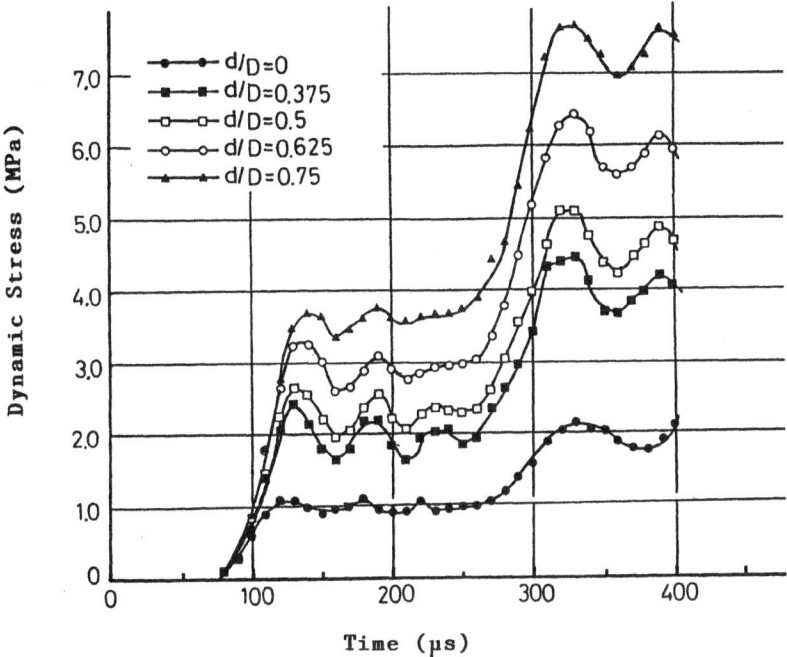

Fig.6 Relation of dynamic stress to time due to change of d/D obtained by FEM

Stress concentration factor

Figure 7 shows the experimental values of dynamic stress concentration factor(α_d).Dividing the maximum stress(σ max) arising on spherical cavity surfaces in cylindrical bars by the maximum stress in a cylindrical bar without cavity(nominal stress σ nom),the dynamic stress concentration factors (α_d)were obtained. They are also determined by numerical calculation (FEM).

For the convenience of comparison, the static theoretical values of Ling in a cylindrical bar with spherical cavity, and the static value by freezing photoelastic experiment of Hui-pih et al.(d/D=0.5) are shown together. As clearly seen in Figure 7 ,the values obtained by resistance wire strain gage method gradually increased as the ratio d/D became larger, and seem to be always larger than the static values of Ling. The values determined by FEM seemed to show somewhat lower values than those obtained by the resistance wire strain gage method, but as a general tendency,both were very similar. However,as d/D increased, the difference became larger. The obtained by dynamic photoelastic experiment showed, generally speaking lower values than those obtained by resistance wire strain gage method and FEM.As the cause of this effect, the difficulty of measuring fractional fringe in dynamic photoelastic fringe measurement, the undesirable effect of an adhesive layer, and the insufficient quantity of light due to the diffused reflection by cylindrical bars,may be mentioned .Anyway, the dynamic stress concentration factor α_d in the cylindrical bars having spherical cavities was in the vicinity of, and above or below the static theoretical and experimental values in the cylindrical bars having spherical cavities, and showed nearly similar change in relation to d/D.

For comparative examination , α_d on the external surfaces of the cylindrical bars having spherical cavities is shown in Figure 8. As clearly seen in Figure 8, in d/D up to about 0.5,values close to the nominal stress were obtained,but as d/D increased further, α_d on the external surfaces of cylindrical bars rose slightly, having been affected by spherical cavities.

Conclusions

The problem of dynamic stress concentration when the cylindrical bars having spherical cavities were subjected to impact load was analyzed by

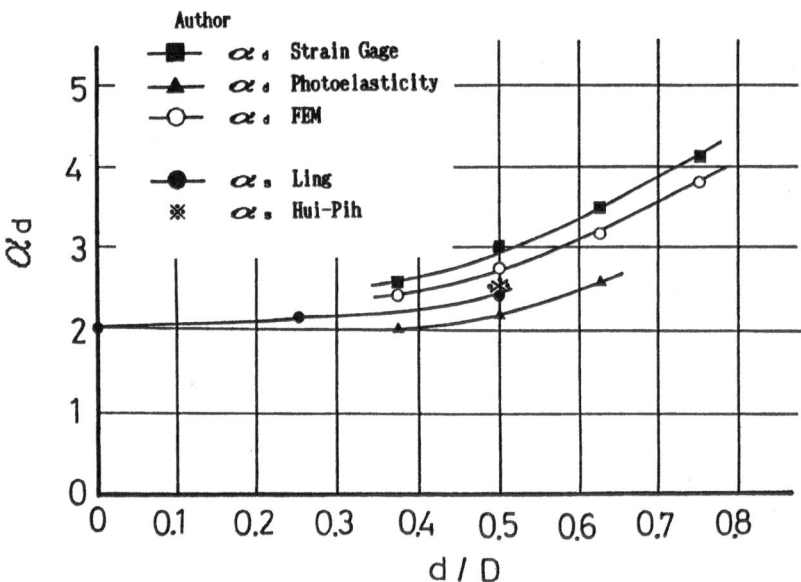

Fig.7 Relation of dynamic stress concentration factor to d/D

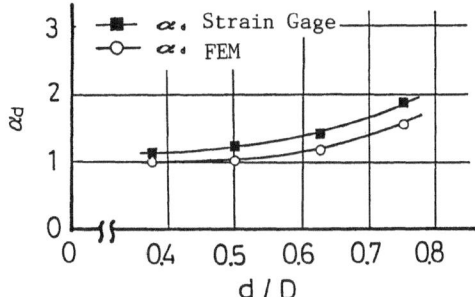

Fig.8 Relation of dynamic stress concenyration factor to d/D

resistance wire strain gage method ,dynamic photoelastic method and FEM.The following conclusions are obtained:

(1) As the ratio d/D of the diameter of spherical cavities to the diameter of cylindrical bars became larger, the dynamic stress concentration factor α_d gradually increased, and its tendency became stronger as the ratio d/D became larger. Above d/D =0.7, α_d rapidly increased.

(2) The dynamic stress concentration factor α_d determined by the resistance wire strain gage method achieved higher values as compared with the static theoretical and experimental values obtained so far, and when the ratio d/D decreased, it converged to the static values.

(3) The dynamic stress concentration factor α_d determined by FEM was close to the dynamic experimental values when the ratio d/D was small, but as the ratio d/D increased, the difference became larger.

(4) The experimental values of dynamic photoelasticity showed the lower values than those obtained by strain gage method and FEM owing to a variety of reasons, but it was found that for the problem in this experiment the difference remained as small as that this method was sufficiently applicable.

References

(1) Atsumi,A.,"Stresses in a Circular Cylinder Having an Infinite Row of Spherical Cavities Under Tension",Trans,ASME,Ser.E,27-1(1960),87-92.

(2) Ling,C.B.,"Stresses in a Circular Cylinder Having a Spherical Cavity under Tension",Q.Q.Appl.Math.,13-4(1956),381-391.

(3) Hui-Pih.,"Three-dimension Photoelastic Investigations of Circular Cylinders with Spherical Cavities in Axial Loading", Exp.Mech., 5-3(1965),90 -96.

(4) Tanimura,S., " Analysis of Stress wave in a semi-infinite elastic/viscoplastic body having a spherical cavity ", Trans   JSME.,(in Japanese),48-434 A(1982),1281-1290.

SHOCK WAVES IN PLATES UNDER ELASTIC AND VISCOELASTIC CONDITIONS -OPTICAL MEASUREMENT BY COMBINING PHOTOELASTICITY AND MOIRÉ

P.R. Dietz, Professor, **G. Wan**, Assistant Professor and **J. Albers**, Assistant Professor
Institut für Maschinenwesen, Technische Universität Clausthal, Clausthal-Zellerfeld, Germany

ABSTRACT

A hybrid optical measuring technique is introduced to measure the whole-field state of stresses and strains under dynamic loads. Using a combined optical bench the displacements are recorded by in plane Moiré technique and synchronously the isochromatics by photoelasticity. To separate the Moiré deformation data white-light processing is applied which also serves to multiply the Moiré fringes and to encode the data for archival storage in color films. The application of the technique is demonstrated for the example of stress waves in plates to simulate solid-sound propagation in machine parts.

INTRODUCTION

With increasing performance and acting speeds dynamic effects in the load and displacement behaviour of machine parts or in the development of new technologies in industrial processing engineering are gaining in importance. Generally, dynamic photoelasticity has proved successful as a basic investigation into the propagation of stress waves in elastic media. The propagation, reflection, superposition and interference of the stress waves are observed using isochromatic images. The repeated single-flash-method developed by Kuske [1] is mostly used in reproducible, non-destructing procedures in which a complete evaluation is achieved by recording fields of isochromatics and isoclinics in continuous time steps.

A complete evaluation using this technique is extremely time consuming due to the repeated isoclinical photographs and is also limited in its reliability due to the delay time action in the microsecond range. In some other problems, e. g. the investigation into the fracture phenomena of impact grinding or the determination of dynamic damping of sound wave propagation in solid materials, inelastic material effects play a major role. Such investigations require an extention of the single-flash-process to enable a complete description of stress and strain relations using an experimentally dynamic stress analysis.

To solve the problem a hybrid experimental technique is required which must supply the complete information necessary for an evaluation. For the investigation of dynamic problems the technique must be able to register separately the results from different measuring processes simultaneously and independently from each other. The hybrid measuring technique presented in this paper consists of a combination of photoelasticity and plane Moiré technique. The developed experimental set-up enables a simultaneous recording of isochromatics (difference of principal stresses) and Moiré isothetics (displacements). The data obtained are registered independently and separately using an appropriate measuring technique during a very short impact procedure taking place in the microsecond range. To separate the U- and V-isothetics from the Moiré crossed-grating fringes the white-light processing technique is applied, which enables furthermore a multiplication and a colour encoding of the isothetics. For the investigation of elastic problems the stresses registered in synchronous photographs can be evaluated experimentally. For viscoelastic problems the complete stress state must be evaluated by combining the isochromatics with isothetics using a creep or relaxation function or by taking additional isoclinical photographs in repeated experiments.

EXPERIMENTAL SET-UP

The experimental set-up for the measurement of dynamic stress states is derived from the single-flash-technique developed by Kuske [1] and is depicted schematically in **Figure 1**.

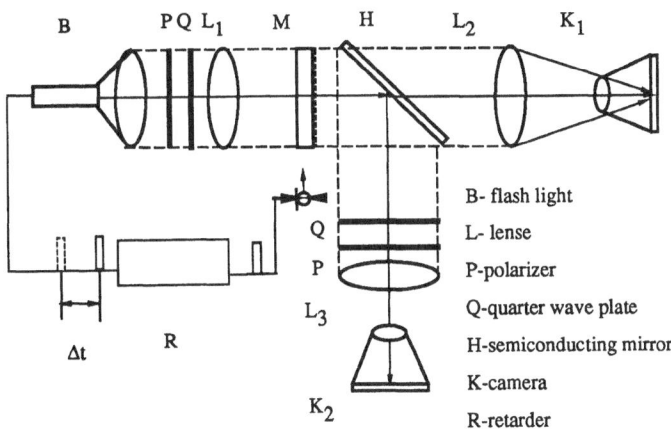

B- flash light
L- lense
P-polarizer
Q-quarter wave plate
H-semiconducting mirror
K-camera
R-retarder

Fig. 1: Combined optical bench for simultaneous photoelastic and Moiré measurement

After the impulse signal the flash is triggered in a certain time lag from the flash gun (B). Using the retarder (R) the time lag is adjustable to an accuracy of 1 μs. The light beam is converted by a polarizer (P_1) and a quarter-wave plate (Q_1) to circulary polarized light. A crossed grating is attached to the surface of the model (M) to act as specimen grating. When the circulary polarized light behind the model passes through the semitransparent mirror (H) it is divided into two beams. Via camera 1, optical path 1 produces the Moiré image, whereas optical path 2 is directed through the second quarter-wave plate (Q_2) and the analyzer (P_2) into camera 2 which records the isochromatics.

Dividing the light beam fulfills the decisive requirement on the combined measuring technique that stresses (isochromatics) and strains (isothetics) are recorded simultaneously and separately.

A direct recording of the isothetics, for which the master grating must be placed directly on the specimen grating, is rejected due to an unavoidable disturbance of the isochromatic image. As a result, the displacement images are recorded by applying the Moiré photography technique [4] ,with which the Moiré fringes are constructed in two stages: First, the deformed specimen grating is projected through the objective at a scale of exactly 1:1 onto a photo-active glass plate in the focus plane of the camera. In the second phase, the deformed specimen grating is superposed by a master grating causing the Moiré fringes to appear.

The advantage of this Moiré recording technique is that the rigid-body displacement, which is unavoidable due to the impact load, can be eliminated by the later pairing with the master grating. The prerequisite for this is the exact mapping of the displaced gratings on the focal plane of camera 1 on a scale of 1:1, which can be managed by using the master grating as an adjustment help. The accuracy of the deformation measurement is limited by the condition that the line density of the specimen grating does not overtax the resolution of the camera. The contrast of the photograph lies at 40% for a line density of 40 lines/mm. An increased line density causes the contrast to fall drastically.

SEPARATION AND MULTIPLICATION OF THE ISOTHETICS

The set-up shown in Figure 1 and the recording technique of Moiré photography allow the simultaneous and sharply divided registration of the stresses and deformations, but two problems remain unsolved. The individual U- and V-Isothetics from the "entire" Moiré fringes, which appear as a result of superposing the crossed specimen and master gratings, must be separated. Secondly, a multiplication of the displacement fringes is necessary because the line density of the gratings used is limited for the reason mentioned.

This problem is solved by employing the optical multiplication process by spatial filtering as developed by D. Post [2]: In principle a Fraunhofer diffraction pattern is produced by pairing the gratings. Using the spatial filtering technique, the useful information is formed by blocking portions of this pattern in the diffraction plane with reference to the orders of diffraction. In reconstructing the image by the unblocked portions of light in a defined diffraction order using an optical bench, the Moiré fringes can be multiplied integrally as required.

Fig. 2 shows a schematical diagram of the optical bench for the separation and multiplication of the isothetics. In the object plane E_1 the pairing of the specimen and master gratings is situated for which the degree of light transmissivity can be described by a transmissivity function $f(x,y)$. A parallel beam of light, which is generated by lense L_1, is shone through the gratings pairing. The light distribution $F(\xi,\eta)$ in the diffraction plane E_2 (focal plane of lense L_2) is then linked to the transmissivity function $f(x,y)$ using a Fourier transformation,

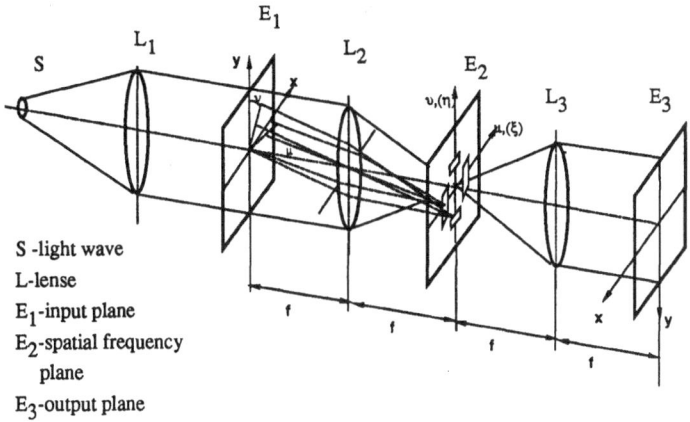

S -light wave
L-lense
E_1-input plane
E_2-spatial frequency
    plane
E_3-output plane

Fig. 2: Optical bench for the separation and multiplication of the isothetics

which corresponds with the phase difference due to the different optical path length in the diffracted beam.

Lense L_3 effects a Fourier transformation which transforms the light distribution $F(\xi,\eta)$ of the diffraction plane E_2 into the image plane E_3 and as such projects the image from E1 into E3 - rotated by 180°.

Multiplication of the Moiré fringes and separation of the isothetics ensue from manipulation by the optical Fourier transformation and spatial filtering in the diffraction plane E_2, whereby the changed information is converted to a photographically registerable linear pattern in the image plane E_3. The experimental procedure is described in the following, the mathematical relations in accordance with the underlying principles given by [9] are to be found in [12].

The diffraction pattern for a pairing of a deformed object grating and the (non deformed) master grating results in the image of a combination of two "connected" gratings. As shown in **Fig. 3**, each order of diffraction of grating 1 participates in that of grating 2 and as such in the whole diffraction order (see also [2, 4]).

The results of the mathematical correlations given in [12] and the consequences for the experimental multiplication are shown in **Fig. 4**. The quantified interference orders are seen in the diffraction plane (Figure 4a). The zeroth order lies in the centre, symmetrically to which the low frequencies are close to the axis, the high frequencies away from the axis. Multiplication ensues from spatial filtering, whereby only the image in the diffraction plane E_2 with the desired order of diffraction is reconstructed in the image plane E3. For this, only the desired order of frequency is allowed to pass through by blocking those remaining.

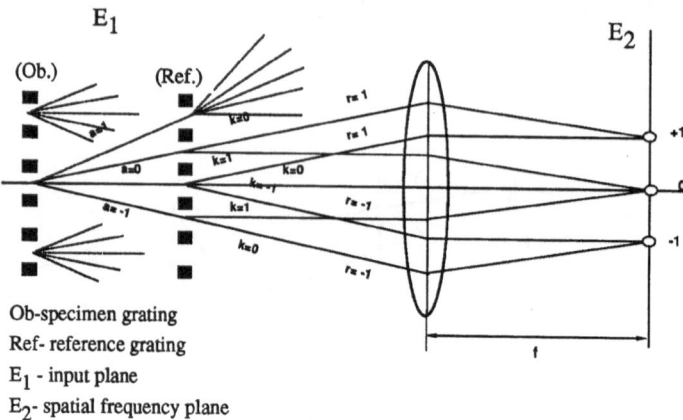

Ob-specimen grating
Ref- reference grating
E_1 - input plane
E_2- spatial frequency plane

Fig. 3: Diffraction of the pairing of Moiré gratings

If the light waves of the order of frequency of +1 and -1 are allowed to pass and all other frequencies are blocked, then an intensity distribution arises due to the retransformation equations as shown in Fig. 4c - the moiré fringes are not multiplied but the contrast of the image is enhanced (Fig. 4d).

In the same manner, the intensity distribution given in Fig. 4e arises if only the light waves of the order of diffraction of +2 and -2 are allowed to pass through. The Moiré fringes are multiplied twice (Fig. 4f). Due to the transformation relations no uniform contrast of the fringes occurs, the multiplied fringes appear brighter (Fig. 4f). This distribution is also to be seen in the intensity diagram (Fig. 4e). According to the same principle, the Moiré fringes can be multiplied n-times by blocking all orders but the n-th order of diffraction. In [5] a multiplication of the Moiré fringes up to eleven times is reported.

THE WHITE-LIGHT PROCESSING TECHNIQUE

In the presented paper the application of white light as light source in place of the hitherto exclusively used laser light for the multiplication technique is described. In addition to multiplication and separation of the isothetics it is also possible to apply white light to encode the fringe pattern for archival storage of simultaneously taken U- and V-isothetics in colour films. Colour variations facilitate the recognition of the plus or minus signs of the isothetics, and for the presented task of a synchronous recording the spatial frequency pseudocolor encoding is of significant advantage.

White light is also diffracted by transmission through line gratings and is interfered whereby the coherent wave length of the light waves increases due to diffraction and interference at the grating.

From Fourier transformation it can be seen that for the application of white light the positi-

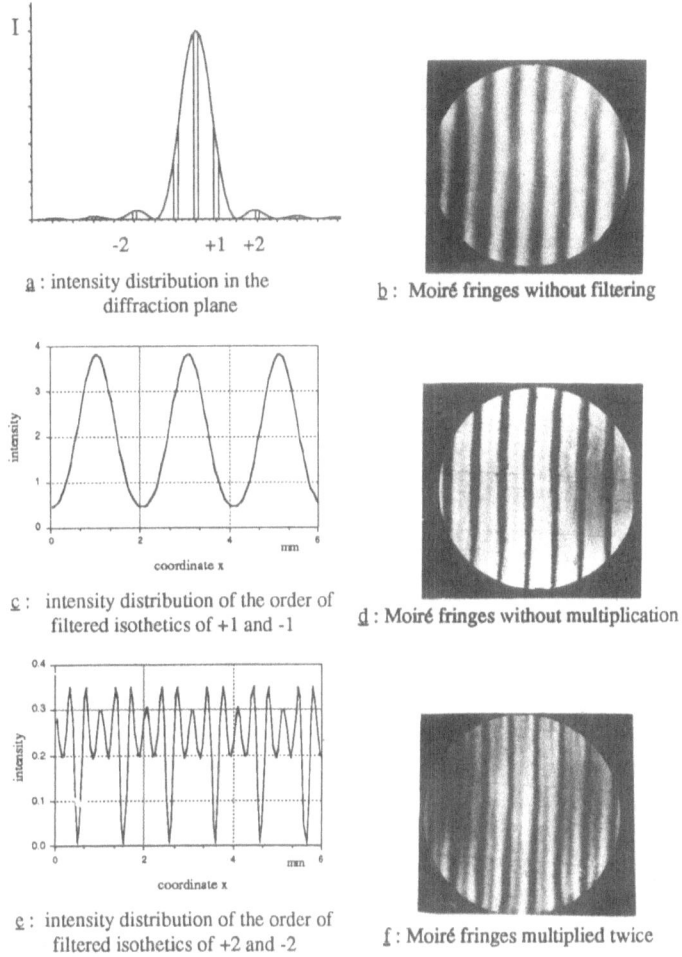

a : intensity distribution in the diffraction plane

b : Moiré fringes without filtering

c : intensity distribution of the order of filtered isothetics of +1 and -1

d : Moiré fringes without multiplication

e : intensity distribution of the order of filtered isothetics of +2 and -2

f : Moiré fringes multiplied twice

Fig.4: Multiplication of Moiré fringes

ons of the orders of interference differ not only according to the angle of diffraction but also with different wave lengths. The final effect is the coloured row of the orders of diffraction (**Fig. 5a**). If the waves of a certain wave length (λ) are to be filtered out of the diffraction plane E_2, whereby a chosen colour is allowed to pass through a small opening at a certain order of diffraction for which the remaining colours are covered, isothetics appear on the image plane E_3 in the colour chosen. This procedure is termed pseudocolour encoding.

If, for example, U-isothetics are encoded in red and V-isothetics in blue, a coloured image of the Moiré fringes appears on the image plane which is, however, differently marked by the colours (**Fig. 5b - 5d**).

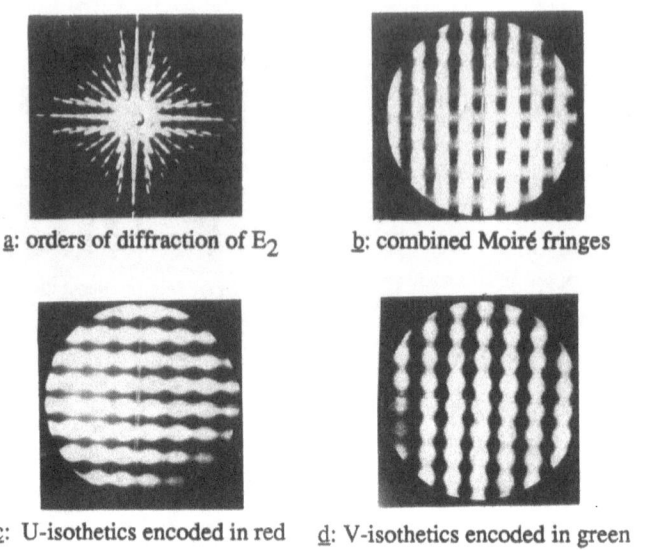

a: orders of diffraction of E_2 b: combined Moiré fringes

c: U-isothetics encoded in red   d: V-isothetics encoded in green

Fig. 5: Separation of U- and V-isothetics by white light processing technique

EVALUATION OF THE COMBINED TECHNIQUE

The evaluation of the coordinate stresses, strains and displacements can be implemented easily using the isothetics and isochromates. **Figure 6** shows the flow chart of the evaluation process in the elastic range. An adaption of spline functions enables the displacement to be reconstructed from the isothetics and the corresponding strain profile to be derived. The two principal strains, which are derived from the displacements, and the principal stress difference give a complete description of the stress and deformation state for the elastic case.

In the viscoelastic range, the stresses and strains are linked by functions of creep or relaxation, which are dependent on time. Evaluation ensues from a continuous measurement series from which the stresses and strains are evaluated step by step by linearisation according to the course of the creep function of the material.

AN EXAMPLE FOR THE APPLICATION OF THE COMBINED TECHNIQUE

As an example for the application of the hybrid measuring technique an experiment is described in which a perforated rectangular plate is loaded by impacts.

The impact load of the model is applied by a steel ball which is pneumatically accelerated in a pipe. During the passage of the ball through the pipe, an impulse emitted from a light barrier triggers an infinitely adjustable time-delay device which releases the flashlight. In accordance with Fig. 1, the deformed specimen grating and the isochromatics are registered simultaneously via optical path 1 on a photo-active glass plate and via optical path 2 in a camera, respectively.

Figure 7 shows the isochromatics and the corresponding isothetics, whereby the U-isothetics are coded with green and the V-isothetics with red. Image processing leads to a digital determination of all isochromatics and the separated isothetics, which are then evaluated by programs such as those described in Fig. 7. The evaluation of the perforated plate under impact loading is given in **Figure 8**. It can be seen that compared with the input waves there is a considerable scattering and damping within the plate due to reflexion and diffrac-

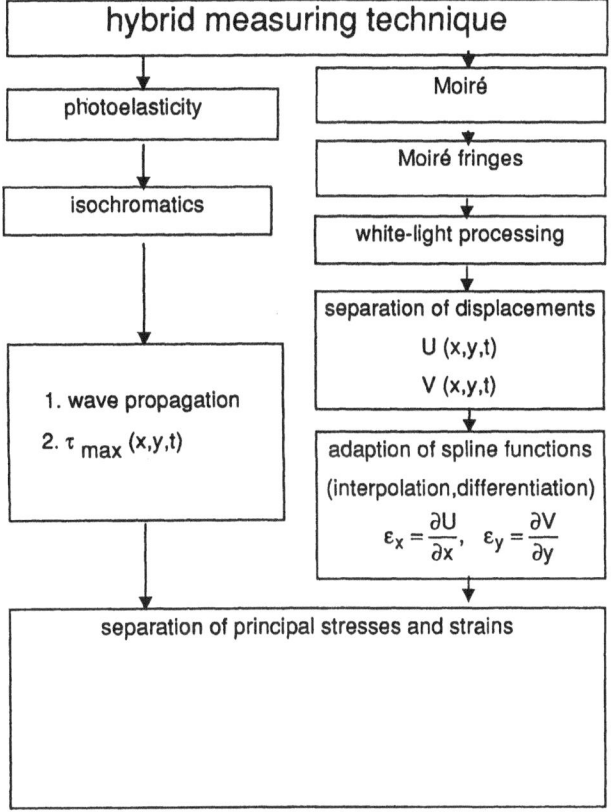

Fig. 6: Flow chart of the evaluation using the hybrid method for the elastic case

tion at the perforations. Waves reaching the upper end are very weak, nearly the whole energy from the impact remains in the plate to be consumed by damping. This example leads

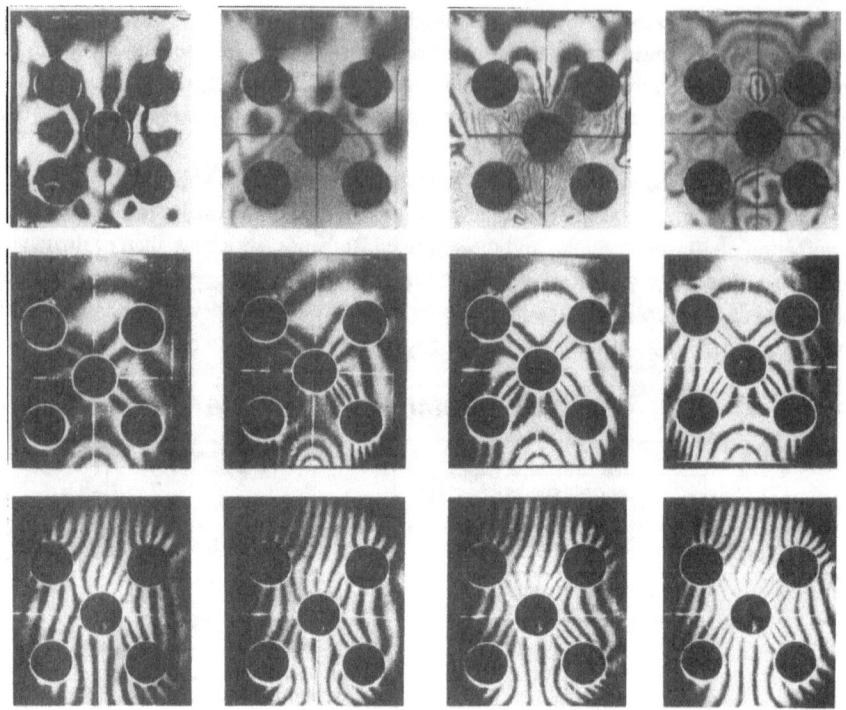

Fig. 7: Isochromatics and Isothetics

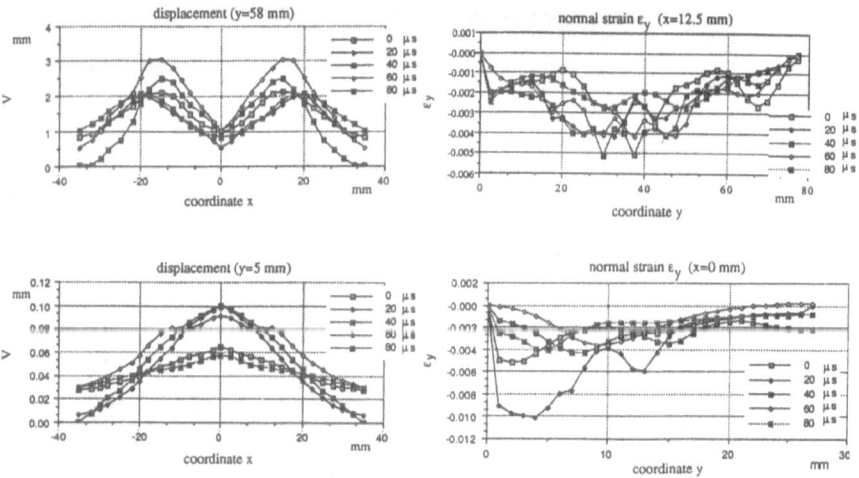

Fig. 8: Evaluation

to basic investigations on solid-sound propagation in machine parts.

SUMMARY

The hybrid measuring technique presented in this paper using the combination of photoelasticity and Moiré offers the possibility to synchronously record the complete information on the dynamic state of time-dependent loading. A complete evaluation of the isochromatics and isothetics using computer programs and image processing enables both the stress and the strain profiles to be determined. As such, this method enables elastic, viscoelastic and local-plastic deformation behaviour to be investigated independently of the law of materials.

REFERENCES

[1]: Kuske A. Robertson G.: Photoelastic   Stress Analysis John Wiley & Sons 1974.

[2]: Post D.: Analysis of moiré fringe multiplication phenomena. Applied Optics 6 (1967) 11, p. 1938-1942

[3]: Jack D.Gaskill: Linear Systems, Fourier Transformation and Optics. John Wiley & Sons 1978.

[4]: Chiang, F.P., Parks V.J. and Durelli: Moiré-fringe interpolation and multiplication fringe shifting. Exp.Mech. 8 (1968) 12, p. 554-560

[5]: Sciammarella, C.A. and Stureon D.L: Digital-filtering techniques applied to the interpolation of moiré fringe data. Exp.Mech. 7.(1967) 11 468-475

[6]: Post, D. and Baracat, W.A.: High-Sensitivity Moiré Interferometry - A simplified Approach. Exp.Mech. 1981, p. 100-104

[7]: Sciammarella, C.A.: Theoretical and Experimental Study on Moiré Fringes. Illinois Institut of Technology, Chicago III, (1960)

[8]: Chiang, F.P.: A method to increase the accuracy of Moiré method. Proceedings of the ASCE, Journal of Eng. Mech. Division 1965

[9]: Sciammarella, C. A.: Basic Optical Law in the Interpretation of Moiré Patterns. Exp. Mech. Mai. 1965.

[10]: Yu, F.T.S.: White light processing technique for archival storage of color films. Applied Optics / Vol. 19. No.14 / July 1980

[11]: Yu, F.T.S., Zhuang, S.L., Chao, T.H. and Dymek, M. S.: Realtime white light spatial frequency and density pseudocolor encoder. Applied Optics / Vol. 19. No.17 / September 1980

[12]: Wan, G. und Albers. J.: Eine hybride Methode zur Bestimmung des vollständigen Spannungs- und Verformungzustandes bei dynamischen Beanspruchungen. (in advance)

EXPERIMENTAL, ANALYTICAL AND COMPUTATIONAL STRESS ANALYSIS USED IN DEVELOPING A MULTI-PURPOSE SEMI-SUBMERSIBLE

Duncan M. Warwick and Douglas Faulkner

Department of Naval Architecture and Ocean Engineering

University of Glasgow

ABSTRACT

In 1988-89 the Department of Naval Architecture and Ocean Engineering at the University of Glasgow was asked to provide R&D support in the analysis of Hydrodynamic and Stability characteristics and in Structural Design for the development of a new modular construction concept of a versatile semi-submersible platform for oil and gas production world-wide. This work was undertaken (for Seaways Engineering Ltd) and the resulting design has undergone a full risk analysis by Lloyd's Register who have awarded their pre-certification certificate to the design which has also received very favourable comment and is now being marketed.

The paper describes the manufacture of a 1/100 scale model of the platform made in PVC (rather than steel) in order to concentrate on the distribution of wave induced primary stresses, especially in way of the column/pontoon corner connections. These stresses were inferred from measurements on 101 strain gauges during dynamic wave loading tests in a wave tank. Analytical and computational modelling of stress distributions for the same wave loads showed good correlation with these measured in-plane primary stresses. The computational models were then used to obtain the total combined primary stress response to regular wave loading for the full-scale structure.

NOTATION

$A(\omega)$: real part of complex transfer function

$B(\omega)$: imaginary part of complex transfer function

$H_c(\omega)$: complex load effect transfer function

$\overline{H_c(\omega)}$: complex conjugate of $H_c(\omega)$

$S_x(\omega)$: response process mean square spectral density function

$S_\eta(\omega)$: wave elevation process mean square spectral density function

$Q(t)$: general internal load effect process

a_o :   incident wave amplitude

q_o :   amplitude of internal load effect process

k   :   wave number = ω^2/g in deep water

x,y,z   :   coordinates of point within wave

α :   wave phase angle ie. position of wave crest relative to wave axis origin, measured in radians

ω :   wave frequency.

INTRODUCTION

The offshore industry is currently considering many different compliant systems for the exploitation of deep water oil and gas reserves. One such concept which aims at achieving maximum constructional simplicilty so that initial capital costs can be dramatically reduced is the Multi-Purpose Semi-Submersible (MPSS) developed by Seaways Engineering Ltd[1,2], Fig.(1). The structural configuration of the MPSS is based on common building blocks of 7.5 metre sided steel cubes themselves pre-fabricated from flat cross-stiffened panels with uniform scantlings. The basic platform is 90 m square and 56.25 m high with an operating draught of 27.5 m.

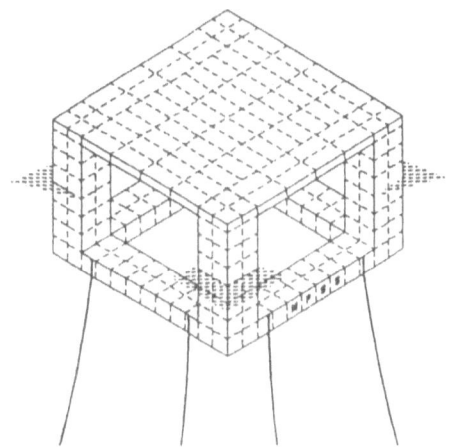

Figure 1. The Multi-Purpose Semi-Submersible (MPSS) concept

In January 1988 Seaways Engineering Ltd with support from Conoco (UK) Ltd approached the Department of Naval Architecture and Ocean Engineering at the University of Glasgow for help in further experimental and computational development of the MPSS structural design. The principal objective of the work was to investigate the primary structural response of the MPSS design. The experimental approach adopted was to construct a 1:100

scale structural model from sheet PVC and to simulate realistic operational and extreme loading conditions in the Departments Hydrodynamics Laboratory[3]. The experimental work was to be supported by computational wave loading and finite element analysis of the MPSS primary structure.

Since it was the elastic structural response which was of interest, the use of PVC material for the model construction was considered quite acceptable, provided it had an adequate elastic range. The PVC structural model was suitable for producing similar primary strain distributions and hence primary field stress distributions to those to be expected in the full scale steel prototype MPSS. No attempt was made to model secondary and tertiary structural behaviour. Non-linear effects were also omitted from the modelling.

MODEL CONSTRUCTION

The 1:100 scale model was constructed using 4 mm PVC sheet throughout. The choice of thickness was controlled by the need to avoid local buckling and minimise secondary and tertiary bending in the unstiffened plate panels. All internal primary structural members were replicated in the PVC model. These included:

(1)   Deck   - all WT longitudinal and WT transverse bulkheads

(2)   Columns   - all WT bulkheads and deck flats

(3)   Pontoon   - all WT longitudinal and WT transverse bulkheads

The model components were connected using a plastic welding technique which produced very little distortion of the PVC panels. The egg-box type internal structure of the deck was bonded to continuous upper and lower deck panels.

To determine the material properties of the PVC tensile test specimens were cut from the same PVC material as used in model construction. The average results obtained from the tests are given in Table 1.0. Since small strains were expected in the tests the elastic range was felt to be acceptable.

TABLE 1.0
Elastic Properties of PVC

E (N/mm2)	Proportional Limit Strain ($\mu\varepsilon$)	Proportional Limit Stress (N/mm^2)	Poissons Ratio
3255	4000	13.02	0.5*

* assumed from literature

The model was strain gauged with a total of twenty-three 45° three-arm rosettes and thirty-two single strain gauges. The gauges were arranged at one corner of the MPSS near the deck/column and pontoon/column intersections, Fig. (2a & b). Rosettes were used close to the deck/column and pontoon/column intersection, while single gauges were positioned more remotely from the corner nodes to confirm the uniform stress fields in the pontoons. All gauges were waterproofed with a flexible sealant.

Figure 2a. 1:100 scale model of MPSS

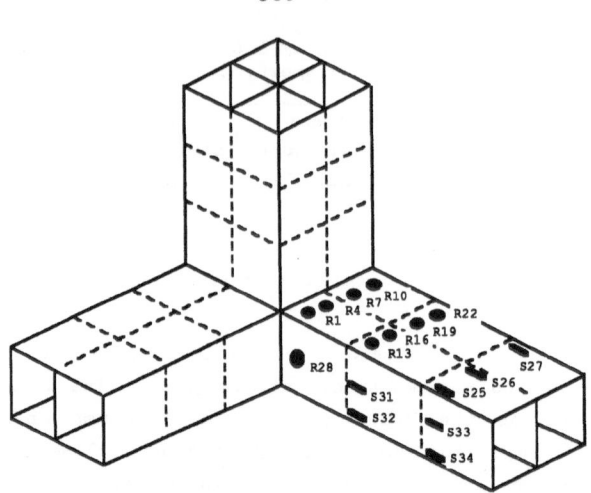

Figure 2b. Strain gauge arrangement on model MPSS

The PVC material was not expected to show large strains under the load levels to be experienced in the 1:100 scale tests so high elongation strain gauges were not considered appropriate. Therefore, standard 10 mm gauge length strain gauges, usually applied to steel structures, were chosen to measure the strains. The choice of a 10 mm gauge length, equivalent to 1.0 metre on the full scale, avoided measurement of local distortions of the primary stress distribution.

The problem of temperature compensation was tackled by preparing two sets of compensating gauges bonded to separate PVC sheets. To compensate the submerged column and pontoon gauges one set of compensating gauges was submerged close by in the tank. For the deck and upper column gauges, which were at ambient air temperature, the second set of compensating gauges were positioned on the deck of the model.

Testing arrangements and procedures

All strain gauge time history responses were logged on a VAX 11/730 mini-computer for further analysis. Wave probe responses were also logged in this way. The strain gauge measurement datum was set at the still water condition so that the gauges measured only the wave stress component of structural response.

Regular waves were applied to the moored MPSS model at two heading angles of 0° and 45° over a wave frequency range of 0.4 Hz to 1.4 Hz. The wave heights represented moderate operating sea conditions on full-scale.

Comparison of theory with experiment

The results from the wave tests conducted on the MPSS 1:100 th scale model were used to validate a finite element model of the MPSS (see next section).

The intention of using three-arm rosette gauges on the pontoon deck was to be able to determine direct and shear principal stresses for example for shear lag effects. Experience in the model tests later showed that it was not possible to satisfactorily measure shear lag effects with the numbers and types of strain gauges used. The theoretical model assumed a uni-axial stress distribution along the pontoons and the resolution of the strain gauge time histories into multiple principal stresses was not necessary in order to obtain results which were directly comparable with predicted stresses. In addition, full-scale preliminary design evaluation is conventionally based upon the assumption of a uni-axial direct stress field in any beam like elements such as pontoons, i.e. simple beam theory. For more detailed work, such as strength analyses of pontoon deck grillages, a bi-axial stress field would be more appropriate. Hence, to obtain a direct comparison between theoretical and test stresses only the stresses derived from the centre gauges of the rosettes, parallel to the longitudinal axis of the pontoon, were used.

Fig.(3) plots the stresses derived from the centre gauges of rosettes R1,R4 and R7, together with the theoretical predictions for the same gauge positions at node 100. It was discovered that gauge R10 was malfunctioning and so was omitted from the comparison.

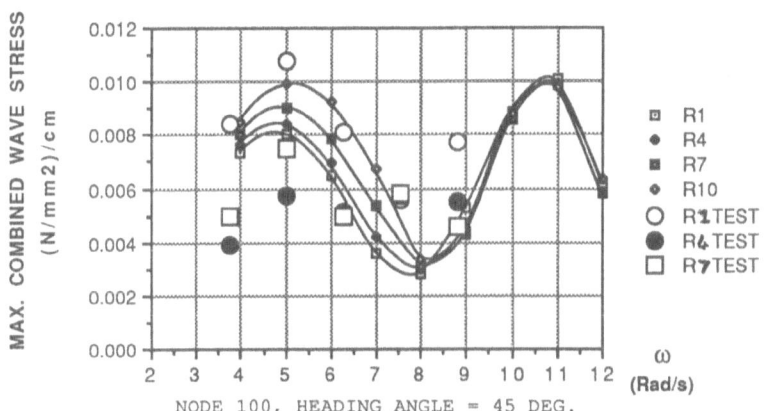

Figure 3. Comparison of test stress results with theoretical predictions

From Fig.(3) it can be seen that the predicted and measured stresses show a similar peak at a wave frequency of 5.0 rad/s. Although there is considerable scatter in the test results the predictions on average very slightly over-estimate the actual values of stress. This statement is quantified in Table (2.0) which presents the predicted and measured stresses averaged across the pontoon deck. The ratio of actual stress to predicted stress has a mean value of 0.97 and a coefficient of variation of 21.5%. These may be interpreted as indicating that the theoretical approach has, on average, a conservative systematic error of about 3% with a random uncertainty of about 21.5%. While individual gauges may show a larger variation from the predicted values the above statistics indicate a satisfactory level of correlation between theory and test when considering average stress levels.

It was concluded[4] that the numerical model exhibited a satisfactor correlation with the actual behaviour of the small scale PVC model. Hence it was used to infer that the numerical model applied to a full-scal analysis would have approximately similar modelling statistics. In doin this it was assumed that non-linear response of the full-scale structur does not occur at full-scale load levels equivalent to those experienced b the PVC model. It was also assumed that non-linear scaling effects ar small and will not produce large differences between small and full-scal actual and predicted behaviour.

Table 2.0:
Comparison of measured and predicted stress
levels at node 100 at 45° heading

Test Average $(N/mm^2) \times 10^3$	Predicted Average $(N/mm^2) \times 10^3$	Test Pred
5.77	7.89	0.731
8.00	8.85	0.904
6.07	7.63	0.796
5.70	4.97	1.149
5.93	4.641	1.278

Modelling bias = 0.972
Coefficient of variation = 21.5%

COMPUTATIONAL STRESS ANALYSIS OF MPSS

The conventional design approach for offshore structures is to apply regular design wave of specified height and period such that the wave models the worst effects of a 100 year storm condition. The wave loading on the structure is evaluated by application of a suitable wave theory and an associated theory of wave/structure interaction eg. Morison's equation or potential theory. In the case of a floating platform the loads are derived such that dynamic equilibrium of the platform is satisfied. The structural analysis then proceeds in the usual static manner. The main advantages of the design wave approach are its relative simplicity, efficiency and ease of standardisation within a design code.

More recent approaches to structural response analysis of offshore structures have favoured the adoption of stochastic analysis methods[5]. These methods have been applied more commonly in random vibration analysis of mechanical systems[6] and in the analysis of random noise in electronics[7]. The methods provide rational tools with which to incorporate the inherent randomness of the environmental wave loading process on offshore structures by introducing response modelling in the probability domain.

For the case of a linear wave/structure system the stochastic approach is based upon the well known frequency domain random response relationship:

$$S_x(\omega) = H_c(\omega) \overline{H_c(\omega)} S_\eta(\omega) \tag{1}$$

Response process statistics can be derived from a knowledge of $S_x(\omega)$ conditioned upon the sea-state characteristics used to determine $S_\eta(\omega)$. However, this paper concentrates on the determination of the load effect transfer function which completely describes the structures response characteristics in the frequency domain.

Wave loading

For a full-scale prototype MPSS a wave loading analysis was carried out based upon three dimensional radiation/diffraction theory. This approach is appropriate for structures whose principal dimensions are such that appreciable scattering of the incident wave field occurs. The program AQWA-LINE was used to carry out the analysis. The fluid forces are assumed to be composed of excitation forces due to the incident wave and reactive forces due to the motions of the structure. The wave excitation forces which induce motion are composed of a diffraction component due to scattering of the incident wave field and a Froude-Krylov component arising from the pressure field within the undisturbed wave. The reactive fluid loading can be estimated by analysing the radiate wave field arising from the motions.

Linear Airy wave theory was used based upon the assumption that the incident wave is harmonic and of small amplitude compared to its length. Hydrostatic forces are computed and together with the structural mass and hydrodynamic forces allow the calculation of the small amplitude rigid body motion response about a free-floating equilibrium position. The program solution technique utilises a distribution of fluid singularities over the mean wetted surface of the body. Fig.(4) shows the diffraction mesh for the MPSS where two-fold symmetry means that modelling of only one quarter of the structure is necessary.

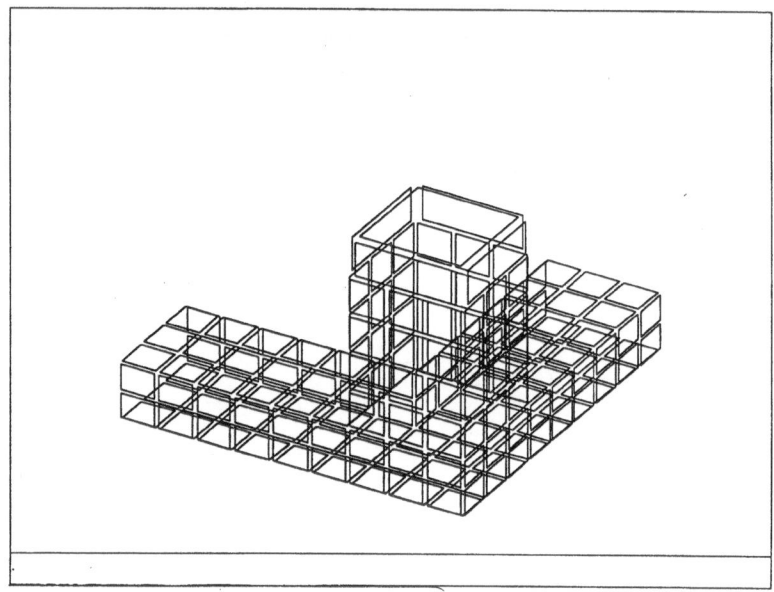

Figure 4. Radiation/Diffraction model of MPSS

Structural response

A global finite element model was developed to assess the primary load effects and stress fields induced by the action of waves upon the MPSS structure. Only the primary field stresses generated in the structure, such as those induced by axial force, vertical and horizontal bending, are modelled and it was not intended to estimate local stress concentrations or any non-linear effects by use of the finite element model. The structural analysis was carried out using the linear finite element program ASAS and the associated suite of post-processing programmes.

The FE model was constructed from a combination of three dimensional beam elements and two dimensional membrane elements, Fig.(5). Three dimensional beam elements were used to model the pontoon and column structure and also the primary deck supporting beams and girders. The average length of the beam elements was 7.5 m. The in-plane shear stiffness of the deck plating was modelled by membrane elements with dimensions 15.0 m by 7.5 m in the central deck area and 7.5 m by 7.5 m at the fore and aft deck edges. The beam element properties were calculated based upon the scantlings of all longitudinally effective material making allowances for loss of effectiveness.

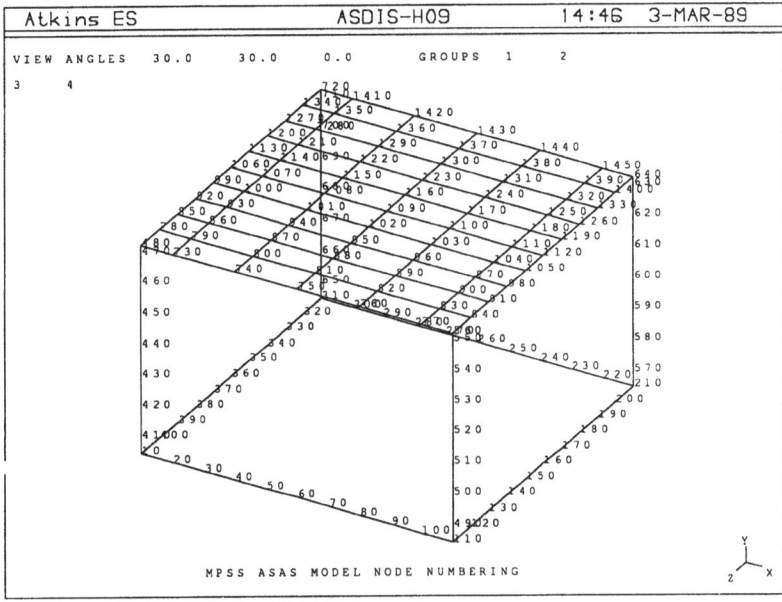

Figure 5. Global finite element model of MPSS structure

The incident, diffracted and radiated velocity potentials and hydrostatic restoring forces generated by the radiation/diffraction analysis were transformed into nodal forces acting on the FE model. The rigid body accelerations calculated in the radiation/diffraction analysis together with the nodal load vector comprised a load system which was in dynamic equilibrium. Hence, all reaction forces at the imposed support points were negligible when compared with the applied nodal loads.

Complex load effect transfer functions
With the aim of carryimg out a stochastic structural response analysis in accordance with equation (1) it was necessary to evaluate the complex load effect transfer functions for any point in the MPSS structure.

Consider a linear structural system acted upon by wave loading only. The input excitation is proportional to the surface elevation given by a cosine wave travelling in the positive x-direction. The output or response to the regular wave excitation will be of the form:

$$Q(t) = q_o \cos(kx - \omega t - \alpha) \qquad (2)$$

In general the phase angle at which the maximum internal load effect occurs will be different from the phase angles of the motion responses and the excitation forces because member load effects are dependent upon both direct member loading and the effects of global structural response whose maxima are not necessarily coincident in time.

Adopting complex exponential notation and noting that it is the real part of the complex function that is of interest, the harmonic output maybe expressed as:

$$Q(t) = a_o \operatorname{Re}\left\{ H(\omega) . e^{-i\omega t} \right\} \qquad (3)$$

where $H(\omega)$ is defined as the complex load effect transfer function and is written in terms of its real and imaginary parts:

$$H(\omega) = A(\omega) - i.B(\omega) \qquad (4)$$

$A(\omega)$ and $B(\omega)$ are functions of ω and position within the structure relative to the wave axis origin. Equation (3) can be shown to be identical to:

$$Q(t) = a_o \sqrt{A(\omega)^2 + B(\omega)^2} \cos(\omega t - \alpha)$$

$$\alpha = \tan^{-1} \frac{B(\omega)}{A(\omega)} \qquad (5)$$

which indicates that the regular wave response amplitude is obtained from the modulus of the complex transfer function and the phase angle is simply the argument of the complex transfer function. The complex transfer function contains all the information required to fully define the frequency response of the load effect of interest.

The real and imaginary parts of $H(\omega)$ are obtained by conducting a structural analysis when the incident wave peak is at the wave axis origin ie. $\omega t = 0$ and when the wave peak is one quarter period advanced ie. at $\omega t = \pi/2$. Hence, two structural load cases per wave frequency must be analysed. The wave height is taken as 2 metres to give $H(\omega)$ directly from equation (3).

Fig.(6) gives the load effect RAO's (modulus of transfer function) for axial load and vertical and horizontal bending moment at node 100 of the structural model for a wave headings of 0°, 22.5° and 45°.

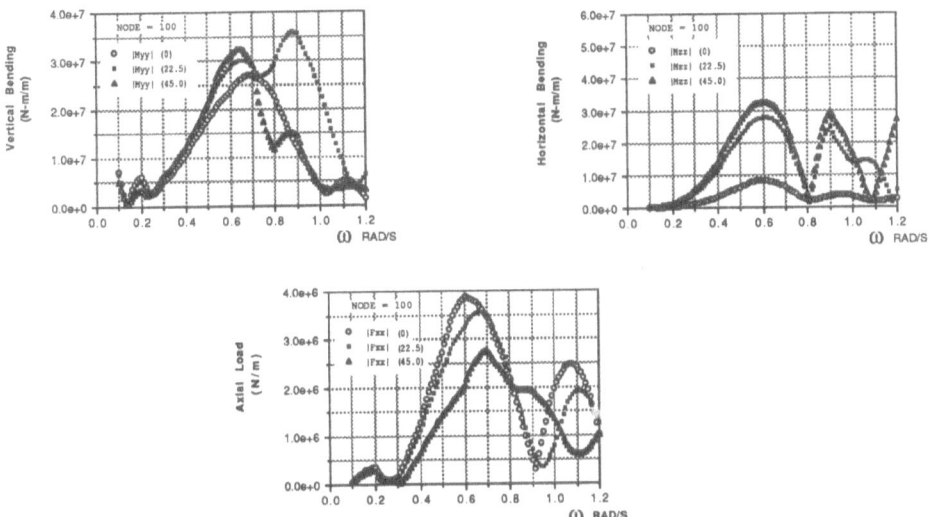

Figure 6. Load effect RAO's for node 100 at 0°, 22.5° and 45° headings

Combining load effects to produce maximum stresses

As noted already the wave phase (position) at which each internal load effect, and hence stress component, reaches its maximum value within a wave cycle will not be the same for all three components. Hence, the approach to the combination of primary stresses must pay due regard to this. The total primary stress at any point in the wave cycle will be given in the time domain as, referring to Fig.(7):

$$\sigma_c(t) = \frac{F_{xx}(t)}{A} - \frac{M_{yy}(t)}{Z_{yy}} - \frac{M_{zz}(t)}{Z_{zz}} \tag{6}$$

In the frequency domain the combined stress transfer function is obtained from[8]:

$$H_c(\omega) = \sum_{i=1}^{n} a_i H_i(\omega) \tag{7}$$

where $H_i(\omega)$ are the individual load effect complex transfer functions and a_i are linear load combination factors given simply by A^{-1}, Z_{yy}^{-1} and Z_{zz}^{-1}

for i = 1,2,3, respectively, for the combination of axial stress with
vertical and horizontal bending stress. The ordinates of $|H_c(\omega)|$ represent
the maxima of the combined stress response in a wave cycle of given
frequency. The phase relationships between component load effects will be
preserved through the real and imaginary parts of $H_c(\omega)$. Hence, the phase
relationship of the combined stress response process will account for phase
differences between component load effects.

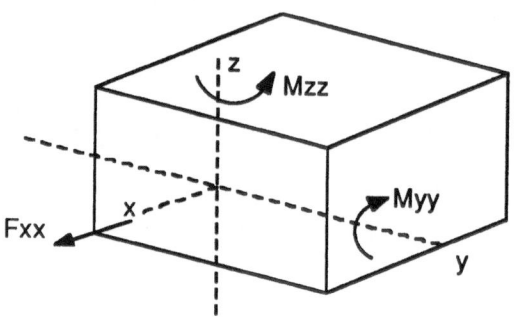

Figure 7. Primary load effects acting on a pontoon cross-section

Fig.(8) shows $|H_c(\omega)|$ for node 100 at 0° and 45° headings. There is a
distinct second peak in the stress response for the 45° heading which does
not occur in the 0° case. By examining the variation of the three primary
stress components throughout one wave cycle at 0.9 rad s^{-1}, as shown in
Fig.(9), it is seen that in the 45° case the stress components are clearly
in phase with one another at this frequency giving rise to a peak in stress
response at this frequency. For the 0° heading the horizontal bending
stress is very much smaller than the vertical bending stress. Since there
is zero direct member loading in the horizontal plane for the pontoons
aligned with the incident wave direction the only source of horizontal
bending is from global frame action of the MPSS and this produces a bending
moment an order of magnitude smaller than in the 45° case.

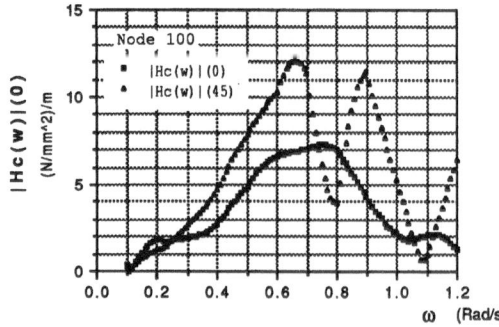

Figure 8. Combined primary stress RAO at node 100 for 0° and 45° headings

The stress components move in and out of phase as wave frequency varies depending upon the contributions of local and global structural response to each stress component. For example, horizontal bending in the pontoons is almost entirely due to direct member loading since the columns are torsionally very stiff. Vertical bending in the pontoon can result from both direct member loading of the pontoon and a "carry over" moment from the vertical columns. At 0.9 rad s^{-1} and 45° the local and global effects are in phase resulting in the second stress peak.

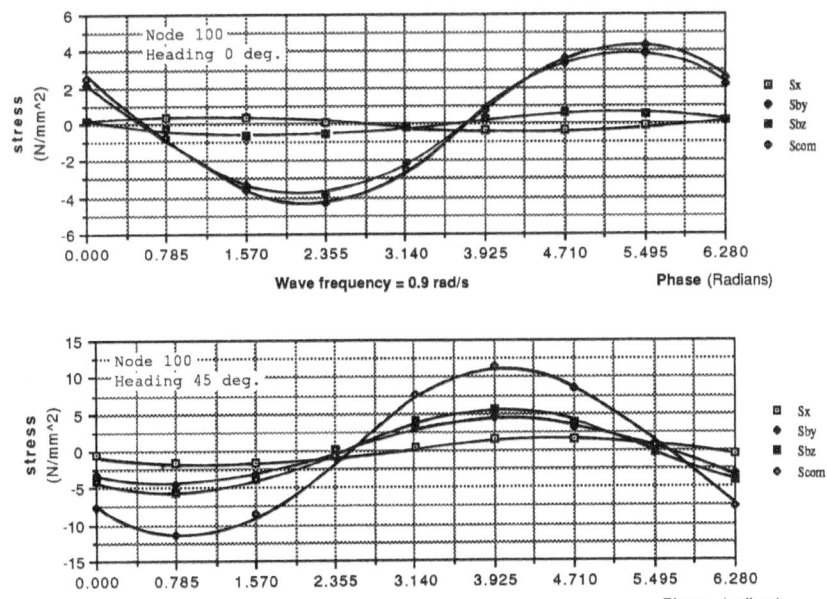

Figure 9. Primary stress component variation with wave position

CONCLUSIONS

The paper has described the construction and testing of a small scale PVC model of a Multi-Purpose Semi-Submersible. The experimental measurements of primary field stresses were found to show a satisfactory correlation with theoretical model predictions.

An approach to obtaining the primary wave stress response of the MPSS in regular waves has been described. The approach implicitly accounts for the different phase relationships between individual load effect responses during the passage of a wave. Despite the common load source, phase differences in the load effects arise out of the differing levels of direct member loading and global platform response which contribute to each load effect.

The most convenient method for determining the combined stress response of a linear wave/structure system utilises a frequency domain solution the essential component of which is the determination of the complex load effect transfer functions. These transfer functions completely define the structural response to wave loading throughout the frequency range of interest.

The idea of using a PVC model with strain gauges was justified but the strain gauge rosettes were not found to justified. The tests supported the design studies undertaken earlier for the sponsors and Seaways Engineering Ltd.

Acknowledgements

The authors and their Department are indebted to Seaways Engineering Ltd and the sponsors of the work described here, Conoco (UK) Ltd.

REFERENCES

[1]   Ryan, B. and Lang, A., C.: "A Breakthrough in Semi-submersible Design", 4th International Conference on Floating Production Systems, IBC Technical Services Ltd, London, 1988.

[2]   Faulkner, D., Incecik, A. and Lang, A., C.: "Design Evaluation of a Multi-Purpose Semi-Submersible", Trans. of Institute of Engineers and Shipbuilders in Scotland, Paper 1492, Vol 132, 1988/89.

[3]   Warwick, D., M.:"Structural Model Testing of a Multi-Purpose Semi-Submersible", Department of Naval Architecture and Ocean Engineering Report NAOE-89-34, University of Glasgow, 1989.

[4]   Faulkner, D. and Warwick, D., M.:"Structural Testing and Analysis of a Multi-Purpose Semi-Submersible", Department of Naval Architecture and Ocean Engineering Report NAOE-89-35, University of Glasgow, 1989.

[5]   Inglis, R., B., Pijfers, J., G., L. and Vugts, J., H.: "A Unified Approach to Predicting the Response of Offshore Structures, Including Extreme Response", Behaviour of Offshore Structures, 1985, Elsevier Science Publishers, Amsterdam.

[6]   Newland, D., E.: "An Introduction to Random Vibration and Spectral Analysis", 2nd Ed., Longman, London, 1984.

[7]   Rice, S., O.: "Mathematical Analysis of Random Noise", Bell System Technical Journal, Vol 23, 1944.

[8]   Mansour, A., E.: "Combining Extreme Environmental Loads for Reliability Based Design", Extreme Loads Response Symposium, SNAME, Arlington, Va., 1981.

RECENT SHELL BUCKLING RESEARCH AT LIVERPOOL

GERARD D. GALLETLY
Department of Mechanical Engineering,
University of Liverpool,
P.O. Box 147, Liverpool L69 3BX.

ABSTRACT

Various shell buckling problems have been studied in the Department of Mechanical Engineering at Liverpool over the past decade. The background to these problems, and the work carried out on them at Liverpool, is discussed briefly in this paper. The shell types which have been investigated are doubly-curved shells (torispheres and hemispheres) and cylinders (mainly unstiffened but some with ring stiffeners). The principal applied loading has been uniform pressure (internal or external) but transverse edge shear and axial compression loads on cylinders have also been studied.

INTERNALLY-PRESSURISED THIN TORISPHERICAL SHELLS

That there might be some problems with thin torispherical shells became evident in 1956 when a large (15 m dia.) fluid coker failed during its hydrostatic proof test in Avon, California [1,2,3]. A number of investigations emanated from a study of that failure, including (i) the experimental verification [4] of predicted elastic internal pressure buckling in torispheres [2], (ii) the limit analysis of torispheres and toricones [5,6] and (iii) the application of the Runge-Kutta numerical method to axisymmetric shell problems using the digital computer (2,7,8].

The plastic buckling of internally-pressurised torispheres was investigated in the 1970's at Nottingham [9,10], Manchester [11,12], Liverpool [13,14] and APV [15]. One of the reasons for the renewed interest in the structural behaviour of internally-pressurised torispheres was because some very thin (i.e. with large D/t-ratios) versions of them occurred in fast breeder reactors (e.g. the Prototype Fast Reactor at Dounreay and the French Superphénix). What one wants to avoid is the occurrence of waves, or wrinkles, around the circumference in the knuckle region - see Fig. 1.

Despite the experimental investigations at APV, Nottingham, Manchester and Liverpool, no suitable design equations to prevent internal pressure

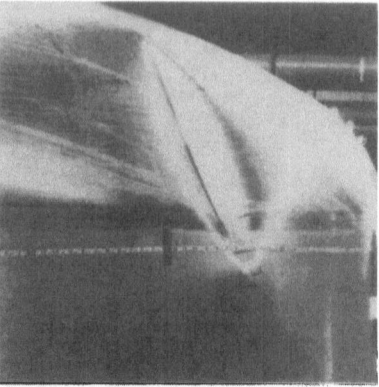

Figure 1. Plastic buckling due to internal pressure in a 3 m diameter
stainless steel torisphere.

buckling in torispheres had emerged prior to 1979. It was, therefore,
decided to undertake some parametric studies on perfect metallic ellipsoidal and torispherical shells using a digital computer. The axisymmetric
plastic collapse mode and the elastic and the plastic bifurcation buckling
modes were studied. 2:1 ellipsoidal shells were investigated first (as
the number of parameters was smaller) and it was possible to derive simple
equations for predicting the buckling and collapse pressures of these
perfect ellipsoidal shells [16,17]. Similar simple equations for perfect
internally-pressurised torispherical shells were obtained shortly afterwards [18,19,20]. The computer program used in the development of these
equations was the finite-deflection, elastic-plastic, shell program BOSOR
5 [21].

Insofar as perfect internally-pressurised steel torispherical shells are
concerned, it was possible to write the plastic buckling pressure as a
function of (r/D), (D/t), (R_s/D) and σ_{yp}, where r = toroidal (knuckle)
radius of the torisphere, R_s = radius of the spherical portion of the
torisphere, t = shell thickness, D = diameter of the cylindrical body, E =
Young's modulus of the shell material and σ_{yp} = yield stress of the shell
material.

Subsequent to the development of this simple equation (in [18]),
additional computer calculations were carried out covering wider ranges of
the parameters [22]. Although different formulae were given in [22], the
numerical differences between the formulae in [18] and [22] were not very
great. Independent numerical verification of the Liverpool internal
pressure buckling formulae was recently presented by Swedish researchers
[23].

Actual dished ends in practice undergo large plastic strains while they
are being formed and residual stresses are produced in the ends. The
plate thickness in the toroidal portion of the ends is also subjected to
considerable thinning in the spinning process. This causes strain
hardening and, for cold-spun austenitic stainless steels, the effect on
the mechanical properties can be substantial. The dished ends are also
geometrically imperfect but internal pressure buckling of dished ends is

one of the few shell buckling problems which is relatively imperfection-insensitive (i.e. the buckling is of the stable kind).

In [24], a discussion is given of the various factors which need to be considered when developing an equation for the buckling pressure of actual fabricated torispherical shells. Clearly, experimental results obtained on full-scale dished ends, and on models of a reasonable size, are important factors in arriving at such a design equation. The results obtained on both small and large tests of internally-pressurised torispherical shells were reviewed in [25] and, a few years later, a simple design equation to prevent internal pressure buckling in fabricated torispherical shells was suggested [26,27]. It is:

$$p_D/YF = \frac{80\ (^r/D)^{0.825}}{(^D/t)^{1.5}\ (R_s/D)^{1.15}} \tag{1}$$

where p_D = design (allowable) internal pressure
  Y = a numerical factor, which depends on how the head is manufactured (crown and segment or pressed and spun) and the type of steel used (austenitic or otherwise).
  F = 0.2% proof stress of the shell wall material.

It should be mentioned that the pressures obtained in the buckling tests of [15] were lower than those of [9,10]. It is not known why the difference occurred, as both series of tests were on large models (3 m dia.) made by the same firm. In deciding on the value of the constant in Eq. (1), the lower test results of [15] were assumed to be controlling.

In comparison with all known test buckling pressures (including the French ones [28]), Eq. (1) gives a minimum factor of safety of 1.5. It was adopted in 1988 by the ECCS in their recommendations on shell buckling [29]. The French had incorporated a version of the simple equation for perfect torispheres in the French pressure vessel code CODAP [30] a few years earlier (they applied a knock-down factor to the equation in [18]).

Most of the work at Liverpool on internally-pressurised torispheres has been on the bifurcation buckling mode. However, numerical studies have also been carried out on the axisymmetric plastic collapse mode. Simple equations for this failure mode were given in [17,19,22,31]. The collapse pressures given by these equations are higher than the corresponding ones given in [6], which uses a small-deflection analysis. However, there has been little experimental verification of the equations in [17,19].

It should also be noted that, for this internal pressure buckling problem, practically all of the large model tests have been conducted on torispherical shells made from high-proof 304 S65 stainless steel. This steel is not used as much nowadays as it was formerly.

RECENT TESTS ON INTERNALLY-PRESSURISED MACHINED TORISPHERICAL SHELLS – DEFORMATION VERSUS FLOW THEORIES OF PLASTICITY

For some torispherical shell geometries, the modes of failure predicted by the deformation and the flow theories of plasticity are different. An example of this occurs for a steel dome with the following geometry: D/t = 500, r/D = 0.15, R_s/D = 1.0, σ_{yp} = 310 N/mm^2.

Deformation theory predicts that bifurcation buckling will occur in the above shell while flow theory predicts that there will be no buckling. The shell program used in the above predictions was BOSOR 5, but its predictions were confirmed by four other independent programs.

The flow theory predictions should, of course, be more correct than those of deformation theory. In order to check whether this was the case or not, it was decided to machine six steel torispherical shells and test them under internal pressure. One of the shells had a geometry which was similar to that mentioned above; the other shells had related geometries.

The results of the tests are given in [32]. Buckles visible to the naked eye appeared in four of the domes and low amplitude waves (picked up by transducers) occurred in the other two. Deformation theory predicted that bifurcation buckling would occur in all six domes whereas flow theory predicted buckling in only two of them. Thus, it would appear that deformation theory is better than flow theory in its prediction of the failure mode for these domes.

However, for axially-compressed cruciform columns, it is known that the inclusion of very small initial imperfections in the analysis was sufficient to bring about reasonably good agreement between the test results and the predictions of flow theory [33]. It is possible that very small initial imperfections could have a similar effect in the present problem. At the author's request, this possibility is being investigated by A. Combescure in France and M. Esslinger in Germany. Both investigators have analysed one of the domes for which flow theory predicted no buckling. By including very small initial imperfections in the analysis, both investigators have obtained buckling predictions using flow theory. Further study of this aspect of the problem is continuing.

CYLINDERS SUBJECTED TO TRANSVERSE EDGE SHEARING LOADS

If an earthquake should occur in the vicinity of a liquid metal fast breeder reactor (LMFBR), the reactor would vibrate and it would subject the cylindrical primary tank to dynamic shears via a stiff horizontal diaphragm. Depending on the magnitude and duration of these shearing forces, the cylindrical tank might buckle. Various nuclear authorities have been concerned about the possibility of shear buckling due to seismic action and tests have been conducted in several countries to study the problem.

The static shear buckling problem is easier to solve than the dynamic one and a solution, in the elastic region, was given by Yamaki et al. [34]. Their work was confirmed at Liverpool using Melinex models. However, the elastic-plastic static buckling of metallic shells subjected to transverse edge shearing loads had not been solved and additional work on this problem (for steel shells) was undertaken at Liverpool. The type of buckling under discussion is shown in Fig. 2.

As may be seen from Fig. 2, the mode of failure is similar to (but not the same as) that which occurs under torsional loading. This fact was utilised by Yamaki when deriving the buckling stress of an elastic cylindrical shell subjected to transverse edge shear forces.

Figure 2.   Buckles in a steel cylindrical model.   The loading was a
transverse edge shearing force.

The torsional analogy can also be used in the plastic buckling case [35].
However, it was found that a quadratic interaction equation would predict
the experimental plastic buckling loads to within 10%. This was as good
as the predictions of some large computer programs.   The interaction
equation was:

$$\frac{1}{\tau_{ET}^2} + \frac{1}{\tau_{yp}^2} = \frac{1}{\tau_p^2} \tag{2}$$

where
τ_{ET} = elastic buckling stress of the cylinder subjected to torsion,
τ_{yp} = yield stress in shear of the shell material, and
τ_p = plastic buckling stress of the cylinder subjected to transverse edge
     shear.

A comparison of the predictions of Eq. (2) with some test results is given
in [36]. An independent evaluation of its utility for predicting static
plastic shear buckling loads may be found in [37].

The buckling of these cylinders under dynamic shear loads has not been
pursued at Liverpool so far. This has been because the test facilities to
apply the necessary forces to models of a reasonable size ($\approx$ 0.6 m dia.)
are not available as yet. However, other groups (e.g. the Japanese) have
tested cylinders under dynamic conditions.

CYLINDERS UNDER SIMULTANEOUS AXIAL COMPRESSION AND EXTERNAL PRESSURE

Insofar as elastic buckling is concerned, this problem has been studied by
aeronautical engineers for the past twenty-five years or so. Recently, it
became of interest to the offshore oil industry, in connection with the
design of the cylindrical legs of oil platforms in the North Sea.   The
problem is also of interest to the nuclear industry in containment shell
structures (partial vacuum plus axial load).

The R/t-ratios of the cylindrical shells used in the offshore field are
much smaller than those used in the aerospace industry.   Thus, plastic,

rather than elastic, failure modes are more likely to occur. The different materials and manufacturing methods in the two industries also lead to different residual stress patterns and initial geometric imperfections. Thus, one cannot assume that the aeronautical experience on this biaxial buckling problem can be transferred directly to the offshore field.

Because of the lack of relevant experimental information on the elastic-plastic buckling of steel cylinders subjected to biaxial compression, the Marine Technology Directorate, SERC, decided to fund research on the problem at several universities in the U.K. The inter-relation of this work is described in [38] and only the Liverpool studies on unstiffened cylinders will be described herein.

Prior to undertaking the combined loading work at Liverpool in 1981, the status of the buckling of cylinders subjected to simultaneous external pressure and axial compression was as follows:

(i)   from the aeronautical work [39], it was known that a linear interaction equation was reasonably safe for design. The equation was:

$$\frac{\sigma_x}{\sigma_{t,x}} + \frac{\sigma_\theta}{\sigma_{t,\theta}} = 1 \tag{3}$$

In the above equation, σ_x and σ_θ are the combined axial and hoop stresses which, acting simultaneously, cause buckling. The quantity $\sigma_{t,x}$ is the test result for buckling due to axial compression alone while $\sigma_{t,\theta}$ is that for buckling due to external lateral pressure alone. These quantities are used in the denominators of Eq. (3) (rather than the corresponding theoretical quantities for perfect cylinders, i.e. $\sigma_{cr,x}$ and $\sigma_{cr,\theta}$) because of the initial geometric imperfections in the models. To utilise Eq. (3) in practice, one needs to know the quantities $\sigma_{t,x}$ and $\sigma_{t,\theta}$. As there are many test results available for single loads, one can use either lower-bound curves or empirical equations to estimate their magnitudes.

(ii) Tennyson and his colleagues at UTIAS (University of Toronto Institute for Aerospace Studies) had recently (1978) published a paper [40] on the elastic buckling of short cylinders subjected to external pressure and axial compression. One of the interesting things they found was that the use of Eq. (3) (for perfect cylinders) might not be safe for small values of the parameter Z($\approx$ $(L/R)^2$ (R/t)). Prior to the UTIAS work, it had been thought that the theoretical interaction curves for perfect cylinders would tend towards a straight-line as Z increased. However, the Toronto work showed that some interaction curves went below the straight-line. This could have serious implications for design.

The Liverpool experimental work on unstiffened steel cylinders included machined and welded models (having $R/t = 100$) and welded models having R/t $\approx$ 330. The L/R-ratios varied from 0.18 to 1.4. The small L/R-ratios are typical of the inter-frame spacing in offshore cylinders. Photographs of the Liverpool failed models appear in [41,42].

In addition to the experimental work, numerical investigations were carried out using the BOSOR 5 program. Some of these calculations were on perfect models (both elastic and elastic-plastic buckling) and some on models with axisymmetric imperfections in them [43].

The results of the Liverpool investigations on combined loads on cylinders are given in Refs. [41,42,43]. Briefly, they are as follows:

(a) The use of a linear interaction equation in design is rather conservative for offshore geometries. A quadratic interaction equation is more economical and is still safe.

(b) For the very short shells ($L/R = 0.18$), the experimental interaction curves had reversed curvature in them. The theoretical interaction curves for these shells were also slightly unusual.

(c) The work of the UTIAS group on the elastic buckling of very short cylinders was confirmed. However, the relevant mode of failure, for the yield points and L/R-ratios of interest to the offshore industry, was shown to be elastic-plastic buckling [43]. The interactive buckling curves for this failure mode are all curved away from the origin (see Figs. 2 and 4 in [42]), i.e. the UTIAS conclusions are not relevant to this case.

EXTERNALLY-PRESSURISED METALLIC HEMISPHERES AND TORISPHERES

The external pressure design rules in BS 5500 [44] concerning torispheres and hemispheres are the same, i.e. the knuckle (or toroidal) radius does not enter into the calculations. In addition, it is not clear how the tolerances in shape affect the allowable pressures. The length of the shell over which the imperfections are measured is smaller in the BS 5500 rules than in those of DnV [45] or DASt 013 [46]. On the other hand, the allowable radial deviation in the DnV rules varies with the R/t-ratio. With the BS 5500 rules, the limitations are on the radii of curvature rather than the radial deviations. However, for increased-radius imperfections (i.e. those which tend towards a flat spot), there is an algebraic relation between the radius of curvature of the imperfection and its radial deviation. The use of BS 5500 with increased-radius imperfections implies that the radial deviation does not vary with R/t.

Thus, there are certain anomalies in the design rules for doubly-curved shells. Such shells are widely used on land as well as in offshore platforms (e.g. flotation tanks) and submersibles. As there had been a failure of a large flotation sphere in the North Sea, it was thought advisable to test some models of externally-pressurised torispheres in order to check the regulations. Several of the torispheres tested had generous knuckle radii (simulating spheres) whereas others had relatively sharp knuckle radii (to establish the effect of knuckle radius on the collapse pressure). A number of parametric studies on the bifurcation buckling and collapse pressures of these shells was also undertaken.

One of the Ph.D. students at Liverpool wrote his thesis [47] on this subject and Fig. 3 shows two failed torispheres from his work. As may be seen, there are two failure modes (at least). One is an axisymmetric mode, occurring at the apex of the spherical part of the shell; the other is an asymmetric mode, occurring in both the knuckle region of the head and the spherical cap.

Additional tests on torispheres were carried out after those in [47] had been completed. The results of all the tests are shown on Fig. 4 (from [48]). It may be seen that several of the test results (torispheres with

(a) Axisymmetric

(b) Unsymmetric

Figure 3. Failure modes in externally-pressurised torispherical shells.

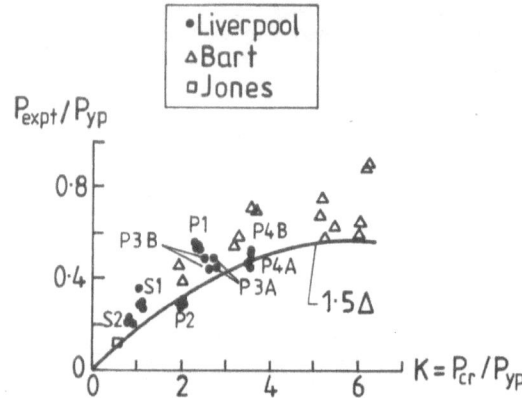

Figure 4. Experimental collapse pressures of as-manufactured tori-spherical domes versus the BS 5500 lower bound (1.5Δ) curve (external pressure).

sharp knuckle radii) are below the BS 5500 predictions (i.e. the BS 5500 values of Δ multiplied by a safety factor of 1.5).

Theoretical work on hemispheres and torispheres with axisymmetric imperfections [49,50] had also shown that one would expect the collapse strengths of externally-pressurised torispheres with sharp knuckle radii to be lower than those with generous knuckle radii. In addition, it was found that initial geometric imperfections sometimes had no effect on the collapse pressures of some torispheres [49,50]. This is unusual for shell buckling problems, in which imperfections are usually very harmful. On the experimental side, this imperfection-insensitivity had been observed by the authors of [47] and [48]. With torispheres which fail plastically, the failure mode can be (i) axisymmetric yielding in the knuckle, (ii) axisymmetric yielding in the region of the apex and (iii) bifurcation buckling near the knuckle/spherical cap junction - see Fig. 9 in [51]. If the axisymmetric imperfections are not too large, then they seem to have little effect on the collapse pressures of torispheres which fail plastically. Imperfections in the shape of the buckling mode, and located in the knuckle, still have to be investigated.

Another unusual aspect of externally-pressurised torispheres is that the normalised collapse curve does not appear to be independent of the yield stress, σ_{yp}. This is illustrated on Fig. 18 in [52], which shows the calculated collapse curves for clamped hemispherical shells and for clamped torispherical shells with sharp knuckle radii (both subjected to external pressure). As may be seen, the collapse curves for the axisymmetric collapse mode in the shallow domes vary with σ_{yp} (however, this was not the case when failure was by bifurcation buckling).

Additional theoretical calculations on torispherical shells, mainly perfect ones, appear in [53]. Some of the collapse curves, plotted as a function of the knuckle radius, are a little unusual. This topic will be discussed in the next section.

For hemispherical shells, attempts have been made to relate the imperfections in shape permitted by Codes to the strength of a given shell [49,50]. To this end, axisymmetric increased-radius imperfections, localised at the apex of the shell, have been used in conjunction with the BOSOR 5 program. The tolerances of the BS 5500 Code were used, as were those of the Norwegian DnV Code.

The results are shown in Fig. 5. Insofar as BS 5500 is concerned, there is a considerable gap between the theoretical predictions using axisymmetric imperfections and the lower-bound curve of the test results. However, the gap is reduced considerably when the template length (over which the imperfections are measured) is increased from $2.4\sqrt{R_{max}t}$ to $4.0\sqrt{R_{max}t}$ (i.e. it becomes similar to the length used by DnV and DASt 013). With the DnV Code, the agreement with the lower bound is good in the elastic region but not so good in the plastic region.

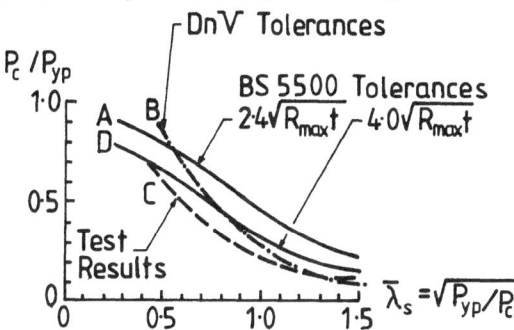

Figure 5. Theoretical predictions for hemispherical shells with axisymmetric imperfections compared with the lower bound of test results (external pressure).

It should be noted that, in the above exercise, no account was taken of residual stresses or of any interaction with nearby imperfections.

With the hemispherical shells, it was possible to generate collapse curves corresponding to given levels of imperfection – these are shown in Fig. 21 in [50]. In these curves, the arc length of the axisymmetric localised imperfection was varied until a theoretical minimum was obtained (this is discussed in [50]). It would be useful to be able to generate such curves for torispheres having various values of r/D (knuckle radius/diameter). It is not clear how easy this task will be, due to some of the collapse curves varying with σ_{yp} (see [53]).

In brief, the studies on externally-pressurised hemispheres and torispheres so far have shown:

(a)   The current design curve in BS 5500 should be lowered a little if it is to be applicable to torispheres with sharp knuckle radii. An alternative solution would be to have separate buckling rules for torispheres having generous knuckle radii and those having sharp knuckle radii.

(b)   That initial geometric imperfections have little effect on the collapse strength of some torispheres under external pressure. This is at variance with the usual harmful effects that imperfections have on the buckling strength of shells. However, in the cases under discussion, the torispheres failed by axisymmetric yielding collapse. This failure mode does not seem to be very sensitive to imperfections.

(c)   The correlation between the shape tolerances in Codes and their allowable strengths does not appear to be very good. There is also disagreement between the tolerances permitted by various national Codes. More work on this problem would be useful.

(d)   In some elevated temperature applications (say, 200°C), the Code rules for torispheres with sharp knuckles may be optimistic - see [54].

MORE TESTS ON EXTERNALLY-PRESSURISED METALLIC TORISPHERES

In the last two years, a further sixteen steel torispherical shells have been tested under external pressure at Liverpool. The yield stress of the steel in these models was 645 N/mm^2 and there was a small amount of strain-hardening after the yield point had been reached (for the models discussed in the previous section, σ_{yp} was in the range 400-485 N/mm^2 and there was no significant strain-hardening). The r/D-ratio varied from 0.06 to 0.18 in these models and the R_s/t-ratio was 62 < R_s/t < 137.

When the test results for these latest models were plotted in the same format as Fig. 4, none of the collapse pressures were below the 1.5Δ curve of BS 5500. It is not known why the results from some of the earlier tests plotted below the 1.5Δ curve from BS 5500 while none of the later ones did.

MACHINED STEEL TORISPHERES SUBJECTED TO EXTERNAL PRESSURE

The variation in the buckling/collapse strength of externally-pressurised perfect steel torispheres with the r/D-ratio is illustrated in Fig. 6 [54]. One might expect the strength to go on increasing as r/D was increased until, finally, the hemispherical shape (r/D = 0.5) was reached. However, the predictions do not follow that pattern. Instead a local maximum is reached, then a local minimum, followed by a rapid increase in strength to that of the hemisphere. It should be noted that the torispherical shells in Fig. 6 were perfect and did not contain any initial radial geometric imperfections.

In order to check the above theoretical predictions, some machined steel models (0.255 m dia.) were manufactured and tested under external pressure. The test results (for R_s/t = 300, R_s/D = 1, σ_{yp} = 304 N/mm^2) are shown in Fig. 8, superimposed on the theoretical predictions. As may be seen, the agreement between theory and test (from [51]) is very good.

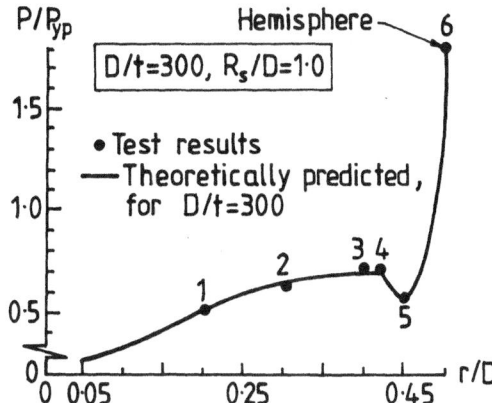

Figure 6. Experimental results and theoretical predictions for externally-pressurised machined steel torispheres (σ_{yp} = 304 N/mm^2).

The big drop in strength in going from r/D = 0.5 to r/D = 0.45 (a factor of 3.3) is due to the deviation from sphericity of the r/D = 0.45 torispherical shell.

OTHER SHELL BUCKLING RESEARCH

Due to space limitations, it has not been possible to describe all the research on shell buckling which has been undertaken at Liverpool during the past decade. However, the topics of the other research investigations, together with the papers published on them, are noted below. Interested readers should consult the references for further details.

(i)   Cylinder/spherical bulkhead combinations under internal pressure [55].
(ii) Cylinders under axial compression [56].
(iii) Ring-stiffened cylinders subjected to external pressure [57].
(iv) Composite torispheres and cylinders [58-64].

REFERENCES

1.   Galletly, G. D., "Stress Failure of Large Pressure Vessels - Recommendations Resulting from Studies of the Collapse of a 68' High x 45' Dia. Pressure Vessel". Tech. Report No. 45-57, Shell Development Corp., Emeryville, California, March 1957.
2.   Galletly, G. D., "Torispherical Shells - A Caution to Designers", Trans. ASME, J. Engng. for Industry, 81, 1959, 51-66. Also published in "Pressure Vessels and Piping Design: Collected Papers 1927-1959", ASME, 1960.

3. Harding, A. G. and Ehmke, E. F., "Brittle Failure of a Large Pressure Vessel", Proc. Am. Petrol. Inst., 1962, 42, Section 3, 107-117.

4. Mescall, J., "Stability of Thin Torispherical Shells under Uniform Internal Pressure", NASA TN D-1510, Washington, December 1962, 671-692.

5. Drucker, D. C. and Shield, R. T., "Limit Analysis of Symmetrically Loaded Shells of Revolution", J. Appl. Mech., 81, No. 1, March 1959, 61-68.

6. Shield, R. T. and Drucker, D. C., "Design of Thin-Walled Torispherical and Toriconical Head Pressure Vessels", J. Appl. Mech., 28, No. 2, 1961, 292-297.

7. Galletly, G. D., "Influence Coefficients for Open-Crown Hemispheres", Trans. ASME, J. Engng. for Power, 82, 1960, 73-81. Also published in Pressure Vessel and Piping Design: Collected Papers 1927-1959", ASME, 1960.

8. Galletly, G. D., "Bending of 2:1 and 3:1 Open-Crown Ellipsoidal Shells", Welding Research Council, Bulletin Series, No. 54, October 1959, 1-9.

9. Stanley, P. and Campbell, T. D., "Very Thin Torispherical Pressure Vessel Ends under Internal Pressure: Test Procedure and Typical Results", J. Strain Analysis, 16, 1981, 171-186.

10. Stanley, P. and Campbell, T. D., "Very Thin Torispherical Pressure Vessel Ends Under Internal Pressure: Strains, Deformations and Buckling Behaviour", J. Strain Analysis, 16, 1981, 187-203.

11. Kirk, A. and Gill, S. S., "The Failure of Torispherical Ends of Pressure Vessels Due to Instability and Plastic Deformation - An Experimental Investigation", Int. J. Mech. Sci., 17, 1975, 525-544.

12. Patel, P. R. and Gill, S. S., "Experiments on the Buckling Under Internal Pressure of Thin Torispherical Ends of Cylindrical Pressure Vessels", Int. J. Mech. Sci., 20, 1978, 159-175.

13. Galletly, G. D., "Internal Pressure Buckling of Very Thin Torispherical Shells - A Comparison of Experiment and Theory", Proc. 3rd Inter. SMiRT Conf., London, Paper G2/3, September 1975, 1-10.

14. Galletly, G. D., "Some Experimental Results on the Elastic-Plastic Buckling of Thin Torispherical and Ellipsoidal Shells Subjected to External Pressure", Proc. 2nd Inter. Colloquium on the Stability of Steel Structures, Liège, April 1977, 619-626.

15. Kemper, M. J., "Buckling of Thin Dished Ends under Internal Pressure", in Vessels Under Buckling Conditions, I. Mech. E., London, Paper C189/72, 1972, 23-32.

16. Galletly, G. D., "Elastic and Elastic-Plastic Buckling of Internally-Pressurized Ellipsoidal Shells", Trans. ASME, J. Press. Vess. Tech., 100, February 1978, 335-343.

17. Galletly, G. D. and Aylward, R, W,, "Plastic Collapse and the Controlling Failure Pressures of Thin 2:1 Ellipsoidal Shells Subjected to Internal Pressure", Trans. ASME, J. Press. Vess. Tech., 101, February 1979, 64-72.

18. Galletly, G. D. and Radhamohan, S. K., "Elastic-Plastic Buckling of Internally-Pressurized Thin Torispherical Shells", Trans. ASME, J. Press. Vess. Tech., 101, August 1979, 216-225.

19. Radhamohan, S. K. and Galletly, G. D., "Plastic Collapse of Thin Internally-Pressurized Torispherical Shells", Trans. ASME, J. Press. Vess. Tech., 101, November 1979, 311-320.

20. Galletly, G. D. and Aylward, R. W., "Elastic Buckling of, and First Yielding In, Thin Torispherical Shells Subjected to Internal Pressure", Inter. J. Press. Vess. and Piping, 7, September 1979, 321-326.

21. Bushnell, D., "BOSOR 5 - A Program for Buckling of Elastic-Plastic Complex Shells of Revolution Including Large Deflections and Creep", Comp. and Struct., **6**, pp. 221-239, 1976.

22. Galletly, G. D. and Blachut, J., "Torispherical Shells Under Internal Pressure - Failure Due to Asymmetric Plastic Buckling or Axisymmetric Yielding", Proc. I. Mech. E., **199**, No. C3, 1985, 225-238.

23. Dillström, P. and Dahlberg, L., "Buckling in Torispherical Pressure Vessel Heads Under Internal Pressure", in 'Stability of Plate and Shell Structures', eds. P. Dubas and D. Vandepitte, 1987, (ECCS Colloquium in Ghent), 367-372.

24. Galletly, G. D., "The Buckling of Fabricated Torispherical Shells Under Internal Pressure", in "Buckling of Shells - Proc. of a State-of-the-Art Colloquium", (ed. E. Ramm), Springer-Verlag, Berlin, 1982, pp. 429-466.

25. Galletly, G. D., "Plastic Buckling of Torispherical and Ellipsoidal Shells Subjected to Internal Pressure", Proc. I. Mech. E., **195**, 26, 1981, 329-345.

26. Galletly, G. D., "Design Equations for Preventing Buckling in Fabricated Torispherical Shells Subjected to Internal Pressure", Proc. I. Mech. E., **200**, No. A2, March 1986, 127-139.

27. Galletly, G. D., "A Simple Design Equation for Preventing Buckling in Fabricated Torispherical Shells Under Internal Pressure", Trans. ASME, J. Press. Vess. Tech., **108**, Nov. 1986, 521-526.

28. Roche, R. L. and Autrusson, B., "Experimental Tests on Buckling of Torispherical Heads and Methods of Plastic Bifurcation Analysis", Trans. ASME, J. Press. Vess. Tech., **108**, May 1986, 138-145.

29. European Recommendations for Steel Construction: Buckling of Shells, ECCS, Av. Louise 326, Brussels, 4th Edition, 1988.

30. CODAP, Code Français de Construction des Appareils à Pression SNCT, AFIAP (10, avenue Hoche, Paris).

31. Galletly, G. D., "Buckling and Collapse of Thin Internally-Pressurized Dished Ends", Proc. I.C.E., **67** (Part 2), September 1979, 607-626.

32. Galletly, G. D., Blachut, J. and Moreton, D. N., "Internally-Pressurised Machined Domed Ends - A Comparison of the Plastic Buckling Predictions of the Deformation and Flow Theories", to be published in Proc. I. Mech. E.

33. Bushnell, D., "Plastic Buckling", in Pressure Vessels and Piping: Design Technology 1982. A Decade of Progress, ASME, New York, 49-117.

34. Yamaki, N., Naito, K. and Sato, E., "Buckling of Circular Cylindrical Shells Under Combined Action of a Transverse Edge Load and Hydrostatic Pressure", Proc. Int. Conf. on Thin-Walled Structures, University of Strathclyde, Glasgow, April 1979.

35. Galletly, G. D. and Blachut, J., "Buckling of a Cantilevered Cylindrical Shell Subjected to a Transverse Shearing Force at Its Tip", Proc. 3rd Inter. Colloquium on Stability of Metal Structures, Paris, November 1983, Preliminary Report, 383-389.

36. Galletly, G. D. and Blachut, J., "Plastic Buckling of Short Vertical Cylindrical Shells Subjected to Horizontal Edge Shear Loads", Trans. ASME, J. Press. Vess. Tech., **107**, May 1985, 101-106.

37. Dostal, M., Austin, A., Combescure, A., Peano, A. and Angeloni, P., "Shear Buckling of Cylindrical Vessels: A Benchmark Exercise", Trans. 9th Inter. SMiRT Conf., Lausanne, Vol. E, 199-208, 1987.

38. Harding, J. E. and Dowling, P. J., "Recent Research on the Behaviour of Cylindrical Shells Used in Offshore Structures", in 'Steel Structures', (ed. M. Pavlovic), Elsevier Appl. Science Pubs., London,

1986, 317-338. Conf. on Steel Structures, Budva, Yugoslavia.

39. Weingarten, V., Morgan, E. and Seide, P., "Final Report on Development of Design Criteria for Elastic Stability of Shell Structures", Space Technology Labs. Inc., STL/TR-60-0000-19425, 1960.

40. Tennyson, R. C., Booton, M. and Chan, K. H., "Buckling of Short Cylinders under Combined Loadings", Trans. ASME, J. Appl. Mech., 45, Sept. 1978, 574-578.

41. Galletly, G. D. and Pemsing, K., "Interactive Buckling Tests on Cylindrical Shells Subjected to Axial Compression and External Pressure - A Comparison of Experiment, Theory and Various Codes", Proc. I. Mech. E., 199, No. C4, 1985, 259-280.

42. Galletly, G. D., James, S., Kruzelecki, J. and Pemsing, K., "Interactive Buckling Tests on Cylinders Subjected to External Pressure and Axial Compression", Trans. ASME, J. Press. Vess. Tech., 109, Feb. 1987, 10-18.

43. Galletly, G. D. and Pemsing, K., "Buckling of Cylinders Under Combined External Pressure and Axial Compression", in "Collapse: The Buckling of Structures in Theory and Practice". IUTAM Symposium (eds. J. M. T. Thompson and G. W. Hunt), Cambridge University Press, 1983, 505-527.

44. BS 5500: 1976 and 1988, Specification for Unfired Fusion Welded Pressure Vessels, Section 3.6, (British Standards Institution, London).

45. DnV (Det norske Veritas): 1982 Buckling Strength Analysis, Classification Note 30.1, Høvik, Norway, July 1982.

46. DASt (Deutscher Ausschuss für Stahlbau), Richtlinie 013, Beulsicherheitsnachweise für Schalen, July 1980, Cologne, Germany.

47. Warrington, B., "The Buckling of Torispherical Shells Under External Pressure", Ph.D. Thesis, University of Liverpool, 1984.

48. Galletly, G. D., Kruzelecki, J., Moffat, D. G. and Warrington, B., "Buckling of Shallow Torispherical Domes Subjected to External Pressure - A Comparison of Experiment, Theory and Design Codes", J. Strain Anal., 22, No. 3, 1987, 163-175.

49. Galletly, G. D., Błachut, J. and Kruzelecki, J., "Plastic Buckling of Externally-Pressurised Dome Ends", "Advances in Marine Structures", (eds. C. S. Smith and J. D. Clarke), 1986, 238-261, Elsevier Applied Science Publishers.

50. Galletly, G. D., Blachut, J. and Kruzelecki, J., "Plastic Buckling of Imperfect Hemispherical Shells Subjected to External Pressure", Proc. I. Mech. E., 201, No. C3, 1987, 259-262.

51. Blachut, J., Galletly, G. D. and Moreton, D. N. "Buckling of Near-Perfect Steel Torispherical and Hemispherical Shells Subjected to External Pressure", to be published in J. AIAA.

52. Błachut, J. and Galletly, G. D., "Clamped Torispherical Shells Under External Pressure - Some New Results", J. Strain Anal., 23, No. 1, 1988, 9-24.

53. Błachut, J. and Galletly, G. D., "Externally-Pressurised Torispheres - Plastic Buckling and Collapse", in Buckling of Structures - Theory and Experiment, ed. I. Elishakoff, J. Arbocz, C. D. Babcock, jr. and A. Libai, Elsevier Science Pubs., Amsterdam, 1988, 29-45.

54. Galletly, G. D., "Buckling of Shallow Dished Ends Under External Pressure - A Caveat", Proc. I. Mech. E., 201, No. C5, 1987, 373-378.

55. Galletly, G. D. and Błachut, J., "Elastic Buckling of Internally-Pressurised Cylinder/Bulkhead Combinations", Proc. I. Mech. E., 201, No. C4, 1987, 259-262.

56. Galletly, G. D. and Błachut, J., "Axially-Compressed Cylindrical

Shells - A Comparison of Experiment and Theory". To be published in the Professor A. Sawczuk Commemorative Volume, 1990.

57. Galletly, G. D. and James, S., "Inter-Ring Buckling of Welded Ring-Stiffened Cylindrical Shells Subjected to External Pressure", Proc. I. Mech. E., J. Proc. Eng., **203**, 1989, 101-114.

58. Galletly, G. D. and Muc, A., "Buckling of Fibre-Reinforced Plastic-Steel Torispherical Shells Under External Pressure", Proc. I. Mech. E., **202**, No. C6, 1988, 409-420.

59. Galletly, G. D. and Muc, A., "Buckling of Externally-Pressurized Composite Torispherical Domes", Proc. I. Mech. E., J. Proc. Eng., **203**, 1989, 41-56.

60. Galletly, G. D., Moreton, D. N. and Muc, A., "Buckling of Slightly Flattened Domed Ends Reinforced Locally with FRP", to be published in Proc. I. Mech. E., 1989.

61. Levy, F., Galletly, G. D. and Mistry, J., "Buckling of Composite Torispherical and Hemispherical Domes". To be presented at CADCOMP 90, Brussels, April 1990.

62. Blachut, J. and Galletly, G. D., "A Numerical Investigation of Buckling/Material Failure Modes in CFRP Dome Closures". To be presented at CADCOMP 90, Brussels, April 1990.

63. Blachut, J., Galletly, G. D. and Gibson, A. G., "CFRP Domes Subjected to External Pressure". To be published in J. Marine Struct., 1990.

64. Galletly, G. D. and Blachut, J., "On The Buckling Strength of Steel and CFRP Dome Closures". To be presented at UDT 90 Conference, London, February 1990.

Stresses in Damaged Circular Cylindrical Shells

Christophe K.W. Tam and James G.A. Croll
Department of Civil and Municipal Engineering
University College London
Gower Street, London WC1E 6BT

ABSTRACT

Circular cylindrical shells containing localised dent type damage are shown to develop high levels of local stress concentration. For a generalised analysis procedure, the dent damage is represented by means of a normal pressure applied to the perfect shell for analysis. For axially loaded cylinders having various end loading conditions, an even more convenient axisymmetric form of analysis procedure is presented; this is suitable for routine damage effect assessment. Predicted stress concentrations are used to provide a simplified calculation of the effects on fatigue life for components containing localised dent damage.

1. INTRODUCTION

As oil and gas exploration moves in deeper and more environmentally hostile waters, the use of weight efficient structural shell components is imposing tighter controls on the engineers and fabricators to ensure that these members are constructed within specified tolerances. But, in addition to fabrication imperfections, these components are also prone to in-service damage such as collision of supply and service vessels. Depending on shell geometry, end fixity and location of the impact, the damage which result are known to exhibit a variety of geometric configurations. Of present interest is how local dent type damage will affect the elastic stress distributions within these components.

Circular cylindrical shells with initial geometric imperfections in the form of a localised dent damage may exhibit considerably reduced static load carrying capacities [1,2]. Surrounding the area of the dent, elastic stress concentrations can reach unacceptably high levels resulting in shorter fatigue life, higher susceptibility to progressive damage growth and a form of ratcheting failure [3,4]. The effects of geometric errors in reinforced cooling towers [5,6] and pressure vessels [7,8,9] have been widely studied. Analytical techniques developed in this context may be usefully applied to the consideration of stress distribution in locally dented shells.

This paper is directed toward the investigation of elastic stress concentrations in locally damaged circular cylindrical shells components under various axial loading conditions. The method is based upon the replacement of dent damage by a statically equivalent distribution of load which for analysis may be applied to the perfect shell. An even simpler procedure based on a 1-dimensional axisymmetric type of analysis is presented and a qualitative assessment of the fatigue implications of local damage is discussed.

2. STRESS ANALYSIS OF CIRCULAR CYLINDERS CONTAINING LOCAL DAMAGE

With the advance of powerful computers, stress analysis in recent years has largely relied upon increasingly sophisticated numerical techniques. For shells with distortional damage, these methods usually require extensive computing effort which is not always convenient for the engineer faced with the need for quick decisions as to consequential action. The following represents a practical alternative.

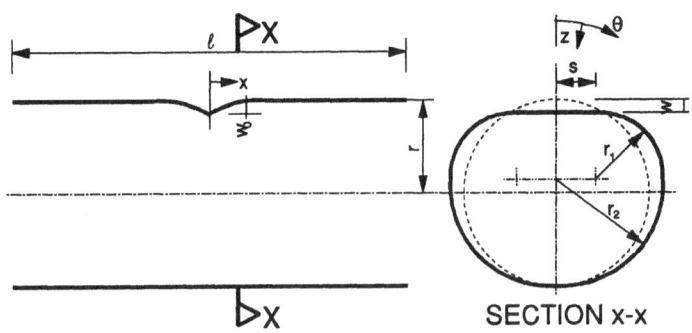

$$w = w_0 e^{-bx/r}, \qquad s = \frac{\pi}{2}w, \qquad r_1 = r - \frac{\pi+2}{4}w, \qquad r_2 = r + \frac{\pi-2}{4}w$$

Figure 1 Idealised damage for analysis

2.1 The 'Equivalent Load' Method

For the analysis of imperfection effects, the 'equivalent load' method, already in use for assessing the elastic stress distributions in reinforced concrete cooling towers containing geometric imperfections, has the advantage of providing a rational and yet convenient means of transforming geometric distortions into a normal pressure distribution applied to the perfect shell for analysis to include imperfection effects. The derivation of the normal pressure is based upon static equivalence. If the imperfect shell is assumed to have the same in-plane stress resultants $(n_x^0, n_\theta^0, n_{x\theta}^0)$ as those of the perfect shell under its design load, then a first order equivalent pressure, p_z^1, may be written as

$$p_z^1 = n_x^0 \chi_x^0 + n_\theta^0 \chi_\theta^0 + 2n_{x\theta}^0 \chi_{x\theta}^0 \qquad -(1)$$

where $(\chi_x^0, \chi_\theta^0, \chi_{x\theta}^0)$ are the errors in axial, circumferential and twist curvatures due to the damage.

When this normal pressure is applied to the perfect shell, a set of incremental stress and moment resultants $(n_x^1, n_\theta^1, n_{x\theta}^1, m_x^1, m_\theta^1, m_{x\theta}^1)$ is obtained. These stresses representing the equilibrium of the out-of-balance pressure p_z^1 which arises when the shell containing the initial imperfection then, allow a first order correction of the stress state in the shell to be written as

$$n_i^* = n_i^0 + n_i^1$$

$$m_i^* = m_i^0 + m_i^1 \qquad \qquad -(2$$

where

$$i = x, \theta, x\theta$$

Subsequent corrections may be set up as an iterative procedure in which the qth order equivalen pressure is obtained on the basis of the $(q-1)^{th}$ order incremental in-plane stress resultants, sucl that

$$p_z^q = n_x^{q-1} \chi_x^0 + n_\theta^{q-1} \chi_\theta^0 + 2n_{x\theta}^{q-1} \chi_{x\theta}^0 \qquad \qquad -(3$$

and

$$n_i^* = n_i^0 + n_i^1 + n_i^2 + \ldots + n_i^q$$

$$m_i^* = m_i^0 + m_i^1 + m_i^2 + \ldots + m_i^q \qquad \qquad -(4$$

A detailed discussion on how the 'equivalent load' method is formulated can be found in ref.[3]

2.2 Stresses in damaged circular cylinder

Depending on impact conditions, circular cylindrical shells can exhibit a variety of dent damage configurations which are in general quite different from those of fabrication geometry errors. Previous analysis [3] based on a doubly-symmetric dent, as shown in Figure 1, has relied upon the use of a 2-dimensional fourier series for the description of the radial distortion profile of the shell. It has been shown that such a fourier series not only provides a compact representation of the dent damage, but also facilitates a convenient solution procedure for a simplified linear shell bending analysis.

For illustration purposes, consider a circular cylindrical shell of radius upon thickness ratic $r/t=100$, having the localised dent damage as that shown in Figure 1. The axial damage profile is taken to have the form

$$w = w_0 e^{-bx/r} \qquad \qquad -(5$$

where w is the radial distortion, w_0 is the maximum distortion amplitude at $(x=0, \theta=0)$ and b is a non-dimensional parameter used to specify the axial extent of the damage localisation. As a specific example, the typical values, $W_0=w_0/r=0.1$ and $b=5$ are considered. It may be of interest to note that the axial extent of damage is dependent upon the level of damage with b decreasing as w_0 is increased [10]

For the damaged shell subject to an axial stress resultant, n_x^0, uniformly distributed around the circumference, the incremental hoop membrane and axial moment stress resultants are shown in Figure 2. The circumferential moment and axial membrane stresses arising basically as consequence of Poisson effects are found to be of little significance and are therefore not shown in the figure. However, they are included in the evaluation of the surface stress concentration factors (SCF) shown in Figure 3. These SCF's are non-dimensionalised by

$$S_i = \frac{\sigma_i^*}{\sigma_x^0} = N_i^0 + N_i^1 \pm 6W_0 M_i^1 \left(\frac{r}{t}\right) \qquad \qquad -(6$$

where

$$N_i^0 = \frac{n_i^0}{n_x^0}, \qquad N_i^1 = \frac{n_i^1}{n_x^0}$$

$$M_i^1 = \frac{m_i^1}{n_x^0 w_0}$$

and $\sigma_x^0 = n_x^0/t$ is the axial membrane stress of the perfect cylinder, σ_i^* is the combined bending and membrane stress on the surface of the cylinder in the circumferential or axial direction.

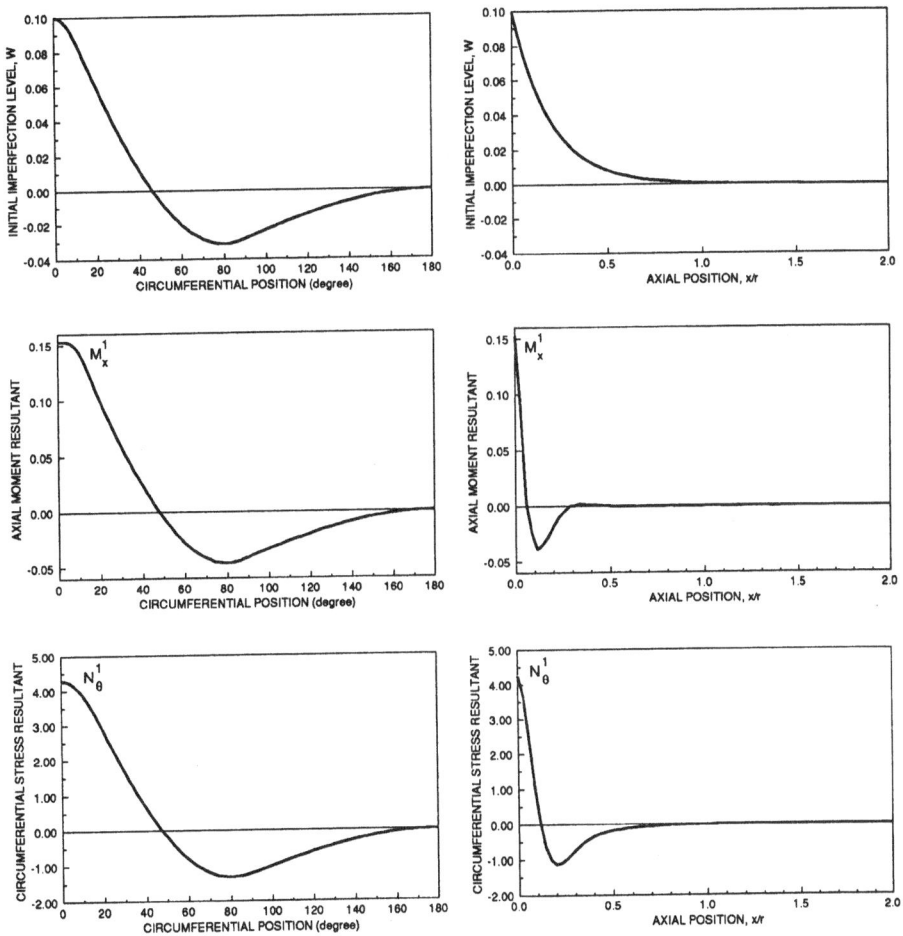

**Figure 2 Incremental stress and moment resultants (at x=0
and θ=0) based on a 2-dimensional fourier representation
of the idealised dent damage shown in figure 1.**

For the present example, Figures 2 & 3 show that significant levels of meridional bending and circumferential membrane actions are developed to counteract the out-of-balance load produced by the dent damage and the applied stress. Axial surface stress concentrations are mainly due to these bending effects with incremental axial membrane stresses being found to be small. I is therefore suggested that higher order corrections according to eq.(3) will be unlikely to produce significant change in the estimations of the final stress state.

Stress concentrations are also found to be sensitive to shell geometry as well as dent configuration [3]. For shells containing the same level and form of damage but with a decreasing wall thickness stress concentrations in the axial direction are found to be more localised around the most damaged position. Even though an increasing proportion of the out-of-balance load is being resisted by the development of hoop membrane forces with a consequent reduction in axial moment, the thinner shell wall is still required to develop higher axial stresses to maintain equilibrium. For shells with constant radius upon thickness ratios, more localised dent damage in which b increases, has been found to yield higher and more localised stress concentrations because of the increased contribution to the resistance from the axial bending moment.

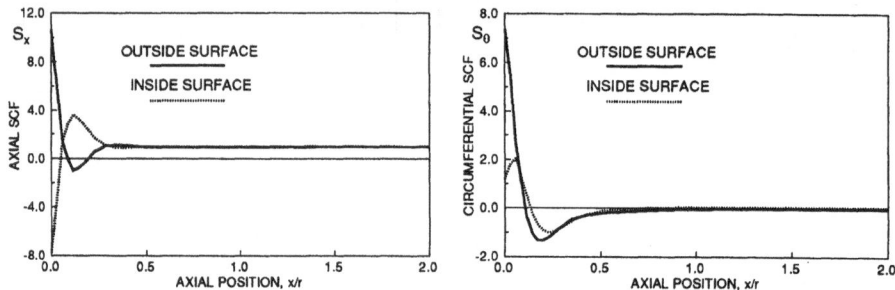

Figure 3 Surface stress concentration factors along the meridian at θ=0

Even under the same impact conditions, dent damage configurations are highly dependent upon damage level and shell geometry, becoming more localised at lower damage levels and with thinner walled shells [3]. For this reason, proper account of their interdependence must be taken into consideration when assessing the effect of variations in shell wall thickness and damage shape. In addition, the use of the present idealised axial damage profile, given by eq.(5), which contains a slope discontinuity at the most damage position, has the effect of over-estimating the stress concentrations. A fuller discussion on how shell geometry, damage configuration, and other parameters can affect elastic stress concentrations can be found in ref.[3,4].

2.3 The Simplified 'Equivalent Load' Method

Although the use of a two-dimensional fourier series to describe dent damage has the ability to cater for a wide range of damage shapes and while it provides a convenient and systematic load function for the subsequent perfect shell analysis, solution procedures still require a fairly high level of computational efforts, especially for damage configuration with sharp curvature changes and/or slope discontinuities. This is due to the damaged shell behaviour being very sensitive to curvature changes for which a fourier series modelling requires a very high number of terms, even though the local dent damage may be closely approximated with a relatively small number of fourier terms.

For an axially loaded cylinder containing a dent damage as shown in Figure 1, it has been shown that meridional moments and circumferential membrane forces are the predominant shell actions offering resistance to the out-of-balance moment produced by the imperfection. It may be observed from Figure 2 and 3 that while the axial stress and moment variations are dependent upon the axial damage profile, their variations in the circumferential direction are approximately proportional to the magnitudes of the radial damage in the circumferential direction. This evidently implies that if the stress variations on a typical meridian can be found, the entire stress state for the damaged shell may be approximated by proportioning the stresses and moments in accordance with the initial imperfection profile in the circumferential direction.

In circumstances where the axial and circumferential displacements (u^1, v^1) may be neglected, the axisymmetric behaviour of the shell may be modelled to contain the all important axial bending and circumferential membrane shell action by the equation

$$D\frac{d^4w^1}{dx^4} + Et\frac{w^1}{r^2} = p^1 \qquad\qquad -(7)$$

where $D = Et^3/[12(1-v^2)]$ is the bending stiffness, E is the Young's modulus, v is the Poisson's ratio, $p^1 = n_x^0 d^2w^0/dx^2$ is the equivalent pressure due to the axial curvature error, w^0 is the radial displacement of the imperfect shell, and w^1 is the radial displacements induced by the equivalent pressure, p^1. The approximate relationships between the incremental radial displacements and the equivalent load induced axial moment and circumferential membrane stress resultants may now be written as

$$m_x^1 = D\frac{d^2w^1}{dx^2}$$

$$n_\theta^1 = -Et\frac{w^1}{r} \qquad\qquad -(8)$$

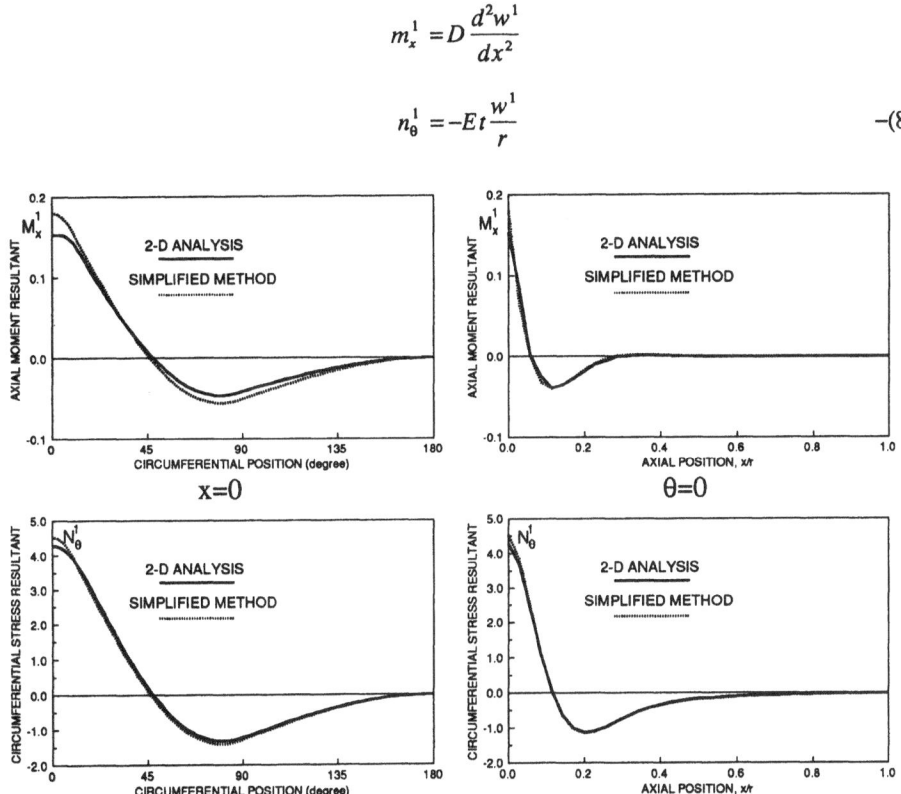

Figure 4 Comparison of results obtained on the basis of the 2-dimensional analysis and the simplified equivalent load method

For the doubly symmetric damage type shown in Figure 1, a representation of the radial damage profile on the meridian at the circumferential position θ=0, will then allow a simple and effective analysis based on eq.(7) to be undertaken. That is, instead of a 2-dimensional fourier series, the idealised damage profile may now be written as

$$w^0(x,\theta) = \overline{w}^0 f(x) g(\theta) \qquad -(9)$$

where $\overline{w}^0$ is the maximum amplitude of the dent at $x=0$ and $\theta=0$, f(x) is the variation of the radial distortion in the axial direction, and $g(\theta)$ is the geometric variations in the circumferential direction at $x=0$. For the damage shape shown in Figure 1, in which the axial deformation is modelled by an exponential function as given in eq.(1), closed form solutions to eq.(7-9) are obtainable [3]. For comparison purposes, the damaged shell considered in the previous section has been studied using the 1-dimensional analysis and results plotted in Figure 4 for comparison with the more elaborated 2-dimensional modelling. In general, the accuracy is sufficiently high to recommend this simplified method as a basis for routine assessment of damage effects in circular cylinders. The small discrepancies observed in Figure 4 are suggested to be largely due to the equivalent load induced in-plane displacements which have been approximated in the 1-dimensional analysis.

3. DAMAGED SHELLS SUBJECTED TO ARBITRARY END LOADING CONDITIONS

The analyses and results (as shown in Figures 2 to 4) presented in §2 have been based upon the shell being subjected to a constant axial membrane stress resultant, n_x^0, uniformly distributed around the circumference at some location far from the dent damage. Under certain circumstances, such a uniform compressive stress may not be achievable, perhaps due to load eccentricity, presence of moment loading or attachment to stiff nodal joints. Although the 2-dimensional analysis described in §2.1 can still cater for varying applied stress situations, solution procedures will require even higher computational efforts. The simplified 1-dimensional 'equivalent load' method, described in §2.3, has the ability of providing a simple and efficient alternative means of assessing the stress state.

For the damaged shell as shown in Figure 1 to be subjected to a varying axial in-plane stress resultant, $\overline{n}_x(\theta)$, at the shell ends, the 1-dimensional equilibrium condition similar to that given by eq.(7) may now be written as

$$D\frac{d^4w^1}{dx^4} + Et\frac{w^1}{r^2} = \overline{n}_x^0(\theta)\frac{d^2w^0}{d^2x} \qquad -(10)$$

It becomes clear that once the original in-plane stress resultant, $\overline{n}_x^0(\theta)$, and the initial imperfection profile are known, the incremental stress and moment results for each meridian at circumferential positions θ may be evaluated in much the same way as previously described. This procedure may be applied to the prediction of response under controlled end displacement and forces.

3.1 Damaged Shells Subjected to Uniform End Displacements

In situations where the damaged shell boundary displacements are controlled, the axial membrane stresses at the shell ends must be distributed in such a way so as to accommodate the reduction in stiffness due to local distortional geometry variations. To obtain these distributions, consider the following procedure. If the shell is subjected to an uniformly distributed stress, end shortening due to axial membrane and bending effects will vary in the circumferential direction. Under displacement controlled situations, the applied stresses on each meridian will have to be adjusted such that the end shortening at each circumferential position is compatible with that of the controlled uniform end displacement.

Based on the simplified 1-dimensional analysis, the axial shortening of a damaged shell of length, l, under uniform axial stress, n_x^0, may be written as

$$e_x = e_x^b + e_x^m = 2 \int_0^{l/2} \varepsilon_x^b dx + 2 \int_0^{l/2} \varepsilon_x^m dx \qquad -(11)$$

and

$$\varepsilon_x^b = \frac{dw^0}{dx}\frac{dw^1}{dx}$$

$$\varepsilon_x^m = \frac{n_x^0}{Et}$$

where ε_x^b and ε_x^m are the equivalent load induced axial strains due to the bending and membrane actions respectively. For the shell to produce a uniform end displacement , the applied in-plane stress resultant variations may be approximated as

$$\overline{n}_x^0 = \frac{1}{\left(1 + \frac{e_x^b}{e_x^m}\right)} n_x^0 \qquad -(12)$$

where n_x^0 is the stress resultant at some circumferential position where the meridian does not contain any geometry errors, e_x^b and e_x^m are end shortenings due to axial bending and membrane effects as defined in eq.(11).

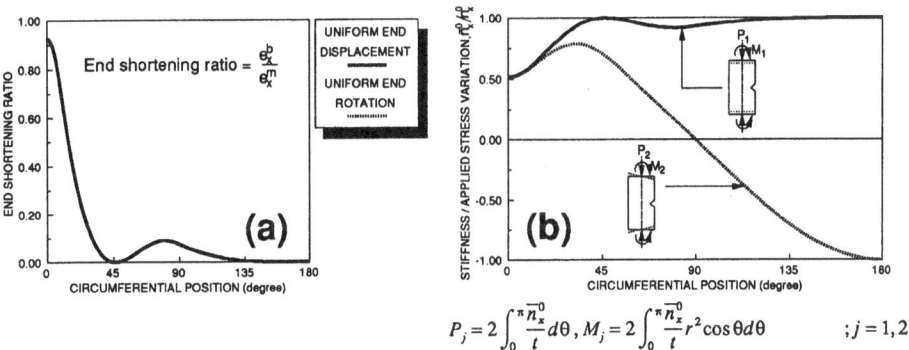

$$P_j = 2 \int_0^\pi \frac{\overline{n}_x^0}{t} d\theta \, , \, M_j = 2 \int_0^\pi \frac{\overline{n}_x^0}{t} r^2 \cos\theta d\theta \qquad ; j = 1, 2$$

Figure 5 (a) Variation of end shortening under stress controlled axial loads and (b) Variation of stiffness reduction to achieve uniform end displacement and end rotation.

For the damaged shell described in §2 to be subjected to a uniformly distributed in-plane stress resultant n_x^0, corresponding with an externally applied pure axial load of $P_0 = 2\pi r n_x^0$, the end shortening based on the 1-dimensional analysis is shown in Figure 5. As expected, the more deformed meridians suffer greater bending deflection under load and therefore exhibit more severe end shortening. In regions where there is little or no initial out-of-straightness, membrane actions dominate and end shortening is minimal. Also plotted in the figure is the adjusted stress distribution required for the same shell to give a uniform end displacement the same as that of the stress controlled situation at $\theta=\pi$ position. This corresponds with an axial load of $P_1=0.92P_0$ and an end moment of $M_1=-0.37P_1r$.

Surface stress concentration factors obtained for such a uniform end displacement loading situation are plotted in Figure 6 for comparison with those of the stress controlled. It may be observed that there is a reduction of the maximum stress concentration at the most damage position. The way in which these incremental stresses vary is directly related to the stiffness reduction as shown in Figure 5b.

3.2 Damaged Shells Subjected to Uniform End Rotations

For a shell subject to uniform end rotations, the stiffness variation around the circumference at the shell ends may be written as

$$\overline{n}_x^0 = \frac{1}{\left(1 + \frac{\epsilon_x^b}{\epsilon_x^m}\right)} n_x^0 \cos\theta \qquad -(13)$$

where the symbols retain the same meanings as those defined in eq.(12). Overall end rotation is assumed to take place about the axis passing through $\theta = \pm\pi/2$; this is taken to produce an axial displacement at $\theta = 0$ the same as the axial displacement at $\theta = \pi$ of the stress controlled case. The modified applied in-plane stress distribution resulting from the stiffness variation is shown in Figure 5b. Surface stress concentration factors resulting from this are included in Figure 6 and the corresponding externally applied load has been found to reduce to $P_2 = -0.055 P_0$ with the end moment $M_2 = -51.2 P_2 r$.

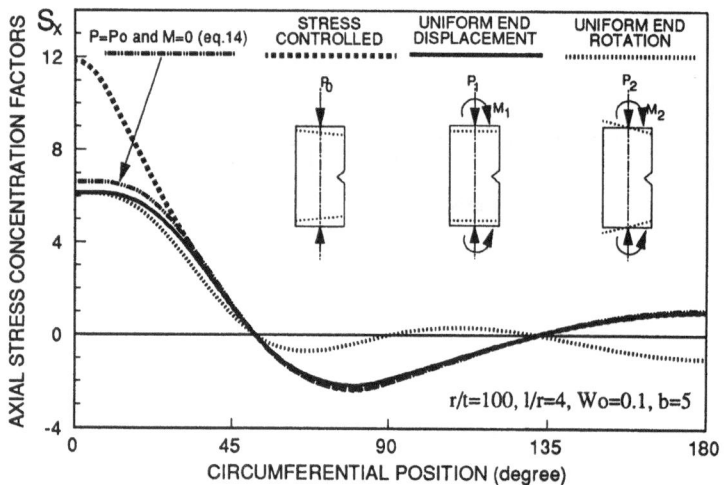

Figure 6 Comparison of surface stresses (at $x=0$) under (i) Stress controlled, (ii) Uniform end displacement, (iii) Uniform end rotation and (iv) Pure applied axial compression

3.3 Damaged Shells Under Combined Axial Compression and End moments

Having established the stiffness distribution for the uniform end-displacement situation, other predefined modes of end loading situation may be achieved by superimposing the corresponding end displacement conditions and uniform rotation. In combined loading situations where the overall axial compression force, P, and the end moment, M, are specified, the final stress state of the damage shell may be obtained by suitably proportioning the load and moment contribution arising from the displacement controlled loading situations, such that

$$P = \alpha P_1 + \beta P_2$$

$$M = \alpha M_1 + \beta M_2 \qquad -(14)$$

where α and β are multiplying factors to be applied to the stress concentration factors obtained on the basis of the uniform end displacement (§3.1) and end rotation loading (§3.2) situations respectively. For the shell being considered to be subjected to a centrally applied axial compression of $P=P_0$ and $M=0$, the stress concentration distribution has been found to be very similar to that of the uniform end displacement controlled case, as shown in Figure 6.

4. FATIGUE IMPLICATIONS OF SHELLS CONTAINING LOCAL DAMAGE

Fabrication defects such as geometric discontinuities at connections, weld irregularities, ... are known to induce high stress concentration regions which have direct implications in so far as the fatigue strength of the member is concerned. Unlike the effects of weld defects, which have received a great deal of attention in the past, the scarcity of information available on the fatigue implication of distortional damage means that we have an as yet incomplete understanding of this problem. Static strength reduction of a structural component is in general dependent upon the nature of stress redistribution occurring in the plastic range. For structural components under low stress high cycle loading conditions, failure of the component is usually governed by the fatigue limit, which has a relationship with the elastic limit of the material. Of current concern is how the knowledge of stress concentration arising from local damage can assist in the estimation of the fatigue endurance of the component.

Cylindrical shells containing local damage, as that shown in Figure 1, have been shown to exhibit high stress concentrations at the most damaged position. The two dimensional nature of these stresses has added implications so far as fatigue analyses are concerned. However, for simplicity, the maximum principal stress theory will be use for the present investigation. This implies that the number of cycles required to cause fatigue failure under combined stresses is the same as that taken by the maximum principal stress acting alone. For this reason, the following fatigue evaluation will be based upon the meridional stress component only.

In the absence of specific design guidance on damaged components, the present investigation will be based on the joint classification system as given in ref.[11,12]. A class B joint is assumed as this deals with materials in the as rolled conditions or with cleaned surfaces. If the stress level always remains elastic, the low stress high cycle SN-curve may be written as

$$log\, N = log\, a - m\, log\, \delta\sigma \qquad -(15)$$

where $\delta\sigma$ is the design stress range. For the component exposed to sea water but provided with adequate cathodic protection, the numerical constants a and m are given by

$$log\, a = 15.01 \quad \text{and} \quad m = 4.0$$

Under cyclic loading conditions, the design stress range for a perfect cylinder will induce a higher stress range in the damaged shell. This has the effect of lowering the number of cycles to failure. A measure of the fatigue strength reduction can be based upon the fatigue cycle reduction factor, N_r, defined as

$$N_r = \frac{N_i}{N_p} = \left(\frac{\delta\sigma_p}{\delta\sigma_i}\right)^m \qquad -(16)$$

where the subscripts i and p are used to denote the imperfect and perfect shell respectively. This compact representation of the fatigue strength reduction can be further simplified by the introduction of the theoretical stress concentration factor, k_t, which may be written as

$$k_t = \frac{\delta\sigma_i}{\delta\sigma_p}$$

$$= \frac{surface\ stress\ level\ of\ the\ damaged\ shell}{surface\ stress\ level\ of\ the\ perfect\ shell} = S_x \qquad -(17)$$

such that

$$N_t = \left(\frac{1}{k_t}\right)^m = \left(\frac{1}{S_x}\right)^m \qquad -(18)$$

where S_x is the axial surface stress concentration factor defined in eq.(6). For the shell being subjected to the stress controlled loading conditions, Figure 7(a) shows how even an extremely small initial dent damage, $W_0=0.01$ and $b=20$, can cause a dramatic reduction in fatigue endurance. Also shown in Figure 7(b) are the variations of the fatigue cycle reduction factor with respect to shell and dent geometry.

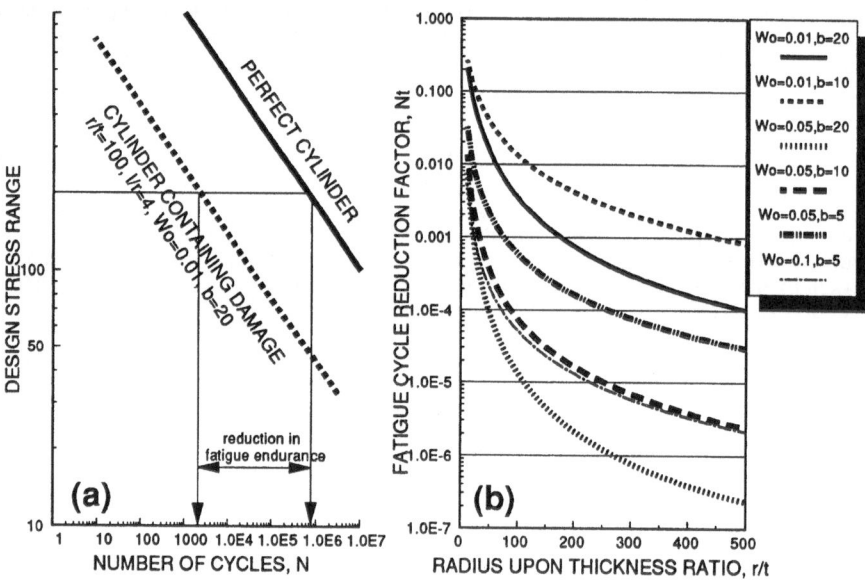

Figure 7 (a) Fatigue life reduction of a cylinder containing local damage; (b) Variations of fatigue cycle reduction factors with shell radius upon thickness ratios for various damage configurations.

5. CONCLUSION

The 'equivalent load' method used for evaluating the extent of elastic stress concentrations in circular cylindrical shells containing localised dent damage has been described. The method is based upon replacing geometric distortions by a statically equivalent normal pressure to be applied to the perfect shell for analysis. Stress changes resulting from an idealised dent damage configuration have been shown to be of significance. Although only axially loaded shells have been considered, the method can be easily extended to cater for a wide range of loading conditions.

A simplified 1-dimensional analysis based on the physical behaviour of an axially loaded damaged shell has also been described. It has been shown, for a certain class of damage, that closed form solutions are possible and that the method is particularly suitable for routine structural integrity monitoring purposes. Shells subjected to various end loading conditions have also been considered. For illustration purposes, a simple fatigue life assessment of a damaged shell has been carried out. It has been found that even very low levels of local damage can dramatically reduce the fatigue endurance of a structural member.

6. REFERENCES

[1] Smith, C.S., Somerville, W.L. and Swan, J.W., "Residual Strength and Stiffness of Damaged Steel Bracing Members", Offshore Technology Conference, OTC paper 3981, 1981, pp.273-282

[2] Ellinas, C.P., "Ultimate Strength of Damaged Tubular Bracing Members", Journ. of Structural Engineering Division, Proc. ASCE, Vol.110, No.2, 1984, pp.245-254

[3] Tam, C.K.W. and Croll, J.G.A., "Elastic Stress Concentrations in Cylindrical Shells Containing Local Damage", *Applied Solid Mechanics -2*, Ed. A.S. Tooth and J. Spence, Elsevier Appl. Sci. Pub., 1988, pp.155-177

[4] Tam, C.K.W. and Croll, J.G.A., "Stress Concentrations in Circular Tubulars Local Damage", Journ. Offshore Mech. and Arctic Eng., Trans. ASME., Vol.111, 1989, pp.278-284

[5] Croll, J.G.A. et al., "A Simplified Approach to the Analysis of Geometrically Imperfect Cooling Towers", Engineering Structures, Vol.1, 1979, pp.92-98

[6] Ellinas, C.P., Croll, J.G.A. and Kemp, K.O., "Cooling Towers with Circumferential Imperfections", Journ. of the Structural Division, Proceedings of the ASCE, Vol.106, ST12, 1980, pp.2405-2423

[7] Haig, B.P., "An Estimation of the Bending Stresses Induced by Pressure in a Tube that is not Quite Circular", Appendix IX, Welding Research Committee 2nd Report, Proc. Inst. Mech. Engrs, 133, 1936, pp.96-98

[8] Carlson R.A., and McKean, J.D., "Cylindrical Pressure Vessels : Stress Systems in Plain Cylindrical Shells and in Plain Pierced Drumheads", Proc. Inst. of Mech. Engrs., 169, 1955, pp.269-293

[9] Steel, C.R. and Skogh, J., "Slope Discontinuities in Pressure Vessels", Journ. of Applied Mech., 37, 1970, pp.587-595

[10] Tam, C.K.W. and Croll, J.G.A., "Analysis of Buckle Initiation in Subsea Pipelines", Journ. Offshore Mech. and Arctic Eng., Trans. ASME., Vol.109, 1987, pp.366-374

[11] Gurney, T.R., *"Fatigue of Welded Structures"*, 2nd edition, Cambridge University Press, 1979

[12] Almar-Næss, A., *"Fatigue Handbook - Offshore Steel Structures"*, published by TAPIR, 1985

AN IMPROVED ANALYSIS FOR CYLINDRICAL VESSELS SUPPORTED ON RIGID SADDLES

FAISAL A MOTASHAR and ALWYN S TOOTH
Department of Mechanical and Process Engineering
University of Strathclyde, Glasgow G1 1XJ, UK

ABSTRACT

The key to understanding the behaviour of support, or contact, problems lies in deriving the interface forces which occur between the support and the component, in this case the vessel. For ease of calculation, the early work by the authors assumed that these pressures were of constant magnitude across the smaller dimension of the support, that is the saddle width. When the supports are of rigid construction, compared to the vessel, such as with a GRP vessel on a steel or concrete support, the above assumption will not be valid. This paper extends the existing analysis to derive the interface pressures across the width of such a support. From these values the stresses, which occur in the critical regions of the vessel, can be derived with more accuracy than previously. An illustrative example of a large storage vessel is presented.

NOTATION

B	Distance of saddle centre profile from vessel end
b_j	Distance of a discrete area centre from vessel end ($x = 0$)
C	Half saddle width
D	Extension rigidity $= Et/(1 - v^2)$
E	Modulus of elasticity
i, j	Discrete areas in the ϕ and x directions
k, l	General discrete areas in the ϕ and x directions
K	Bending rigidity $= Et^3/12(1 - v^2)$
L	Length of vessel (tan/tan length)
$N_x, N_\phi, N_{x\phi}$	Stress resultants (see Figure 1)
$M_x M_\phi M_{x\phi}$	
NA, NC	Total number of discrete areas in the x direction and ϕ direction
P_x, P_ϕ, P_r	Externally applied loading in the x, ϕ and radial directions
$P_{xmn}, P_{\phi mn}, P_{rmn}$	Loading and displacement coefficients, in the x, ϕ and radial
u_{mn}, v_{mn}, w_{mn}	directions employed in the Fourier series
q	Surcharge internal pressure
R	Mean radius of the cylindrical vessel
t	Wall thickness of vessel

u, v, w	Mid surface displacements in the x, ϕ and radial directions (Figure 1)
x, ϕ, z	Coordinates in the axial, circumferential and radial directions (Figure 1)
α	Angle from bottom of vessel ($\phi = 0°$) to surface of contained fluid
β, γ	Half discrete area size in the ϕ and x directions
Δ	Vertical upward rigid body displacement of saddle with respect to end of the vessel
λ	$m\pi R/L$
v	Poisson's ratio
ρ	Specific weight of contained fluid
ρ_v	Specific weight of vessel material
ϕ_i	Angular distance of a discrete area centre from ($\phi = 0°$)

INTRODUCTION

The design of horizontal vessels supported on twin saddles has been dealt with by several authors over the years. However, the approach given in the Pressure Vessel Standards BS5500 and ASME are essentially the work of L.P. Zick [1], who used a modified beam and ring analysis so that the mathematical model for the vessel, predicted values which agreed with the experimental results he had available. More recent work by Tooth, et al [2,3] had indicated that Zick's treatment for the vessel full of fluid predicts stresses which are in reasonable agreement with the experimental values when a flexible saddle is employed. However, when the saddle is rigid, compared with the vessel, Zick's treatment underestimates the peak stresses which actually occur at the horn.

The key to understanding the behaviour of such problems lies in deriving the interface forces which occur between the support and the vessel. The magnitude and distribution of these forces depends upon the vessel flexibility and the rigidity of the support. In the earlier analytical work by Tooth et al [5], the configuration of the support was found to have a crucial effect on the stress in the vessel - primarily in the 'horn' region of the saddle. For example, when a flexible saddle is employed, the vessel stresses can be reduced by up to 50%.

In order to determine the interface pressures between the saddle and the vessel, the saddle contact area is divided into a number of discrete areas, each of which is subject to **unknown** uniformly distributed pressures in both the radial and tangential direction. For ease of calculation, the early work [2-5] assumed that these pressures were of constant magnitude across the saddle width. That is, the saddle had an element of radial flexibility across the width to avoid pressure high spots.

When the saddles are of rigid construction (i.e. no radial or tangential flexibility) compared to the vessel, such as with a GRP vessel on a steel or concrete support, the above assumption will not be valid. It is proposed, therefore, to extend the existing analysis to derive the interface pressures which occur at points across the width of such a rigid support. From these values the stresses in the vessel can be obtained.

GOVERNING DIFFERENTIAL EQUATIONS

If we assume that a linear elastic small displacement situation exists for the support problem, the behaviour of thin-walled circular cylindrical shells can be described using the set of differential equations proposed by Sanders [6]. Rearranging these equations they can be expressed in terms of the mid-surface displacements, u,v,w and the applied loadings, P_r, P_x and P_ϕ, as follows:-

$$\begin{bmatrix} L_1 & L_2 & L_3 \\ L_2 & L_4 & L_5 \\ L_3 & L_5 & L_6 \end{bmatrix} \begin{bmatrix} w \\ u \\ v \end{bmatrix} = \frac{R^2}{D} \begin{bmatrix} P_r \\ - & P_x \\ - & P_\phi \end{bmatrix} \tag{1}$$

The expressions for the L's are given in the Appendix.

The stress resultants, N_ϕ, N_x, M_ϕ, M_x, etc (see Figure 1) can be related to the mid-surface displacements u, v and w by using the constitutive relations and the strain displacement relations.

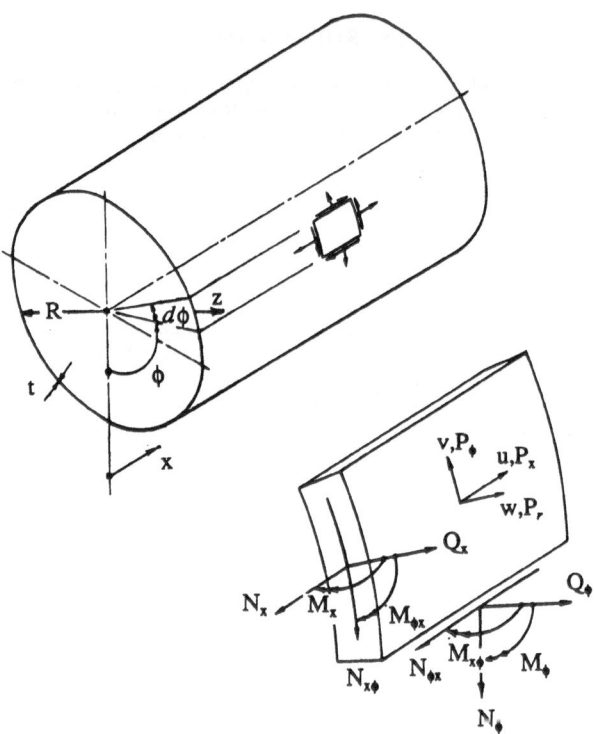

Fig 1   Positive directions of mid-surface displacements, stress resultants and loading.

FOURIER EXPANSION SOLUTIONS

The double Fourier Series expansions method is employed to solve equations (1). All the known and unknown force and moment resultants and displacements can be represented in this series form. The following are the Fourier expansions for displacements and surface loading with symmetry about $\phi = 0°$.

$$
\begin{bmatrix} w \\ u \\ v \\ P_r \\ P_x \\ P_\phi \end{bmatrix} = \sum_{m=0}^{\infty} \sum_{n=0}^{\infty} \begin{bmatrix} w_{mn} & \cos & n\phi & \sin & (\lambda x/R) \\ u_{mn} & \cos & (n\phi) & \cos & (\lambda x/R) \\ v_{mn} & \sin & (n\phi) & \sin & (\lambda x/R) \\ P_{rmn} & \cos & (n\phi) & \sin & (\lambda x/R) \\ P_{xmn} & \cos & (n\phi) & \cos & (\lambda x/R) \\ P_{\phi mn} & \sin & (n\phi) & \sin & (\lambda x/R) \end{bmatrix} \tag{2}
$$

where $\lambda = m\pi R/L$

The origin of the coordinate system is taken at one end of the vessel. This implies that the cylinder is supported at the ends, through rigid diaphragms.

By substituting equation (2) into equation (1) the following matrix equation, involving the coefficients of displacements and loadings, is obtained

$$
\begin{bmatrix} Z_{11} & Z_{12} & Z_{13} \\ Z_{12} & Z_{22} & Z_{23} \\ Z_{13} & Z_{23} & Z_{33} \end{bmatrix} \begin{bmatrix} w_{mn} \\ u_{mn} \\ v_{mn} \end{bmatrix} = \frac{R^2}{D} \begin{bmatrix} P_{rmn} \\ P_{xmn} \\ P_{\phi mn} \end{bmatrix} \tag{3}
$$

where

$$
Z_{11} = 1 + k(n^2 + \lambda^2)^2
$$

$$
Z_{12} = \lambda \left[\frac{1}{2}(1-v)kn^2 - v \right]
$$

$$
Z_{13} = n(1 + n^2\lambda) + \frac{1}{2}(3-v)kn\lambda^2
$$

$$
Z_{22} = \lambda^2 + \frac{1}{8}(1-v)(4+k)n^2
$$

$$
Z_{23} = -\frac{1}{8}n\lambda[4(1+v) - 3k(1-v)]
$$

$$
Z_{33} = n^2(1+k) + \frac{1}{8}(1-v)(4+9k)\lambda^2
$$

The matrix equation (3) can be evaluated using Cramer's rule or the Co-factor method of matrix inversion, from which u_{nm}, v_{nm} and w_{nm} are expressed as functions of P_{rmn}, P_{xmn} and $P_{\phi mn}$. Using the stress resultant - displacement relations the values of N_x, N_ϕ, M_x and M_ϕ can be obtained for each of the loading cases (radial P_{rmn}, tangential $P_{\phi mn}$ and longitudinal P_{xmn}).

FOURIER SERIES REPRESENTATION OF THE LOADING

The only unknowns in the double series expression which arise from the above approach, for w,v,u and the stress resultants N_x, N_ϕ, M_x, M_ϕ, etc., are the loading coefficients P_{rmn}, $P_{\phi mn}$ and P_{xmn}. These can be related to the applied loading P_r, P_ϕ and P_x by multiplying both sides of the loading terms contained in equations (2) by suitable orthogonal functions such that integration over the surface of the cylinder eliminates all but one of the terms in each Fourier expansion. This procedure results in the following expressions for P_{rmn} and $P_{\phi mn}$.

$$P_{rmn} = \frac{1}{L\pi}\int_o^L \int_o^{2\pi} P_r \sin(m\pi x/L)dxd\phi \qquad (n=0, m=1,2,3)$$

$$P_{rmn} = \frac{2}{L\pi}\int_o^L \int_o^{2\pi} P_r \cos n\phi \sin(m\pi x/L)dxd\phi \qquad (n=m=1,2,3) \qquad (4)$$

$$P_{\phi mn} = \frac{2}{L\pi}\int_o^L \int_o^{2\pi} P_\phi \sin n\phi \sin(m\pi x/L)dxd\phi \qquad (n=m = 1,2,3)$$

Radial and tangential loading of any form can be represented by the use of equation (4). A detailed outline of the above method and compendium of these solutions are provided by Duthie and Tooth [7]. The convergence of the series solution depends on the types of loading considered as well as the vessel dimensions. A discussion of this can be found in Refs. [5,8].

THE RIGID SADDLE SUPPORTED VESSEL

The cylindrical vessel of length L, mean radius R and of constant wall thickness t is supported on twin rigid saddles which are located on equal distance B, from the vessel ends. It is assumed that the loading on the cylinder is symmetrical about the vertical plane through the cylinder centre line. The saddle is welded to the vessel.

The Interface Force System

The saddle/vessel contact area is assumed to be divided into a number of equal size discrete areas. $2\beta \times 2\gamma$ in size, as shown in Figure 2. The discrete areas in the axial direction are identified as 'j' with a total number equal to NA, and in the circumferential direction as 'i' with a total number of NC on both sides of the vessel nadir ($\phi = 0°$).

Each discrete area is loaded with a uniform radial pressure and tangential shear. For example, on area 'ij' a radial pressure of p_{ij} and tangential shear of t_{ij} is assumed to act. The radial and tangential displacements of the vessel at a general point 'kl' due to p_{ij} and t_{ij} are given by:-

$$w_{kl} = t_{ij} (w_t)_{ij,kl} + p_{ij} (w_r)_{ij,kl}$$

$$v_{kl} = t_{ij} (v_t)_{ij,kl} + p_{ij} (v_r)_{ij,kl} \qquad (5)$$

where $(w_t)_{ij,kl}$ and $(v_t)_{ij,kl}$ are the radial and tangential displacements of point 'kl' due to unit tangential shears applied over area 'ij' and $(w_r)_{ij,kl}$ and $(v_r)_{ij,kl}$ are the radial and tangential displacements of point 'kl' due to unit radial pressures applied over area 'ij'.

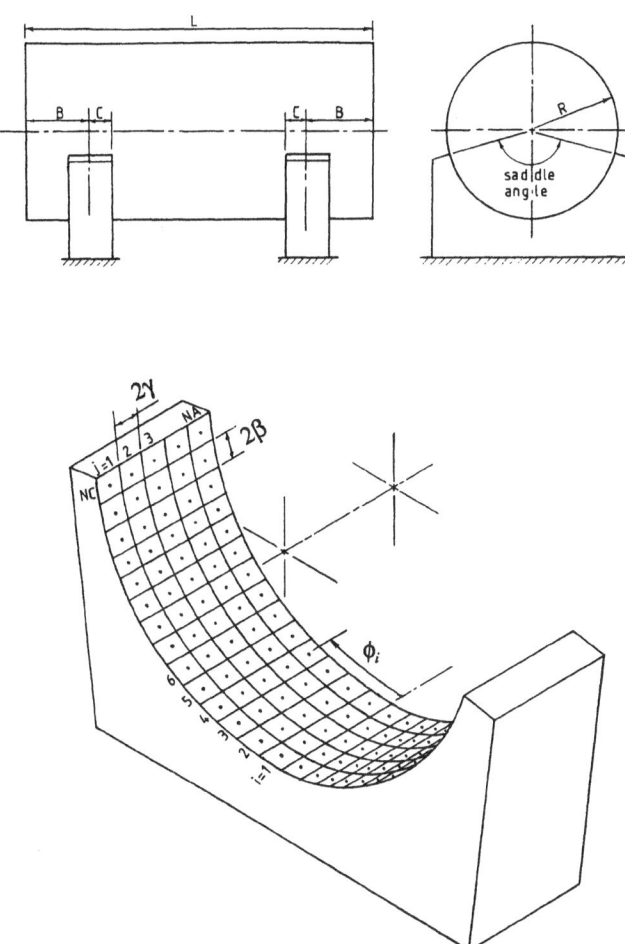

Fig 2   The saddle/vessel contact surface of a twin saddle supported cylinder.

The total radial and tangential displacements of point 'kl' on the surface of the vessel due to all the interface forces are then given by:-

$$W_{kl} = \sum_{j=1}^{NA} \sum_{i=1}^{NC} t_{ij}(w_t)_{ij,kl} + \sum_{j=1}^{NA} \sum_{i=1}^{NC} p_{ij}(w_r)_{ij,kl}$$

$$V_{kl} = \sum_{j=1}^{NA} \sum_{i=1}^{NL} t_{ij}(v_t)_{ij,kl} + \sum_{j=1}^{NA} \sum_{i=1}^{NC} p_{ij}(v_r)_{ij,kl} \tag{6}$$

These expressions are valid for the NA discrete areas along the saddle width and the NC areas around each half attachment about $\phi = 0°$. They can be rewritten in matrix form:-

$$[W] = [WT][T] + [WR][P]$$

$$[V] = [VT][T] + [VR][P] \tag{7}$$

The elements of the flexibility matrices [WR], [VR], [WT] and [VT] are given by the series forms of the displacements w and v in terms of the loading functions. For example [WR] is the radial displacement w at the centre of the general element 'kl' due to a unit radial pressure applied at the four areas 'ij'. These areas are located at equal distances from the vessel ends and at equal angles from the vessel nadir. It is expressed as follows:-

$$(w_r)_{ij,kl} = \frac{R^2}{kD} \sum_{n=o}^{\infty} \sum_{m=1}^{\infty} f_1(\lambda, n) P_{rmn} \cos n\phi_k \sin(\lambda b_l/R) \tag{8}$$

where f_1 is a function of λ and n.

The loading coefficient P_{rmn} for the four areas detailed above, is obtained from equations 4 where

$P_r = 1$ in the region

$$(\phi_i - \beta) \leq \phi \leq (\phi_i + \beta) \text{ and } \{ \begin{array}{ccccc} (b_j - \gamma) & \leq & x & \leq & (b_j + \gamma) \\ (L - b_j - \gamma) & \leq & x & \leq & (L - b_j + \gamma) \end{array}$$

$$(-\phi_i - \beta) \leq \phi \leq (\phi_i + \beta) \text{ and } \{ \begin{array}{ccccc} (b_j - \gamma) & \leq & x & \leq & (b_j + \gamma) \\ (L - b_j - \gamma) & \leq & x & \leq & (L - b_j + \gamma) \end{array}$$

$P_r = 0$ otherwise

The resulting values for P_{rmn} are:-

$$P_{rmn} = \frac{16\beta}{m\pi^2} \sin(m\pi b_j/L) \sin(m\pi\gamma/L) \qquad (n=0, m=1,3,5,...)$$

$$P_{rmn} = \frac{32}{mn\pi^2} \sin(m\pi b_j/L) \sin(m\pi\gamma/L) \cos n\phi_i \sin(n\beta) \qquad (n=1,2,3,...; m=1,3,5,...) \tag{9}$$

Gravity and Pressure Loading on the Vessel

In addition to the reactive interface forces the vessel is subjected to the applied loading which is a combination of hydraulic pressure, internal pressure surcharge, q, and the self weight of the vessel (specific weight ρ_v). The loading coefficients for a partially filled vessel (filled to height α) are:

$$P_{mn} = \frac{4}{m\pi}\left[q + \frac{\rho R}{\pi}(\sin\alpha + \alpha\cos\alpha)\right] \qquad (n=0, m=1,3,5...)$$

$$P_{mn} = \frac{4}{m\pi}\left[\rho_v t + \frac{\rho R}{\pi}(2\alpha - \sin 2\alpha)\right] \qquad (n=1, m=1,3,5,...)$$

$$P_{mn} = \frac{8\rho R}{m n\pi^2(n^2-1)} \quad [\sin(n\alpha)\cos\alpha - \cos(n\alpha)\sin\alpha] \qquad (n=2,3,4,...; m=1,3,5,...)$$

$$P_{\phi mn} = -\frac{4}{m\pi}\rho_v t \qquad (n=1, m=1,3,5...) \quad (10)$$

$$P_{\phi mn} = \text{otherwise}$$

Using equations (10) the radial and tangential displacements at the centres of the discrete areas of the support can be derived. These can be written in matrix form, [WHSW] for the radial and [VHSW] for the tangential displacement.

Compatibility

In the series forms for the displacement, the displacement of the ends of the vessel are assumed to be zero. In reality it is the supports which remain fixed. To compensate for this the vessel is given a rigid body movement Δ in the vertical direction. The corresponding values of the radial and tangential displacement at the centres of the saddle discrete areas, due to this movement are respectively expressed as:

$$\Delta[CS] \text{ and } - \Delta[SN]$$

where [CS] and [SN] are the vectors of elements $CS_i = \cos\phi_i$, $SN_i = \sin\phi_i$ and ϕ_i is the angle to the centre of the discrete 'ij'.

The unknown interface forces 'p_{ij}' and 't_{ij}' which act at the various discrete areas will in general cause radial and tangential displacements of the saddle. The elements of the saddle flexibility matrices can be obtained from say a Finite Element method. However, in the treatment presented here the saddle is rigid and thus these are set to zero.

Compatibility is enforced by the following:

$$[WR][P] + [WT][T] + \Delta[CS] + [WHSW] = 0$$

$$[VR][P] + [VT][T] + -\Delta[SN] + [VHSW] = 0 \qquad (11)$$

Equilibrium

In the saddle region it is possible by using the sign convention of Figure 1 to write:

$$[SN]^T[T] = S + [CS]^T[P] \tag{12}$$

where:

$$S = \text{Total weight supported/4xDiscrete area size}$$
$$S = W/16R\beta\gamma$$

and $[\ ]^T$ is the transpose of $[\ ]$

Determination of Vessel Stresses and Displacements

Using the compatibility and equilibrium equations (11) and (12) it is possible after some manipulation to determine the unknown interface forces [T] and [P] in terms of the known applied loading. Combining these with the gravity and pressure loadings in the vessel the total loading coefficients P_{rmn}, $P_{\phi mn}$ can be obtained. These are used to determine the displacements and stress resultants at all points in the vessel. These are shown in detail in Ref [8]. Certain techniques have been devised to reduce the computing time required to generate the vessel flexibilities and to derive the subsequent distribution of stresses. Details of these area given in Refs [8,9].

ILLUSTRATIVE EXAMPLE

The analysis presented is capable of predicting the displacements and stress resultants in cylindrical vessels which are supported on rigid saddles. To illustrate the way in which a non-uniform interface pressure across the saddle width influences the stressing in the vessel a selection of typical theoretical results are presented for a large gas receiver. The vessel is 3658mm mean diameter, 54860mm barrel length, 26.6mm constant wall thickness. The saddles are assumed to be rigid and welded to the vessel. They are 762mm wide, of at angle 162° and located 6805mm from the ends of barrel length.

To examine the distributions of interface forces and stresses across the saddle width, the width was progressively divided into, 3,5,7 and 9 areas, while the number of areas round the saddle arc was kept the same (NC = 20). A typical set of results for the radial and tangential interface pressures is shown in Figure 3. This shows high tangential shear at both saddle corners J=1 (nearest the end) and J=5 (nearest the vessel centre). High radial pressures also occur nearest the vessel centre, j=5. The signs of the radial pressures indicate the presence of a circumferential moment, which is required to enforce compatibility in the immediate horn region.

The magnitude of both the radial and tangential pressure are much higher than when the uniform pressure assumption is prescribed. Although, as one would expect when they are averaged across the width they are equal to the uniform pressure case. The large positive and negative tangential shear forces, at j=1 and 5, form a reactive longitudinal moment across the saddle resisting the applied loading in the vessel.

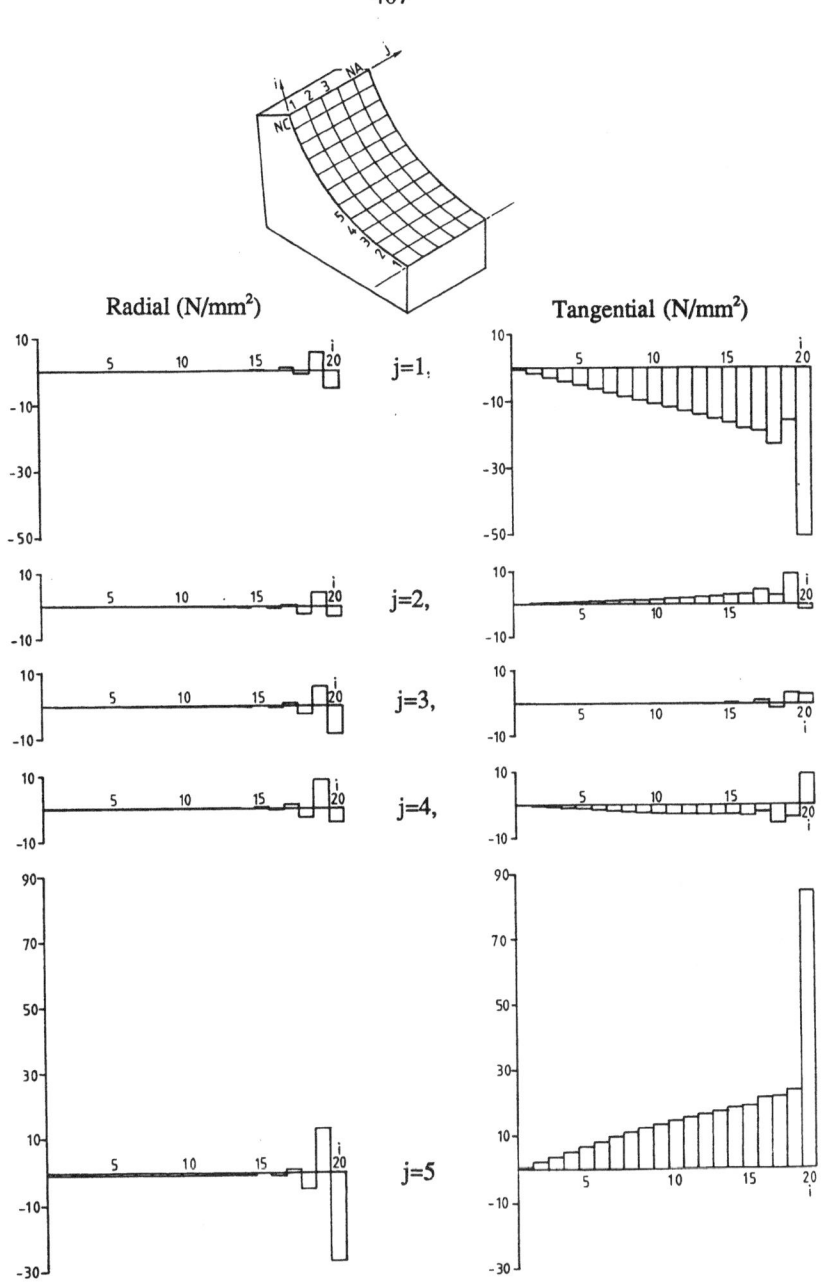

Fig 3   Typical distribution of radial and tangential interface pressures over the surface of a
rigid saddle NC=20, NA=5.

It is found that the peak stress value, in the saddle horn area, increases in magnitude and its location shifts from the saddle centre profile towards the edge nearest the vessel centre. The distribution of the outside circumferential stress across the saddle width in the horn area for NA values from 1 (i.e. uniform pressure), 3,5,7 and 9 for the case NC = 20, are presented in Figure 4. The variation of the peak stress magnitude with the number of discrete areas both across the saddle and round the half saddle width (NC = 30,20,10) is shown in Figure 5. It is clear that high values of NC are required to obtain a converged result. It was found that when NA = 1, an ideal was reached when NC = 90.

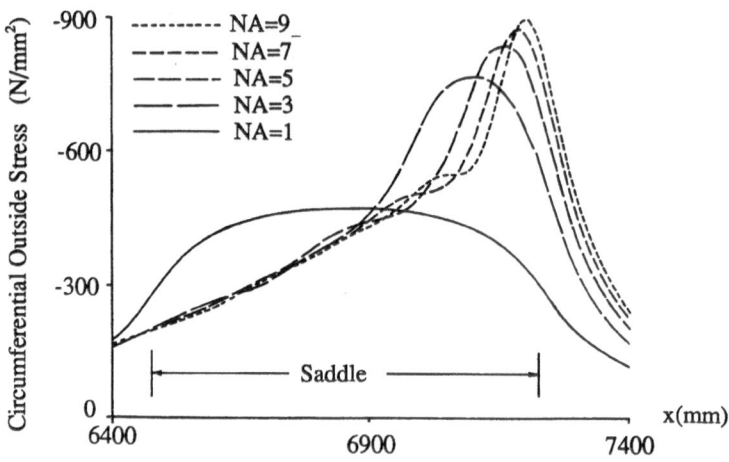

Fig 4   Distribution of the outside surface circumferential stress across the saddle width, in the horn region for NC=20 and NA = 1,3,5,7 and 9.

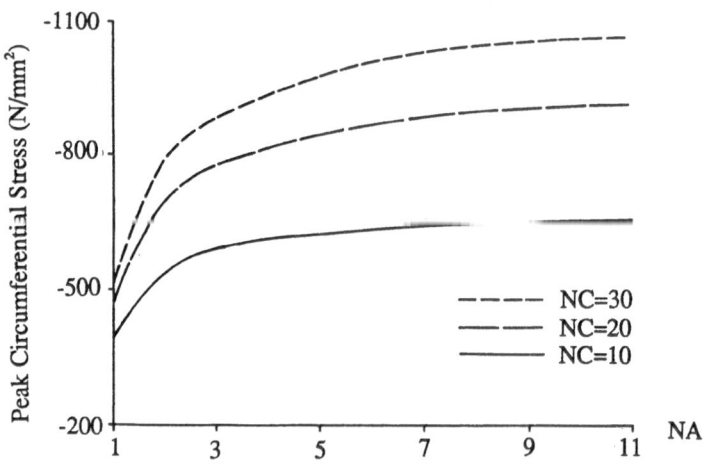

Fig 5   Variation of the peak stress with the number of discrete areas across and round the saddle.

CONCLUDING COMMENTS

The analysis presented enables an assessment to be made of the peak stresses which occur in a vessel in the region of a saddle support which is rigid compared to the vessel. The approach has wide application and can be used to derive the magnitude and distribution of interface forces in other contact problems. The results present the clear message that, where possible, supports should contain an element of flexibility so that the interface pressures should be more uniformly distributed and the resulting stresses thereby reduced.

REFERENCES

1.   Zick, L.P., Stresses in large horizontal cylindrical pressure vessels on two saddle supports. Welding Res Supplement, 1951, Vol 30, No. 9, 435-445s.

2.   Duthie, G. and Tooth, A.S. The analysis of horizontal cylindrical vessels supported by saddles welded to the vessel - a comparison of theory and experiment, 3rd Int. Conf. on Pressure Vessel Tech, Tokyo, 1977, pp 25-38.

3.   Tooth, A.S., Duthie, G., White, G. C. and Carmichael, J., Stresses in horizontal storage vessels - a comparison of theory and experiment, J. of Strain Analysis, 1982, 17, No. 3, 169-176.

4.   Wilson, J. D. and Tooth, A. S., The support of unstiffened cylindrical vessels, 2nd Int. Conf. on Pressure Vessel Tech, San Antonio, 1973, pp 67-83.

5.   Duthie, G., White, G. C., and Tooth, A. S., An analysis for cylindrical vessels under local loading - application to saddle supported vessel problems, J of Strain Analysis, 1982, 17, No. 3, 157-167.

6.   Sanders, J. L., Jr, An improved first approximation theory for thin shells. NASA, TR-R24, 1959.

7.   Duthie, G. and Tooth, A. S. Local loads on cylindrical vessels; Fourier series solution, In Behaviour on Thin-Walled Structures, ed. J. Rhodes and J. Spence. Elsevier Applied Science, London 1984, pp 235-272.

8.   Motashar, F.A. The analysis of horizontal cylindrical vessels - supports, local attachments and diaphragms. Ph.D thesis, University of Strathclyde, Glasgow, 1988.

9.   Ong, L. S. The effect of high internal pressure on cylindrical vessels with initial geometric imperfection Ph.D thesis, University of Strathclyde, Glasgow, 1985.

APPENDIX : COEFFICIENTS OF MATRIX [L]

$$L_1(\ ) = 1 + k\left[\frac{\partial^2}{\partial\phi^2} + R^2\frac{\partial^2}{\partial x^2}\right]^2(\ )$$

$$L_2(\ ) = vR\frac{\partial}{\partial x}(\ ) + \frac{1}{2}(1-v)Rk\frac{\partial^3}{\partial x\partial\phi^2}(\ )$$

$$L_3(\ ) = \frac{\partial}{\partial\phi}(\ ) - \frac{1}{2}(3-v)kR^2\frac{\partial^3}{\partial x^2\partial\phi}(\ ) - k\frac{\partial^3}{\partial\phi^3}(\ )$$

$$L_4(\ ) = R^2\frac{\partial^2}{\partial x^2}(\ ) + \frac{1}{8}(1-v)(4+k)\frac{\partial^2}{\partial\phi^2}(\ )$$

$$L_5(\ ) = \frac{1}{8}[4(1+v) - 3k(1-v)]R\frac{\partial^2}{\partial x\partial\phi}(\ )$$

$$L_6(\ ) = (1+k)\frac{\partial^2}{\partial\phi^2}(\ ) + \frac{1}{8}(1-v)(4+9k)R^2\frac{\partial^2}{\partial x^2}(\ )$$

where $k = t^2/(12R^2)$

STRESSES AND DEFLECTIONS DUE TO EXTERNAL LOADS ON THE NOZZLE
BRANCHES OF CYLINDRICAL PRESSURE VESSELS

A.C. WORDSWORTH
Lloyds Register, London

ABSTRACT

The paper examines the use of simple parametric equations which were
developed for calculating stress concentration and stiffness factors at
the joints between tubular members of offshore structures for
calculating the corresponding factors at cylindrical pressure vessels
with external loads applied via nozzle branches. Stress concentration
factors calculated in this way are compared with those obtained from
published pressure vessel experimental data. Generally good agreement
is shown. They are also compared with those obtained by methods
commonly applied in the design of pressure vessels.

SYMBOLS

a	=	Radius of vessel in inches.
d	=	Branch outside diameter
D	=	Vessel outside diameter
E	=	Young's Modulus
F_{CAX}	=	Axial flexibility coeff.
		= axial flexibility x E.D.
F_{CIB}	=	In-plane bending flexibility coeff.
		= in-plane flexibility x $E.D^3$
F_{COPB}	=	Out-of plane bending flexibility coeff.
		= out-of plane flexibility x $E.D^3$
K_C	=	Out-of-plane spring constant (in.lb.f/radian)
K_L	=	In-plane spring constant (in.lb.f/radian)
L	=	Distance between vessel supports
R	=	Vessel radius (R=a if R is in inches).
S.C.F	=	Stress concentration factor
	=	$\dfrac{\text{Stress being considered}}{\text{maximum relevant nominal stress in branch}}$

Suffixes C and R refer to component of stress
circumferential and radial to the branch

S.C.F$_B$	=	Nominal vessel bending stress/nominal branch axial stress.
S.N.C.F	=	Strain concentration factor
	=	$\dfrac{\text{strain being considered}}{\text{maximum relevant nominal strain in branch}}$
t	=	Branch thickness
T	=	Vessel thickness
Z	=	Branch section modulus.

INTRODUCTION

With the exploitation of the hydrocarbon resources beneath the North Sea over the last twenty years, considerable research has been done on the stress analysis of the tubular joints, or nodes, of the steel jacket structures which may be used to support the topsides of offshore platforms.

In the absence of significant relevant data available at the time, attention was initially directed to theoretical work which had been done on the analysis of the stresses and deflections due to external loads on the nozzle branch pipes in cylindrical pressure vessels. However, the simplifying assumptions made in this work together with its restricted validity ranges limited its value. A number of research projects were therefore initiated which have yielded simple semi-empirical parametric equations for calculating the stress concentration factors (S.C.F's) and stiffnesses of nodes.

Simultaneously, but separately, some further work has been done on pressure vessels with externally loaded branches.

It is now appropriate to examine and compare the results of these two strands of research with the object of capitalising on the large expenditure of effort on node analysis to improve the accuracy, extend the validity range, and simplify the methods used on the analysis of pressure vessels.

This paper describes such a comparison as follows:-

(i) To identify inherent differences between nodes and pressure vessels a comparison is made of the strains and S.C.F's measured on a tubular 'T' joint tested first as a node and then as a pressure vessel.

(ii) The published methods of calculating cylindrical pressure vessel response to external branch loads are reviewed.

(iii) Parametric equations developed for calculating S.C.F's and stiffnesses at nodes are presented.

(iv) S.C.F's obtained from published experimental data are compared with those obtained from the methods described in (ii) and (iii) above.

(v) S.C.F's and stiffnesses obtained by the methods described in (ii) and (iii) above are presented graphically to highlight areas of discrepancy.

INHERENT DIFFERENCES BETWEEN T NODES AND CYLINDRICAL PRESSURE VESSELS WITH EXTERNALLY LOADED BRANCHES

The basic differences between an offshore 'T' node and a cylindrical pressure vessel with a branch are that:

a)   The shell of the pressure vessel is subject to significant differential pressure while the node is not.

If it is assumed that the deflections of a pressure vessel due to differential pressure are small it can be concluded that its response to externally applied branch loads will be independent of such pressure and, to this extent, the response will be no different from those of an equivalent node.

b)   The portion of the shell within the nozzle is absent from a pressure vessel but present in a node where it is sometimes referred to as the 'plug'. In order to investigate the effect of the presence or otherwise of this plug, data from a strain gauged acrylic tubular 'T' joint specimen, tested first with the plug and then without it, are presented. A small scale acrylic tubular 'T' joint specimen was fabricated, strain gauged, and tested using techniques [1] whose validity had been thoroughly verified by comparative tests on full scale steel specimens.

The specimen dimensions, in mm, were D=150.6; d=73.7; T=t=5.23; L=1067. The horizontal and vertical elements of a 'T' joint would be identified as the chord and brace of a node and as the vessel and branch of a pressure vessel. The latter terminology is adopted hereafter.

The vessel was simply supported at either end and the 120 strain gauge elements connected to a computer controlled data logger.

Axial load, out of plane (O.O.P) and, in-plane (I.P) bending moments were applied separately to the free end of the branch and the resulting strains recorded. This was done initially with the plug present and then with it removed to represent node and pressure vessel conditions respectively.

The measured strains were converted to strain concentration factors (S.N.C.F's) by dividing them by the maximum nominal strains in the branch. 'Hot-spot' S.C.F's were obtained by extrapolating the S.N.C.F's into the branch/vessel junction and by then taking account of the orthogonal S.N.C.F's as appropriate.

This 'hot-spot' S.C.F specifically excludes the notch concentration effect, which arises due to the sharp corner at the junction, and is the factor normally required by designers of welded fabrications to, for example, BS5500 Enquiry Case 79.

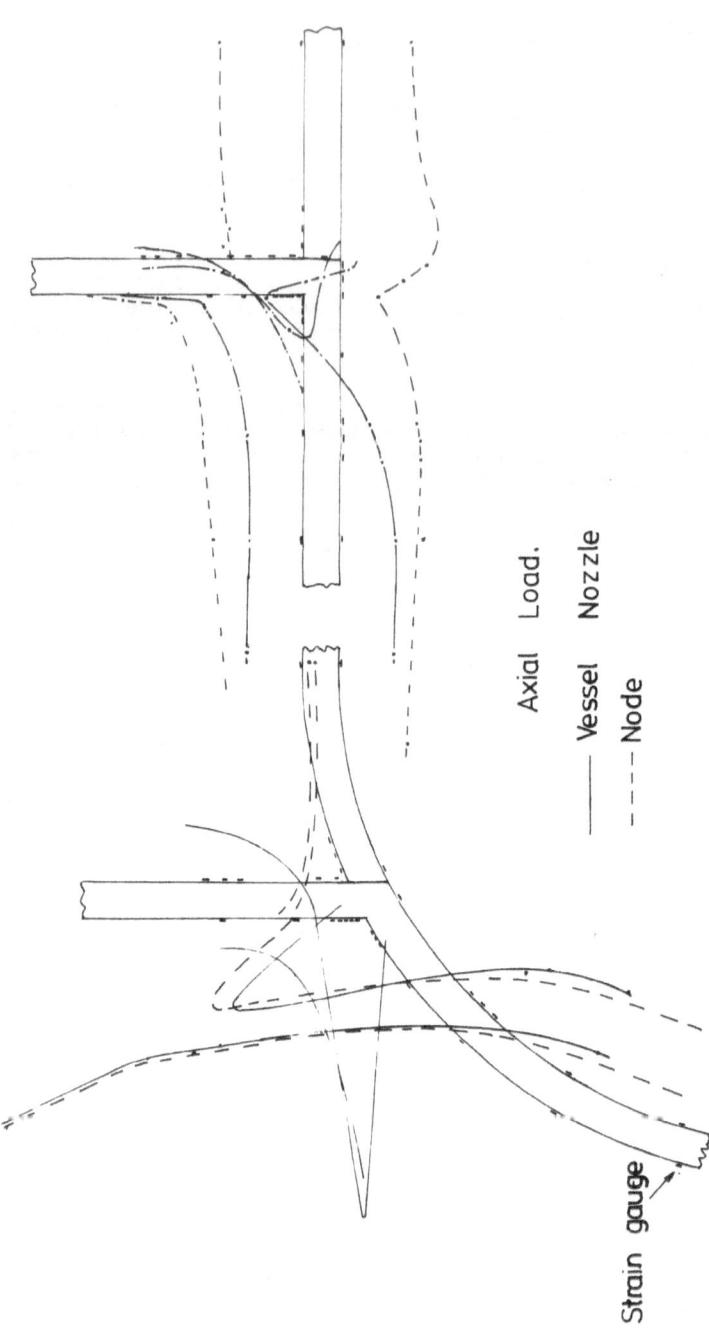

Figure 1. Comparison of Strain Concentration Factors at Acrylic Vessel Nozzle and at Node.

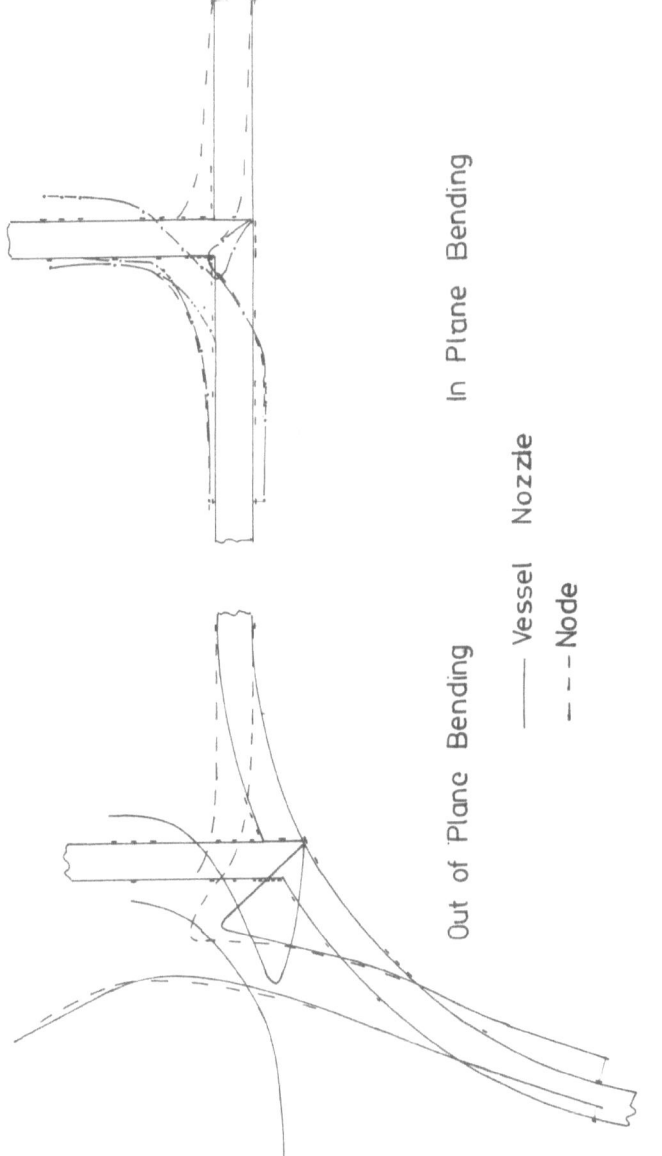

Out of Plane Bending   In Plane Bending

—— Vessel Nozzle
– – Node

Figure 2.   Comparison of Strain Concentration Factors
at Acrylic Vessel Nozzle and at Node.

The S.N.C.F's corresponding to the strains in the planes of the sections are plotted in Figs 1 & 2. From these figures it is apparent that, in general, the plug is subjected to relatively small strains and that its removal has corresponding small effect on the strains on the outer surfaces of the vessel and nozzles. The exception to this generalisation occurs on the longitudinal plane when axial load is applied to the branch. In this case the continuity of the path of the overall bending load is interrupted by its removal. This results in the redistribution and concentration of the stress due to overall bending of the vessel.

The S.C.F's obtained on the transverse and longitudinal centrelines are given in Table 1.

Table 1.
Acrylic Specimen Hot Spot S.C.F's

	Axial Load				O.O.P	I.P
	Position 1		Position 2		Position 1	
	Trans. C.L	Long. C.L	Trans. C.L	Long. C.L	Trans. C.L	Long. C.L
Node S.C.F$_R$	17.2	7.9	----	----	11.4	4.7
Node S.C.F$_C$	10.5	3.1	----	----	5.4	2.6
Vessel S.C.F$_R$	17.8	6.4	0	0	11.3	4.8
Vessel S.C.F$_C$	13.6	3.5	7.6	-3.4	5.8	3.1

Position 1 refers to positions at the junction between the branch and the vessel.

Position 2 refers to positions on the surface of the hole which is created by the removal of the plug.

Table 1 shows that the removal of the plug has minimal effect on the maximum S.C.F's.

Under branch axial load S.C.F$_C$ is increased by 30% and 13% on the transverse and longitudinal centrelines respectively.

The S.C.F's of 7.6 and -3.4 on the edge of the hole arise from the overall bending of the vessel and represent S.C.F's of 2.1 and -0.95 when based on the more appropriate gross nominal bending stress in the vessel.

The results from the tests therefore tend to indicate that, with the exception of S.C.F$_C$ at the saddle under branch axial load, the differences between node and pressure vessel response to external branch loading are small.

CYLINDRICAL PRESSURE VESSEL BRANCH LOAD CALCULATION METHODS

The three commonly used methods of calculating the response of cylindrical pressure vessels to external branch loads are given in.

(i)   BS5500.   Appendix G [2]. Here an analysis of cylinders subjected to load distribution over rectangular areas is modified to cover the case of a cylindrical branch. Membrane and bending stresses at crown and saddle positions (ie on the longitudinal   and transverse centrelines) are computed and superimposed on the pressure stresses to determine maximum stresses.

Disadvantages of the method are that the theory assumes uniform load distribution, which has been shown to be invalid, and that the d/D validity range is restricted to 0.25 maximum. The method is tedious to use, involving interpolation of a number of graphs.

TEIXEIRA et al [3] present a simplified version of this method but it can only be used to calculate the maximum stresses due to external loads alone.

(ii)   Welding   Research   Council   Bulletin   107   [4] was first published in 1965 and it presents a "cook book" simplification of work by Bijlaard which has been modified in the light of some experimental data and extrapolated to cover d/D ratios up to 0.5 maximum.

Membrane and bending stresses are again computed and superimposed.

This method also involves interpolation of graphs.

Murad and Sun [17] present spring constants, also derived from work by Bijlaard, which can be used to calculate branch deflection.

(iii)   Welding Research Council Bulletin 297 [5] was published in 1984 as a supplement to WRC.107 Validity ranges and general procedures are similar to those for WRC.107 but, unlike the latter, account is taken of the branch stiffness.

As well as allowing vessel stresses to be calculated, methods are also presented for computing branch stresses and deflections.

NODE CALCULATION FORMULAE AND THEIR APPLICATION TO
PRESSURE VESSELS

In contrast to the methods used for pressure vessels which rely on graphical presentation of the data from which the individual stress components are computed the node methods use semi—empirical parametric equations to calculate S.C.F's.   These are applied to branch maximum nominal stresses to obtain the maximum stresses due to the various loadings.

There are several alternative sets of equations for calculating these S.C.F's, but it has been concluded that the widely used Wordsworth/Smedley formulae, which have been developed from acrylic specimens [6], are the most suitable in terms of reliability [7]. These equations include the non-dimensional ratios d/D, R/T etc which are represented by the symbols beta, gamma etc. The equations are presented here in terms of d, D, T etc. This enables the variation of S.C.F with variation of the different parameters to be more readily appreciated.

In the vessel.

Axial load. At saddle

$$S.C.F._R = \frac{t.d}{T^2} \left(3.39 - 3.21\left(\frac{d}{D}\right)^2\right) \tag{1}$$

O.O.P.

$$S.C.F._R = \frac{t.d}{T^2} \left(.8 - .575\left(\frac{d}{D}\right)^5\right) \tag{2}$$

I.P.

$$S.C.F._R = \left(\frac{D}{T}\right)^{.6}\left(\frac{t}{T}\right)^{.8} \left(.792\left(\frac{d}{D}\right)^{.25} - .346\left(\frac{d}{D}\right)^2\right) \tag{3}$$

In the Branch

$$S.C.F. = 1 + .63 \; S.C.F._R \tag{4}$$

In all the above cases, provided L/D is not excessive, $S.C.F._R$ will be greater than $S.C.F._C$.

The equations were developed from data for joints with .13 < d/D < 1 ; 8 < R/T < 32 ; .25 < t/T < 1.5.

However, their general form, at least for the saddle S.C.F's which are proportional to td/T^2, is such that some extrapolation beyond these ranges is not unreasonable.

On pressure vessels the maximum stresses due to pressure, which are superimposed on those due to branch loading, are maxima in a direction circumferential to the branch. The acrylic node test specimen data were therefore re-analysed to obtain the following equations for vessel $S.C.F._C$ with $\frac{d}{D} < 1$.

O.O.P.

$$S.C.F._C = .47 \; S.C.F._R \tag{5}$$

I.P.

$$S.C.F._C = .71 \; S.C.F._R \tag{6}$$

The data in Table 1 imply that these $S.C.F_C$ are increased by 7% and 19% respectively by the removal of the plug. However, the stresses represented by them occur on the outer surfaces of the vessels while the maximum stresses due to pressure occur on the inside. Therefore, to avoid excessive conservatism it is suggested that the above increases be discounted.

S.C.F$_C$'s for the axial load case may be composed of two components, those due to the overall bending of the vessels, which are amplified by the removal of the plug, and those due to local bending of the vessel shells. At the crown position on nodes the S.C.F$_R$ corresponding to the latter stress is given in [6] by

$$S.C.F_R = 0.7 + .97 \left(\frac{D}{T}\right)^{.5} \frac{t}{T} \left(1 - \frac{d}{D}\right) \tag{7}$$

If S.C.F$_B$ = nominal vessel bending stress/nominal branch axial stress then, for a vessel simply supported on either side of the branch and with no additional loads.

$$S.C.F_B = \frac{L.d.t}{D^2 T} \tag{8}$$

The following equations, obtained from limited data, are tentatively proposed.

At saddle S.C.F$_C$ = (S.C.F$_B$ x 2.1) + (S.C.F$_R$ x .4) (9)

At crown S.C.F$_C$ = (S.C.F$_B$ x -.95) + (S.C.F$_R$ x .9) (10)

The example given in the Appendix illustrates the use and relative significance of these equations.

Equations for calculating node stiffnesses, which have been developed by Fessler from tests on epoxy specimens and compared with data from steel specimens by Tebbett, are recommended by the Underwater Engineering Group. [8]

They are

Axial load $F_{CAX} = 2.3 \left(\frac{R}{T}\right)^{2.3} e^{-3.3 \, d/D}$ (11)

O.O.P $F_{COPB} = 48.1 \left(\frac{R}{T}\right)^{2.5} e^{-3.7 \, d/D}$ (12)

I.P. $F_{CIPB} = 171 \left(\frac{R}{T}\right)^{1.65} e^{-4.6 \, d/D}$ (13)

No validity ranges are imposed on these equations.

Ref [7] also quotes the following equation from DNV

I.P. $F_{CIPB} = 18.6 \left(\frac{T}{R} - .01\right)^{-(2.35 - 1.5 \, d/D)}$ (14)

limits of .33< d/D < .8 ; 10 < R/T < 30 are prescribed for this equation.

TABLE 2

Comparison of S.C.Fs obtained from published experimental data and calculated values.

2A

JOINT IDENTIFICATION GEOMETRIC PARAMETERS						AXIAL LOAD S.C.F'S					OUT-OF-PLANE BENDING S.C.F'S					IN-PLANE BENDING S.C.F'S				
NO	REF	SPECIMEN	d/D	R/T	t/T	Measured	Node Eq'n	WRC 297	WRC 107	BS 5500 AppG	Measured	Node Eq'n	WRC 297	WRC 107	BS 5500 AppG	Measured	Node Eq'n	WRC 297	WRC 107	BS 5500 AppG
1	9	Mehringer 1	.124	47.0	.612	20.3	16.1	21.6	17.5	19.5	8.7	5.7	7.4	6.5	11.0	5.3	4.8	5.1	4.8	6.4
2	10	O.R.N.L. 3	.129	25.0	.840	17.2	12.1	18.5	14.8	16.5	5.0	4.3	4.9	4.1	8.4	3.7	4.3	3.8	3.4	4.7
3	11	Cranch	.155	39.4	.448	9.9	10.5	10.4	12.1	11.7	4.4	3.8	4.2	4.5	7.1	3.2	3.4	3.0	3.3	4.3
4	9	Mehringer 2	.182	47.0	.72	23.2	25.0	31.0	22.3	25.0	14.6	9.9	13.2	10.3	15.6	6.6	6.0	6.7	6.1	8.7
5	12	Schoessow 2	.202	14.0	.421	4.6	4.6	5.0	5.8	7.3	3.5	1.9	1.9	2.0	3.8	2.0	1.9	1.4	1.6	2.1
6	12	Schoessow 1	.208	21.8	.673	9.8	11.7	14.9	12.6	15.8	5.8	4.9	5.8	5.0	8.7	3.8	3.6	3.8	3.6	5.1
7	13	Decock 1	.207	25.2	1.00	24.2	20.0	28.8	20.3	22.6	3.2	8.3	10.6	8.0	13.2	1.7	5.5	6.1	5.4	8.1
8	13	Decock 5	.207	41.9	1.00	41.0	33.5	40.4	28.9	30.9	6.9	13.9	18.8	13.5	20.1	3.5	7.4	9.2	8.0	13.1

2B

NO	REF	SPECIMEN	d/D	R/T	t/T	Measured	Node Eq'n	WRC 297	WRC 107	BS 5500 AppG	Measured	Node Eq'n	WRC 297	WRC 107	BS 5500 AppG	Measured	Node Eq'n	WRC 297	WRC 107	BS 5500 AppG
9	13	Decock 6	.373	25.2	.500	16.5	13.5	12.6	10.3	11.2	9.8	7.6	7.4	7.7	8.9	2.9	3.4	3.4	3.8	5.7
10	13	Decock 2	.373	25.2	1.00	34.2	26.9	33.3	19.7	22.4	15.5	15.1	19.2	14.1	17.2	6.1	6.0	7.4	7.1	11.2
11	14	Acrylic	.480	14.4	1.00	17.8	16.1	24.9	13.1	17.3	11.3	11.0	13.4	9.8	12.8	4.8	4.4	5.4	4.7	8.3
12	14	O.R.N.L. 1	.500	50.0	.500	36.3	28.0	23.3	9.2	12.1	35.3	19.6	16.4	16.4	15.5	10.0	5.3	4.4	5.3	7.5
13	15	Riley C-1*	.500	115	.980	81.7	127	145	24.2	25.5	74	88.5	94.8	63.5	51.8	15.1	14.9	19.5	14.3	18.5
14	16	HardenBergh R*	.633	10.0	.687	8.9	7.2	12.8			6.1	6.5	7.3	8.1		2.5	2.6	2.9		
15	13	Decock 7	.701	25.2	.600	20.2	14.9	21.5			13.3	14.9	16.6			4.9	3.9	4.5		
16	13	Decock 3	.701	25.2	1.00	37.4	24.8	42.2			24.5	24.8	30.9			8.0	5.8	8.2		

2C

NO	REF	SPECIMEN	d/D	R/T	t/T	Measured	Node Eq'n	WRC 297	WRC 107	BS 5500 AppG	Measured	Node Eq'n	WRC 297	WRC 107	BS 5500 AppG	Measured	Node Eq'n	WRC 297	WRC 107	BS 5500 AppG
17	16	HardenBergh S*	1.00	10.0	1.00		3.6	20.4			4.4	4.5	16.1			3.7	2.7	4.6		
18	13	Decock 4	1.00	25.2	1.00	15.6	9.1	37.5			5.9	11.3	39.7			5.7	4.7	8.7		
19	13	Decock 8	1.00	41.9	1.00	13.8	15.1	83.3			7.3	18.9	63.4			7.0	6.4	12.0		

* Data taken from reference [4]

TABLE 3

Average percentage differences between S.C.F's obtained from
experimental data and from calculation methods. Percentage Standard
Deviations given in brackets.

3A

JOINTS	CALCULATION METHOD	AXIAL LOAD	OUT OF PLANE BENDING	IN PLANE BENDING	ALL LOAD CASES
1 to 8	Node Eq'ns	-7 (17)	13 (75)	41 (84)	16 (66)
ie d<.25 D	WRC 297	16 (18)	41 (102)	48 (104)	35 (82)
	WRC 107	0 (22)	15 (70)	37 (87)	17 (65)
	BS5500 App G	13 (31)	91 (107)	100 (141)	65 (110)

3A'

1 to 8	Node Eq'ns	-3 (19)	-26 (13)	-1 (10)	-10 (18)
excluding	WRC 297	19 (19)	-13 (17)	-6 (12)	0 (21)
7 & 8	WRC 107	8 (20)	-21 (15)	-8 (7)	-7 (19)
	BS5500 App G	23 (30)	37 (27)	26 (11)	28 (23)

3B

1 to 13	Node Eq'ns	-5 (23)	4 (60)	22 (70)	7 (55)
ie d<.5 D	WRC 297	14 (31)	25 (83)	32 (85)	23 (69)
	WRC 107	-19 (33)	1 (57)	23 (71)	1 (57)
	BS5500 App G	-8 (40)	50 (99)	81 (115)	41 (96)

3B'

1 to 13	Node Eq'ns	-1 (25)	-16 (19)	0 (10)	-6 (20)
excluding	WRC 297	20 (30)	-3 (23)	4 (17)	7 (25)
7,8 & 12	WRC 107	-13 (33)	-19 (12)	-1 (15)	-11 (22)
	BS5500 App G	0 (40)	21 (31)	43 (30)	21 (38)

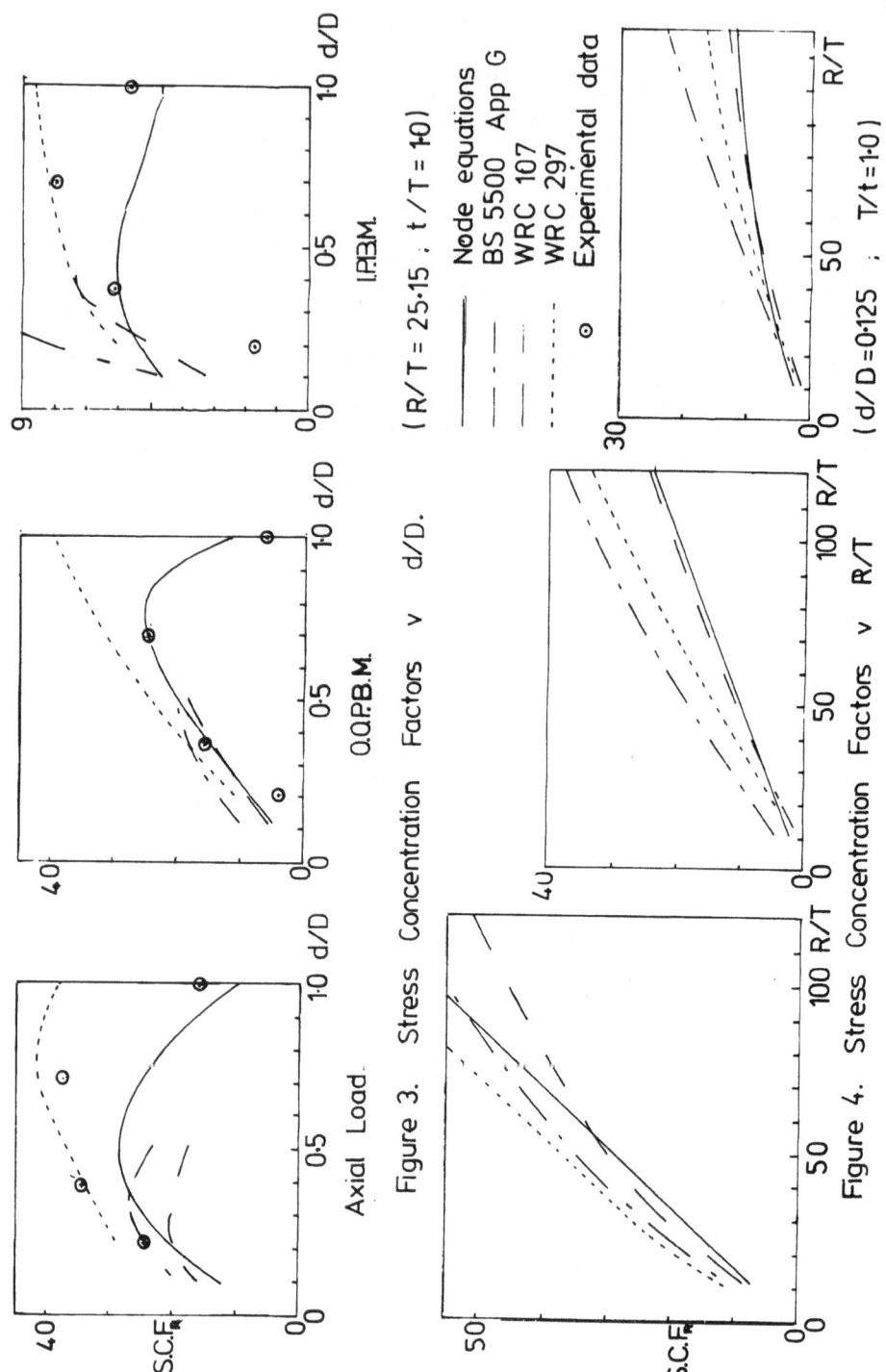

Axial Load.   O.Q.P.B.M.   I.P.B.M.

Figure 3.   Stress Concentration Factors v d/D.

$(R/T = 25.15 ; t/T = 1.0)$

Node equations
BS 5500 App G
WRC 107
WRC 297
⊙ Experimental data

Figure 4.   Stress Concentration Factors v R/T.

$(d/D = 0.125 ; T/t = 1.0)$

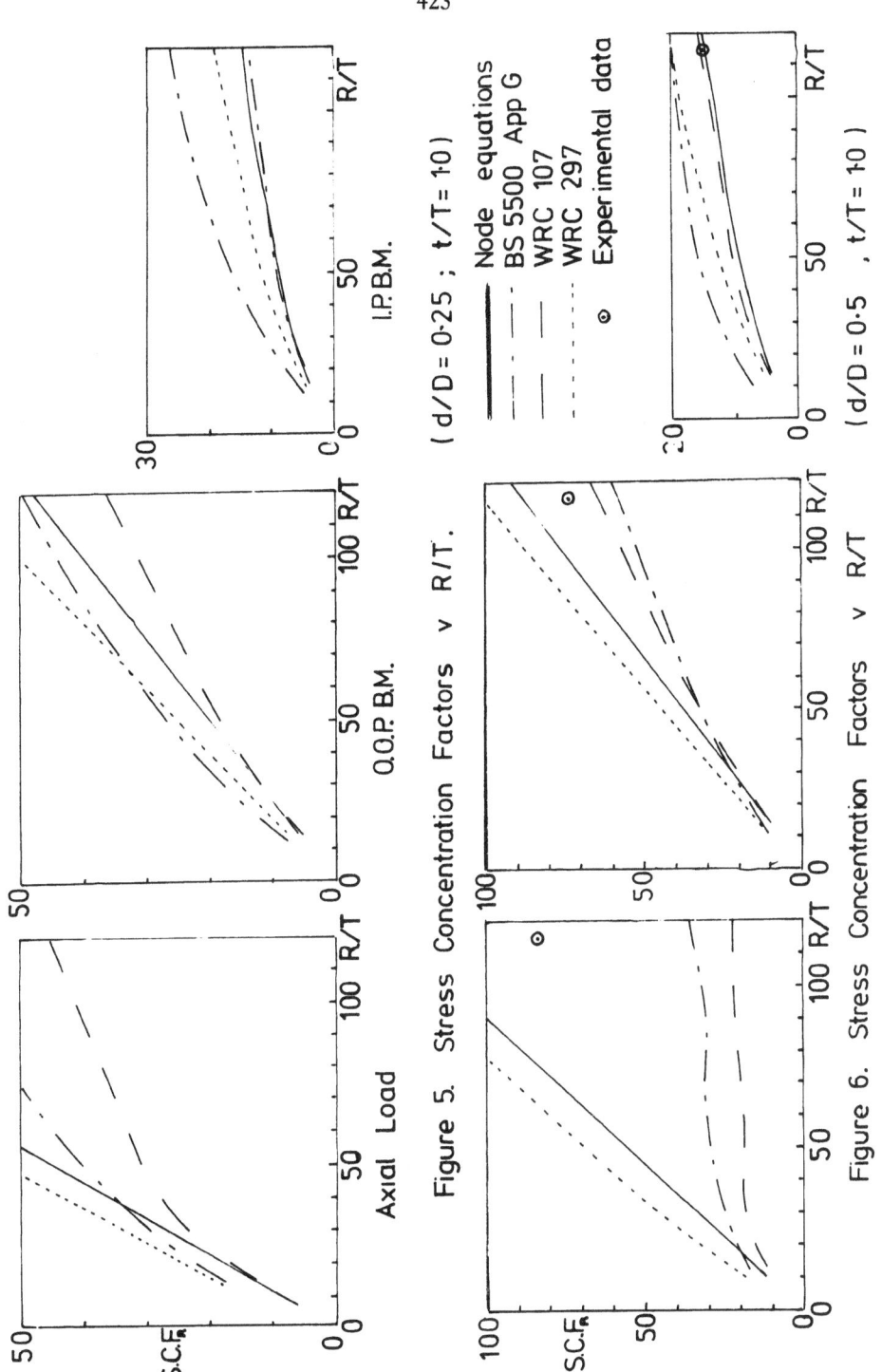

Figure 5. Stress Concentration Factors v R/T.
(d/D = 0·25 ; t/T = 1·0)

Figure 6. Stress Concentration Factors v R/T
(d/D = 0·5 , t/T = 1·0)

424

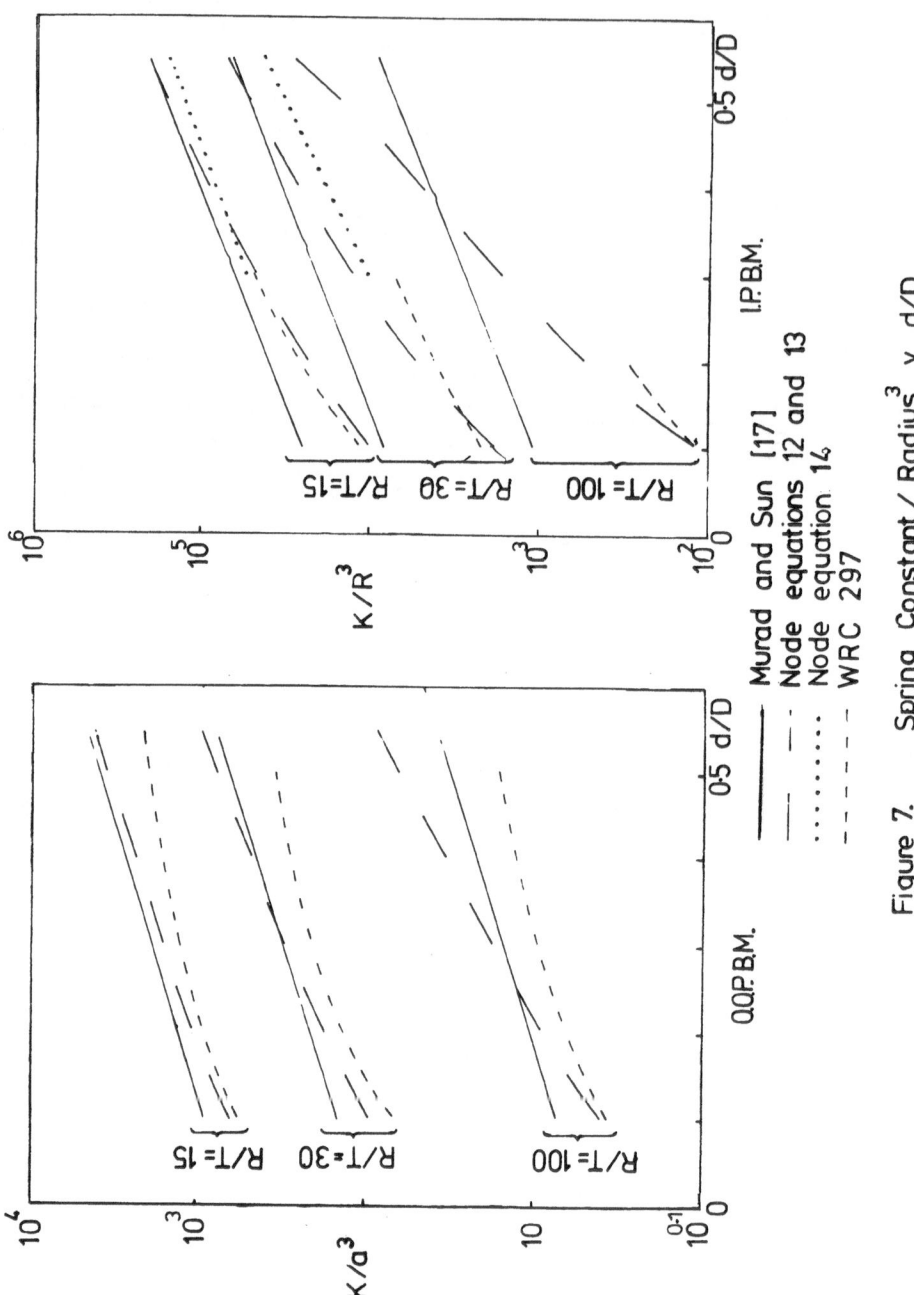

Figure 7. Spring Constant / Radius³ v d/D

COMPARISON OF CALCULATED VESSEL S.C.F'S WITH THOSE FROM EXPERIMENTAL DATA

A salutary conclusion to come from the many tests on large scale steel nodes has been that differences of 25% have been found in the S.C.F's obtained for nominally identical specimens [7]. There are no obvious reasons for concluding that similar discrepancies would not occur in pressure vessels but, despite this, a method of assessing the validity of methods for calculating S.C.F's is to compare them with those obtained experimentally from steel test specimens. Vessel S.C.F's from published data and those obtained by the various calculation methods are compared in Table 2. From a cursory examination it can be seen that, with the exception of specimens 7, 8 and 12, the experimental data fall within the general scatter bands of the calculated values. For specimens 7, 8 and 12 the experimental values are significantly outside those bands and it is tempting to discount them.

Average percentage differences and standard deviations between experimental and calculated S.C.F's are given in Table 3. Table 3A and 3A' include specimens within the d/D validity range for BS5500 App G, the latter with specimen 7, 8 and 12 excluded. Tables 3B and 3B' include the specimens within the d/D validity range of WRC 107 and WRC 297.

COMPARISON OF VESSEL S.C.F'S AND STIFFNESSES OBTAINED BY DIFFERENT CALCULATION METHODS

To highlight the differences between S.C.F's and stiffnesses obtained by the different calculation methods they are shown plotted against either d/D or R/T in Figs 3 to 6. For the cases where branch or vessel are outside the validity ranges it is common practice for designers to indulge in some extrapolation and the figures include data obtained in this way.

Experimental data from specimen Nos 7, 10, 16 and 18 are included in Fig 3 and from specimen no 13 in Fig 6.

Fig 7 is taken from [17] with data from other stiffness calculation methods superimposed. It should be noted that the spring constant units in this figure are in lb.f. and inches, as in the original reference.

CONCLUSION

Reference to Tables 2 and 3 indicates that, where experimental data are available, the node equations, WRC 107 and WRC 297 predict the maximum S.C.F's with similar levels of consistency. The methods given in BS5500 App G appear to be less reliable.

The areas where the S.C.F's and stiffnesses calculated by the different methods diverge and where additional theoretical or experimental data are required are self-evident from Figs 3 to 7.

Calculations of the S.C.F's and stiffnesses using the node equations are more straightforward, less tedious, and therefore less error-prone, than those using the methods for pressure vessels which are described herein. They also cover vessels with d/D ratios of greater than 0.5, which are outside the validity range of the pressure vessel methods. Comparison with experimental data from specimen no 13 [15] indicates that the equations are not unconservative when extrapolated from a maximum R/T ratio of 32, for which they were originally developed, to a ratio of 115.

REFERENCES

1. Wordsworth. A C, Experimental determination of stresses at at unstiffened tubular T and X joints. Joint Australasian Welding and Testing Conference. Perth 1977.

2. British Standard Specification for unfired fusion welded pressure vessels. 1988.

3. Teixeira. M.A, McLeish. R.D, Gill. S.S, A simplified approach to calculating stresses due to radial loads and moments applied to branches in cylindrical pressure vessels. Journal of Strain Analysis. No 4. 1981.

4. Wichman. K.R, Hopper. A.G, Mershon. J.L, Local stresses in spherical and cylindrical shells due to external loadings. W.R.C. Bulletin 107 (Aug 1965 revised March 1979)

5. Mershon. J.L, Mokhtarian. K, Ranjan. G.V, Rodabaugh. E.C, Local stresses in cylindrical shells due to external loadings on nozzles – supplement to WRC Bulletin 297 August 1984.

6. Wordsworth. A.C, Smedley. G.P, Stress concentration at unstiffened tubular joints. European Offshore Steels Research Seminar. Cambridge. Nov 1978.

7. Tebbett. I.E, Lalani.M, A new approach to stress concentration factors for tubular joint design. Offshore Technology Conference. Houston 1984.

8. Underwater Engineering Group. Design guidance on tubular joints in steel offshore structures.

9. Mehringer. F.J, Cooper. W E, Experimental determination of stresses in the vicinity of pipe appendages to a cylindrical shell. S.E.S.A. Proceedings 1957 XIV. 159-174.

10 Gwaltney. R.C, Bolt. S.E, Corum. J.M, Bryson.J.W, Theoretical and experimental stress analyses of ORNL Cylinder-to-cylinder model 3. Oak Ridge National Laboratory Report ORNL 5020. June 1975.

11. Cranch. E.T, An experimental investigation of stresses in the neighbourhood of attachments to a cylindrical shell. W.R.C. Bulletin 60. May 1960.

12. Schoessow. G.J, Kooistra. L.F, Stresses in a cylindrical shell due to nozzle or pipe connections. J. Appl. Mech. 1945. A107-A112.

13 Decock. J. External loadings on nozzles in cylindrical shells. 4th International conference on pressure vessel technology. Vol 2 p. 127-134. I. Mech. E. 1980.

14. Corum. J.M, Holt, S.E, Greenstreet. W.L, Gwaltney. R.C, Theoretical and experimental stress analysis of ORNL thin-shell cylinder - to - cylinder model No 1. Oak Ridge National Laboratory Report 4553, Oct 1972.

15. Riley. W.F, Experimental determination of stress distributions in thin-walled cylindrical and spherical pressure vessels with circular nozzles.

16. Hardenbergh. D.E, Zamrik. S.Y, Effects of external loadings on large outlets in a cylindrical pressure vessel. WRC Bulletin 96. May 1964.

17. Murad. F.P, Sun. B.C, On radial and rotational spring constants on piping nozzles. 5th International Conference on pressure vessel technology. San Francisco 1985.

APPENDIX

This appendix presents an example of the use of the node equations for calculating stresses at a loaded nozzle. Loadings have been chosen to illustrate the significance of the component stresses.

Support span	= 3600mm.	$D = 900$
Pressure	= .75N/mm^2.	$T = 16$
Axial tension	= 15000 N.	$d = 323.9$
O.O.P. = I.P.	= 4550 NM	$t = 9.52$

Branch C.S.A. = 9402mm^2; Branch modulus = 717924mm^3

Branch nominal axial stress = $\dfrac{15000}{9402}$ = 1.6N/mm^2

Branch nominal bending stress = $\dfrac{4550 \times 10^3}{717924}$ = 6.34N/mm^2

Vessel nominal pressure stress = $\dfrac{.75 \times 900}{2 \times 16}$ = 21.1N/mm^2

Axial Load

Bending S.C.F (8) = $\dfrac{3600 \times 323.9 \times 9.52}{900^2 \times 16}$ = .857

Saddle S.C.F$_R$ (1) = $\dfrac{9.52 \times 323.9}{16^2}\left(3.39 - 3.21\left(\dfrac{323.9}{900}\right)^{.5}\right)$ = 17.64

S.C.F$_C$ (9) = (.857 × 2.1) + (.4 × 17.64) = 8.86

Crown S.C.F_R (7) $= .7 + .97\left(\dfrac{900}{16}\right)^{.5} \dfrac{9.52}{16} (1- \dfrac{323.9}{900})$ $= 3.47$

S.C.F_C (10) $= (.857 \times -.95) + (.9 \times 3.47)$ $= 2.31$

O.O.P S.C.F_R (2) $= \dfrac{9.52 \times 323.9}{16^2} (.8 - .575\left(\dfrac{323.9}{900}\right)^5)$ $= 9.59$

S.C.F_C (5) $= .47 \times 9.59$ $= 4.51$

I.P S.C.F_R (3) $=\left(\dfrac{900}{16}\right)^{.6}\left(\dfrac{9.52}{16}\right)^{.8}$

$[.792\left(\dfrac{323.9}{900}\right)^{.25} -.346\left(\dfrac{323.9}{900}\right)^2] = 4.21$

S.C.F_C (6) $= .71 \times 4.21$ $= 2.99$

Pressure S.C.F_C (BS5500 Code case 19) $= 3.87$
Although BS5500 App G implies that this S.C.F_C occurs at
all positions around the nozzle published data and logic
indicate that at the saddle it will reduce to half the
value at the crown ie. $= 1.94$

Total stress
At saddle f_R $= (17.63 \times 1.6) + (9.59 \times 6.34)$
         $= 28.21 + 60.80$ $= 89.01 N/mm^2$
    f_C $= (8.86 \times 1.6) + (4.51 \times 6.34) + (21.1 \times 1.94)$
         $= 14.18 + 28.59 + 40.93$ $= 83.7\ N/mm^2$
At crown f_R $= (3.47 \times 1.6) + (4.21 \times 6.34)$
         $= 5.55 + 26.69$ $= 32.24 N/mm^2$
    f_C $= (2.31 \times 1.6) + (2.99 \times 6.34) + (21.1 \times 3.87)$
         $= 3.7 + 18.96 + 81.66$ $= 104.31 N/mm^2$

NOTE:
     If the axial load is a thrust rather than a tension it will
produce compressive and tensile membrane stresses at the saddle and
crown respectively rather than vice-versa. These are superimposed on
the tensile membrane stresses due to pressure. Tensile shell bending
stresses will still be produced but these will occur on the inner rather
than the outer surface of the vessel.
     To take account of the above changes the addition signs between
the pairs of brackets in equations 9 and 10 above would be changed to
subtraction signs. The branch axial nominal compresive stress would
then be entered as a negative value when computing the total stresses.

OPTIMUM DESIGN OF COMPOSITE–REINFORCED PRESSURE VESSELS

S. J. HARDY & N. H. MALIK
Department of Mechanical Engineering,
University of Wales, Swansea.

ABSTRACT

The paper presents an analytical study of composite wound metallic pressure vessels with cylindrical geometry. The variables are the section thicknesses of liner and composite, with a defined winding pattern. An attempt is made to obtain optimal pressure vessel efficiencies by varying the thickness ratios. A further comparative analysis is performed for identical geometry metal vessels.

NOTATION

l	length of fillet	R	inner radius
r	central opening radius	$\overline{S}$	inplane composite shear strength
t_i	thickness of ith section	T_{max}	max. section thickness
t_{ci}	minimum collapse thickness	V	internal volume
$x, \underline{X}$	variable, variable vector	V_f	fibre volume fraction
E_x	tensile modulus	W	total vessel weight
E_y	transverse modulus	$\overline{X}$	longitudinal tensile strength
FS	factor of safety (1.5)	$\overline{\overline{X}}$	„ compressive strength
G_j	j–th constraint	$\overline{Y}$	transverse tensile strength
G_{xy}	shear modulus	$\overline{\overline{Y}}$	„ compressive strength
L	length of cylinder	η	pressure vessel efficiency
P	applied internal pressure	$\eta c, \eta_m$	composite, metal vessel efficiencies
P_{cr}	buckling pressure	ν, ν_{xy}	Poisson's ratios
P_L	limiting design pressure	ρ	density
P_w	winding pressure on liner		

INTRODUCTION

During the last two decades the use of composite materials in the automobile and aerospace industry has increased significantly. The high strength, stiffness, low specific weight and good fatigue properties of these materials has made them a viable alternative to traditional engineering materials particularly for fabricated products. This increased usage of composites may or may not depend on their economic potential, but is generally due to an engineering necessity.

Composite pressure vessels and tubing with high strength to weight ratios may be readily produced by various fabrication means. Using filament winding techniques, a variety of winding patterns can be generated to meet the performance demands. Major high performance applications of wound vessels include those of rocket motor cases in the aerospace industry, but applications such as fibre reinforced plastic[FRP] storage tanks, pressurized pipes, electrical insulators and pressure vessels are emerging, leading to progress in filament winding technology as well as analysis–synthesis capabilities for filament wound products. Typical advantages gained when compared to purely metal products are weight savings, corrosion resistance, low thermal conductivity, low notch sensitivity and high impact/shatter resistance. The patterns generated are based on relatively complex geometry and are not easily analysed. Design and analysis methods for cylindrical and spherical FRP vessels have been reported by the Society of Plastic Industry[e.g.1–3].

Vessel designs are usually developed assuming netting or orthotropic laminate theory. Netting analysis came into use during early filament winding development and has continued to be used. Although quite limited in its applicability, it serves a useful purpose in approximating hoop and meridional stresses in order to establish the geometry of the vessel. An underlying assumption is the absence of matrix or resin as a load carrier. On the other hand, orthotropic laminate theory has provided a basis for better modelling of the process. In addition, analytical techniques such as the finite element method provide a means of examining the geometry, the material elastic response and the stress/strain behaviour of the structure.

In the present study a mathematical simulation is used to investigate the influence of winding pattern and material response on the performance of a typical internally pressurized vessel. An attempt is made to obtain optimal dimensional and material specifications. It is assumed that the vessel is subjected to mechanical loads only (cold testing), the overwinds consist of balanced in–plane and hoop windings and the composite is wound so as to achieve as near uniform thickness and isotropic conditions as possible.

BACKGROUND

Winding of Vessels

Two basic winding methods have been adopted in the past[4], helical and polar(or planar) winding, each of which generates a distinctive winding pattern. In the former case, the pattern formed due to continuous mandrel rotation and traverse of the carriage is essentially a multi–circuit helix. In polar winding, due to indexed movement of the mandrel and longitudinal motion of the feed arm, the fibres lie adjacent to each other and the fibre path forms a plane of intersection with the mandrel. Hoop windings form reinforcements for the cylindrical portion of the vessel(fibre angle approaching 90 deg. to axial direction) and longitudinal windings are placed at low angles for similar reasons.

For high performance components, the selection of the optimum winding pattern is mainly influenced by the structural efficiency of the head. The earliest vessels were wound at the classical 54.75 deg., giving the 2 : 1 balanced ratio of hoop to meridional stress required for internally pressurized cylinders. As a consequence, most failures occured in the head regions, which acted as soft spots. To counteract this effect, additional low angle windings were used for head reinforcement and hoop windings for the cylindrical part. One problem encountered was that of slippage of tape or filament during manufacture. To resolve this problem, geodesic helical winding and planar winding types were developed together with supplementary techniques(e.g. human expertise, friction controlled trajectory) for NC winding machines, assisted by CAD methods[5].

Geodesic or isotensoid implies that the fibre path describes the shortest distance between two points on a curved surface and this leads to uniform fibre tension. In the head region the geodesic path is taken as tangential to the axial opening. In the planar case, the filament path describes a line of intersection of an inclined plane with the head. For any given axial opening and vessel diameter, the geodesic contour yields only one solution and this determines the winding angle in the cylindrical section of the vessel. Over the head, however, the angle is constantly changing. For the planar wind, the angle is established by the L/D ratio as well as axial openings. The requirements to generate both of these contours via netting theory are described elsewhere[4] and results exist for a parametric study of these windings[6]. The fundamental winding patterns can then be reduced to three types (1) multi–circuit helical winding,(2) single circuit polar with hoop reinforcements and (3) low angle helical windings in conjunction with hoop reinforcements.

For filament wound composites, the laminate topology is usually balanced and the pairs of laminates at + and – angles also meet the condition of zero shear strain. Moreover the corresponding stress–strain relationship becomes simplified due to the absence of shear distortion terms. The generation of elastic constants and their orientation to transformed planes such as for the axisymmetric case, has been discussed by several authors[7,8] and relationships for a balanced multi–layer composite have been given[9], using the condition of strain compatibility between adjacent laminates. The general features of a typical internally lined FRP vessel are shown in Fig.1. The liner is employed as a mandrel during the manufacture, ensures gas tightness and also provides an improvement in mechanical strength.

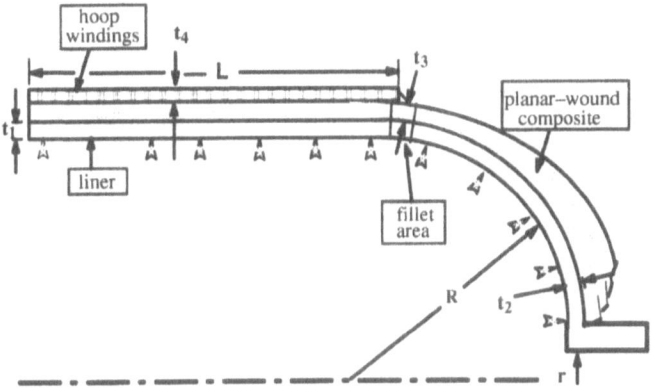

FIGURE 1. Sketch of a typical internally lined composite pressure vessel

Optimization techniques

The optimization of supporting structures made from composites began in the early seventies, following the provision of accurate information on the material behaviour of FRP s. A number of references and papers exist [eg 10,11], in which composite panels and structural components are optimized with regard to various objectives. In addition much information on the optimum design of conventional pressure vessels is available [e.g.12,13], taking virtually every design aspect into account. Optimization techniques themselves vary depending on the nature of the investigation and are based on mathematical programming or structural optimality considerations, in order to obtain improvements in structural efficiency[14]. Within the design–synthesis of composite structures, a certain degree of optimism is usually assumed, in order to demonstrate the superiority of a composite component over its metal counterpart. An example of this is the wide variety of results which are based on optimum fibre orientation in composite panels and components, where no account is taken of the fabrication constraints or practicalities. Early studies on the design optimization of composite pressure vessels were based on simplified hoop–axial stress relationships in order to obtain a reasonable fibre trajectory in the various regions of the vessel. More recently FEM has been adopted for analysis and improvement in such design exercises[9,15].

The utilization of existing mathematical optimization software (written with isotropic materials in mind) to solve composite structural design problems has limitations, owing to the large number of design variables and related constraints. Consequently attention is being focused on the application of multi–objective techniques[16,17] and multi–level optimization[18,19].

OPTIMIZATION ANALYSIS

Finite element model

Two data preprocessors are used to help prepare the finite element input. A mesh generation program calculates the node and element data for the model. A winding simulation routine produces a composite thickness profile and fibre angles throughout the pattern assuming a smooth distribution of layers. Laminates in the geometry are assigned the local thickness and fibre orientations of the specified latitudes. The accumulated thickness obtained provides an overall thickness profile. Classical axisymmetric 6 and 8 noded isoparametric displacement elements are used to model the final structure. Over an element, the fibre orientation angle is assumed to be constant within a laminate. Moreover the inaccuracy in the polar region, due to 3–dimensional stress–strain effects and material placement has to be a compromise. The fillet region is needed to provide continuity between differing thicknesses. The axial opening is a passive region and is not included in the redesign. The flow scheme in Fig.2 shows the role of different routines in the program for mesh generation, property assignment, analysis and post–processing. The programming language adopted throughout is FORTRAN–77.

Optimization model

The optimum design of such vessels can be presented in the classical form of a system objective function that is minimized or maximized subject to a set of applied

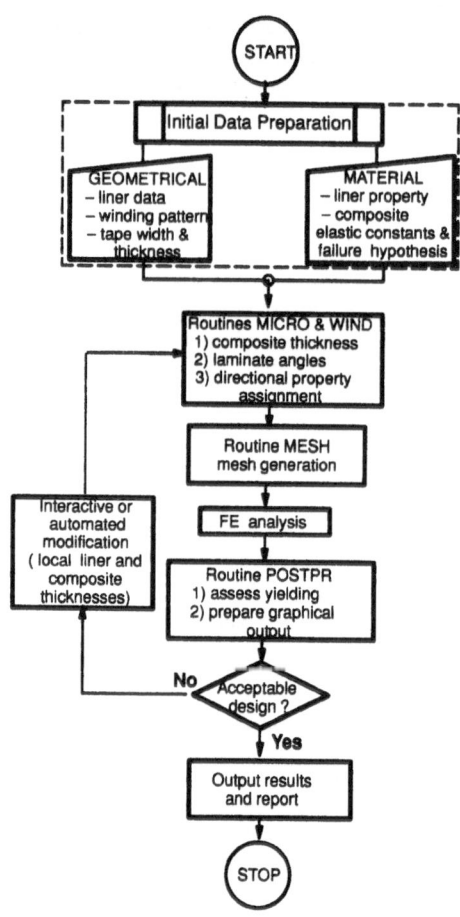

FIGURE 2. Outline for the design procedure

constraints. This mathematical process requires the development of a model in the form of explicit equations to describe the objective and constraint functions. Here it should be noted that analysis modelling such as FEM and design modelling are distinct but interrelated aspects of the procedure. The design model description is based on predictions from the analysis model.

The design criterion used for the objective function formulation is the maximization of the vessel efficiency ($P_L V/W$). In discretized form, this can be written as :–

$$MAX \ \eta(\underline{x}) \ \text{subject to constraints} \ \ G_j(\underline{x}) > 0, \quad j = 1,2...m$$

where $\underline{x}$ is the vector of design variables and $G_j(\underline{x})$ is the set of linear and nonlinear constraints imposed on the problem. The constraints have a dual role in that they control the boundary description within prescribed limits and also provide a means of imposing upper and lower limits in order to control the design model.

The routine which is employed here(E04VDF), uses a sequential quadratic programming(SQP) algorithm in which the search direction is the solution of a QP problem. The algorithm treats bounds on variables, linear constraints and nonlinear constraints separately. Firstly an iterative procedure determines a feasible point by evaluating the constraints and secondly generates a sequence of feasible solutions in order to extremize the objective function. The details of the routine are given in [20].

Analysis procedure

The procedure adopted for a complete vessel analysis is shown in the form of a flowchart in Fig.3. There is an inner loading loop which, based on a given combination of thicknesses determines the design pressure(P_L) and the vessel efficiency(η). The outer optimization loop modifies the thicknesses in order to optimise η. The solution to the finite element investigation proceeds in a step–by–step manner. The loading is applied incrementally and the following two conditions are used to determine maximum load:–

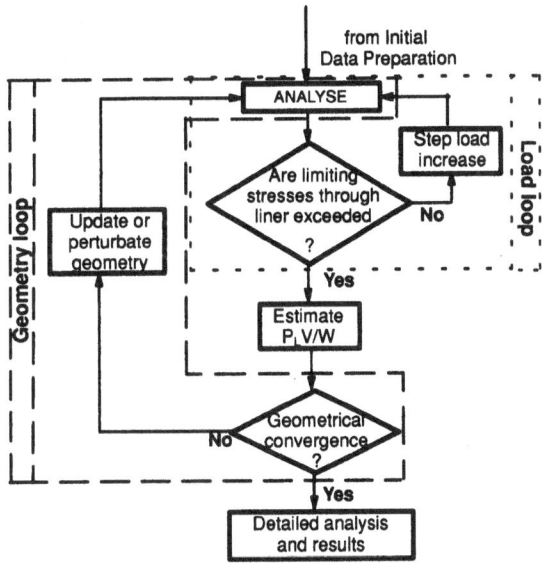

FIGURE 3. Outline of design algorithm

(1) Tresca criterion for the liner.
(2) Either the Tsai–Hill or maximum fibre equivalent stress criterion for the composite[7].
The explicit form of these criteria is given in the Appendix. In both cases, a factor of safety of 1.5 is assumed for the limiting stress. In order to obtain maximum benefit, the load loop terminates only when stresses throughout the liner thickness are above the limiting stress using an elastic analysis. If either the metal or composite limiting criteria is violated, then the last load step is replaced by a

modified load step, using an interval halving procedure. Convergence is achieved when the current load increment is less than a specified fraction of the original load step. This design pressure is then used to calculate the vessel efficiency. The optimization loop then modifies the design parameters sequentially within the specified constraint envelope. This new set of design parameters is then used as input to the loading loop and η is recalculated. Global optimization is achieved when a particular set of design variables generates the highest value of the efficiency.

DESIGN CASE STUDY

A cylindrical pressure vessel with a simple hemispherical head and an axial opening has been considered. The vessel is made up of a metal liner, with planar and hoop windings. Although the investigation is restricted to the above parameters, the application can be extended to other more complex head shapes (eg. ellipsoidal or torospherical) and winding patterns.

The geometry is defined by fixed parameters **R,L,r, and l**. The design variables are four thicknesses of the profile t_1 to t_4 respectively, see Fig.1. Constraints on the solution are :–

(i) The wall and head liner thicknesses have to be greater than the minimum buckling or collapse thicknesses, t_{c1} and t_{c2} respectively, for the specified external winding pressure P_w (based on the internal pressure[4]) on the liner. See Appendix.

(ii) In order to formulate a finite problem, the overall thickness at the equator is restricted to a maximum value, T_{max}.

The overall vessel geometry is R = 0.25, L/R =0.5, r/R = 0.2 , l/L = 0.1 and the following cases are considered :–

P_w/P = 1%, 5% , R/T_{max} = 6.0, 9.0, 12.0, in conjunction with three basic pairs of materials:–
1) E–Glass Epoxy + Steel (EGS)
2) Kevlar–49 Epoxy + Titanium (KET)
3) T300–5208 Graph. Epoxy + Aluminium (GRA)
for which the properties are given in Tables 1 and 2. Fig.4 shows the relative merits of these material pairs in terms of modulus and strength.

Material	Steel	Titanium	Aluminium
Property			
E (GPa)	209.	350.	70.
ν	0.3	0.3	0.33
$\overline{X}$ (GPa)	0.689	1.033	0.344
ρ (kg/m³)	7800	4000	2695

TABLE 1. Properties assumed for the liner.

FIGURE 4. E/ρ vs $\overline{X}/\rho$ for composites and metals.

Material	EGS	KET	GRA
Property			
V_f	0.72	0.6	0.7
E_x	60.7	76.0	153.0
E_y	24.8	5.5	10.9
ν_{xy}	0.23	0.34	0.3
G_{xy}	11.99	2.3	5.6
$\overline{X}$	1.2994	1.40	0.6895
$\underline{X}$	0.821	0.235	0.7585
$\overline{Y}$	0.0459	0.012	0.0276
$\underline{Y}$	0.1744	0.053	0.0965
$\overline{S}$	0.0448	0.034	0.0621
ρ	2169	1419	1699

TABLE 2. Properties assumed for the composite[7]. Moduli and strengths in GPa. Density in kg/m³

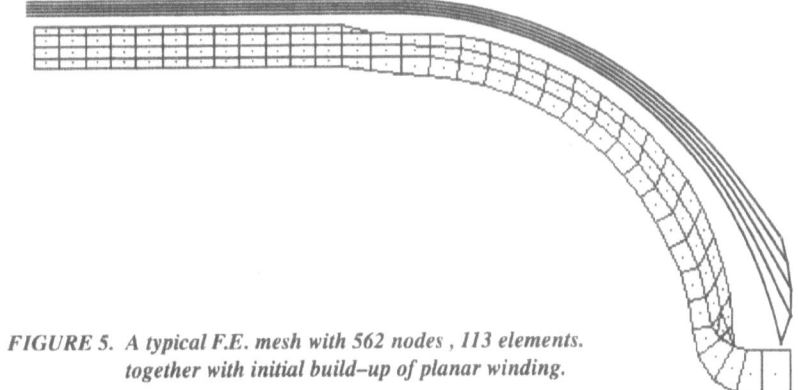

FIGURE 5. A typical F.E. mesh with 562 nodes , 113 elements. together with initial build–up of planar winding.

A laminate thickness and tape width of 0.00264 and 0.011 respectively is assumed. Given the above data and the initial design vector $\mathbf{x_n}^{1,}$ the finite element mesh is generated. An example of such a mesh is shown in Fig. 5 along with a typical arrangement of the planar windings generated by the routine, where the attempt to keep a uniform thickness implies the termination of successive winds near the poles, due to the local build–up.

RESULTS

The variation of optimized vessel efficiency with vessel geometry is shown in Fig.6(a) and 6(b) for winding pressures of 1% and 5% of the design pressure respectively. In all cases the Kevlar–Titanium material pair results in efficiencies which are a factor of ~2 greater than the nearest rivals(Graphite–Aluminium). This is because Kevlar49–Epoxy, being a compliant material, is supported by a liner with higher modulus and moderately better strength than the two other liner materials. Fig.7 gives an indication of the levels of improvement in efficiency that can be achieved when composite materials are added to augment the metal liners. Optimized composite vessel efficiency is divided by the optimized efficiency of its purely metal (liner) counterpart and it is clear that factors of improvement of upto 3 are possible. In most cases, the GRA and EGS material pairs exhibit similar levels of improvement with KET, for reasons already discussed, being less beneficial.

Figures 6 and 7 show that the results are both geometry and winding pattern dependent. In the first case, greater efficiencies and benefits are achieved with thicker walled vessels. This is because, as R/T_{max} increases, the composite becomes less effective due to winding imperfections and the liner tends to make up for this deficiency. Subsequently there is a relative increase in weight with a substantial decrease in efficiency. Secondly, increasing the winding pressure results in a marked decrease in the efficiency and benefits that can be expected. The primary cause of this effect is the higher content of metal in the structure as a consequence of the constraint on buckling thickness. This deterioration in efficiency becomes more pronounced at higher R/T_{max} levels, where liner thickness is the major contributor to the total weight, and the expected benefits from raising design pressures are counteracted by the resulting heavier vessel.

The optimized distribution of metal liner, planar and hoop composite thicknesses are based on a compromise between the elastic properties of the metal and the composite material. As an illustration, the results for the Graphite–Aluminium material pair are presented in Fig.8(a) and 8(b) for the cylinder and head region respectively. Accumulated thicknesses are normalised with respect to the total thickness in order to observe their relative contribution to the overall thickness. Considering the wall region, the liner : total thickness ratio stays almost constant at low winding pressure, but increases with reducing thickness at higher winding pressure due to the buckling constraint imposed on it. Correspondingly the hoop composite thickness ratio shows a slight decrease. The planar composite thickness ratio reduces with increasing R/T_{max}, since partially it is responding to the load compensation being made by the liner, but is mainly influenced by the requirement in the head region(see Fig.8(b)) where again Aluminium, being the weaker material, has to satisfy the requirements on strength as the total thickness goes down.

Finally the variation of maximum design pressure with geometry for the optimized vessel is shown in Fig.9(a) and 9(b) for the two winding pressures. Comparing these results with those in Fig.6 it is seen that the Kevlar–Titanium pair still provides the best solution. However the EGlass/Epoxy–Steel combination is superior to Graphite/Epoxy–Aluminium, when maximum design pressure is used as the criterion. This feature is not only evident from Fig.4, but was also identified by Gerstle[8], where certain material pairs were found advantageous for overall efficiency, whereas others had superior burst pressure capabilities.

DISCUSSION

The paper presents the development of an analysis and design model for applications in the area of composite reinforced metallic pressure vessels. The given examples indicate that the F.E. method is capable of predicting behaviour of such vessels and furthermore that a design model can be developed to help optimize the vessel response, especially to investigate the optimum geometry and materials. Prestressing or induced winding pressure helps to create higher pressure capabilities, although the vessel weight increases considerably. These possibilities could prove invaluable to the designer, if the predictions obtained from the study can be verified against experimental test data. In this case parametric studies and extensive mechanical testing can be substantially reduced. The resulting vessel efficiencies illustrate the significant advantages of composite reinforcement over pure metals for low to medium pressures and cold conditions.

REFERENCES

1. Foerster A.F. & Boiler T.J, Structural characteristics and optimization of filament–wound cylindrical rocket motor cases , SPE Transactions, April, 1964,pp. 103–106.

2. Brown J. Jr., Simplified stress analysis of filament reinforced plastic pressure vessels , SPE Journal, Vol.17,9, Sept. 1961, pp. 989–991.

3. Chiao T.T., Design for commercial filament winding , SPE Journal, Vol.22, April, 1966, pp. 43–47.

4. Shibley A.M., Filament winding , ed. George Lubin Handbook of fibre glass and advanced plastics composites , SPE Polymer Technology Series, Van Nostrand Reinhold Company, 1969., pp. 438–484.

5. Eckold G.C., Thomas D.G.L. and Wells G.M., Computer aided filament winding , Proceedings of the I.Mech–E Conference on Design in Composite Materials , March,1989, pp. 127–136.

6. Rosato D.V. and Grove C.S., Filament winding : Its development, manufacture, applications and design , Interscience Publishers, 1964. pp. 171–248.

7. Vinson J.R. and Seirakowski R.L, The behaviour of structures composed of composite materials , Martinus Nijhoff Publishers, 1986.

8. Gerstle G.P. Jr., High performance advanced composite spherical pressure vessels , ASME paper 74–PVP–42, June, 1974.

9. Chen M.C. & Chewlow L.N.O., Computer analysis of filament–reinforced metallic spherical pressure vessels

, Computers and Structures, Vol.7., 1977 pp. 93–102.

10. Venkayya V.B., Structural optimization : a review and some recommendations ,Int. J. Num. Meth. Engng.,Vol.13, 1978. pp. 203–228.

11. Sobeiski J.,–NASA symposium on recent experiences in multidisciplinary analysis and optimization, NASA–CP–2327, Hampton, VA, April 1984.

12. Middleton J. and Petruska J., Optimal pressure vessel shape design to maximize limit load , Engineering Computations, Vol.3, 4, Dec. 1986.

13. Maksimov L. Y., Design of cylinders to withstand high internal pressures , Russian Engng. J., Vol.44, 1964, pp. 5–6.

14. Hardy S.J. & Malik N.H., Optimum design of laminated structural members , Proceedings of the IMech.E.,Conference on Design in composite materials , London, 1989. pp. 49–60.

15. Nyssen C., Nonlinear incremental analysis up to failure of aeronautical structures , Computers and Structures, Vol.12, 1980, pp. 593–605.

16. Eschenauer S. and Fuchs W., Modelling, structural analysis and optimization of composite structures , Zeitschrift fur Flugwissenschaften und Weltraumforschung , BD.11, 1987, pp. 201–210.

17. Malik N.H. and Hardy S.J., Multi–objective design of composite laminated structures , NUMETA–90, Swansea, Jan. 1990, pp. 500–509.

18. Kam T.Y. & Chang R.R, Optimal design of laminated composite plates with dynamic and static considerations , Computers and structures, July 1989, pp. 387–393.

19. Weiji L. and Baohua S., Multilevel optimization procedure of composite structure , ed. Marshall I.H., Proceedings of 4th Int. Conf. on Composite Structures , Paisley, 1987, pp. 1.357–1.367.

20. NAG Fortran Library Manual–Mark 12, Vol.4., The Numerical Algorithms Group Ltd., 1987.

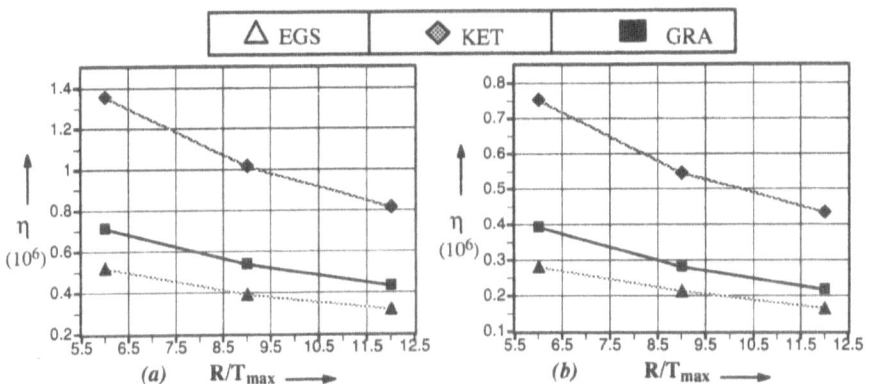

FIGURE 6 Variation of efficiency with R/T_{max} for (a)$P_w=1\%P$ (b)$P_w=5\%P$

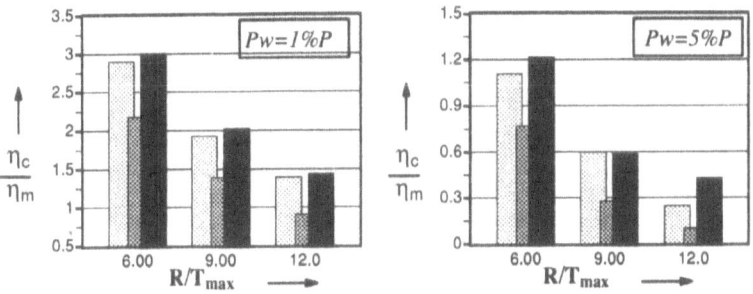

FIGURE 7 Efficiency factor for composite to metal vessels

438

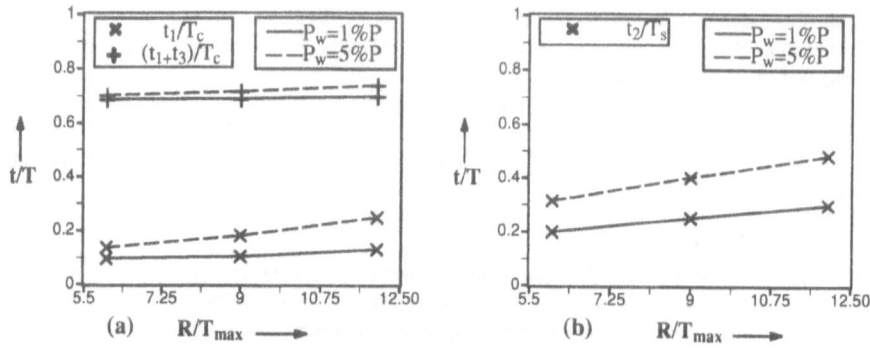

FIGURE 8. Thickness ratios in (a)Wall, (b)Head for GRA.
$(T_c = t_1 + t_3 + t_4, \ T_s = t_2 + t_3)$

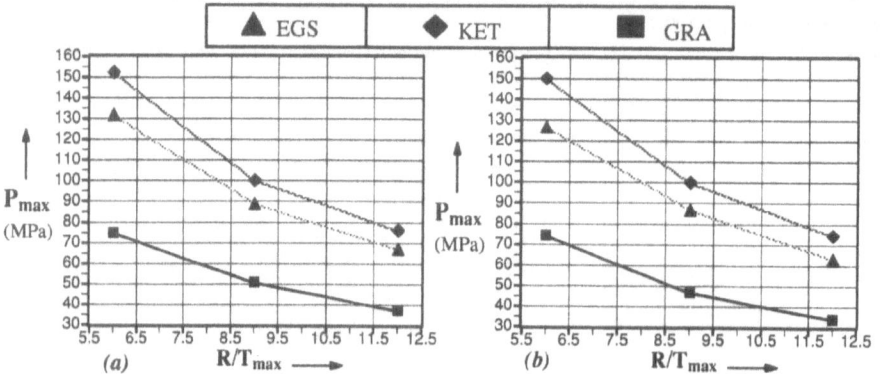

FIGURE 9 Variation of limiting pressure with R/T_{max} for (a)$P_w=1\%P$ (b)$P_w=5\%P$

APPENDIX

(a) Tsai–Hill criterion is:

$$\left[\frac{\sigma_{11}}{X}\right]^2 + \left[\frac{\sigma_{22}}{Y}\right]^2 + \left[\frac{\sigma_{11}\sigma_{22}}{XY}\right] + \left[\frac{\sigma_{12}}{S}\right]^2 - FS \leq 1$$

(b) The fibre breaking criterion is:

$$\left[\frac{\sigma_{11}}{X}\right]^2 - FS \leq 1$$

where requisite values are implied for yield strengths $\overline{X}$ and $\overline{Y}$ in tension or compression. σ_{11}, σ_{22} and σ_{12} are respectively the directional material stresses in longitudinal transverse and shear.

(c) External collapse pressures for

(1) A short cylinder with clamped ends :

$$P_{cr} = 0.807 \frac{Et^3}{LR^2} \sqrt{\left(\frac{1}{1-v^2}\right)^3}$$

(2) A spherical part with clamped ends

$$P_{cr} = \frac{Et^3}{4R^3}\left(\frac{1}{1-v^2}\right)$$

A TWO-DIMENSIONAL STRESS ANALYSIS OF ADHESIVE BUTT JOINTS

SUBJECTED TO TENSILE LOADS

TOSHIYUKI SAWA
Department of Mechanical Engineering, Yamanashi University
4-3-11 Takeda, Kofu, Yamanashi, 400 Japan

KATSUHIRO TEMMA
Kisarazu National College of Technology, Japan

HIROHISA ISHIKAWA
Mitsubishi Heavy Industry Co.Ltd., Japan

ABSTRACT

Replacing dissimilar adherends and an adhesive with finite strips, stress distributions in adhesive butt joints subjected to external tensile loads were analyzed strictly as a three-body contact problem using the two-dimensional theory of elasticity. The effects of stiffness and thickness of adhesive bonds on the stress distributions were demonstrated by numerical calculations. In addition, the stress singularity caused at the edge of the interface was discussed. The analytical result was fairly consistent with the experimental one concerning the strains produced on the adherends.

INTRODUCTION

Recently, due to the improvement in the properties of adhesive bonds adhesive joints have become increasingly used in mechanical structures . Adhesive joints have some advantages over mechanical joints in such areas as bolting and riveting, but structures which consist of adhesive joints have previously been designed empirically. However, data is now available for the establishment of an optimal design method. In establishing such a design method for adhesive joints, accurate knowledge of the stress distributions of joints is essential. Up to now, many investigations(1)-(10) have been carried out on lap, scarf and butt adhesive joints subjected to tensile, bending and shear loads. However, few investigations(11) have been done on butt joints in which dissimilar finite strips are joined by an adhesive.

This paper deals with a stress analysis of adhesive butt joints ,in which dissimilar finite plates are joined by an adhesive,subjected to external tensile loads. Replacing dissimilar adherends and the adhesive with finite strips respectively,the stress distributions are analysed strictly as a three-body contact problem using a two-dimensional theory of elasticity. The effects of the ratios of Young's modulus among the adherends and the adhesive on the stress distributions are clarified by numerical calculations,and the stress singularity at the edge of the interface is examined. The analytical result is compared with the experimental result concerning strains produced on the adherends.

THEORETICAL ANALYSIS

Figure 1 shows the case where an adhesive butt joint with two adherends which have dissimilar dimensions and materials, is subjected to an external tensile load. The adherends are replaced with finite strips [I] and [III],and the adhesive with finite strip [II] as shown in Figure 2. The length of strips [I] and [III] are represented by $2l$,their height by $2h_1$ and $2h_3$,their Young's modulus by E_1 and E_3,and their Poisson's ratio by ν_1 and ν_3,respectively. The characteristics of strips [II] are designated by $2l,2h_2,E_2$ and ν_2,respectively. It is assumed that the tensile loads acting on the upper and the lower ends of the joints are replaced with symmetrically distributed stresses $F(x)$ and $G(x)$ with respect to the y_1 and the y_3 axes over ranges of $x \leq e_1$ on the upper surface and $x \leq e_2$ on the lower surface of the strips [I] and [III]. Expanding the distributions $F(x)$ and $G(x)$ into Fourier series,the boundary conditions shown in Figure 2 ,where the displacement in the x-

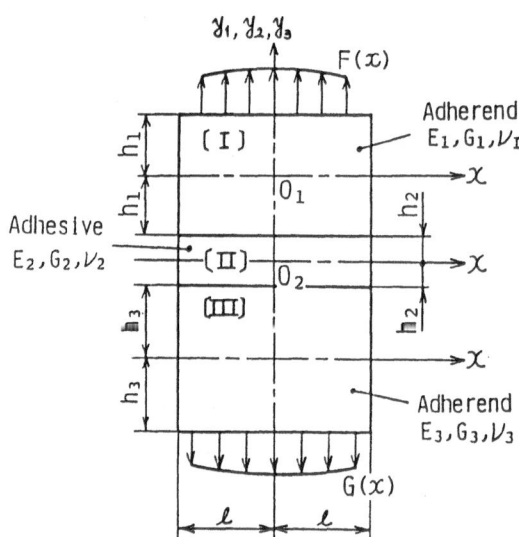

Figure 1. A butt adhesive joint of dissimilar adherends subjected to an external tensile load

direction is denoted by u,the displacement in the y-direction by v,and upper suffixes I,II and III correspond to the finite strips [I],[II] and [III],respectively, are expressed as follows.

(i) on the finite strip [I](adherend)

$$x = \pm \ell : \sigma_x^I = \tau_{xy}^I = 0$$

$$y_1 = h_1 : \sigma_y^I = F(x) = a_0 + \sum_{s=1}^{\infty} a_s \cos\left(\frac{s\pi x}{\ell}\right) \quad (1)$$

$$\tau_{xy}^I = 0$$

(ii) on the finite strip [II](adhesive)

$$(2)$$

$$x = \pm \ell : \sigma_x^{II} = \tau_{xy}^{II} = 0$$

(iii)on the finite strip [III](adherend)

$$x = \pm \ell : \sigma_x^{III} = \tau_{xy}^{III} = 0$$

$$y_3 = -h_3 : \sigma_y^{III} = G(x) = b_0 + \sum_{s=1}^{\infty} b_s \cos\left(\frac{s\pi x}{\ell}\right)$$

$$\tau_{xy}^{III} = 0 \quad (3)$$

(iv) at the interface between the finite strips [I] and[II]

$$(\sigma_y^I)_{y_1=-h_1} = (\sigma_y^{II})_{y_2=h_2}$$

$$(\tau_{xy}^I)_{y_1=-h_1} = (\tau_{xy}^{II})_{y_2=h_2}$$

$$(u^I)_{y_1=-h_1} = (u^{II})_{y_2=h_2}$$

$$(\partial v^I / \partial x)_{y_1=-h_1} = (\partial v^{II} / \partial x)_{y_2=h_2} \quad (4)$$

(v) at the interface between the finite strips [II] and [III]

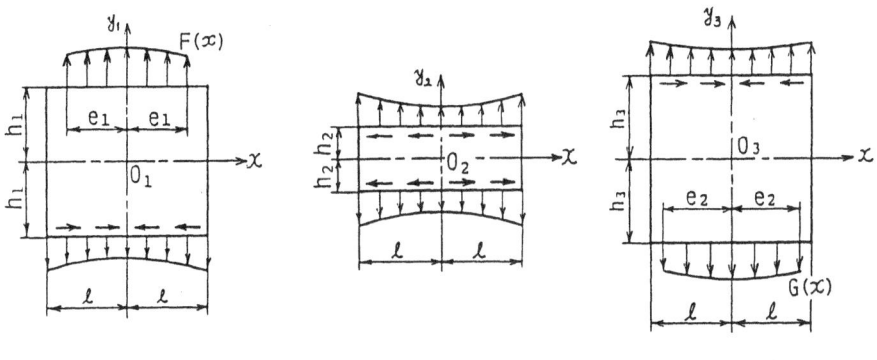

(a)adherend[I]   (b)adhesive[II]   (c)adherend[III]
Figure 2. A model for analysis

$$(\sigma_y^{II})_{y_2=-h_2} = (\sigma_y^{III})_{y_3=h_3}$$

$$(\tau_{xy}^{II})_{y_2=-h_2} = (\tau_{xy}^{III})_{y_3=h_3}$$

$$(U^{II})_{y_2=-h_2} = (U^{III})_{y_3=h_3}$$

$$(\partial V^{II}/\partial x)_{y_2=-h_2} = (\partial V^{III}/\partial x)_{y_3=h_3} \tag{5}$$

where

$$a_0 = \frac{1}{2\ell} \int_{-e_1}^{e_1} F(x)dx \ , \quad a_s = \frac{1}{\ell} \int_{-e_1}^{e_1} F(x)\cos\left(\frac{s\pi x}{\ell}\right)dx$$

$$b_0 = \frac{1}{2\ell} \int_{-e_3}^{e_3} G(x)dx \ , \quad b_s = \frac{1}{\ell} \int_{-e_3}^{e_3} G(x)\cos\left(\frac{s\pi x}{\ell}\right)dx$$

Each finite strip is analysed as a three-body contact problem using Airy's stress functions which are selected from solutions for the method of separation variables.

EXPERIMENTAL METHOD

Figure 3 shows the dimensions of the specimens used in the experiments. They are manufactured from steel (S45C,JIS) and aluminum,and both of their thicknesses are 8mm. Two specimens were bonded by an adhesive(Scotch-Weld 1838,Sumitomo 3M Co. Ltd.) and a tensile load was applied using pins. Strains were measured at positions situated 2 mm from the interface with strain gauges (KFC-C1-11,Kyowa Dengyo Co.Ltd.Japan).

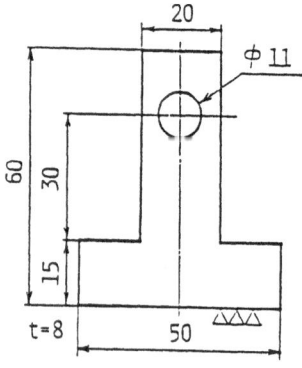

Figure 3. Dimensions of specimens used in experiment

ANALYTICAL RESULTS AND COMPARISON WITH EXPERIMENTAL RESULT

Analytical results

In the numerical calculations, the number, N, of the terms in the series was put as 50. Figure 4 shows the stress distributions at the interfaces $y_2 = h_2$ and $y_2 = -h_2$, where the ratio E_1/E_3 was held constant at 3 and the ratio E_1/E_2 was 5, 10 and 100. The abscissa is the ratio of the distance x to the half length 1 of the adhesive and the ordinates indicate the ratio of each stress to the mean normal stress σ_{ym} at the interface. In Figure 4 , it was assumed that tensile stresses act uniformly on the upper($y_1 = h_1$) and the lower($y_3 = -h_3$) surfaces in the region $x \leq 1$. From the results , it is seen that singular stresses occur at the edge of the

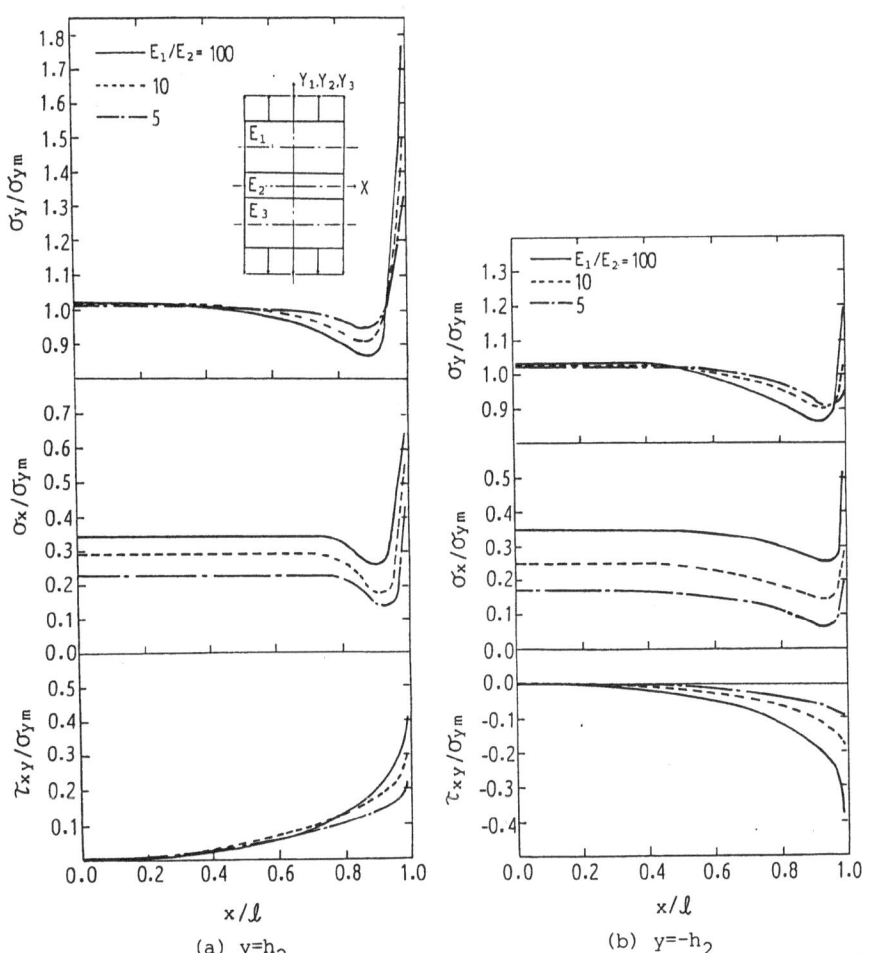

(a) $y = h_2$

(b) $y = -h_2$

Figure 4.   Effect of the ratio of Young's modulus of adherend to that of adhesive on stress distribution

interfaces and each stress becomes larger at $y_2=h_2$ than at $y_2=-h_2$. The absolute values of stress σ_y and the shear stress τ_{xy} become larger near the edge $x/l=1.0$ of the interface with an increase of the ratio E_1/E_2.

Figure 5 shows the effect of the ratio E_1/E_3 of Young's modulus. The ratio E_1/E_2 was held constant at 10 and the ratio E_1/E_3 was 3 and 6. Each stress near $x/l=1.0$ ($y_2=h_2$) increases with an increase of the ratio E_1/E_3 and decreases near $x/l=1.0$ ($y_2=-h_2$). Figure 6 shows the effect of the ratio of Poisson's ratio. The ratio E_1/E_3 of Young's modulus was put as 3, the ratio ν_1/ν_3 of Poisson's ratio as 1.0 and the ratio ν_1/ν_2 was put as 0.75, 0.86 and 1.00. Each maximum stress increases at $y_2=h_2$ than at $y_2=-h_2$.

Figure 7 shows the effect of the thickness $2h_2$ of the adhesive layer, where h_1 equals h_3 and the value h_1/h_2 was 2 and 5, and E_1/E_3 was

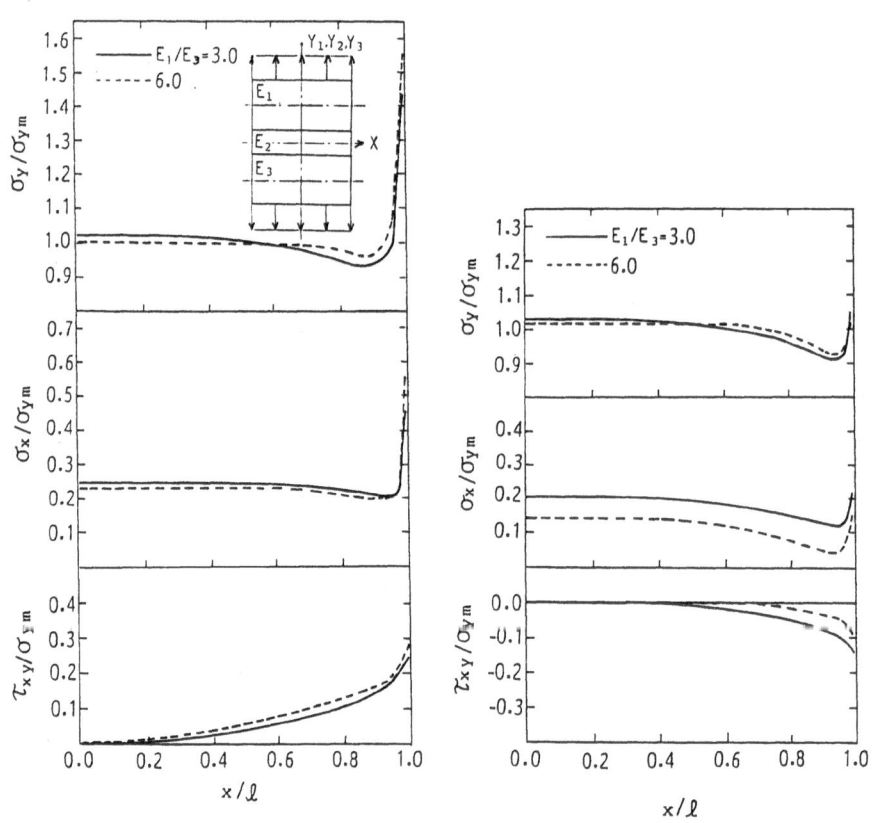

(a) $y=h_2$ (b) $y=-h_2$

Figure 5. Effect of the ratio of Young's modulus
between adherends

1 and 5 ,and $E_1/E_2=3.0$.

Figure 8 shows the relationship between the ratio of the normal stress σ_y to the mean stress σ_{ym} and the distance r from the edge in logarithmic scales in order to examine the property of the stress singularity at the edges. This case corresponds to Figure 4. In general, the singular stress is expressed approximately by Eq.(6),where K is the intensity of the stress singularity and λ is the order of the singularity. The distance from the edge is denoted by r and expressed by the equation r=(1-x)/1. The parameter λ is determined by the method which was demonstrated in the references(12)(13). The singular stresses are expressed in the figure.

$$\sigma = K\, r^{-\lambda} \tag{6}$$

Comparison between analytical results and experimental ones

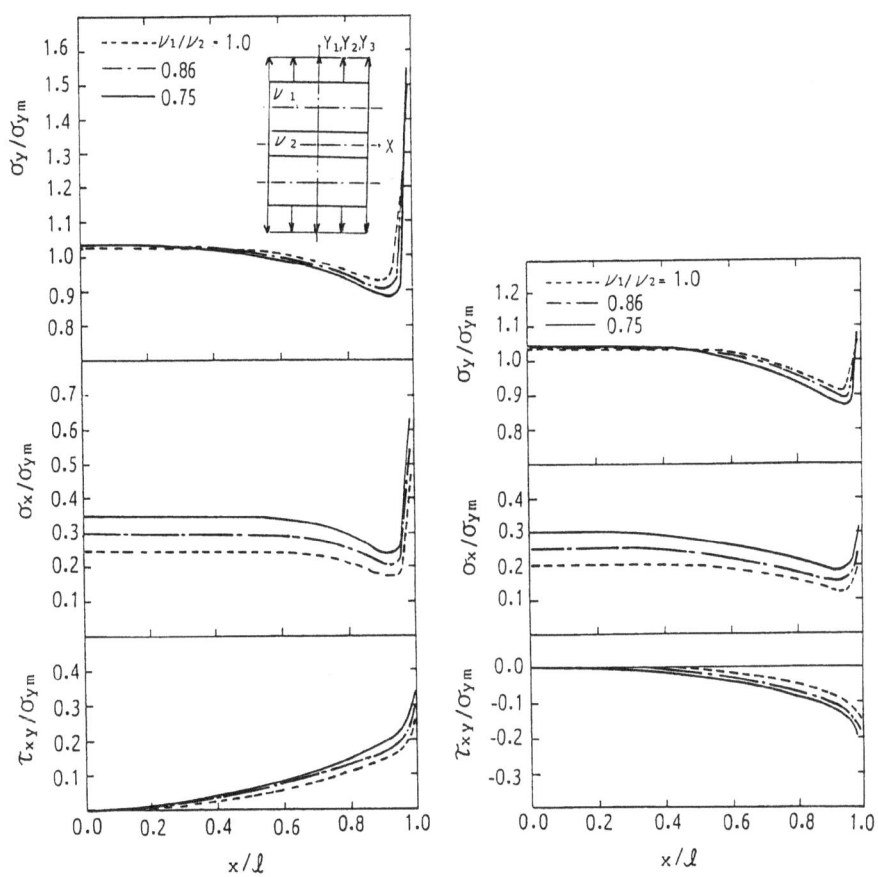

(a) $y=h_2$ (b) $y=-h_2$

Figure 6. Effect of the ratio of Poisson's ratio

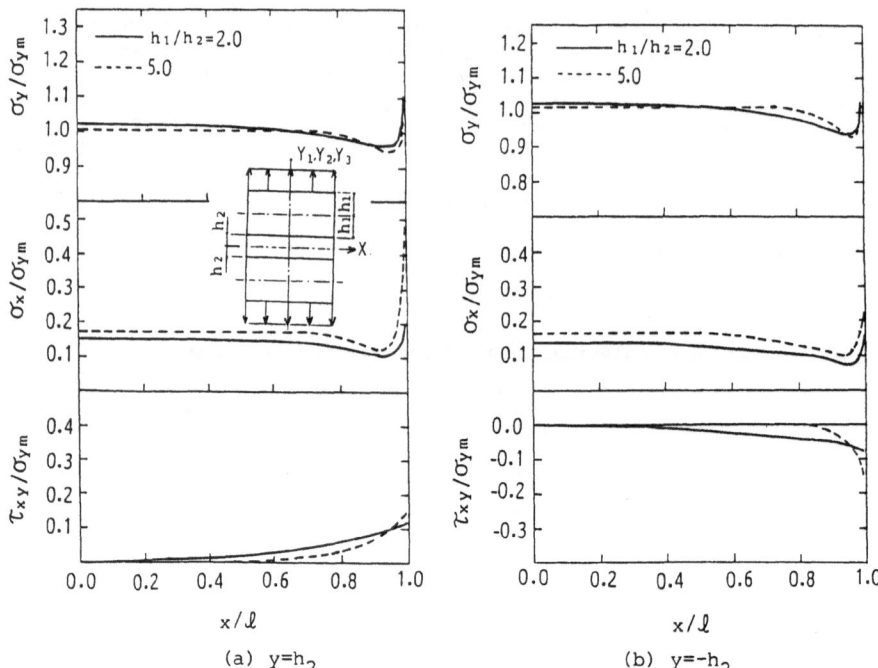

(a) $y=h_2$ (b) $y=-h_2$

Figure 7. Effect of the thickness of adhesive layer

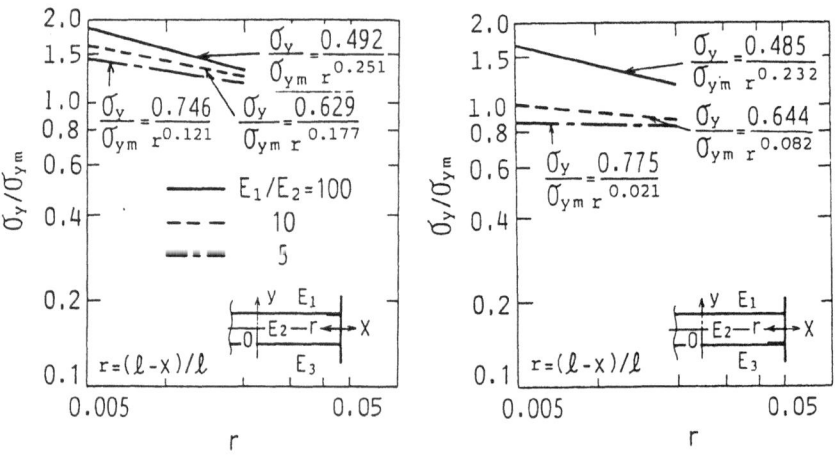

(a) $y=h_2$ (b) $y=-h_2$

Figure 8. Singular stress at edge of interface(in the case of Figure 4)

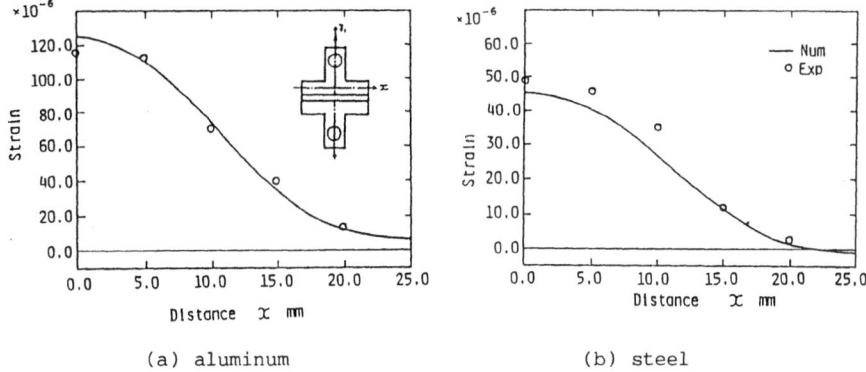

(a) aluminum   (b) steel

Figure 9. Comparison between analytical result and experimental one
concerning strains produced on adherends

Figure 9 shows a comparison between the analytical result with the
experimental one concerning the strains in the y-direction produced at
the positions (y_1=-5.5mm,y_3=5.5mm) situated 2mm from the interface. In
the numerical calculation, the material constants E_1 and ν_1 were put as
206Gpa and 0.3(steel),E_3 and ν_3 as 68.7 Gpa and
0.3(aluminum),respectively. Those of the adhesive layer were put as 3.6
Gpa and 0.38 ,and $2h_2$ was measured as 0.055mm. It is seen that the
analytical result is fairly consistent with the experimental one.

CONCLUSIONS

This paper dealt with a two-dimensional stress analysis of butt adhesive
joints of dissimilar adherends subjected to external tensile loads. The
following results were obtained.
(1)Replacing adherends and an adhesive layer with finite strips ,a method
of analysis of stress distributions in butt adhesive joints , where
adherends are dissimilar, was demonstrated as an elastic contact problem
by using a two-dimensional theory of elasticity.
(2)Stress distributions at the interfaces were analyzed by the numerical
calculations. The effects of the ratios of Young's moduli and Poisson's
ratio among adherends and an adhesive,the thickness of the adhesive layer
and the external load distributions on the stress distribution at the
interfaces were clarified.
(3) An experiment was performed concerning the strains on the adherends.
The analytical result was fairly consistent with the experimental one.

REFERENCES

1. Suzuki,Y.,Three-dimensional Finite Element Analysis of Adhesive Scarf
   Joints of Steel Plates Loaded in Tension, Bull JSME,1984,27,No.231,

pp1836-1845.

2. Chen,D. and Cheng,S.,An Analysis of Adhesive-Bonded Single-Lap Joints,Jour. Appl. Mech.,1983,50,pp109-115.

3. Renton,W.J. and Vinson,J.R.,On the Behavior of Bonded Joints in Composite Material Structures,Eng. Fracture. Mech.,1975,7,pp41-60.

4. Kaplevatsky,Y. and Raevsky,V.J.,On the Theory of the Stress-Strain State in Adhesive Joints,Jour. Adhesion,1976,6,pp65-77.

5. Wah.T. ,Stress Distribution in a Bonded Anisotropic Lap Joint, Jour.Eng. Materials. Technology.,1973,pp174-181.

6. Sawa,T.,Nakano,Y. and Temma,K.,Stress Analysis of T-type Butt Adhesive Joint Subjected to External Bending Moments, Jour. Adhesion,1987, 24,pp1-15.

7. Nakano,Y.,Temma,K. and Sawa,T.,A Two-Dimensional Stress Analysis of Butt Adhesive Joints Having a Circular Hole in the Adhesive,JSME Int,Jour.,1988,31,No.3,pp507-513.

8. Alwar,R.S. and Nagaraja,Y.R.,Elastic Analysis of Adhesive Butt Joints, Jour. Adhesion,1976,7,pp279-287.

9. Adams,R.D. and Peppiet,N.A.,Stress Analysis of Adhesive-Bonded Lap Joints Jour. Strain Analysis,1974,9,No.3,pp185-196.

10.Sawa,T.,Iwata,A. and Ishikawa,H.,Two-dimensional Stress Analysis of Adhesive Butt Joints Subjected to Tensile Loads,Bull JSME,1986, 29,No.258,pp4037-4042.

11.Sawa,T.,Temma,K. and Tsunoda,Y.,Axisymmetric Stress Analysis of Adhesive Butt Joints of Dissimilar Solid Cylinders Subjected to External Tensile Loads, Int. Jour. Adhesion and Adhesives, 1989,9,No.3,pp161-169.

12.Hattori,T.,Sakata,S. and Watanabe,T.,A Stress Singularity Parameter Approach for Evaluating Adhesive and Fretting Strength,The Winter Annual Meeting of ASME, Advances in Adhesively Bonded Joints, 1988,pp43-50.

13.Bogy,D.B.,Two Edge-Bonded Elastic Wedges of Different Materials and Wedge Angles Under Surface Tractions,Jour. Appl. Mech.,1971,38, pp377-386.

449

A STRESS ANALYSIS OF A BAND-ADHESIVE BUTT JOINT SUBJECTED TO A TORSIONAL LOAD

Yuichi NAKANO
Department of Mechanical Engineering
Shonan Institute of Technology
1-1-25 Tsujido Nishikaigan Fujisawa, Kanagawa, 251 JAPAN

Toshiyuki SAWA
Department of Mechanical Engineering
Yamanashi University
4-3-11 Takeda Kofu, Yamanashi, 400 JAPAN

ABSTRACT

The stress distribution and the displacement are examined when an adhesive butt joint, in which two dissimilar tubular or solid shafts are bonded at the interface partially, are subjected to a torsional load. In the analysis, general representations of the stress and the displacement are given as a torsional problem. Then, the effects of the ratio of the shear modulus of an adhesive to that of the shafts, the bonded portion of band-adhesive, i.e., partially bonded and the thickness of the adhesive on the stress distribution at the interface are clarified by the numerical calculations. In the numerical calculation of the band-adhesive joint, it is seen that the shear stresses at the inner and the outer circumferences of the interface become steep so that the stresses are singular at the points.

INTRODUCTION

Recently, adhesive bonding is used widely in many manufacturing fields, i.e., from daily necessities to cars, electrical devices, and space instruments. At present, in joining machine parts or machine elements with an adhesive, it has not yet been established how bonding strength of the joints can be estimated and how the quality of bonding states can be measured, especially by non-destructive tests. Consequently, mechanical structures with adhesive joints are not always used with full confidence. Adhesive joints are often used as a supplemental aids in conventional joining such as bolted, riveted and welded joints. So fundamental data, such as the strength of various static and dynamic loadings, thermal strength, and durability have not been determined so that design of adhesive joints and an effective test method should be developed. Up to now, some experimental and theoretical studies have been carried out on the stress distributions of adhesive joints under such types of loadings as tension, bending and shearing [1]-[5]. On the other hand, studies on adhesive bonded joints subjected to torsional loads are few.

The purpose of this study is to obtain fundamental data for the strength design of adhesively joined transmitting shafts. An adhesive butt joint with two solid or tubular shafts are subjected to a torsional load and its strength is examined as a torsion problem. Moreover, the effects of the ratio of the shear modulus of the shafts to that of an adhesive, the thickness of the adhesive and the bonded portion of band-adhesive, i.e., partially bonded at the outer part and at the inner part of the interface, on the stress distribution at the interface are clarified by the numerical calculations.

THEORETICAL ANALYSIS

Figure 1 shows a model for analysis of a butt adhesive joint. Two semi-infinite dissimilar tubular shafts, called adherend(II) and adherend(III) hereinafter, are joined by adhesive(I) and the joint is subjected to a torsional load T at the ends. The inner and the outer diameters of the adherend(II) and (III) are denoted by 2c, 2d, 2e and 2f, the inner and the outer diameters of the adhesive band(I) are also denoted by 2a and 2b, respectively. Their shear moduli and Poisson's ratios are $G_2, \nu_2, G_3, \nu_3, G_1$ and ν_1 respectively. By using cylindrical co-ordinates(r, θ, z), the stresses and the displacement are expressed as follows[6];

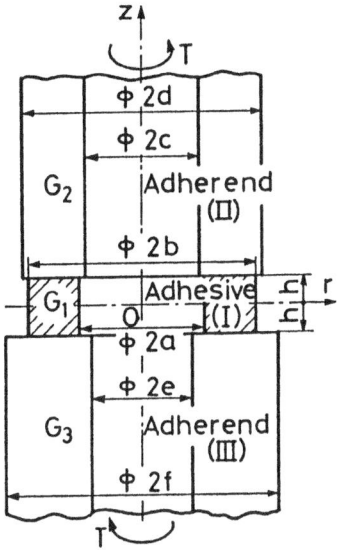

Figure 1. Model of adhesive joint of shafts subjected to a torsional load

$$\frac{\partial^2 V_\theta}{\partial r^2} + \frac{1}{r}\frac{\partial V_\theta}{\partial r} - \frac{1}{r^2}V_\theta + \frac{\partial^2 V_\theta}{\partial z^2} = 0 \tag{1}$$

$$\tau_{\theta z} = G\frac{\partial V_\theta}{\partial z} \tag{2}$$

$$\tau_{r\theta} = Gr\frac{\partial}{\partial r}\left(\frac{V_\theta}{r}\right) \tag{3}$$

In these equations, V_θ denotes the displacement in the circumferential direction, and $\tau_{\theta z}$ and $\tau_{r\theta}$ denote shear stresses. Boundary conditions of the joint are expressed by the following equations(4)-(13).

$$(V_{\theta 1})_{z=h} = (V_{\theta 2})_{z=h} \qquad (a \leq r \leq b) \tag{4}$$

$$(V_{\theta 1})_{z=-h} = (V_{\theta 3})_{z=-h} \qquad (a \leq r \leq b) \tag{5}$$

$$(\tau_{\theta z1})_{z=h} = (\tau_{\theta z2})_{z=h} \qquad (a \leq r \leq b) \tag{6}$$

$$(\tau_{\theta z1})_{z=-h} = (\tau_{\theta z2})_{z=-h} \qquad (a \leq r \leq b) \tag{7}$$

$$(\tau_{\theta z2})_{z=h} = 0 \qquad (c \leq r < a),(b < r \leq d) \tag{8}$$

$$(\tau_{\theta z3})_{z=-h} = 0 \qquad (e \leq r < a),(b < r \leq f) \tag{9}$$

$$(\tau_{r\theta 1})_{r=a} = (\tau_{r\theta 1})_{r=b} = 0 \qquad (|z| \leq h) \tag{10}$$

$$(\tau_{r\theta 2})_{r=c} = (\tau_{r\theta 2})_{r=d} = 0 \qquad (z > h) \tag{11}$$

$$(\tau_{r\theta 3})_{r=e} = (\tau_{r\theta 3})_{r=f} = 0 \qquad (z < -h) \tag{12}$$

Torsional load T=const. $\qquad (z \leq \infty) \tag{13}$

The subscripts 1,2 and 3 correspond to the adhesive(I), adherend(II) and adherend(III), respectively. Eq.(1) is solved by a method of separation of variables taking Eqs.(10)-(12) into consideration, and then the shear stresses $\tau_{\theta z1}$ and $\tau_{r\theta1}$ of the adhesive(I) are expressed from Eqs.(2) and (3) as follows[7];

$$G_1 V_{\theta1} = A_0' r + A_0 zr + \sum_{n=1}^{\infty} C_1(\alpha_n r)(A_n \sinh(\alpha_n z) + B_n \cosh(\alpha_n z)) \qquad (14$$

$$\tau_{\theta z1} = A_0 r + \sum_{n=1}^{\infty} \alpha_n C_1(\alpha_n r)(A_n \cosh(\alpha_n z) + B_n \sinh(\alpha_n z)) \qquad (15$$

$$\tau_{r\theta1} = -\sum_{n=1}^{\infty} \alpha_n C_2(\alpha_n r)(A_n \sinh(\alpha_n z) + B_n \cosh(\alpha_n z)) \qquad (16$$

where, $\quad C_j(\alpha_n r) = J_j(\alpha_n r) - J_2(\alpha_n a)Y_j(\alpha_n r)/Y_2(\alpha_n a)$
$\quad\quad J_j(\alpha_n r)$: Bessel function of 1st kind of order j
$\quad\quad Y_j(\alpha_n r)$: Bessel function of 2nd kind of order j
$\quad\quad \alpha_n$: the n-th positive root satisfying $C_2(\alpha_n b)=0$

$C_2(\alpha_n a)$ becomes 0 too, so that Eq.(10) is satisfied under these definitions. Similarly, Eqs.(17)-(19) for the adherend(II) and Eqs.(20)-(22) for the adherend(III), are expressed.

$$G_2 V_{\theta2} = D_0' r + D_0 zr + \sum_{m=1}^{\infty} D_m C_1(\beta_m r)\exp(-\beta_m z) \qquad (17$$

$$\tau_{\theta z2} = D_0 r - \sum_{m=1}^{\infty} D_m \beta_m C_1(\beta_m r)\exp(-\beta_m z) \qquad (18$$

$$\tau_{r\theta2} = -\sum_{m=1}^{\infty} D_m \beta_m C_2(\beta_m r)\exp(-\beta_m z) \qquad (19$$

$$G_3 V_{\theta3} = F_0' r + F_0 zr + \sum_{l=1}^{\infty} F_l C_1(\gamma_l r)\exp(-\gamma_l z) \qquad (20$$

$$\tau_{\theta z3} = F_0 r - \sum_{l=1}^{\infty} F_l \gamma_l C_1(\gamma_l r)\exp(-\gamma_l z) \qquad (21$$

$$\tau_{r\theta3} = -\sum_{l=1}^{\infty} F_l \gamma_l C_2(\gamma_l r)\exp(-\gamma_l z) \qquad (22$$

β_m and γ_l are the m-th and l-th positive roots satisfying $C_2(\beta_m d)=0$ and $C_2(\gamma_l f)=0$, respectively and $C_2(\beta_m c)$ and $C_2(\gamma_l e)$ become 0 too.

$A_0', A_0, \{A_n\}, \{B_n\}, D_0', D_0, \{D_m\}, F_0', F_0$ and $\{F_l\}$ in Eqs.(14)-(22) are undetermined coefficients and among them $A_0, \{A_n\}, \{B_n\}, D_0, \{D_m\}, F_0$ and $\{F_l\}$ are obtained from the boundary conditions expressed in Eqs.(4)-(13), where the shear stresses and the displacements at each interface are the same each other. Besides, A_0', D_0' and F_0' cannot be determined independently but they are concerned with the displacement V_θ only. So they are determined in this analysis such that the displacement becomes 0 at the outer circumference of the adherends. By using these determined coefficients, the stresses and the displacements of the joint are obtained.

NUMERICAL RESULTS

Band-adhesive joint of tubular shafts

Numerical calculations were carried out in the case where the two adherends (II) and (III) were of the same dimensions (c=e, d=f, c/d=0.5) but of the different materials($G_2 < G_3$). To investigate the effect of the position of the adhesive band on the stress distributions and the displacements, two types of adhesive joints were examined numerically, one bonded at the outer interface, called type(A) and the other bonded at the inner interface, called type(B), also the bonded area of each type was the same. Moreover, the effect of the ratio of the shear modulus of the adhesive to that of the adherend and the effect of the thickness of the adhesive were examined in the case of the two types mentioned above. In the calculations, the number of terms, N, of the series was taken as 200.

Figure 2 shows the effects of the ratios of the shear modulus of the adhesive to that of the adherends on the shear stresses(a) and the displacements(b) at each interface between the adhesive and the adherend, in the case of type(A). In Figure 2, the shear stresses and the displacements are normalized using $\tau_N = 2Tf/\pi(f^4 - e^4)$ and $V_N = 2Tf^2/\pi G_3 (f^4 - e^4)$.

Also, these effects in the case of type(B) are shown in Figure 3. From these figures, it is seen that the maximum stresses become steep in the case when the ratio of the shear modulus $G_1/G_2 = 1$, at the inner circumference of the interface, i.e. r=a, in type(A) and at the outer circumference of the interface, i.e. r=b, in type(B). When the ratio of the shear modulus is small, i.e., $G_1/G_3 = 0.1$, the shear stresses distribute linearly according to the radial position of the interface. Also it is seen that the maximum displacements become larger when $G_1/G_2 = 1$, in both cases.

454

Moreover, comparing the results of type(A) with those of type(B), th maximum stresses and the displacements become smaller in type(A), the join where tubular shafts are bonded at the outer interface of the adherends.

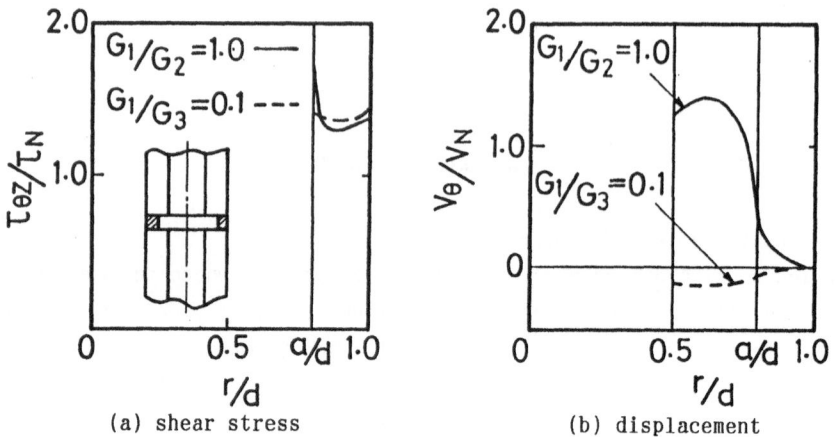

(a) shear stress   (b) displacement

Figure 2. Effects of the ratios of shear modulus of adherends to that of an adhesive on the shear stress distribution and on the displacement type(A): $(G_1/G_2=1$, $G_1/G_3=0.1$, $2h/d=0.1$, $z=\pm h$, $a=\sqrt{(c^2+d^2)}/2$, $c=e=0.5d$, $b=d=f$

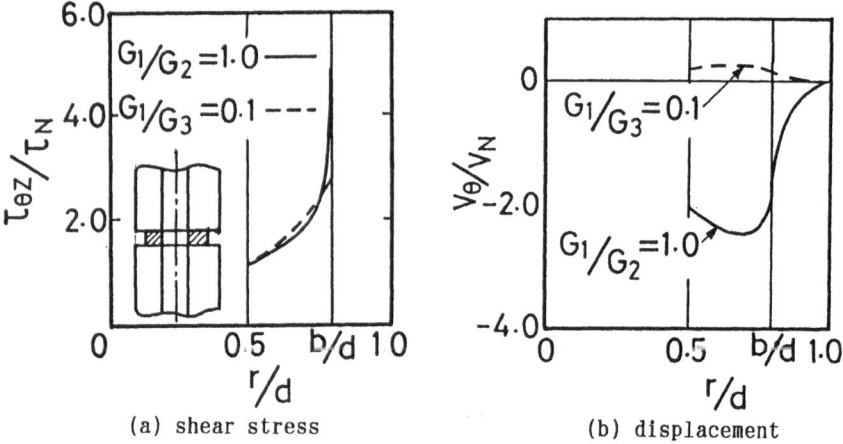

(a) shear stress   (b) displacement

Figure 3. Effects of the ratios of shear modulus of adhesive to that of adherend on the shear stress distribution and on the displacement type(B): $(G_1/G_2=1$, $G_1/G_3=0.1$, $2h/d=0.1$, $z=\pm h$, $b=\sqrt{(c^2+d^2)}/2$, $a=c=e=0.5d$, $d=f)$

Joints of tubular vs. solid shafts

As a special case, numerical calculations were also carried out for the two types of joints of the same material$(G_2=G_3)$. One is a joint of a tubular and a solid shafts, and the other is a joint of two solid shafts.

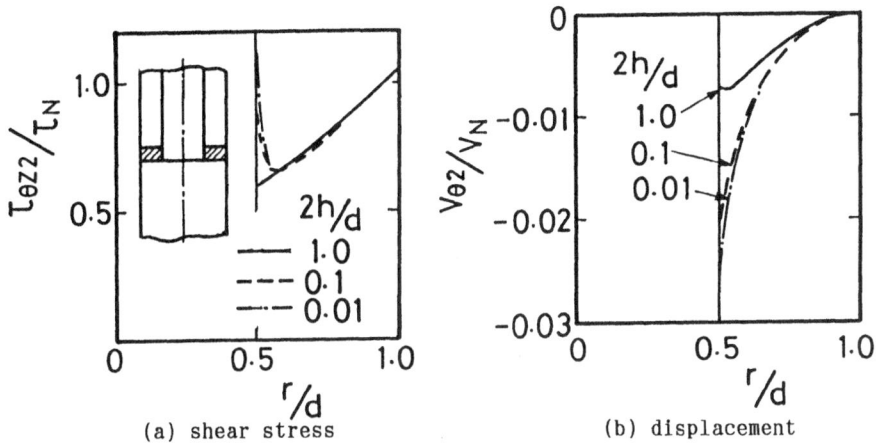

(a) shear stress (b) displacement

Figure 4. Effects of the thickness of the adhesive at a joint of a tubular and a solid shaft
$(G_1=G_2=G_3$, $z=h$, $a=c=0.5d$, $e=0$, $b=d=f)$

Figure 4 shows the effect of the thickness of the adhesive on the stress distributions at each interface of the joint of a tubular (adherend(II)) and a solid shaft (adherend(III)). At the joint, the outer diameters of both the adherends are the same($d=f$) and the inner diameter of the adherend(II) is half of the outer diameter of the adherends($c=0.5d$).

From these figures, it is seen that the the shear stresses and the displacements become steeply large with a decrease of the thickness $2h$, at the inner circumference of the interface between the adhesive and the adherend(II). Also it is clarified numerically that the effect of the thickness on the shear stress distribution and on the displacement at the interface between the adhesive and the adherend(III) is negligibly small.

Joints of solid vs. solid shafts

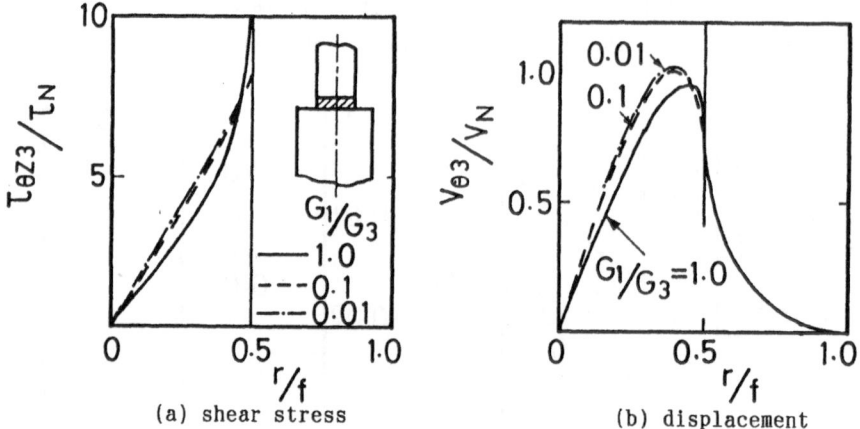

(a) shear stress   (b) displacement

Figure 5. Effects of the ratio of shear modulus of adhesive to that of an
adherend at a joint of two solid shafts
$(G_2=G_3,\ 2h/d=0.1,\ z=-h,\ a=c=e=0,\ b=d=0.5f)$

Figure 5 shows the effect of the ratio of the shear modulus of an
adhesive to that of an adherend on the shear stress distributions and on
the displacements at the interface between the adhesive and the
adherend(III) of larger diameter. At the joint of two solid shafts, outer
diameter of the adherend(II) is half of that of the adherend(III) (d=0.5f).

From the figures, it is seen that the shear stress becomes singular at
the edge of the interface between the adhesive and the adherend(III), i.e.
r=a, when $G_1/G_3=1$. On the other hand, the maximum displacements decreases
with an increase of the ratio of the shear modulus.

CONCLUSIONS

This paper dealt with a stress analysis of a butt adhesive joint subjected
to a torsional load and some mechanical characteristics of the joint were
examined from theoretical analyses.

The results obtained are as follows;

1)In the case of adhesive butt joint of tubular shafts, the singularity of the stress increases at the inner and the outer circumferences of the interface with an increase in the ratio of the shear modulus of an adhesive to that of the shafts or with a decrease of the thickness of the adhesive.

2)The torsional strength of the band-adhesive joint in which tubular shafts are bonded at the outer interface of the adherends is larger than that for tubular shafts bonded at the inner interface.

3)In the case of solid shafts of different size, the shear stress becomes singular at the edge of the interface between the adhesive and the adherend of large diameter, with an increase of the ratio of the shear modulus of the adhesive to that of the shafts.

REFERENCES

1. Erdogan,F. and Ratwani,M., Stress distribution in bonded joints. J. Composite Mater., 1971, 5, 378-393.
2. Sen,J.K. and Jones,R.M., Stresses in double-lap joints bonded with a viscoelastic adhesive. Part 1: Theory and experimental corroboration. AIAA Journal, 1980, 18, 1237-1244.
3. Delale,F. and Erdogan,F., Viscoelastic analysis of adhesively bonded joints. J. Appl. Mech., 1981, 48, 331-338.
4. Rao,N.B., Rao,Y,V,K,S., and Yadagiri,S., Analysis of composite bonded joints. Fibre Sci. Tech., 1982, 17, 77-90.
5. Adams,R.D., Coppendale,J. and Peppiatt,N.A., Stress analysis of axisymmetric butt joints loaded in torsion and tension. J. Strain Analysis, 1978, 13, 1-10.
6. Sneddon,I.N., Fourier Transforms, McGraw-Hill, New York, 1951, pp.501.
7. Sawa,T., Nakano,Y. and Temma,K., A stress analysis of butt adhesive joints under torsional loads. J. Adhesion, 1987, 24, 245-258.

STRENGTH EVALUATION OF SCARF AND STEPPED LAP JOINTS BONDED WITH ADHESIVE RESIN

TOSHIO SUGIBAYASHI
KIYOMI MORI
Faculty of Engineering,
Takushoku University
815 Tate-machi, Hachioji, Tokyo 193, Japan

HIDEKI KYOGOKU
Eastern Industrial Research Institute
of Hiroshima Prefecture
3-232-6 Higashifukatsu-cho, Fukuyama, Hiroshima 721, Japan

KOZO IKEGAMI
Research Laboratory of Precision Machinery and Electronics,
Tokyo Institute of Technology
4259 Nagatsuta-cho, Midori-ku, Yokohama 227, Japan

ABSTRACT

The deformation and strength of adhesively bonded scarf and stepped lap joints were investigated both analytically and experimentally. The joint consisted of adherends of carbon steel and adhesive of epoxy resin. The strain distributions of both joints under tensile loadings were found out by the finite element method, and their validity were confirmed by the experiments. The strength of both joints was predicted by applying the strength criterion corresponding to the adherends, adhesive layer and adhering interfaces on the stress distributions in the both joints. The difference in the initial failure strength and the final fracture strength between scarf and stepped lap joints was examined by the obtained analytical and experimental results. It was found that the initial failure strength of scarf joints was better than that of stepped lap joints because of the difference in stress distribution of both joints.

INTRODUCTION

Adhesively bonded joints have many advantages over mechanical fastenings for jointing structural materials and mechanical elements, and bonding electric or electrical parts, since they do not decrease the strength of adherends and are scarecely increasing the weight of the adhesive. There are many kinds of joints, the strength of which are affected greatly by the shape of adherend. A butt joint and a single lap joint have great stress concentration under loadings on the adhesive layer at the end of its butt section and its lap section. To improve the joint strength by decreasing the stress concentration, the stepped lap joint having adherends with both butt and lap sections and the scarf joint with tapered adherends are proposed. Concerning stepped lap joints, Erdogan et al.[1] and few other researchers [2-3] have reported the results of stress analysis. Lubkin [4] and many researchers [3-7] have reported on their stress analysis of scarf joints. The research on stress analysis by means of the finite element method has been reported by Suzuki et al.[8] and other researchers.

In the present paper, the deformation and strength of stepped lap joints and scarf joints are investigated both analytically and experimentally. Furthermore, the effect of the difference in deformed states between those joints on their joint strength are examined.

STRESS ANALYSIS

A Model of Analysis

Scarf Joint: The coordinates and dimensions for the scarf joints are shown in Fig.1. The longitudinal and thickness directions correspond to the x and z directions, and the parallel and vertical directions to the adhesive layer to the s and n directions. The joint length, the lengths of left and right adherends and the overlap length are represented by notations 1, 1_1 and 1_2, respectively. The thicknesses of adherends and the adhesive layer are represented by notations t_1 and t_2, respectively. The acute angle of the adherends is shown by the scarf angle θ. The position in the s direction is shown by the non-dimensional $\overline{S}$ ($\overline{S}=s\cdot\sin /t_1$). $\overline{S}=-0.5$, 0 and 0.5 indicate the upper side surface, central position and lower side surface of the joint, respectively. Carbon steel is used as adherends, and epoxy resin as adhesive. The elastic constants of these materials are given in Table 1.

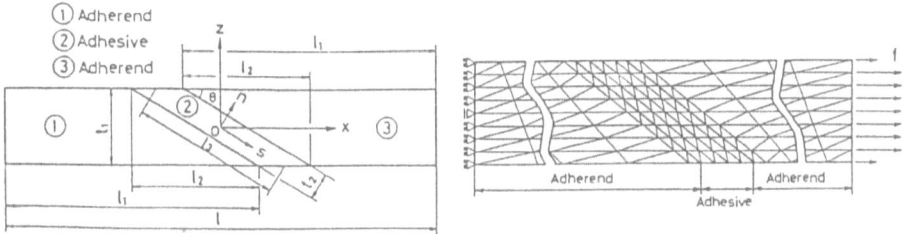

Fig.1. Coordinates and dimensions   Fig.2. Finite element configuration
        of a scarf joint.   and boundary condision of
                                                   a scarf joint.

TABLE 1
Material constants.

	Adherend (Carbon steel)	Adhesive (Epoxy resin)
Young's modulus GPa	206	3.33
Poisson's ratio	0.33	0.34

The stress analysis is conducted with elastic finite element method on the assumption of a plane strain condition. The finite element configuration and boundary conditions are shown in Fig.2.

Stepped Lap Joint: The coordinates and dimensions for the stepped lap joints are shown in Fig.3. This joint has both butt and lap sections, and the number of steps N_s is defined by that of butt sections. The joint length, the lengths of left and right adherends, and the overlap length are represented by notations l, l_1, l_2 and l_3, respectively. The thickness of the adherends and the steps on the butt sections are represented by notatons t_1 and t_3, respectively. The thickness of adhesive layer on both butt and lap sections is shown by t_2. Each step is S_1, S_2, S_3, S_4 and S_5 downward from the upper right of stepped lap section, and the butt section is called S_{iB}(i=1,$\cdots$,5) and the lap section S_{iL}(i=1, $\cdots$,4). The positions in the x and z directons is shown by the non-dimensional $\bar{X}$ ($\bar{X}=x/l_3$) and $\bar{Z}$ ($\bar{Z}=z/t_3$), respectively, and the overlap length on the lap section by $\bar{L}o$ ($\bar{L}o=l_2/t_1$). The stress analysis is carried out by applying elastic finite element method on the assumption of a plane strain condition. The finite element configuration and boundary conditions are shown in Fig.4.

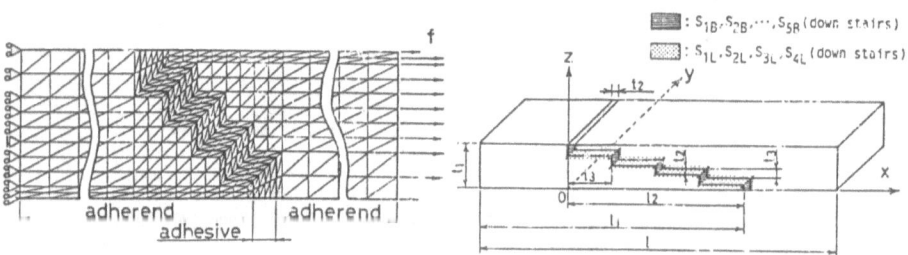

Fig.3. Coordinates and dimensions of a stepped lap joint.

Fig.4. Finite element configuration and boundary condition of a stepped lap joint.

Strain Distribution

Scarf Joint: The stress analysis in a scarf joint of carbon steel adherends having the dimensions of l=178mm, t_1=10mm and 25mm wide was carried out, the overlap length l_2 being varied. Figure 5 represents the strain distributions

along the s direction in the adherends, to which an average
stress σ_a of 10 MPa is applied, near the adhesive layer
(n≑1.0mm). The average stress σ_a is defined as a load divided
by the cross sectional area of the joint. The ordinate and
abscissa show, respectively, the strain and non-dimensional
position $\overline{S}$ in the s direction. The strain ε_x changes slightly
in the vicinity of $\overline{S}=-0.5$, but the strains ε_z and γ_{xz} are
constant over almost the whole range of $\overline{S}$. Thus, the strains of
the scarf joint scarcely change. The experimental results of
strain distributions in the joints having carbon steel
adherends under the average stress σ_a of 10MPa are shown in
Fig.5. They are in good agreement with the analytical results.
This indicates that the boundary conditions used in this
analysis are validated.

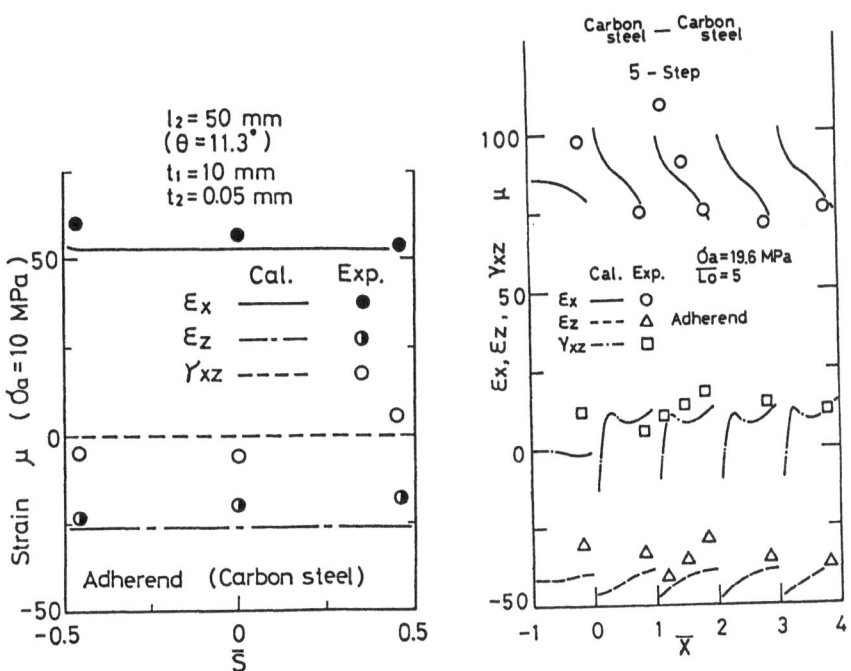

Fig.5. Distribution of strains in   Fig.6. Distribution of strains in
       adherends of the scarf joint.   adherends of the stepped
                                                    lap joint.

Stepped Lap Joint: The stress analysis of a stepped lap
joint with t_1=10mm, t_2=0.05mm, t_3=2mm, $\overline{L}_o$=5 and 25 mm wide was
caried out. Figure 6 shows the strain distributions along the x
direction near the overlap of the lower adherend, to which an
average tensile stress σ_a of 19.6 MPa is applied. These
strains of positions of $\overline{Z}$=4.74, 3.74, 2.74, 1.74 and 0.74 are
shown, respectively, at each step of $\overline{X}$=-1 to 4 along the

overlap range. These strain distributions of $\overline{X}$=0.0, 1.0, 2.0, 3.0 and 4.0 are distinctive because these positions are edges of the steps. The greater the non-dimensional position $\overline{X}$ is on each step, the smaller the strain ε_x is, but the greater the strain ε_z is. The strain γ_{xz} is negative value at the corner of the step and becomes positive at the middle. This strain distribution is a similar on each step. The experimental results in the adherend near the adhering interface under the average tensile stress σ_a of 19.6 MPa are shown in Fig.6. They coincides well with the analytical results. This gives the good indication that the boundary conditions used in this analysis are validated.

STRENGTH PREDICTION OF JOINTS

Strength Criterion

The authors have previously proposed [9] the adhesive strength criterion under combined states of stress that is determined experimentally by a butt joint of two thin-walled tubes. This criterion is applyed to the adhering interface. The strength of each part of the joint is calculated by applying the adhesive strength criterion and von Mises criterion to the adhering interface and the adherend or the adhesive layer, respectively. The joint strength is obtained by the minimum value among the calculated results.
The von Mises criterion is applied to the adherends and adhesive layer. They are represented as follows:
For the adherend,

$$F_1=(\sigma_z^2-\sigma_x\sigma_z+\sigma_z^2+3\tau_{zz}^2)^{1/2}/\sigma_{01}=1 \tag{1}$$

For the adhesive layer,

$$F_2=(\sigma_z^2-\sigma_x\sigma_z+\sigma_z^2+3\tau_{zz}^2)^{1/2}/\sigma_{02}=1 \tag{2}$$

where σ_{01} and σ_{02} correspond, respectively, to the yield strength of adherend and the fracture strength of adhesive resin. For the material in this research, σ_{01}=343MPa and σ_{02}=64MPa.
The adhesive strength criterion is applied to the adhering interface and represented as follows:
For the scarf joint,

$$F_3=|\sigma_n/\sigma_{04}|^m+|\tau_s/\tau_{01}|^m=1 \tag{3}$$

where σ_n and τ_s denote, respectively, the tensile and shear stresses on the adhering interface,

$$\sigma_n=\frac{1}{2}(\sigma_x+\sigma_z)+\frac{1}{2}(\sigma_x-\sigma_z)\cos 2\theta+\tau_{xz}\sin 2\theta$$

$$\tau_s=-\frac{1}{2}(\sigma_x-\sigma_z)\sin 2\theta+\tau_{xz}\cos 2\theta$$

For the stepped lap joint,

$$F_3=|\sigma_z/\sigma_{04}|^m+|\tau_{xz}/\tau_{01}|^m=1 \tag{4}$$

In equations (3) and (4), the constants σ_{04}, τ_{01} and m are obtained as follows; for the adhesive thickness 0.05mm σ_{40}=31.5 MPa, τ_{01}=30.8 MPa and m=8.66.

Analytical and Experimental Results

Scarf Joint: The strength of the scarf joint is calculated on the basis of the strength criterion, equations (1), (2) and (3). The strength distributions of the joint having adherends of carbon steel adherends with t_1=10mm and l_2=50mm are calculated by applying this criterion to the stress distributions in the corresponding parts of the joints, and these are illustrated in Fig.7. In this figure, σ_a indicates the evaluated stress, when the stress in each part of the joint satisfies the strength criterion F_i=1 (i=1,2,3). It is found from Fig.7 that each strength distribution of adherend, adhesive layer and adhering interface is approximately constant but that of both the adhesive and interfaces are smaller. This fact indicates that the onset of initial crackings occurs at those positions.

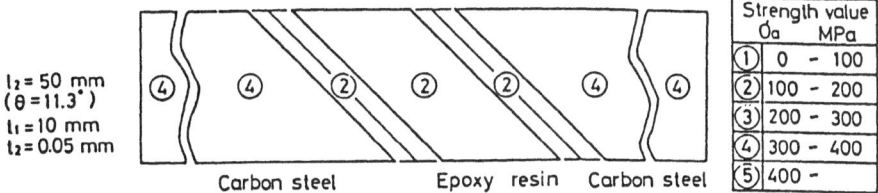

Fig.7. Strength distributions in the scarf joint.

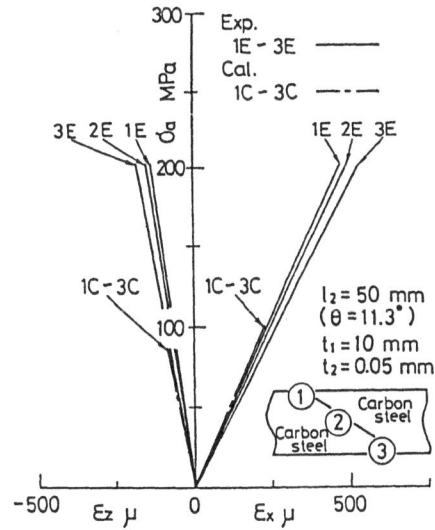

Fig.8. Variations of strains in the scarf joint with an increasing tensile load.

The strains ε_x and ε_z at three positions of the joint illustrated in Fig.8 are measured with increasing tensile loadings on the joint until it fractures. Figure 8 shows the variations of ε_x and ε_z with increasing tensile loadings. The ordinate and abscissa show the average stress σ_a and the strain, respectively. Both strains ε_x and ε_z vary linearly. This fact suggests that the scarf joint having adherends of the same metal fractures in a brittle manner.

Typical fractured surfaces of the joint ($t_1=10mm$) are given in Fig.9. In the case of the butt joint ($l_2=0mm$), the cohesive fracture exists over the whole fractured surface. But in the cases of a non-zero length of l_2, a slight interfacial fracture exists at edges of adhesive layer, and a cohesive one exists on most of the fractured surface. This fact coincides with the calculated result, as shown in Fig. 7, that the adhering interface tends to fracture.

Fig.9. Fractured surface of the scarf joint

Stepped Lap Joint: The strengths of both interfaces and adhesive layer in a stepped lap joint depicted in Fig.6 are given in Table 2. The strengths of interfaces on butt sections are smaller in this joint, particularly that of position I_2 is minimum.

TABLE 2
Predicted strength values at
each position in figure 10.

i	Interface I_i σ_c MPa	Adhesive A_i σ_c MPa
1	37.6	84.3
2	29.0	104.0
3	45.0	112.0
4	30.4	115.0
5	46.4	114.2
6	30.7	116.3
7	46.0	113.1
8	30.4	114.8
9	37.1	101.6
10	35.2	83.9

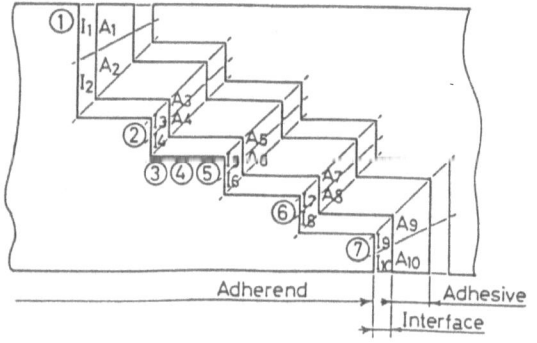

Fig. 10. Positions of strain gauges and
parts for strength values in
Table 2.

The strain ε_x measured with increasing tensile loadings
at the positions of 1 to 7 on the joint is shown in Fig.10. The
ordinate in this figure give the average stress σ_a. The
strain ε_x at the positions 1, 5 and 7 vary linearly with an
increasing tensile stress σ_a, whereas beyond the region of
the tensile stress $\sigma_a=30$ MPa it decreases. Further, with
increasing tensile stress the strain ε_x at the position of 2
varies non-linearly, and finally the joint fractures at $\sigma_a=86$
MPa. This fact suggests that the stepped lap joint fails
initialy at the positions 1, 5 and 7, the crack propagates with
increasing tensile stresses and finally it fractures.

To compare the analytical results with experimental ones,
both results are shown in Fig.11. Since the strength of
adhesive layer at position 1 is smaller, the strain measured
until it fractures and the calculated strength of adhesive are
shown in Fig.11(a).

The fractured surface of the joint is shown in Fig.12. It
is clear from this figure that the joint fractures at the
interface between the adherend and the adhesive layer.

Comparison of Scarf Joint with Stepped Lap Joint

The variations of the joint strength against overlap
length are illustrated in Fig.13. The ordinate shows the
average failure stress σ_c which is given by the failure load
divided by the cross sectional area of the joint. The abscissa
shows the non-dimensional length, $\overline{Ls}$, i.e. the overlap length
l_2 divided by the adherend thickness t_1. It is found that σ_c
changes linearly above $\overline{Ls}=1.0$. The experimental results of the
joint strength are also illustrated in Fig.13. Since the
experimental results agree well with the analytical ones. The
experimental result of the initial failure strength
approximately coincides with that of the final fracture,

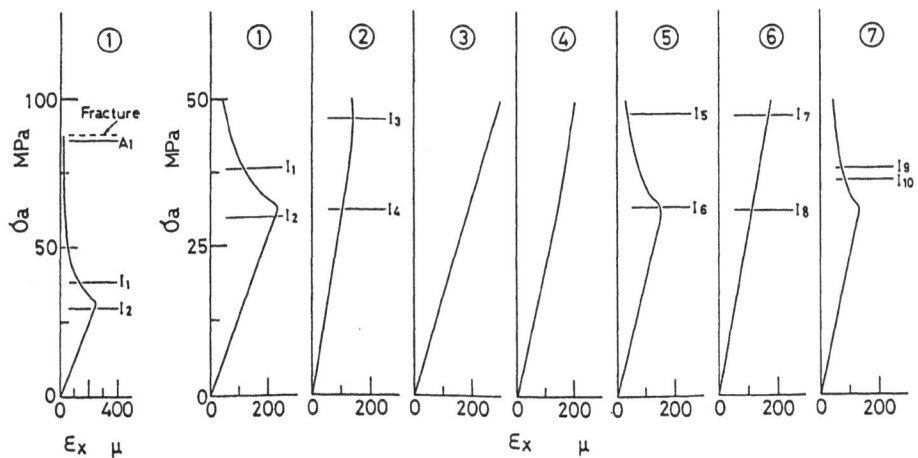

(a) Variation of ε_x (b) Variations of ε_x corresponding
    until fracture.   parts of the joint.
Fig.11. Variations of strain ε_x corresponding to parts of the
stepped lap joint.

Fig.12. Fractured surface of the stepped lap joint.

because the initial failure and the final fracture in a scarf joint takes place almost simultaneously.

The calculated initial failure strength against the number of steps N_s are shown in Fig.14. The following three results among the initial failure strength shown in Table 2 are chosen. Ahs and I_n are the calculated strength of edge of adhesive layer and interface, i.e. the element of the upper edge of left adherend on the butt section in Fig.3. Ahr_1 is the calculated strength of weak part of adherend at the corner of the butt section S_{1B} and the lap section S_{1L}. The strength is in order of the edge of interface I_n, the edge of adhesive layer Ahs and the corner of adherend Ahr_1. Since Ahr_1, Ahs and I_n scarcely increased with increasing number of steps, they are hardly affected by them.

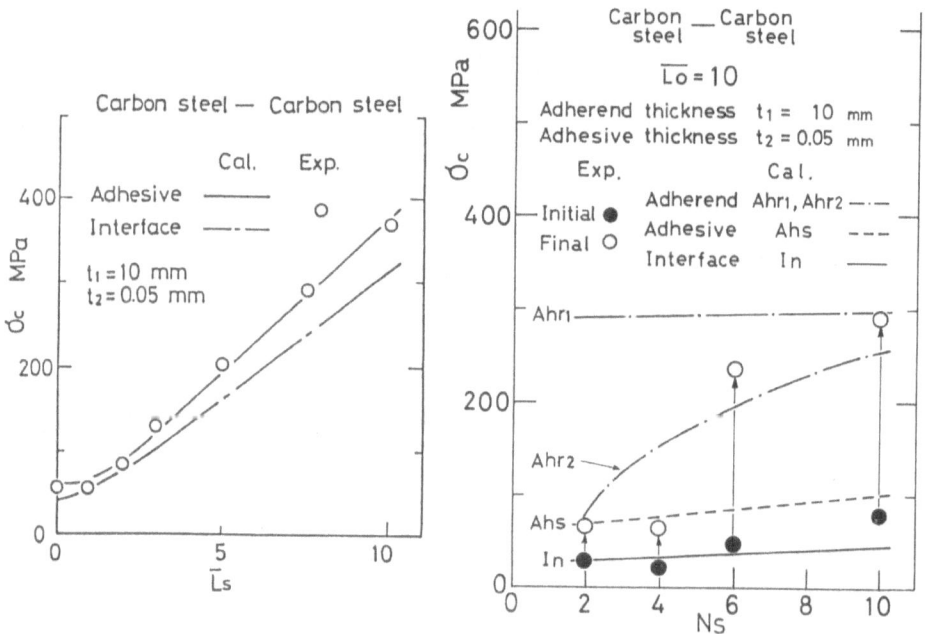

Fig. 14. Variations of strength stepped lap joint as a function of number of steps.

Fig. 13. Variations of strength of the scarf joint as a function of adhesive length.

The calculated final fracture strength Ahr_2 is shown in Fig.14. This is the calculated strength of element on the right edge of lap section S_{1L} obtained by applying the stress redistribution after the initial failure calculated on the assumption that it takes place on the lap section to the strength criterion (1). The strength of adherend Ahr_2 after initial failure is near that of Ahs and I_n in the case of $N_s=2$, but it increases with increasing number of steps. In the case of $N_s=10$, it approaches the strength Ahr_1 of the corner in the adherend. This fact suggests that the initial failure takes place at the edge of interface, and its strength is not affected by number of steps, while the final fracture strength is affected by them.

CONCLUSIONS

The deformation and strength of adhesively bonded scarf and stepped lap joints were investigated both analytically and experimentally. The results are as follows:
(1) Comparing the strain distributions of scarf joints with stepped lap joints, it was found that the former varies slightly, while the latter changes greatly at the edge of each lap section.
(2) Comparing the initial failure strength of scarf joints with stepped lap joints, the former was found to increase linearly with increasing overlap length, while the latter increases slightly with increasing number of steps. Thus, the initial failure strength of the scarf joint is better than that of the stepped lap joint.
(3) Comparing the final fracture strength of scarf joints with stepped lap joints, the former was found to agree well with the initial failure strength, while the latter increases greatly with increasing number of steps.

REFERENCES

1. Erdogan, F. and Ratwani, M., Stress distribution in bonded joints. J. Composite Materials, 1971, 5, pp. 378-93.
2. Cushman, J.B., McCleskey, S.F. and Ward, S.H., Test and analysis of Celion 3000/PMR-15 graphite/polyimide bonded composite joints. NASA CR-3602, 1983.
3. Hart-Smith, L.J., Adhesive-bonded scarf and stepped lap joints. NASA CR-112237, January 1973.
4. Hart-Smith, L.J., Analysis and design of advanced composite bonded joints. NASA CR-2218, April 1974.
5. Lubkin, J.L., A theory of adhesive scarf joint. J. Appl. Mech., 1957, 24, pp. 2555-60.
6. Wah, T., Plane stress analysis of a scarf joint. Int. J. Solids Structures, 1976, 12, pp. 491-500.
7. Wah, T., The adhesive scarf joint in pure bending. Int. J. Mech. Sci., 1976, 18, pp. 223-28.
8. Suzuki, Y., Adhesive tensile strengths of scarf and butt joints of steel plates. Bull. Japan Soc. Mech. Eng., 1985, 28, pp. 2575-84.
9. Sugibayashi, T. and Ikegami, K., Deformation and strength of single lap joints by tensile-shear load. Trans. Japan Soc.Mech. Eng., 1984, 50 C, pp. 17-24.

A NEW CONCEPT IN MECHANICAL EVALUATION OF ADHESIVE TEST SPECIMENS: STIFF VERSUS FLEXIBLE ADHERENDS

VICTOR WEISSBERG* AND MIRCEA ARCAN**

ABSTRACT

An analytical - numerical study of the influence of adherend stiffness on adhesive bonded joints and of the related test specimens provided an insight into some of their fundamental properties.

The stress field in the vicinity of singular locations, in stiff and flexible adherend joints was investigated. Linear elastic finite element analysis was used to compare the different systems. Parametric finite element analysis allowed to develop closed form expressions for the stress distribution, in the vicinity of the singular points.

A general relationship between the stress intensity factor SIF and the stress concentration factor SCF, unique for adhesively bonded joints was developed. The SIF/SCF relationship is a system of data characterizing a wide range of stiff to flexible adherend joints; it also provided an unified approach to strength and fracture of adhesive bonded joints.

The subsequent definition of a physical factor, the Relative Adherend Stiffness, which was derived by a fracture mechanics approach, permitted a more general characterization of adhesive bonded joints.

A comparative test program comprising both stiff adherend (butterfly system) and flexible adherend (DCB, ENF, MMF) test specimens was performed. Analysis of test results using the SIF/SCF relationship, provided a compatible systematic characterization of the different testing specimens.

* Tel Aviv University, Tel Aviv 69978, Israel
** ILLINOIS INSTITUTE OF TECHNOLOGY, CHICAGO, IL., 60616, on sabbatical leave from Tel Aviv University

INTRODUCTION

Many researchers (1,2) agree that the key factor which governs the behaviour of adhesive bonded joints, is the relative stiffness of the adherends.
It should be pointed out that an apparently stiff adherend does not necessarily provide a stiff adherend joint.
Recently, it has been shown by the Authors (3) that the physical parameter which characterises the test specimen is the derivative of the compliance with respect to the crack length. This parameter, F has been called by the Authors: The Relative Adherend Compliance.

$$F = \frac{dc}{da} \bigg/ \frac{dc}{da} \bigg|_S \qquad\qquad [1]$$

where a - the crack length

$\frac{dc}{da}$ compliance change rate

$\frac{dc}{da}\bigg|_S$ compliance change rate of an infinitely stiff adherend joint

It was shown that the relative adherend compliance F is directly related to the stress concentration factor (SCF) in the adhesive layer :

$$F = C^2 \qquad\qquad [2]$$

where C is the stress concentration factor
It is obvious that for infinitely stiff adherend joint, F=C=1.

A general relationship between the stress intensity factor SIF and the stress concentration factor SCF, unique for adhesively bonded joints, was developed, References (4) and (5).

$$K_I = K_{IS}\, C_t \qquad\qquad [3]$$

and

$$K_I = K_{IIS}\, C_s \qquad\qquad [4]$$

where $K_{I/II}$ are SIFs, $K_{IS/IIS}$ are the SIFs of infinitely stiff adherend joints and $C_{t/s}$ are the SCFs in tension and shear respectively.

It was shown by numerical analysis that the SIF/SCF relationships are accurate for adherend/adhesive elastic modulus ratios higher than 20.

The ideal infinitely stiff adherend joint concept has proved to be useful for a simplification of the analysis.

A stiff adherend (butterfly) test specimen was analyzed in detail and shown to behave almost as an ideal infinitely stiff adherend joint.

A representative flexible test specimen was analyzed, and a comparative stiff versus flexible adherend test program was performed.

THE STIFF ADHEREND (BUTTERFLY) TEST SPECIMEN

An adhesively bonded joint made of finite stiffness adherend material, tailored so that the parameter F=1, and consequently the joint will behave like an ideal infinitely stiff one.

The stiff adherend (Butterfly) test specimen and its circular loading system are presented in Figures 1 and 2. Dimensions are in millimeters. This is a versatile stiff adherend test specimen proposed by ARCAN et al [6+9]. The versatility of the specimen consists in the fact that by loading the specimen in a rotated position, a mixed mode $(M_I + M_{II})$ may be induced. The stiff adherend specimen was attached with two pins on each side in order to make it statically determinate; in the finite element model, boundary conditions were set accordingly.

The stiff adherend test specimen was analysed in Mode I and Mode II loading, for a various crack lengths.

It can be seen that the deep symmetrical cracked specimen $(\frac{a}{w} = .7)$

is practically an infinitely stiff adherend joint in tension as well as in shear loading (Figure 3).

FLEXIBLE ADHEREND TEST SPECIMEN

The end notch flexural test specimen (ENF) was chosen as the representative flexible adherend test specimen.

The ENF specimen is essentially a three point flexural specimen with a crack, as shown schematically in Figure 4.

This simple yet efficient test specimen induces a pure Mode II condition as the crack tip.

RUSSELL AND STREET (10) obtained a closed form solution for the strain energy release rate (G_{II}) of the homogeneous ENF test specimen.
To this expression a correction factor C_{ENF} due to the presence of the adhesive layer was added by the authors. The correction factor was obtained numerically by finite element analysis, see Figure 5.

$$G_{II} = \frac{9}{16} (\frac{P}{b})^2 \frac{a^2 C_{ENF}}{E_A (t+h)^3} \qquad [5]$$

471

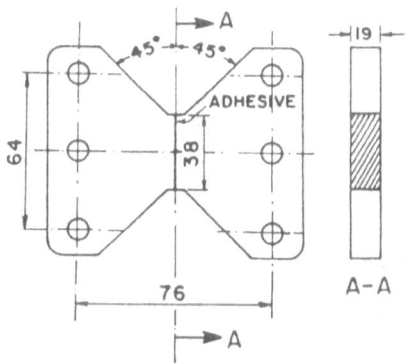

FIGURE 1   STIFF ADHEREND SPECIMEN

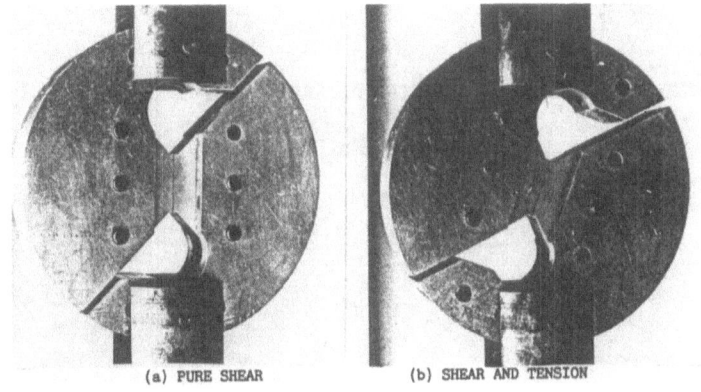

(a) PURE SHEAR   (b) SHEAR AND TENSION

FIGURE 2   ASSEMBLY VIEW OF TESTING SYSTEM

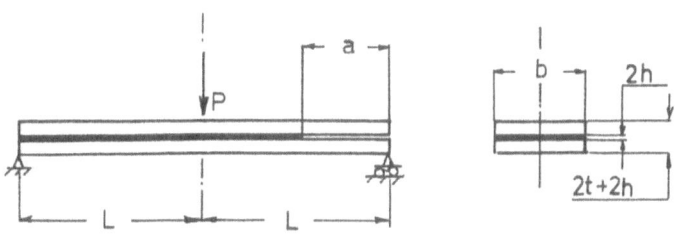

FIGURE 4   END NOTCHED FLEXURAL (ENF) TEST SPECIMEN

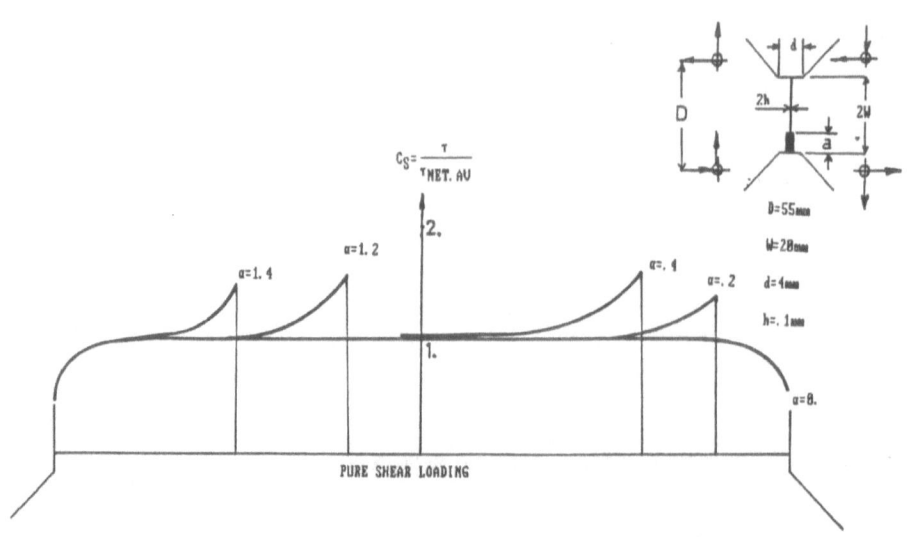

FIGURE 3   STRESS DISTRIBUTION ON STIFF ADHEREND TEST SPECIMEN

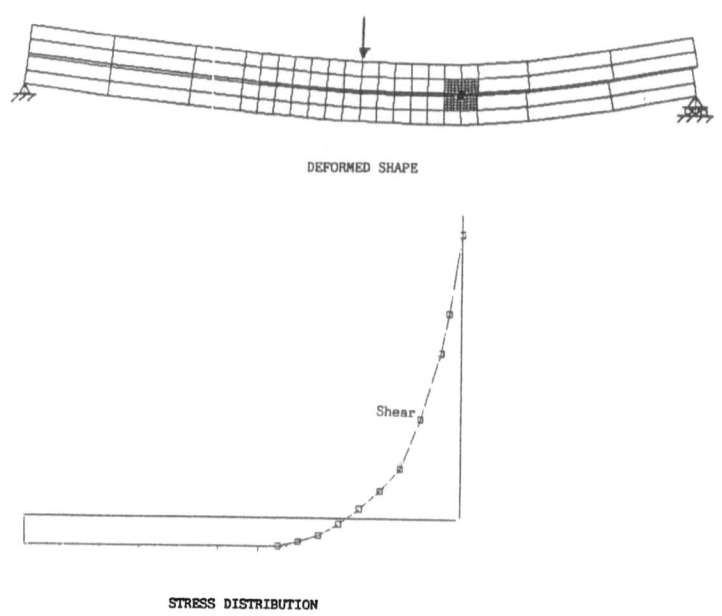

FIGURE 5   END NOTCHED FLEXURAL (ENF), FINITE ELEMENT ANALYSIS

where a - The crack length
      b - Specimen width
      t - Adherend thickness
      h - Adhesive thickness
      E_A - Adherend modulus
      C_{ENF}^A - The correction factor

For adhesive adherend elastic modulus ratios higher than 20. C_{ENF} = 1.5

The maximum shear stress in the adhesive layer was obtained by means of the SIF/SCF relationship :

$$G_{II} = \frac{K_{II}^2(1-\nu_a^2)}{E_a} \quad \text{and} \quad G_{IIs} = \frac{K_{IIs}^2(1-\nu_a^2)}{E_a} \qquad [6]$$

where E_a and ν_a are the adhesive elastic modulus and Poisson ratio respectively.

Substituting in the SIF/SCF relationship [4], the maximum shear stress was obtained :

$$\tau_{max} = \sqrt{\frac{G_{II}}{G_{IIs}}} \qquad [7]$$

Substituting G_{II} and G_{IIs} in [7] a closed form expression for τ_{max} is obtained :

$$\tau_{max} = .56 \, P \, \frac{a}{b} \, \sqrt{\frac{E_a}{E_A(t+h)^3 h}} \qquad [8]$$

TEST RESULTS AND DISCUSSION

Typical load deflection diagrams were obtained with the stiff adherend (Butterfly) specimen, and with flexible ENF test specimens, are shown in Figure 6.

Figure 6 illustrates the difference between the stiff adherend and flexible adherend behaviour.

In the flexible adherend test specimen, the adhesive yields, and the load continues to increase in a nonlinear way, due to the elasto-plastic stress redistribution, until the failure strain is reached.

In the stiff adherend test specimen the stress can not be redistributed, being uniformly applied. Therefore, all the adhesive layer yields simultaneously. The strain increases at a constant load, up to a maximum strain at which failure occurs.

474

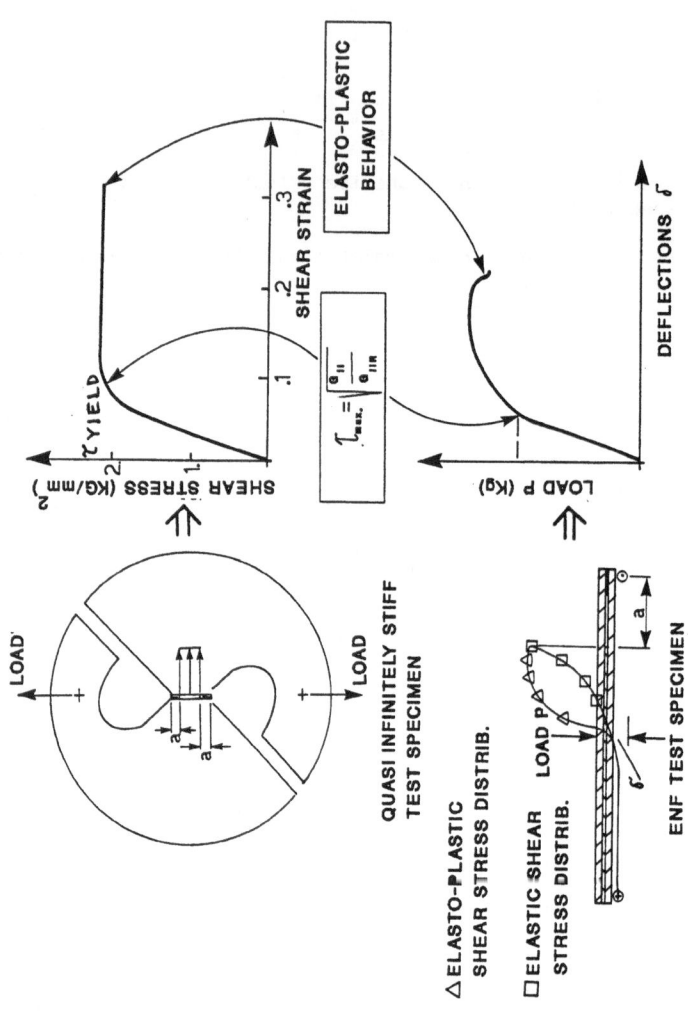

FIGURE 6 COMPARISON BETWEEN THE BEHAVIOUR OF THE STIFF
ADHEREND (BUTTERFLY) END ENF TEST SPECIMEN

It was demonstrated that the same failure mechanism occurred in both stiff and flexible adherend test specimens.

Flexible test specimens are generally considered to provide the fracture characteristics, and the stiff adherend the strength characteristics, of the adhesive bonded joints.

Analysis of the test results using the SIF/SCF relationship confirmed that a Unified Fracture and Strength Characterization of Adhesive Bonded Joints is possible.

REFERENCES

1. Adams D.A., Wake C.W. Structural Adhesive Joints in Engineering Elsevier Applied Science Publishers, London 1984.

2. Hart-Smith L.J., "Difference between Adhesive behavior in Test Coupons and Structural Joints" Presented to ASTM Adhesive Committee D-14 Meeting, Phoenix, Arizona, March 1981.

3. Weissberg V., Arcan M. "Stiff versus Flexible Adherend Adhesive Bonded Joints" to be published in Journal of Adhesion.

4. Weissberg V., Arcan M. "Stress Intensity - Stress Concentration Factor Relationship for Inhomogeneous Bodies with the Application to Adhesive Bonded Joints" Procedures of Numerical Methods in Fracture Mechanics V 23-27 April 1990.

5. Weissberg V., Arcan M. "Invariability of Singular Stress Fields in Adhesive Bonded Joints" to be published in the Int. Journal of Fracture.

6. Arcan M., Hashin Z., Voloshin A., "A Method to Produce Uniform Plane-Stress States with Application to Fiber - Reinforced Materials", Experimental Mechanics, Vol. 18, 1978, pp 141-146.

7. Weissberg V., Arcan M., "A Uniform Pure shear Testing Specimen for Adhesive Characterization" Adhesively Bonded Joints : Testing, Analysis, and Design ASTM STP 981, W. S. Johnson, Ed., Philadelphia, 1988, pp 28-38.

8. Arcan M., Bank-Sills L., "Mode II Fracture Specimen - Photoelastic Analysis and Results" Proceedings, Seventh International Conference on Experimental Stress Analysis, Israel, 1982, pp 187-201.

9. Bank-Sills L., Arcan M., Bortman Y., "A Mixed-Mode Fracture Specimen for Mode II Dominant Deformation", Engineering Fracture Mechanics, Vol. 1, 20, 1984, pp 145-157.

10. Russell A.Y., Street K.N. "Moisture and Temperature Effects on the Mixed-Mode Delamination Fracture of Unidirectional Graphite Epoxy" in Delamination and Debonding of Materials ASTM. STP876 W.S. Johnson ed 1985.

THE APPLICATION OF THE FINITE ELEMENT METHOD TO THE STUDY OF CRACKING IN MASONRY ARCH BRIDGES

B.S. CHOO, M.G. COUTIE and N.G. GONG
Department of Civil Engineering
University of Nottingham
Nottingham, NG7 2RD

ABSTRACT

In this paper, the behaviour of masonry arch bridges is examined using the finite element method and the results are compared with experimental data from recently conducted large scale testing of brick arch bridges. The no-tension characteristic and the non-linear stress-strain relationship of the arch materials is considered. Both one dimensional tapered beam elements and two dimensional elements are used to model the arch. In the 1-d analysis, the depths of the tapered beam elements are varied to represent the depths of the arch ring as radial cracks develop and crushing of the material due to high compressive stress occurs in the arch ring. For the 2-d analysis, nodal separation of the elements is used to represent both radial cracking of the arch ring and also tangential cracking which causes separation of the arch ring in multi-layered brick arch bridges. Reasonable agreement between the two methods and experimental data is achieved.

INTRODUCTION

The masonry arch bridge, one of the oldest forms of bridge in the world, under load produces mainly compressive forces in its construction material. It was a popular form of bridge before the development of material with high tensile strength and has proved to be durable and reliable with limited maintenance. In the last century, with the development of rail transport, more than 40,000 masonry arch bridges were built for the railway system in UK. There are presently over 30,000 masonry arches in service with British Railways (BR) alone. However, over the years these bridges deteriorate, resulting in some structural damage (for example ring separation, loss of bricks and cracking). It is expensive and unnecessary to replace all the masonry arch bridges on the road and railway systems with modern bridges. Hence, it is desirable to be able to assess the actual strength of the arch bridge in its present deteriorated condition. Both the Transport and Road Research Laboratory (TRRL) and British Rail Research (BRR) are undertaking research programmes to re-examine currently available methods of assessing the load carrying capacity of arch bridges. These programmes include the development of theoretical models and full/large scale testing.

The mechanism and 'MEXE' (Military Engineering Experimental Establishment) [1,2] methods are in common use in the UK, but their usefulness is limited. When analysing an arch by the mechanism method, a suitable failure pattern must be assumed before proceeding with the calculations. Based on this approach it is possible to find [3] a load carrying capacity for an arch bridge.

With the development of computer based numerical methods, the finite element method has also been used to simulate the complex behaviour of masonry arch bridges [4,5,6]. The arch ring has been represented by a string of short beam elements. The fill material above the arch is treated simply as dead load on the arch ring. The one dimensional (1-d) finite element approach using tapered beam elements, developed by the authors [6] as part of a larger study by BRR into the behaviour and deterioration of masonry arches, has been verified against data from TRRL and BRR large scale tests for both load-deflection and cracking characteristics. However the main disadvantage of 1-d finite element models is their inability to examine the effects of tangential cracking which leads to ring separations. To do this it is necessary to use 2-d elements to simulate multi-layered brickwork and to permit nodal separation which represents cracking when the nodal tensile stresses exceed the tensile capacity of the arch material. In this paper, the cracking characteristic of masonry arches and its influence on the ultimate strength of the arch are examined using 1-d tapered beam elements and 2-d finite elements.

ANALYTICAL APPROACH

The model

A masonry arch bridge is composed of the arch ring and fill material in between spandrel walls which act to retain the fill material over the arch ring. For the purposes of structural analysis, it is conventional to consider the arch ring as the main structural component and the fill material as dead loading above the arch ring, as shown in Fig. 1. The fill also serves as the medium through which the surface load is dispersed to the arch ring. The total load on the bridge is then transferred to the abutments by the arch ring. In this paper, it is assumed that only the arch ring has structural stiffness. The total loading on the arch ring consists of the self-weight of the bridge, the weight of the fill above the arch and the imposed load. Most load tests on arches are conducted by applying a line load across the fill over the width of the arch. To obtain the loading on the arch ring due to the line load, it is assumed that the load is distributed on to the arch ring through the fill as though both the fill and the arch are part of a semi-infinite elastic medium.

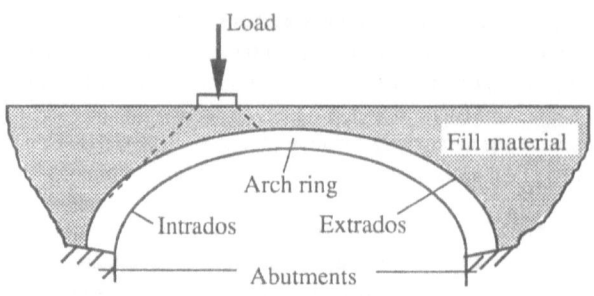

Figure 1. Arch bridge.

Although the fill above the arch ring is not considered as part of the structure, it may affect the structural behaviour of the arch ring by providing lateral passive resistance as a result of arch deformation. To allow for this, one dimensional horizontal elements are attached at the arch ring. These elements are activated only when the arch ring moves into the fill.

1-d analysis

For the 1-d analysis, a series of straight tapered beam elements is used to simulate the arch rings. It is assumed that the arch material has no tensile capacity, and under compression has the bi-linear stress-strain relationship shown in Fig. 2. Therefore, any tension in the arch would lead to radial cracking. In the analysis load is applied incrementally. At each load step the stresses between elements are checked for the occurrence of tension. When this is found, the material is deemed to have cracked. As the cracks develop, the depth of the tapered beam is reduced to model the uncracked portion of the arch ring which is in compression. In addition, when the compressive stress reaches the crushing (ie yielding) value of the material, it results in zones which do not contribute to the stiffness of the arch. Therefore, the 'effective arch ring' comprises only that uncracked and uncrushed portion of the arch, as shown in Fig. 3. The required effective depth of the arch ring can be calculated [6], for the three possible cases shown in Fig. 4, based on conditions of equilibrium of the nodal forces (P and M) and the internal stresses.

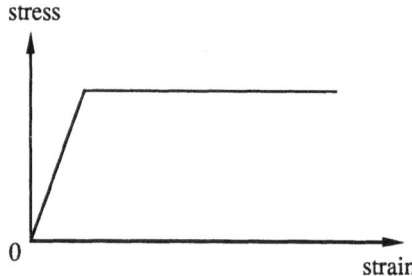

Figure 2. Stress-strain relationship of the arch material.

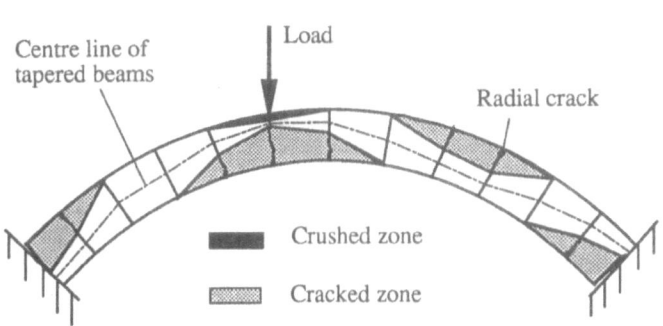

Figure 3.   Arch ring comprising tapered beam elements.

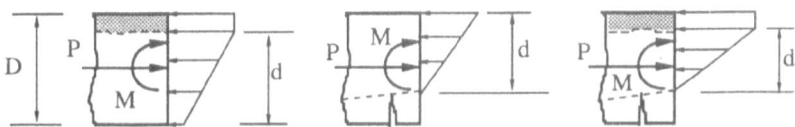

Figure 4. Possible stress distribution and corresponding effective arch depth d.

2-d analysis

In order to simulate radial cracking and ring separation in a brickwork arch bridge, it is necessary to discretise the arch ring using quadratic plane stress elements capable of being separated from each other when the stresses exceed the tensile strength of the material, as shown in Fig. 5. This is a direct way of simulating the effects of tensile cracking of the brickwork. It should be noted that cracks may close up with further application of load. For example, due to dead load only, radial cracks appear at the springing. With the application of imposed loading at quarter span, the crack at the springing further away from the load will begin to close up. Hence, those elements adjacent to the crack should be able to come into contact with each other at nodal points but should not overlap. This is achieved by re-connecting the affected elements at nodal points. This approach has been used in this study to examine the development of both radial cracking and ring separation of masonry arch bridges and their influence on the ultimate strength of the arch. However, this approach does not allow for any shear deformation between elements.

Older arch bridges that have seen service for a period of time may develop both radial and tangential cracks resulting in some separation of the brickwork. When assessing the present load carrying capacity of such bridges, it is necessary to model these existing radial cracks and ring separations which may be considered as initial defects. With further increase in load on the arch the affected elements may close up without being in alignment. To facilitate this in the model, additional joint elements with high axial stiffness in compression only but with no shear stiffness are employed at the tangential crack interface, as shown in Fig. 6. The use of these joint elements overcomes the limitation described above. To examine the influence of ring separation on the ultimate strength of masonry arches this method of modelling has been employed.

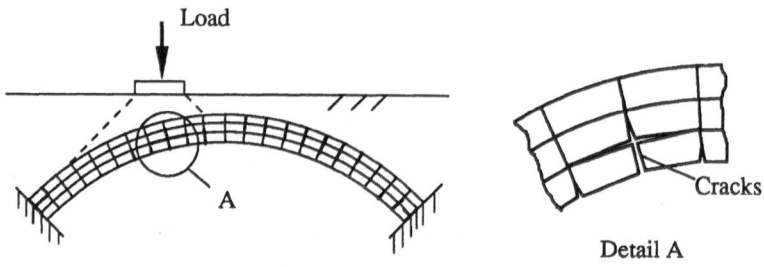

Figure 5. 2-d mesh and nodal separation between elements to represent cracking.

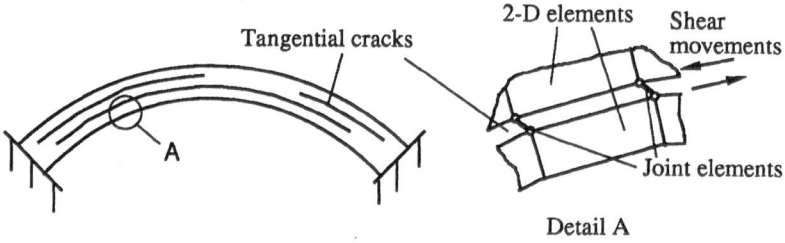

Figure 6. Separated arch rings and joint elements.

Solution procedure

The incremental loading procedure adopted for both 1-d and 2-d analyses has been discussed already. Additionally, at each new load level allowance is made for the change of ring geometry and the active pressure loading generated by deflection of the arch ring into the fill. The stress-strain diagram of Fig. 2 is followed in both cases.

NUMERICAL EXAMPLES

Numerical studies have been conducted, using both the 1-d and 2-d models, to determine the minimum number of elements required to obtain reasonably accurate solutions and to provide load-deflection and crack depth data for comparison with experimental results.

The accuracy of a finite element solution is affected by the mesh density adopted. Using Bridgemill bridge (Fig. 7) as an example [7], the variation of collapse load with number of elements used for both 1-d and 2-d analyses is shown in Fig. 8. It should be noted that three rings of 2-d elements were used for this comparison. Clearly, a three fold increase in the number of elements (ie corresponding to a seven-fold increase in the number of degrees of freedom) is required to achieve the same level of accuracy using 2-d elements. Nevertheless, it is shown that the collapse load obtained using both types of element converges with an increase in the number of elements used.

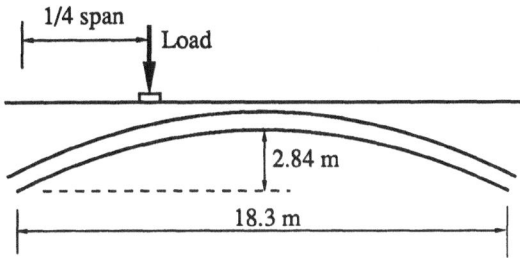

Figure 7. Geometry of the Bridgemill arch bridge.

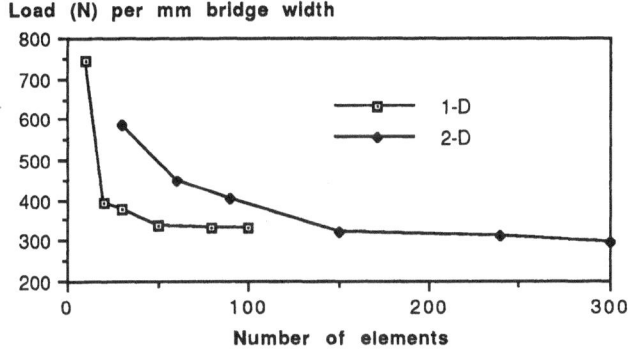

Figure 8. Variation of computed failure load with number of elements.

Numerical results using both 1-d and 2-d analyses have been compared with data obtained from load tests on a full size brick arch bridge and a large scale laboratory model. The full size test was conducted by the TRRL [8] at Torksey on a three ring brick arch bridge that was in poor condition. A large tangential crack separating the lower ring from the upper two rings could be clearly seen on the spandrel wall of the arch bridge, even before testing of the bridge started, as shown in Fig. 9. It was not obvious if this ring separation existed in the arch barrel. The maximum load attained by the bridge was 1080 kN and at collapse the arch disintegrated into individual bricks. Hence it may be assumed that the mortar in the arch was in a very weak state.

The results of finite element analyses of the Torksey bridge using both 1-d and 2-d elements are presented in Fig. 10 and Table 1. The calculated collapse load using the 1-d tapered beam elements was 1050 kN. For the 2-d model, using only 60 elements, four cases were considered. Firstly the arch was modelled without using joint elements: in case 1 only radial cracking was permitted while in case 2 both radial and tangential cracks were permitted. In both cases, the calculated collapse load was 8% above the test value. In the third case an initial defect in the form of ring separation of the lower layer of brickwork from the upper two layers of the arch ring was considered. This corresponded to the above mentioned observed ring separation before the bridge was tested. The computed collapse load of 1072 kN is much closer to the test collapse load. Finally, the extreme case where all the three rings were modelled as separated resulted in a relatively brittle failure at a much lower collapse load of 412 kN. This is probably because the structure very quickly became unstable as a mechanism. Case three clearly modelled the actual condition of the bridge most closely.

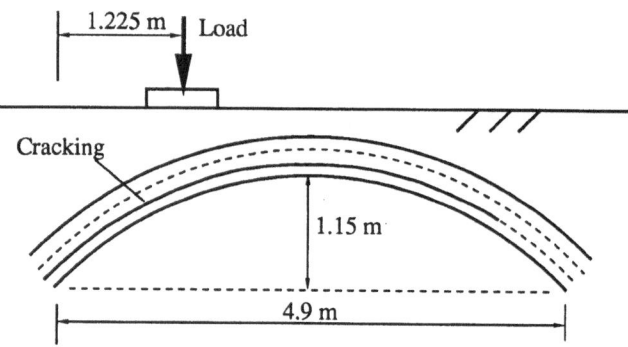

Figure 9. Geometry of the Torksey brick arch bridge.

As part of British Rail's masonry arch bridges assessment programme a 5-metre span model brickwork arch bridge was recently tested to collapse at Bolton Institute of Higher Education. The dimensions and loading of the arch model is shown in Fig 11. Using 50 tapered beam elements in a 1-d analysis the predicted collapse load of 1748 kN compares well with the actual collapse of 1720 kN. The development of crack lengths at 1/4 and 3/4 span (ie hinge locations) was also recorded. The results of crack development at the corresponding positions are plotted in Fig. 12. A good agreement is obtained. Using the same input data in the 2-d model (case 1) the computed collapse load is 1611 kN, that is 6% lower than the test result.

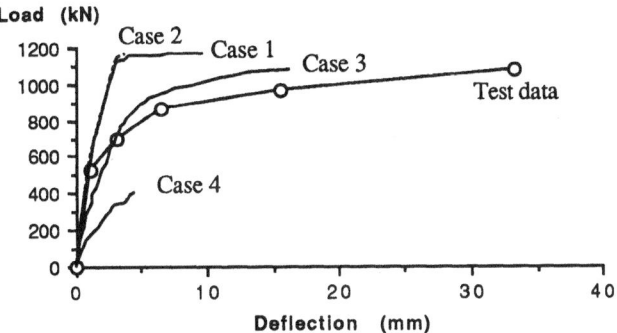

Figure 10. Load deflection plots for Torksey bridge.

Table 1

Comparison of test ultimate strength with FE results for Torksey Bridge

Test strength	1-d result	2-d results			
		without joint elements		with joint elements	
		Case 1 R.C. only	Case 2 R.C.+T.C.	Case 3 1 R.S..	Case 4 2 R.S
1,080 kN	1050 kN	1,163 kN	1,163 kN	1,072 kN	402 kN
CPU time*	5 mins	14 mins	27 mins	40 mins	21 mins

Note: R.C. = radial crack; T.C. = tangential crack; R.S. = ring separation
* all computing was carried out on Apple MacIntosh II

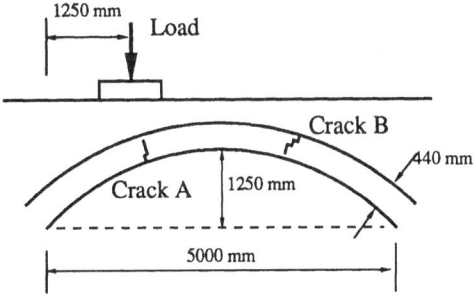

Figure 11. Dimensions of the large scale arch bridge tested at Bolton.

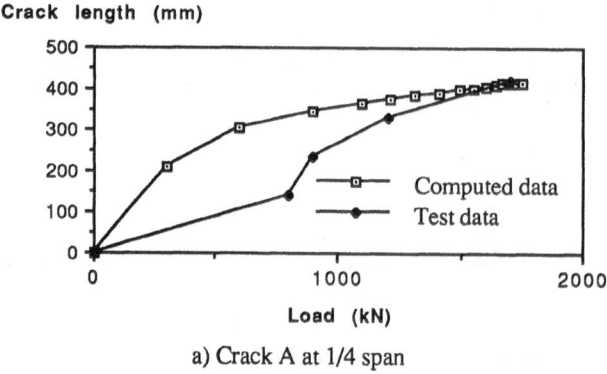

a) Crack A at 1/4 span

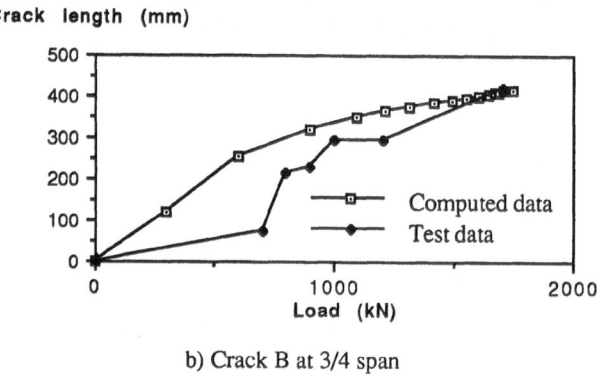

b) Crack B at 3/4 span

Figure 12. Increase of the crack length with load - Bolton test.

CONCLUSIONS

The one and two dimensional finite element models described here have been shown to provide a satisfactory method of predicting the behaviour under load and collapse of some brick masonry arch bridges. A good accuracy has been achieved. The 1-d model is cheaper to run but the 2-d model must be used if ring separation is to be considered.

ACKNOWLEDGEMENT

This project is sponsored by British Rail. The authors would like to thank Mr. C. Lemmon and Mr. S. Broomhead of British Rail Research in Derby for their assistance and for providing the authors with relevant information.

485

REFERENCES

[1] Department of Transport, The assessment of highway bridges and structures, Departmental Standard BD 21/84, DTp, London.

[2] Department of Transport, The assessment of highway bridges and structures, Advice Note BA 16/84, DTp, London.

[3] Towler, K. and Sawko, F. Limit state behaviour of brickwork arches, 6th Int. brick masonry conference, Rome, May 1982

[4] Crisfield, M.A., A finite element computer program for the analysis of masonry arches, TRRL laboratory report 1115, 1984

[5] Crisfield, M.A. and Packham, A.J., A mechanism program for computing the strength of masonry arch bridges, TRRL research report 124, 1987.

[6] Choo, B.S., Coutie, M.G., and Gong, N.G., Finite element analysis of masonry arch bridges using tapered elements, to be published.

[7] Hendry, A.W., Davies, S.R., and Royles, R., Test on stone masonry arch at Bridgemill - Girvan, TRRL contractor report 7, 1985.

[8] Page, J., Load test to collapse on two arch bridges at Torksey and Shinafoot, TRRL research report 159, 1988.

STRESS ANALYSIS IN MASONRY WALLS
BY ULTRASONIC MEASUREMENTS

RENATO S. OLIVITO

Dept. of Structures - University of Calabria - Cosenza (ITALY).

ABSTRACT

Non-destructive methods play an important and active role in estimating the state of preservation of brick masonry in buildings of historical interest.

This paper describes a method of measuring the compressive strength of brick masonry walls by means of simple ultrasonic measurements carried out on a selection of the bricks themselves.

A series of preliminary ultrasonic tests was conducted on individual bricks and on brick masonry models, measuring pulse velocity with a commercially available ultrasonic unit.

After processing this experimental data a correlation was obtained between the compressive strength and the indirect ultrasonic velocity of brick.

INTRODUCTION

The study of materials which do not withstand tension has recently seen a singular convergence of interests: on one hand the natural interest of structural engineers [1-....-7], while on the other, that of mathematicians attracted by the possibility of studying the phenomenology of fracture in masonry structures using the continuum theory [8-9-10].

Moreover, the structural behaviour of masonry is of considerable importance in the renovation and consolidation of monumental and architectonic property.

But such consolidation is only one of the reasons behind this renewed interest in masonry structures.

Through finite element discretization, many diversified studies, from a numerical point of view, have been developed in this field. These studies principally regard the formulation of the static equilibrium problem and the dynamic behaviour of masonry structures [11-12].

On the contrary, there are far fewer experimental contributions, mainly because of the complexity of modelling and the difficulties of measuring characteristic quantities, in particular measuring the compressive strength of a masonry wall without damaging it [11-12].

The refinement of calculation techniques, developed in recent years, requires a more accurate definition of input data, among which the mechanical properties of materials assume particular importance.

The simplest and most immediate approach to the problem of characterizing materials is the use of destructive mechanical tests carried out on non-modified samples taken from structures.

In the case of buildings and monuments of historical interest, it is normally impossible to take samples even of minimal dimensions. Hence the importance of experimentation and the need to carry out on site investigations by means of non- destructive methods.

This paper presents a simple non-destructive procedure for measuring the compressive strength of a brick masonry wall by identifying the relationship between compressive strength and ultrasonic velocity measurements on a selection of bricks in the wall.

The experimental results obtained with this method were supported by those obtained from destructive tests on brick masonry samples taken from both real masonry structures and others prepared in the laboratory.

MATERIALS AND EXPERIMENTAL TECHNIQUES

Masonry is a typical composite material, with two fundamental components,mortar and brick, which have different mechanical properties: brick presents brittle fracture behaviour while mortar presents ductile fracture behaviour characterized by large strains (Fig. 1).

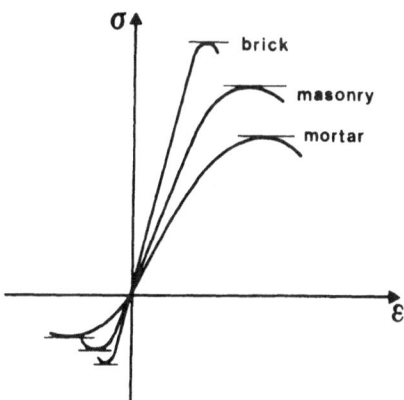

Figure 1. Uniaxial tensile and compression test on mortar and brick.

TABLE I.

Geometrical and ultrasonic measurements on bricks.

Sample	Dimensions Cm			Weight Kg	Density Kg/Cmc	Ultrasonic Velocity m/s			Maximum Load KN	Maximum Stress N/mm^2
	a	b	c			Va	Vb	Vc		
1	5,2	12,3	25,4	3,06	1883	2418	1434	1936	610,00	19,50
2	5,0	12,5	25,4	3,14	1977	3190	1900	2424	1200,00	37,79
3	5,0	12,5	25,4	2,80	1763	3030	1705	2238	890,00	28,03
4	5,0	12,6	25,0	2,92	1854	3030	1831	2370	1180,00	37,46
5	5,1	12,4	25,2	2,88	1807	2589	1521	1961	640,00	20,48
6	5,3	12,3	25,3	2,94	1782	3193	1946	2338	1080,00	34,70
7	5,4	12,3	25,6	3,02	1776	2727	1629	2124	820,00	26,04
8	5,2	12,1	25,1	2,95	1867	2796	1748	2140	790,00	26,01
9	5,2	12,4	25,4	3,00	1831	2587	1570	1945	700,00	22,22
10	5,3	12,5	25,1	2,93	1831	2787	1768	2229	935,00	29,80
11	5,3	12,4	25,2	3,00	1811	2961	1832	2362	1075,00	34,40
12	5,4	12,4	25,3	3,00	1771	2700	1570	2155	875,00	27,89
13	5,4	12,3	25,2	3,00	1792	3051	1730	2200	890,00	28,70
14	5,0	12,3	25,4	2,80	1792	2688	1565	2120	760,00	24,65
15	5,1	12,3	25,3	3,02	1903	3018	1904	2416	1050,00	33,74
16	5,0	12,2	25,5	2,76	1774	2688	1646	2300	800,00	25,71
17	6,0	12,2	25,5	3,36	1800	2985	1650	2100	795,00	25,55
18	5,1	12,3	25,5	2,88	1814	3187	1892	2507	1073,00	34,23
19	5,3	12,3	25,5	3,04	1829	3099	1801	2372	1023,00	32.60
20	5,0	12,3	25,5	2,80	1785	2732	1589	2070	730,00	23,27
21	5,0	11,0	24,7	2,72	1852	2976	1747	2256	770,00	26,20
22	5,5	12,1	24,9	3,08	1917	2910	1709	2190	830,00	27,55
23	5,4	12,0	24,8	3,08	1917	2842	1657	2012	770,00	25,87
24	4,9	12,1	24,8	2,7	1836	3025	1626	2175	815,00	27,18
25	5,2	12,0	24,8	3,06	1978	2826	1744	2189	990,00	33,27
26	6,3	11,8	25,0	3,00	2059	2880	1671	2168	910,00	30,85
27	5,0	12,2	24,8	2,86	1890	3048	1650	2309	840,00	27,76
28	5,5	11,9	24,7	3,18	1968	2723	1732	2065	850,00	28,92
29	5,3	12,0	24,7	3,12	1987	2849	1783	2176	890,00	30,02
30	5,5	11,9	24,8	3,09	1907	2835	1635	2132	800,00	27,10
31	5,1	11,8	24,8	3,04	2013	2948	1735	2214	905,00	30,90
32	4,8	12,0	24,8	2,70	1888	2743	1890	2340	1050,00	35,62
33	5,2	12,2	25,0	2,96	1866	2873	1694	2183	915,00	30,00
34	5,0	12,1	25,0	3,00	1986	2809	1819	2285	900,00	29,75
35	5,2	12,0	25,0	3,10	1987	2841	1641	2174	810,00	27,00

Both materials (mortar, brick) present a higher compressive strength than tensile strength. Therefore it seems reasonable to assume that the mechanical properties of the masonry material are intermediate between the component's properties, as shown in Fig. 1.

In order to correlate the mechanical characteristics of the masonry components with those of the masonry as a whole, a series of tests on sample runs of 35 bricks was carried out.

All the bricks used in the experimental investigation were subjected to grinding in order to eliminate, as fully as possible, any errors resulting from surface flaws.

The geometrical dimensions and the ultrasonic velocities of a typical run of bricks, measured according to the three directions, are shown in Tab. I.

The ultrasonic velocity measurements were carried out by using piezo-electric transducers with a nominal frequency of 150 kHz and an outside diameter of 25 mm.

A commercially available ultrasonic unit was used to measure the transmission time. Silicone vacuum grease served as a coupling medium between the transducer and the brick surfaces, in order to obtain an adequate contact. The pulse velocity was determined as the direct path length divided by the transmission time. All ultrasonic velocities and other measurements were made by the same operator.

The influence of the transducer frequency was examined on 10 bricks, by comparing the results obtained using two different sizes of piezo-electric transducers: 150 kHz, outside diameter 25 mm, and 1 MHz, outside diameter 7.5 mm (Tab. II).

TABLE   II.
Ultrasonic measurements using 150 kHz and 1 MHz transducers.

Sample	Weight Kg	Dimensions Cm			Transducers					
					150 KHz			1 MHz		
		a	b	c	Va	Vb	Vc	Va	Vb	Vc
1	2,83	5,3	12,3	25,3	2960	1585	2340	3011	1767	2343
2	3,10	6,0	12,3	25,0	3141	1730	2304	3141	1757	2304
3	2,82	5,3	12,2	25,4	2994	1856	2343	3045	1871	2345
4	3,17	5,9	12,2	25,3	3041	1859	2356	3105	1874	2356
5	2,90	5,6	12,3	25,3	3077	1949	2353	3146	1919	2358
6	3,13	6,0	12,3	25,4	2913	1653	2193	3030	1722	2243
7	2,96	5,4	12,3	25,1	2967	1919	2442	3016	1943	2446
8	2,69	4,9	12,3	25,3	2753	1701	2194	2882	1723	2201
9	2,95	5,3	12,3	25,3	2677	1777	2247	2760	1775	2285
10	3,16	5,8	12,2	25,4	2749	1685	2230	2871	1748	2216

A further 15 bricks were subjected to a uniaxial compression test, perpendicular to the largest surface, measuring the transmission times at 100 kN load intervals up to maximum load.

The compression tests were carried out using the testing apparatus illustrated in Fig. 2; the transmission times were determined using 150 kHz transducers.

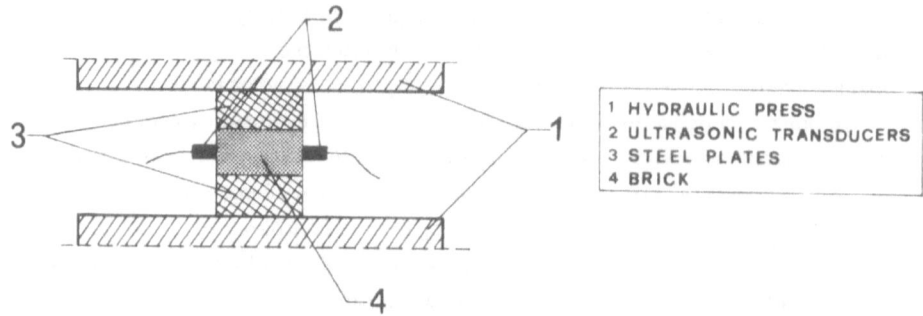

Figure 2. Testing apparatus for the compression test on bricks.

Finally, in order to study the mechanical behaviour of the masonry as a whole, 10 brick masonry models with dimensions of 40x25x42 cm were constructed, utilizing cement mortar (M1) with an average compressive strength not inferior to 12 N/mm^2 (Fig. 3).

The brick masonry models were subjected to compression tests with measurements of direct (Fig. 3a) and indirect (Fig. 3b) ultrasonic velocity at 100 kN load intervals up to maximum load.

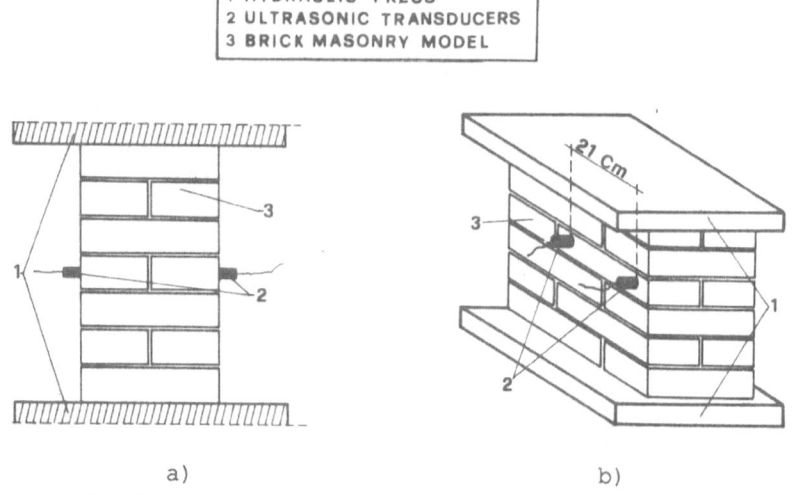

a)   b)

Figure 3. Brick masonry model and experimental arrangement.

EXPERIMENTAL RESULTS

The number of samples (n > 10) of Tab. I enables us to develop a statistical relationship between the compressive strength of bricks and the ultrasonic velocities when measured according to the three directions.

This processing was carried out by means of the least square method applied to the experimental data of Tab. I.

The correlations found between compressive strength (R) and ultrasonic velocity (V_b, V_c) are:

$$R = 0.0343 \, V_b - 29 \tag{1}$$

$$R = 0.0281 \, V_c - 32 , \tag{2}$$

which are plotted in Figs. 4 and 5.

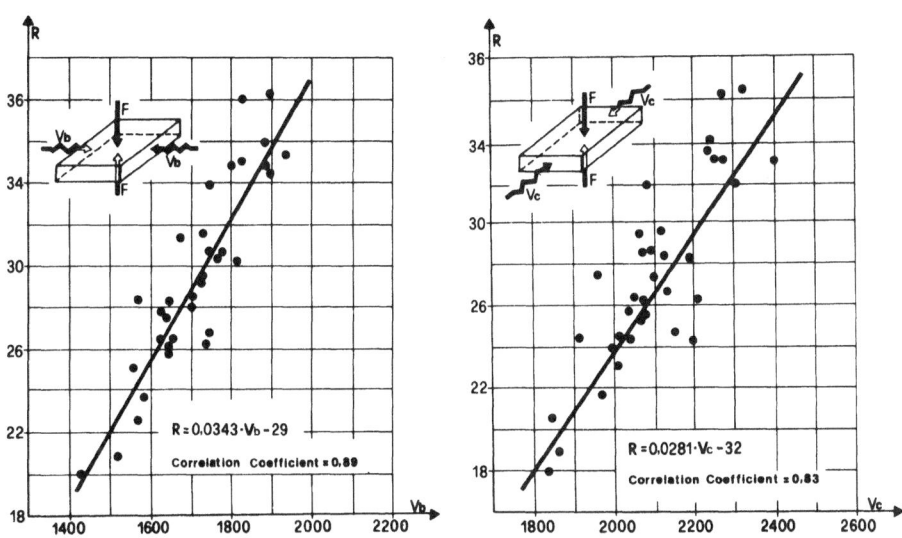

Figure 4. R-V_b straight line   Figure 5. R-V_c straight line
       correlation.   correlation.

The correlations obtained offer satisfactory reliability for bricks of compressive strength superior to 20 N/mm^2.

The uniaxial compression tests carried out both on single bricks and on brick masonry models, enabled us to obtain useful information regarding the behaviour of these elements when subjected to load.

In order to study also the fracture behaviour, a damage coefficient was defined as a function of pulse velocity :

$$D = 1 - (V / V_0) \tag{3}$$

where V_0 is the value of the ultrasonic velocity for an unstressed element and V for a stressed element [14-15].

Figs. 6-7 show the outline of the damage coefficient with respect to the non-dimensional ratio K= σ/σ_r, where σ is the stress corresponding to a certain load level and σ_r is the maximum stress.

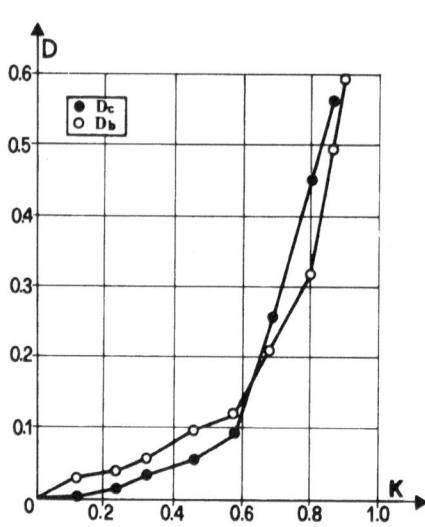

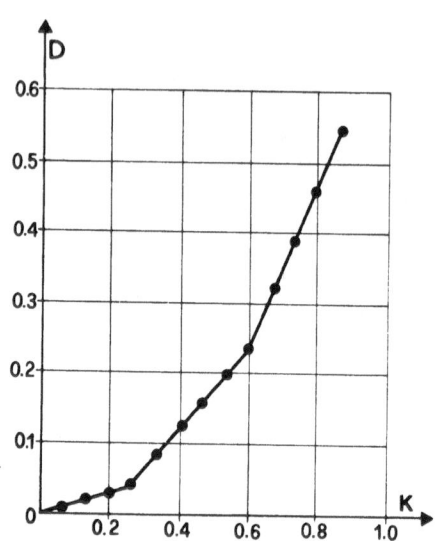

Figure 6. Damage coefficient of bricks according to b and c directions.

Figure 7. Damage coefficient of masonry models.

It should be noted that Fig. 6 shows that D_c values are lower than D_b ones up to 60% of maximum stress. These results confirm that the direction of cracking is predominantly orthogonal to the length of the brick.

The damage curve of the brick masonry models shown in Fig.7 describes three different sections. The first (k< 0.25) indicates a slight increase of damage, caused by the settlement of the masonry material subjected to load.

In the second part, by increasing the compressive stress (0.25 < k < 0.6), cracks appear, located on bricks and along the contact zones between brick and mortar.

The third part (k > 0.6) is characterized by considerable increases of the damage coefficient for small increases of the compressive stress. The masonry models present vertical cracks from top to bottom.

A PROPOSED ULTRASONIC PROCEDURE

In order to obtain an experimental relationship between the brick compressive strength of a masonry wall, measurements of indirect ultrasonic velocity on 21 cm of brick path were carried out.

The average of these indirect ultrasonic velocities was:

$$\overline{V}_i = 2350 \text{ m/s.}$$

Now let ρ_1 and ρ_2 be the following ratios:

$$\rho_1 = \overline{V}_b / \overline{V}_i \tag{4}$$

$$\rho_2 = \overline{V}'_i / \overline{V}_i \tag{5}$$

where $\overline{V}_b$ is the average of the direct ultrasonic velocities of Tab. I, $\overline{V}_i$ the average of the indirect ultrasonic velocities of the bricks and $\overline{V}'_i$ the average of the indirect ultrasonic velocities of the brick masonry models.

The values of ρ_1 and ρ_2 are 0.725 and 1.160 respectively.

Eqs. (4) and (5) offer

$$\overline{V}_b = 0.725 \, \overline{V}_i \tag{6}$$

$$\overline{V}_i = \overline{V}'_i / 1.160 \tag{7}$$

By substituting Eq. (7) with Eq. (6) we obtain

$$\overline{V}_b = 0.625 \, \overline{V}'_i \tag{8}$$

Eq. (8) can be utilized in Eq. (1) to obtain

$$R = 0.0214 \, \overline{V}'_i - 29 \tag{9}$$

Eq. (9) enables us to estimate the maximum strength of the bricks set in a masonry wall.

Therefore, given the mechanical properties of brick and mortar shown in Tab. III [13], the maximum strength of the masonry is easily determined by means of measurements of indirect ultrasonic velocity on bricks with a measuring distance of 21 cm.

494

TABLE III.
Mechanical properties according to the Italian Standards.

Compressive Strength of Brick	Compressive Strength of Masonry			
	Mortar			
	M1	M2	M3	M4
N/mm²	N/mm²	N/mm²	N/mm²	N/mm²
1,5	1,2	1,2	1,2	1,2
3,0	2,2	2,2	2,2	2,0
5,0	3,5	3,4	3,3	3,0
7,5	5,0	4,5	4,1	3,5
10,0	6,2	5,3	4,7	4,1
15,0	8,2	6,7	6,0	5,1
20,0	9,7	8,0	7,0	6,1
30,0	12,0	10,0	8,6	7,2
40,0	14,3	12,0	10,4	

This proposed procedure was tested on real brick masonry structures; the experimental results obtained were compared to those obtained by uniaxial compression tests of samples drawn from the same structures.

The comparison showed a good agreement with an approximation of ± 3 % .

CONCLUSIONS

The proposed ultrasonic procedure, based on measuring the time travel of a pulse or train of waves through a measured path length, furnishes a substantial contribution to the stress analysis of brick masonry structures.

The experimental investigation conducted on single bricks and on brick masonry models, enabled us to find an interesting correlation (Eq. 9) between the compressive strength and the indirect ultrasonic velocity of brick.

The correlation found was tested on real brick masonry structures. The results obtained are satisfactory when compared to those obtained by destructive tests of samples drawn from the same structures, since they show a discrepancy of only ± 3 % .

REFERENCES

1. Heyman, S., The Masonry Arch, Ellis Horwood Series in Engineering Science, John Wiley & Sons Ltd, Chichester, 1982.

2. Villaggio, P., Stress diffusion in masonry walls, Journal of Structural Mechanics, 9, 1981, p. 439.

3. Como, M. and Grimaldi, A. , A unilateral model for the limit analysis of masonry walls, Unilateral Problems in Structural Analysis, G. Del Piero & F. Maceri Eds, Springer, Wien, 1985.

4. Del Piero, G., Constitutive equation and compatibility of the external loads for linear elastic masonry-like materials, Meccanica, Vol. 24, 1989, pp. 150-162.

5. Di Pasquale, S., Questioni concernenti la meccanica dei mezzi non reagenti a trazione, Atti VII Congresso Nazionale AIMETA, Trieste, 1984, Vol. 5, pp. 227-238.

6. Page, A.W., The biaxial compressive strength of brick masonry, Proc. Inst. Civ. Eng., 2, 71, 1981, pp. 893-906.

7. Shalin, S., Structural Masonry, Prentice-Hall, Inc., Englewood Cliffs, New Jersey, 1971.

8. Signorini, A., Un teorema di esistenza ed unicita' nello studio dei materiali poco resistenti a trazione, Rend. Accademia Nazionale dei Lincei, Vol. 2, 1925, pp. 401-406.

9. Anzellotti, G., A class of non-coercive functionals and masonry-like materials, Amm. Inst. Henri Poicare', Vol. 2, 1985, pp. 261-307.

10. Giaquinta, M. and Giusti, G., Researches on the equilibrium of masonry structures, Arch. Rational Mech. Analysis, Vol. 88, 1985, pp. 358-392.

11. AA. VV., Comportamento statico e sismico delle strutture murarie, G. Sacchi Landriani & R. Riccioni Eds, Clup, Milano, 1982.

12. AA. VV., Studi italiani sulla meccanica delle murature, Atti del Convegno Stato dell'Arte in Italia sulla Meccanica della Muratura, A. Grimaldi & A. Giuffre' Eds, Roma, Ottobre 1985.

13. La Tegola, A., Consolidamento degli edifici in muratura.Progettazione e tecniche d'intervento, Dept. of Structures, University of Calabria, Rep. N. 112, Oct. 1988.

14. Daponte, P and Olivito, R.S., Crack detection measurements in concrete, to appear on Proc. of Microcomputer Applications Conference, Dec. 14-16, 1989, Los Angeles, CA U.S.A.

15. Daponte, P. and Olivito, R.S., Metodi ultrasonici per il rilevamento del danno nel calcestruzzo: Analisi nel dominio del tempo e della frequenza, Convegno Nazionale in ricordo di Riccardo Baldacci e Michele Capurso, Roma-CNR, 25-26 Ott. 1989, pp. 297-306.

EXPERIMENTAL RESULTS ON MOMENT REDISTRIBUTION IN REINFORCED CONCRETE CONTINUOUS BEAMS

GIUSEPPE SPADEA

Department of Structures - University of Calabria
87030 Arcavacata di Rende (CS) - Italy

ABSTRACT

An experimental investigation of moment redistribution in hyperstatic reinforced concrete beams is described. The experimental equipment, using
a mechanical worm gear actuator, allows strain controlled tests in order
to determine the moment–curvature relationship, including the range of
large curvatures from the first yield up to nearly the ultimate load.
A relatively simplified calculation model, based on the average reduction
of the bending stiffness in cracked zones is described. The tension stiffening effect is also taken into account.

   In the light of these assumptions the present work develops a numerical analysis for the evaluation of moment redistribution in reinforced
concrete continuous beams in the cracked regime, and also presents a parametric investigation checked against experimental results which confirm
the validity of the analytical model outlined.

INTRODUCTION

The flexural behaviour of statically indeterminate reinforced concrete
beams, after first cracking, is influenced by the variation in flexural
rigidity caused by crack growth in zones where bending moments exceed the
first crack moment [1, 2]. Thus a moment redistribution must be considered
in the analysis. This gives the advantage of a reduction in the congestion
of reinforcement at the supports of continuous members, or in the nodes of
frames, and also produces a reduction of peaks in bending moment envelopes.
In addition a more realistic assessment of the behaviour of beams is gained in calculating both deformations and bending moments in the midspan
sections [3].

Various parameters, which are difficult to determine, influence the variation of flexural rigidity and bending moment redistribution. For this reason, modelling is rather complex and creates severe difficulties in numerical applications. Through the use of suitable moment–curvature relationships, which take into account both the finite increment of curvature due to the crack opening and the tension stiffening effect, and by neglecting the stress diffusion around crack tips, it is possible to develop easier calculation models which furnish results sufficiently close to the experimental ones available in the literature [4, 5, 6].

The model outlined for calculating the redistribution of bending moments is not based on discrete cracking but instead takes into account an evaluation of the overall pattern of cracking directly taken from the diagrams of the elastic bending moments.

The effect of cracking on the increase of curvature and reduction of bending moment cannot be neglected, especially for high quality concrete beams. In addition experimental results have revealed significant differences between global and local moment–curvature relationship. In fact the global relationship is usually determined by considering a constant bending moment distribution and the relative rotation of the initial and final sections of the beam element under consideration. This approach can be adopted when a uniform distribution of the external load occurs because, in this case, the actual elastic bending moment distribution in the midspan zone can be approximated by a constant distribution. This is not admissible when the bending moment gradient is high, as in the support zones or in the case of concentrated loads. The calculation of these cases requires the consideration of local moment–curvature relationships involving parts of the beam including a crack [7, 8].

When the bending moment distribution is almost uniform, because of the dishomogeneity of concrete, the opening of each crack occurs successively and (for the same reason) the spacing of cracks is stochastic. Since the tension stiffening effect depends on the spacing of cracks, this effect is therefore different in each beam element between two adjacent cracks. Differences between local and global moment–curvature relationship become significant in the transition stage from the uncracked (I) to the cracked (II) condition, because each crack produces a sudden increase of curvature and a marked redistribution of bending moment. The above phenomena cause differences between local and global moment–curvature diagrams which, as will be shown, must be considered in the analysis.

THE OUTLINE OF THE PROBLEM

Fig. 1 shows, for a practical beam in which the tension steel yields, a typical moment–curvature diagram that can be well idealized to the trilinear relationship represented in the same Fig. 1. In many cases it is sufficiently accurate to consider only the simplified bilinear relationship shown in Fig. 2.

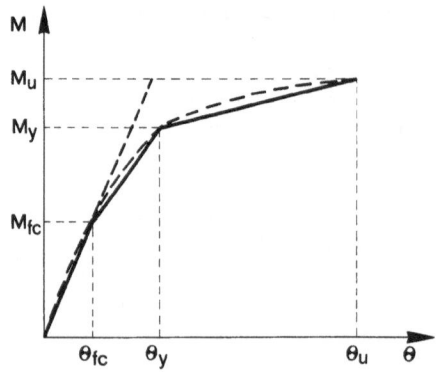

Figure 1. Global moment–curvature relationship of a beam section falling in tension.

Figure 2. Idealized moment–curvature curve of a beam section falling in tension.

In the case both of lightly reinforced rectangular sections (or T sections subject to negative bending moments) and of high gradient in bending distribution, we must examine more thoroughly the transition phase from the first (I) to the second stage (II) and consider the local moment-curvature relationship.

The local moment-curvature relationships, experimentally obtained show unstable branches (Fig. 3); the slight differences mainly attributable to the material non-homogeneity, allow us to assume an M-θ average diagram (Fig. 4).

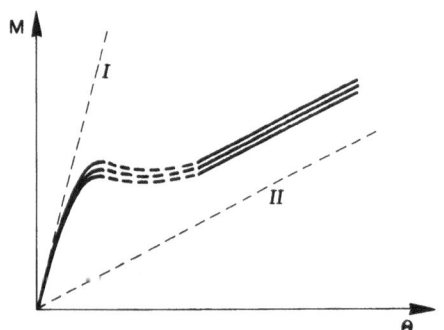

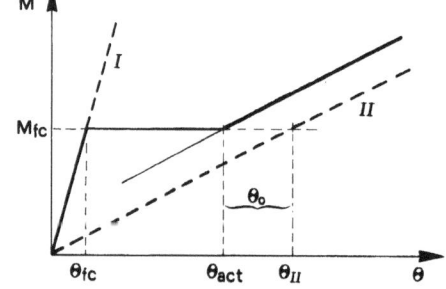

Figure 3. Local moment-curvature relationships.

Figure 4. Idealized local moment-curvature curve.

It can be seen that the first part (uncracked stage, $M < M_{fc}$) is well represented by the straight line of the classical theory; similarly the second part also assumes linear behaviour (cracked stage, $M > M_{fc}$) although the average curvature values are influenced by the stiffness of the stretched concrete between two adjacent cracks (tension stiffening).

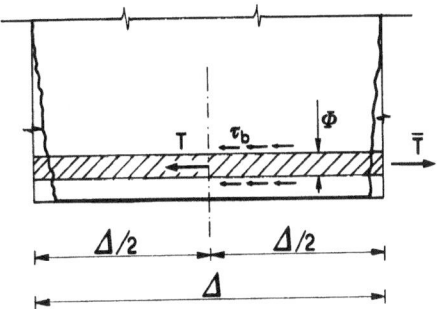

Figure 5. Valuation of the tension stiffening effect.

This effect can easily be determined as follows without considering closely the transition phase between stages I and II [7, 8], Fig. 5.

Let us consider a portion of a beam including two adjacent cracks. Bond stresses τ_b are assumed to have a constant distribution along the element. Isolating the reinforcement bar (diameter Φ), from equilibrium considerations we can write:

$$T = \bar{T} - \Delta\pi\phi\tau_b/2 \tag{1}$$

$$\varepsilon_s = T/(E_s A_s) = \varepsilon_s - \Delta\pi\phi\tau_b/(2A_s E_s)$$

replacing ε_s with its average value $\varepsilon_{sm} = (\varepsilon_s + \bar{\varepsilon}_s)/2$, and given the actual curvature of the cracked zone

$$\Theta_{act} = \Theta_{II} - \Theta_o = (\bar{\varepsilon}_c + \varepsilon_{sm})/d \tag{2}$$

(where d is the effective depth of the cross section), we have:

$$\Theta_{act} = [\bar{\varepsilon}_c + \bar{\varepsilon}_s - \Delta\pi\phi\tau_b/(4E_s A_s)] \tag{3}$$

from which:

$$\Theta_o = \Theta_{II} - \Theta_{act} = \Delta\pi\phi\tau_b/(4dE_s A_s) \tag{4}$$

The crack spacing can be evaluated by the classical relation:

$$\Delta = K f_r A_c/(\pi\phi\tau_b) \tag{5}$$

where $1 < K < 2$ in the case of members subject to axial tension force, and $K = 0.5$ in the case of bending. In addition f_r and A_c are the tensile strength and the tensed area of concrete, respectively. Therefore:

$$\Theta_o = K f_r A_c/(4dE_c A_s) \tag{6}$$

is the quantity to be determined.

The assumption of the simplified moment-curvature relationship described above allows us to establish a methodology for evaluating the effect of cracking on bending moment redistribution of hyperstatic reinforced concrete structures. Fundamentals of this methodology can be found in [4]: both general and particular equations can be derived for continuous beams, which are the most widely used reinforced concrete structures.

This method permits us to determine the relations between bending moments and curvatures at the extremities of the members. With reference to a beam subject to any given load distribution and to end moments M_a and M_b the cracking in those parts in which $M>M_{fc}$ produces an increase of rotations Θ_{*c} at the extremes. Therefore we obtain:

$$\Theta_a = \Theta_{ae} + \Theta_{ac} \tag{7}$$

where:

$$\Theta_a = \Theta_{aa}M_a + \Theta_{ab}M_b + \Theta_{ao} \tag{8}$$

is the elastic rotation produced by the external loads.

For small increases in the moments at the extremes we can write (in matrix form):

$$\Theta X + [\Theta + \Theta'_c]\delta X + \Theta_c + \Theta_o = 0 \tag{9}$$

Θ being the deformation matrix, X the hyperstatic moments vector, Θ the contribution of cracking to the rotations and with the position $\Theta'_c = \{\partial\Theta_c/\partial X\}$. Since both Θ_c and Θ'_c depend on the unknowns an iterative method must be used.

With reference to the diagram of Fig. 4, the curvatures in the first and second stage are given by:

1st stage $(M<M_{fc})$ $\qquad \Theta^I = M\,\Theta_{fc}/M_{fc}$

2nd stage $(M>M_{fc})$ $\qquad \Theta^{II} = M\,\Theta_2/M_{fc} - \Theta_o$
$$\tag{10}$$

Therefore the increase of curvature due to the cracking is:

$$\Delta\Theta = (\Theta_2 - \Theta_{fc})\,M/M_{fc} - \Theta_o \tag{11}$$

The increments of rotations in a and b, caused by cracking of that part of the beam in which $M>M_{fc}$, are:

$$\Theta^i_{ac} = \int_{\xi^i_1}^{\xi^i_2} [\Delta\Theta(\xi)(1-\xi/L)]\,d\xi$$

$$\Theta^i_{bc} = \int_{\xi^i_1}^{\xi^i_2} [\Delta\Theta(\xi)\,\xi/L]\,d\xi \tag{12}$$

where ξ^i_1 and ξ^i_2 are the abscissae corresponding to the two extremities of the cracked part under consideration.

The total effect, of the various cracked parts is therefore:

$$\Theta_{ac} = \Sigma\,\Theta^i_{ac}\;;\quad \Theta_{bc} = \Sigma\,\Theta^i_{bc} \tag{13}$$

Thus, the analysis of a structure requires the determination of the moments at the extremes in the elastic condition at the assessment of the cracked parts along the members. In the case of a continuous beam on three supports $M_b = 0$; with reference to the central support and the midspan, the cracked parts are identified in the following figure:

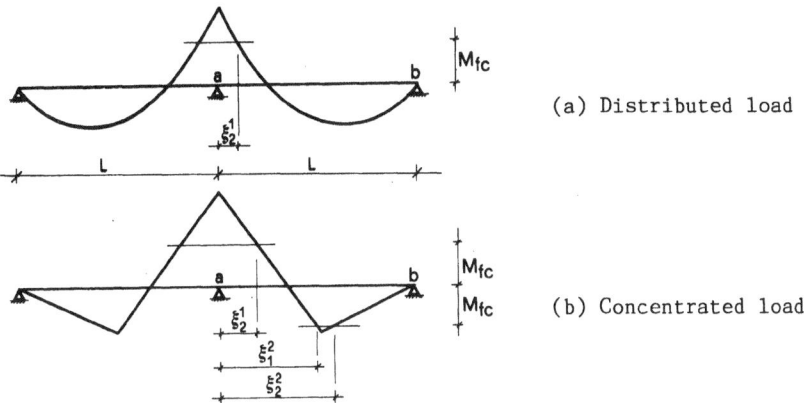

(a) Distributed load

(b) Concentrated load

Figure 6. Cracked parts of a two span reinforced concrete continuous beam.

NUMERICAL ANALYSIS

The identification of geometric and mechanical parameters which mainly influence the crack pattern and moment redistribution of continuous beams has been developed by a parametric investigation. With reference to a two span continuous beam the parameters considered in the analysis were:
- cubic compressive strength of concrete R_{ck};
- content of tension and compression steel μ and μ';
- depth-to-width ratio of rectangular sections d/w ;
- flange thickness-to-web width ratio of T-beams sections f/w.

For the sake of brevity only the results pertaining to the most relevant cases are given. They can be discussed on the basis of the relationship between the moment at center support and the applied load (distributed q or concentrated F) and are shown in Figs. 7, 8, 9, 10.

The analysis of these figures suggests the following considerations:
- the size of the cross section has considerable influence on redistribution: this is more pronounced in the case of a T section with respect to a rectangular section;
- the two load conditions taken into consideration (distributed or concentrated) exhibit a different influence on redistribution. In fact, considering the case of concentrated forces, we observe that, in connection with the reinforcing steel content on the top or the bottom of the section, after the first cracking at the support, cracking can occur also in the midspan sections. This influences the type and the amount of redistri-

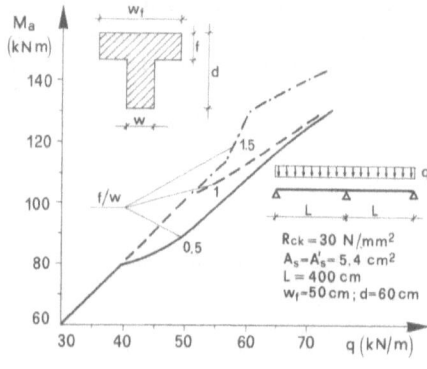

Figure 7. Size effect of the cross T section.

Figure 8. Size effect of the cross rectangular section.

bution;
- provided that the percentage of reinforcing steel is the same in the midspan as in the support section, redistribution is not influenced by the overall amount of reinforcing steel. On the contrary, such redistribution is strongly affected by significant differences in the steel content of the midspan and support sections;
- the increasing compressive strength produces a higher redistribution at a greater load level, while for low grade concretes a smaller redistribution occurs earlier.

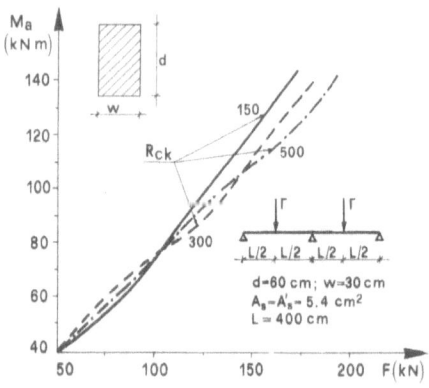

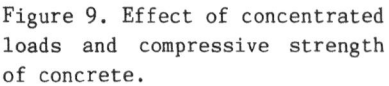

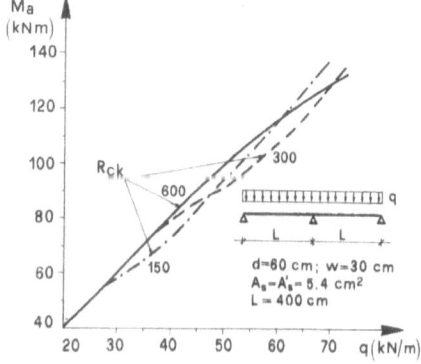

Figure 9. Effect of concentrated loads and compressive strength of concrete.

Figure 10. Effect of compressive strength of concrete.

EXPERIMENTAL INVESTIGATION AND CONCLUDING REMARKS

The proposed modelling of the nonlinear behaviour of reinforced concrete continuous beams and the numerical results shown above can be validated against experimental results. Full scale tests were carried out on three reinforced concrete beams each with rectangular cross sections. The dimensions and the mechanical properties of the beams are given in detail in Fig. 11.

The amount of steel reinforcement and the grade of concrete were chosen in order to allow a higher moment redistribution. In fact cracking is more pronounced with a lower tensile steel content in the support section than in the midspan section, even with higher concrete strength.

The beams were manufactured from the same batch of concrete, cast in plysteel forms, and compacted on a vibrating table. Casting and curing were particularly accurate in order to avoid shrinkage cracks. The beams were tested on a steel frame. In order to minimize the settlement of the central support the steel frame apparatus was strongly stiffened in the central zone; in any case the displacement of the central support was accurately measured by two inductive displacement transducers LVDT.

The forces were applied by a worm gear actuator which allowed controlled displacement tests. The applied load and the reactions of the supports were measured by four high class accuracy load cells (mod. C1, 0.1%, HBM). The curvatures were measured in the midspan section and in the central support section by two couples of LVDT.

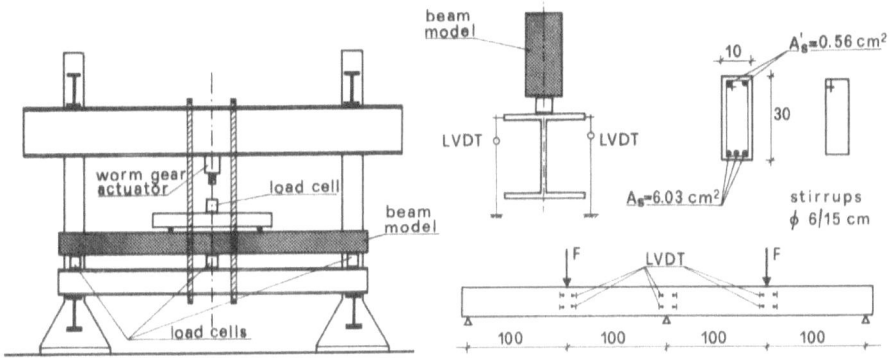

Figure 11. Testing apparatus: loading scheme and instruments.

The complete scheme of the instruments used is given in Fig. 11. All instruments (load cells and displacement transducers) were connected to an automatic scanning and measuring station.

The results of the experimental investigation can be examined in the following diagrams, relative to the first of the beams tested, which represent, respectively:
- the reactions in the lateral and central supports R_b and R_a (Figs. 12 and 13);

– the bending moment in the central support M_a (Fig. 14);
versus the external applied force F.

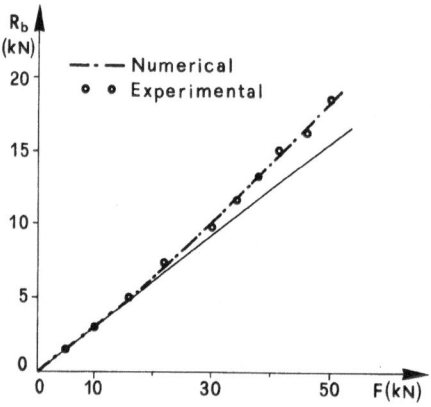

Figure 12. Reaction of the lateral
supports versus the external force.

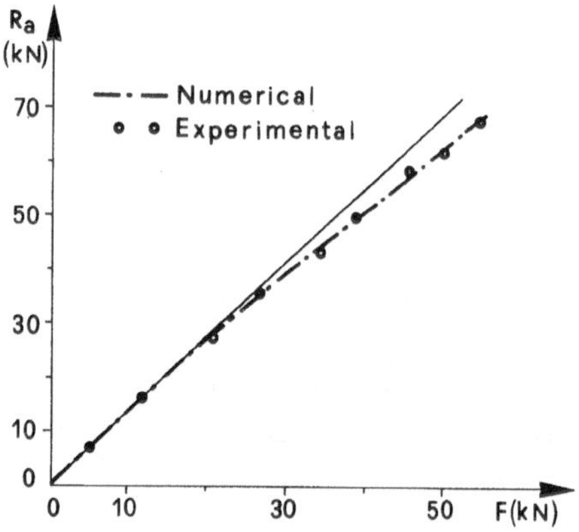

Figure 13. Reaction of the central
support versus the external force.

In the diagrams above the deviation from linearity represents the
amount of redistribution; it should be noted that the results of Figs.
13 and 14 are a confirmation of the results given in Fig. 12. A clearer
picture of the behaviour of the beam can be seen in Fig. 15 were the
moment-curvature diagram of the central support section is plotted.

The agreement found between the experimental and numerical results
in each diagram is close enough to be satisfactory. In addition the results
relative to the second and third beams tested completely confirmed the
results given here.

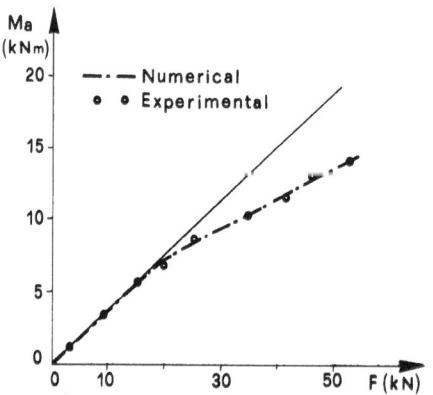

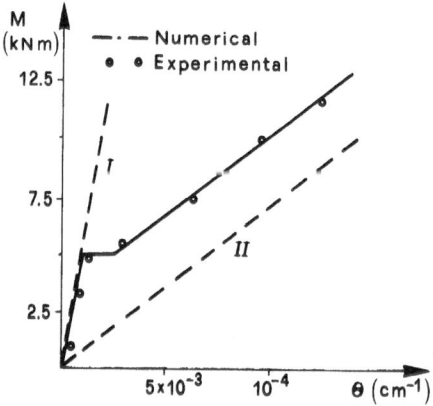

Figure 14. Bending moment in the
central support versus the external force.

Figure 15. Moment-curvature diagram
of the central support section.

It can be concluded that the behaviour of hyperstatic reinforced concrete members as estimated by the mechanical model outlined shows a good agreement both with previous results available in the literature and with the test results obtained in the present investigation. Thus a calculation scheme which takes into account both the increment of curvatures consequent to the crack opening in finite parts of the members and the tension stiffening effect, enable us to determine the redistribution of stresses and strains in any type of statically indeterminate concrete structure.

REFERENCES

1. Burnett, E.F.P., Rotation capacity of reinforced concrete flexural elements. In Inelasticity and Non-Linearity in Structural Concrete, ed. M.Z. Cohn, SMD, University of Waterloo, Ontario, Canada, 1973, **8**, pp. 181-210.

2. Ghosh, S.K. and Cohn M.Z., Non-linear analysis of strain-softening structures. In Inelasticity and Non-Linearity in Structural Concrete, ed. M.Z. Cohn, SMD, University of Waterloo, Ontario, Canada, 1973, **8**, pp. 315-32.

3. Park, R. and Paulay, T., Reinforced Concrete Structures, John Wiley & Sons, New York, 1975, pp. 195-269.

4. Giuriani, E., Gli effetti della fessurazione sulla ridistribuzione dei momenti nelle strutture iperstatiche in c.a.: fondamenti teorici. In Studi e Ricerche, Corso di Perfezionamento per le Costruzioni in Cemento Armato F.lli Pesenti, Politecnico di Milano, 1982, **4**, pp.163-87

5. Giuriani, E., Studio sulla ridistribuzione dei momenti provocata dalla fessurazione nelle travi continue di c.a.. In Studi e Ricerche, Corso di Perfezionamento per le Costruzioni in Cemento Armato F.lli Pesenti, Politecnico di Milano, 1982, 4 pp. 189-209.

6. Gelfi, P. and Giuriani E., Effetti della fessurazione sugli appoggi ed in campata nelle travi continue di c.a.. La Prefabbricazione, 1986, n. 1, pp. 19-30.

7. Giuriani, E., On the effective axial stiffness of a bar in cracked concrete. In Bond in Concrete, ed. P. Bartos, Applied Science Publishers, London, 1982, pp. 107-26.

8. Giuriani, E. and Rosati, G., Deformabilità degli elementi inflessi di c.a.soggetti ad una singola fessura. Effetti nelle strutture iperstatiche. In Studi e Ricerche, Corso di Perfezionamento per le Costruzioni in Cemento Armato F.lli Pesenti, Politecnico di Milano, 1984, **6**, pp. 119-49.

A PRELIMINARY INVESTIGATION OF REINFORCEMENT DUCTILITY
IN REINFORCED CONCRETE SLABS

R. H. Scott
Lecturer,
School of Engineering and Applied Science,
University of Durham, South Road, Durham, DH1 3LE.

P. A. T. Gill
Senior Lecturer,
School of Engineering and Applied Science,
University of Durham, South Road, Durham, DH1 3LE.

ABSTRACT

The development of a technique for internally strain gauging the wires in
fabric reinforcement is described, with 31, 41 and 75 electric resistance
strain gauges being installed in wires of 8, 10 and 12 mm diameter
respectively. The performance of these gauged wires was evaluated in
four slab specimens subjected to three point bending. Results from the
tests are discussed and details are given of a test programme currently
in progress to investigate the associated phenomena of reinforcement
ductility and moment redistribution.

INTRODUCTION

BS8110, the current British Standard for the Structural Use of Concrete
(1), permits the bending moments in the members of a continuous frame at
the ultimate limit state to be derived using an elastic analysis.
However, the ductility of the tension reinforcement in a reinforced
concrete member allows plastic hinges to form which modify the elastic
moment distribution and BS8110 recognises this by permitting a process of
moment redistribution. Subject to maintaining equilibrium between
internal and external forces, moments at a hinge may be reduced by up to
30% in braced frames, or by up to 10% in unbraced frames. EC2, the new
Eurocode for reinforced concrete (2), also permits moment redistribution
to occur, but distinguishes between high ductility and low ductility
reinforcement, with 30% redistribution being allowed in the former case,
but only 15% in the latter. No redistribution at all is permitted in
unbraced frames.

During the drafting of EC2, discussion arose concerning the related areas of moment redistribution, reinforcement ductility and allowable rotation of a reinforced concrete section, largely prompted by the work of Eligehausen and Langer (3). This prompted a re-examination of the ductility of UK reinforcement, with particular concern being raised with regard to fabric reinforcement used in slabs. As a consequence, the authors are undertaking a three year research programme, funded by the Science and Engineering Research Council, to examine the associated problems of reinforcement ductility and moment redistribution. A series of two-span beams and slabs are to be tested, using internally strain gauged reinforcing bars to give very detailed data concerning reinforcement strain distributions. Bars of this type have already been used with success in a number of investigations and up to the time of undertaking the work described in this paper the technique had been developed to the point where 100 strain gauges (gauge length 3 mm) could be installed in a 4 × 4 mm duct running longitudinally through the centre of each bar. References 4-9 describe the technique and application in detail but, briefly, each gauged rod is formed by milling two solid bars to a half round and then machining a longitudinal groove in each to accommodate the strain gauges and their lead wires. After installation of the gauges and wiring, the two halves are glued together to give the appearance of a normal reinforcing bar which may then be concreted into a test specimen. However, it was considered that a 12 mm diameter bar was the smallest size able to accommodate a 4 × 4 mm duct. Consequently, further development work was required in order to suit the technique for the smaller diameter wires commonly used in fabric reinforcement. This work was undertaken in advance of the SERC funded investigation and forms the main part of this paper. Brief details of the main investigation are also given and these will be elaborated on during the conference presentation.

FABRIC GAUGING DEVELOPMENT WORK

Ribbed wires, 8, 10 and 12 mm diameter, were used. It was considered that the "standard" 4 × 4 mm duct was suitable for the 12 mm wire, but this was reduced to 3.2 × 3.2 mm for the 10 mm wire and 2.5 × 2.5 mm for the 8 mm wire, these dimensions being selected on the basis of keeping the ratio of the cross-sectional area of the machined wire to that of the solid wire as constant as possible. Actual values were 88%, 87% and 86% for the 8, 10 and 12 mm wires respectively.

The form of fabric reinforcement, longitudinal wires welded to cross-wires, imposed an additional stage in the strain gauging procedure not previously encountered with bars. Machining and gauging of finished fabric was not attempted, instead main and cross-wires were welded together in the School's workshops using a purpose-made jig. Each mat had three longitudinal wires at 100 mm centres with cross-wires at 200 mm centres. The central longitudinal wire in each mat was gauged, so the "bottom" half of a machined wire was initially welded in place, the "top" half being glued on once gauging was complete. The practice established with bars of using epoxy resin for this operation was continued with the wires with complete success, the two halves being clamped together for three or four days while the adhesive cured. Gauging a half-wire which had cross-wires attached was slightly more awkward than working on one in isolation but presented no real problems.

As a start three mats were produced, one each with 8, 10 and 12 mm longitudinal wires, cross-wires being 8 mm in all cases. 31 strain gauges were installed in the 8 mm wire, 51 in the 10 mm wire, and 75 in the 12 mm wire, spaced at 12.5 mm centres over the central 250 mm, and at 25 mm elsewhere, giving gauged lengths of 500, 1000 and 1600 mm for the 8, 10 and 12 mm wires respectively. The 8 and 12 mm wires were completed satisfactorily, but there were five failures in the 10 mm wire, which was disappointingly high. Consequently, an additional 10 mm wire was machined and gauged to form a fourth mat which was far more successful with one failure only out of the 51 gauges installed. Strain gauges with a gauge length of only 2 mm had to be used in the smaller grooves, and fifteen high elongation gauges were provided at the centre of each rod to measure the very high central strains anticipated in the test specimens. Details of the gauges used are given in Table 1.

TABLE 1
Strain Gauge Specifications

Gauge Length (mm)	Gauge Width (mm)	Base (mm × mm)	Strain Limit (%)	Location
3	1.8	9 × 3.5	3	12 mm wire
2	0.9	5.5 × 1.5	3	8 & 10 mm wire
2	1.8	7.5 × 4*	10-15	All wires

* this dimension trimmed to suit duct width

Gauge resistance : 120 ohm ⎫
Gauge factor   : 2.12   ⎬ all gauges
Sensing element : foil   ⎭

TEST SPECIMENS

To assess the performance of the gauged wires, the finished mats were cast into four slab specimens, each 1800 mm long, 300 mm wide and 115 mm deep. Slab 1 had 8 mm longitudinal wires, Slabs 2 and 2A 10 mm wires (the latter being the additional specimen) and Slab 3 12 mm wires. The gauged wire in each mat was load cycled before being cast into a slab specimen in order to bed-in the gauges and check the strain gauge installation.

Concrete for the test specimens used 10 mm aggregate, had an aggregate/cement ratio of 5.5 and a water/cement ratio of 0.6. Test cubes and cylinders were cast with each specimen for the determination of compressive and indirect tensile strengths respectively.

TESTING

Specimens were loaded in three-point bending with supports being 1600 mm apart and the central point load applied using a hydraulic ram. Slabs were cast and tested in the order 2, 3, 1, 2A. Load was applied in increments of 1 or 2 kN and the development of flexural cracks noted. This procedure continued until gross yield of the wire at the central crack occurred, whereupon load was held constant whilst the specimen continued to deflect. This required very delicate load control. Slabs 1, 2 and 2A were tested to failure, Slab 3 was loaded until gross yield of the wires occurred and was then unloaded to observe the residual strains.

A full set of strain gauge readings was recorded at each load increment using a computer controlled data acquisition system, and further scans were taken during the post-yield constant load phase. Central deflections were recorded during the early stages of each test using a dial gauge. When deflections exceeded the travel of this gauge, approximate values were recorded with a steel rule, but this ceased during the late stages of each test for reasons of safety. Average surface strains were measured at early stages in each test using a 200 mm Demec gauge. These were compared with the corresponding strain gauge readings. This enabled a check on strain gauge performance to be effected at these early stages when, before cracking, surface strains and wire strains would be comparable. Such comparisons were not possible once cracking had led to bond breakdown and consequent loss of strain compatibility between the wires and surrounding concrete.

RESULTS AND DISCUSSION

The gauged wires performed well and gave very detailed data concerning strain distributions in the specimens, indicating that the purpose of the work, the development of a technique for internally strain gauging fabric reinforcement, was successfully achieved.

Figure 1 illustrates the changes that occurred in the early stages of a test as cracks developed, with strains peaking at cracks where the wires had to carry virtually all the tensile force in the cross section, and declining between cracks as load was shared between the wires and the surrounding concrete due to the action of bond. As would be expected, the number of cracks in a specimen increased with increasing reinforcement percentage, an implication of this being that cracks did not necessarily coincide with cross-wire positions. Bond stresses were typically in the range 1.0 to 2.5 N/mm^2 whilst the wires were behaving elastically, but decreased rapidly once the inelastic zones of the stress-strain relationships were reached. These values may be compared with those for 12 mm Torbar in earlier work (4) which were typically 4.0 to 5.0 N/mm^2.

The post-yield behaviour of the wires was most interesting, with plastic hinges being developed at the centre of all specimens, as shown in Figures 2 and 3. Peak strains of 5.4% in the 8 mm wire and 3.3% in the 10 mm wire were recorded although, with strains increasing very rapidly at this late stage in a test, actual failure strains were

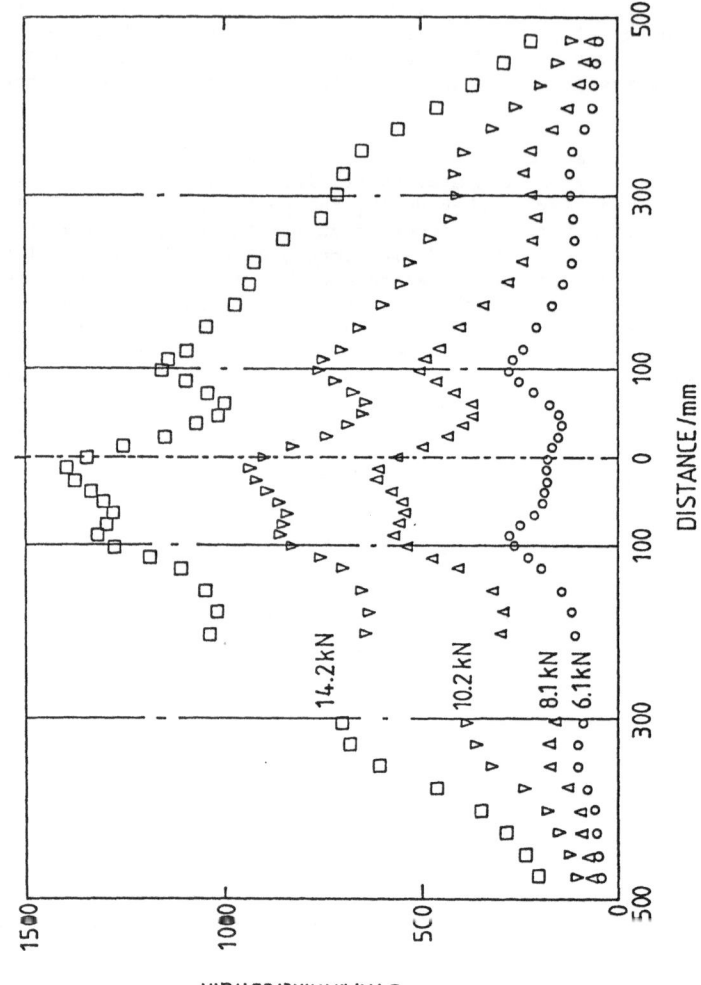

FIG. 1 : SLAB 2A STRAIN DISTRIBUTIONS : 1

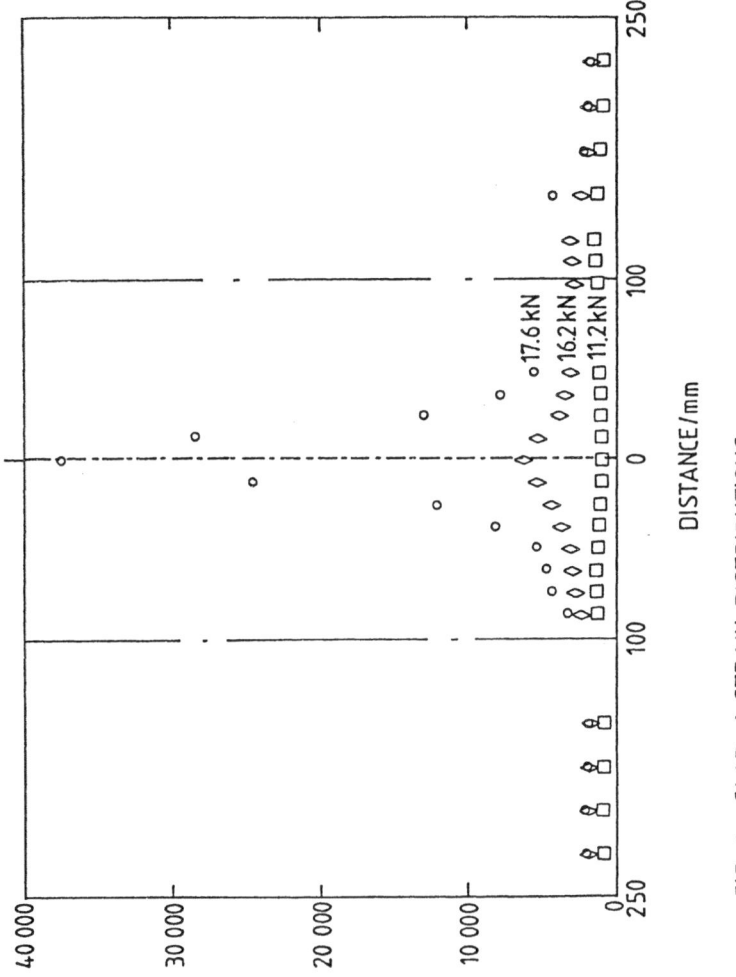

FIG. 2 : SLAB 1 STRAIN DISTRIBUTIONS

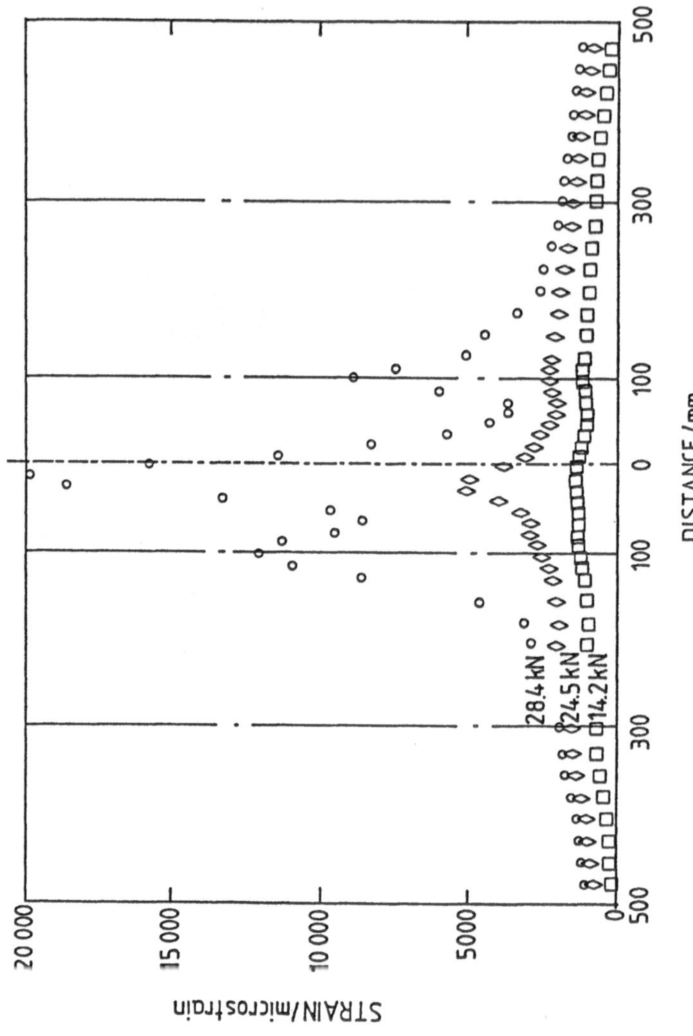

FIG. 3 : SLAB 2A STRAIN DISTRIBUTIONS: 2

believed to have been much higher. This is supported by the values for percentage reduction in area, calculated from measurements on the solid wires (as measurements on the gauged wires were complicated by the indeterminate distortion around the ducts). Average values were 66.3% for the 8 mm wire (standard deviation 2.5%) and 41.0% for the 10 mm wire (standard deviation 5.5%). The width of the plastic hinge zone varied, being 100 mm Slab 1, 200 mm in Slab 2 and 450 mm in Slab 3. This increase with increasing reinforcment percentage is a likely result of the closer crack spacings already referred to. Strains in the plastic hinge zone recovered by about 2000 microstrain when Slab 3 was unloaded.

DEVELOPMENTS

The experience gained in these tests suggests that it should be possible to increase the number of gauges installed in the 8 and 10 mm wires in future work. Gauging 6 mm wire may also be practicable, though with a duct size of 2 × 2 mm (at best) this would mean the use of strain gauges with a 1 mm gauge length. Up to 20 could be installed in a duct of this size.

These developments will be particularly useful in the SERC funded work. A series of two-span beams and slabs up to 5 m long are to be tested to investigate moment redistribution from the central support into the adjacent spans. Each span will be loaded with a central point load. A variety of reinforcement arrangements will be used at the central support for comparison of the effects of bar diameter and reinforcement percentage. Those for the first series of tests, which will be for beams, are shown in Figure 4. Span reinforcement layouts will permit up to 30% moment redistribution, but these can be adjusted in the light of early test results. The use of gauged bars in the tension and compression reinforcement both at the support and the centre of one span will enable very detailed measurements of strains and bond stresses to be made. Surface concrete strains will be measured with a Demec gauge and surface mounted strain gauges. The data from these tests will be used to calibrate a numerical model of the specimens' behaviour, work on which will commence shortly. Early test results will be presented at the conference.

CONCLUSIONS

The work has demonstrated the practicability of internally strain gauging fabric reinforcement with 31, 41 and 75 gauges being installed in 8, 10 and 12 mm wires respectively. Four slab specimens tested in three point bending clearly showed the development of plastic hinges at the central load positions. The success of the tests suggested that more gauges could be installed in future specimens and that consideration could be given to the gauging of 6 mm wire.

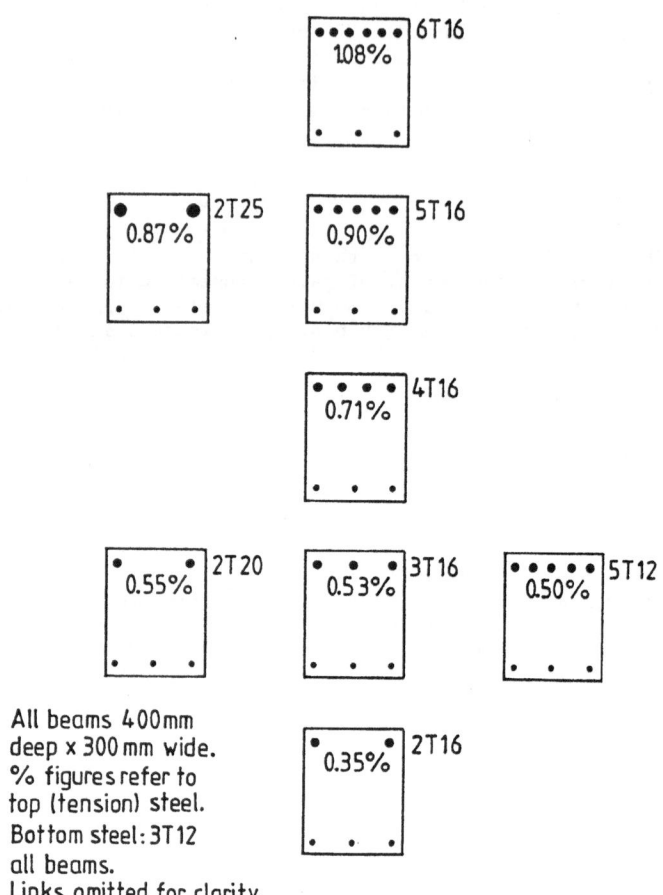

All beams 400mm
deep x 300 mm wide.
% figures refer to
top (tension) steel.
Bottom steel: 3T12
all beams.
Links omitted for clarity.

Fig. 4: <u>SECTIONS THROUGH PROPOSED TEST BEAMS</u>
<u>AT CENTRAL SUPPORT</u>

515

ACKNOWLEDGEMENTS

Funding for the fabric gauging development work was provided by the Fabric Reinforcement Development Association whilst that for the main investigation is coming from the Science and Engineering Research Council. Mr. S.P. Wilkinson was responsible for the installation of the strain gauges.

REFERENCES

1. BS8110, Structural use of concrete. British Standards Institution, 1985.

2. EC2: Eurocode No. 2, Design of concrete structures. Part I, General Rules and Rules for Buildings, Final Draft, December 1988.

3. Eligehausen, R. and Langer, P., Rotation capacity of plastic hinges and allowable degree of moment redistribution. University of Stuttgart, January 1987. (English translation)

4. Scott, R.H. and Gill, P.A.T., Short-term distributions of strain and bond stress along tension reinforcement. The Structural Engineer, Vol. 65B, No. 2, June 1987, pp.37-43, 47.

5. Scott, R.H. and Gill, P.A.T., Measurement of internal concrete strains using embedment strain gauges. Magazine of Concrete Research, Vol. 39, No. 139, June 1987, pp.109-112.

6. Judge, R.C.B., Scott, R.H. and Gill, P.A.T., Force transfer in compression lap joints in reinforced concrete. Magazine of Concrete Research, Vol. 41, No. 146, March 1989, pp.27-31.

7. Judge, R.C.B., Scott, R.H. and Gill, P.A.T., Strain and bond stress distributions in tension lap joints in reinforced concrete, to be published in Magazine of Concrete Research, March 1990.

8. Scott, R.H. and Gill, P.A.T., Time dependent distributions of strain and bond stress along tension reinforcement, to be published in Proc. I.C.E., June 1990.

9. Scott, R.H. and Gill, P.A.T., An experimental investigation into the performance of reinforced concrete connections". Applied Solid Mechanics 3, University of Surrey, April 1989, Proceedings published by Elsevier Science Publishers Ltd., pp 128-142, (ISBN 1-85166-435-1).

A NEW COMPUTER-AIDED SYSTEM FOR PHOTOELASTIC STRESS ANALYSIS WITH STRUCTURE-DRIVEN TYPE IMAGE PROCESSING

Masahisa TAKASHI, Shizuo MAWATARI, Yoshiaki TOYODA
College of Science and Engineering
Aoyama Gakuin University
6-16-1, Chitosedai, Setagaya-ku, Tokyo, Japan
and
Takeshi Kunio
Department of Mechanical Engineering
Kanto Gakuin University
4834, Mutsuura-cho, Kanazawa-ku, Yokohama, Japan

ABSTRACT

Two major difficulties in the current computer-aided photoelastic analysis, namely automatic determination of principal stress direction and of fringe order over the whole area of specimen, were successfully overcome with single-valued representative functions for principal stress direction with the diagonal summation theorem and for fringe order with the concept of transversality of intersection in Catastrophe Theory. Utilizing a structure-driven type image processing, not only the stress trajectories but also the stress distribution on an arbitrarily designated axis are successfully calculated on the basis of theories of differential geometry and ordinary differential equations, respectively. A new system developed in this study is useful for two dimensional analysis as well as for the stress-frozen technique in three dimensional analysis, in which the isochromatics and the isoclinics are inevitably superimposed.

INTRODUCTION

Although a number of studies, most of which are developed on the basis of the so-called Fringe Diagram Method, have been published up to date as introduced in a recent review paper[1], current computer-aided techniques for the photoelastic analysis are not always successful. A major problem here is the lack of investigation into theories and algorithms suitable for digitized brightness data of image which involves unavoidable noise.

It has been pointed out that the following problems are still unsolved in development of an automatic system for two dimensional photoelastic stress analysis.
a) Determination of principal stress direction by means of isoclinics which often show extreme broadness and poor contrast in image,

b) Automatic determination of fringe order from a single piece of isochromatic pattern over the whole area of specimen,
c) Reduction of analytical error caused by unavoidable noise and inappropriate algorithms in numerical solution of differential equation.
   It would be necessary to review the basic relationships in optical
equation of photoelasticity and to construct better representative functions for both principal stress direction and fringe order, taking the
fundamental relationship in elasticity into account.
   In this paper, the authors propose a new computer-aided system for
two dimensional photoelastic stress analysis over the whole area of specimen. From a view point of structure-driven type image processing with
principles and algorithms suitable for digitized image data, the system can
surmount the difficulties mentioned above.

BACKGROUND FOR COMPUTER-AIDED PHOTOELASTICITY

Three Types of Digital Image Processing
The recent developments in practical digital image processing are classified into three types[2] as follows;
1) Data driven processing:
   This type processes input data with transforming, binarizing, sharpen-
   ing and smoothing techniques in addition to contrast adjustment/en-
   hancement, noise reduction and geometrical compensation, in order to
   extract characteristic features of object image.
2) Model driven processing:
   This group selects an optimum pattern after several steps of preproc-
   essing such as noise reduction and normalization.
3) Structure driven processing:
   This type reconstructs an appropriate image by considering brightness
   distribution of object image as realization of a certain function and
   by solving the corresponding known structural equations taking its
   physical background into consideration.
   For photoelastic analysis, the first type seems to be fruitless,
while the second one inappropriate. The structure driven type processing
is promising, since image data of isochromatics and/or isoclinics are
definitely representable with the structural equations in optics.

Current Trends of Computer-Aided Photoelastic Stress Analysis
The following two major types of computer-aided photoelastic techniques are
currently in practical use for experimental data handling.
  i) Fringe diagram method[1]: In this method, isochromatics and isoclinics
     are usually obtained through a circular and/or a plane polariscope
     separately. Since only the streaks of them are extracted with the
     data driven type image processing, large part of data between adjacent
     fringes is discarded. Omitted information has to be reproduced by
     proper inter/extrapolation techniques.
 ii) Function construction method[2][3]: Brightness distribution of image
     data over the whole area of specimen is effectively utilized in order
     to construct single-valued functions of principal stress direction and
     difference, respectively. The system proposed by the authors has been
     developed on the basis of a concept which belongs to the second cate-
     gory.

Mathematical Preparation for Structure Driven Image Processing
Approximation of Representative Functions: When representing experimental

data as realization of a certain function, its form is usually unknown. An arbitrary continuous function can be, however, substituted in broader aspects with a set of cubic spline functions, according to a functional approximation theory in the field of numerical analysis. For the problem of automatic placement of knots in the spline function theory, a new successful algorithm has already been developed by the authors[4]. Thus, a set of cubic spline functions is, if necessary, applicable to construction of a representative function for experimental data.

Foliation and Parameter Family: Consider a two-dimensional rectangular image domain M_x x M_y which includes specimen area S and a real value function F on S, which represents the brightness distribution of experimental data. Assume F is sufficiently differentiable. The domain S and the function F are divided into a set of leaf as follows;

$$S_x = \{(x,y) \in S: y \in M_y\}, \qquad S_y = \{(x,y) \in S: x \in M_x\} \qquad (1)$$

$$F_x = F \mid S_x, \qquad\qquad F_y = F \mid S_y. \qquad (2)$$

Namely, the leaf S_x or S_y is the coordinate line along y- or x-axis in the area S obtained by fixing the value of one of the variables x or y. And the F_x and F_y are defined on S_x and S_y, respectively. Resolution of S into a set of S_x or S_y is called 'Foliation[5].' Each set of F_x or F_y is a single variable parameter family.

Concept of Unfolding and Photoelastic Data: Photoelastic data depend not only on magnitude of load and thickness of specimen but also on wave length of incident light and photoelastic constant of the material. Accordingly, a piece of photoelastic data, f, is in general measured under a specified combination of test condition, $u \in R^e$. In order to extend the result obtained under a specific condition to more general circumstances, it is convenient to adopt the concept of 'Unfolding[6].'

A Newly Developed Formula of Numerical Differentiation: Calculation of numerical differentiation is extremely important in study of image processing in which digitized data involving noise have to be handled. In such cases, the derivative of the Lagrange interpolation formula is usually used in order to find the numerical value of the derivative $f'_p = f'(x_p)$ at a point x_p from the tabulated values $f_k = f(x_k)$. Accuracy of this type formula is susceptible to loss of significant digits due to cancellation. Its effect on accuracy of analysis is detrimental especially when the data involve unavoidable noise.

In order to construct a numerical differentiation formula which stands for representative functions of experimental data involving noise, it must be appropriate to take differentiation of a smoothing polynomial into consideration. The new formula[7] obtained is given as;

$$g'(x_0) = \frac{\sum_{k=1}^{n} (f_k - f_{-k})}{hn(n+1)} \qquad (3)$$

where $h = x_{k+1} - x_k$.

Relative Boundary Value and Solution of Differential Equation: Equilibrium equations in two-dimensional elasticity are still valid even when a constant is added to all of stress components, σ_x, σ_y and τ_{xy}.

On resolving two-dimensional image data into foliation and on solving ordinary differential equation in a framework of initial value problem, one can start from a relative boundary value selected arbitrarily, instead of an actual ones. Solution of the differential equation will be easily corrected by comparison of both boundary values, namely the assumed and the actual one. This method gives us advantages not only of easiness in solution but also of reduction of both occurrence and propagation of errors.

Diagonal Summation Theorem

The well-known equation of isochromatics and isoclinics observed through a plane polariscope is given as;

$$I = a^2 \sin^2 2\phi_1 \sin^2(\delta/2) \tag{4}$$

where a is the amplitude of incident polarized light, $\phi_1 (=\phi-\theta)$ is the angle between the principal axes of polarizer and stress, and δ is the phase difference arisen after passage of light through specimen. Also, the relationship between phase difference δ and principal stress difference $(\sigma_1-\sigma_2)$ is written as $\delta = (2\pi c/\lambda)d(\sigma_1-\sigma_2)$, where, c is a photoelastic constant and d the thickness of specimen.

Brightness of isochromatic fringe is, as well known, written as;

$$L = a^2 \sin^2(\delta/2) = a^2 \sin^2(N\pi), \tag{5}$$

because of the relationship between fringe order N and phase difference,δ that is, $N = \delta/(2\pi)$.

The 'Diagonal Summation' theorem[8][9], which regards separation of isoclinics and isochromatics from several images obtained through a plane polariscope under the same loading condition, is derived by the following manner.

Four different image patterns which have brightness distributions $I(\theta_i)$, i=1,2,3,4, are measured under the conditions of $\theta_1 - \theta_2 = \theta_3 - \theta_4 = \pi/4$. For an ideal case with no noise, the following relation is easily obtained,

$$I(\theta_1) + I(\theta_2) = I(\theta_3) + I(\theta_4) = a^2 \sin^2(N\pi). \tag{6}$$

This equation implies that summation of brightness distribution of only two different images through a plane polariscope under different angles separated by $\pi/4$ can make an isochromatic image. Since effects of unavoidable noise cannot be neglected in a practical situation, four images are necessary to separate isochromatics and isoclinics. Assuming that effects of noise do not depend on θ, the following values are designated as follows inorder to facilitate calculation for noise elimination,

$$A = I(\theta_2) - I(\theta_1), \qquad B = I(\theta_4) - I(\theta_3)$$
$$C = [A\cos 4(\theta_1 - \theta_3) - B] / \sin 4(\theta_1 - \theta_3).$$

Thus, the diagonal summation theorem is derived as follows,

$$(A + C)^{1/2} = a^2 \sin^2(N\pi) = L, \tag{7}$$

and an image pattern of isochromatics is separated from the original one measured through a plane polariscope. Dividing the brightness distribution of original image Eq.(4)), by the separated one of isochromatics L, isoclinics over the whole area can be also obtained.

According to this theorem, there is no need to take a number of image patterns for construction of an isoclinic image pattern. Also highly accurate data of both isochromatics and isoclinics are easily obtainable only from four pieces of image pattern under a plane polariscope without the use of a circular polariscope.

Theorem for Increase/Decrease Reversal of Fringe Order
Useful and effective methods for this problem have not been established yet because of complicated features of isochromatic patterns dependent on stress distribution.

Referring to Eq.(7) in the previous section, it will be expected that fringe order over the whole area of specimen could be determined by inversely solving the equation from only a single isochromatic image pattern. It is not, however, so easy because inverse cosine function of Eq.(7) is multivalued as,

$$N = \{2k\pi \pm Cos^{-1}(1-2L/a^2)\}/2\pi, \quad k = 0, 1, 2, \ldots \quad (8)$$

Relations between the first- and second-derivatives of fringe order N and brightness L with respect to x, a spatial variable limited within S_y, are given by,

$$D^1L = (D^1N)\pi a^2 \sin 2N\pi,$$

$$D^2L = \pi a^2\{(D^1N)^2 2\pi \cos 2N\pi + (D^2N)\sin 2N\pi\},$$

where, $D^k = \partial^k/\partial x^k$, (k = 1, 2). Investigating the above two equations on the basis of the concept of "Transversality" in Catastrophe Theory[10], the following useful theorems applicable to determination of fringe order can be obtained. The increase/decrease reversal of fringe order[11][12] along a scanning line, for example along x-axis, takes place at points as below;
(a) When 2N is not an integer, at the extreme points of brightness distribution,
(b) When 2N is an integer, i) at points of the extremal intersection of isochromatics with x-axis, and ii)at points of the inflectional intersections of isochromatics with x-axis, under the condition of $\partial^2 L/\partial y^2 \neq 0$.

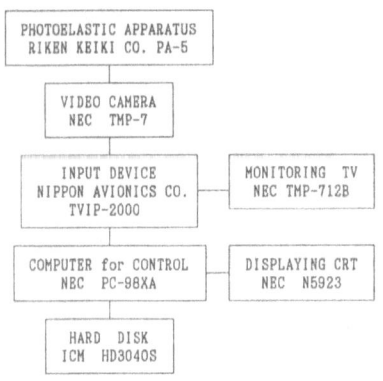

Figure 1. Hardware system[13].

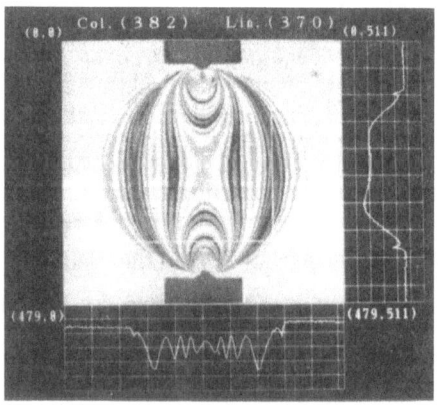

Figure 2. Displaying tool for brightness distribution[13].

EXPERIMENTS

Figure 1 shows an illustration of hardware system used in this study. In order to take high quality image data, attentions are paid to several points as follows;

(a) Amplitude of random noise generated by fluctuation of light source is minimized by increasing the intensity of incident light.

(b) Brightness of image is adjusted so that is kept from 0 to 255(in the case of 8-bits data) in gray level at any observed point.

(c) Unnatural decline of brightness in the vicinity of the extrema is avoided as possible, by monitoring the brightness distribution with a displaying tool[13][14] as shown in Figure 2.

As mentioned in the previous sections, four image patterns under a plane polariscope are stored in computer memory.

ALGORITHMS AND RESULTS ANALYZED

Principal Stress Direction

Succeeding the discussion in the section on Diagonal Summation theorem, the algorithm for determination of principal stress direction can be easily composed as follows. Suppose $A^2 + C^2 \neq 0$, then putting $D = \cos 4(\phi - \theta_1)$ and $E = \sin(\phi - \theta_1)$, D and E are computed. The principal stress direction ϕ is obtained as,

$$\phi = \theta_1 + n\pi + (-1)^n \sin^{-1}(E), \quad (n = 0, \pm 1, \pm 2, \ldots). \quad (9)$$

Here, if $\theta_1 = 0$, one obtains $n = 0$ from Eq.(9) under under the condition

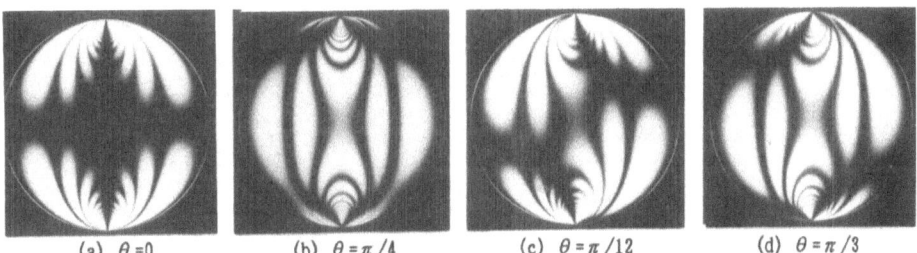

(a) $\theta = 0$ (b) $\theta = \pi/4$ (c) $\theta = \pi/12$ (d) $\theta = \pi/3$

Figure 3. Four images through a plane polariscope[8, 9]

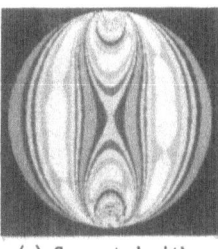
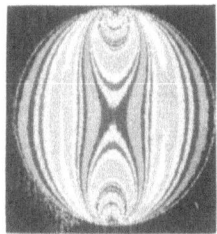

(a) Separated with Diagonal Summation   (b) Theoretical values   (c) Separated with Diagonal Summation   (d) Obtained through a circular polariscope

Figure 4. Displayed images of principal stress direction (a) and (b), and principal stress difference (c) and (d)[8, 9]

$|\phi| \leqq \pi/2$. Thus, the principal stress direction ϕ against the reference axis is calculated over the whole area of specimen.

If $A^2 + C^2 = 0$, additional considerations are needed because of the division by zero. For this case, the direction is calculated by the linear inter-/extrapolation with values of ϕ of surrounding points.

Figure 3(a)-(d) show four example images obtained through a plane polariscope at the angles $\theta_1 = 0$, $\theta_2 = \pi/4$, $\theta_3 = \pi/12$, $\theta_4 = \pi/3$, respectively. The results separated with the algorithm mentioned above is also shown in Figure 4(a)-(d).

Fringe Order and Principal Stress Difference

Fringe order N and/or principal stress difference $(\sigma_1 - \sigma_2)$ are calculated using Eq.(8) over the whole area of specimen, since the value of k can be decided from boundary conditions. Also, selection of sign, namely positive

(a) Original isochromatics separated with Diagonal Summation

(b) Fringe order automatically asigned with the method proposed

Figure 5. Displayed images of a three-points bending specimen[11, 12]

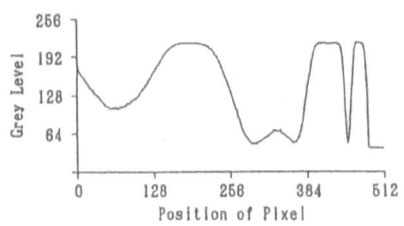

(a) Brightness distribution of isochromatics

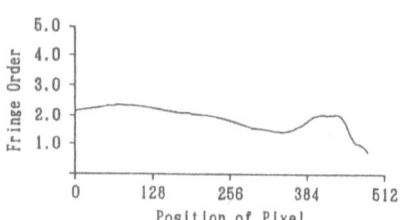

(b) Fringe order distribution

Figure 6. Graphical expressions of the results on S_y[11, 12]

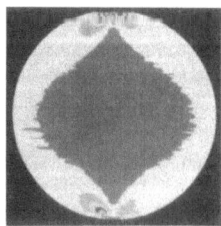

(a) Calculated values of σ_x

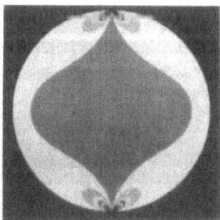

(b) Theoretical values of σ_x

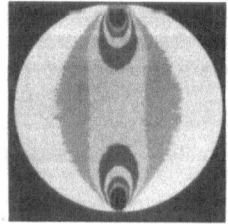

(c) Calculated values of σ_y

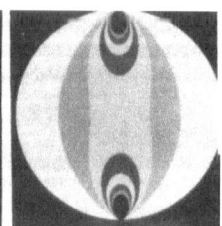

(d) Theoretical values of σ_y

Figure 7. Comparisons of the results with the theoretical values

Table 1. Evaluation of errors

Components of Stress	y / r	① Theoretical Values (×100)	② Shear Differ- ence Method (×100)	③ Our Method (×100)	④ Errors in ② (%)	⑤ Errors in ③ (%)
σ_x	0.0	26.560	27.502	26.556	23.451	0.025
	0.2	25.509	28.803	25.507	11.056	0.019
	0.4	22.463	23.378	22.472	38.767	1.942
	0.6	19.271	22.526	19.287	51.097	0.835
	0.8	35.273	63.918	34.915	107.487	1.590
σ_y	0.0	88.565	98.391	87.006	18.773	1.761
	0.2	91.019	99.194	92.200	12.870	1.298
	0.4	99.974	111.522	101.343	16.242	1.369
	0.6	120.922	127.806	122.392	9.060	1.216
	0.8	177.540	197.123	173.320	15.461	2.377

or negative, is performed with the theorem of 'Increase/decrease reversal' of fringe order on a scanning line. Figure 5(a) and (b) show displayed images an original isochromatics and a result obtained with the method poroposed, respectively. Furthermore, the brightness and the fringe order distribution analyzed on Sy are shown in Figure 6(a) and (b).

Separation of Stress Components in Cartesian Coordinates

Equilibrium equations of two-dimensional stress state in a Cartesian coordinates is generally written as;

$$D_1 \sigma_x + D_2 \tau_{xy} + f_1 = 0, \text{ and } D_2 \sigma_y + D_1 \tau_{xy} + f_2 = 0, \qquad (10)$$

where $D_1 = \partial/\partial x$, and $D_2 = \partial/\partial y$. The stress components σ_x, σ_y, and τ_{xy} have to satisfy a set of given boundary conditions. According to the classical method of shear difference, Eq.(10) can be rewritten as;

$$\sigma_x = \sigma_{x0} - \int_{x0}^{x} (D_2 \tau_{xy} + f_1) dx, \text{ and } \sigma_y = \sigma_{y0} - \int_{y0}^{y} (D_1 \tau_{xy} + f_2) dy.$$

Since the current status of discretization for the differential/integral approximation in the shear difference method is still quite primitive, generation and propagation of errors due to discretization can not be avoided. In order to achieve a higher accuracy in the calculation, it is necessary to establish more rigorous discretization for those equations. Thus, in this paper, one of Eq.(10) is rewritten as;

$$D_1 \sigma_x = -D_2 \tau_{xy} - f_1. \qquad (11)$$

Here, using the principal stress direction ϕ and the difference $(\sigma_1-\sigma_2)$ calculated before in addition to the relationship of $\tau_{xy} = -0.5(\sigma_1-\sigma_2)\sin(2\phi)$, the values of $D^2\tau_{xy}$ are accurately obtained with the better numerical differentiation method in Eq.(3). Moreover, it is convenient to utilize Euler- or Heun-method in order to solve Eq.(11) numerically, considering it as a initial-value problem of ordinary differential equation on the leaf S_y together with a concept of the so-called relative boundary. The results obtained are shown in Figure 7(a)and (c) together with theoretical values (b) and (d), respectively. Evaluation of errors of the method proposed are compared with those by other methods in Table 1.

Stress Trajectories and Separation of Principal Stresses

Regarding a principal stress trajectory in two-dimensional elasticity as a plane curve s ? C(s), the length of tangent vector at C(s) with respect to the arc length s is shown to be unity in differential geometry. A structural equation of principal stress trajectory with the principal stress direction θ at a point C(s) is, therefore, derived as,

$$dC(s)/ds = (\cos\theta, \sin\theta). \tag{12}$$

Starting from an arbitrary point inside the area of specimen and solving Eq.(12) numerically with Euler- or Heun-method, stress trajectory is easily drawn.

Prior to solving Lamé-Maxwell equation numerically on each leaf S_y, curvature of a stress trajectory has to be obtained. Denoting the principal stress direction and the curvature at a point C(s) on a stress trajectory as $\phi(s)$ and $\rho_c(s)$, respectively, the following relation is obtained,

$$\rho_c(s) = d\phi(s)/ds. \tag{13}$$

Accordingly, the curvature $\rho_c(s)$ is readily obtained through a simple calculation of the right hand side in the above equation using Eq.(3), after determining $\phi(s)$ at each point on the stress trajectory drawn with the method mentioned above.

Coming to this stage, components of principal stress σ_1 and σ_2 are easily calculated in a manner similar to that in the previous section. It should be emphasized that the method proposed is not only simple but conformable to the definition of stress trajectory and the basic elastic relation. Typical examples of the results obtained are shown in Figure 8 and Figure 9.

CONCLUDING REMARKS

For the purpose of developing a better computer-aided system for two dimensional photoelastic stress analysis, several concepts, theorems and techniques from a wide variety of mathematical fields are utilized and integrated. Not only improvement in each step of analysis, such as numerical differentiation /integration, but also consistency in the flow of analysis on digital image involving unavoidable noise is attained. The application

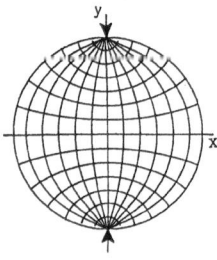

Figure 8. Expression of
principal stress
trajectories

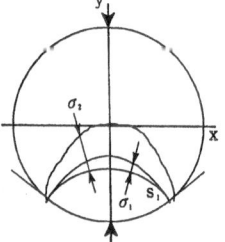

(a) From calculated
values

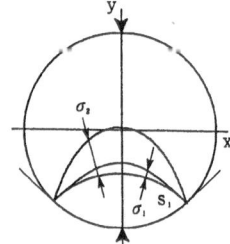

(b) From theoretical
values

Figure 9. Distribution of principal stresses
σ_1 and σ_2 on a obtained trajectory

of the system to experimental data gives good agreement with the results obtained from theoretical and/or numerical analyses such as by FEM or BEM. Using the system, components of strain and displacement can also be obtained as needed.

Moreover, it should be emphasized that the new system developed in this study must be useful and effective not only for two dimensional photoelastic analysis but also for the stress-frozen technique in three dimensional one, in which isochromatics and isoclinics are inevitably superimposed.

REFERENCES

1. Patterson, E.A., Automated photoelastic analysis, Strain, 1988, 24, 15-20.
2. Takashi, M., Toyoda, Y., Mawatari, S., Hanai,K. and Shimasaki, K., Structure driven processing for image understanding and its application to photoelastic analysis. Proc. of 6th Int. Cong. on Exp. Mech., Soc. for Exp. Mech., USA 1988, pp.189-194.
3 Mawatari, S., Takashi, M. and Toyoda, Y., A generalization method for assignment of isochromatic fringe order using structure driven-type image processing. Trans. of JSME, 1989, 55A, pp598-607(in Japanese).
4. Mawatari, S., Takashi, M., Toyoda,Y., A simplified algorithm for determining knots for smoothing with spline functions. Denshi jouhou tuushin gakkai ronbunshi, 1989, J72-D-II, 18-6-1823(In Japanese).
5. Lawson, JR., H.B., Foliations, Bull. Amer. Math. Soc. 1974, 80, 369-418.
6. Poston,T. and Stewart, J., Catastrophe theory and its applications, Pitman, London, 1978, pp.143-150.
7. Mawatari, S., Takashi, M., Toyoda, Y., Differentiable structure of a representative function for experimental data and treatment of errors due to noise. Trans. of JSME., 1990, 56A, (to be published in Japanese).
8. Mawatari, S., Takashi, M, and Toyoda, Y., Whole-area photoelastic analysis by image processing on the principal stress direction and separation of isochromatics from isoclinics, Trans. of JSME., 1989, 55A, pp1423-1428(in Japanese).
9. Mawatari, S. Takashi, M., Toyoda, Y. and Kunio, T., A single-valued representative function for determination of principal stress direction in photoelastic analysis. In Proc. 9th Int. Conf. on Exp. Mech., Copenhagen Denmark, 1990.
10. Poston,T. and Stewart, J., Catastrophe theory and its applications, Pitman, London, 1978, pp.66-72.
11. Mawatari, S., Takashi, M., Toyoda, Y., Automatic assignment of isochromatic fringe order using a single-valued function in plane photoelasticity. Trans. of JSME., 1990, 56A, 902-908(in Japanese).
12. Mawatari, S., Takashi, M., Toyoda and Kunio, T., A new method of Computer-aided fringe order determination of isochromatics in two dimensional photoelasticity. In Proc. of 9th Int. Conf. on Exp. Mech., Copenhagen,Denmark, 1990.
13. Mawatari, S., Shimasaki, K., Hanai, K., Takashi, M. and Toyoda Y., Effective input, display and storage of fringe pattern data using a microcomputer. Proc. of Jpn. Soc. Photoelasticity, 1989, 9, pp15-22(in Japanese).
14. Mawatari, S., Takashi, M. and Toyoda, Y., Computer definition of photoelastic fringes using spline smoothing method. Proc. of Jpn. Soc. Photoelasticity, 1989, 9, 23-29(in Japanese).

WHOLE-FIELD MEASUREMENT OF PRINCIPAL STRESS DIRECTIONS
FROM PHOTOELASTIC EXPERIMENT USING IMAGE PROCESSING SYSTEM

EISAKU UMEZAKI, TAMOTSU TAMAKI
Nippon Institute of Technology, Miyashiro, Saitama 345, Japan

AKIRA SHIMAMOTO
Saitama Institute of Technology, Okabe, Saitama 369-02, Japan

and
SUSUMU TAKAHASHI
Kanto Gakuin University, Mutuura, Yokohama 224, Japan

ABSTRACT

A method for measuring automatically isoclinic parameters (i.e. principal stress directions) in the whole field of a model from photoelastic experiment has been developed. A basic idea for obtaining the directions of principal stresses was to utilize that a time series curve of light intensity changes sinusoidally at each point in the model using a plane polariscope in which a polarizer and analyzer were rotated, and the minimum location on the curve corresponds to the direction of principal stresses. The minimum location was detected by using the Fourier-series expansion of the curve. The above method was realized on an automatic polariscope with a TV camera, a personal computer, an image processing equipment etc. Using this system, the directions of principal stresses in the whole field could be automatically determined with a high accuracy and without any interpolation.

INTRODUCTION

Recently, several systems for analyzing the whole photoelastic fringe pattern in a model and separating stress components have been developed [1]-[4], but not fully automated yet. One of the cause is that the completely automatic acquisition of photoelastic data (i.e. isochromatic fringe orders and isoclinic parameters) is not achieved in the whole field of a model. Redner[5] and Allison and Nurse[6] have developed similar systems for the completely automatic acquisition of photoelastic data at a point, but been unsuitable for the acquisition in the whole field. Nisida and Ohi[3] have obtained absolute isochromatic fringe orders automatically at each pixel in digitized images by use of a lineary-increasing load method or a shifting light wavelength method called by them.

On the one hand, for obtaining isoclinic parameters Muller and Saackel[1] and Seguchi et al.[2] have utilized techniques for extracting the center lines of fringes in binary images. Umezaki et al.[4] have adopted a technique for searching minimum light intensities corresponding to isoclinic parameters. However, their techniques were not very accurate because changes in isoclinic light intensity are considerably small compared with isochromatic one, the light intensity contains many noises, hence isoclinic fringes are not clear, and interpolation was required for determining isoclinic parameters over the entire domain. Mawatari et al.[7] have used a technique for computing the phase of the wave front from four isoclinic patterns, applying the four-phase shift method. This technique was based on the assumption that light intensity changes sinusoidally without containing noises, and hence was sensitive to noises.

In this study, a method was developed which can obtain automatically isoclinic parameters, that is, the directions of principal stresses over the entire domain without any interpolation and is insensible of noises, and its effectiveness was investigated.

PRINCIPLE OF MEASUREMENT

A basic idea for obtaining the direction of principal stresses was to utilize a time series curve at each point in a photoelastic model using the plane polariscope set in the dark field. As shown in Fig.1, the time series curve of light intensity at a noted point of the model changes sinusoidally with the rotation of the polarizer and analyzer. The location of minimum light intensity on the curve corresponds to the direction of principal stresses at the noted point.

Plane polariscope
The light intensity emerging from the plane polariscope of crossed

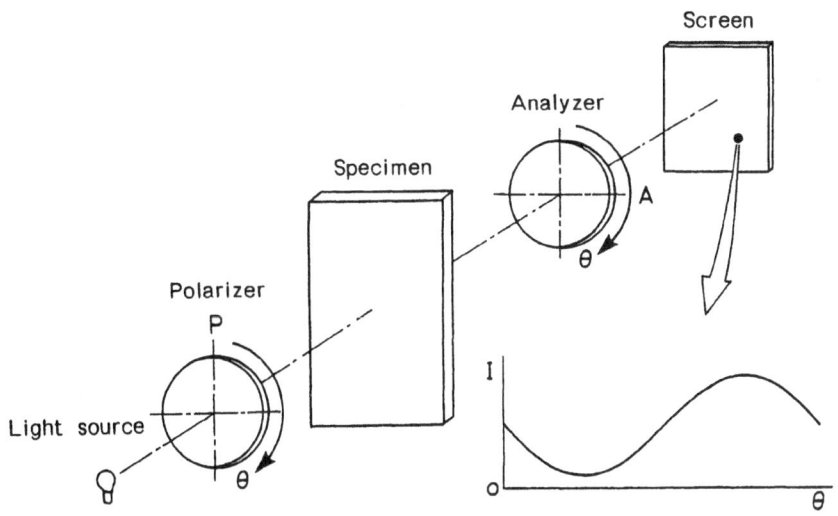

Figure 1. Plane polariscope and measurement of direction of principal stresses.

polarizer and analyzer can be described by

$$I=a^2\sin2\pi N\sin^2 2\phi_p \qquad (1)$$

where a is the light amplitude, N is the fringe order and ϕ_p is the angle between the position of polarizer assembly and the direction of one of principal stresses.

Change of light intensity due to polarizers rotation
As shown in Fig.2, when crossed polarizer (P) and analyzer (A) are rotated by θ from a selected reference (R), equation (1) can expressed as:

$$I=I_0\sin^2 2(\phi-\theta) \qquad (2)$$

where $I_0(=a^2\sin^2\pi N)$ is a constant and ϕ is the direction of one of principal stresses to the reference axis (R). Adding a background intensity (b) to equation (2), we can obtain

$$I=I_0\sin^2 2(\phi-\theta)+b \qquad (3)$$

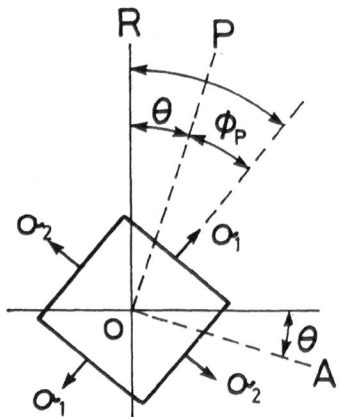

Figure 2. Relation between direction of principal stresses, polarizer (P) and analyzer (A).

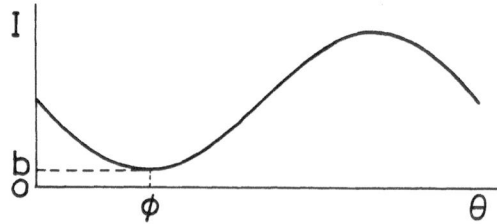

Figure 3. Light intensity curve obtained from rotation of crossed polarizers.

The equation (3) shows that the light intensity changes sinusoidally, and becomes minimum (b) when the polarizer is parallel to one of the principal stresses, as shown in Fig.3. Therefore, we can know the direction of principal stresses by finding the location of minimum light intensity on the curve.

Determination of minimum location on light intensity curve

Equation (3) can be rewritten as:

$$I=-(I_0/2)\sin(4\theta-4\phi+\pi/2)+(I_0/2)+b \tag{4}$$

Substituting $C_1=-I_0/2$, $\beta_1=4\theta$, $\phi_1=-4\phi+\pi/2$ and $a_0=I_0/2+b$ into equation (4), we can obtain

$$I=C_1\sin(\beta_1+\phi_1)+a_0 \tag{5}$$

C_1, ϕ_1 and a_0 in equation (5) are computed from the Fourier-series expansion of equation (5) by use of discrete light intensity data, I_0, I_1, I_2, $\cdots$, I_{n-1}, $I_n(I_0=I_n)$ obtained by rotating polarizer and analyzer at equal intervals of θ from 0 to 90 degrees as follows.

$$C_1=-\sqrt{a_1^2+b_1^2}$$

$$\phi_1=\tan^{-1}(a_1/b_1) \tag{6}$$

$$a_0=\frac{1}{n}\sum_{m=0}^{n-1}I_m$$

where

$$a_1=\frac{2}{n}\sum_{m=0}^{n-1}I_m\cos(m\frac{2\pi}{n})$$

$$b_1=\frac{2}{n}\sum_{m=0}^{n-1}I_m\sin(m\frac{2\pi}{n}) \tag{7}$$

Therefore, the direction of principal stresses is obtained as

$$\phi=(-\phi_1+\pi/2)/4 \tag{8}$$

Correction of principal stress directions

ϕ calculated from equation (8) was adapted as the principal stress direction when true one was between 0 and 45 degrees, but was corrected by adding 45 degrees to it when between 45 and 90 degrees. Note that the directions of either maximum or minimum principal stresses are obtained from the above mentioned procedure. In order to change minimum principal stress direction (ϕ') into maximum one (ϕ), we can use the equation

$$\phi=\phi'-\pi/2 \tag{9}$$

By use of this method, the principal stress directions are almost determined automatically over the entire domain. However, this method is not effective on integer order isochromatics.

EVALUATION OF THE METHOD

The method proposed can measure the direction of principal stresses with a high accuracy. However, noises contained in the light intensity may reduce the accuracy. So, the effect of noises on the accuracy of the measurement of principal stress directions was mainly investigated as follows.

Fig.4 shows the effect of the number of data (N) on the principal stress direction (ϕ) using artificial light intensity with noises which are given considering their appearance in actual one. A light intensity in case of ϕ=10 degrees was investigated. The number of noises was 13, and their magnitude (ΔI) was 0.083I_0. The values of ϕ shown in these figures were obtained from the present method. These results indicated that the principal stress directions obtained depend on the number of data, and its accuracy is about 0.003 degrees at (N) above 100.

Fig.5 shows the effect of noise distribution on the principal stress direction using artificial light intensity with N=128 and $\Delta I/I_0$=0.083. CASE1 simulated the light intensity with few noises whose signs all were plus, CASE2 one with ten noises whose signs all were plus and CASE3 was

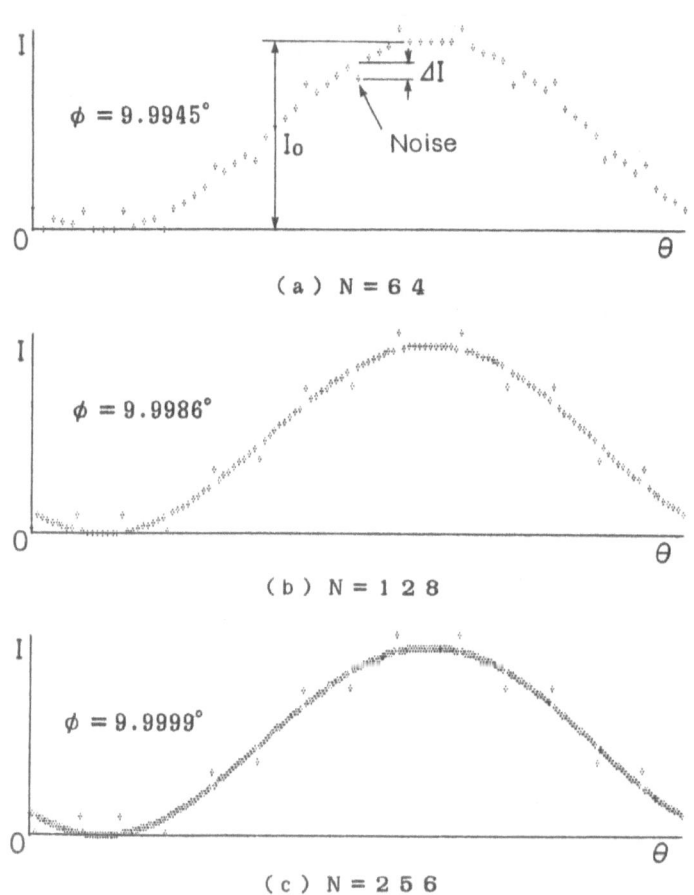

(a) N = 6 4

(b) N = 1 2 8

(c) N = 2 5 6

Figure 4. Effect of number of data (N) on direction of principal stresses (ϕ). $\Delta I/I_0$=0.083.

the same light intensity as in Fig.4. The accuracy of CASE1, CASE2 and CASE3 was 0.0001, 0.008 and 0.002 degrees, respectively. These results showed that the accuracy depends on how noises are distributed rather than the number of noises. The accuracy became low with increase in the magnitude or number of noises, but alternate noises such as CASE3 reduced a fall of the accuracy.

Fig.6 shows that the effect of light intensity constant I_0 on the principal stress direction using artificial light intensity with N=128 and the same noise distribution as CASE3 in Fig.5. The magnitude of noises always was constant. From this figure, it was found that the accuracy becomes low with decrease in light intensity constant I_0. Then, it is probable that the accuracy becomes low near integer order isochromatics be-

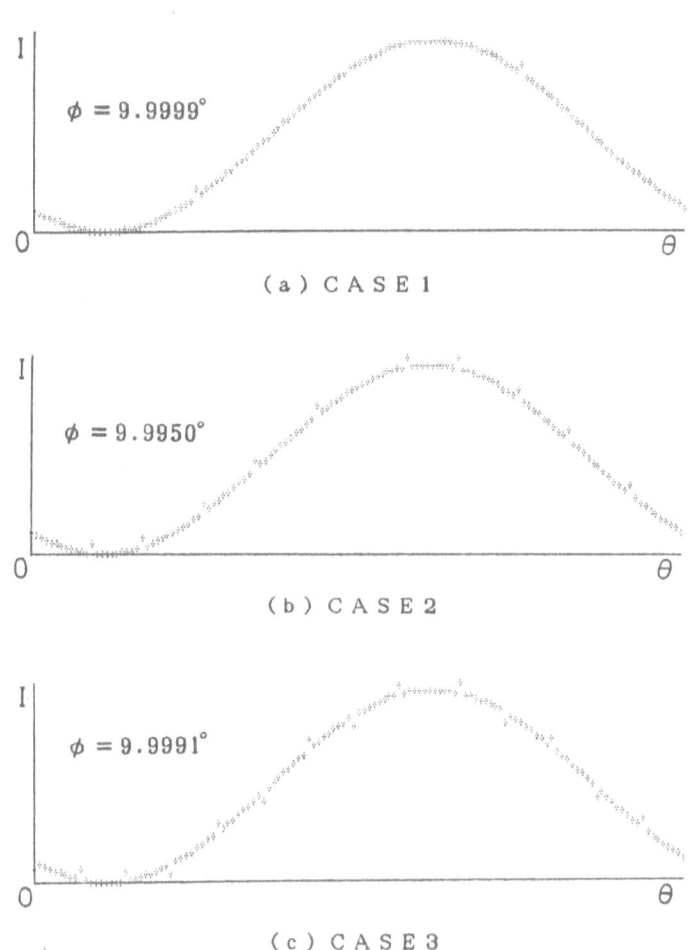

(a) CASE 1

(b) CASE 2

(c) CASE 3

Figure 5. Effect of noise distribution on direction of principal stresses (ϕ). N=128 and $\Delta I/I_0$=0.083.

cause the light intensity constant is small.

Judging generally the above results, the principal stress directions obtained by use of the Fourier-series expansion of the light intensity distribution were high accuracy except those near integer order isochromatics.

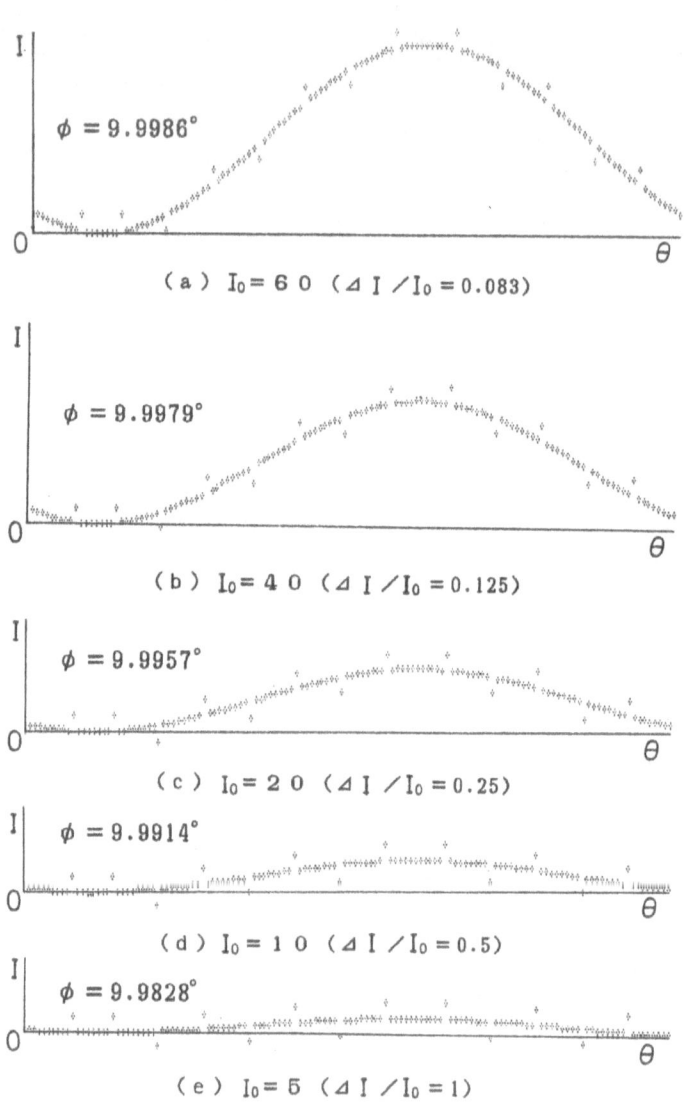

(a) $I_0 = 6\,0$ ($\Delta I / I_0 = 0.083$)

(b) $I_0 = 4\,0$ ($\Delta I / I_0 = 0.125$)

(c) $I_0 = 2\,0$ ($\Delta I / I_0 = 0.25$)

(d) $I_0 = 1\,0$ ($\Delta I / I_0 = 0.5$)

(e) $I_0 = 5$ ($\Delta I / I_0 = 1$)

Figure 6. Effect of light intensity constant (I_0) on direction of principal stresses (ϕ). N=128 and ΔI=5.

FULL-FIELD AUTOMATIC POLARISCOPE

Fig.7 shows the system constitution for measuring automatically the direction of principal stresses in the whole field. The polarizer and analyzer were rotated by a stepper motor driven according to the number of pulses transmitted from a personal computer. the isoclinic images were taken by a TV camera, digitized by an image processing equipment (256Vx256H pixels, 6bits), and stored into a hard disk at each step of the rotation. The images stored were the average of five images taken continuously at each step. As a result of the average, noises were reduced.

After all isoclinics images intended between 0 and 90 degrees were stored, the time series curves of light intensity at each pixel were formed. These curve data were stored into the hard disk, and analyzed for determining the directions of principal stresses by using the Fourier-series expansion method. The procedure of input of the images to determination of the principal stress directions was completely automated.

APPLICATION

This full field automatic polariscope was applied to a circular disk of 90.5mm in diameter subjected to a concentrated load of 78.5N. Figs.8 and 9 show the light intensity at points (A) and (B) locating approximately on an isoclinic of 0 and 10 degrees, respectively. The values of principal stress directions at pixels before and behind point (A) on a horizontal line of y/a=0.6 were -1.2300 and 2.1148 degrees, respectively, and at point (B) 8.4360 and 11.6607 degrees, respectively. Fig.10 shows the directions of principal stresses along a line of y/a=0.2 and 0.6 obtained by the present system, shown by symbols "O" and "△", compared to those by the theory, shown by solid lines. Both results almost agreed with the theory at the range of x/a values from -0.4 to 0.4, but did not at the range of x/a<-0.4 and x/a>0.4 because wide zero-order isochromatics

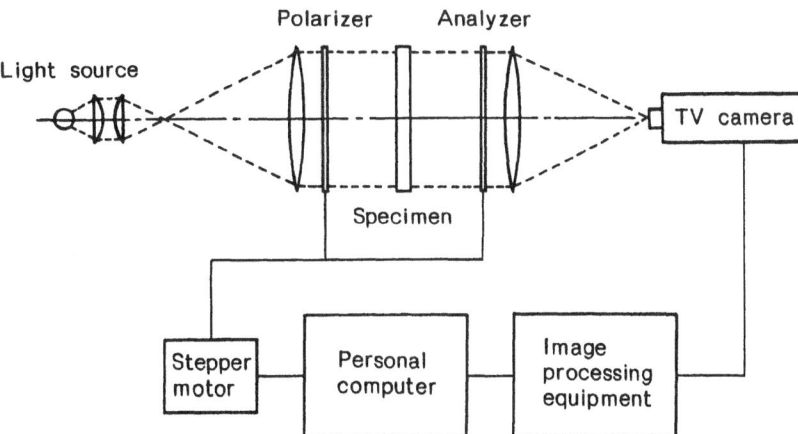

Figure 7. Diagram showing layout of the automatic polariscope for measuring direction of principal stresses.

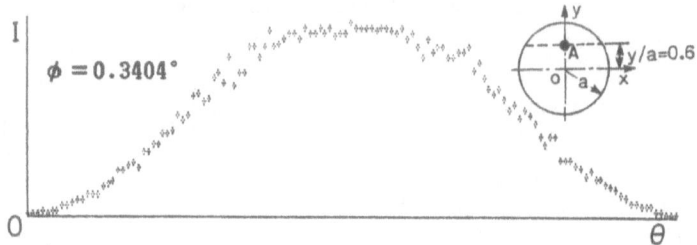

Figure 8. Light intensity at point (A) in circular disk under concentrated loads.

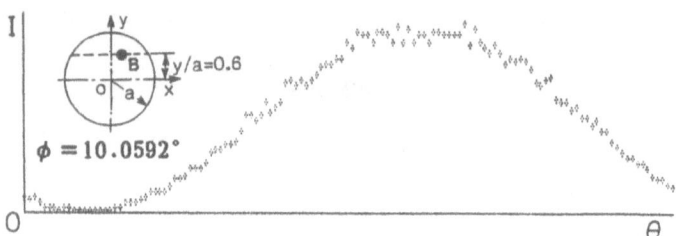

Figure 9. Light intensity at point (B) in circular disk under concentrated loads.

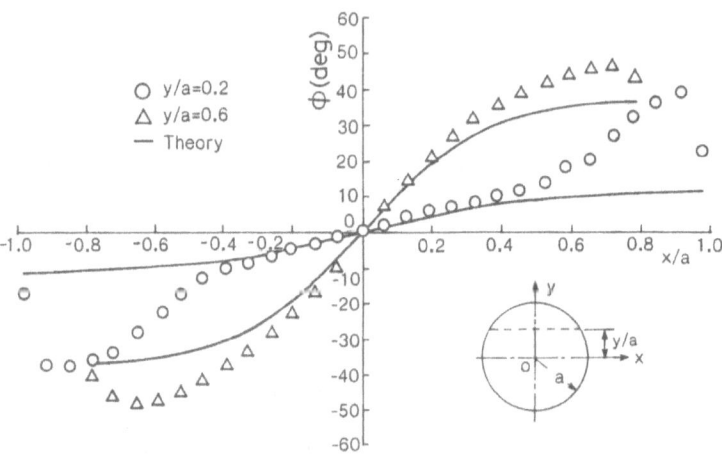

Figure 10. Directions of principal stresses along a line of y/a=0.2 and 0.6 obtained from the proposed method , shown by symbols "O" and "△".

appeared. Therefore, in addition to the proposed method a technique will
be desired for measuring accurately the principal stress directions on and
near integer order isochromatics.

CONCLUSIONS

The system described was capable of measuring automatically the directions
of principal stresses in the whole field of the model without any
interpolation, and with high accuracy. From now on, a technique will be
desired for measuring accurately the principal stress directions on and
near integer order isochromatics.

REFERENCES

[1]Muller, R.K. and Saackel, L.R., Complete automatic analysis of photoelastic fringes, Exp. Mech., 1979, 19(7), 245-251.
[2]Seguchi, Y., Tomita, Y. and Watanabe, M., Computer-aided fringe-pattern
analyzer-A case of photoelastic fringe , Exp. Mech., 1979, 19(10), 362-
370.
[3]Nisida, M. and Ohi, S., On a fully automatic system of two-dimensional
photoelastic stress analysis. I. Fundamental study of the method (in
Japanese), Proc. of the Japan Society for photoelasticity, 1988, 8(2),
1-6.
[4]Umezaki, E., Tamaki, T. and Takahashi, S., Automatic stress analysis of
photoelastic experiment by use of image processing, Exp. Tech., 1989,
13(12), 22-27.
[5]Redner, S., New automatic polariscope system, Exp. Mech., 1974, 14(12),
486-491.
[6]Allison, I. M. and Nurse, P., Interactive control for an automatic
polariscope, Strain, 1979, 15(3), 90-94.
[7]Mawatari, S., Takashi, M. and Toyoda, Y., Whole-area photoelastic analysis by image processing on the principal-stress direction and separation of isochromatics from isoclinics (in Japanese), Trans. of the
Japan Society of Mechanical Engineering, 1989, 55(514)A, 1423-1428.

THE BENEFITS AND PITFALLS OF AUTOMATIC PROCESSING FOR PHOTOELASTIC TEST DATA

I. M. ALLISON,
Department of Civil Engineering,
University of Surrey,
Guildford, Surrey, GU2 5XH.

P. NURSE,
Polytechnic of the South West,
Plymouth, Devon.

ABSTRACT

The essential features of an automatic polariscope are listed. Special attention is given to the problem of boundary location, which is required for making reliable measurements of the critical values at stress concentrations. A description is given of point-by-point measuring systems which provide unambiguous values for both the principal inclination and stress difference with sufficient precision to permit successful evaluation of the separate stress components through difficult regions. The results of tests used to quantify the resolution and accuracy of the system have been discussed. It is suggested that similar criteria should be used in evaluating the performance of any automatic fringe processing system.

INTRODUCTION

Automatic methods for recording and processing photoelastic test data are particularly attractive because the special expertise and high level of concentration needed to make satisfactory manual measurements can be reduced. However loss of the discrimination possessed by human operators presents difficulties in ensuring the reliability of data recorded automatically from models or stress frozen slices which are liable to surface defects and unwanted optical disturbances.

The practical difficulties of making successful automatic measurements can be illustrated by the wide variety of different devices which have been developed during the last four decades. These range from sensitive photometers [1], [2], used to improve the precision of individual observations of isoclinic angle or isochromatic fringe number, to sophisticated systems for incorporating fully automatic recording and processing of the optical data [3], [4].

More recently the availability of hardware in the form of a video camera linked to a digital image store [5],[6], together with software which allows well established image processing techniques to be employed appears to make the construction of a full field automatic polariscope an attractive and feasible proposition. A comprehensive survey of these developments has been made by Patterson [7] who concluded that the potential of photoelastic stress analysis had been considerably increased by the availability of video and image processing equipment, and that full field analysis can now be attempted. However in discussing the relative merits of full field and point-by-point measuring systems insufficient attention appears to have been given to a number of factors which will have an adverse influence on the performance of existing full field polariscopes. These include;

(1)   the characteristics and quality of the optical system

(2)   the difficulty of determining the isoclinic angle with sufficient precision, which can involve the introduction of additional independent experimental or numerical data.

(3)   the need for logical intervention to identify the isochromatic fringe number and gradient.

(4)   the problem of measuring isochromatic fringe numbers in the areas of most interest where the fringe gradients are high.

(5)   the lack of qualitative information about the accuracy and reliability of observations made using existing full field systems.

(6)   justification in terms of output information for the large amounts of data processing involved.

It is the purpose of this paper to highlight the critical features of some of these issues, and to indicate how they can be resolved using a particular point-by-point measuring system. It is also hoped to demonstrate the advantages of an alternative approach to the problem of deriving stress data from photoelastic models by focussing attention upon the following practical requirements;

(1)   precise location of model boundaries and internal points.

(2)   accurate determination of boundary stress values, particularly at points of stress concentration where the fringe gradients are high.

(3)   determination of the basic data needed to evaluate the separate stress components across selected sections.

Automatic Point-by-Point Measurement

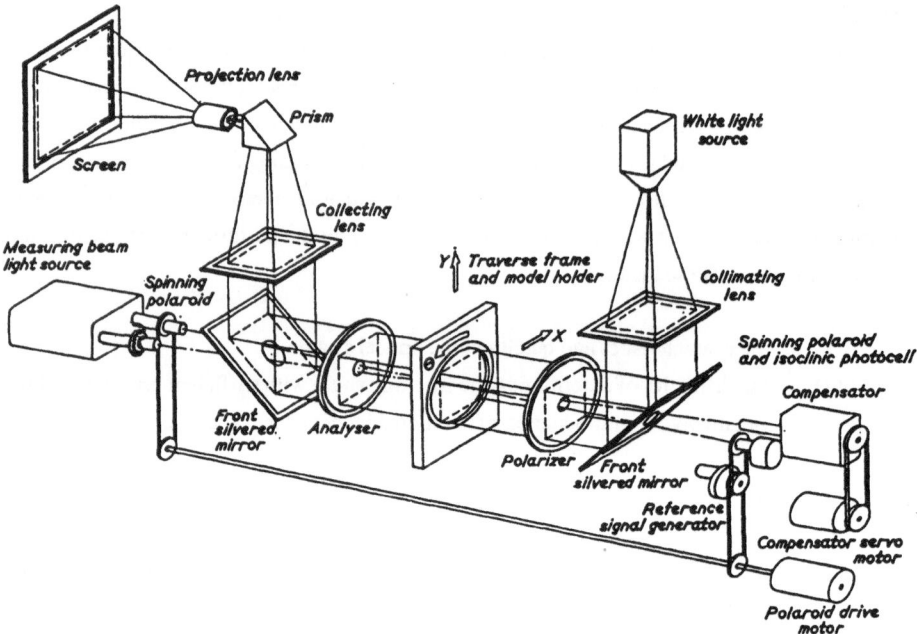

Figure 1. General arrangement of automatic polariscope optical systems.

Figure 1 shows the latest version of the instrument, built at the University of Surrey, which incorporates three independent optical systems. Two of these are similar using a high intensity mercury arc source to generate beams with a cone angle of 2.2×10^{-4} radians which are aligned to converge at the datum of the traversing frame in a spot which is 0.1 mm in diameter. The first converging beam is used to measure the isoclinic angle whilst the second measures the isochromatic fringe number. The third system is of lower optical quality and provides a general view of the fringe pattern in the model which is projected onto a screen set in the polariscope control console.

The model support consists of a frame having digitised orthogonal traverses driven by servo motors which permit the location of spatial co-ordinates within a 250mm x 250mm field to an accuracy of ±0.01mm. This frame also provides for rotation of the model about the optical axis with the angular position being displayed to an accuracy of ±0.1 degrees. Traverse movements may be initiated by programmed commands or for convenience can be controlled manually from the central console. The x,y and θ co-ordinates of the current position of the model with respect to the polariscope optical datum are displayed visually and are also available in digital form for input to the control computer.

Since the prototype was constructed, refinements have been made to improve the precision and reliability of both the isoclinic and isochromatic measurement systems. These have reduced the incidence of anomalous observations to a negligible level and increased the speed of the machine by using more direct processing of the original observations.

Measurement of Principal Inclination

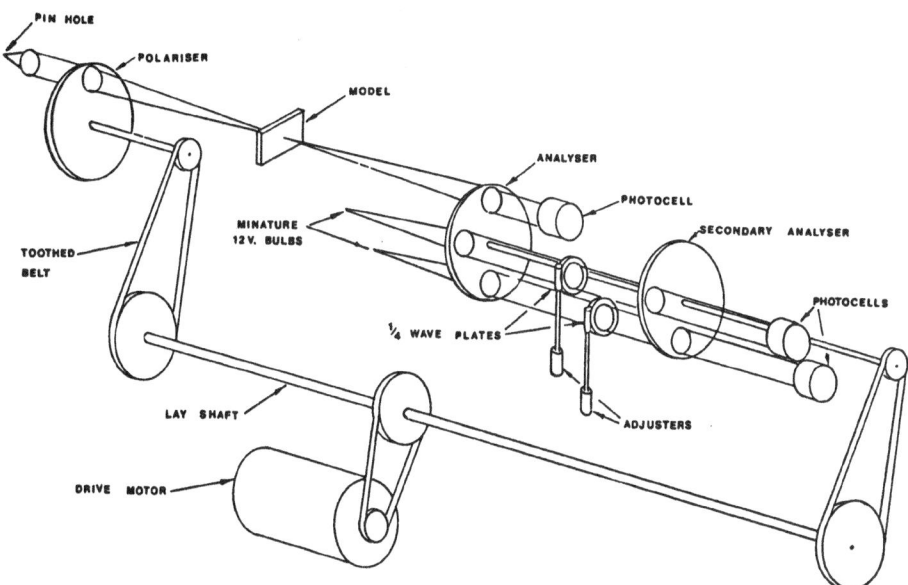

Figure 2. Details of isoclinic angle measuring system.

The revised arrangement for determining the inclination of the principal stress which lies in the first quadrant is shown in figure 2. The light output from the polariscope, and hence the signal generated by the photodetector, is modulated at four times the rotational frequency of the coupled polariser and analyser, with the amplitude at time t being given by:

$$I_1 = \frac{a}{2}(1 - \cos 4\omega t)$$

where   a   maximum amplitude of output signal

               ω circular frequency of rotating polaroids.

The two reference beams are modulated at the same circular frequency, and are arranged so that a constant phase difference of π/2 is maintained between them. All three analogue signals are converted into square waves of constant amplitude, which allows a pulse counting technique to be used in determining the relative phase difference of the polariscope signal to a precision of ± 0.1 degrees. This system has been found to give completely consistent results even at singular points when the principal inclination is given a default value of zero degrees.

Boundary Location

When a photoelastic model is uniformly illuminated by depolarised light precise location of the boundary is a relatively simple matter, and the specification of an optical system which will provide excellent definition in these circumstances has been given by Jessop and Harris [8]. The difficulties of obtaining similar results when the model is observed in a conventional polariscope are well documented, and the apparent shift in the boundary position which occurs when the field is changed from light to dark can be particularly disconcerting.

If information is required at a stress concentration where the fringe gradient normal to the boundary is very high the accuracy with which the extreme value of the isochromatic fringe number can be determined becomes a function of the boundary location. For making accurate photographic records of fringe patterns Coutts [9] has advocated the use of datum marks scribed at a known distance from the boundary. These allow both the true scale and boundary position to be plotted on the enlarged image.

The procedure for using the isoclinic beam to locate the model boundary in a point-by-point polariscope is illustrated in figure 3.

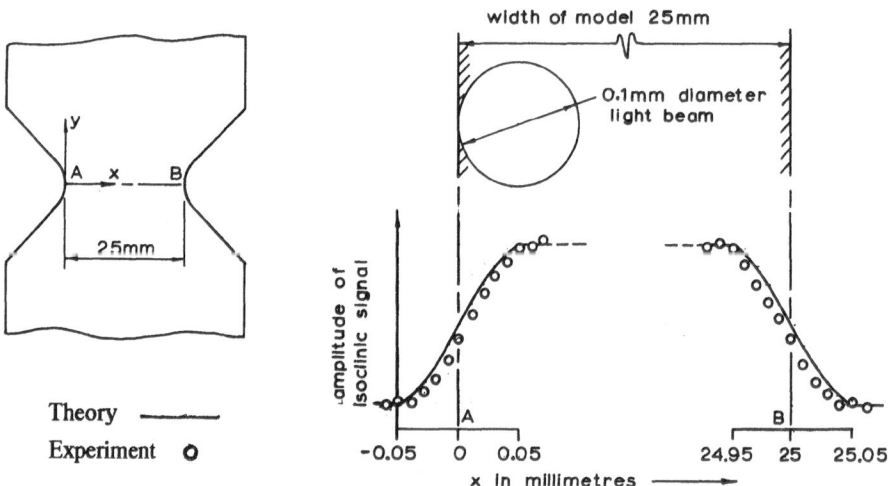

Figure 3. Variation in signal amplitude as the isoclinic beam crosses the edge of the model.

The model is first oriented so that a traverse may be made along the normal to the boundary at the point where the measurement is required. With the coupled crossed polaroids spinning the photodetector output remains at a constant minimum level whilst the beam is in the zone outside the model boundary. As the beam crosses the boundary the signal amplitude increases and reaches a maximum when the entire beam just passes through the model. Although flare and diffraction effects cause an apparent image shift, the deviation is small and consistent. Accordingly it is possible to process experimental observations of the photodetector output so that the boundary may be located within an accuracy of ±0.02 mm.

Measurement of Isochromatic Fringe Order

To facilitate automatic measurement of both the integral and fractional parts of the isochromatic fringe number it has been decided to use a Soleil-Babinet compensator in preference to the more widely used methods which only yield the fractional part of the observed value. The arrangement of the measuring system in current use is shown in figure 4. Light from a high pressure mercury arc is focussed upon a 75 micron diameter pinhole, and is then transmitted as a narrow cone of rays through the model before being refocussed through a circular compensator, which is rotated at 250 rpm, onto a photodetector.

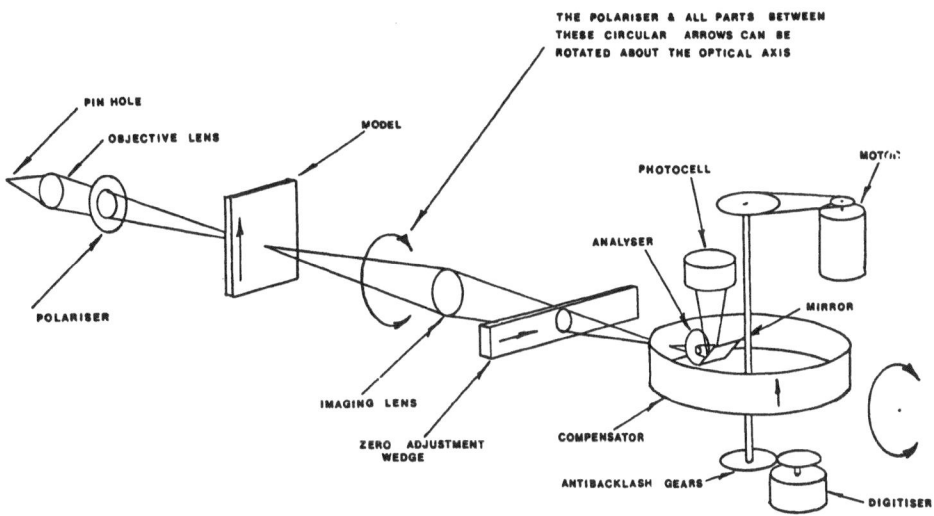

Figure 4. Details of isochromatic fringe measuring system

A servo mechanism, controlled continuously by the output from the isoclinic measuring system, is used to maintain the correct alignment of the compensator with the principal stress directions in the model. If the algebraically greater principal stress lies in the first quadrant it follows that the output will be positive. Conversely the output will be negative if the algebraically least principal stress falls in the first quadrant. It follows that determination of the stress difference at any point in the field is completely unambiguous.

Choice of the bandwidth for the light source is a matter of compromise. If a monochromatic source is used extinction occurs for every integral fringe variation in the compensator, thus making it impossible to identify the integral fringe order. When a white light source is used the difference between the minimum intensities of the zero and first order fringes is a maximum. However dispersion effects will prevent compensation being effected above the fifth order fringe. A mercury arc has particularly strong emissions in the yellow (5780A), green (5460A) and violet (4360A) wavelengths and accordingly provides adequate discrimination between the minimum intensities of the zero and first order fringes whilst allowing reliable compensation over a range of 22 fringes.

The output from the compensator will be of the form;

$$I_2 = \sum_{\substack{r=1 \\ n}} \frac{a_r}{2} \left[1 - \cos 2\pi \frac{\lambda}{\lambda_r} (N - k\psi) \right]$$

where a_r maximum intensity emitted for illumination by wavelength λ_r

           N birefringence measured in model for illumination at wavelength λ

           k compensator calibration factor.

           ψ angular displacement of compensator from the datum position

and I_2 is the overall intensity obtained by summation over the n transmitted wavelengths.

Since the light source is not monochromatic it is necessary to manufacture the compensator from the same material as the photoelastic model.

Whilst observations made in a uniform stress field demonstrate that this compensator ha a measuring accuracy of ±0.005 fringes, the precision will be affected by the fringe gradien across the 0.1mm aperture of the measuring beam.

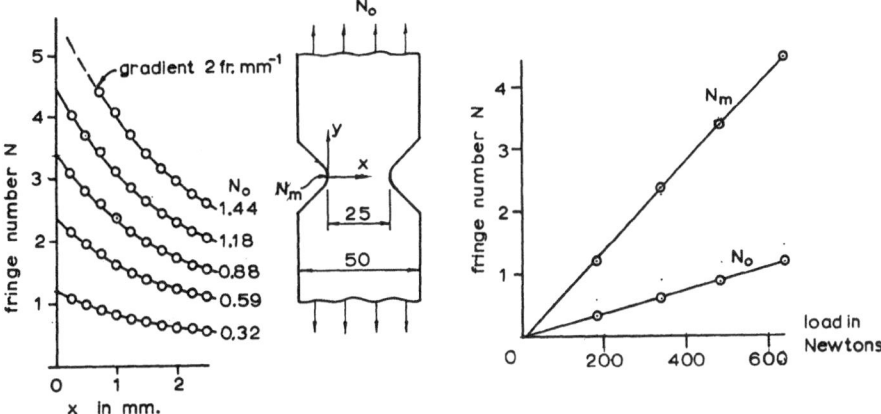

Figure 5. Observations of fringe values in the fillet radius of a notch as the applied load is increased.

Figure 5 shows the results of measurements made at the fillet stress concentration in a notched plate as the load, and hence the fringe gradient, is increased. Observations made at internal points were extrapolated using a polynomial curve fitting routine to obtain the boundary values. Except in the case of the highest tangential load, the resulting tangential stresses at the boundary vary linearly with the applied load. In the case of the highest load the fringe gradient reached a value of 2 fringes per mm at a distance of 0.75mm from the boundary and reliable compensation of the fringe order could no longer be achieved. Detailed examination of this effect revealed that reliable observations could be made in a field with a fringe gradient of 1.75 fringes per mm, whilst unsatisfactory results were obtained when the gradient increased to 2 fringes per mm.

This performance is sufficient to meet the conditions encountered in practical engineering components when easily measured nominal stress or strain levels are applied. Measurements of the optical parameters and the spatial co-ordinates with the reliability and precision provided by this point-by-point system has been shown [10] to yield satisfactory values of the separate stress components in both live loaded models and sections cut from stress frozen components. It seems likely however that some difficulty might be encountered in designing a full field system with the range and discrimination to provide comparable performance.

CONCLUSIONS

1. Stress data is normally required at points of stress concentration or across selected sections. Automatic point-by-point measurements to satisfy this requirement.

2. The ability to locate the boundaries of a deformed model within an accuracy of ±0.02mm is an important attribute if reliable values of a fillet stress concentration are to be obtained.

3. Measurement within an accuracy of ±0.01 fringes in the presence of gradients up to 1.5 fringes per mm is essential to the satisfactory measurement of fillet stresses.

4. Unambiguous measurement of the principal inclination within ±0.1° is required if individual stress components are to be derived using integration of the equilibrium equations.

5. Unless full field methods can offer commensurate accuracy there appears to be little merit in undertaking the large amount of additional data processing which is implied.

REFERENCES

1. Brown, A.F.C. and Hickson, V.M. 'Improvements in photoelastic technique obtained by using a photometric method' Brit. Jnl App Phys. 1 (1950), pp 29-44.

2. Frocht , M.M., Hui Pih and Landsberg, D. 'The use of photometric devices in the solution of the general three dimensional photoelastic problem'. Opt.Comm. XII(1954) 181-190.

3. Allison, I.M. and Nurse, P. 'Optical data acquisition for an automatic polariscope' Proc. 7th All Union Conference on Photoelasticity, Tallinn (Nov. 1971) Part 1. 93-105.

4. Redner, S. ' A new automatic polariscope system'. Exp. Mech. 14 (1974) 486-491.

5. Mueller, R.K. and Saackel, L.R., 'Complete automatic analysis of photoelastic fringes' Exp. Mechs. 19 (1979) 245-252.

6. Umezaki, B., Tamaki, T. and Takahashi, S., 'Automatic stress analysis from photoelastic fringes due to image processing using a personal computer'. Proc. Soc. Photo 504 (1984) 127-134.

7. Patterson, E.A., 'Automated photoelastic analysis. Strain Vol.24, No.1 (1988) 15-20.

8. Jessop, H.T. and Harris, F.C., 'Photoelasticity: Principles and Methods'. Cleaver Hume Press Ltd., London 1949.

9. Coutts, J.A., 'Stresses around multiple holes in plates, with special reference to water tube boiler drums' PhD thesis. University of London, 1956.

10. Allison, I.M. and Nurse, P., 'Optimisation of photoelastic stress separation procedures' Proc. 6th Int. Conf. Exp. Stress Analysis, Munich, VDI Verlag, (1978) 41-45.

3-D PHOTOELASTICITY AND TV-HOLOGRAPHY FOR THE ANALYSIS OF ROTATING COMPONENTS

Dr.R.W.T.PREATER

Thermo-Fluids Engineering Research Centre,
City University,
Northampton Square, London, ECIV OHB, UK.

ABSTRACT

Two and three dimensional photoelasticity is an extremely useful stress analysis tool for the design and development of complex engineering structures. Confirmation by analysis of the prototype may become more difficult however for economic reasons. The development of a non-contact method of analysis requiring the minimum surface preparation and plant shut-down is therfore doubly attractive. Electronic Speckle Pattern Interferometry (ESPI) or Tv-Holography using a pulsed laser is a technique which displays considerable industrial potential in this way.

INTRODUCTION

Experience in the analysis of large rotating structures used in the mining industry such as drum winders (1) using 3-D photoelastic model techniques revealed a certain reluctance on the part of the operators to allow extensive prototype analysis for economic reasons. Often in such cases either the model structure or the loading configuration is simplified in order to satisfy the sort of empirical design procedures adopted and the final operating dynamic conditions remain unknown. Experimental stress analysis of the prototype using conventional strain gauge techniques would require costly shut-down periods and production delays which may only be contemplated in the event of subsequent failure. A non-contact method of approach requiring only minor surface preparation, painting with matt white paint and little or no interruption to the production line is therefore extremely attractive. Electronic Speckle Pattern Interferometry (ESPI) and now sometimes called Tv-Holography, pioneered at Loughborough University by Butters and Leendertz (2) is a technique which displays this potential. In-plane strain measurement using ESPI was demonstrated by Denby and Leendertz (3) on static components in the laboratory using an argon-ion c/w laser and a conventional closed-circuit television system but with electronic processing. Further developments have now produced a well established technique for the measurement of component deformation. Both out-of-plane and in-plane displacement measurements have been made under dynamic conditions using a pulsed laser. Development for the analysis of in-plane displacements on rotating components has been carried out at City University.

For rotating components the pulse width of 20 ns for the pulsed laser freezes the component motion. Two beam oblique illumination of the component surface gives a displacement sensitivity direction in the plane of the illuminating beams. The use of a high resolution speckle tv-camera gives fine quality speckle images of the surface, yielding clear high contrast interference fringe patterns on image subtraction.

Early researchers using this technique of pulsed ESPI, suggested a limiting component tangential velocity of 2 ms^{-1} (4). This was not considered to be a major drawback for rotating components since for many applications the maximum strains occur close to the axis where tangential velocities are low. However, as the technique development has proceeded the limiting conditions have become more clearly defined. If there are velocity components in the system sensitivity direction which exceed 5 ms^{-1} then fringe "blurring out" occurs, or no interference fringes are observed. Where the velocity direction is normal to the sensitivity direction fringe information is retained. In the latter case, providing an initial state speckle image may be recorded and subsequent live load images are available for electronic substraction, interference fringes may be produced over a very wide range of component rotational speeds. Simple plane mirror optics give satisfactory fringe information for speeds up to 5000 rpm and tangential velocities of 60 ms^{-1}, and twin cylindrical mirror optics have extended this to speeds in excess of 12500 rpm and 150 ms^{-1} (5).

Both the above optical systems require clear optical access to the component surface for illumination and tv-viewing. This would eliminate the analysis of confined components or those totally enclosed by casings. Fibre optics, however, offer the possible penetration of shrouding or casings as well as the development of a very much more compact system. Preliminary tests using fibre optics have shown fibre optic ESPI to be feasible (6). Their use with pulsed lasers has shown the vulnerability to damage but with the introduction of diode lasers this may prove a suitable alternative technique.

ELECTRONIC EQUIPMENT

In order to achieve good high contrast interference fringe patterns in the analysis of rotating components, high precision triggering of the pulsed laser is required to give the correct register of speckle images for image subtraction. The present system gives satisfactory results for a wide range of component speeds but where the deformation is predominantly speed dependent some modification may still be required to suit the particular application in order to avoid timing errors. Load, torque or temperature effects where the component speeds remain relatively constant are less likely to be affected by such errors.

The use of fibre optics in place of conventional optical components would not require any major changes to laser triggering techniques except where modulated diode lasers are employed. Here the laser modulation would require to be synchronised with the component rotation and with similar precision to that of the pulsed laser. The modulated diode laser would have the added advantage of producing continuous interference effects rather than single shots. Tv-viewing would require use of a fibrescope or borescope through casings and in regions of limited access.

OPTICAL SYSTEM

Simple plane mirror optics in the conventional system, Figure 1, gives a sensitivity direction in the plane of the illumination and provides interference information for in-plane displacements for component speeds of 5000 rpm and tangential velocities of 60 ms^{-1}. At higher speeds where the components of the tangential velocity in the sensitivity direction exceed a critical value of 5 ms^{-1} "blurring out" occurs and causes a loss of fringe information. With increasing component speed this produces progressive reduction in the area of useful information to a narrow band along the component diameter. The introduction of the twin cylindrical mirror system, Figure 2, produces radial illumination and hence radi al sensitivity throughout. This removes the velocity component in the sensitivity direction and restores full field fringe information for component speeds tested up to 12500 rpm and tangential velocities of 150 ms^{-1}. Both the plane and cylindrical system however require clear optical access for illumination and tv-recording of the speckle images.

Recent advances in the design of fibre optics and the increasing power of diode lasers which may be modulated, present possible alternatives to the more conventional optical systems. Single mode fibres of 200 μm in diameter, requiring only a simple means of support in a hypo dermic needle tube of 0.75 mm diameter, may be used to guide the illumination into areas of limited access and penetrate casings or shrouds to illuminate enclosed rotating components, Figure 3.

Although diode laser powers are still relatively low in comparison with the pulsed laser, output powers of 30 and 40 mW may certainly be used for ESPI and without causing fibre optic damage. The modulation frequencies of 50 MHz are now approaching those of rotation frequencies of the higher speed components. The compactness of the laser source and the flexibility of a fibre optic system are added attractions to this form of development.

EXPERIMENTAL RESULTS

The pulsed ESPI experimental results that have been achieved have far exceeded the original suggested limitation of 2 ms^{-1} for dynamic in-plane displacement and strain measurements. In the early stages of development even this was considered worthwhile for the analysis of the slower rotating components and in regions close to the component axis. Figure 4 shows the surface radial stress distribution for a Koepe winder drum cheek-plate obtained from early photoelastic model results. The maximum tangential velocity achieved at the cheek-plate/flange connection would be of the order of 3 ms^{-1}. Starting from rest, or snatch loading, would be the severest loading condition, and this was certainly within the velocity range. The present analysis capability at velocities of 150 ms^{-1} and above has opened up a vast range of components which may now be included.

During the course of development of the system, the plane mirror optics were found to cause fringe "blurring out". Figure 5 shows the full field fringe information on a 250 mm disc rotating at 4000 rpm. At speeds above 5000 rpm fringe "blurring out" occurs, causing the fringe area to be progressively reduced to a narrow band coinciding with the component horrizontal diameter Figure 6 and 7. With the introduction of the twin cylindrical optics, Figure 2, with radial sensitivity throughout, full field fringe information is restored for the higher rotational speeds also, Figure 8, at tangential velocities of the order of 150 ms^{-1}. Figures 9 and 10 show more recent test results at 9150 rpm and 110 ms^{-1} with improved fringe contrast. Tests at even higher speeds between 12500 and 20000 rpm will be carried out shortly with the use of a larger air motor drive with a free running capacity of 27000 rpm.

The cylindrical mirror optics also provide more uniform illumination for unpainted component surfaces. Vented brake disc tests, where the brake pad path could not be painted with matt white paint, in Figures 11 and 12, show cooling contraction displacements after a series of braking cycles for the disc rotating with vented cooling at 800 rpm. Specular reflections from the unpainted surface had previously caused apparent additional hor izontal fringes with the plane mirror optics.

The tv-camera fitted with a macro-zoom lens gives a significant apparent depth of focus so that non-plane surface components may be included. Figure 13 shows test results for a model aircraft propeller blade rotating at 2700 rpm, a component which displays considerable out-of-plane configuration.

Preliminary tests using two beam oblique fibre optic illumination of the component surface have been carried out using a HeNe c/w laser. Laser Diode tests have also been carried out under c/w and modulation modes of operation. Figure 14 and 15 show results of these tests. To date these results look promising, but high frequency tests using this system await the delivery of a quality pulse generator which can provide a good square wave output over the full frequency range.

CONCLUSION

Although the pulsed laser ESPI technique was initially likely to be confined to the analysis of relatively slow rotational speed components it was proposed for confirmation of prototype deformations. In the event the capability has far exceeded initial expectations. Plane mirror optics were found to be a little restricted but development of the twin cylindrical mirror system now extends the speed range to include high speed components at rotational speeds in excess of 12500 rpm and tangential velocity of 150 ms^{-1}. Current development tests now in progress, using a larger capacity air motor, may now extend the speed range still further.

Fibre optics in conjunction with laser diodes may provide a suitable alternative optical system for regions of limited optical access. The flexibility and compactness of such a system lends itself to use in an industrial environment.

ACKNOWLEDGEMENTS

Initial work in the development of pulsed laser ESPI suitable for the analysis of high speed components was supported by the Department of Trade and Industry and British Rail. Advanced design high performance brake disc components were provided by BL Technology Ltd. The current work on high speed components is supported by an SERC/Rolls-Royce Co-founded Research Programme.

REFERENCES

(1) PREATER R.W.T.and ATKINSON L.T.J.
Theoretical and Experimental Techniques used in the Design of Winding Engine Drum Structures.
Proc.Int.CONF.S.A.I.MECHE Hosting .Men.Materials.Minerals. P.24 pp. 253-271, 1973.

(2) BUTTERS J.N.and LEENDERTZ J.A.
Holographic and Video Techniques applied to Engineering Measurement.
Trans.Inst.Measurement and Control V4, N12, pp. 349-354, 1971.

(3) DENBY D.and LEENDERTZ J.A.
Plane Surface Strain Examination by Speckle Pattern Interferometry using Electronic Processing.
Jnl.Strain Analysis. V9, N1, pp. 17-25, 1974.

(4) COOKSON T.J.,BUTTERS J.N.and POLLARD H.C.
Pulsed Lasers in Electronic Speckle Pattern Interferometry.
Optics and Laser Technology, pp. 119-124, 1978.

(5) PREATER R.W.T.
A Novel Optical System for the Analysis of In-plane Strain on High Speed Rotating Components.
Electro-Optics/Laser UK'86 I.P.C.Sc.and Tech.Press.Ed.H.G.Jerrard, 1986.

(6) PREATER R.W.T.
Measuring Rotating Component In-plane Strain using Conventional Pulsed ESPI and Optical Fibres.
Conf.Laser Technologies in Industry,Portugal.
SPIE.Vol 952. P.150, 1988.

550

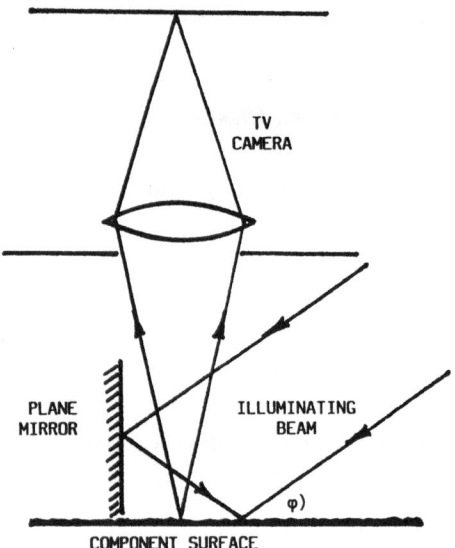

Figure 1. Plane mirror optics.

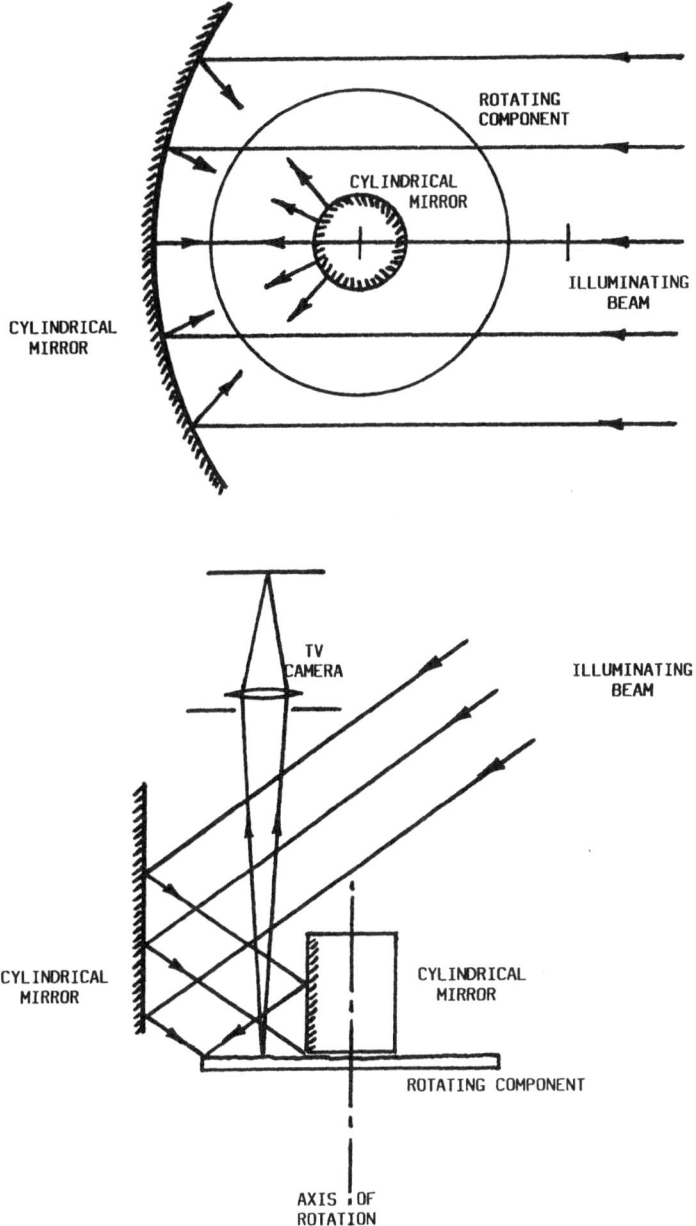

Figure 2. Twin-cylindrical mirror optics.

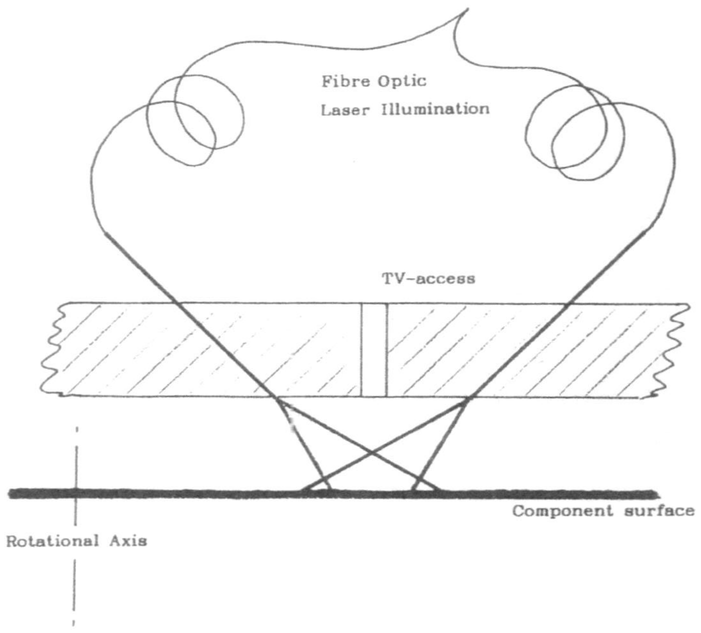

552

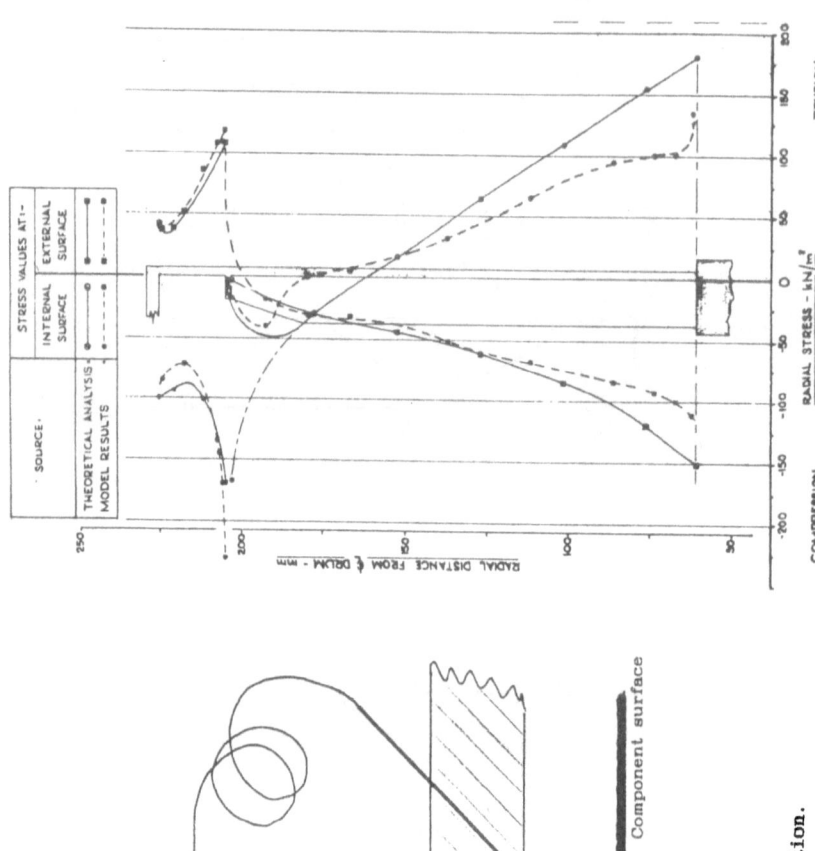

Figure 4. Surface radial stress distribution for a Koepe
Winder drum cheek-plate.

Figure 3. ESPI using Fibre Optic Illumination.

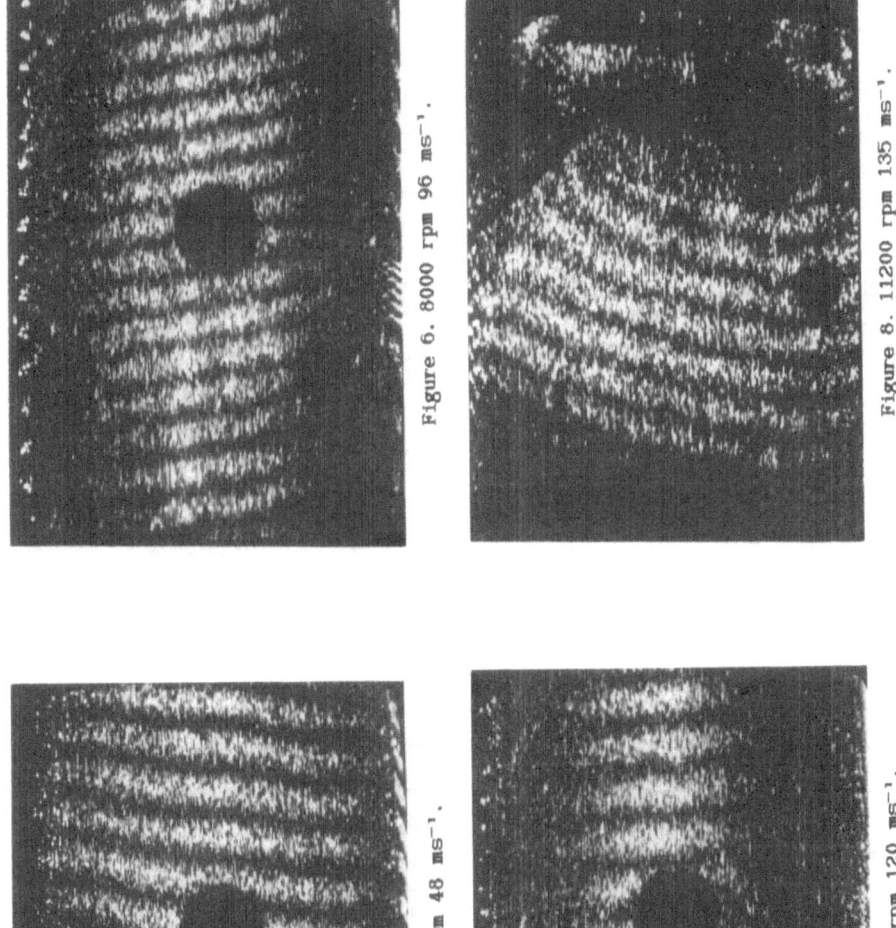

Figure 5. 4000 rpm 48 ms⁻¹.

Figure 6. 8000 rpm 96 ms⁻¹.

Figure 7. 10000 rpm 120 ms⁻¹.

Figure 8. 11200 rpm 135 ms⁻¹.

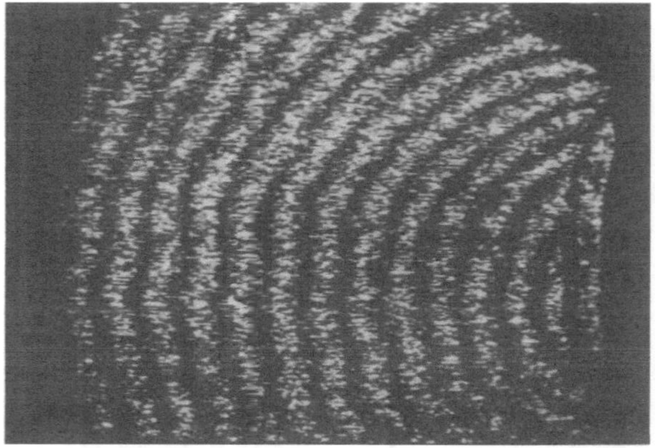

554

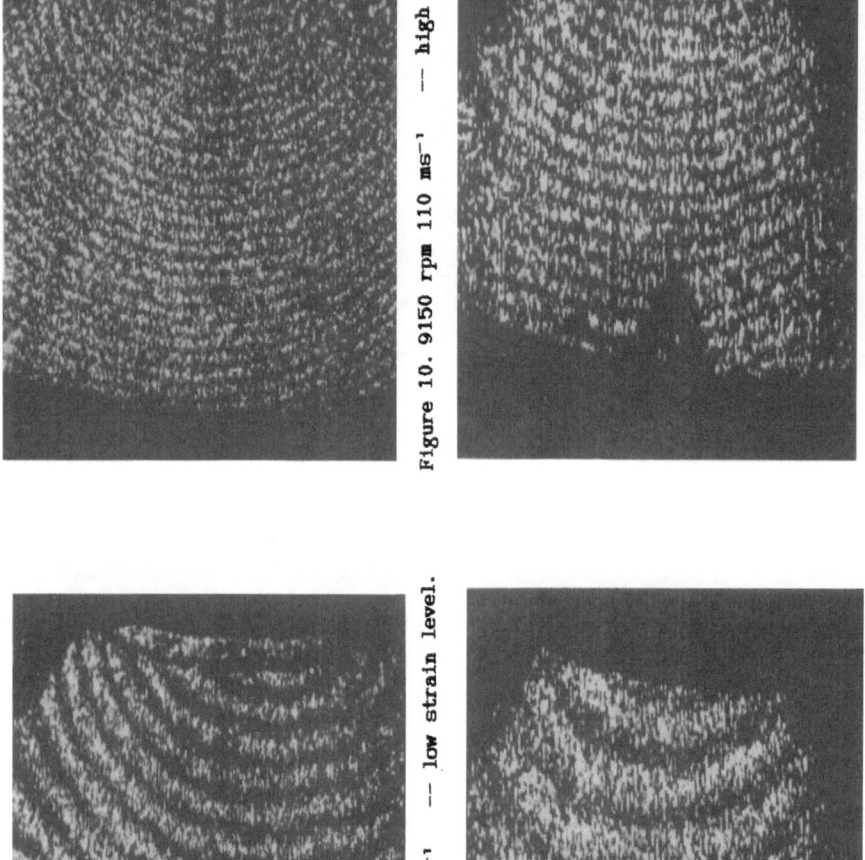

Figure 9. 9150 rpm 110 ms⁻¹ -- low strain level.   Figure 10. 9150 rpm 110 ms⁻¹ -- high strain level.

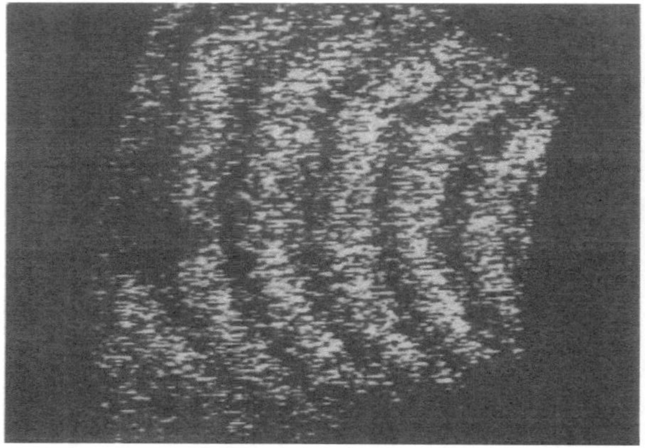

Figure 11. Vented brake disc displacements after 15 secs.   Figure 12. Vented brake disc displacements after 45 secs.

555

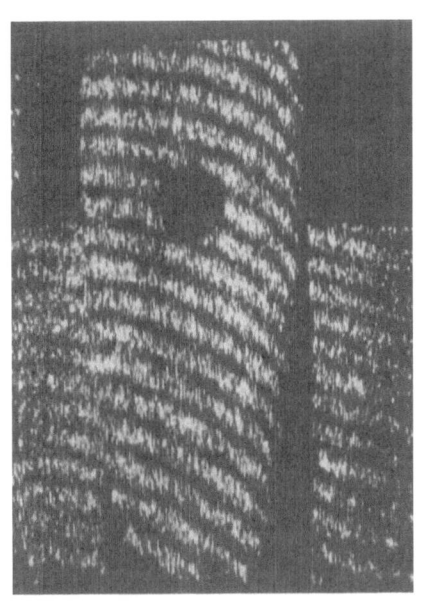

Figure 13. Model Aircraft propeller blade at 2700 rpm.

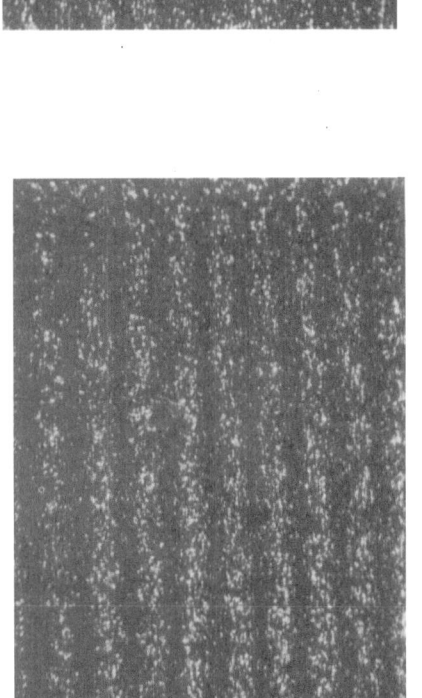

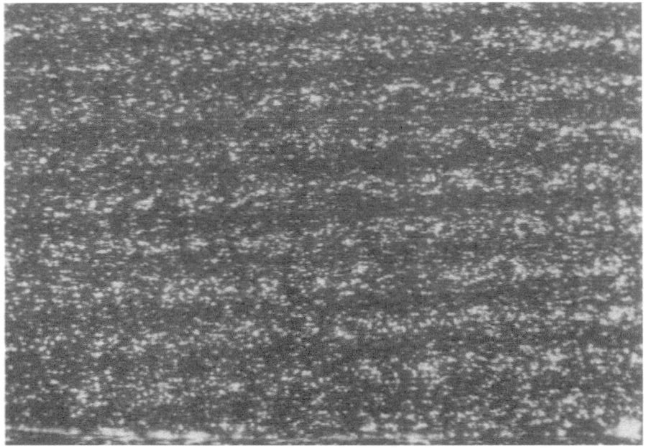

Figure 15. Infrared Diode Laser ESPI.

Figure 14. HeNe Fibre Optic ESPI.

LEAST SQUARES METHODS FOR PHOTOELASTICITY

Donald G. Berghaus
Georgia Institute of Technology

ABSTRACT

Least-squares methods are presented as having many advantages for photoelastic stress analysis. Excess information can be included to obtain an overdetermined solution of improved accuracy. It is also possible to increase the influence of data in which there are high confidence or of relationships which are considered to have high accuracy. Least-squares opens the door to the inclusion of other linear stress relationships and, specifically, enables photoelasticity to be combined with the finite element method. Methods are demonstrated in example problems.

INTRODUCTION

The essential stress-optical equations of photoelasticity relate the principal stress difference and principal directions to the isoclinic and isochromatic fringe parameters. Solution for the separate stress component requires the use of additional information. The elastic field differential equations have provided a ready source for this information together with boundary conditions.

Finite difference approximations to the field equations provide linear equations. When a system of such equations is written over a solution region, and when static equilibrium and boundary conditions are also added, an excess of information is produced [1]. A solution which uses all of the information may be obtained using least-squares. Such solution will give stress results over the entire field and will permit weighting more heavily those items of information in which there is more experimental confidence. This paper shows how such solutions may be constructed and solved. Least-squares is also shown to be useful for a solution method which combines photoelasticity and the finite-element method.

A GENERAL METHOD FOR PLANE PROBLEMS

A solution region is used which extends across the model as shown in Figure 1. A network of solution stations is superposed on the region. The stations are along lines parallel to the x and y coordinate axes. In the equations, subscript i represents the ith row while j represents the jth column [1].

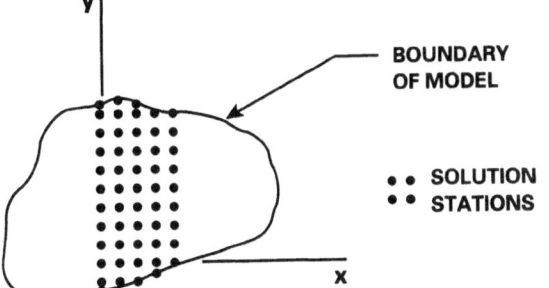

BOUNDARY
OF MODEL

Figure 1.
Solution Region

•• SOLUTION
•• STATIONS

The Photoelastic Effect

If the isoclinic angle is denoted (θ) and the compensated isochromatic
fringe order is designated (N), then

$$(\tau_{xy})_{i,j} = 1/2 \ (\tfrac{f}{t}) \ (N)_{i,j} \ \sin 2(\theta)_{i,j} \tag{1}$$

$$(\sigma_x)_{i,j} - (\sigma_y)_{i,j} = \tfrac{f}{t}(N)_{i,j} \ \cos 2(\theta)_{i,j} \tag{2}$$

For the isoclinic angle, (θ) is measured positive counter-clockwise to the
direction of greatest principal stress. The stress-optical coefficient is f
and the model thickness is t. In eqs. 1 & 2, the stress components are
unknown at all solution stations.

Stress-Equilibrium Relations

For plane problems, the stress equilibrium equations may be written:

$$\frac{\partial \sigma_x}{\partial x} + \frac{\partial \tau_{xy}}{\partial y} = 0 \quad \text{and} \quad \frac{\partial \sigma_y}{\partial y} + \frac{\partial \tau_{xy}}{\partial x} = 0 \tag{3}$$

The numerical equations are written in finite difference form. For the
solution station network shown, if the rows and columns do not have
identical spacing, or if some of the stations do not lie along the rows or
the columns (e.g., at free boundaries), the first of eqs. (3) may be
written:

$$(\sigma_x)_{i,j+1} - (\sigma_x)_{i,j} + \frac{1}{2} \ (\frac{x_{i,j+1} - x_{i,j}}{y_{i+1,j} - y_{i-1,j}}) \ [(\tau_{xy})_{i+1,j} - (\tau_{xy})_{i-1,j}]$$

$$+ \frac{1}{2} \ (\frac{x_{i,j+1} - x_{i,j}}{y_{i+1,j+1} - y_{i-1,j+1}}) \ [(\tau_{xy})_{i+1,j+1} - (\tau_{xy})_{i+1,j-1}] = 0 \tag{4}$$

A similar equation is written for the second of eqs. 3. These equations may
be simplified for equal station spacing, which is the case, except adjacent
to region boundaries. Each of the finite difference equations relates
stress components at six points and each may be written for each "six-point-
nest" within the solution region.

Boundary Conditions

Shear and normal stresses are zero at locations on free boundaries. For
angle, α, to the normal, from the horizontal x axis; for the zero normal
stress:

$$\sigma_x \cos^2\alpha + \sigma_y \sin^2\alpha + 2\tau_{xy} \sin\alpha \cos\alpha = 0 \qquad (5)$$

For the zero shear stress:

$$(\sigma_x - \sigma_y) \cos\alpha \sin\alpha + \tau_{xy} (\sin^2\alpha - \cos^2\alpha) = 0 \qquad (6)$$

Static Equilibrium Conditions

If a free body is identified which consists of the portion of the model to the right or left of a section line which passes along one of the columns o: solution stations, then the following equilibrium equations may be written:

$$\Sigma F_x = 0; \qquad \Sigma F_y = 0; \qquad \Sigma M_o = 0 \qquad (7)$$

Along the section line the shear and normal stresses contribute to the equilibrium requirement. If the summations are written in incremental form along this line, the following equations are obtained:

$$t \left[\sum_{i=1}^{n} (\sigma_x)_{i,j} (\Delta y)_{i,j} \right] = \Sigma F_{x(ext)} \qquad (8)$$

$$t \left[\sum_{i=1}^{n} (\tau_{xy})_{i,j} (\Delta y)_{i,j} \right] = \Sigma F_{y(ext)} \qquad (9)$$

$$t \left[\sum_{i=1}^{n} (\sigma_x)_{i,j} (y_o)_{i,j} (\Delta y)_{i,j} \right] = -M_{o(ext)} \qquad (10)$$

In these equations, there are n solution stations along the section line. The effective area at a station is $t(\Delta y)$ and (y_o) is measured from (y) to a convenient pivot point. The subscript (ext) denotes external load or moment.

Weighting

When a least-squares solution is employed, the solution is biased toward th equations which have the numerically highest coefficients for the unknown quantities. This effect may be used to increase or decrease the relative influences of those equations in which there is higher or lower experimenta (or numerical) confidence. This procedure is often referred to as "weighting". For example, comparatively lower confidence may be held in th photoelastic equations near the boundaries and in the finite difference approximations to the stress-equilibrium equations in the same regions, for a general boundary. Comparatively high confidence may be found in the boundary conditions. For a problem in which the same confidence is held fo: all of the equations, the coefficients should be of the same order.

Inadvertent weighting must be avoided. For the preceding equations, the equilibrium equations, as written, are nominally weighted lower by a factor $(t\Delta y)$ for eqs. 9 and 10 and by a factor $(y_o\ t\Delta y)$ for eq. 11. This may be eliminated by dividing both sides of equations 9 and 10 by a factor $(t\overline{\Delta y})$ where $\overline{\Delta y}$ is the average value of $(\Delta y)_{i,j}$. A similar factor, $(y_o\ t\overline{\Delta y})$, is used for equation 11.

Example Problem

Figure 2a shows an isochromatic fringe pattern with a solution station grid
for a solution region in a beam with a deep notch under the influence of a
simple bending moment. The overall height, H, of the cross-section is 4.76
cm, and the notch is 1.74 cm deep with a radius of 0.48 cm. The nominal
station spacing is 3.25 mm. For this problem there are 59 solution stations
and 178 unknowns. There is a total of 219 equations to solve for these
quantities.

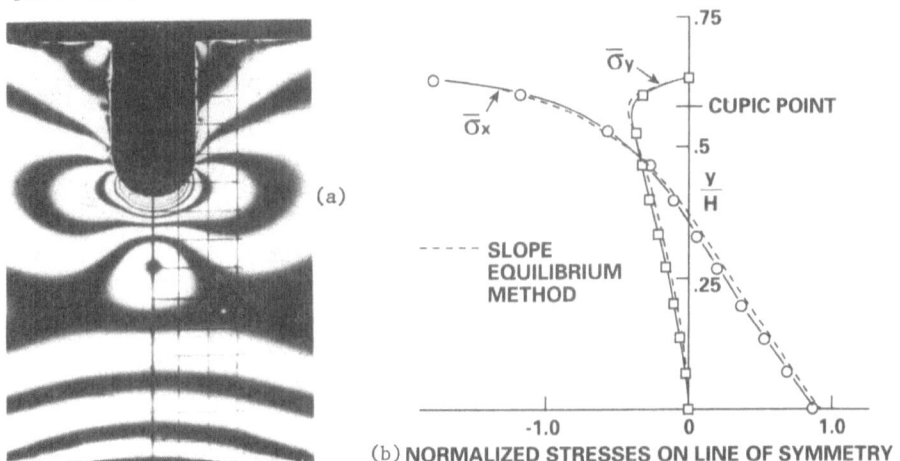

(a)

Figure 2. Beam with Deep Notch, Isochromatic Fringes (a)
For (b), $\bar{\sigma} = \sigma/S_m$, $S_m = Mh/(2I)$, $I = t\,h^3/(12)$, height h at notch

The results obtained along the line of symmetry are shown in Fig. 2b,
along with results obtained by slope-equilibrium, a method of superior
accuracy along lines of symmetry. Table 1 shows the effects of weighting
the different equations. The effect of the relatively crude finite-
difference approximations to the stress-equilibrium equations in the notch
region is apparent.

ADAPTING THE LEAST-SQUARES METHOD TO THE PERSONAL COMPUTER

The small desk-top personal computer (PC) is an ideal tool for experimental
stress analysis, providing rapid economical computation and graphical
display. It is possible to revise the least-squares approach to provide a
better fit of the problem to the small computer, both in terms of program
complexity and storage requirements. There are four changes [2].

A smaller solution region is presumed. It consists of a number of rows
nominally parallel to a free model boundary and a number of columns
perpendicular to them. The region is essentially rectangular, except at the
single free boundary (presumed to be at the top of the region) with all
station spacings the same in both directions, except adjacent to the
boundary.

A single unknown, the tangential stress, is used at boundary stations.
All of the co-ordinate stress components are expressed in terms of the
tangential stress using transformation equations.

TABLE 1
Effect of equation weights: notched beam in bending

$(W_1)^{be}$	$(W_1)^{bf}$	$(W_1)^{se}$	e_s
1.0	1.0	1.0	0.050
1.0	0.75	1.0	0.052
10.0	1.0	1.0	0.046
1.0	1.0	0.2	0.032
10.0	0.75	0.2	0.030
10.0	1.0	0.2	0.031
10.0	0.75	0.1	0.031

$(W_1)^{be}$ = weight for boundary condition equation
$(W_1)^{bf}$ = weight for photoelastic equation at boundary station
$(W_1)^{se}$ = weight for stress-equilibrium equatons at notch
e_s = root-mean-square error with respect to slope-
equilibrium results for normal stresses along line
of symmetry.

Four point finite difference approximations are used for the stress equilibrium equations (eqs. 3). For example, in place of eq. 4, the following equation is used:

$$\frac{\Delta y}{\Delta x} [(\sigma_x)_{i+1,j+1} + (\sigma_x)_{i,j+1} - (\sigma_x)_{i+1,j} - (\sigma_x)_{i,j}]$$
$$+ (\tau_{xy})_{i+1,j+1} + (\tau_{xy})_{i+1,j} - (\tau_{xy})_{i-1,j} - (\tau_{xy})_{i,j} = 0 \quad (11)$$

Four point approximations are more accurate than those they replace, but they cannot be used adjacent to general free boundaries (such as the deep notch of Fig. 2).

A single static equilibrium equation, for moments, is used. If the lower left corner of the region is designated i=1, j=1, this equation, which includes contributions from stress components on all interior boundaries of the solution region, is written for a region consisting of I rows ($1 \leq i \leq I$) and J ($1 \leq j \leq J$) columns:

$$(\frac{1}{\Delta x})^2 \{ \sum_{i=1}^{I} (y_0)_{i,1} (\sigma_x)_{i,1} (\Delta y)_{i,2} + \sum_{i=1}^{I} (x_0)_J (\tau_{xy})_{i,j} (\Delta y)_{i,J}$$
$$- \sum_{i=1}^{I} (y_0)_{i,J} (\sigma_x)_{i,J} (\Delta y)_{i,J} - \sum_{j=1}^{J} (x_0)_{1,j} (\sigma_y)_{1,j} (\Delta x)_{1,J} \} = 0 \quad (12)$$

The static force equilibrium equations can be shown to be redundant with stress equilibrium considerations for this solution region and they are not used.

Compatibility Equation in Terms of Stress

The compatibility equation was introduced to this problem [3]. For plane problems, the compatibility equation may be written:

$$\left(\frac{\partial^2}{\partial x^2} + \frac{\partial^2}{\partial y^2}\right)(\sigma_x + \sigma_y) = 0 \tag{13}$$

This is the familiar LaPlace equation. For a solution station i,j within the region, not adjacent to the free boundary, this equation may be approximated using finite differences:

$$-4\left[(\sigma_x)_{i,j} + (\sigma_y)_{i,j}\right] + (\sigma_x)_{i,j+1} + (\sigma_y)_{i,j+1} + (\sigma_x)_{i,j-1} + (\sigma_y)_{i,j-1}$$

$$+ (\sigma_x)_{i+1,j} + (\sigma_y)_{i+1,j} + (\sigma_x)_{i-1,j} + (\sigma_y)_{i-1,j} = 0 \tag{14}$$

For stations adjacent to the boundary the equation must be modified to include the varied station spacing. This equation is traditionally used separately to find the sum $\sigma_x + \sigma_y$ at locations within a closed region when all boundary values are known. The requirement to know all boundary values is relaxed because the equation is used only as an additional source of information in an otherwise completely defined problem.

Example Problem

Figure 3a shows an isochromatic fringe pattern for a disk (5.1 cm diam.) in diametral compression, together with a solution station grid. A solution region was identified consisting of the line of symmetry and the two columns immediately adjacent for the five rows. There are 15 solution stations, three of which are on the free boundary. A least-squares solution was sought and the values for normal stresses along the center line of the solution region are compared with those produced by incremental integration

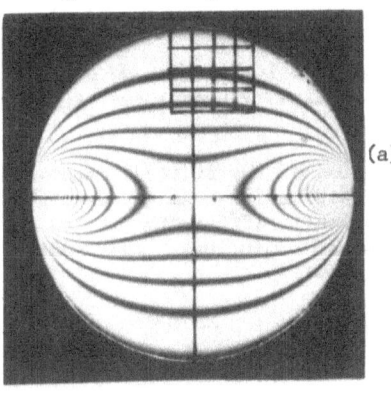

(a)

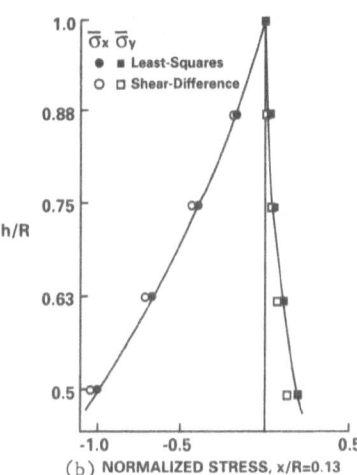

(b) NORMALIZED STRESS, x/R=0.13

Figure 3. Disk in Compression
Solution Grid and Isochromatics (a)
For (b), $\bar{\sigma} = \sigma\, t\, D/P$

of the stress equilibrium equation, using more data (shear difference). For this region, there are 39 unknown stress values and 47 equations, including compatibility. When compared with theoretical stress values for this problem (Figure 3b) the advantage of least-squares in avoiding error accumulation is apparent.

A second solution region was identified consisting of stations on the top four rows and to the right of and on the line of symmetry; 16 solution stations (four on the free boundary). A solution was obtained using theoretical values for isochromatic and isoclinic fringe parameters to examine the effect of the compatibility equation. For this region there are 40 unknowns and 51 equations when all are used. Table 2 shows that the compatibility equation has a small effect on the solution and in the interest of efficient use of storage and simple programming (especially with the small computer) it may be omitted.

TABLE 2
Effect of removing information from stress solution

Information Removed	Number of Equations	δ(percent)
None	51	0.23
Static Equilibrium	50	0.41
Compatibility	47	0.25
Static Equilibrium and Compatibility	46	0.46

δ = RMS error with respect to theoretical solution
$\delta = \delta/S_m$; $S_m = P/(td)$ where P is the load on the disk and d is the diameter.

COMBINING PHOTOELASTICITY WITH THE FINITE ELEMENT METHOD

This method is inserted at this point because it has been developed for plane problems and the stress optical equations are the same as previously used [4].

The method is essentially an expansion of the finite-element method to include photoelastic data as additional information. An excess of information is then available which permits simplifications in the problem definition and which leads to a least-squares solution. The problem is posed such that isoclinic data is not needed.

To briefly review the finite element method, the region of interest is subdivided into triangular elements (Figure 4) which are connected and transmit forces at their nodes. Each element experiences constant strain. The element geometry permits the element strain components to be expressed in terms of the node displacements. Linear elastic material properties are used to express stress in terms of strain. For the element:

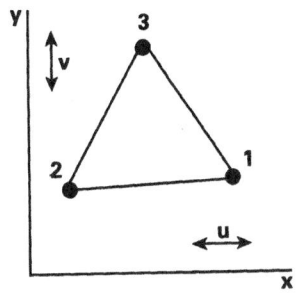

Figure 4.
Triangular Element

$$\{\sigma\} = [D] [B] \{^u_v\} \qquad (15)$$

Matrix B relates the element strain to node displacements, while D relates element stress to element strain.

For the element, the virtual work produced by the forces at the element nodes is equated to the strain energy experienced by the element. For the strain energy:

$$\delta U = \int_V \epsilon^T \sigma \, dV \qquad (16)$$

where V is the volume of the element, V = At, with the plane area and thickness equal to A and t. For the virtual work:

$$\delta W = \sum_{i=1}^{3} (F_{xi} u_i + F_{yi} v_i) = \{^u_v\}^T \{^{F_x}_{F_y}\} \qquad (17)$$

Equating these two quantities and remembering the constant element strain, a final element equation is produced:

$$[K] \{^u_v\} = \{^{F_x}_{F_y}\} \qquad (18)$$

where K is the element stiffness matrix; $K = [B]^T[D][B](V)$. Using eq. 18 it is possible to specify a consistent combination of six node displacements and/or forces and solve for the remaining forces and displacements.

For stress analysis over the region, two force equilibrium equations are written at each node with contributions from each element at the node; but they are written in terms of node displacements using eqs. 18. The loading system of applied forces is placed at certain of the nodes with displacement constraints also at certain nodes. The system of equations is solved for node displacements which then are used to solve for stresses using eqs. 15.

Inclusion of Photoelastic Data

The solution region is configured such that there are a majority of boundaries which are either free or which are lines of symmetry. Along such boundaries, principal directions are known and solution for isoclinics is not necessary.

Consider points (1) and (2) to be along a free boundary of a rotated element, with $x_2 > x_1$ and $y_2 > y_1$. Thence for the tangential stress: $\sigma_T = Nf/t$. An expression for the strain may be written in terms of the displacements (u_1, v_1) and (u_2, v_2):

$$(u_2 - u_1) \cos\phi + (v_2 - v_1) \sin\phi = \frac{s}{E} \frac{Nf}{t} \qquad (19)$$

where $\phi = \arctan (y_2 - y_1)/(x_2 - x_1)$ and $s = [(x_2 - x_1)^2 + (y_2 - y_1)^2]^{1/2}$.

For an element along a line of symmetry, consider points (1) and (2) to lie along such line, parallel to the x axis. Let the element be an isosceles triangle with base, $\ell = x_1 - x_2$ and height h to point (3). Thence a displacement equation, similar to eq. 19 may be written:

$$u_1 - u_2 - \frac{\ell}{h} v_3 = \frac{\ell(1+\nu)}{E} \frac{Nf}{t} \qquad (20)$$

A similar equation may be written for elements along a vertical line of symmetry. In eqs. 19 and 20, E is Young's modulus and ν is Poisson's ratio.

Applied Loads

Because of the excess of information provided with the photoelastic data, it is not necessary to use applied node forces. The affected node equations can be removed. Loading may be more reasonably introduced using static equilibrium in the vertical and/or horizontal directions. For equilibrium in the y direction, with n elements along the line of symmetry:

$$\Sigma F_y = \sum_{i=1}^{n_i} \sigma_{yi} \ell_i t \qquad (21)$$

In terms of the node displacements:

$$(\frac{1-\nu^2}{Et}) \Sigma F_y = \sum_{i=1}^{n_i} [\frac{v_{3i}}{h_i} + \nu \frac{(u_{2i} - U_{1i})}{\ell_i} \qquad (22)$$

For stress solution, the photoelastic equations and the static equilibrium equations are combined with the node equilibrium equations in an overdetermined system of equations which is solved using least-squares. It is necessary beforehand to adjust the coefficients to prevent inadvertent weighting. As they are written, the coefficients differ by several orders of magnitude. It is also necessary to assure that the boundary tangential stresses are tensile and that $\sigma_x > \sigma_y$ along the line of symmetry and to change the signs of the appropriate photoelastic equations if required.

Example Problem

The combined method was applied to a portion of the disk (5.1 cm diam.) in diametral compression. The isochromatic fringe pattern with the finite element grid superimposed on it are shown in Figure 5a. For this problem, there are 40 elements from 28 nodes. The solution obtains 47 displacement values from 51 equations. The stress results along the second row of elements above the horizontal diameter (x axis) are compared with the theoretical values in Figure 5b.

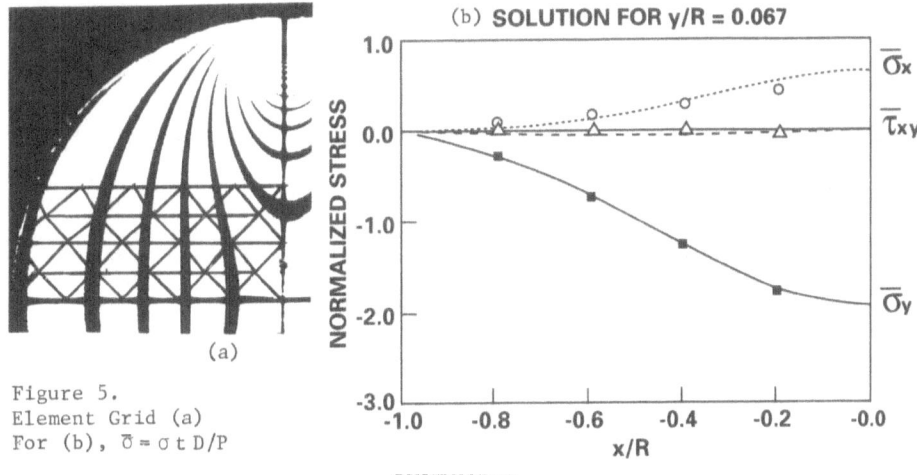

Figure 5.
Element Grid (a)
For (b), $\bar{\sigma} = \sigma \, t \, D/P$

CONCLUSION

The purpose of this paper has been to show how least-squares can be a valuable tool for application with photoelasticity, and to show how it can extend photoelasticity for use with the finite element method. The method has been employed in plane problems. Least-squares is applied to three dimensional scattered light photoelasticity in Reference 5.

The least-squares reduction of the overdetermined system of equations is given in the first two references. An algorithm for large systems of equations is listed in the first reference. A simpler algorithm for smaller problems is given in Reference 2.

REFERENCES

1. Berghaus, D.G., "Overdetermined Photoelastic Solutions Using Least-Squares", Experimental Mechanics, 13(3), 1973, pp. 97-104.

2. Berghaus, D.G., "Calculations for Experimental Stress Analysis - Using the Personal Computer", Publication #S-029, Soc. for Exp. Mech., Bethel, CT, USA (1987).

3. Berghaus, D.G., "Adding the LaPlace Equation to Least-Squares Photoelastic Stress Solutions", Experimental Techniques, 13(12), 1989, pp. 18-21.

4. Berghaus, D.G., "Combining Photoelasticity and Finite Element Methods for Stress Analysis Using Least-Squares", Proc. 1989 SEM Spring Conference on Experimental Mechanics, Cambridge, MA., USA, May 29-June 1, pp. 231-237.

5. Berghaus, D.G., and Aderholdt, R.W., "Photoelastic Analysis of Interlaminar Matrix Stresses in Fibrous Composite Models", Experimental Mechanics, 15(11), 1975, pp. 409-417.

PHOTOELASTIC ANALYSIS OF DOVETAIL JOINTS FOR TURBINE BLADES

B. KENNY, E. A. PATTERSON and K. S. S. ARADHYA
Department of Mechanical and Process Engineering
University of Sheffield,
Mappin Street,
Sheffield, S1 3JD

ABSTRACT

Both room temperature and frozen stress techniques of photoelasticity were used to study the stress distributions in two-dimensional models of an axially loaded dovetail type of joint. The geometry of the joint and the location of the loading were varied. Stress separations were carried out in the interior of the models as close to the loaded boundaries as was found to be feasible, using the Shear Difference method.

Steep principal stress gradients existed immediately below the surface at the end of the contact zone in the area adjacent to the fillet, which could give rise to severe fatigue conditions. The load distribution along the flank did not appreciably affect the maximum principal stress values below the contact surface. Reduction of the dovetail flank angle appeared to cause an increase in the principal stresses below the contact area.

INTRODUCTION

The Dovetail or the derivative Fir-tree type of turbine blade fixing is widely used for attaching blades to compressor or turbine discs. A consideration of the typical geometry and loading conditions for such a joint indicates that a maximum principal tensile stress occurs on the free surface of the fillet radius at the root of the projection in both blade and disc [1], [2], [3], [4]. Not all failures of such components originate from cracks in this region however and it is necessary to consider the stress distributions internal to the contact areas of the blade root and disc.

The aerodynamic loads on turbine blades are small in comparison to the radial loads due to rotation and many turbines have blades which are located in axial dovetail slots in the periphery of the discs. It was therefore considered that a two-dimensional model of the blade-disc joint would be appropriate and that the photoelastic method was the most suitable experimental technique to employ to consider the internal stress distributions.

EXPERIMENTAL DETAILS

Initially, two-dimensional photoelastic models were tested and analysed at room temperature, but it was discovered that the loading conditions on the contact surfaces of the models were not reproducible, which prevented all the necessary checking of the experimentally obtained photoelastic data. It was therefore decided to stress freeze the models so that the analysis could be checked where appropriate.

Fig. 1 shows the details of the loading rig used and a typical model geometry. The model and its surrounding loading frame were profiled from a sheet of CT200 epoxy resin using a high speed router. The models were encased in an aluminium box which acted as the axial load and as a constraint to prevent out of plane distortion of the models and to also provide constraint to the dovetail slot in what would be the circumferential direction of the slot. Thus "opening out" of the dovetail slot was prevented by this means.

Consideration was given to the nature of the contact loading for the model. Examination of real turbine blades and discs which have seen service provides evidence of a non uniform type of contact loading where a surface stress singularity situation is a possibility. To model the contacting surfaces in the same photoelastic material would result in two conforming surfaces at the stress freezing temperature of the model material. In the event it was decided to investigate the two extreme contact conditions by having a resin to resin contact on one side of the model and a resin to aluminium contact on the other side of the model. Both contact areas were provided by separate pads which were bonded to the slot part of the model in the appropriate location. Several models were used to find the effect of different load locations and two different flank angles namely 45° and 35° were used. Results were obtained for three different loading positions on the 45° flank angle model and one loading position for the 35° flank angle model.

568

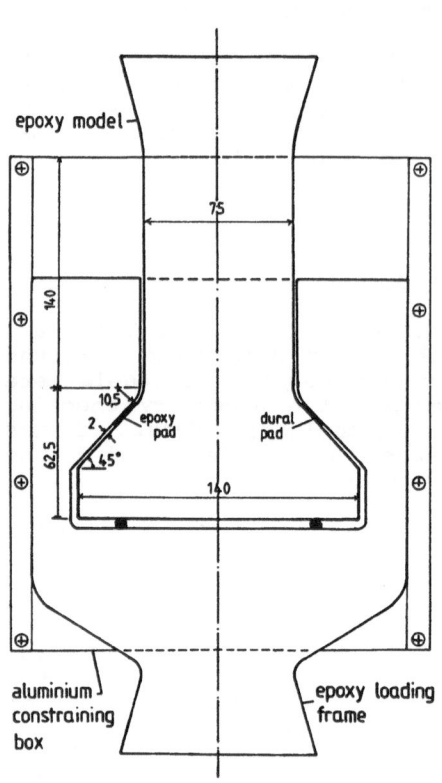

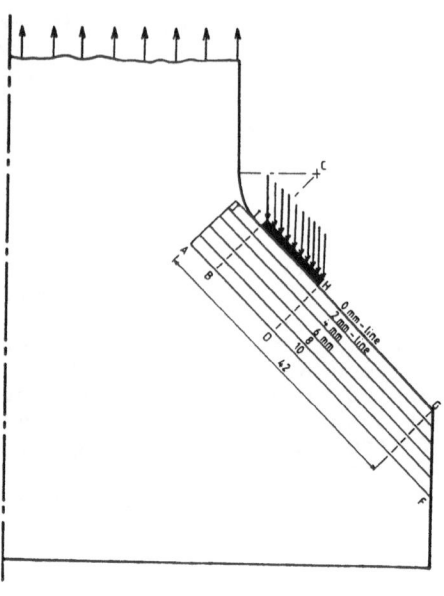

Figure 2   Detail of model and
               the lines of
               stress integration.

Figure 1   The model and loading
               frame.

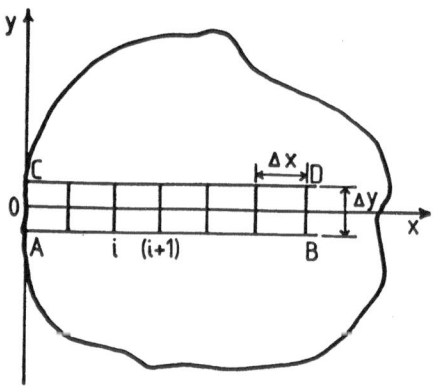

Figure 3   Reference system used
               for stress
               separation procedure.

STRESS DETERMINATION

Frocht's [5] shear difference method was used for separating stresses. To aid this procedure a grid of lines 2 mm apart was drawn on the two-dimensional models below the contact flanks, as shown in Figure 2. The isoclinic and isochromatic parameters along the lines parallel to the contact surface were obtained at 2 mm intervals. The means of at least three readings at each point were used to obtain these parameters.

The isoclinic parameter, "θ", was measured with respect to the flank angle, so that the x-axis was parallel to the contact flank with the origin being situated at the end of the flank furthest from the fillet radius.

The distribution of the cartesian stress component σ_x along the line of interest was obtained by integrating the differential equations of equilibrium, which in two-dimensions is given in simplified form by:

$$(\sigma_x)_{i+1} = (\sigma_x)_i - \left[\frac{\Delta \tau_{xy}}{\Delta y}\right]_{i+\frac{1}{2}} \Delta x \qquad (1)$$

with the notation of Figure 3.

The use of standard two-dimensional stress relationships enabled all the stress components to be separated.

DISCUSSION OF RESULTS

Figure 4 shows the separated principal stresses for the 45° flank angle case with the loading adjacent to the fillet. The principal stresses have been normalised with respect to the nominal average axial stress in the shank of the model.

On observing the resulting isochromatic fringe patterns in the loaded frozen models it was decided that the nearest one would expect to get to the flank boundary in order to obtain reliable photoelastic data was on a line at a distance of 2 mm from the flank boundary. Consequently, a line which was 3 mm from the loaded flank boundary was the nearest on which one could obtain derived separated stress data.

The significant features are the high gradient of positive stress at the fillet end of the contact region and the double peak of the compressive principal stress distribution. In practice, the loading of the contact area will not be constant since there is clearance between the mating surfaces of the blade and disc. This may have the effect of causing the curves of Figure 4 to move closer to or further

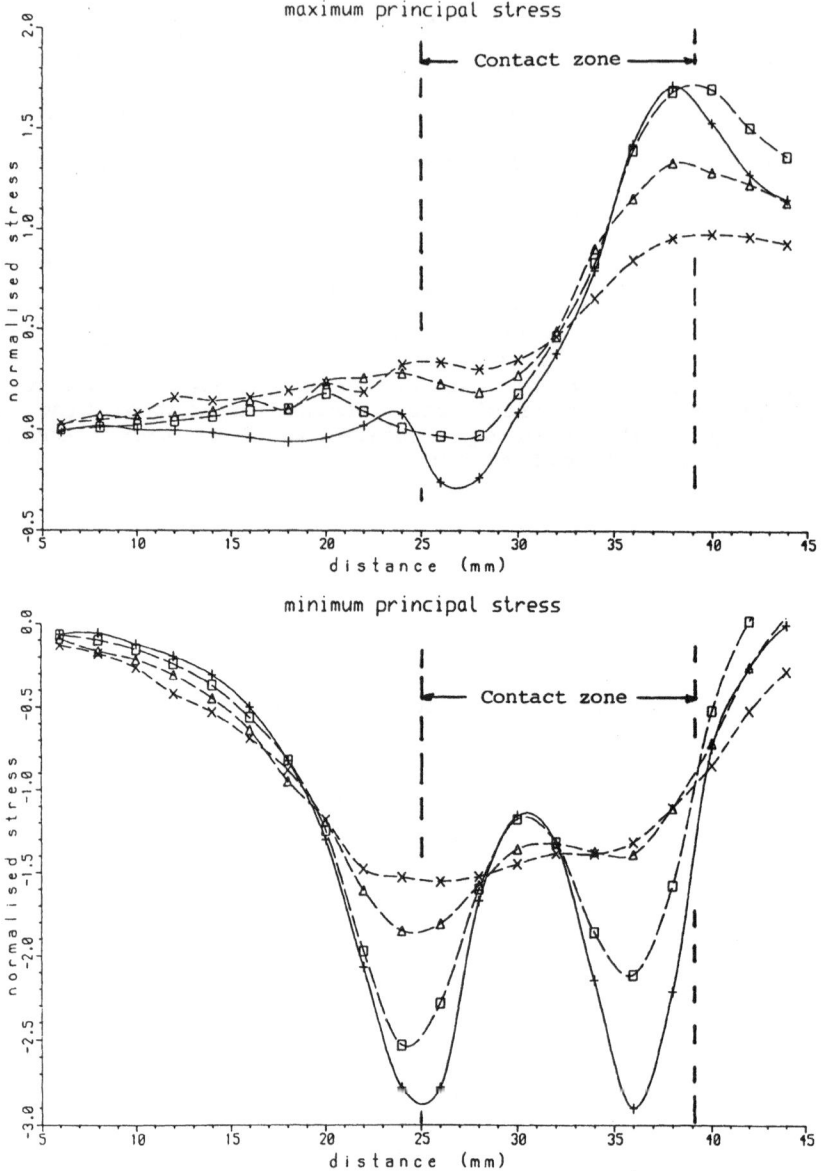

Figure 4   Distribution of principal stresses along lines
            parallel to the contact surface when the flank is
            loaded at the fillet end.   Depths below the surface:
            + 4; □ 6; △ 8; x 10 mm.

away from the origin of the graphs. Thus any fluctuation in the contact area conditions would give rise to a significant change in stress magnitude and even a sign reversal, leading to fatigue conditions at this location. It should be noted that as one approaches the boundary, the rate of increase of the positive principal stress becomes greater and also the gradient of this stress with distance along the flank increases. Thus, the nearer to the loaded boundary one gets, the more severe is the tendency to fatigue conditions.

It was found for all cases studied that the maximum principal stress occurred on the fillet boundary. The stress at this location however is only slightly affected by a change of the contact conditions and so it is likely that fatigue is not a problem in this locality.

The distributions of normalised principal stresses for the four model cases studied are shown in Figure 5. It is observed that when the contact area is near to the fillet region, the gradient of the positive principal stress is greatest. As the loading position moves away from the fillet region, a more uniform positive principal stress distribution is obtained. Since the load in each case was the same, the bending stress at the fillet region increases as the loading location moves away and so there is a maximum principal stress value at the fillet end of the flank when the load is located at the tip of the flank.

The effect of the dural-epoxy and the epoxy-epoxy contact conditions may be observed in Figure 6 where the principal stress differences for the 45° flank angle models are presented. It is seen that the general trends of the distributions are the same for both types of contact conditions and that away from the areas of contact, there are no significant differences in the magnitudes. Some differences do exist at a depth of 2 mm below the surface however, but it was concluded that these were mainly associated with experimental accuracy and could have been due to variations in loading pad finish or fixing, or to errors in fringe determination close to the singularity on the surface at the edge of the contact. The major difference occurs for the case of loading adjacent to the fillet where the peak stress difference under the dural pad is 2 times that under the epoxy pad and $1\frac{1}{2}$ times greater than the peaks at the tip end of contact. (This result was not in accordance with a similar loading case for a 35° flank angle model and was therefore attributed to misalignment of the loading pad in the case of the 45° flank angle model).

It was observed that when the load was not adjacent to the fillet, the maximum fillet stress occurred at an angular position of 34° from the shank of the model whereas this changed to an angular position between 21° - 26° for loading adjacent to the fillet, depending on the flank angle used.

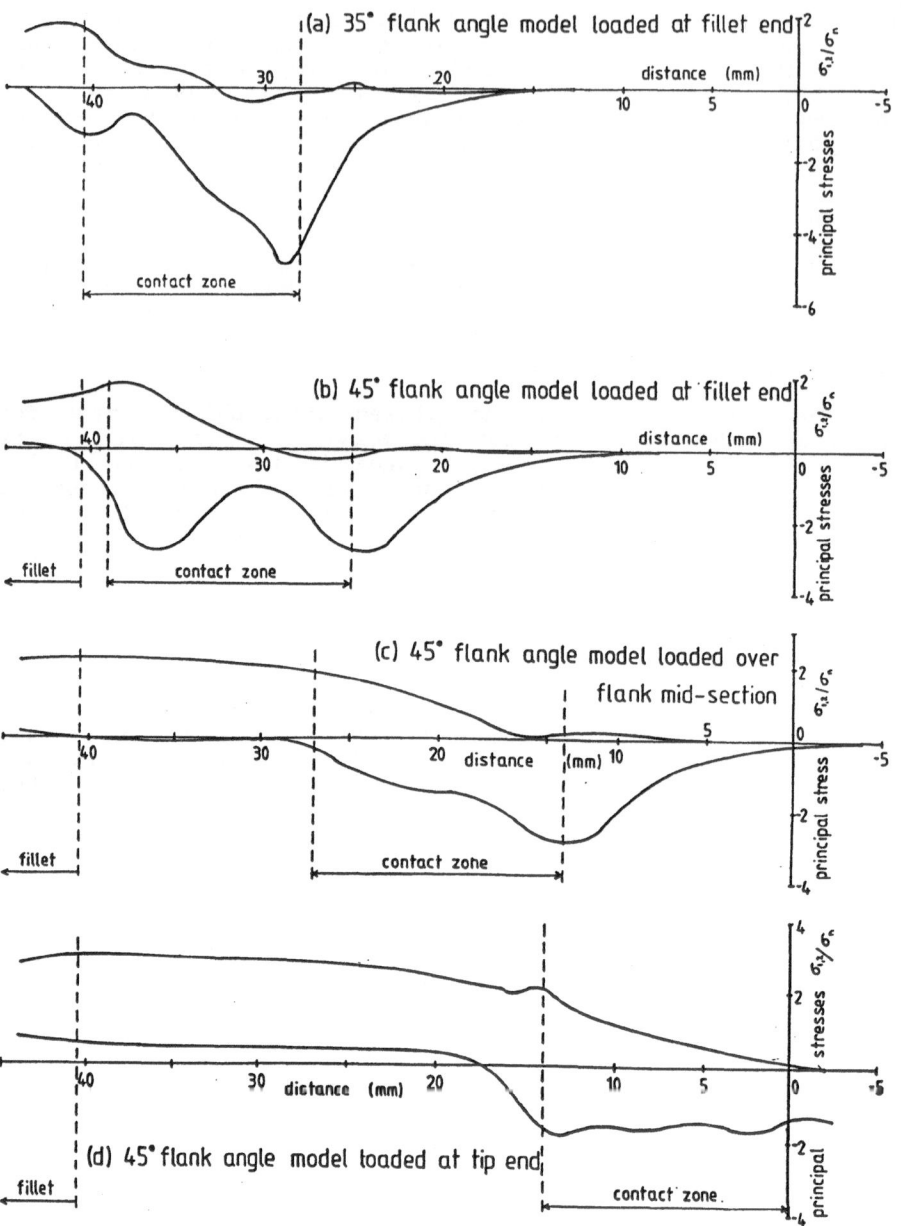

Figure 5   Distribution of normalised principal stresses
along a line 3 mm below the contact flank.

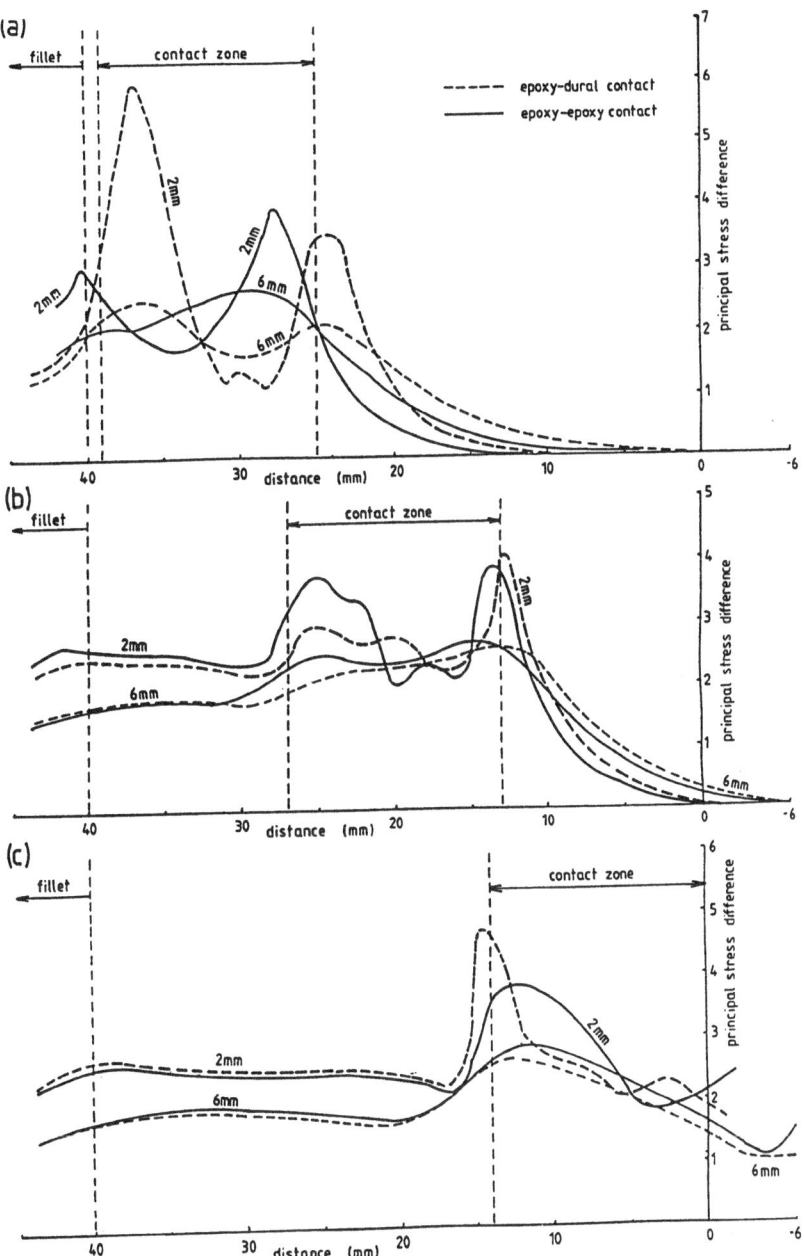

Figure 6   Comparison of principal stress difference below the surface of both sides of the model when loaded (a) toward the fillet and (b) the middle (c) the tip of the flank.

ACCURACY OF RESULTS

The separated stress components obtained by the Shear Difference method were used to determine a load balance for a typical section of the model. The maximum error in such a load balance was found to be 7%.

CONCLUSIONS

In the 45° flank angle model, the stress concentration in the fillet radius increased as the load approached the fillet region.

The tensile principal stresses tend to decrease with depth from the contact surface on the fillet side of contact and increase with depth on the tip side of contact. The variation of principal stress magnitude (and direction) was greatest as the flank surface was approached.

The maximum tensile principal stress was found to be in the fillet root.

Steep principal stress gradients could result in fatigue conditions on the flank surface at the fillet end of contact.

It was unlikely that fatigue conditions would be critical for the fillet region.

REFERENCES

1. Hetenyi, M., Some applications of photoelasticity in turbine generator design, J. Appl. Mech., 1939, 61:A151-A155.

2. Heywood, R. B., Tensile fillet stress in loaded projections, Proc. Instn. Mech. Engrs., 1948, 159:384-391.

3. Durelli, A. J., Dally, J. W., Riley, W. F., Stress and strength studies on turbine blade attachments, Proc. SESA., 1958, 16(1):171-186.

4. Fessler, H., Woods, P. J., Stress concentrations at axially loaded projections of flat bars, J. Strain Anal., 1980, 15(3):137-143.

5. Frocht, M. M., Photoelasticity, vol. 1, Chapman Hall, New York, 1941, 262-264.

INACCURACIES of PHOTOELASTIC MEASUREMENTS

Jan Cernosek
president
PhotoStrain, 11166 Ables Ln.,Unit C,
Dallas, TX 75225, USA

Abstract.

Three-dimensional photoelasticity is a powerful but complicated method requiring a great number of steps on the path from the blueprint to the information related to the stress distribution. The presentation attempts to discuss the inaccuracies which are usually omitted in literature because they cannot be submitted to rigorous mathematical analysis.

Introduction.

Three-dimensional photoelasticity is an exceptionally powerful, but complicated method of an experimental stress analysis. It is the only method which offers the information permanently stored in a "stress-frozen" model. This information which is related to stress at each point of the model can be decoded solely at the discretion of the analyst who can change the decoding strategy or even the goal of the study any time during decoding procedure.

Despite of this indisputable advantage of stress-freezing photoelasticity over any other experimental method, its use declined in accord with the development of powerful numerical methods of stress analysis based on rapidly improving computers. The reason was simple. The stress-freezing, three-dimensional photoelasticity, as it was originally established, was a very complicated, time consuming, and therefore expensive method. Only recently, new techniques and materials were developed which made photoelasticity by stress-freezing once again cost-effective.

But three-dimensional photoelasticity still remains a complicated method requiring a great number of steps on the path from the blueprint to the information related to the stress distribution and its magnitude. A general question - What is the reliability, accuracy and repeatability of photoelasticity - does not have a single and simple answer.

In order to gain at least some insight into this problem, this presentation attempts to discuss sources of the inaccuracies which are usually omitted because they cannot be submitted to rigorous mathematical analysis.

Methodology of stress-freezing three-dimensional photoelasticity.

The accuracy and reliability of stress-freezing, three-dimensional photoelasticity can be affected by any or all steps/aspects of its methodology. They are:

1) Preparation of the model:
a) Fabrication of the pattern using information supplied from the blueprint.
b) Fabrication of the mold for casting the model.
c) Casting the model.
d) Solidification or primary curing of the material.
e) Machining of the model to conform specifications of the blueprint (effect of machining stresses).
f) Secondary or final curing of the model material. Effect of material and geometrical dissimilarities.
g) Calibration of the optical and mechanical properties of the material.

2) Stress-freezing and slicing of the model:
a) Fabrication of the loading fixture.
b) Selection of the level of the load applied to the model (effect of large deformations).
c) Stress-freezing cycle, selection of a thermal regime (heating and cooling rates, length of the soaking period to achieve uniform temperature through the mass of the model).
d) Selection of slices and their location.
e) Slicing of the model and preparation of slices (thickness of slices, smoothing their surfaces, enhancing their transparency).

3) Measurement of birefringence (fringe order and isoclinic parameter) and evaluation of stress:
a) Accuracy of the measurement of fractional fringe order (compensation methods).
b) Effect of the gradient of stress through the thickness of the slice.
c) Measurement of the stress at the edge of the slice. Effect of the time-effect and rind-effect.
e) Relationship between stress and birefringence (material fringe constant).
f) Accuracy of the methods for evaluation of out of slice plane stress components (oblique incidence, sub-slicing).

4) Relation of model and prototype stress:
a) Effect of boundary conditions.
b) Distribution of contact stresses and effect of the size of contact areas.
c) Effect of Poisson's ratio on the stress distribution.

It is quite obvious that this presentation cannot address all of these aspects of the stress-freezing procedure. While some of them are frequently discussed in literature there are others which were conspicuously ignored. Some of those will be addressed in this presentation.

Preparation of the model.

The traditional way of fabrication of a photoelastic model was by machining from a block of fully cured model material. Preparation of this block alone was an expensive and time consuming process. Because the curing of epoxy based materials is a strong exothermic reaction the low reactivity curing agents (mostly anhydrides) have to be used. Thus, the curing of such a block could take two or even more weeks. The machining itself is an expensive and tedious process. The development of rind-free materials together with the development of an "on shape" casting technique greatly enhanced the effectiveness of photoelasticity. Fast curing materials could be used because the volume, and therefore the amount of heat evaluated, was greatly limited. The materials which are now available enable complicated models to be cured overnight.

The curing of any epoxy based material is accompanied with shrinkage which is restricted by the mold. The natural question therefore arises: What is a residual birefringence in the model and what is therefore the base-line of the analysis? In order to answer this question the three-dimensional model of the connecting rod was prepared using a standard procedure and submitted to a stress-freezing cycle with no load applied. The model was then cut into slices which were evaluated in the polariscope. The model material was epoxy resin EPON 828 cured by the addition of 3 percent of tris(dimethylaminomethyl) phenol tri(2-ethyl hexoate). This is a liquid curing agent which permits curing of the model at relatively low temperature of 150°F. The model was cast into the rigid mold which consisted of an inner lining made from condensation cure silicon elastomer and an outside rigid shell made of high temperature resistant epoxy. The model was solidified overnight, machined and submitted to a typical stress freezing cycle: heating with a gradient of 20°F/hour from room temperature to temperature of 275°F, soaking at this temperature for 8 hours, and cooling to the room temperature with a gradient of 5°F/hour. The model was then sliced using the high speed band saw and slices were kept overnight in the drying oven at temperature of 150°F before they were evaluated in the transmission polariscope.

Fig. 1 depicts the photoelastic model together with the pattern which was used for fabrication of the mold. Fig. 2 depicts the distribution of the residual birefringence in typical slices. The fringe order was normalized to the typical thickness of a slice of 0.1 inch. This residual birefringence includes all possible contributions from individual steps of the model preparation, stress freezing and slice preparation procedures. It is the "noise" which is superimposed onto the useful signal which would be the birefringence induced by a mechanical stress.

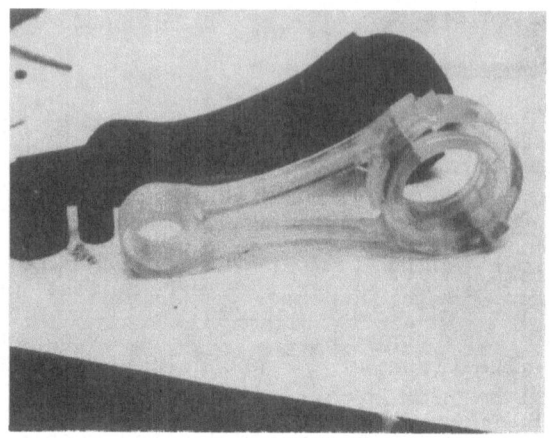

Fig. 1 - Photoelastic model of connecting rod.

Taking into consideration that the typical fringe order in a 0.1 inch thick slice is in the range from 2 to 6 orders, this "noise" represents a typical error of 1.5 to 5 percent which is quite acceptable for a typical design support study.

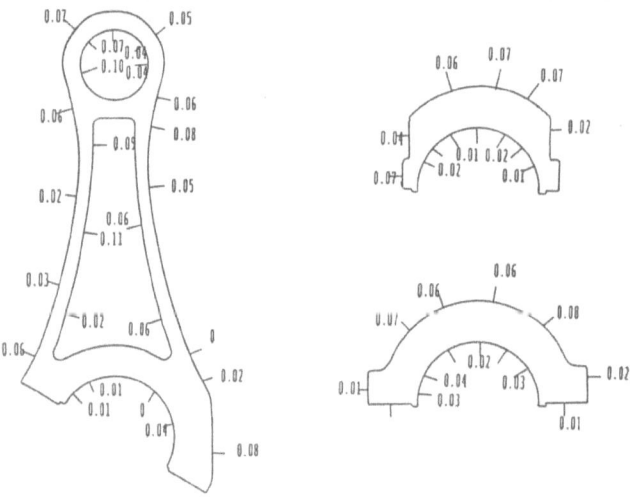

Fig. 2 - Residual birefringence. Thickness of slice 0.1 inch.

But this level of residual birefringence prevents the use of some methods based on the light intensity distribution measurement as for example "half fringe photoelasticity".

Effect of large deformation.

The modulus of elasticity of typical photoelastic material for stress-freezing is within 1600 to 4500 psi range. Thus, the strain corresponding to a typical fringe order (normalized to thickness of 1 inch) is within 2 to 6.5 percent. Frequently, this fact was claimed to be a major setback of stress-freezing method because the analysis of the large strain field was generalized onto a small strain field in the prototype.

The study of the effect of large strain on the stress concentration in the fillet of the shoulder of a cylindrical bar provided some insight into this problem. The bar with parameters of D/d=1,66 and R/d=0.33 (see Fig. 3) was manufactured from special, very soft (modulus of elasticity of 42 psi) material which exhibits a strain-freezing phenomenon (the deformation remains frozen-in if the material is heated at 210°F for at least 6 hours). The strain of various levels in the cylindrical section "d" was introduced, bar was "strain-frozen", sliced, profile of a distorted fillet was copied into a three-dimensional photoelastic model and analyzed. The results of this study are depicted in Fig. 3. It can be observed that the strain up to 10% brings the change of a stress-concentration factor up to 5%, which is not a negligible but an acceptable error if we realize that this level of strain is seldomly achievable.

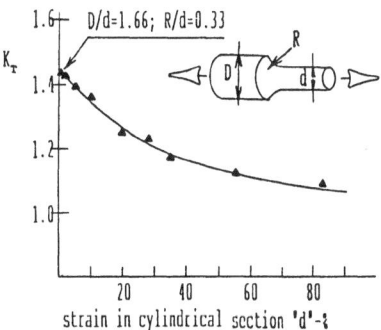

Fig. 3 - Effect of large strain.

Measurement of birefringence.

Much can be found in the literature about accuracy of various

methods of measurement of parameters of birefringence. Most
analyses limited themselves to monochromatic (single light
wave) light because these analyses then can be performed
within rigorously determined conditions. The use of white
light in two dimensional photoelasticity was limited to the
analysis of the field of isoclinics because the isochromatics
of higher order appear to be washed out in white light. Only
recently some interest in the methods of analysis of the
spectral content could be indicated.

There is an indisputable advantage in using white light in
three dimensional photoelasticity because of the nice
definition of fringes up to the fourth order, easily
recognizable order of the fringe and its nice and sharp
definition (if the fringe is for example defined as a line
dividing red and green colors). The fringe pattern of white
light contains more information than the monochromatic fringe
pattern. The difference can be compared to the difference of
single and parallel processing in computer.
Fig. 4 depicts the intensity of the light (adjusted for the
sensitivity of the average human eye) for the light length in
the range of 450 to 650 nm for fringe orders within range of 1
to 4. Fig. 5 depicts the light intensities during the course
of the Tardy's compensation for a fractional fringe order of
2.25. It can be seen that the intensity distribution
approaches that one for fringe order 2 when compensation angle
(angle of rotation of an analyzer from its "dark field"
position) is increasing from zero to its nominal value of 45°.

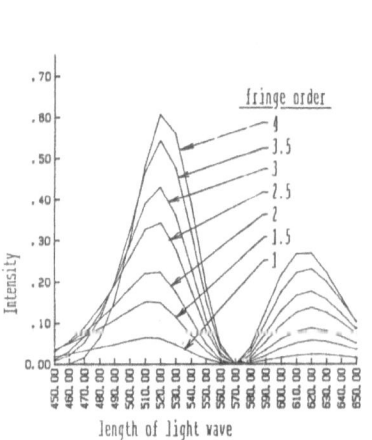 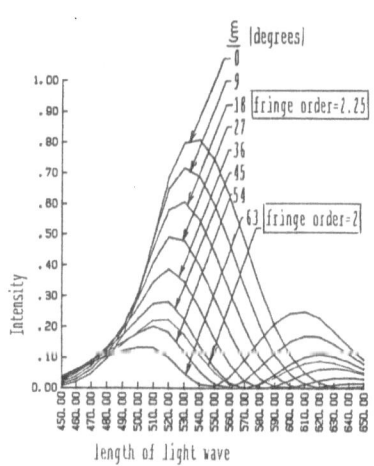

Fig. 4 - White light.   Fig. 5 - Intensity of light.
         Intensity distribution   Tardy's compensation

Fig. 6 depicts the distribution of the intensities of light
when the compensation angle approaches 45°. For comparison,
there is in this figure also the distribution of light
intensities for the fringe order 2. It can be seen that
the spectral distribution for the fringe orders 2 and 2.25 are

not identical but Fig. 7 indicates that their ratio remains nearly constant for lengths of light wave within a range of 500 to 650 nm.

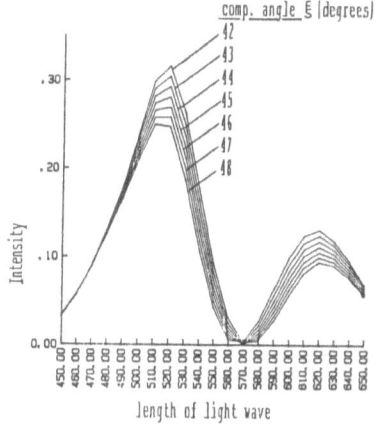

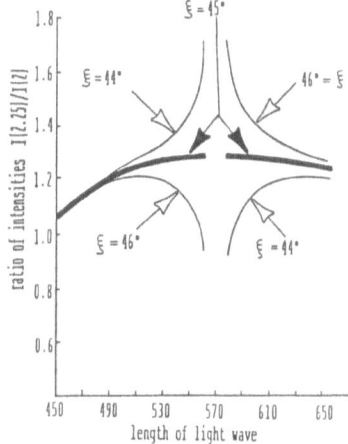

Fig. 6 - Intensity of light. Compensation angle approaches 45°.

Fig. 7 - Ratio of intensities. Fringes (2) and (2.25)

These results indicate that the Tardy' compensation in white light is feasible and deserves more study. It is much more convenient to determine the location of the fringe as the boundary between the red and green colors then to try determine where the center-line of the fringe is. This is especially convenient for determination of the fractional fringe order at the edge of the slice (and this is the only location where the fringe order has to be determined in three-dimensional photoelasticity).

Fringe order at the edge of the slice.

Three-dimensional photoelasticity is capable of determining the stress at the point of the free surface of a three-dimensional, stress-frozen model. Thus, the measurement of the fringe order (the direction of polarization is known for slices cut in such a way that their plane is perpendicular to the tangential plane to the free surface) is of upmost importance.

Unfortunately, the fringe order at the boundary of the slice can be affected not only by residual birefringence discussed above but also by parasitic birefringence because of an absorption or desorption of moisture. To protect slices from such an effect it is recommended to keep them in a drying oven at temperature of 140 to 170°F and remove them just before measurement. The persistent questions arise: How long will it take to cool a slice to room (say 75°F) temperature? How long can a slice be kept out of the oven before the fringe order at the boundary is affected by moisture?

582

Fig. 8 illustrate a somehow surprising finding that it took nearly fours hours to stabilize the fringe at the edge of 0.123 inch thick slice. Fig. 9 depicts the fringe pattern in this slice.

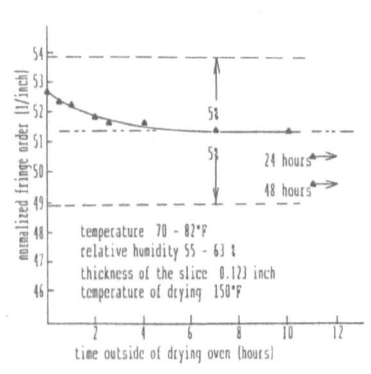

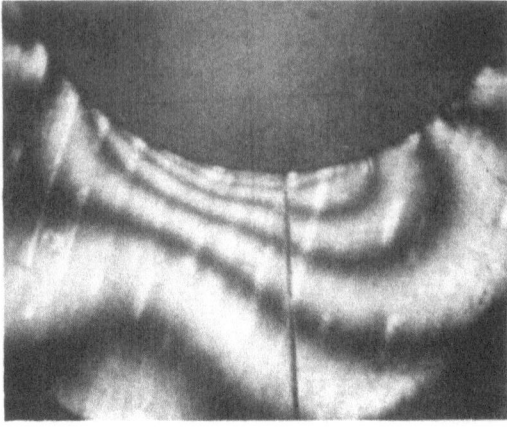

Fig. 8 - Stabilization of fringe after drying.

Fig. 9 - Fringe pattern after drying. T=0.1 inch.

As the way of coping with the problem of time-effect, the extrapolation of "the fringe order-distance from the boundary" function to the edge of the slice was suggested. This approach is sometimes also favored by those who prefer the automatic fringe pattern processing using digitizing cameras in order to avoid difficulties at or in the vicinity of the edge of the slice.

In order to asses the validity of such extrapolation the position of the fringes was determined with the accuracy of ± 0.001 in the projection polariscope (10x magnification) equipped with a x-y translating table. The least square method was employed to determine the coefficients of the polynom of the fourth order. The extrapolated value of the fringe to the boundary was 5.89 which is 7% less than the value measured at the edge of the slice (6.32). The procedure was repeated after 48 hours at 75°F and approximately 52% relative humidity. The extrapolated value was found to be 6.4 much closer to the originally measured fringe order than for measurement immediately after stabilization of fringe after removal from the drying oven. Fig. 10 depicts the results of the extrapolation after the slice was immersed for 24 hours in the water. The fringe pattern in the vicinity of the edge is in Fig. 11. The absorption of the moisture into the slice evidently affects the fringe pattern only in the vicinity of the edge. Thus the method of extrapolation using the least square approach is quite efficient if the accuracy of ±5% is acceptable.

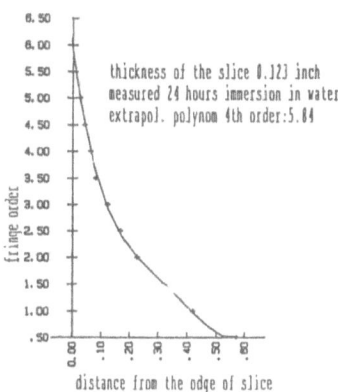

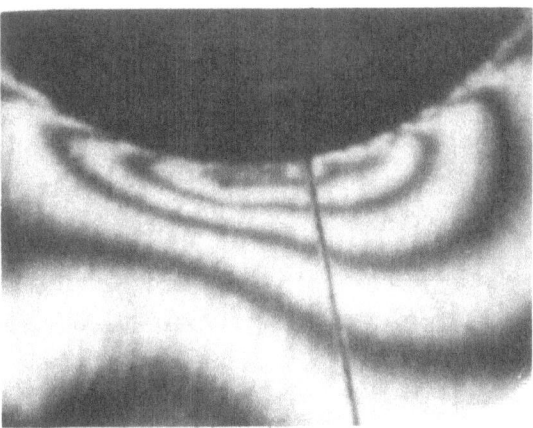

Fig. 10 - Extrapolation
to the edge.

Fig. 11 - Fringe pattern after
24 hours immersion.

Evaluation of stress.

The final goal of any photoelastic analysis should be the
determination of stress, its distribution and its magnitude.
If the slice is sufficiently thin for assumption of constant
stress through the thickness to hold, then the normalized
fringe order, at the point of free surface, is directly
proportional to the component of stress in the plane of the
slice. The other two components of the stress tensor have to
be determined by additional procedure, sub-slicing or by the
oblique incidence method.

The disadvantage of sub-slicing is obvious. It is tedious,
time consuming and requires very careful "book-keeping"
because the handling of small (say 0.1 x 0.1 x 0.1 inch)
elements is difficult.

Thus, the oblique incidence method seems to be the only useful
method capable of providing the desired information in a
reasonably cost-effective way.

What is its accuracy? The principles of the oblique incidence
methods have been firmly established. At the point of free
edge the stresses can be determined from following
relationships:

$$S_{xx} = (DELTA_n)K/T \; ; \; S_{xy} = (DELTA_{o1} - DELTA_{o2})K/(4Tsin \Psi)$$

$$S_{yy} = (DELTA_n - (DELTA_{o1} + DELTA_{o2})/(2cos \Psi))K/(tan \Psi)^2$$

where ψ is an oblique incidence angle. The rotation with respect to the normal to the tangential plane (axis "z") was considered. It is obvious that in such point the directions of polarization are normal and tangential to the edge of the slice. But this is not true if the slice was cut with feathered edge. The stress components can be then determined from the following equations:

$$S_{xx}-S_{yy}(\sin \alpha)^2=DELTA_n K/T;$$

$$S_{xy}=-(DELTA_{o1}\cos(2\phi_1)-DELTA_{o2}\cos(2\phi_2)K/T(\sin(2\psi)\cos(\alpha)$$

$$S_{xx}=((DELTA_{o1}\cos(2\phi_1)+DELTA_{o2}\cos(2\phi_2)K-S_{yy}\cos^2(\psi))/$$
$$(\sin^2(\psi)\cos^2(\alpha)-\sin^2(\alpha))$$

where α is the angle of the feathered edge and ψ is the oblique incidence angle. Even if the point is located in the plane of symmetry (shear stress equals to zero) the direction of polarization is not perpendicular and tangential to the edge. This fact is depicted in Fig. 12.

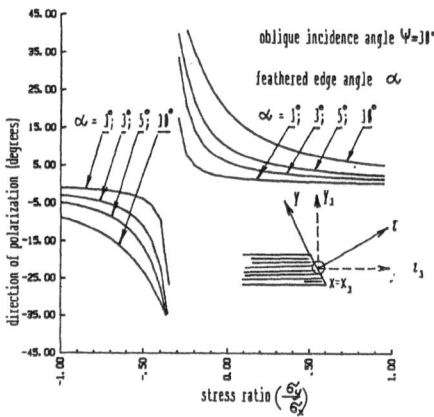

Fig. 12 - Isoclinic angle. Slice with feathered edge. Plane of symmetry.

Even if the slice is cut in such a way that the edge of the slice is perpendicular to the plane of slice the oblique incidence method can provide quite erroneous results.

Let us, for example, assume that the slice is cut from the plate containing the circular hole. The plate is stretched by the uniformly distributed tensile stress "S" and also loaded by tensile stress "S" applied in the direction parallel to the axis of the hole. Let us also assume that the point under consideration is located in the plane of symmetry. There is a rotation of directions of polarization along the path of the light for oblique incidence.

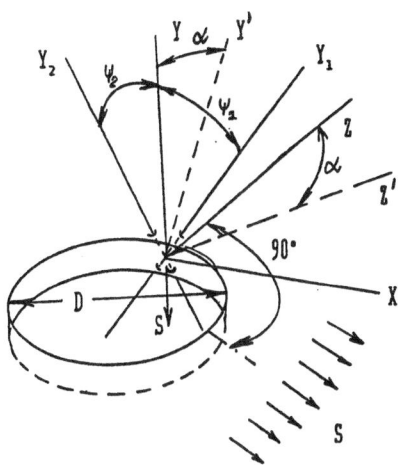

Fig. 13 - Oblique incidence method. Definitions.

In order to gain some idea what inaccuracy can be expected the
problem was analyzed in the reversed direction: Knowing the
stress, the parameters of birefringence were calculated in
every point of the light paths. Then the characteristic
quantities (characteristic directions and charecteristic phase
retardation) of the equivalent system of retarders were
calculated and using the arrangement for Tardy's compensation
the apparent fringe order was calculated from the condition of
the minimum flux from the analyzer. These results were then
used to determine the apparent stress perpendicular to the
plane of the slice. The results are listed in Table 1.

Table 1.

Dia. of hole	Thickness of slice	Angle of incidence	Stress "S"	Error S_{yy} %
0.5	0.1	20	10	11.7
0.5	0.1	30	10	11.0
0.5	0.1	45	10	13.1
0.5	0.1	20	5	12.6
0.5	0.1	30	5	12.1
1.5	0.1	45	5	11.8
0.75	0.1	20	10	5.3
0.75	0.1	30	10	5.1
0.75	0.1	45	10	7.2
0.5	0.1	20	1	24.5
0.5	0.1	30	1	24.9
0.5	0.1	45	1	27.1
0.375	0.1	30	10	20.0
0.375	0.075	30	10	12.5
0.375	0.050	30	10	5.8
0.375	0.025	30	10	1.0

Theoretical magnitude of $S_{yy}=S$.

The need for thin slice and for the high fringe order in the
plane of the slice can be clearly seen.

Repeatability of photoelastic experiment.

The last question which should be asked is: If the experiment
is repeated, if the identical model is analyzed again, perhaps
at different level of loading, how would the results of these
two experiments agree?

In the effort to find the answer to this question a relatively
simple study was performed twice - at a different level of
load -and the results were normalized to one load. The
models (simple flexure with shoulder, circular fillet and
hole through the shoulder) were prepared from different
batches of material, cast "on shape" into different molds,
stress-frozen in separate thermal cycles and loaded by loads
which ratio was equaled to 2. Thus, the experiment included
the inaccuracies introduced by possible geometrical
dissimilarity, errors in material calibration, differences in
stress concentration because of changes of shape of fillet on
the tensile and compression side of the flexure, inaccuracies
introduced during slicing and preparation of slices and errors
of fringe measurement. The results are depicted in Fig. 14.
It can be seen that the overall repeatability of
three-dimensional photoelasticity is quite acceptable.

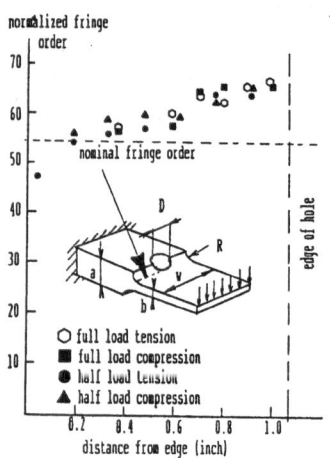

Fig. 14 - Repeatability. Maximum fringe in fillet.

Conclusion.

The statement that three-dimensional photoelasticity is quite reliable, accurate and cost-effective method seems not to be exaggeration. One has to realize that the most important problem of engineering is not what accuracy of engineering analysis can be achieved but rather what inaccuracy can be accepted and still produce reliable and useful information. The unnecessary accuracy is a luxury which nobody can afford.

FUNDAMENTALS OF PHOTOVISCOELASTIC TECHNIQUE FOR ANALYSIS OF

TIME AND TEMPERATURE DEPENDENT STRESS AND STRAIN

TAKESHI KUNIO
Professor Emeritus of Keio University, Yokohama &
Professor of Department of Mechanical Engineering,
Kantoh Gakuin University,
4834 Rokuura Kanazawa-ku, Yokohama 236, JAPAN

YASUSHI MIYANO and SUGURU SUGIMORI
Professors of Materials System Research Laboratory,
Kanazawa Institute of Technology,
7-1 Ohgigaoka Nonoichi, Kanazawa 921, JAPAN

ABSTRACT

Polymeric materials show remarkable time and temperature dependent visco
elastic behaviors, which are quite different from elastic ones, under cer
tain conditions, for example, at the temperature ranging from the glass
state to the rubbery.   Although photoelasticity is an useful experimenta
method for the analysis of elastic stress, it is not applicable for analy
sis of time and temperature dependent viscoelastic stress and strain.
    In this paper, fundamentals of photoviscoelasticity, by which th
above viscoelastic stress and strain can be analyzed, are described briefl
together with two characteristic examples of viscoelastic problems; one i
a problem of non-proportional loading, which is out of question in elasti
problem, the other is a problem of thermoviscoelastic stress and strain du
to rapid cooling of viscoelastic body.   These problems are closely relate
to practical use of polymeric materials, in particular, the latter of whic
is one of the essential problems for clarification of generation mechanis
of residual stress and strain.

INTRODUCTION

It is well known that photoelasticity is very useful for elastic stres
analysis.   However, for the purpose of analyzing experimentally time an
temperature dependent stress and strain in a viscoelastic material, phc
toviscoelasticity should be used instead of ordinary photoelasticity
This paper intends to describe fundamentals of the photoviscoelastic tech
nique [1]-[6] as well as its application to practical use.

Prior to making a practical photoviscoelastic analysis, mechanical and optical characterizations of the materials [1]-[3] concerned are required (These correspond to procedure of material calibrations in photoelastic experiment). In other words, it is necessary to determine the time and temperature dependent viscoelastic and photoviscoelastic coefficients represented as functions of these variables. However, the description of characterization for visco- and photovisco-elastic coefficients of polyurethane rubber and epoxy resin employed are neglected here.

In this paper, examples of the practical photoviscoelastic analyses are demonstrated on two characteristic problems. The first is the analysis of a viscoelastic square plate having a central hole, to which the non-proportional load is subjected under a constant temperature. [4] In this case, the principal axes of polarized light, principal stress and strain do not coincide with each other. The second is the analysis of an epoxy strip which is subjected to rapid cooling from both sides. [5],[6] This is a so-called thermoviscoelastic problem in a wide range of temperature change.

A SQUARE PLATE HAVING A CENTRAL HOLE
LOADED NON-PROPORTIONALLY ON ADJACENT SIDES

In case of ordinary two-dimensional photoelasticity, the directions of principal stress, principal strain and polarization of light generally coincide completely at any point, while they do not always in photoviscoelasticity. However, under the special conditions, for example, the proportional loading at constant temperature, all these three have the same direction. In such a case, the photoviscoelastic situations at any time with respect to stress and strain are just similar to those in standard photoelasticity. Therefore, the time-dependent principal stress and strain differences can be analyzed photoelastically, using the characterization master curves of the inverse-relaxation stress-birefringence coefficient $C^{-1}{}_{\sigma r}(t)$ and the inverse-creep strain-birefringence coefficient $C^{-1}{}_{\varepsilon c}(t)$ determination of the fringe order at any time.

The experiment demonstrated here is concerned with the case of disagreement of these three. As an example, the case of non-proportional loading is employed. In this case, the principal stress difference, principal strain difference and their directions at any time are analyzed from recording the time-dependent photoviscoelastic isoclinics and isochromatics.

Experimental Procedures

Figure 1 shows the geometry of the specimen employed, loading method and loading conditions. The dead load W_1 is applied to the specimen over all the time and the step load W_2 is applied after t = 0. In this way, the non-proportional loading could be realized. The constant temperature during the experiment is as low as T = -41.3 °C, because the polyurethane rubber shows decisive viscoelastic behavior at this temperature.

Standard linear- and circular polariscopes are used for taking time-dependent photoviscoelastic pictures of isoclinic and isochromatic patterns, respectively.

Some typical pictures of isoclinic and isochromatic fringe patterns at several times, measured from the instant (t = 0) of W_2 loading, are demonstrated in Fig. 2. Isoclinic parameter α, i.e. the direction of polariza-

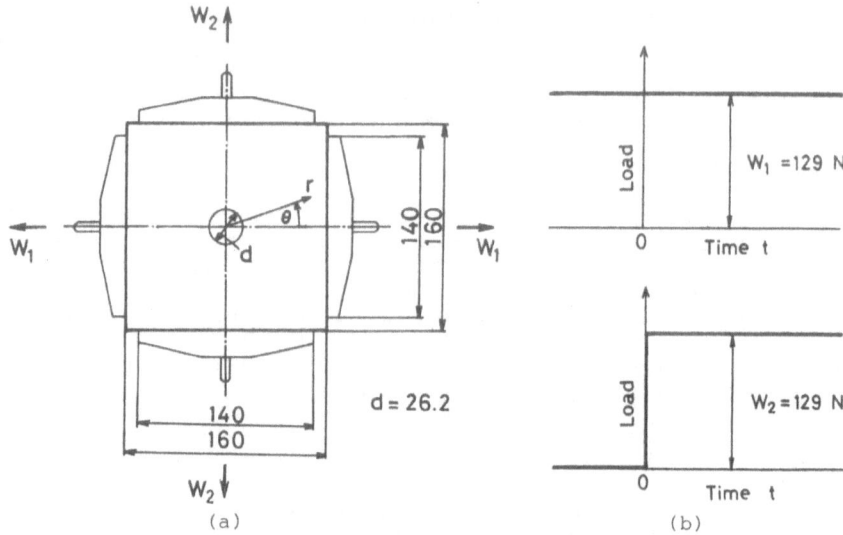

Figure 1. Specimen geometry, loading method and non-proportional loadinc condition. (unit : mm)

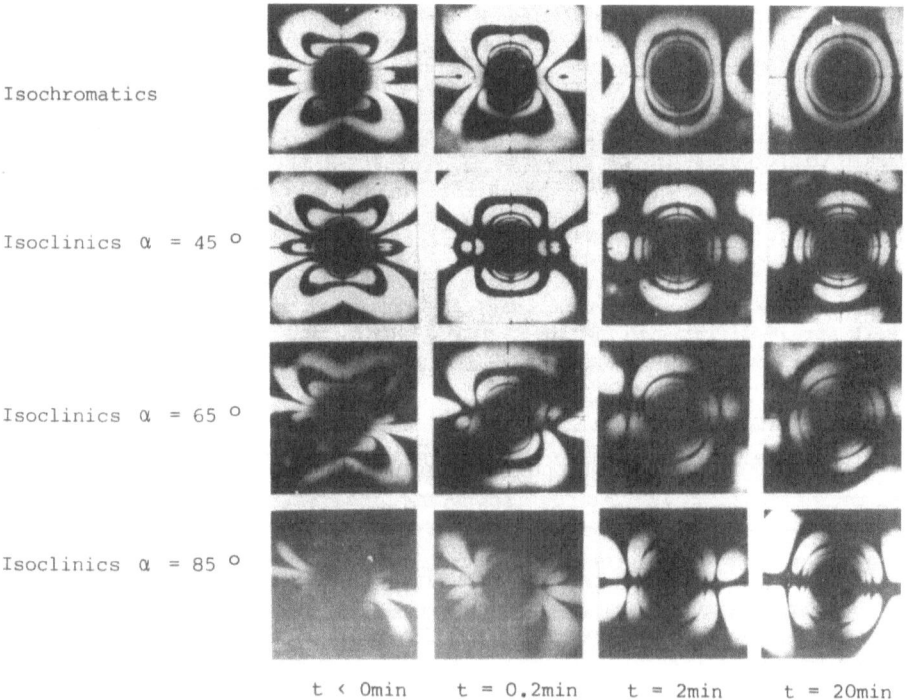

Figure 2. Typical fringe patterns due to non-proportional loading at con-stant temperature T = -41.3 °C (polyurethane resin).

tion of light is measured as the angle to the horizontal line.

Calculations of Stress and Strain

From the fringe patterns in Fig. 2, the variation of fringe order $N(t)$ with time at a point $P(r = 0.75d, \theta = 45^{\circ})$ in Fig. 1 (a) can be read as shown in Fig. 3 (a). The principal stress difference $\sigma_1 - \sigma_2$ and the angle β of principal stress direction to horizontal line can be calculated by introducing the above data $N(t)$ and $\alpha(t)$ into the following equations:

$$(\sigma_1 - \sigma_2)\cos(2\beta) = \frac{1}{h} \int_{-\infty}^{t} c^{-1}{}_{\sigma r}(t-\tau) \frac{dN(\tau)\cos(2\alpha)}{d\tau} d\tau$$

$$(\sigma_1 - \sigma_2)\sin(2\beta) = \frac{1}{h} \int_{-\infty}^{t} c^{-1}{}_{\sigma r}(t-\tau) \frac{dN(\tau)\sin(2\alpha)}{d\tau} d\tau \tag{1}$$

where h is the thickness of the specimen. Also, as far as the calculations of the principal strain difference $\varepsilon_1 - \varepsilon_2$ and the angle γ of the principal strain direction, the following equation are used:

$$(\varepsilon_1 - \varepsilon_2)\cos(2\gamma) = \frac{1}{h} \int_{-\infty}^{t} c^{-1}{}_{\varepsilon c}(t-\tau) \frac{dN(\tau)\cos(2\alpha)}{d\tau} d\tau$$

$$(\varepsilon_1 - \varepsilon_2)\sin(2\gamma) = \frac{1}{h} \int_{-\infty}^{t} c^{-1}{}_{\varepsilon c}(t-\tau) \frac{dN(\tau)\sin(2\alpha)}{d\tau} d\tau \tag{2}$$

The variations of principal stress difference together with its direction and principal strain difference together with its direction are shown in Fig. 3 (b) and Fig. 3 (c), respectively. It can be found from these figures that the experimental and theoretical results agree fairly well with each other, and also that, because the input to the specimen is given by loads at the boundary, the stress responds immediately after loading, while the strain response shows a remarkable time-dependent viscoelastic behavior. Also, the directions of the principal stress, the principal strain and the polarization of light at the point P are shown in Fig. 4. Disagreement of these directions is characteristic feature of viscoelastic behavior.

A EPOXY STRIP SUBJECTED TO RAPID COOLING FROM BOTH SIDES

This section is dealt with a photoviscoelastic analysis of the thermal stresses and strains in a viscoelastic strip having a circular hole subjected to rapid cooling from both sides. The analysis of the stress and strain concentrations at the edge of the circular hole in the strip is a problem of technical interest.

After keeping an epoxy strip having a circular hole at a holding temperature which is sufficiently higher than the glass transition tempera-

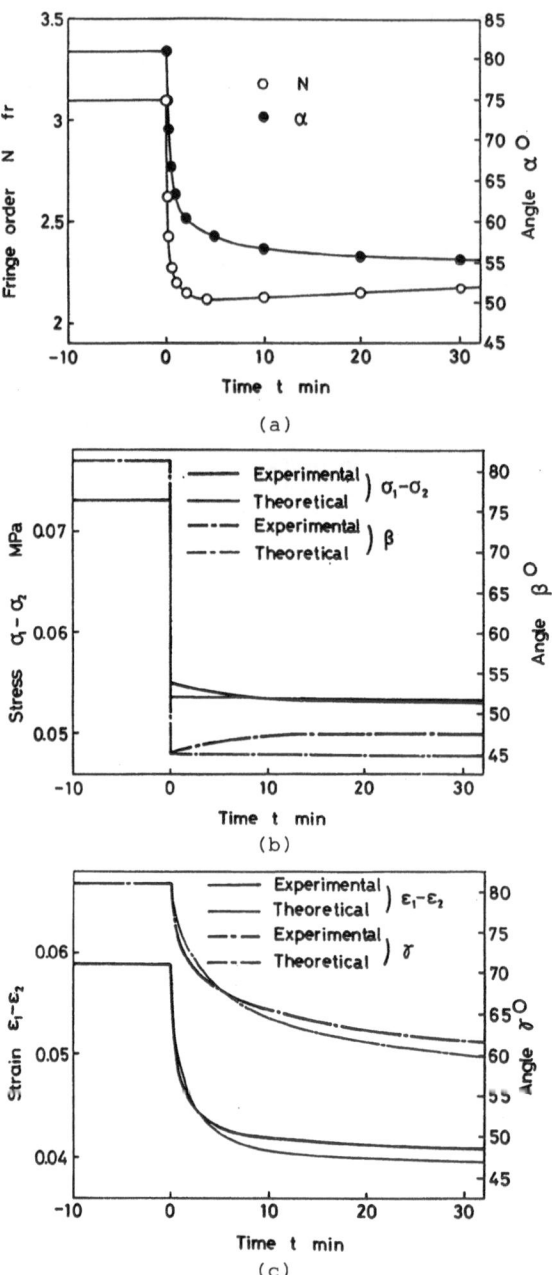

Figure 3. Transient fringe orders N(t), angle $\alpha(t)$ of polarization axis, principal stress difference $\sigma_1(t) - \sigma_2(t)$, angle $\beta(t)$, principal strain difference $\varepsilon_1(t) - \varepsilon_2(t)$ and angle $\gamma(t)$ at point P at T = -41.3 $^\circ$C.

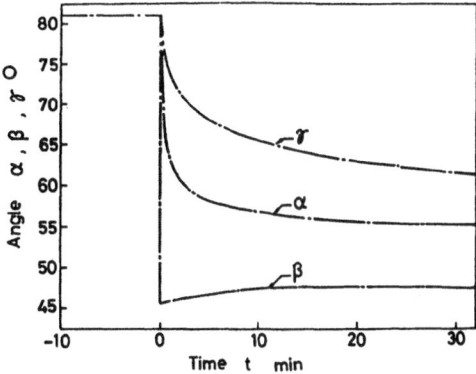

Figure 4.   Transient angles $\alpha(t)$, $\beta(t)$ and $\gamma(t)$ at point P   at   T = -41.3 °C.

ture, this strip is subjected to rapid cooling from both sides in water. Both the transient birefringence and temperature are measured during the rapid cooling.   From these measurements, the transient thermal stresses and the transient strains are determined using hereditary integrals based on the two-dimensional linear photoviscoelastic theory.

Experimental Procedures

The dimensions of the specimen and the rapid cooling method are shown in Fig. 5.   The specimen is a strip having a central hole.   The width (the x axis) is sufficiently larger than the thickness (the z axis) and the length (the y axis) is sufficiently larger than the width.

The temperature of 180 °C, at which the material is in rubbery state, is selected as the holding temperature T_h, because it is sufficiently higher than the glass transition temperature of epoxy resin T_g = 125 °C.   The temperature of 10 °C is selected as the cooling temperature T_c at which the material is in glassy state.

The insulating materials made of transparent type silicone rubber are put on both surfaces of the specimens for preventing the cooling of both surfaces as shown in Fig. 5.   The strain sensitivity of this material is very much smaller than that of epoxy resin.   The specimen is put in a thermostatic oven so that the specimen could be kept at the holding temperature T_h before rapid cooling.   At time t = 0 , the specimen is rapidly cooled from both sides, put into running water at the cooling temperature T_c, then it is kept in water until the specimen attains T_c.

Through direct measurement of specimen's temperature during rapid cooling, the transient temperature distribution is confirmed to be approximately one dimensional in the x direction and is symmetric with respect to the z axis, in spite of the specimen having a circular hole at the center. Using the actual values of temperature measured at two or three points on the x axis of each specimen during cooling by thermocouples, the transient temperature distribution T(x,t) on the x axis is theoretically calculated by means of the heat conduction equation.

On the other hand, the transient isochromatic fringe pattern at any time during rapid cooling is recorded by video camera using an ordinary circular polariscope.   The transient fringe order distribution on the x axis of specimen is obtained from the isochromatic fringe pattern.

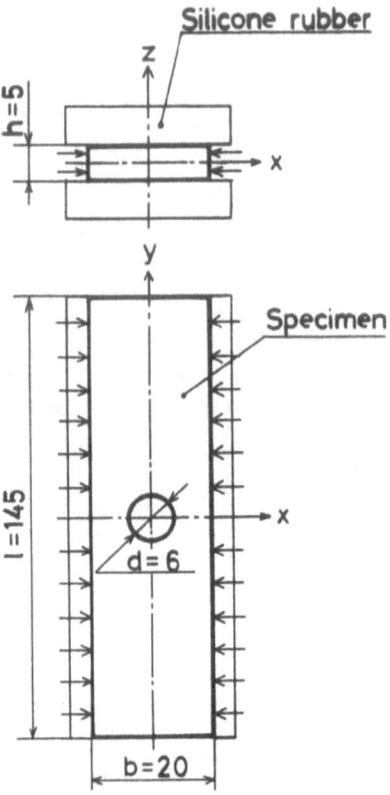

Figure 5. Specimen geometry and cooling method. (unit mm)

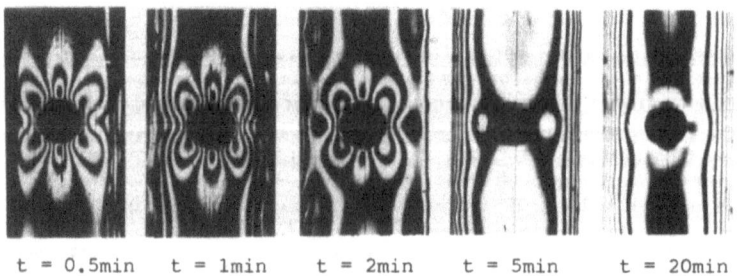

t = 0.5min   t = 1min   t = 2min   t = 5min   t = 20min

Figure 6. Isochromatic fringe patterns due to rapid cooling(T_h = 180 $^\circ$C, T_c = 10 $^\circ$C, epoxy resin).

Calculation of Thermal Stress and Strain

Figure 6 shows the photographs of the isochromatic fringe patterns of specimen in dark field at time $t = 0.5, 1, 2, 5$ and 20 min after the specimen is subjected to rapid cooling. These fringe patterns changes remarkably with time. A considerable change of the fringe patterns is observed along the hole after rapid cooling. The fringe pattern immediately after loading is very similar to the fringe pattern of the photoelastic strip having a circular hole which is subjected to the longitudinal loading. However, this characteristic fringe pattern vanishes gradually as time passes.

Both distributions of fringe order and temperature on the x axis of specimen during rapid cooling are shown in Fig. 7.

The temperature distribution of specimen at any time is symmetric with respect to the x axis. Therefore, the axes of principal stresses, principal strains and polarization of light on the x axis of specimen coincide with each other. Since the time-temperature equivalent law holds for the relation of stress-strain-birefringence, the principal stress difference $\sigma_1 - \sigma_2$ and the principal strain difference $\varepsilon_1 - \varepsilon_2$ can be determined by substituting $T(t)$ and $N(t)$ into the following equations:

$$\sigma_1 - \sigma_2 = \frac{1}{h} \int_0^t C^{-1}\sigma_r(t'-\tau',T_0) \frac{dN(\tau)}{d\tau} d\tau$$

$$(3)$$

$$\varepsilon_1 - \varepsilon_2 = \frac{1}{h} \int_0^t C^{-1}\varepsilon_c(t'-\tau',T_0) \frac{dN(\tau)}{d\tau} d\tau$$

with the relation

$$t' = \int_0^t \frac{du}{a_{T_0}(T(u))} \qquad (4)$$

where T_0 = reference temperature
         a_{T_0} = time-temperature shift factor with respect to T_0
         t' = reduced time with respect to T_0

The thermal stress distributions on the x axis of the rapidly cooled specimen are shown in Fig. 8 (a). The thermal stresses are very small in the entire region of x axis and show a slightly tensile one in the vicinity of the cooled sides just after the start of cooling. The high tensile stress concentrations are generated suddenly at the edge of circular hole after considerable elapsed time. Simultaneously, the thermal stresses become compressive in the vicinity of cooled sides. The distribution of strains on the x axis during rapid cooling are shown in Fig 8 (b). The high compressive strain concentration is generated at edge of hole just after the start of cooling, then it becomes slightly low with time. Such behavior of thermal stresses and strains at the edge are very characteristic of a viscoelastic body.

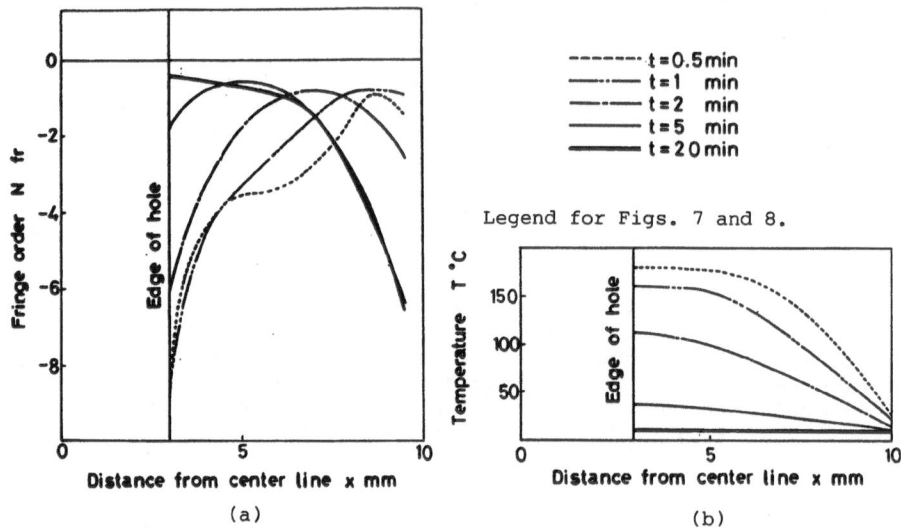

Figure 7. Distribution of fringe order N(x,t) and temperature T(x,t) during rapid cooling.

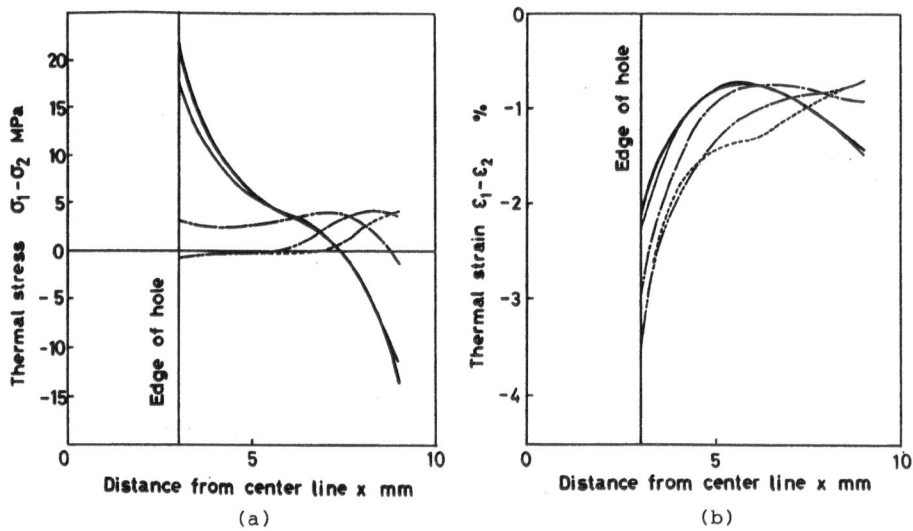

Figure 8. Distribution of principal stress difference $\sigma_1(x,t) - \sigma_2(x,t)$ and principal strain difference $\varepsilon_1(x,t) - \varepsilon_2(x,t)$ during rapid cooling.

597

CONCLUDING REMARKS

The photoviscoelastic technique is one of the experimental techniques for analysis of time and temperature dependent viscoelastic stress and strain.

In this paper, the practicality of this techniques was demonstrated, divided into two parts.

The first part was dealt with viscoelastic analysis of a square plate of polyurethane rubber having a circular hole, which is subjected to non-proportional loading. In this problem, there contains a characteristic feature that the three principal directions of stress, strain and polarized light do not, in general, coincide with each other. The effectiveness of photoviscoelasticity to such a problem was shown through the experimental results.

In the second part, thermoviscoelastic analysis using photoviscoelastic techniques together with time-temperature equivalent law was described. As an example, thermal stress in an epoxy strip subjected to rapid cooling were analyzed. Generally speaking, since thermal stress and strain vary with time, the residual stress and strain can be obtained as the values of viscoelastic ones after a long lapse of time. Therefore, this example provides a very interesting problem applicable to practical use for estimation of residual stress and strain of polymeric materials.

ACKNOWLEDGEMENTS

The authors wish to express their thanks to Profs. M. Takashi and A. Misawa for many helpful advises and discussions.

REFERENCES

1. Arenz, R.J., Ferguson, C.W. and Wiliams, M.L., The Mechanical and optical characterization of a Solithane 113 composition, Experimental Mechanical, 1957, 7 (4), 183-8.
2. Arenz, R.J., Ferguson, C.W., Kunio, T. and Williams, M.L., The mechanical and optical characterization of Hysol 8705 with application to photoviscoelastic analysis, GALCIT SM 63-31, California Institute of Technology, 1963.
3. Miyano, Y., Tamura, T. and Kunio, T., The mechanical and optical characterization of polyurethane with application to photoviscoelastic analysis, Bulletin of JSME, 1969, 12, 26-31.
4. Kunio, T. and Miyano, Y., Photoviscoelastic analysis by use of polyurethane rubber, Applied Mechanics (Proceedings of the Twelfth International Congress of Applied Mechanics), 1968, 269-76.
5. Sugimori, S., Miyano, Y. and Kunio, T., Photoviscoelastic analysis of thermal stress in a quenched epoxy beam, Experimental Mechanics, 1984, 24 (2), 150-6.
6. Miyano, Y., Sugimori, S. and Kunio, T., Photoviscoelastic analysis of thermal stress in an epoxy strip with a circular hole by rapid cooling, Proceedings of the 1985 SEM Spring Conference on Experimental Mechanics, 1985, 191-8.

STRESS MEASUREMENTS BY STRAIN GAGES MADE OF NICKEL FOIL

Masaichiro SEIKA[*] and Kikuo HOSONO[**]

[*] Professor, Department of Mechanical Engineering, Daido Institute
of Technology, 2-21 Daido-cho, Minami-ku, Nagoya 457, Japan

[**] Research Associate, Department of Mechanical Engineering, Faculty
of Engineering, Nagoya University, Furo-cho,
Chikusa-ku, Nagoya 464, Japan

ABSTRACT

Strain gages made of nickel foil are developed for measuring the elastic
surface stress of machine parts in operation. Sticking a piece of nickel
foil about 10 μm thick on the surface of a specimen subjected to repeated
loads, the elastic stress is measured by observing slip-bands in the foil
resulting from repeated strains. Calibration studies with rotating-bending
tests are performed on round steel bars with nickel foil to determine the
applicable test temperatures for the gage, and to establish the relation
between the threshold stress for the first appearance of slip-bands and the
number of stress cycles. The peak stresses in grooved shafts under bending
are measured with the nickel foil gages, and the accuracy of the results is
examined.

INTRODUCTION

Strain gages made of nickel foil have been developed for measuring the elastic surface stress of machine parts in operation [1], [2], [3]. Sticking a
piece of nickel foil about 10 um thick on the surface of a specimen subjected to repeated loads, the elastic stress has been measured by observing
slip-bands in the foil resulting from repeated strains.

A basic study on the measurement of elastic stresses using strain gages
made of nickel foil is described in this paper. Rotating-bending tests are
performed on round steel bars with nickel foil to examine the applicable
test temperatures for the nickel foil gages and variation of the density of
slip-bands due to the magnitude of cyclic stress, and to establish the relation between the threshold stress for the first appearance of slip-bands
and the number of stress cycles. The peak stresses in grooved shafts subjected to bending at several definite test temperatures are measured with
the nickel foil gages and the accuracy of the results is examined.

EXPERIMENTAL PROCEDURE

Testing machine and test specimen

A rotating-beam-type machine (3600 rpm, 98 N•m) was used for fatigue tests. Tapered calibration specimens shown in Fig. 1 and grooved specimens shown in Fig. 2, the dimensions of which are tabulated in Table 1, were made of round bars of carbon steel.

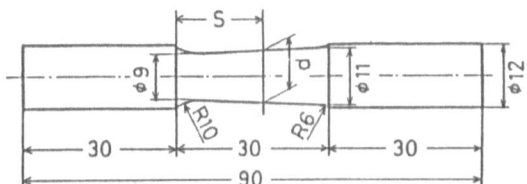

S: Distance to the position where
slip-bands begin to appear

Figure 1. Tapered calibration specimen

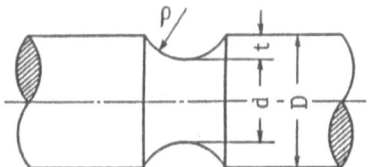

Figure 2. Circular shaft with a circumferential groove

TABLE 1

Dimensions of shafts with a circumferential groove

Specimen	D mm	d mm	ρ mm	t mm	D/d	ρ/d	t/ρ
(a)	15.0	10.0	2.5	2.5	1.5	0.25	1.0
(b)	12.0	8.0	4.0	2.0	1.5	0.5	0.5
(c)	12.0	8.0	8.0	2.0	1.5	1.0	0.25

Nickel foil and its adhesion

A stainless steel plate polished by buffing was electroplated with nickel, and then a sheet of plating foil about 10 μm thick was made by stripping the deposited layer from the stainless steel plate. By annealing the plating foil at 950 °C for 60 seconds in a vacuum, crystal grains of 30 μm in mean diameter could be produced in the nickel foil. The nickel foil heat-treated was cut into a number of small rectangular pieces (2 mm × 25 mm) and squares (1 mm × 1 mm, 2 mm × 2 mm). They were stuck on the tapered part of the calibration specimen and the root of the groove as the foil gages. A thermosetting-type adhesive based on silicone was used to stick the foil gages. The adhesive layer to stick the foil was about 2 μm in thickness.

Measurement of threshold stress
The threshold stress of the foil gage was measured with the tapered calibration specimen by the rotating-bending test under a constant temperature and a specified number of stress cycles. Two strips of the rectangular foil were stuck on the tapered part so that they might be arranged symmetrically along the axis of the specimen. If the slip-bands in the attached foil began to appear at the position of the distance S (Fig. 1) under a bending moment M, the threshold stress σ_p in tension-compression was calculated by the following formula.

$$\sigma_p = 32M/\pi d^3 \tag{1}$$

where d was the diameter at the position. The slip-bands were observed with an optical microscope at a magnification of 100×. The distance S was determined as the average of the measured values in the attached foil. The deviation of each measured value of S from the average value was within 1 mm and it was a small value less than 2.0 MPa in terms of the stress in tension-compression.

Standard photographs
Using the tapered calibration specimen tested under the conditions of a constant temperature and a specified number of stress cycles, photomicrographs showing the density of slip-bands were taken at several positions along the attached foil. Such a series of photographs, together with the corresponding repeated stresses acting on the specimen, has been used as the 'standard photographs' to estimate the stress value in an object, by comparing the slip-bands in the foil attached to the object with those in the standard photographs.

Measurement of peak stress based on threshold stress
The peak stresses in grooved shafts under bending were measured with the specimens shown in Fig. 2 and Table 1. Two pieces of the square foil were stuck at the symmetrical positions on the periphery of the root of groove. Using the specimens thus prepared, rotating-bending tests at a constant test temperature were carried out to generate the slip-bands in the attached foil. The slip-bands were observed with an optical microscope at a magnification of 100×. When more than three zones of slip-bands were found in the field of the microscope, the appearance of slip-bands was judged 'dense'. When two or three zones of slip-bands were found, it was judged 'thin', which was quite similar to the appearance of slip-bands at the critical position of the tapered calibration specimen. Thus the standard stress σ_0 was determined by the nominal stress $\sigma_d = 32M/\pi d^3$ corresponding to 'thin'. When the specimen was subjected to the standard stress σ_0, the stress at the root of the groove had become a threshold stress σ_p, and the stress-concentration factor k was obtained by the formula, $k = \sigma_p/\sigma_0$.

RESULTS AND DISCUSSIONS

Calibration curves
Figure 3 shows the relation between the threshold stress σ_p and the number of stress cycles N in the range from room temperature (R. T.) to 350°C. The

σ_p - log N curve, i. e., the calibration curve of the foil tested at a constant temperature T, descends obliquely with an increase in N and then approaches to a horizontal line for $N > 2.0 \times 10^6$. The σ_p -value at a constant N reduces surely with a rise in T. At temperatures near 400 °C, however,the observation of slip-bands was rather difficult because the foil was discolored to dark blue by oxidation. Furthermore the σ_p -value was not affected by the test-interruption accompanied with a temperature change, provided that a specified test tempersature below 350 °C was unchanged throughout the total number of stress cycles. Hence the nickel foil gage can be used satisfactorily for the tests in the range from room temperature to 350°C.

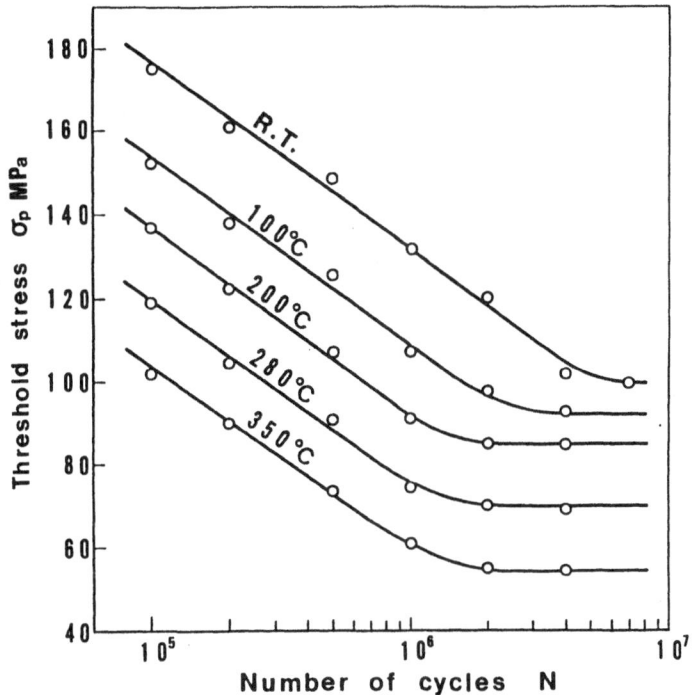

Figure 3. Calibration curves at definite temperatures

Two kinds of techniques based on the calibration curve were applied to measure the peak stresses in the grooved specimens subjected to rotating-bending.

The method based on a specified repetition-number: Several specimens with grooves of the same shape were used, and they were tested under a specified number of stress cycles N,where the threshold stress σ_p had been known by the calibration curve. The specimens were subjected to separate nominal stresses, σ_d; as a result, the standard stress σ_o corresponding to the appearance of 'thin' slip-bands at the root of the groove could be found. Hence the stress-concentration factor k was obtained by the ratio of σ_p/σ_o.

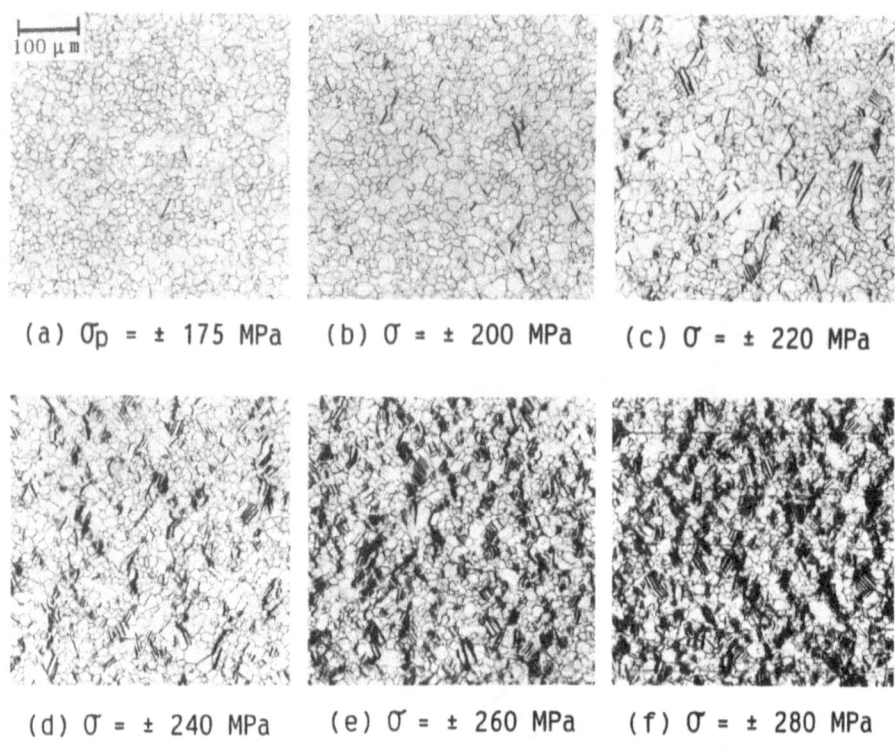

(a) $\sigma_p = \pm 175$ MPa   (b) $\sigma = \pm 200$ MPa   (c) $\sigma = \pm 220$ MPa

(d) $\sigma = \pm 240$ MPa   (e) $\sigma = \pm 260$ MPa   (f) $\sigma = \pm 280$ MPa

Figure 4. Standard photographs (T = R.T., $N = 10^5$)

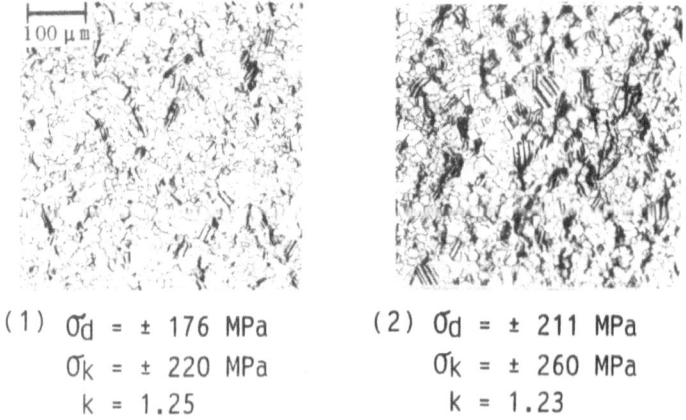

(1) $\sigma_d = \pm 176$ MPa
$\sigma_k = \pm 220$ MPa
$k = 1.25$

(2) $\sigma_d = \pm 211$ MPa
$\sigma_k = \pm 260$ MPa
$k = 1.23$

Figure 5. Appearance of slip-bands at groove-root of specimen (b)
(T = R.T., $N = 10^5$)

The method based on a specified nominal stress: One grooved specimen was used,and it was tested under a specified nominal stress σ_d. The test was interrupted at short intervals to examine the appearance of slip-bands at the root of the groove; as a result, the critical number of stress cycles N_c corresponding to the appearance of 'thin' slip-bands could be found. Hence both the threshold stress σ_p and the standard stress σ_0 at the number of cycles N_c were determined, and the stress-concentration factor k was obtained. In the present experiment, however, this method could be applied in the range of $N < 2\times10^6$, where the calibration curves fell obliquely.

Measurement of peak stresses using standard photographs

Figure 4 shows the standard photographs at T = R.T. and $N = 10^5$. The density of slip-bands is thicker gradually with an increase in stress amplitude. The slip-bands in Fig. 4(a) correspond to those at the threshold stress σ_p.

Figure 5 shows the appearance of slip-bands at the root of the groove of the specimen (b) subjected to rotating-bending under T = R.T. and $N = 10^5$. In the figure, σ_d is the nominal stress and σ_k is the stress amplitude attached to the standard photograph with a similar density of slip-bands. Hence the stress-concentration factor k is estimated by the ratio of σ_k/σ_d.

The results for three kinds of grooved specimens are shown in Table 2. For comparison, the measured values of k by the nickel-electroplating method at 180 °C [4], those by electric-resistance strain gages at room temperature [5], the calculated values of k by Neuber's formula [6], and those by the finite-element method [7] are also shown in the table. The present results are in good agreement with the previous ones.

TABLE 2

Measured values of stress-concentration factor
(T = R.T., $N = 10^5$)

Specimen	Nominal stress σ_d (MPa)	Standard photograph σ_k (MPa)	Stress-concentration factor $k = \sigma_k/\sigma_d$	Previous results
(a)	148	220	1.49	1.55[4] 1.50[5]
	174	260	1.49	1.47[6] 1.53[7]
(b)	176	220	1.25	1.22[4] 1.28[5]
	211	260	1.23	1.25[6]
(c)	190	220	1.16	1.10[4] 1.14[5]
	229	260	1.14	1.14[6]

Measurement of peak stresses using calibration curves

Table 3 shows the values of k measured by the method based on a specified repetition-number under T = 200 °C and $N = 5\times10^5$, where the threshold stress σ_p is equal to 106 MPa as shown in Fig. 3. Table 4 shows the values of k measured by the same method under T = 280 °C and $N = 4\times10^6$,where σ_p = 70 MPa. Table 5 shows the values of k measured by the method based on a specified nominal stress at T = 100 °C.

In Table 3~ Table 5, the values of k obtained by the previous studies are also shown. It is found that the present results are in good agreement with those obtained previously.

TABLE 3

Measured values of stress–concentration factor
$(T = 200°C, N = 5 \times 10^5, \sigma_p = 106$ MPa$)$

Specimen	Nominal stress σ_d (MPa)	Appearance of slip-bands	Standard stress σ_0 (MPa)	Stress–concentration factor $k = \sigma_p/\sigma_0$	Previous results
(a)	76 70 65	Dense Thin None	70	1.51	1.55 [4] 1.50 [5] 1.47 [6] 1.53 [7]
(b)	91 85 82	Dense Thin None	85	1.25	1.22 [4] 1.28 [5] 1.25 [6]
(c)	96 92 88	Dense Thin None	92	1.15	1.10 [4] 1.14 [5] 1.14 [6]

TABLE 4

Measured values of stress–concentration factor
$(T = 280°C, N = 4 \times 10^6, \sigma_p = 70$ MPa$)$

Specimen	Nominal stress σ_d (MPa)	Appearance of slip-bands	Standard stress σ_0 (MPa)	Stress–concentration factor $k = \sigma_p/\sigma_0$	Previous results
(a)	50 47 45	Dense Thin None	47	1.49	1.55 [4] 1.50 [5] 1.47 [6] 1.53 [7]
(b)	61 55 53	Dense Thin None	55	1.27	1.22 [4] 1.28 [5] 1.25 [6]
(c)	67 64 58	Dense Thin None	64	1.09	1.10 [4] 1.14 [5] 1.14 [6]

TABLE 5

Measured values of stress-concentration factor (T = 100°C)

Specimen	Nominal stress σ_d (MPa) = σ_0	Critical number of cycles $N_c \times 10^6$	Threshold stress σ_p (MPa)	Stress-concentration factor $k = \sigma_p/\sigma_0$	Previous results
(a)	74	0.9	111	1.50	1.55[4] 1.50[5]
	73	0.8	113	1.55	1.47[6] 1.53[7]
(b)	78	1.7	99	1.27	1.22[4] 1.28[5]
	80	1.4	102	1.28	1.25[6]
(c)	95	1.2	105	1.11	1.10[4] 1.14[5]
	93	1.2	105	1.13	1.14[6]

CONCLUSIONS

Strain gages made of nickel foil have been studied to measure the elastic stress on the surface of a specimen subjected to repeated loads at high temperatures. The following conclusions are drawn from the results obtained.

(1) Nickel foil is made by annealing the nickel-plating foil at 950°C for 60 seconds in a vacuum. It is about 10 μm thick and composed of crystal grains of 30 μm in mean diameter.

(2) When the threshold stress σ_p is plotted as the ordinate and the number of stress cycles N as the abscissa in a logarithmic scale, the σ_p - log N curve, i. e., the calibration curve of the nickel foil gage, at a constant test temperature, descends obliquely with an increase in N and then approaches to a horizontal line for $N > 2\times10^6$.

(3) The applicable test temperatures of the nickel foil gage are in the range from room temperature to 350°C.

(4) The σ_p - value is not affected by the test-interruption accompanied with a temperature change, provided that the specified test temperature is unchanged throughout the total number of stress cycles.

(5) The nickel foil gage has sufficient accuracy as demonstrated by measuring the peak stresses in grooved shafts under bending.

REFERENCES

1. Seika, M. and Hosono, K., A basic study on a strain gauge made of nickel foil.(Calibration tests and stress measurement by standard photographs). Trans. Japan Soc. Mech. Engrs.,(A),(in Japanese), 1989, **55**, 917-22.

2. Seika, M. and Hosono, K., A basic study of a strain gauge made of nickel foil. (Application of the methods using the proper stress and the proper number of cycles to stress measurement). Trans. Japan Soc. Mech. Engrs., (A), (in Japanese), 1989, **55**, 2335-41.

3. Seika, M. and Hosono, K., A basic study of a strain gage made of nickel foil. EXPERIMENTAL MECHANICS, 1989, **29**, 388-91.

4. Seika, M., Hosono, K., and Ōta, H., Measurement of cyclic stresses in high temperature range using nickel electroplating. 3rd Report: Application of a sulfamate bath. Bul. Japan Soc. Mech. Engrs., 1979, **22**, 793-800.

5. Kikukawa, M. and Sato, Y., Stress concentration in notched bars under tension or bending. 2nd Report: U-grooved shafts. Trans. Japan Soc. Mech Engrs., (in Japanese), 1972, **38**, 1673-80.

6. Neuber, H., Kerbspannungslehre, 2nd Ed., Springer-Verlag, Berlin, 1958, p. 12.

7. Sato, Y., Kikukawa, M., and Matsui, T., Stress concentration in notched bars under tension or bending. 4th Report: Finite element analysis of semicircular grooved shafts. Trans. Japan Soc. Mech. Engrs.(in Japanese), 1976, **42**, 3701-9.

APPLICATION OF CAUSTIC METHOD TO STRESS ANALYSIS IN PLATES WITH DISCONTINUOUSLY DISTRIBUTING LOAD

KIYOSHI ISOGIMI
Department of Mechanical Engng.,Faculty of Engng.,
Mie University
Kamihamacho 1515, Tsu, 514, JAPAN

ABSTRACT

In this paper, the properties of caustics necessary for apply-ing the caustic exprimental method to stress analysis are clar-ified. Under the simplified two-dimensional condition, the ob-jects of study are a half plane and a wedge-shaped plane with a discontinuously distributing oblique load divided into two or three parts over a loading region. Specimens are made of an a-crylate resin plate having a thickness of about 3 mm. The caus-tics formed by the reflected light rays at both surfaces of the plates are observed and those characteristics with loading con-ditions are investigated in detail. On the other hand, the com-putational caustic curves based on the optical theory are plot-ted. These two caustics coincide accurately enough. With the results obtained, convenient simple formulas for evaluating one end value of the external load are proposed.

INTRODUCTION

The caustic experimental method has proved to be very useful in analysis of contact problems. Author has been investigating about the characteristics of caustics for a plane subjected to various kinds of external loads in de-tail[1]. The final purpose of these research is to analyze the stresses caused by cutting operations and clarify the relationships between the properties of caustics and tool wear propagation and tool life.

The first object is a half plane with discontinuously distributing load over the definite length along the boundary. The second is a wedge-shaped plane with the same load along one of the boundaries. That is, un-der independent oblique loads on every region divied into two or three parts over the loading width, the caustic formed by the reflected light at the front and rear surfaces of plate are compared with the computational caustic curves. Moreover, very convenient simple procedures for evaluating one end value of the external load by measuring the dimensions of the caus-tics obtained are proposed.

THEORETICAL CAUSTIC EQUATIONS

Equation for Half Plane

Notations are shown in Fig.1(a). Consider an elastic, homogeneous and iso-tropic half plane subjected by an arbitrarily discontinuously distributing load inclined with angle ϕ to the normal of the boundary over loading width 2a. $P_1, P_2, \ldots, P_n$ are the load intensity coefficients to be multiplied with standard external force σ. The loading function can be represented without depending on the divided number of odd or even.

$$F(t) = \frac{\sigma}{2}\left(\cos\phi - i\sin\phi\right)\sum_{j=1}^{n} P_j\left[1 - \frac{\left|t - \left(\frac{n-2j+2}{n}\right)a\right|\left|t - \left(\frac{n-2j}{n}\right)a\right|}{\left\{t - \left(\frac{n-2j+2}{n}\right)a\right\}\left\{t - \left(\frac{n-2j}{n}\right)a\right\}}\right] \tag{1}$$

The stress condition at any point Z in a plate are given by using following complex stress function[2],

$$\Phi(z) = \frac{1}{2\pi i}\int_{-a}^{a}\frac{F(t)}{t-z}\,dt \tag{2}$$

$$\Psi(z) = \frac{1}{2\pi i}\int_{-a}^{a}\frac{\overline{F(t)}}{t-z}\,dt - \frac{1}{2\pi i}\int_{-a}^{a}\frac{F(t)}{(t-z)^2}\,t\,dt \tag{3}$$

Caustic curves can be calculated by the formulas based on the optical theory[3],

$$|4C\Phi''(z)/\lambda| = 1 \tag{4}$$

$$W = \lambda z + 4C\overline{\Phi'(z)} \tag{5}$$

where C is the optical constant of the employed material, and λ is dimensional ratio of the optical layout.

Substitutuing Eq.(2) into Eqs.(4) and (5), the equations of the caus-tic curves are derived,

$$\left|Cm\sum_{j=1}^{n}\left\{P_{n-j+1}\left(z_0 + \frac{n-j+1}{n}\right)\Big/\left(z_0 + \frac{n-j+2}{n}\right)^2\left(z_0 + \frac{n-2j}{n}\right)^2\right\}\right| = \sum_{j=1}^{n}P_j \tag{6}$$

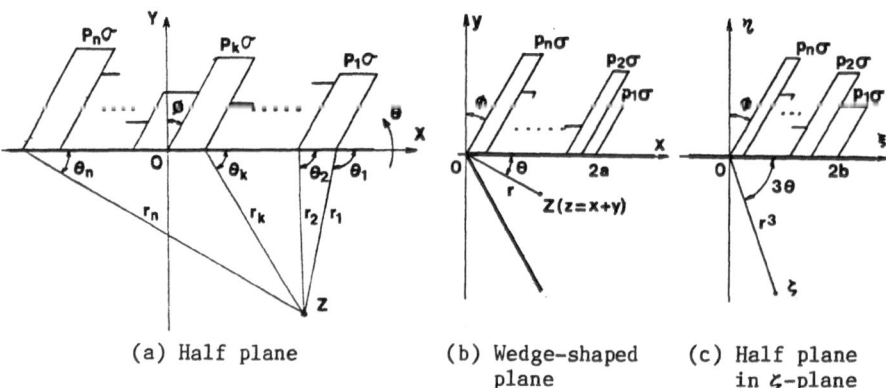

| (a) Half plane | (b) Wedge–shaped plane | (c) Half plane in ζ-plane |

Figure 1. Notations for each plane with general discontinuous load.

$$W = \lambda a \left[r_0 (\cos\phi - i \sin\phi) - \left(Cm/2\sum_{j=1}^{n} P_j\right)\left\{\sum_{j=1}^{n} \frac{P_j}{r_j r_{j+1}} \sin(\phi + \theta_j + \theta_{j+1})\right.\right.$$

$$\left.\left. - i \sum_{j=1}^{n} \frac{P_j}{r_j r_{j+1}} \cos(\phi + \theta_j + \theta_{j+1})\right\}\right] \tag{7}$$

where $Cm = (8C\sigma/\pi \lambda a^2 n) \sum_{j=1}^{n} P_j$.

Equation for Wedge-shaped Plane

Figure 1(b) shows a plane with a wedge angle of 60° subjected to a load on the divided n-parts. The loading function is given by replacing (n-2j+2) and (n-2j) in Eq.(1) with 2(n-j+1) and 2(n-2j).

A state at any point Z in the wedge-shaped plane is mapped on the lower half plane of plane $\zeta (= \xi + i \eta)$ with function $w(\zeta)$ as shown in Fig.1(c).

$$Z = w(\zeta) = \zeta^{1/3} \tag{8}$$

where $-\pi/3 \le a r g (z) \le 0$, $-\pi \le a r g (z) \le 0$.

As Eqs.(2) and (3) are formed in ζ-plane, considering the relations of Eq.(8), the equations of caustic curves are

$$\left|\frac{3a^5 \sigma rCm}{4Sm} \sum_{j=1}^{n} P_j \left(\frac{2a}{n}\right)^3 [(n-j+1)^3 - (n-j)^3]\right.$$

$$\left.\times \left[4r^6 e^{6i\theta} - \left(\frac{2a}{n}\right)^3 [(n-j+1)^3 - (n-j)^3] r^3 e^{3i\theta} - 2\left(\frac{2a}{n}\right)^6 (n-j+1)^3 (n-j)^3\right] \middle/ r^2 r'^2 \right| = 1 \tag{9}$$

$$W = \lambda r \left[(\cos\theta + i \sin\theta) - \frac{6a^5 r \sigma |Cm|}{n^3 Sm} \sum_{j=1}^{n} P_j \frac{[(n-j+1)^3 - (n-j)^3]}{r r'}\right.$$

$$\left.\times [\sin(\phi - 2\theta + \Theta + \Theta') + i \cos(\phi - 2\theta + \Theta + \Theta')]\right] \tag{10}$$

where $Cm = 8CSm/\lambda \pi a^2$ \tag{11}

$$Sm = (\sigma/\pi)\sum_{j=1}^{n} P_j \tag{12}$$

$$r = r^3 - 2[2(n-j+1)ar/n]^3 \cos3\theta + [2(n-j+1)a/n]^6$$

$$r' = r^3 - 2[2(n-j)ar/n]^3 \cos3\theta + [2(n-j)a/n]^6 \tag{13}$$

$$\Theta = \tan^{-1}[r^3 \sin3\theta / [r^3 \cos3\theta - 2^3 (n-j+1)^3 a^3 /n^3]]$$

$$\Theta' = \tan^{-1}[r^3 \sin3\theta / [r^3 \cos3\theta - 2^3 (n-j)^3 a^3 /n^3]] \tag{14}$$

EXPERIMENTAL POROCEDURE

The equipment with a lever-type loading device for every region is used to subject the predecided loads at each region of the loading width divided into two or three parts. The direction of the external loads is always kept vertically downward. The dimension ratio of the optical layout is constant at $\lambda = 2.78$. Specimens are made of PMMA plates with a thickness of 3 mm. The optical constant of employed PMMA plates is $C \doteqdot 0.195$ mm^2/MPa.

Experiments are performed under the following loading conditions. In the case of loading divided into two parts, the load intensity ratios are 1:3, 1:2, 2:3, 1:1, 3:2, 2:1, 3:1 under a constant total load, and load inclined angles are $\phi = 0°$, 15° and 30°. On the other hand, in the case of loading divided into three parts, the load intensity coefficients are selected as $P_t = 0.33$, 0.5, 1, 2 and 3, and the load inclined angles are $\phi = 0°$, 15° and 30°.

RESULTS AND DISCUSSIONS

Properties of Caustics
As an example, considerations are performed in only one case of loading divided into three parts under Cm = 0.638 and 2a = 6.2 mm.
Figures 2(a)∼(c) and Figs.3(a)∼(c) show computational and practical caustics in half plane for load intensity coefficients at the central region P_2 = 0.3, 1 and 2 under ϕ =15° and P_1 and P_3 = 1 constant.

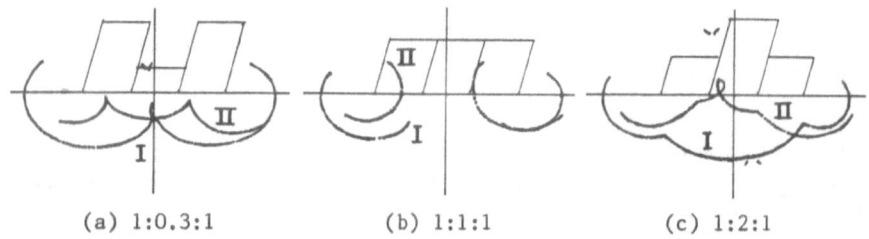

(a) 1:0.3:1   (b) 1:1:1   (c) 1:2:1

Figure 2. An example of computational caustic curves in a half plane.

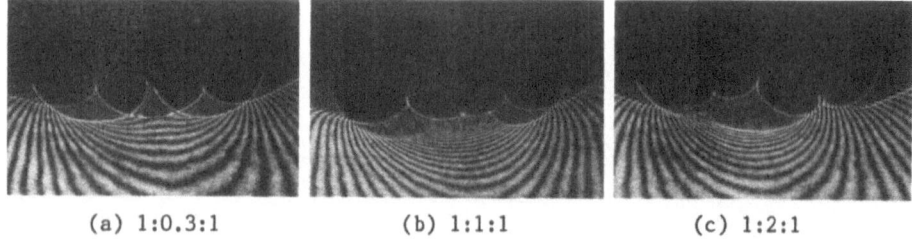

(a) 1:0.3:1   (b) 1:1:1   (c) 1:2:1

Figure 3. An example of practically obtained caustics in a half plane.

Figs.4 and 5 show the caustics in wedge–shaped plane under ϕ =0° and P_1 and P_2=1 constant. The characteristics of the computational curves is in good agreement with the practical caustic.
At first, the properties of the caustics formed at front surface (Curve I) are investigated. Though their whole widths are kept about constant for any value of P_2, the configurations show the following characteristic deformations. In case of the half plane, with an increase in the P_2 value, the pointed peak at the center becomes lower and lower and it separates into two branches. In case of the wedge–shaped plane, the value of P_2 coming near to P_1 and P_3, two branches of the caustic are more distant from each other. Furthermore, when P_2> 1 in both type planes, the third new branch of being downward convex generates and it grows gradually together with an increase in the P_2 value.
On the other hand, the rear caustic (Curve II) changes in both type planes as follows. For P_2<1, the caustics are constructed with two upward pointing continuous branches. And for P_2>1, the similar third branch generates between the two.
It appears that load inclined angle ϕ gives few effects. That is, the whole width of the caustics is almost completely constant without depending on angle ϕ.

611

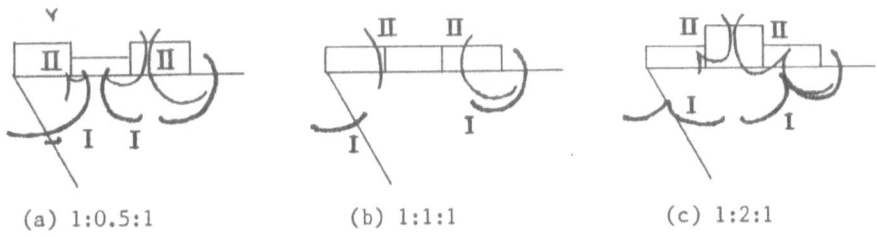

(a) 1:0.5:1   (b) 1:1:1   (c) 1:2:1

Figure 4. An example of computational curves in a wedge shaped plane.

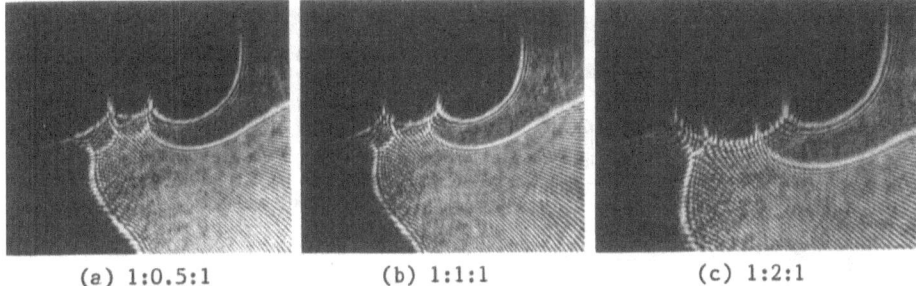

(a) 1:0.5:1   (b) 1:1:1   (c) 1:2:1

Figure 5. An example of practical caustics in a wedge shaped plane.

Figures 6(a),(b) show the caustic curves for change in load intensity coefficient P_1 under $\phi=0°$ only especially in case of the wedge-shaped plane. The very characteristic appearance of the front caustic curve can be found. If the load subjected on a wider region is stronger than that on a narrower one, the caustic consists of three branches, and if the load is weaker, the caustic consists of two branches.

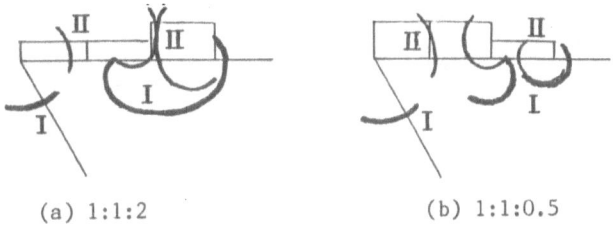

(a) 1:1:2   (b) 1:1:0.5

Figure 6. Relationship between front caustic and load intensity ratio.

In the caustic experimental method, it is necessary to clarify the states of the external loadings by using the caustics generated in practical stress fields. For this purpose, the relationships between loading conditions and dimensions of the front caustics are investigated. As measuring objects, whole maximum width L and depth D from the origin are chosen.

Figures 7(a),(b) show the relationships between P_2 and L, D.

The relations in these figures can be described as

$L \cong m_1 P_2 + m_2$

$$D \cong \sum_{i=1}^{5} n_i P_2{}^{5-i} \quad : \text{half plane} \tag{15}$$

$$D \cong n_1 P_2 + n_2 \quad : \text{wedge-shaped plane} \tag{16}$$

The values of parameters m_i, n_i are listed in Tables 1(a),(b).

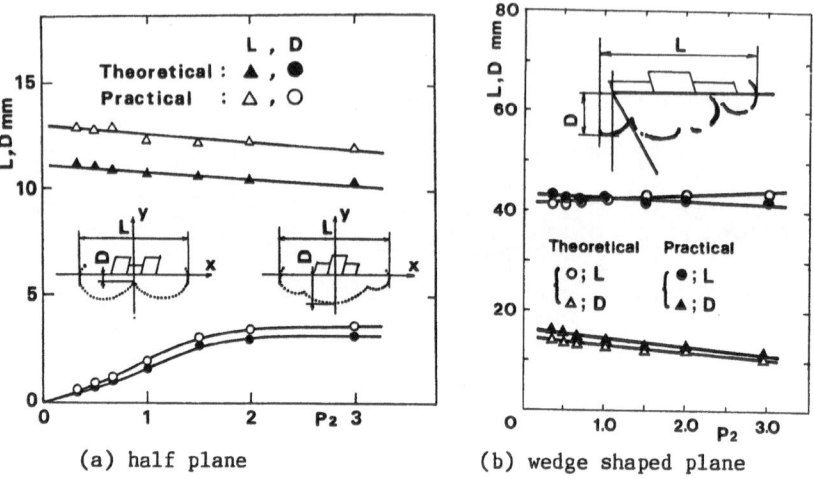

(a) half plane   (b) wedge shaped plane

Figure 7. Relationships between a coefficient P_2 and caustics' dimensions.

TABLE 1(a)
Values of parameters m_i and n_i for half plane.

Load inclined angle ϕ	L		D				
	m_1	m_2	n_1	n_2	n_3	n_4	n_5
0°	−0.28	10.58	0.15	−1.00	1.93	0.27	0.03
15°	−0.31	11.12	0.15	−1.05	1.84	1.04	0.03
30°	−0.34	12.22	0.20	−1.26	2.02	1.21	0.02

TABLE 1(b)
Values of parameters m_i and n_i for wedge shaped plane.

Dimensions	$C^*=0.3, \quad \phi=15°$		
mm	Parameters	i=1	i=2
L	m_i	0.957	41.313
D	n_i	−1.334	14.337

Evaluation of External Loads

Consider only one case in which the external load at the right side region changes. Figure 8 shows the relations between load intensity coefficient

P_1 and maximum width L (for the half plane :Fig.(a)) or distance Lx (for the wedge-shaped plane :Fig.(b)) concerned with every of inclined angle ϕ.

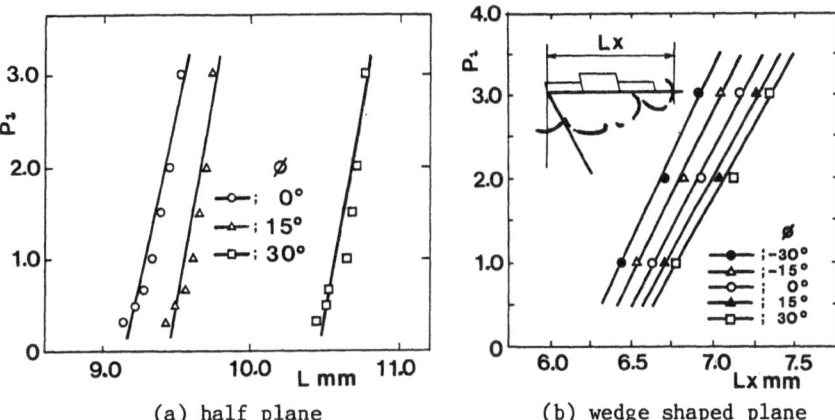

(a) half plane   (b) wedge shaped plane

Figure 8. Relationships between load intensity coefficient P_1 and dimensions L or Lx.

By measuring dimension L or Lx of practically generated caustics we can determine corresponding load intensity coefficient P_1 univocally. Then, there is a very simple expression between P_1 and L or Lx,

$$L, L_x = a_1 P_1 + a_0 \tag{17}$$

The values of parameters a_1 and a_0 are shown in Table 2.

TABLE 2
Values of parameters a_1 and a_0.

Load inclined angle $\phi °$	half plane		wedge shaped plane	
	a_1	a_0	a_0	a_1
30	10.44	0.122	−24.40	4.09
15	9.45	0.105	−23.25	3.85
0	9.15	0.133	−22.76	3.71
−15			−22.11	3.57
−30			−21.79	3.48

The load values obtained by Eq.(18) are compared with the practical load values on the right-side region under various kinds of loading conditions in Table 3. Both values coincide well within an error of 9 % for the half plane and 12 % for the wedge-shaped plane; therefore, it is thought that this experimental method is very useful.

STRESS DISTRIBUTIONS

The stress components raised in planes in cases of the loading conditions

TABLE 3
Examples of evaluation of external load values.

| | Dimensions L,Lx mm | Loading angles ϕ° | External load values | | |
			Calculated values MPa	Practical values MPa	Ratios
half plane	43.2 39.2 42.7	-15 0 15	15.58 7.04 17.58	15.4 7.2 16.4	1.01 0.98 1.07
wedge shaped plane	9.45 10.49	0 30	22.4 104.9	20.0 100.0	1.12 1.05

shown in Figs.1(a) and (b) can be derived by the following procedures.

Components σ_x, σ_y and τ_{xy} at any point z in a plane are generally related to one another,

$$\sigma_x + \sigma_y = 4 R e \, \Phi(z)$$

$$\sigma_y - \sigma_x + 2 i \tau_{xy} = 2[z\Phi'(z) + \Psi(z)] \tag{19}$$

Substituting Eqs.(2) and (3) into the above formulas, we can obtain that for the half plane,

$$\sigma_x = (\sigma / \pi) \sum_{j=1}^{n} P_j [(\Theta_j - \Theta_{j+1})\cos\phi - 2\sin\phi \, l \, n(r_j/r_{j+1})$$

$$+ [\sin(\phi + 2\theta_j) - \sin(\phi + 2\theta_{j+1})]/2] \tag{23}$$

$$\sigma_y = (\sigma / \pi) \sum_{j=1}^{n} P_j [(\Theta_j - \Theta_{j+1})\cos\phi - [\sin(\phi + 2\theta_j) - \sin(\phi + 2\theta_{j+1})]/2] \tag{24}$$

$$\tau_{xy} = (\sigma / \pi) \sum_{j=1}^{n} P_j [(\Theta_j - \Theta_{j+1})\sin\phi - [\cos(\phi + 2\theta_j) - \cos(\phi + 2\theta_{j+1})]/2] \tag{25}$$

that for the wedge-shaped plane,

$$\sigma_x = (\sigma / \pi) \sum_{j=1}^{n} P_j [(\Theta - \Theta')\cos\phi - 2\sin\phi \, l \, n(r/r')$$

$$+ \frac{24a^3 r^3 [(n-j+1)^3 - (n-j)^3]\sin\theta}{n^3 \, r \, r'} \cos(\phi - 2\theta + \Theta + \Theta')] \tag{20}$$

$$\sigma_y = (\sigma / \pi) \sum_{j=1}^{n} P_j [(\Theta - \Theta')\cos\phi$$

$$+ \frac{24a^3 r^3 [(n-j+1)^3 - (n-j)^3]\sin\theta}{n^3 \, r \, r'} \cos(\phi - 2\theta + \Theta + \Theta')] \tag{21}$$

$$\tau_{xy} = (\sigma / \pi) \sum_{j=1}^{n} P_j [(\Theta - \Theta')\sin\phi$$

$$+ \frac{24a^3 r^3 [(n-j+1)^3 - (n-j)^3]\sin\theta}{n^3 \, r \, r'} \sin(\phi - 2\theta + \Theta + \Theta')] \tag{22}$$

Figures 9(a)~(c) and 10(a)~(c) show each example of the stress distributions for both planes.

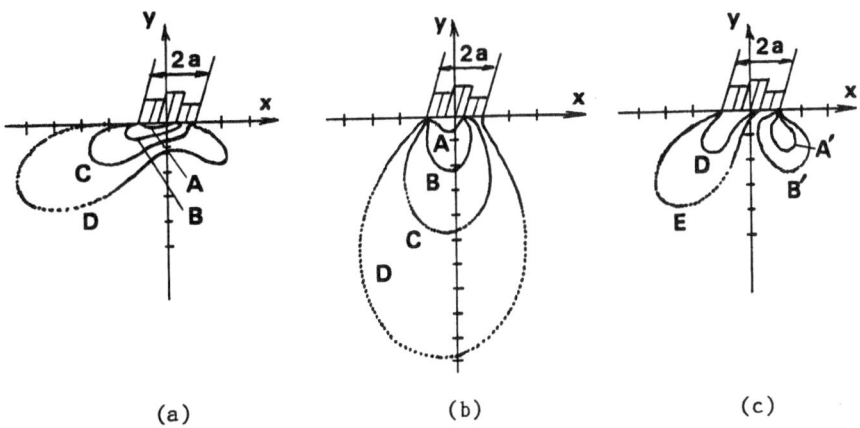

A:−27.0, B:−13.5, C:−6.75, D:−3.38, E:−1.69, A':3.38, B':1.69 MPa
2a=6 mm, Total load=180 N, Load intensity ratio=4:5:3, ϕ =15°.
Figure 9. An example of stress distributions in a half plane.

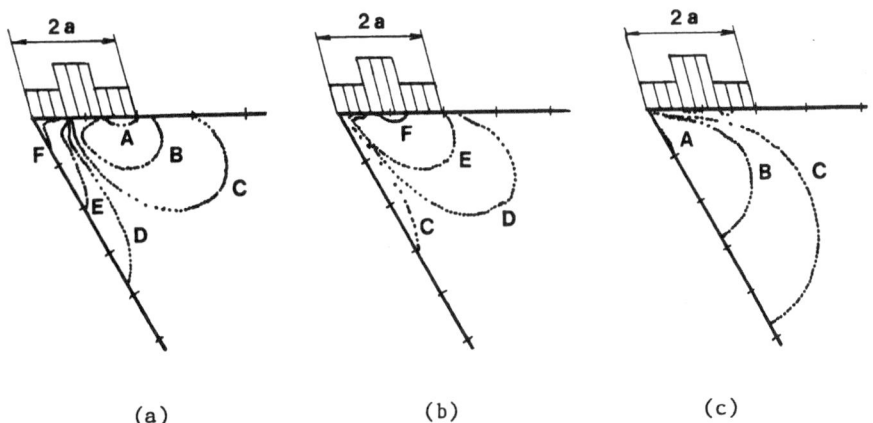

A:−27.0, B:−6.75, C:−1.69, D:1.69, E:6.75, F:27.0 MPa
Total load=180 N, Load intensity ratio=1:2:1, ϕ=15°, 2a=6.2 mm
Figure 10. An example of stress distributions in wedge shaped plane.

CONCLUSIONS

The conclusions obtained in this research are as follows.
(1) The characteristics of computational curves based on the optical theory
are in good agreement with the corresponding practical caustics.
(2) According to the front caustic for the loading divided into three
parts, when the load value at one of both side regions is smaller than that
at the central region, it consists of three branches, and when the load
value is bigger than that at the central region vice versa, it consists of
two branches.

(3) The relations between the load intensity coefficients and the dimensions of caustics are expressed by very simple approximate formulas. As a whole, the width of a caustic is decided only by the load value subjected at the end region regardless of the divided number, we can evaluate the load value at the corresponding region by measuring the dimensions of the practically generated caustics.

REFERENCES

1. For an example, Isogimi, K. and Kitagawa T., Applications of caustic experimental method to stress analysis in cutting tools. Journal of the Japan Soc. of Precision Engng., 1988, 54, 390–395.

2. Muskhelishvili, N.I., Some basic problems of the mathematical theory of elasticity, Nordhoff Int. Pub., Netherland,1977.

3. For an example, Theocaris, P.S., Stress singularities at concentrated load, Exp. Mech., 1973, 12, 511.

OPTICAL AND MECHANICAL INTERACTION OF STRUCTURALLY INTEGRATED OPTICAL FIBER SENSORS

J. S. Sirkis, C. Mathews, and H. Singh
Department of Mechanical Engineering
University of Maryland
College Park, Maryland 20742   USA
301-454-8859

ABSTRACT

The mechanical and optical interaction of structurally integrated optical fiber strain sensors are explored with fiber and moire' interferometry. A single mode optical fiber is embedded in a four point bend specimen to investigate both the phase shift induced in the optical fiber by the specimen strain field, and the stress concentrations induced in the specimen by the fiber acting as a stiff inclusion. The phase shift measured with a Mach-Zehnder fiber interferometer is under predicted by the combination of simple beam theory and the three-dimension optical phase-strain relation. This result is expected since, unlike surface mounted optical fiber sensors, embedded sensors significantly perturb the strain state in the immediate vicinity of the fiber. A hybrid analysis using moire' interferometry and finite element methods is used to better estimate the true average strain state in the optical fiber, which is in turn used to calculate the optical phase shift. However, the phase shifts calculated by this method over predicts the measured phase shifts. Possible reasons for the over prediction include the material property choice for the optical fibers, and the use of a plane stress finite element analysis. Both problems are discussed below. Finally, the hybrid stress analysis technique is used to investigate the micromechanical stress state in a 500μm square region local to the embedded optical fiber. The most significant stress concentration attains a value of 1.72, and the fiber induced perturbation in the stress state is restricted to within three fiber diameters from the fiber center.

INTRODUCTION

The recent emphasis in "intelligent structures" has been to identify and evaluate structurally integrable sensors and actuators. The emphasis has been on embedding these devices without regard to the resulting degradation in the sensor/actuator performance or the structural integrity. Optical fiber sensors have been identified as prime candidates for use with intelligent structures. This is mainly due to fiber's filamentary geometry, light weight, chemical inertness, and high bandwidth. This paper addresses both the performance of an embedded optical fiber sensor and the stresses induced in the structure by the presence of the fiber sensor. The performance issues mainly center on the theoretical basis for inferring the strain induced optical phase shift in an embedded fiber. The phase shift in an embedded fiber is a function of the pointwise state of strain along the fiber path; therefore, a one-to-one map between a single strain component and the optical phase shift does not exist as it does with surface mounted optical fiber sensors [1].

Czarnek et. al. [2] first established that embedded optical fibers can cause significant strain concentrations in laminated composite host structures. Salehi et. al. [3] reproduced Czarnek et. al.'s experiments which they then augmented with a finite element analysis. It is by now well known that the observed strain concentrations are due to the formation of a lenticular resin rich zone between the adjacent plies which surround

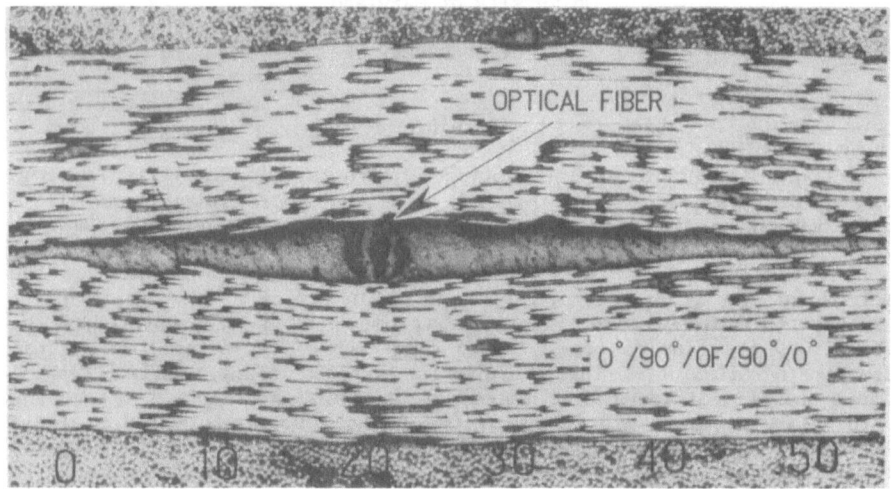

Figure 1. Single Mode Optical Fiber Embedded in a
Graphite-Epoxy Laminated Composite.

the embedded fiber (Figure 1). Salehi et. al. [3] have pointed out that even moire' interferometry lacks the spatial resolution to experimentally record the details of mechanical interaction in area shown in Figure 1. One way to overcome this resolution limitation is to combine moire' interferometry and finite element methods to form a hybrid technique, provided the host material is isotropic on the length scales of interest. The region encompassing an optical fiber embedded in an isotropic host material is investigated with just such a method. The interaction between an isotropic host and embedded fiber sensors is important in its own right since it is projected that host materials for intelligent structures will be both monolithic and heterogeneous. This investigation also lends insight into the performance issues related to optical fiber sensors embedded in heterogeneous hosts. Where appropriate, comparisons with existing information about the interaction mechanics of optical fibers embedded laminated composite hosts are made.

OPTICAL FIBER PHASE SHIFT

Interferometric optical fiber strain sensors use strain induced phase shifts as a means of monitoring strain [1]. The phase shift results from a change in length, birefringence, and higher order dependencies on dispersion and non-linear material and optical response [4]. The dominant properties, however, are the change in length and linear birefringence. The surface mounted sensor theory is significantly simplified by the experimentally verified assumption that only the strain component tangent to the fiber path is transmitted from the structure to the fiber. The fiber therefore is always in a spatially varying uniaxial state of strain. No such assumption has been justified for embedded sensors. The relation between the retardation in an embedded optical fiber and the strain state along the fiber is given in Reference [4] as

$$\Delta\phi = \frac{2\pi n_o}{\lambda}\int_{\Gamma}\left(\varepsilon_1 - \frac{1}{2}n_o^2\left[P_{12}(\varepsilon_1 + \varepsilon_2 + \varepsilon_3) + \frac{1}{2}(P_{11} - P_{12})(\varepsilon_2 + \varepsilon_3)\right]\right)d\Gamma. \tag{1}$$

In this equation, ε_1 is the strain tangent to the fiber path (Γ), ε_2 and ε_3 are the secondary principal strains as defined by Theocaris and Gdoutos [5], n_o is the refractive index in the unstrained state, λ is the wavelength of the propagating light, and P_{11} and P_{12} are Pockels' strain optic coefficients. The first term in the integral represents the fiber length; whereas the second term represents the average refractive index [1] and directly follows from three-dimensional photoelasticity [5]. The strain state is a continuous function of the position along the fiber path which implies that the fiber remains unbroken.

Equation (1) is a function of the average strain state in the fiber, but the strain state in the fiber is most likely not the strain state that would be present in the host material if the fiber were removed. A four point bend beam is used to both verify the phase-strain model embodied in Equation (1) and to access the perturbation in the strain field induce by the presence of the optical fiber. The procedure used here is to apply known load history to beam and simultaneously measure the resulting phase history in fiber interferometer. The load history can then be used with simple beam theory and Equation (1) to calculate the phase history assuming that the fiber does not perturb the local strain state. Finally moire' interferometry will be combined with finite element methods and Equation (1) to calculate the strain induced phase shifts assuming the fiber does perturb the local strain field. The state of stress averaged through the thickness of the beam is assumed to act equivalently to plane stress in order to facilitate the use of simple beam theory and finite element methods. The implications of this assumption are discussed in the concluding section.

Figures 2a and 2b depict the experimental arrangement. The beam (30.5 cm. long and 2.54 cm. square) is cast from a polyester resin (Adhesive Products Corp.'s ALPLEX) with a stripped single mode fiber (EOTec's FS-SC-3611) embedded at the beam's midpoint and 6.88 mm above the neutral axis. ALPLEX is transparent after curing so visual inspection of the embedded fiber is possible. A very soft ultraviolet setting adhesive (Desolite 950X132) is used to protect the embedded fiber at the entrance and egress points. The load cell and phase meter signals are recorded with a personal computer equipped with a 12 bit analog-to-digital converter. After the load and phase history are recorded, the load is returned to zero. The U and V moire' displacement fields are subsequently recorded for an applied load of 890.0 Newtons.

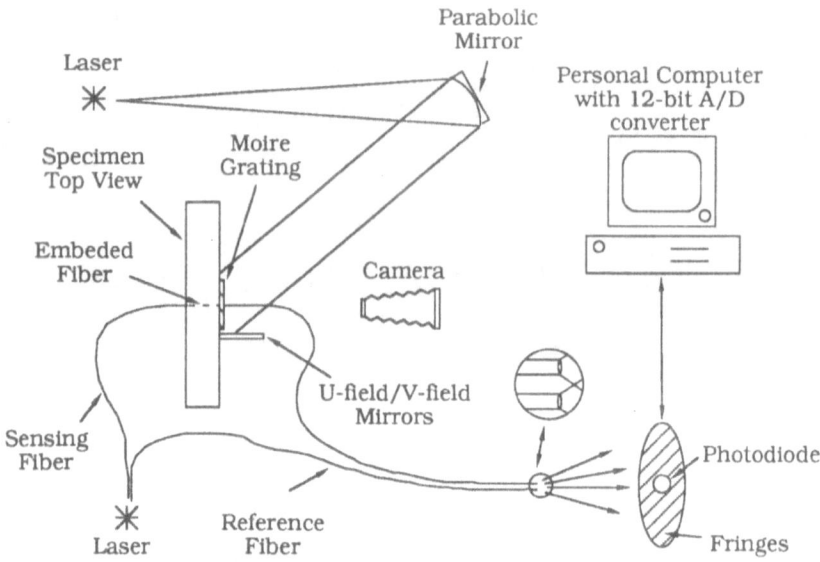

Figure 2a.   Experimental Arrangement Showing Fiber and Moire' Interferometry.

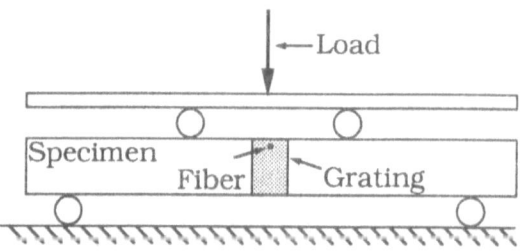

Figure 2b.   Four Point Beam in Pure Bending.

The optical fiber is embedded transverse to the beam length so that Euler beam theory provides the unperturbed strain along the fiber path as

$$\varepsilon_2 = \frac{Pay}{2EI} \quad \text{and} \quad \varepsilon_1 = \varepsilon_3 = -\nu\frac{Pay}{2EI}; \tag{2}$$

where y is the distance of the fiber from the neutral axis, E and ν are respectively the Young's modulus and Poisson's ratio of the beam, a is the separation of the load bearing rollers (Figure 2b) and P is the applied load. In this case, the strain state is constant along the fiber path so that Equation (1) becomes

$$\Delta\phi = - \frac{\pi n_o}{\lambda} \frac{ay}{EI}\left(\nu + \frac{1}{2}n_o^2\left[P_{12}(1 - 2\nu) + \frac{1}{2}(P_{11} - P_{12})(1-\nu)\right]\right)wP. \tag{3}$$

The material properties of the ALPLEX beam (E and ν) are deduced from the 890.0 N moire displacement fields shown in Figures 3a and 3b. The ratio of the two normal strains is calculated at five locations and then averaged to find Poisson's ratio. The x-strain component is compared to ε_2 in Equation (2) at each of the five points to calculate the average Young's modulus. The Young's modulus and Poisson's ratio found with this method are 6.17 GPa and .417, receptively. Figure 4 shows the recorded load-phase response and the load phase response found by using Equation (3). Both curves are linear, but the beam theory under predicts the true load-phase response by 21%. Pure fused silica Pockels' constant were used in these calculations (P_{11} = 0.121 and P_{12} = 0.27). The lack of correlation between the two curves suggests that either Equation (3) is an inadequate model of the strain induced optical retardation, or the strain state in the fiber is significantly different from the strain state which would be present if the fiber were not.

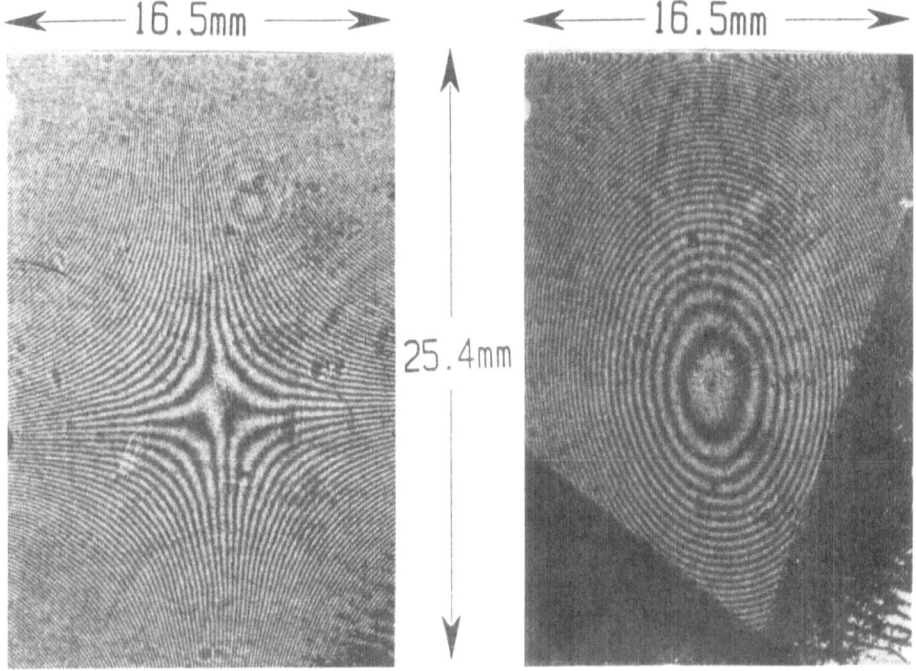

Figure 3a. U-field (890.0 N).   Figure 3b. V-field (890.0 N).

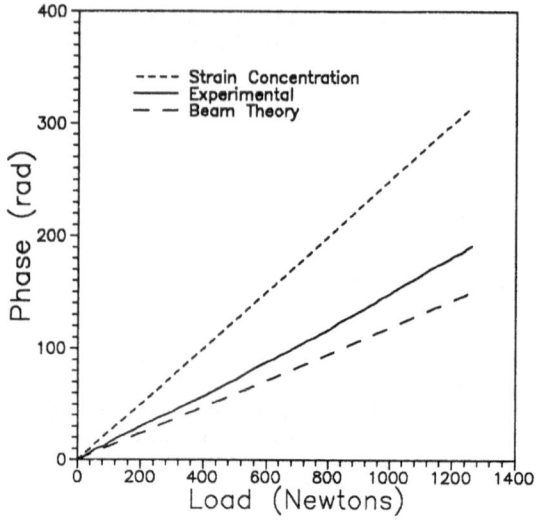

Figure 4.   Phase–Strain History.

FIBER INDUCED STRESS STATE

The hybrid technique used to investigate the mechanical interaction between the embedded optical fiber and the ALPLEX host consists of using the U and V fields from 2400 lpmm moire' interferometry to determine displacement boundary conditions for a finite element stress analysis code.   A plane stress finite element model of a 1600μm by 1600μm square region surrounding the fiber is developed and shown in Figure 5.   The fiber diameter is 80μm (d in Figure 5) and the fiber material properties are taken as those of fused silica (E = 70 GPa and ν = .17).   The beam material properties which were calculated in the previous section are used in the finite element model.

Enlargements of the U and V fields for the 890 N load case are shown in Figures 6a and 6b.   The fiber is darkened in these figures for better visualization.   It is apparent from Figures 6a and 6b that the optical fiber induced perturbations in the displacement fields are not detected by the moire' interferometry.   This result is in contrast to the noticeable perturbations in the U and V fields shown for the laminated composite hosts studied by both Czarnek et. al. [2] and Salehi et. al. [3].   All three sets of U and V fields (Czarnek et. al., Salehi et. al., and in this paper) are obtained for roughly the same far field nominal strains (about 3000μ). The difference in displacement perturbation between monolithic and laminated composite hosts is due to the difference in inclusion characterization.   The optical fiber embedded in a monolithic material represents a round and stiff inclusion.   Where, as seen from Figure 1, the inclusion in a laminated composite is soft (resin filled) and resembles a crack.   The aspect ratio of the inclusion shown in Figure 1 is approximately 13.

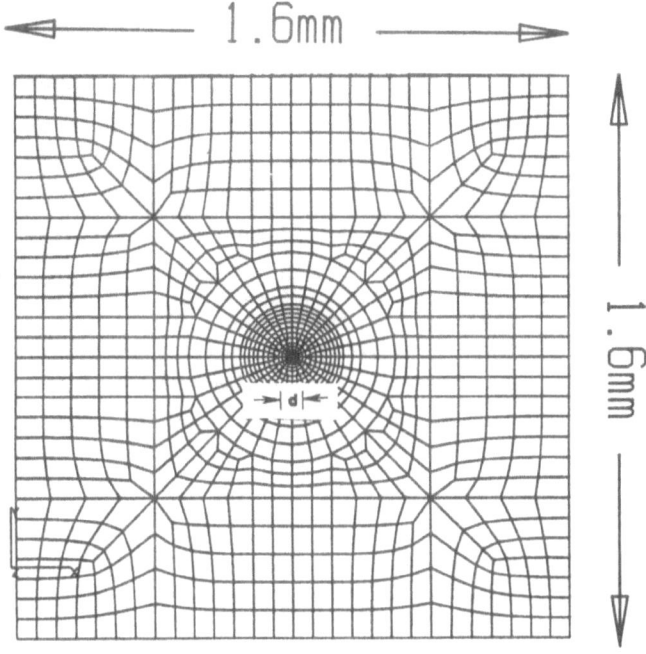

Figure 5. Finite Element Mesh Used in the Hybrid Analysis.

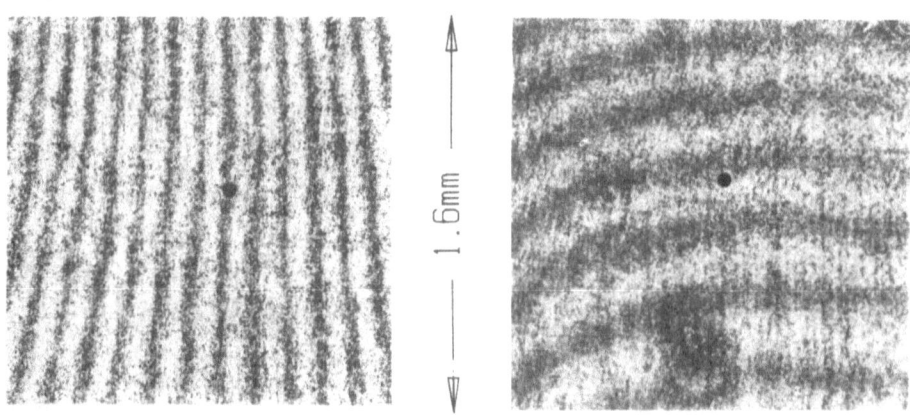

Figure 6a. Enlargement of the
890.0 N U-field.

Figure 6b. Enlargement of the
890.0 N V-field.

The hybrid finite element-moire interferometric method is used to perform a detailed stress analysis of the 500μm by 500μm square region centered on the embedded optical fiber. Figure 7a and 7b show the σ_{xx} and σ_{yy} stress distributions (the optical fiber is darkened). In these figures, the σ_{xx} and σ_{yy} contour increments are 1.17 MPa and .81 MPa, respectively. Figures 7a and 7b show that the optical fiber induced stress fields are concentrated within three fiber diameters of the inclusion; and in this case, the x-stress components are an order of magnitude larger than the y-stress components. The σ_{xx} stress concentration factor is 1.72 and is located on the left and right boundaries of the optical fiber. The stress concentration factor is the maximum stress divided by the stress component ten fiber diameters directly to the right of the optical fiber. The τ_{xy} and σ_{yy} stress concentration factors are 1. and 11.28, respectively. The σ_{yy} stress concentration is deceivingly high because the far field σ_{yy} are not significant. The ratio of the maximum σ_{yy} to the maximum σ_{xx} is .22.

⟵————— .5mm —————⟶   ⟵————— .5mm —————⟶

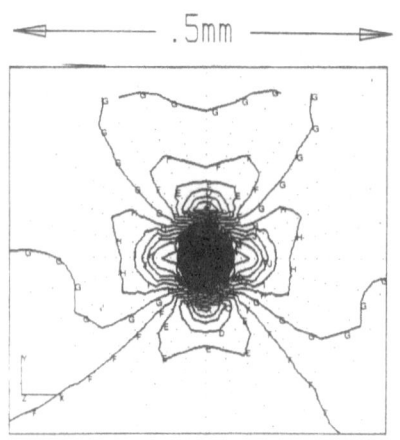   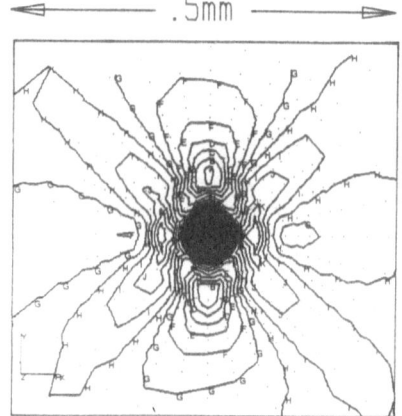

Figure 7a. σ_{xx} Stress Field.   Figure 7b. σ_{yy} Stress Field.

The hybrid stress analysis technique is used to calculate the strain concentration factors so that the true strain induced phase shift can be calculated. The strain concentration factors are defined as the average strain in the fiber (averaged over all the fiber finite elements) divided by the strain 10 fiber diameters directly to the right of the fiber, and are 1.92, 1.55, and 2.07 for ε_{xx}, ε_{yy}, and ε_{zz}, respectively. These strain concentrations are assumed to be correct for all elastic loads. The strain induced phase shift as a function of the applied load can then be calculated by multiplying the strain components in Equation (2) by the corresponding strain concentration factors and then using Equation (1). The phase shift found by this method is shown in Figure 4 along with the simple beam theory and the experimental phase response. As can be seen in this figure, the hybrid stress analysis results significantly over predict the measured phase history.

CONCLUSIONS

Simple beam theory and the hybrid stress analysis results combined with the retardation integral (Equation (1)) for embedded optical fiber sensors respectively under and over predicted the measured load-phase response by more than twenty percent. The simple beam theory is not expected to agree well with the measured phase response since this theory does not account for the inevitable perturbation induced by the fiber's presence. The poor phase response predictions provided by the hybrid stress analysis is somewhat surprising. The assumption that each plane normal to the fiber responds identically to the surface plane where the moire' measurements are made may be inappropriate. A state of plane stress certainly does not exist near the fiber on interior planes. Further the fiber material properties (both mechanical and optical) used in this analysis were those of fused silica. This common practice has increasingly drawn criticism. Efforts are underway to address these concerns.

The hybrid finite element-moire' interferometry analysis provides an experimental interrogation of the interaction micromechanics between the fiber and its host material. The local stress anomalies in this monolithic host material are shown to be confined to a 240μm radius (three fiber diameters) region centered around the embedded optical fiber. Further, the micromechanical analysis shows that stress concentrations are developed as a result of the fiber's presence and that for this loading, the most significant stress concentration factor (1.72) is associated with σ_{xx}. Czarnek et. al. [2] showed that strain concentrations as high as 16 can be produced as a result of an optical fiber being embedded in a laminated composite. As shown above, the strain concentration developed by embedding an optical fiber within a monolithic host is about ten times smaller. This leads to the conclusion that monolithic host materials may represent a safer host for embedded optical fiber sensors

ACKNOWLEDGEMENTS

This work was partially supported by the National Science Foundation under Grant No. ECS-8914865. The U.S. Government has certain rights to this material.

REFERENCES

1.   Sirkis, J. S., and Haslach, H. W., Interferometric Strain Measurement by Arbitrarily Configured, Surface Mounted, Optical Fibers," to appear in Jou. Lightwave Technology.

2.   Czarnek, R., Guo, Y.F., Bennett, K. D., and Claus, R. O., "Interferometric Measurements of Strain Concentrations Induced by an Optical Fiber Embedded in a Fiber Reinforced Composite," Proc. SPIE Vol. 1170, Boston, MA., pp. 43-54, 1988.

3.   Salehi, A., Tay, A., Wilson, D., and Smith, D., "Strain Concentration Factors Around Optical Fibers by FEM and Moire' Interferometry," Proc. 5th ASM/ESD Adv. Comp. Conf., pp. 11-19, Dearborn Mich., Sept., 1989.

4.   Sirkis, J. S., and Haslach, H. W., "Full Phase-Strain Relation for Structurally Integrated Interferometric Optical Fiber Strain Sensors," Submitted to Fiber Optic Smart Structures and Skins III at EO/Fibers 90.

5.   Theocaris, P. S., and Gdoutos, E. E., MATRIX THEORY OF PHOTOELASTICITY, Springer Verlag, New York, 1979.

ACCURACY AND PRECISION IN THE THERMOELASTIC STRESS ANALYSIS
TECHNIQUE

P.STANLEY and J.M.DULIEU
Department of Engineering
Simon Building
University of Manchester
Oxford Road, MANCHESTER M13 9PL

ABSTRACT

The thermoelastic stress analysis technique is based upon the use of the SPATE (Stress Pattern Analysis by the measurement of Thermal Emission) equipment for the radiometric monitoring of the temperature changes induced by cyclic loading in the elastic range. In this paper the factors affecting the accuracy and precision of the technique, as used for quantitative stress studies, are reviewed and assessed.

INTRODUCTION

The thermoelastic stress analysis technique (1) has been developed and applied successfully over the last few years to a wide range of problems including crack-tip stress fields (2), pressure vessel stresses (3), composite materials (4) and modal analysis (5). The SPATE equipment consists essentially of a sensitive infra-red detector, together with facilities for control, signal correlation and processing, and signal display, including the necessary post-processing software. The detector is a cadmium-mercury-telluride device operating at liquid nitrogen temperature. In operation, a selected "frame" of the surface of the cyclically loaded specimen or component is marked out using an optical system incorporated in the detector unit. The frame is "scanned" point-by-point and the detector output is amplified, correlated with a reference signal and displayed on a high-resolution video monitor. Hard copy versions of the display or samples from it are also available. To meet a broad variety of practical restraints and to accommodate a wide range of signal level, the operator has the choice of (i) stand-off distance, (ii) size, position and shape of the scanned frame, (iii) density of the point array within the frame from which signals are received, (iv) sampling time per point, (v) display scale factor, (vi) time constant for the signal processing and (vii) display mode.

The underlying physical basis of the technique is the thermoelastic effect. Beginning with Kelvin's formulation (6), it is readily shown (1) that

$$\sigma_1 + \sigma_2 = AS \qquad (1)$$

where σ_1 and σ_2 are the changes in the principal surface stresses at a particular point, S is the displayed signal from that point and A is a constant of proportionality or calibration constant. In the present appraisal of the technique it is assumed that equation (1) is valid. It is therefore important that the assumptions in its derivation and the limitations on its validity are recognised if potentially serious systematic errors in derived stress values are to be avoided.

It is assumed that the temperature changes occur adiabatically as the stress is cycled. This implies that for a particular material there is a limiting minimum cyclic frequency below which signal losses related to temperature gradients (and therefore to stress gradients) become increasingly important. In an experimental study (1) it has been shown that this limiting frequency (which is proportional to the thermal diffusivity of the material) is approximately 2Hz for steel; at 0.6Hz in the specimen studied there was a signal loss of 33%. A further important assumption made in deriving equation (1) is that the relevant material properties are independent of temperature and stress. In explaining the "mean stress effect" (7), Wong et al (8) set this assumption aside and re-worked the theory leading to equation (1). They showed that $\partial E/\partial T$ (the temperature derivative of Young's modulus) was an important detail, and that for materials in which this derivative was significant in relation to the quantity $\alpha E^2/\sigma_m$ (α, coefficient of thermal expansion; σ_m mean stress in stressing cycle), then equation (1) was deficient. The effect is not significant for steel but could be for other materials, especially the non-metallics. It is also assumed that the detector is a linear device with response directly proportional to input. Oliver and Webber (9) claim that for a typical detector the departure from linearity over the thermoelastic temperature range (0-0.5K) is less than 1%.

A standard procedural detail in thermoelastic stress studies is to apply a matt black coating to the surface to be examined, usually by means of an aerosol spray. This standardises and enhances the surface emissivity. The possible attenuating effect of this coating on the thermoelastic response of the substrate has been studied in detail by McKelvie (10). This work confirms the results of earlier empirical work (11) to the effect that for the usual coating thicknesses then, at typical loading frequencies (e.g. less than 30Hz), these attenuation effects are negligible. Some further work on coatings is described in ref.(9), and Mackenzie (12) has reported a detailed recent study. A further coating-related consideration is the extent to which the surface emissivity decreases as the surface viewing angle increases from zero (i.e. normal viewing) to the "glancing" angle of 90^0. Stanley and Chan (11) have shown that the emissivity of the RS Matt Black Heat Radiator surface coating is independent of viewing angle up to angles of approximately 55^0, but that there are serious signal losses for higher viewing angles. An unspecified surface paint (paint A) is referred to in ref.(9) which has a uniform emissivity value for viewing angles up to 73^0. Clearly, the possibility of serious signal impairment due to this effect can be disregarded in the great majority of applications.

The SPATE system is designed specifically for a sinusoidal loading, with a reference signal input range from 300mV to 10V.

The foregoing brief review covers potential sources of systematic error associated with the conditions necessary for the validity of equation (1); the rest of the paper deals with possible errors in the thermoelastic signal S attributable to the measuring system itself and examines a number of alternative techniques for the determination of the calibration constant, A.

THE THERMOELASTIC SIGNAL S

The following factors are considered, which may affect the quality and reliability of the thermoelastic signal S:-

i) noise, i.e. that inherent in the system and that due to electromagnetic acoustic or mechanical pick-up,
ii) phase matching,
iii) time constant and sampling time,
iv) scanning spot size and density.

A minimum level of inherent noise is an inevitable feature of any electrical/electronic system and puts a lower limit on the attainable accuracy. In the SPATE system the predominating inherent noise source is the infra-red detector; inherent noise contributions from the correlator, the processing computer and connecting cables are relatively small. The system noise envelope is directly related to the minimum resolvable temperature difference (MRTD) of the system, which is in turn dependent upon the detector unit construction. The quoted MRTD for a typical system is 1mK, corresponding to stress changes of approximately 1 MPa in steel and 0.4 MPa in aluminium. The MRTD is specific to a particular system and it is noteworthy that the value for different systems may vary considerably; Machin et al (7) refer to a threefold increase in the signal/noise ratio of one system relative to another. The system noise level is dependent largely upon the effectiveness of the RF shielding incorporated in the detector unit. The noise level can be estimated quantitatively from the recorded variations in a nominally constant signal. Another indication of the noise level can be obtained from the "zero stress" signal, i.e. that obtained with the detector unit shutter closed or the detector scanning an unstressed black card.

A film pellicle is included in the detector unit lens system to provide protection from dust and inadvertent intrusions. A noise contribution has been attributed to this insertion (13).

The phenomenon of acoustic noise pick-up has not been systematically studied but is certainly real. The effect is particularly important when the noise emanates from the test specimen itself as opposed to more general environmental noise. Bream et al (13) were unable to eliminate a specimen noise level of 108 dB, which was equivalent to a significant signal, but with the noise reduced to 90 dB they were able to introduce an offset into the system which eliminated the noise effect. The 108 dB specimen noise level will be uncommon and sustained environmental noise at this level in a laboratory would be a cause for concern. However, inadvertent transients may approach this magnitude and the SPATE user must be aware of an acoustic noise sensitivity in the equipment. The importance of transmitted mechanical vibrations is illustrated in Figs. 1 and 2. The former shows the received signal from an unstressed black card with an average level of 33 mK when one leg of the tripod supporting the detector unit was touching the test machine base. The latter shows the effects of isolating the unit. The authors concluded that transmitted mechanical vibrations and acoustic noise have a similar effect on the measured signal.

Phase-matching in which the phase of the reference signal is shifted to match that of the received signal, is an important step in the setting-up of a SPATE scan. This is done by adjusting the in-phase component (the x-component) and the out-of-phase component (the y-component) of the received signal to maximum and zero values respectively. Small finite y-component values are generally unavoidable and often unimportant. If, however, the phase-matching has been carried out at a high signal location, then, as the scanning covers regions of lower signal level, a finite y-component can result in a pronounced systematic phase-banding effect. This is illustrated in the Argand plots of Fig.3, in which the phase-matched signal is represented by the point A; the noise envelope around A is indicated. The finite y-component persists over the signal range and it is clear that as the signal decreases the phase angle θ increases systematically and practically regardless of signal variations within the noise envelope. Phasing-in at a low signal point (e.g. point B in Fig.3) should be avoided also because, although in absolute terms the finite y-component may be reduced, the effects of signal variations

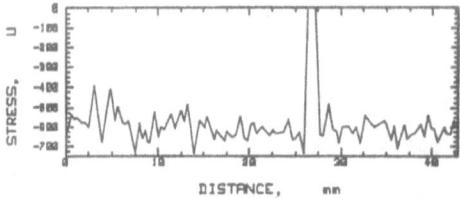

Fig. 1 "Black card" response with
spurious mechanical
vibration pick-up

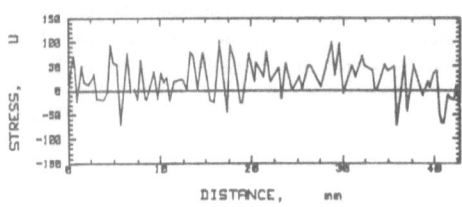

Fig. 2 "Black card" response with
pick-up eliminated

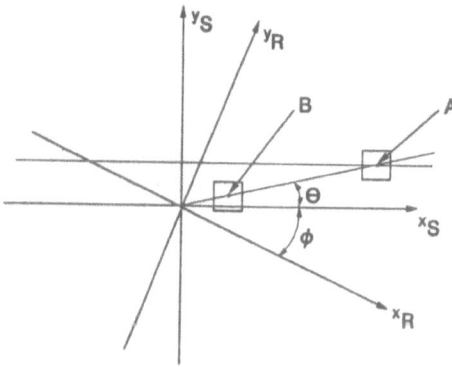

x_R y_R reference signal axes
x_S y_S thermoelastic signal axes
θ phase angle
φ phase shift angle

Fig. 3 Argand diagram showing
influence of noise after
phase matching

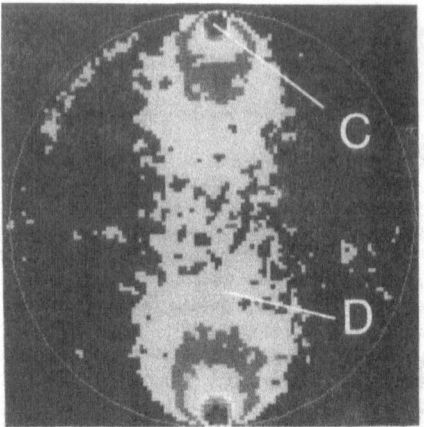

a) signal contours

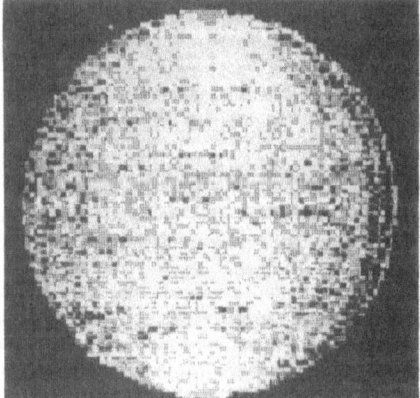

b) phase banding in low
stress region

c) phase banding eliminated

Fig. 4 Circular steel disc in
diametral compression

within the noise envelope may be such as to prevent satisfactory matching. Phasing–in is best carried out at a point of average, or typical, signal level. Fig.4a shows a signal contour plot for a circular disc in diametral compression. The phase angle plot (Fig.4b) displays the banding effect in the low stress region after phase–matching at point C where the stresses are relatively high. The banding effect is significantly reduced in Fig.4c, which was produced after phase–matching at point D.

The SPATE user also has a choice of time constant (i.e. the characteristic response time of the system) and sampling time (i.e. the signal acquisition time at a particular point). The SPATE manual (14) recommends that the time constant should be 10–30% of the sampling time. The time constant does not affect the signal magnitude but, with high values, the system is less responsive to variations in successive signal peaks and less noisy signals are obtained for a given sampling time. Large time constant values appear to be desirable, therefore, but they can give rise to a significant display defect in cases where there are high stress gradients or step–changes in stress within the scanned frame. This particular defect (known as "shuttering") can be seen in Fig.5, a hole–in–a–plate scan, around the edge of the hole. The time constant and sampling time in this case were 100ms and 1s respectively. For the step–change in signal at the hole edge the system is "over–damped" with this time constant. Reducing the value of the time constant eventually eliminates this effect but the system may become "under–damped" with a significantly higher noise level. The effect of increasing the sampling time is to reduce the noise content of the time–averaged signal in direct proportion to $1/\sqrt{t}$. The choice of optimum values of time constant and sampling time is a matter for detailed consideration and will depend, in general, on the particular application. Both are to be seen in relation to the cyclic load frequency. Reducing the frequency has the same effect as reducing the time constant or sampling time, and vice versa.

Signal detail is affected by the size of the scanning "spot" (i.e. the area of the specimen from which the signal is received) and the spatial distribution of the spots within the scanned frame. The spot size is proportional to the distance of the detector from the specimen and, clearly, the bigger the spot the more significant will be the averaging effect over the spot area and the consequent "rounding off" of localised signal peaks. No significant problems of this kind have been experienced with typical working distances between 250 mm and 750 mm (and corresponding spot sizes of 0.5 mm and 1.0 mm). A related potential problem concerns the displacement amplitude of a test specimen with a non–uniform stress distribution. If this is comparable with the spot diameter, then, depending upon the stress gradient, significant signal "smudging" may result. Order–of–magnitude calculations for steel indicate that this effect is unlikely to be significant in typical cases. The choice to be made in selecting the spacing of the scanning spots is between i) a sparse coverage for widely spaced spots, and ii) a proportionately increased total scan time for closely spaced (possibly overlapping) spots.

Positional errors in signal location (i.e."sighting" errors) can occur as a result of any misalignment between the optical axis of the system and the detector axis, leading to misinterpretation of the displayed scan data. Some signal "drift" (equivalent to a few millikelvin per hour in steel) may also occur and it is advisable that some check on this possibility should be made when long scan periods are contemplated.

THE CALIBRATION CONSTANT A

Three essentially different methods are available for the determination of the calibration constant A (see equation (1)):

    i)   from a formula given in ref.(14),
   ii)   calibration against a measured stress,
  iii)   calibration against a calculated stress.

These are reviewed and compared below.

The SPATE manual (14) gives the following formula for a factor F which is identical to the constant A:

$$A = \frac{DGR}{Te\ 2048K} = \frac{DGR\rho\ C_\sigma}{Te\ 2048\ \alpha} \tag{2}$$

where
D = temperature responsivity of the system,
G = correlator sensitivity setting,
R = surface temperature correction factor,
C_σ = specific heat of material at constant stress,
ρ = material density
T = mean absolute temperature at measuring point,
e = emissivity of surface,
K = thermoelastic constant of the material (i.e. $\alpha/\rho C$),
α = coefficient of linear thermal expansion.

The derivation of this expression is covered in the Appendix. Seven of the eight quantities involved in it (i.e. all except G, the correlator sensitivity which is pre-set by the user) are experimental in nature. D is specific to the detector, K (i.e. α, ρ and C_σ) is specific to the material and R and e relate to the surface condition. Errors in any of these seven quantities will accumulate to give an error in A; for example, a probable random error of 1% in each of them will result in a probable random error (or "uncertainty") of 2.6% (i.e. $\sqrt{7 \times 1^2}$) in A, and pro rata for greater errors.

D, the temperature responsivity of the equipment is determined by the manufacturers (9) for a surface temperature of 20°C. The quantity may differ by an order of magnitude for different equipments. It is usually given to two significant figures; a particular value cited in the literature (15) is 9.0 °K/V. The random error is probably of the order 1% (9); no estimate can be made of possible systematic error. The correction factor R is introduced in order to make allowance for the fact that D is temperature dependent. The graph of R versus temperature provided by the manufacturers (14) is in the form of a slightly curved line with a negative slope of approximately 0.01/ °K. It is not clear how this plot was derived and the probable errors associated with it cannot be assessed, but it is clear that an error of 1 °K in the specimen surface temperature T will result in an error of at least 1% in R, and therefore in A. T appears explicitly in the denominator of the A expression. In an ordinary laboratory environment, errors of at least 1 °K in inferred values of specimen surface temperature cannot be considered unlikely. The surface emissivity e is that of the paint coating; values quoted by Oliver and Webber (9) for each of three unspecified paints vary by 1-3% over a viewing angle range 0 to 45°. The thermoelastic constant of the material K depends upon the density ρ, the specific heat C_σ and the coefficient of thermal expansion α. These in turn depend, to some extent, upon material composition (even for nominally identical materials) and whilst nominal two-figure values are readily found for pure substances and most alloys, good-quality values for materials under investigation are not always easily obtained. The difficulty is illustrated by three values of K for "steel", obtained from three independent publications –

ref (14) 3.5×10^{-12} m²/N
ref (15) 3.0×10^{-12} m²/N   (for En 8)
ref (16) 2.8×10^{-12} m²/N

These show a spread of 10% about the mean value.

The correlator "sensitivity" G is a display scaling factor which allows the received signal to be spread over the full available DAC range of ± 2048, so that full advantage can be taken of the display facilities and signal overloads are avoided. The choice of G is best made when the maximum signal values in a scan have been identified. The error in a particular value of G is that associated with any high–quality electronic device, i.e. 1–2%.

When the film pellicle is used in the lens system, an appropriate attenuation factor should be included in the numerator of equation (2). The value of this factor is quoted as 1.16 (15); the origin and accuracy of this figure are not known. The calibration method outlined above is "absolute" in that it depends on the characteristics of the measuring system and on the physical properties of the material and the surface. A thermoelastic signal is measured in the determination of D, but no direct stress measurement is required. Harwood and Cummings (15) have used the method to provide a numerical value of the stress in a tensile specimen of En 8 steel, which differed by 2.3% from an independent value. Nevertheless, because of the difficulties referred to above, particularly the likely uncertainty in the relevant material properties, the present authors do not see this as the best approach where good quantitative stress values are required. Brown (17) comes to a similar conclusion and gives a possible calibration error of 20% for this method.

The second calibration method ((ii) above, referred to in the SPATE manual (14) as the "known stress" method) requires an independent measure of the stress field responsible for the thermoelastic signal. Strain gauges are bonded to a suitable specimen (e.g. a tensile specimen) in an area of reasonably uniform signal and the SPATE signal, S, is read from an adjacent point on the specimen. The stress change $(\sigma_1 + \sigma_2)$ is inferred from the gauge response and the calibration constant is derived from the relationship

$$A = \frac{\sigma_1 + \sigma_2}{S} = \frac{G\,(\sigma_1 + \sigma_2)}{2048\ V} \tag{3}$$

where V is the voltage output of the detector.

The accuracy of the derived value does not depend on an accurate knowledge of the applied load; loading machine calibration and specimen misalignment, etc., are therefore unimportant. The accuracy is dependent upon gauge alignment if a single gauge is used (e.g. on a tensile specimen), but not if a rosette is used since the strain sum is an invariant of the system. There may be a significant systematic error in the derived A value if the stress sum at the gauge point differs from that at the reading point, unless the ratio of the signal values at the two points is introduced as a factor in equation (3). A preliminary scan is therefore an important requirement in using this calibration approach.

If a high quality is assumed in the strain gauge circuitry and associated instrumentation, accuracy limits will probably be dictated by the noise level in the signal, S, and by the degrees of uncertainty in the gauge factor and the relevant material properties, E (Young's modulus) and ν (Poisson's ratio). Gauge cross–sensitivity may be an important factor in some cases. Young's modulus values for steel is usually quoted within the range 200–210 kN/mm^2; an assumed value may therefore be in error by several percent. Single figure errors in the second decimal place of Poisson's ratio (i.e. 0.30 instead of 0.29) will produce a 1.5% error in derived stress values.

The present authors have used the third calibration method ((iii) above) in which a measured SPATE signal is related to a known stress value or stress distribution produced in a sample specimen by a known applied load. It is essential that the load value is known accurately and this may entail a separate test machine calibration using, for example, a proving ring. It is also essential that the dimensional tolerances in the specimen, the test jig and/or the loading attachments are such that the relevant stresses can be calculated with confidence. With these provisos, the accuracy of the derived calibration constant is dependent only upon signal noise level and possible sighting errors. A possible disadvantage of the method is that for large prototype studies it may be necessary to have available a suitable material sample for testing, unless a "known stress" region can be arranged on the test piece itself.

Suitable specimens for this form of calibration are i) the simple tensile specimen, ii) the beam in 4-point bending, iii) the Brazilian disc. Depending on the material and test facilities available, the beam or disc may be more convenient than the simple tensile specimen. The 4-point beam specimen is particularly attractive since, if line scans are taken transversely across the central part of the beam where the bending moment is uniform, the calibration is not effected by sighting errors, the linearity (or otherwise) of the thermoelastic response is evident over a wide range of both positive (i.e. tensile) and negative (i.e. compressive) stress, and the relative noise is spatially displayed in the signal plot. An area scan over the central part of a mild steel beam (200mm long, 31 x 31mm cross-section) is shown in Fig. 6; a transverse line scan through the data (i.e. from top to bottom) is shown in Fig. 7. It is readily shown that the calibration constant A (equation (1)) can be obtained from such a scan as (M/I) divided by the gradient of the line plot where M and I are the bending moment and second moment of area respectively. The value obtained for the case illustrated was 0.125 N/mm^2/signal (for G=100mV) with an estimated standard deviation of 0.002.

The circular disc specimen (Fig. 4) requires no end grips or shackles but since surface stress readings are involved it is most important that the resultant diametral load is applied in the mid-plane of the specimen if spurious bending effects are to be avoided. (This contrasts markedly with the case of a comparable photoelastic specimen in which the isochromatic fringes are related to the average through-thickness stress.) From the available analytical stress solution for this form of specimen, a linearised form of the signal variations along the vertical and horizontal diameters can be derived and thence values of the calibration constant A. (Sighting errors are not automatically eliminated in this specimen and the position of the signal readings relative to the centre of the disc must be accurately known.) A signal plot along the vertical diameter of a mild steel disc loaded in cyclic diametral compression is shown in Fig.8. The calibration constant (A) value derived from a linearised version of this plot was 0.255 N/mm^2/signal for a G value of 5mV. (N.B. Different detectors with different D values were used in the calibration work on the beam and the disc.)

CONCLUDING REMARKS

Within certain limits, the accuracy of an experimental technique is not always the most important of the factors which might influence the choice of technique for a particular purpose. Nevertheless, for the confident and reliable use of a technique, particularly where quantitative results are required, it is important that the principal error sources are identified and their nature understood. In this paper the possible errors in thermoelastic stress data obtained by means of the SPATE equipment have been reviewed under three broad headings:-

i)   The equation relating the thermoelastic temperature change (and
       therefore the SPATE signal) directly to the peak-to-peak stress change only is
       an approximation, to the extent that the temperature dependence of Young's
       modulus is ignored. Within the resulting approximation limits, the validity of
       the equation requires that certain experimental and procedural conditions are
       observed. These conditions are generally well understood and do not
       significantly limit the range of use of the equipment.

ii)   The SPATE signal is a measure of a very small physical quantity. Reliable

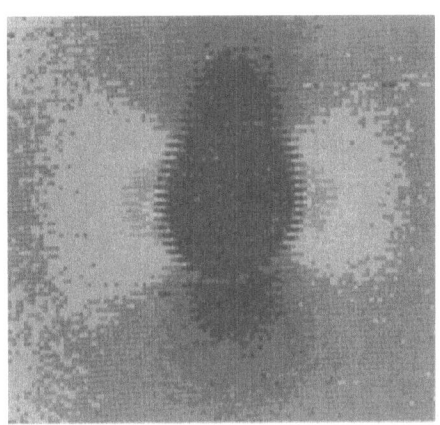

Fig. 5 "Shuttering" effects at the
signal discontinuity at the
edge of hole in a stressed plate

Fig. 6 Area scan over central
part of steel beam in
4-point bending

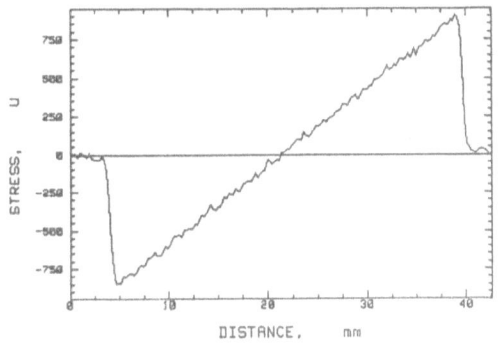

Fig. 7 Line scan (from top
to bottom) through
data displayed
in fig. 6

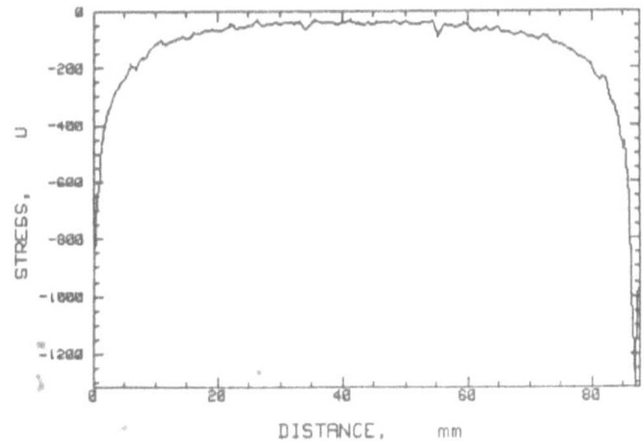

Fig. 8 Line scan (along
vertical diameter)
through data
displayed in fig. 4a

quantitative measurements depend on efficient noise suppression and signal processing, and sound operational practice.

iii) Several alternative calibration techniques are avaliable, each with particular possible error characteristics. In choosing a technique it is important that both systematic and random errors are considered and effectively minimised.

ACKNOWLEDGEMENTS

The authors are grateful for helpful talks with P.Hunt, A.Barker, G.Everett and R.Owens of Ometron Limited, and W.Cummings and K.Mackenzie of NEL, East Kilbride.

REFERENCES

1. STANLEY, P. and CHAN, W.K., "Quantitative stress analysis by means of the thermoelastic effect", J.Strain Anal., 1985, 20, 129-137.

2. STANLEY, P. and CHAN, W.K., "The determination of stress intensity factors and crack-tip velocities from thermoelastic infra-red emissions", Proc. International Conf. on Fatigue of Engineering Materials and Structures, (I.Mech.E.), Sheffield, Sept.1986, 1, 105-114.

3. STANLEY, P., "Stress separation from SPATE data for a rotationally symmetrical pressure vessel", Proc. International Conf. on Stress and Vibration (Incorporating 3rd Int.Conf. on Stress Analysis by Thermoelastic Techniques), London, March 1989, SPIE Vol 1084, 72-83.

4. STANLEY, P. and CHAN, W.K., "The application of thermoelastic stress analysis techniques to composite materials", J.Strain Anal., 1988, 23, 137-143.

5. CUMMINGS, W.M. and HARWOOD, N., "Thermoelastic stress analysis under broad-band random loading", Proc. SEM Conf. on Experimental Mechanics, Las Vegas, June 1985, 740-746.

6. TODHUNTER, I. and PEARSON, K., "A History of the Elasticity and Strength of Materials", Vol 2, 1893, Cambridge Press.

7. MACHIN, A.S., SPARROW, J.G. and STIMSON, M.G., "The thermoelastic constant", Proc. 2nd International Conf. on Stress Analysis by Thermoelastic Techniques, London, February 1987, SPIE Vol 731, 26-31.

8. WONG, A.K., JONES, R. and SPARROW, J.G., "Thermoelastic constant or thermoelastic parameter?", J.Phys.Chem.Solids, 1987, 48, 749-753.

9. OLIVER, D.E. and WEBBER, J.M.B., "Absolute calibration of the SPATE technique for non-contacting stress measurement", Proc. Vth International Cong. on Experimental Mechanics, Montreal, June 1984, 539-546.

10. McKELVIE, J., "Consideration of the surface temperature response to cyclic thermoelastic heat generation", Proc. 2nd International Conf. on Stress Analysis by Thermoelastic Techniques, London, February 1987, SPIE Vol 731, 44-53.

11. STANLEY, P. and CHAN, W.K., "SPATE stress studies of plates and rings under in-plane loading", Experimental Mechanics, 1986, 26, 360-370.

12. MACKENZIE, A.K., "Effects of surface coatings on infra-red measurements of thermoelastic responses", Proc. International Conf. on Stress and Vibration (Incorporating 3rd Int.Conf. on Stress Analysis by Thermoelastic Techniques), London, March 1989, SPIE Vol 1084,

13. BREAM, R.G., GASPER, B.C., PAGE, S.W.T. and LLOYD, B.E., "Operational experiences with the SPATE 8000 dynamic stress measurement system", Proc. 2nd International Conf. on Stress Analysis by Thermoelastic Techniques, London, February 1987, SPIE Vol 731, 181-204.

14. SPATE 8000 and SPATE 9000 Manuals, Ometron Ltd., London.

15. HARWOOD, N. and CUMMINGS, W.M., "Calibration of the thermoelastic stress analysis technique using sinusoidal and random loading conditions", Strain, 1989, 25, 101-108.

16. BAKER, L.R. and WEBBER, J.M.B., "Thermoelastic stress analysis", Optica Acta, 1982, 29, 555-563.

17. BROWN, K. "Calibration studies: collaborative report by the United Kingdom Stress Pattern Analysis by Thermal Emission (SPATE) Users'Group.", Proc. 2nd International Conf. on Stress Analysis by Thermoelastic Techniques, London, February 1987, SPIE Vol 731, 205-211.

APPENDIX

DERIVATION OF CALIBRATION FORMULA

The scaling of the 12-bit digital-to-analogue conversion device (DAC) incorporated in the equipment depends on the setting of the correlator "sensitivity", G, which determines the signal in millivolts which corresponds to the full scale range of $\pm$ 2048 DAC units.

i.e. 2048 DAC units $\equiv$ G mV

or 1 DAC unit $\equiv \dfrac{G}{2048}$ mV (A1)

The DAC unit is the unit of the "uncalibrated" signal. This stage of the derivation is illustrated in the first quadrant of Fig. A1.

The detector responsivity, D, is the ratio of the black-body temperature change "seen" by the detector to the resulting signal. Equation (A1) can therefore be developed to give

1 DAC unit $\equiv \dfrac{DG}{2048}$ mK (A2)

which, with allowance for different surface temperatures and imperfect surface emissivity, becomes

$$1 \text{ DAC unit} = \frac{DGR}{e2048} \text{ mK} \qquad (A3)$$

where R and e are the surface temperature correction factor and the surface emissivity respectively. Allowance for the film pellicle can also be made at this stage if necessary. These steps are represented in the second quadrant of Fig. A1.

The thermoelastic constant of the material, K, is now introduced to relate the temperature change to the stress change, giving

$$1 \text{ DAC unit} = \frac{DGR}{Te2048K} \text{ stress units}$$

$$= \frac{DGR\rho C_\sigma}{Te2048\alpha} \text{ stress units} \quad (A4)$$

This is the constant of proportionality or calibration constant, A, in equation (1).

The route through Fig. A1 in developing equation (A4) is marked in the figure. The more direct approach to equation (1) is represented in the fourth quadrant of the figure. The diagram could also be constructed by proceeding clockwise from the stress axis in the sequence:–

$$1 \text{ stress unit} \equiv KT \text{ mK} \equiv \frac{KTe}{DR} \text{ mV} \equiv \frac{KTe2048}{DGR} \text{ DAC units}$$

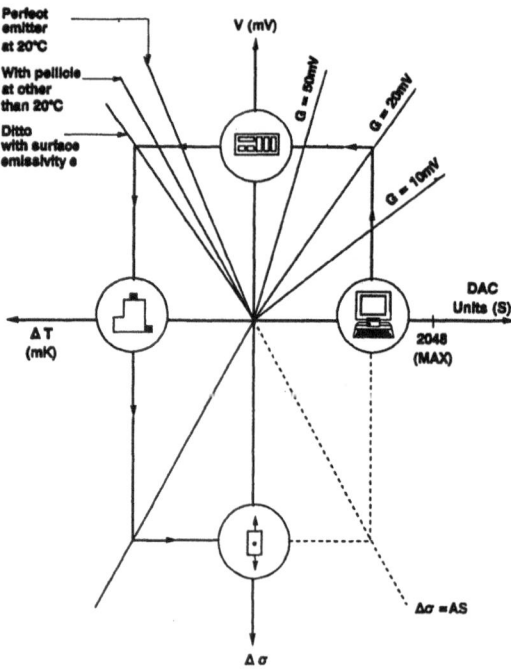

Fig. A1   Derivation of calibration formula

HYBRID TECHNIQUE TO ANALYZE 3-D STRESS - STRAIN STATES

KARL-HANS LAERMANN
BUGH Wuppertal / FB 11
Pauluskirchstr. 7, D-5600 Wuppertal 2, F.R.G.

ABSTRACT

Because of the development of hybrid techniques in experimental mechanics, i.e. the combination of experimental analysis with computer-oriented processes of evaluation, the advantages of the high resolving power of the presently available measuring - and recording techniques together with the possibility to evaluate the increasing amount of data can be utilized meaningfully only when the physical phenomena are described by advanced theories. Then new fields will be opened to apply methods of experimental analysis to as yet unsufficiently solved problems in solid mechanics. As one of those problems in experimental analysis, the determination of the internal stress - strain - state in threedimensional solids will be considered.
Based on the boundary-integral method, a reconstruction algorithm will be derived, by which the displacements in arbitrary points inside a threedimensional, simply connected solid can be determined from experimentally given boundary values u_i^r. The experiments are carried out by means of the multi-hologram-analysis combined with digital image processing.

INTRODUCTION

Recent developments in the techniques of holographic interferometry combined with digital image processing and computer-oriented evaluation procedures enable the experimental/numerical analysis of many problems in solid mechanics, which could not be investigated with appropriate reliability as yet. According to the principle of "hybrid techniques" (1), (2), (3), it is possible now to analyze the internal stress - strain - state in threedimensional solids in case of elastic or even viscoelastic response of material in a non-destructive process. In a multiple-hologram-analysis (4), (5) with at least three non-coplanar holograms combined with automatic digital image processing of the reconstructed interferograms, the total displacement vector $\underline{u}^{\Gamma}$ in discrete points on the surface Γ of a threedimensional, simply-connected solid can be determined. The components of these vectors and their first derivatives as well are taken as input data to calculate the displacements inside the solid by a reconstruction procedure derived from the boundary integral method (6), (7).

NUMERICAL SOLUTION OF THE LAPLACE-DIFFERENTIAL EQUATION

Under the condition of elastic, isotropic and homogeneous response of material, the state of displacement in a threedimensional object can be described by Lamé-Navier's differential equation

$$\nabla^2 u_i + \frac{1}{1-2\mu}\Phi_{,i} = 0, \quad \Phi \equiv u_{k,k} ; \ i,k \in [1/3]. \tag{1}$$

As it can be proved (8), that

$$\nabla^2(x_i\phi) \equiv 2\phi_{,i}, \tag{2}$$

eq. (1) can be formulated as

$$\nabla^2[u_i + \frac{1}{2}\frac{1}{1-2\mu}(x_i\phi)] \equiv \nabla^2 f_i = 0, \tag{3}$$

i.e. f_i denotes a harmonic function.

If two harmonic functions f_i and $1/r$ as a fundamental solution are given in a threedimensional domain G, which are steady in G and show steady derivatives to the 2nd order including the boundary Γ, the value of the function f_i in an arbitrary point P_o inside G holds with reference to Green's formula (9)

$$f_i(P_0) = \frac{1}{4\pi}\iint\limits_{(\Gamma)} [\frac{1}{r} f^{\Gamma}_{i,n} - f^{\Gamma}_i(\frac{1}{r})_{,n}] \, d\Gamma. \tag{4}$$

Considering Dirichlet's problem of the first kind, i.e. the boundary values f^{Γ}_i of the function f_i are given, the derivative $f^{\Gamma}_{i,n}$ in the direction of the boundary normal in P on the boundary is to eliminate. Assuming a Green's function G_1 (P, P_o) with the boundary values $1/r$ on Γ to be existing, eq. (4) yields

$$f_i(P_0) = \iint\limits_{(\Gamma)} f^{\Gamma}_i(P) \cdot G(P,P_0) \, d\Gamma, \tag{5}$$

where the correlation function

$$G(P,P_0) = -\frac{1}{4\pi} [\frac{1}{r} - G_1(P,P_0)]_{,n} \tag{6}$$

has been introduced. In order to determine this correlation function, an algorithm for discrete numerical evaluation has been derived (see (6),(7)).

In an arbitrary point P_0 inside G, the origin of a local carthesian coordinate system x'_j, the harmonic function f_i is expanded in Taylor-series in the directions of the coordinate axes. These series are truncated after the 4th term and introduced into the Laplace differential equation $\nabla^2 f_i = 0$. Introducing the distances $\pm\xi_j$ between the intersection points of the coordinate axes with the object surface Γ and the origin of the local coordinate system (Fig. 1) instead of $\pm\delta x'_j$ and consequently introducing the boun-

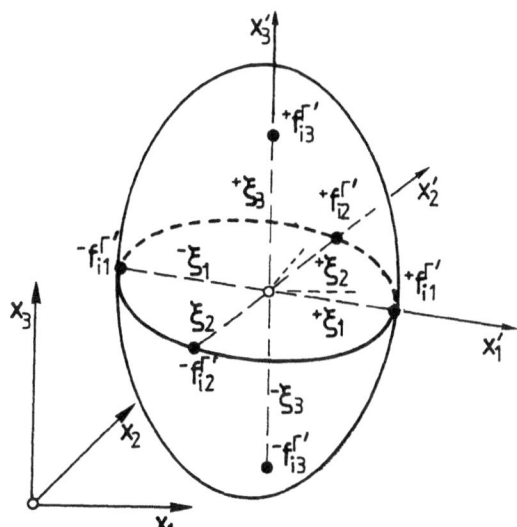

dary values $f_i^{\Gamma'}$, an appropriate value of $f'_i(P_0)$ will be obtained and thus an approach of the correlation function $G(P,P_0)$. Considering the definition of f_i and Einstein's summation convention, the relation between u'_i and the boundary values of the displacement vector holds

Figure 1. Principle and denotation of the reconstruction algorithm.

$$u'_i[w_{jj}+m\,w_i] \approx w_j\,(^+u_{ij}^{\Gamma'}{}^+h_k + {}^-u_{ij}^{\Gamma'}{}^-h_k)\,\delta_{jk} + m\,w_i\,(^+u_{ii}^{\Gamma'}{}^+h_i + {}^-u_{ii}^{\Gamma'}{}^-h_i) +$$
$$+ m\,(^+u_{ji,j}^{\Gamma'} - {}^-u_{ji,j}^{\Gamma'})(1-\delta_{ij})/l_i \tag{7}$$
$$\text{(not summing up over } i\,)$$

In eq. (7), w_j denotes a weighting factor

$$w_j := [|{}^+\xi_j| \cdot |{}^-\xi_j|]^{-1} \tag{8}$$

and h_j a linear form function

$$^{\pm}h_j := ^{\mp}\xi_j \cdot l_j^{-1} \; , \quad l_j := |^+\xi_j| + |^-\xi_j| \; . \tag{9}$$

For abbreviation

$$m = \frac{1}{2} \cdot \frac{1}{1-2\mu} \tag{10}$$

has been introduced.

Discrete integration of eq. (7) over $\mathbb{R}^3$, i.e. taking the weighted mean of the approximate values u_i' over an integer number ν of coordinate triples x_i' transformed onto a global coordinate system x_k by the transformation matrix $[a_{ki}]$, yields the components $u_k(P_0)$ of the interior displacement vector in point P_0

$$u_k(P_0) = \sum_\nu [a_{ki}] \cdot (K_i') \cdot \left(\sum_\nu W_i \right)^{-1} . \tag{11}$$

RECONSTRUCTION ALGORITHM FOR AXISYMMETRIC PROBLEMS

How to proceed, will be demonstrated by considering an axisymmetric cylindrical solid, axially loaded. With the twodimensional displacement vector $\underline{u}^T = (u_r, u_3)$, the evaluation of eq. (11) is split up into two phases, considering equidistant cross sections x_3 = const (Fig. 2).

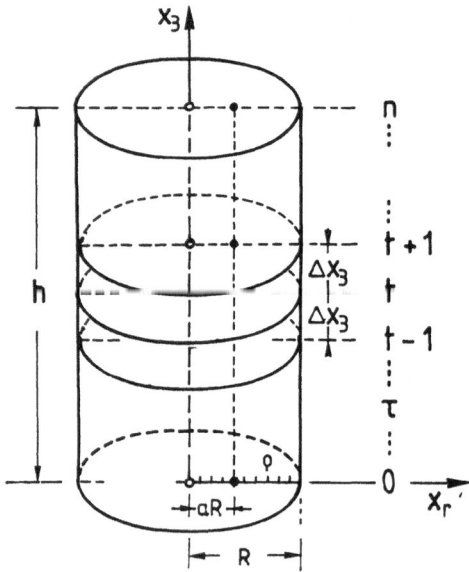

Figure 2. Cross sections and their denotations in the reduction process.

In phase 1, the radial displacement in cross section t holds

$$u_r(aR,t) = K_{rt} + L_r[u_r(aR,t-1) + u_r(aR,t+1)];\qquad(12)$$

$$K_{rt} := \frac{1}{W_r}\,\frac{(2+m)a}{R^2(1-a^2)}\,u_r^\Gamma(t),$$

$$L_r := (2W_r \cdot \Delta x_3^2)^{-1},$$

$$W_r := \frac{1+m}{R^2(1-a^2)} + \Delta x_3^{-2}.$$

This component of the displacement vector along a = const, $0 \le t \le n$, will be calculated in a reduction process:

$$u_{rt} = y_{rt} + \sum_{\tau=t}^{n-2} U_{r\tau} \cdot y_{r(\tau+1)} + U_{r(n-1)} u_{rn}^\Gamma\qquad(13)$$

$$y_{rt} = \sum_{\tau=1}^{t} U_{r\tau} \cdot L_r^{-1} \cdot K_{r\tau} + U_{r1} \cdot u_{r0}^\Gamma$$

$$U_{r\tau} = U_r(t) \cdots\cdots U_r(\tau)$$

$$U_r(\tau) = L_r Q_{r\tau};\quad Q_{r\tau} = [1 - L_r^2 Q_{r(\tau-1)}]^{-1};\quad Q_{r1} = 1\ .$$

Having determined u_r in discrete points ρ in equidistant intervals Δr along the radius R, the derivatives $u_{r,r}(\rho)$ in the cross sections t are given also; cubic spline approximation is recommended.

In phase 2, the axial component u_3 of the internal displacement vector will be determined in an analogous procedure as afore described, based on the following equation:

$$u_3(aR,t) = K_{3t} + L_3[u_3(aR,t-1) + u_3(aR,t+1)]\qquad(14)$$

$$K_{3t} := \frac{1}{W_3}\Big\{\frac{2}{R^2(1-a^2)}\,u_3^\Gamma(t) + \frac{m}{2\Delta x_3}[u_{r,r}(aR,t+1) - u_{r,r}(aR,t-1)]$$

$$L_3 := (1+m)\cdot(2W_3\Delta x_3^2)^{-1},$$

$$W_3 := [R^2(1-a^2)]^{-1} + (1+m)\Delta x_3^{-2}.$$

Thus the displacement vector $\underline{u}^T = (u_r, u_3)$ in discrete points P_o as well as the first derivatives $u_{\alpha,\beta}, \alpha, \beta \in [r,3]$ and consequently the strain tensor $\underline{\varepsilon}$ can be calculated inside the domain G from experimentally given boundary values $\underline{u}^\Gamma$. Regarding Hooke's law of elasticity, the stress tensor $\underline{\sigma}$ can be determined also.

MULTIPLE-HOLOGRAM-ANALYSIS

For an axisymmetric cylindrical solid, axially loaded (Fig. 3)

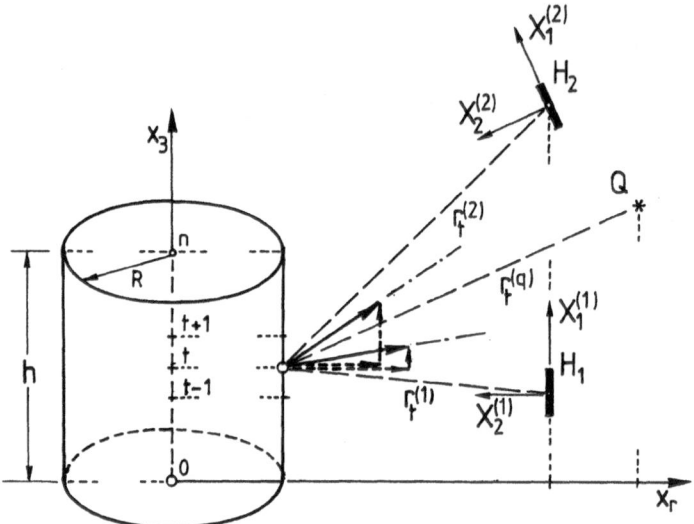

Figure 3. Principle of the holographic setup.

two holograms $H_\rho, \rho \in [1,2]$ are necessary only to determine the twodimensional displacement vector $\underline{u}^\Gamma$ in discrete points t along the generatrix of the surface of revolution. (Another set of holograms might become necessary to get the fringe data on the end planes). The reconstructed holograms are evaluated by means of digital image processing (10), (11); the principle of the electronic recording and processing system is shown in Fig. 4. The position of the integer fringe orders are determined along the generatrix. By cubic spline approximation, a function $N^{(\rho)}(x_3)$ will be obtained, which enables determination of the correct fringe order in the discrete points t, the position of which in the image plane is given considering the image scale

$$A^{(\rho)} = h / \bar{h}^{(\rho)},$$

(15)

where h denotes the real length of the generatrix, $\bar{h}^{(\rho)}$ the respective length in the reconstructed hologram.

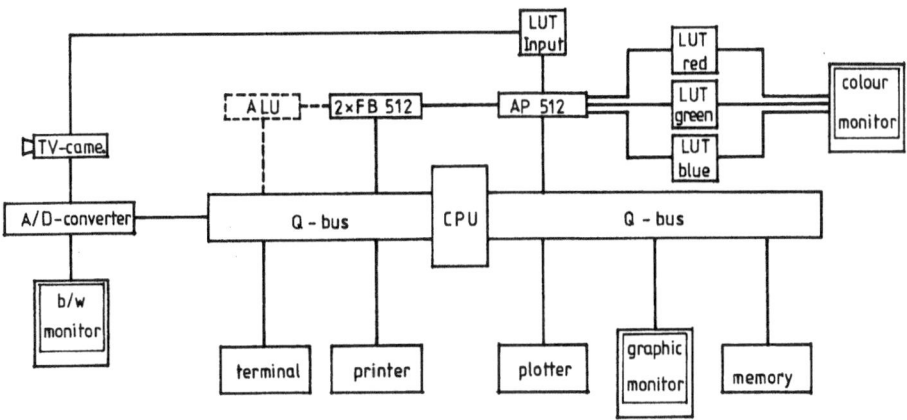

Figure 4. Principle of the electronic recording and processing system.

Finally the displacement vector in point t holds (4)

$$\underline{u}_t^{\Gamma} = \pm \lambda \cdot \underline{\underline{G}}_t^{-1} \cdot \underline{N}_t \; ; \quad \underline{N}_t = (N_t^{(\rho)}) \; ; \quad \underline{u}_t^{\Gamma} = (u_{\alpha t}^{\Gamma}) \; ; \quad \rho \in [1,2] ; \alpha \in [r,3] \qquad (16)$$

with the matrix of geometry

$$\underline{\underline{G}}_t = [a_{\alpha\rho} + a_{\alpha q}]_t \qquad (17)$$

$$\text{with} \quad a_{\alpha\rho} = (x_{\alpha\rho} - x_{\alpha t})/|r_t^{(\rho)}|$$
$$a_{\alpha q} = (x_{\alpha q} - x_{\alpha t})/|r_t^{(q)}|$$

The components of $\underline{u}_t^{\Gamma}$ are obtained in the direction of the sensitivity vectors and therefore to transform onto the coordinate axes x_r, x_3 (see Fig.3).

EXAMPLES OF APPLICATION

The application of the described method is demonstrated by two examples. Fig. 5, a/b shows the holograms of a cylinder between two rigid steel plates, axially pressed, Fig. 6, a/b shows the interference fringe pattern of a cylinder, centrally loaded by a load P applied to a rigid stamp. In Figures 5c and 6c, results of evaluation are given.

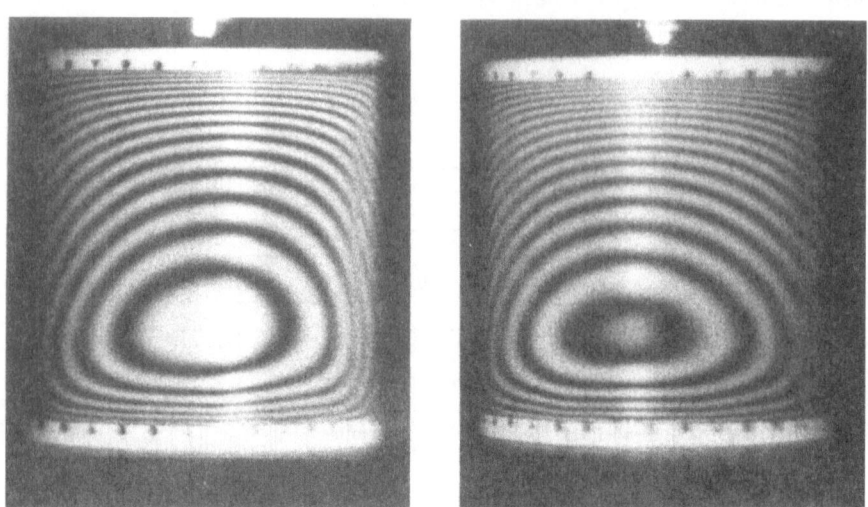

Figure 5, a/b. Interferograms of a cylinder between rigid plates and displacement Δ.

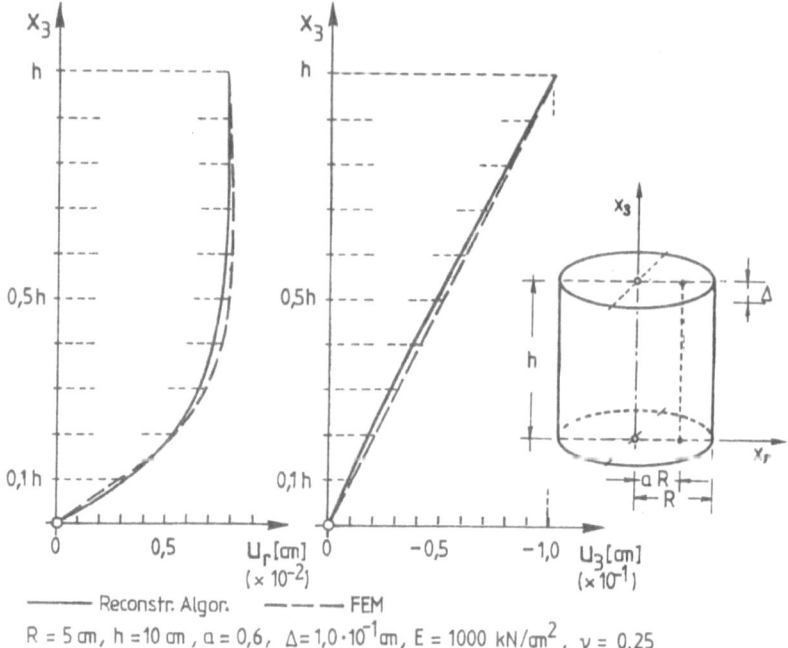

Reconstr. Algor. ——— ———— FEM

$R = 5 \, cm$, $h = 10 \, cm$, $a = 0,6$, $\Delta = 1,0 \cdot 10^{-1} cm$, $E = 1000 \, kN/cm^2$, $\nu = 0,25$

Figure 5, c. Results of evaluation.

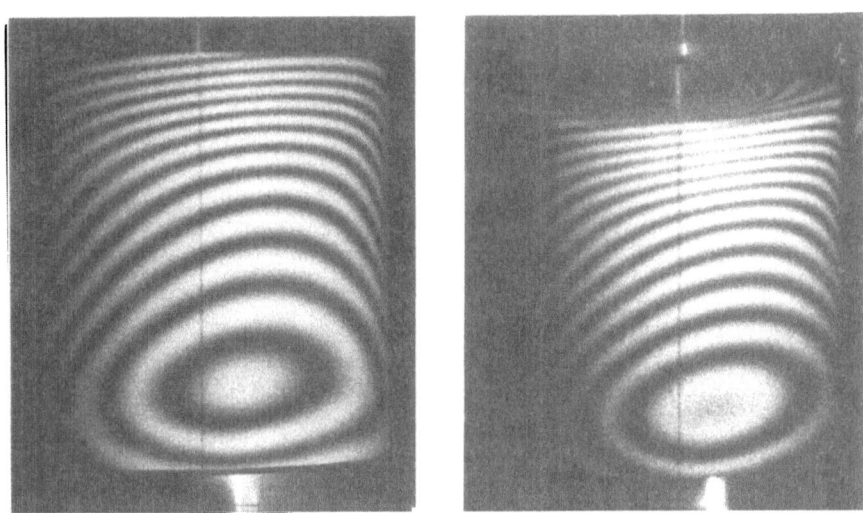

Figure 6, a/b. Interferograms of a cylinder under axial load P.

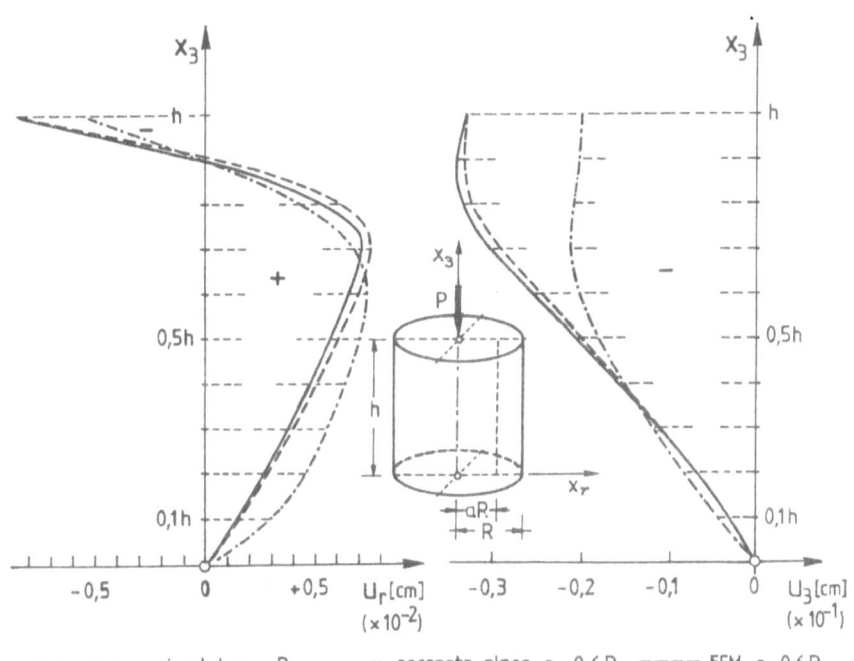

--·--· experim. data r = R, ———— reconstr. algor., r = 0,6 R, ———— FEM, r = 0,6 R

R = 10 cm, h = 20 cm, a = 0,6, P = 500 kN, E = 1000 kN/cm⁻², ν = 0,25

Figure 6, c. Results of evaluation.

648

To prove the reliability and accuracy of the derived method, results of
FEM-analysis are compared with the hybrid results. The coincidence seems
to be quite sufficient.

It must be pointed out, that the method can be applied to arbitrarily
shaped two- and threedimensional objects.

REFERENCES

1. Laermann, K.H., Recent Developments and Further Aspects of Experimen-
    tal Stress Analysis in the Federal Republic of Germany and Western
    Europe. Experimental Mechanics, Vol. 21, No. 2, II/1981.

2. Laermann, K.H., Über das Prinzip der hybriden Technik in der experimen-
    tellen Spannungsanalyse. Messen + Prüfen/Automatik, April 1983.

3. Laermann, K.H., Recent Trends in Experimental Analysis. Mechanika Teo-
    retyczna i Stosowana 2/3, 21, Warszawa 1983.

4. Wernicke, G. and Osten, W., Holographische Interferometrie, VEB-Fach-
    buchverlag Leipzig 1982.

5. Laermann, K.H., Dreidimensionale Analyse des Verformungszustandes
    eines Festkörpers mit Hilfe der holografischen Interferometrie. Laser-
    Optoelektronik in der Technik, W. Waidelich (Herausg.), Springer-Verlag
    Berlin 1987.

6. Laermann, K.H., Reconstruction of the Internal State of Displacement in
    Solids from Boundary Values. Mech. Res. Com., Vol. 11(2), Pergamon-
    Press 1984.

7. Laermann, K.H., Rekonstruktionsalgorithmus zur numerischen Lösung der
    1. Randwertaufgabe. ZAMM, Bd. 85, Hft. 5, 1985.

8. Leipholz, H., Einführung in die Elastizitätslehre, Wiss. u. Technik,
    Verlag G. Braun, Karlsruhe 1968.

9. Smirnow, W.I., Lehrgang der Höheren Mathematik, Teil II, 8. Auflage,
    VEB-Deutscher Verlag der Wissenschaften, Berlin 1968.

10. Alms, K., Erfassung und Lokalisation von Interferenzstreifenmustern
    mit Hilfe der digitalen Bildverarbeitung. Dipl.-Arb. BUGH Wuppertal,
    1987.

11. Laermann, K.H., Testing Installation for Automated Analysis of the In-
    ternal Stress-Strain-State in Threedimensional Solids. IMEKO-TC 15-
    Conference in Moscow, Oct. 1989, to be published by Nova Scientia Pu-
    blisher, New York.

DEFORMATIONS AND STRAINS SURROUNDING WELDMENTS USING MOIRE INTERFEROMETRY

J.S. EPSTEIN
The Department of Civil Engineering
The Georgia Institute of Technology
Atlanta, GA, 30332-0355, USA

ABSTRACT

Research on surface deformation and strain fields surrounding flawed and intact weldments
will be reviewed. The principal tool of investigation will be moire interferometry. Experimental results will be compared with existing computational solutions. Three material systems
will be presented. A 304 stainless steel butt weld with a nickel filler containing diffusion
bond line cracks, a 9-2 1/4 chrome K weldment containing a chromium depletion zone and a
commercially pure (CP) / 6Al-4V titanium diffusion bond with and without cracks. Issues
such as field mixity due to heterogeneous materials, asymptotic field analysis for elasto -
plastic interface cracks and slip band instability emanating from a diffusion bond line will be
addressed.

INTRODUCTION

The mechanical response of weldments are in many ways more complicated than two phase
metal matrix or polymeric composite systems. Weldments typically contain at least four
phases of materials, the upper and lower base materials, the filler material and the heat affected zone separating base from filler. Further, unlike two phase composites, weldments
contain graded interfaces. All these factors yield fundamental scientific questions on first, the
basic understanding of elastic and elastic plastic interfaces, second, the fundamental validity
of elastic and elasto-plastic fracture parameters where extreme material heterogeneity exists
and third, the extent that local mixity of the crack tip field is induced due to heterogeneity of
the materials across the bond line.

Computational studies of the deformation and strain fields of weldments emphasizing material bond line heterogeneity are found by Blume[1], Chavez[2], Matic[3], and Prinaris[4]. In addition, Prinaris addressed the fields surrounding a bond line crack. Idealized crack/interface problems of two substrates with a zero thickness bond line have been conducted under elasto plastic conditions, by Shih[5] and Zywicz[6]. Both authors found that the shape of the asymptotic crack tip stress and displacement field scaled with the exterior far field loading and material properties across the interface raising questions regarding mathematical separability of the stress and displacement asymptotic fields. For explicit ratios of shear to tensile remote loading, crack closure was revealed deep within the crack tip plastic zone. This closure phenomena is similar in concept to the elastic interface crack problem. In addition, Shih found that the J parameter was path independent for selected regions local to the crack tip.

However, little experimental investigation has taken place to ascertain the elastic and elasto-plastic deformation and strain fields local to a weldment. Recent work was performed on elastic cracks by Chaing[7] using photoelasticity and Liechti[8] using Michelson interferometry. Liechti confirmed under pure shear loading that crack closure occurs. This paper addresses experimental fracture issues of the weldment under elasto plastic strains using moire interferometry.

EXPERIMENTAL RESULTS

MOIRE INTERFEROMETRY

The principal experimental tool employed is moire interferometry. Post[9] has given a thorough review of this method. Moire interferometry relies on two beam interference with a phase diffraction grating replicated on the specimen surface. Typically aluminized, this grating is brought to a null state before deformation by collecting the +1 and -1 diffraction orders from the respective impinging beams which emanate perpendicular to the grating and travel collinear in space. As deformation occurs, the +1 and -1 diffraction orders develop a mutual angle of intersection resulting in two beam interference. The interpretation of the resulting interferogram is the same as found in amplitude moire:

$$U = N/F \qquad (1)$$

where:

U is the displacement in the direction perpendicular to the grating lines and is constant along a fringe relative to a known boundary condition,

N is the fringe order assigned to a fringe,

F is the grating frequency usually in lines/mm (l/mm) or inch (l/in).

Typically the phase grating is crossed in orthogonal directions permitting observation of the U (X direction) and V (Y direction) fields. The information can be recorded on film or video.

The next three sub-sections outline work on three different material systems, (i), a 304 stainless steel butt weldment with nickel filler containing diffusion bond line cracks (ii), a 9 - 2 1/2 chrome K weldment containing a chromium depletion zone and (iii), a commercially pure (CP) / 6Al-4V titanium diffusion bond with and without cracks.

304 STAINLESS STEEL WITH BOND LINE CRACKS

Computational work by Shih[5] for small and large scale yield indicates that even with large material heterogeneity across the diffusion bond, the effect of plasticity tends to relax the mixed mode behavior approaching the crack tip. These results are novel when compared to elastic composite systems; for example, Liechti has found that elastic crack tip initiation and propagation occurs under highly mixed mode conditions.

Figure 1 details a loading geometry for a 304 stainless steel (304 SS) - Nickel filler metal butt weld utilized by Chavez, Deason and Epstein[10]. The weld was formed by a gas tungsten metal arc process. Strain gages were bonded to the nickel filler and 304 stainless base metal in addition to 600 l/mm gratings. Figure 2 details the strain gages verses moire interferometry for the stainless steel and nickel. Figure 2 reveals that the nickel yields before the stainless steel and has an overall lower strain hardening response thus leading to considerably more deformation in the nickel. Figure 3 details the Y displacement field for the weldment under large scale yield. Evident in this figure are three diffusion bond line cracks. Crack 1 is essentially an edge crack. Crack 2 lies 2 mm to the right of crack 1. Crack 3 lies 0.75 mm from crack 2. Figure 4 details the ratio of fringe orders at a set radius for the nickel and 304 SS taken at 90 degrees from the crack tip. For cracks 1 and 2, the mixity ratio approaches 1 i.e. mode I loading occurs in the limit approaching the crack tip. For crack 3 which lies close to crack 2, the mixity ratio actually increases indicating an interaction between the asymptotic fields of cracks 2 and 3. The results for cracks 1 and 2 confirm the work of Shih. The results for crack 3 remain unconfirmed computationally.

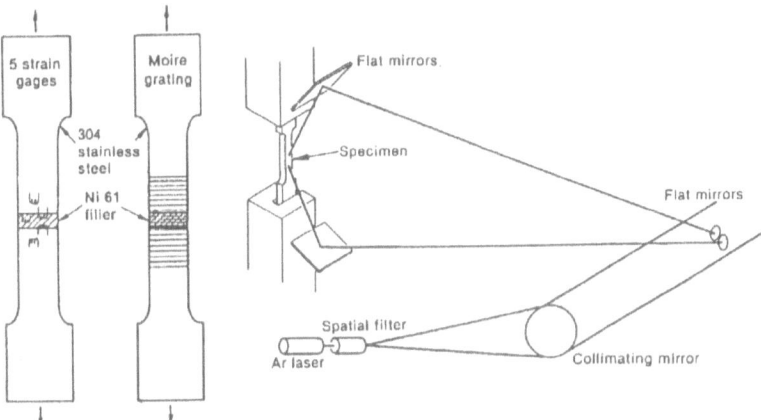

Figure 1. The 304 SS/ Nickel tension test.

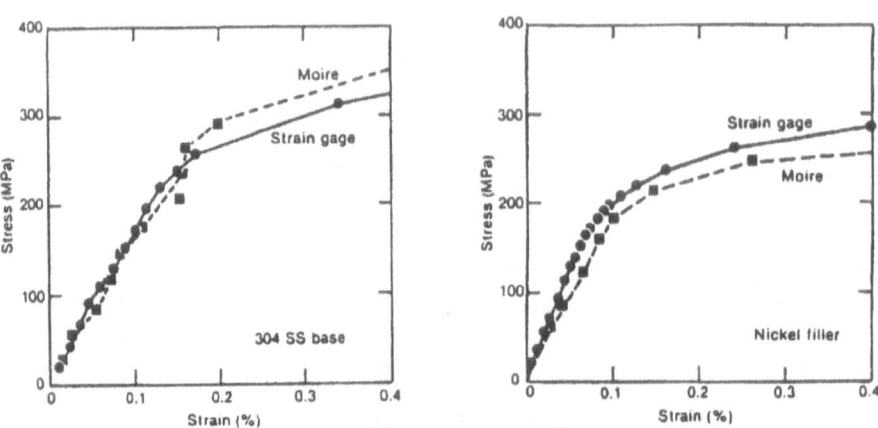

Figure 2

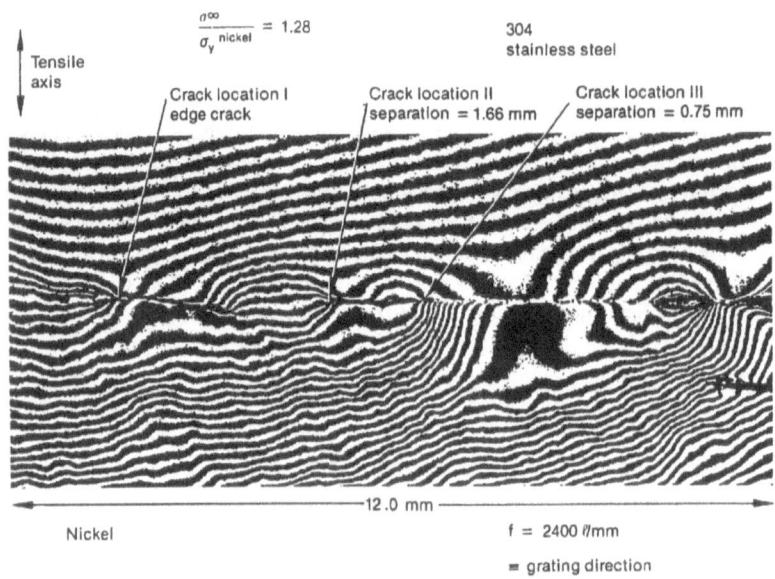

Figure 3

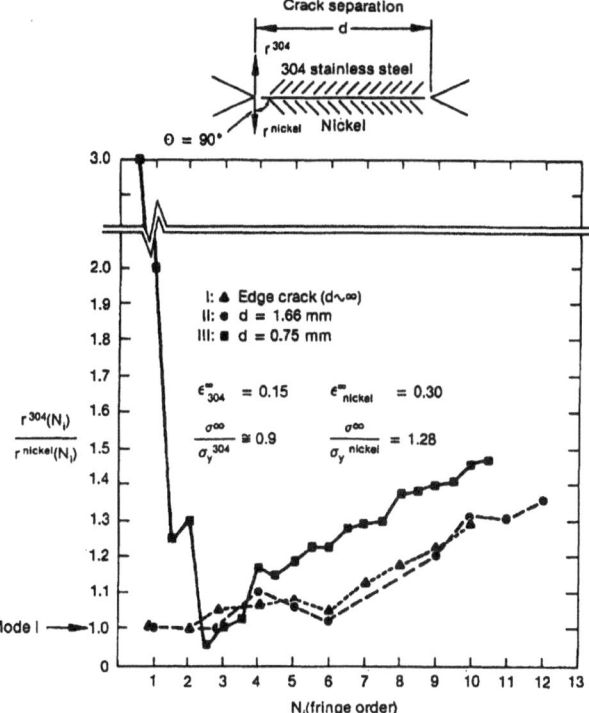

Figure 4. The ratio of fringe orders for the nickel and 304 SS .

9 - 2 1/4 CHROME K WELD

Figure 5 details a enlargement of a 9 - 2 1/4 chrome K weld whereby the chromium gradient between base metals results in a "soft" or a carbon depletion zone. This zone is unique to this material combination and cannot be reproduced in standard Gleeble testing which only simulates the thermal cycle found during a weld operation. Hardness testing of this zone indicates that it yields early relative to the base materials and subsequently strain hardens before the base metals. The result of this relatively wide specimen width to zone thickness is a rigid strip phenomena while the other materials yield. The light bands surrounding the filler metal (which is deposited at an angle) in figure 5 are heat affected zones (HAZ). Figure 6 details the X displacement field for this geometry under tension. Figure 6 details the truly complicated nature of the deformation in a typical weldment whereby 5 materials are undergoing simultaneous deformation. The 2 1/4 chrome yields early showing high amounts of uniform X contraction. The heat affect zone above the 2 1/4 chrome material shows little deformation

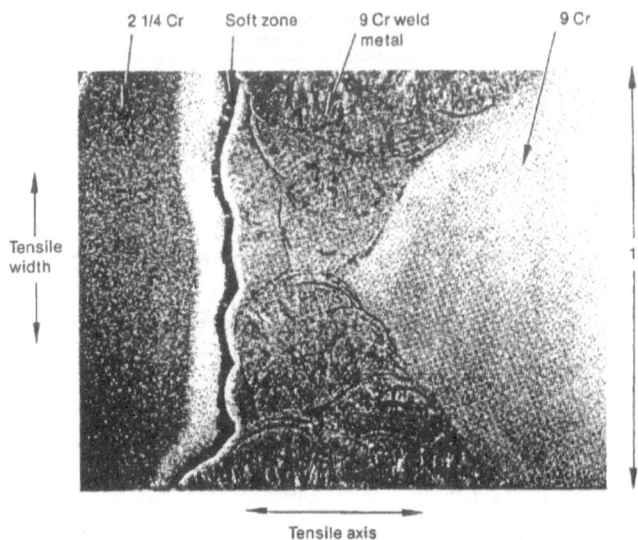

Figure 5. An enlarged section of the 9 - 2 1/4 Chrome K weld.

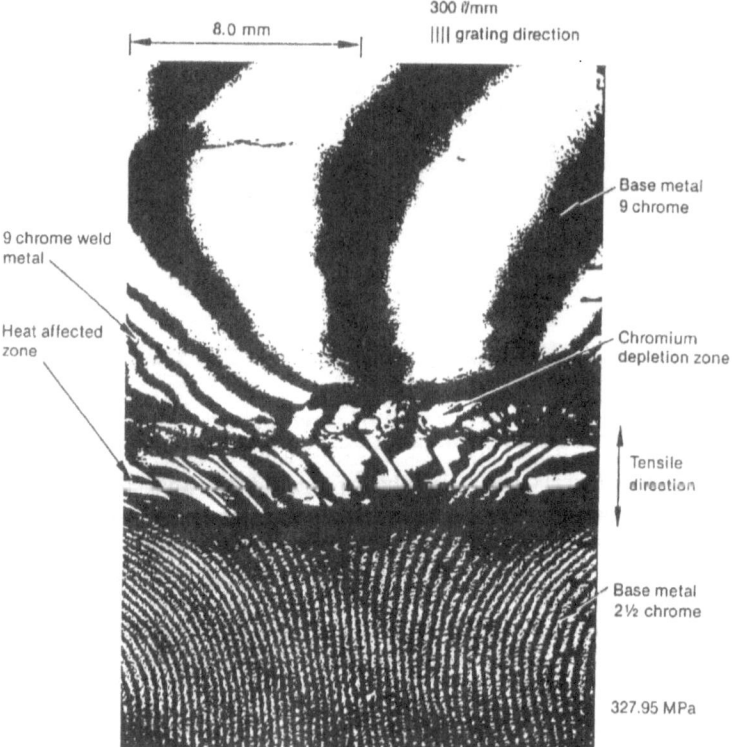

Figure 6. The X displacement field for the K weld.

with high amounts of inter - granular shear by the sharp angular deviations of the bulk fringes. The carbon depletion zone reduces even further in contraction essentially acting as a rigid strip barrier to the 9 chrome upper base metal. Later work by Chavez[2] combined moire interferometry with continuum finite elements to iteratively determine the uniaxial stress strain response of the carbon depletion zone. As noted standard thermal simulation gleeble testing cannot extract the soft zone uniaxial stress strain curve as it results solely on the relative chromium concentration difference across the weldment.

COMMERCIALLY PURE TITANIUM DIFFUSION BONDED TO 6AL-4V TITANIUM

The last two examples were determined from essentially as received weldments. Recent work has concentrated on fundamental studies of cracked and non - cracked bimetallic systems whereby on one material yields while the other remains elastic. Diffusion bonding was chosen to essentially form a mathematical interface eliminating the third filer material. The system of commercially pure titanium (CP Ti) diffusion bonded to 6Al-4V titanium (6Al-4V Ti) was chosen to eliminate processing reduced residual stresses. The CP Ti yields at 403.6 MPa while the 6Al-4V yields at 837 MPa. For this work the CP Ti yields while the 6Al-4V remains essentially elastic.

Figure 7 details a generalized plane stress sample of the outlined diffusion bond system. Cross 600 l/mm gratings were replicated with an effective multiplication of 2. The specimen was mounted in a specially designed interferometric loading fixture to eliminate rigid body rotation. Specimen loading occurred under fixed displacement with an extensometer monitoring overall elongation and a load transducer monitoring external load. Figure 8 details a series X displacement fields for this specimen. At higher ratios of remote load to CP yield, shear bands can be observed emanating from the diffusion bond region. The middle of the bond remains relatively shear free as evidenced by the straightness of the X displacement fringes. Figures 9 a,b and c detail the X,Y and shear strain fields determined at a remote stress to CP yield stress ratio of 0.7. This data was obtained by full field digitization and

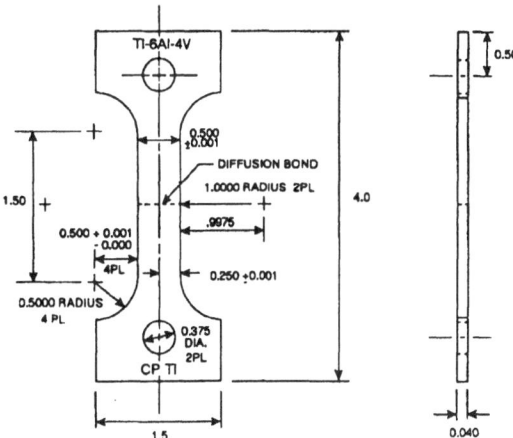

Figure 7. The CP/6Al-S4V diffusion bond specimen.

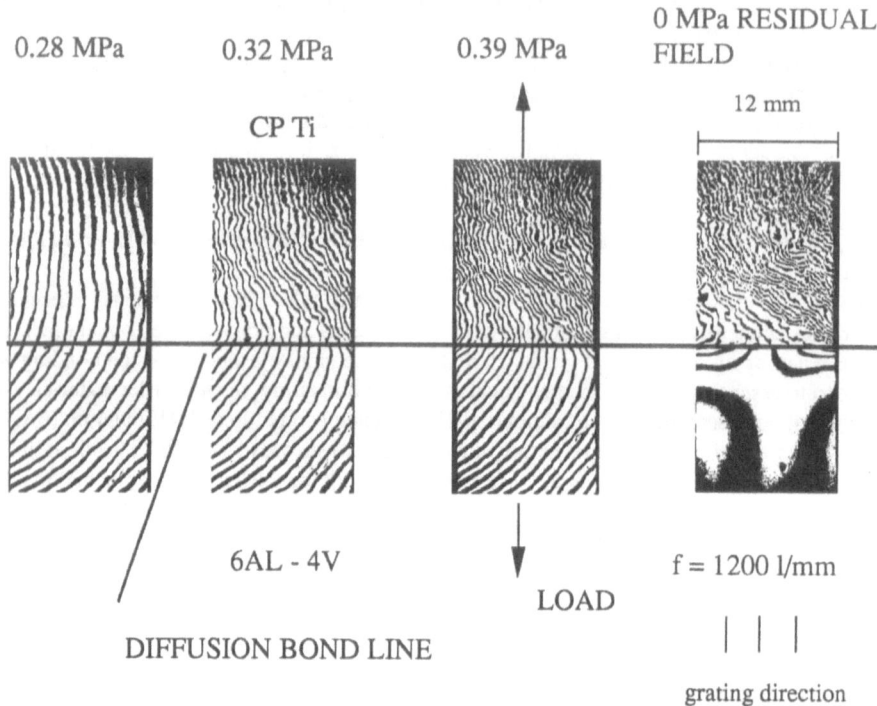

Figure 8. The X displacement field for the CP Ti / 6Al -4V diffusion bond specimen.

splining of the interferogram data. The resulting splines are surfaces of fringe order as height verses X and Y position. Derivatives are then taken of these surfaces to obtain surfaces of strain. Topographic contours are then generated to yield figures 9. The shear strain topographic contour is the addition of the cross derivative of the surface in the X and Y directions. The X strain, figure 9a details good symmetry indicating uniform contraction. Some bending of the X contours occurs due to shear bands crossing the specimen causing bending even under uniform tension. Figure 9b details the Y strain field showing a gradual transition at the diffusion bond line to the uniform far field strains. Figure 9c details the shear strain field showing strong shear band instabilities emanating form the right hand corner of the diffusion bond. The center portion of the bond line is essentially shear free due to symmetry. These results indicate that if a crack were to inserted in the central portion of the bond line, then high triaxial conditions would prevail with essentially mode I loading. Further it is evident that bond failure would start at the corner due to the differing yield properties of the two materials.

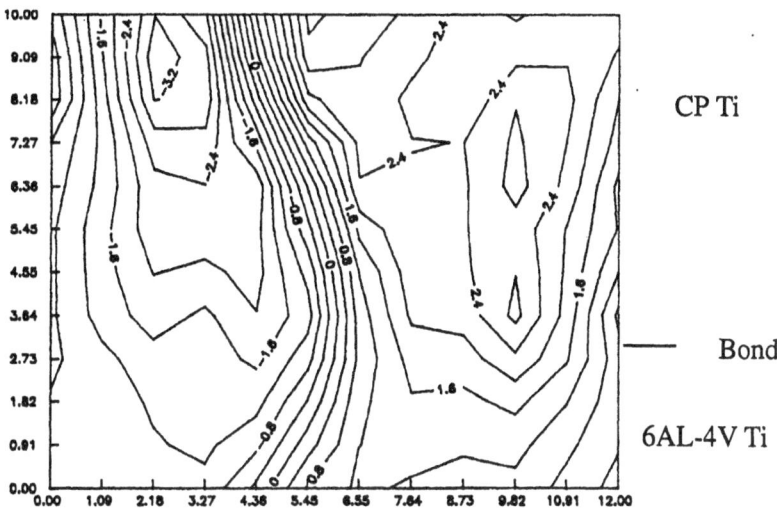

Figure 9a. The X strain field for 0.39 Mpa (all dimension in mm).

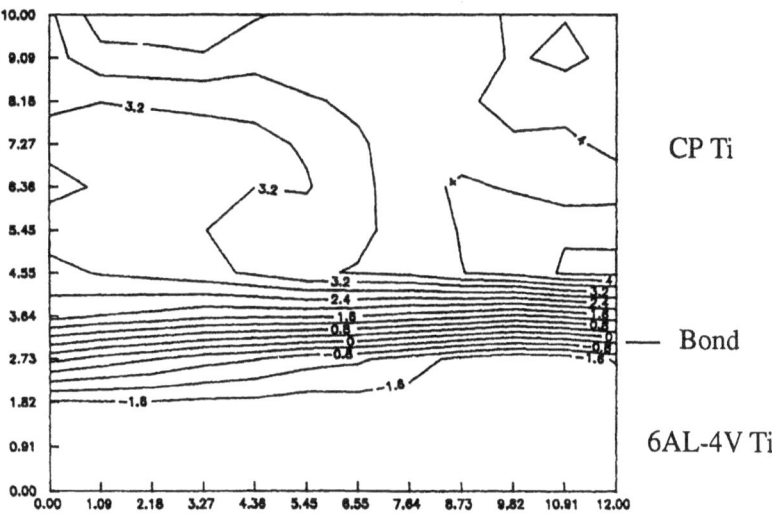

Figure 9b. The Y strain field for 0.39 Mpa.

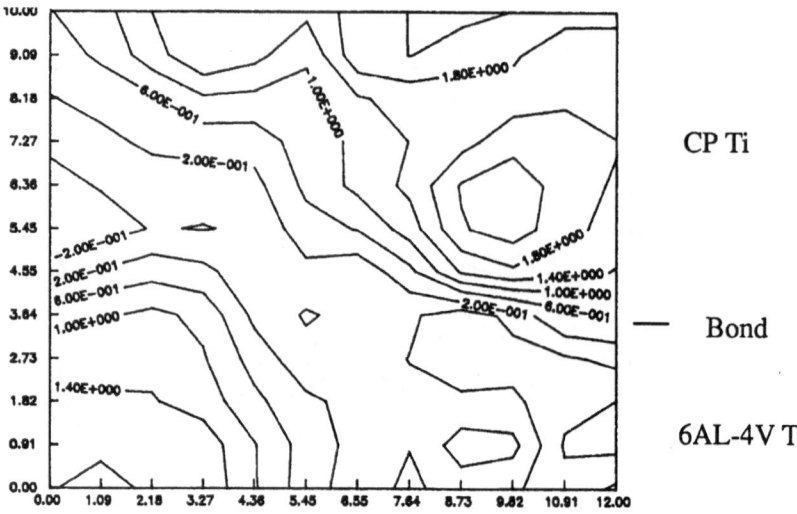

Figure 9c. The shear strain field for 0.39 Mpa.

Figure 10 details the X and Y displacement field for a diffusion bond specimen of the same material systems with a central through crack (Epstein[11]). In this case, the specimen is 3 mm thick instead of the 1 mm generalized plane stress specimen. The X and Y displacement fields were analyzed for three aspects of the near tip asymptotic field, (i), the plastic zone emanating form the crack tip, (ii), the local strain field in terms of polar coordinates from an X - Y cartesian grating and the near tip singularity order. Using a J_2 yield criterion, figures 11 and 12 detail the plastic yield zones for a far field stress to CP yield stress ratio of 0.77 and 0.83. The lower cusp portion of the CP yield zone lying at the diffusion bond line is unique to the interface crack and qualitatively agrees with the results of Zywicz[6]. Figure 13 details the radial and theta displacement field for the CP and 6Al-4V titanium by converting the carte- sian X - Y fields numerically. The angular variation of the fields is in qualitative agreement with the work of Shih[5]. Figures 14 and 15 detail the radial, theta and radial-theta strain fields for two different radii from the crack tip. For the inner radius, figure 14, peak strains exceed 2% close to 90 degrees while the 6Al-4V strain reaches 0.5%.

X Field   CP Ti   Y Field

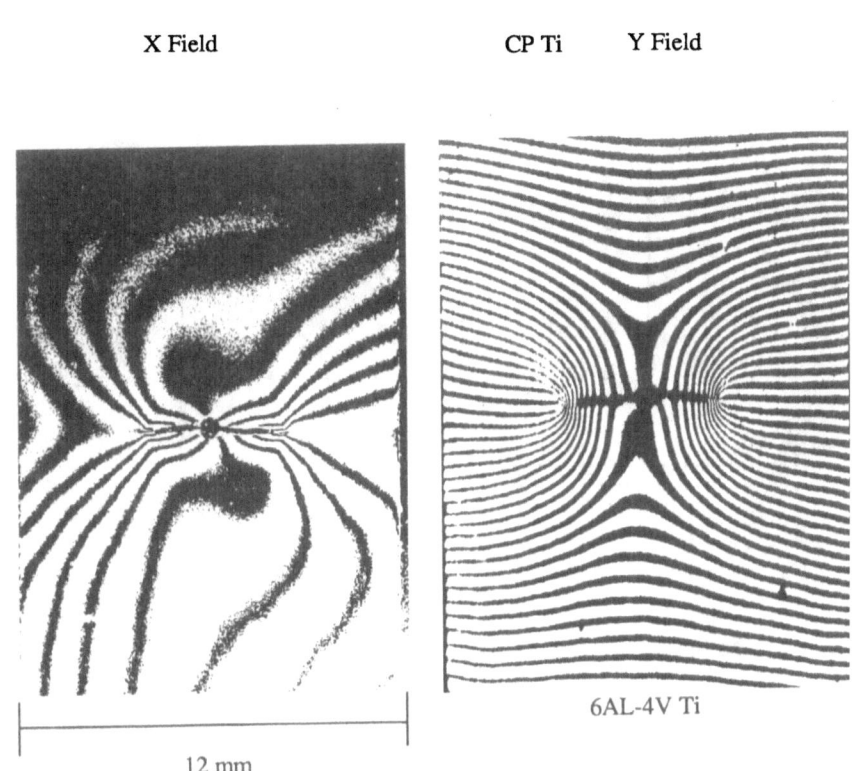

12 mm

f = 2400 l/mm

Figure 10. The X and Y displacement fields for a through crack diffusion bonded plate.

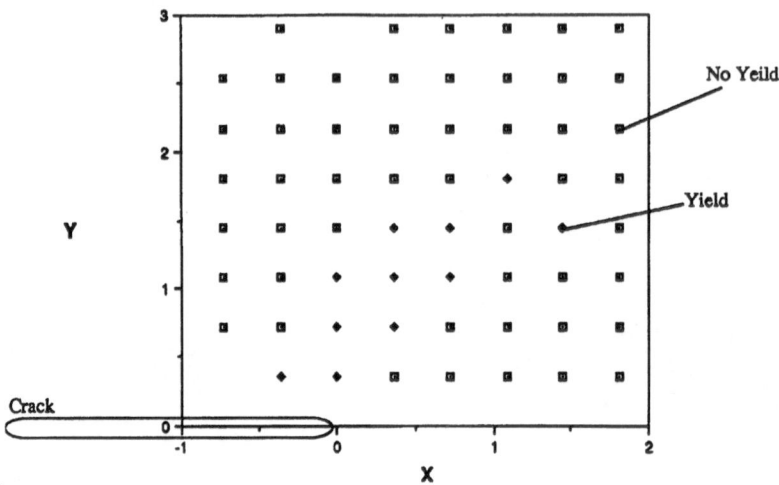

Figure 11. The interfacial yield zone at a remote stress to CP Ti yield stress ratio of 0.77.

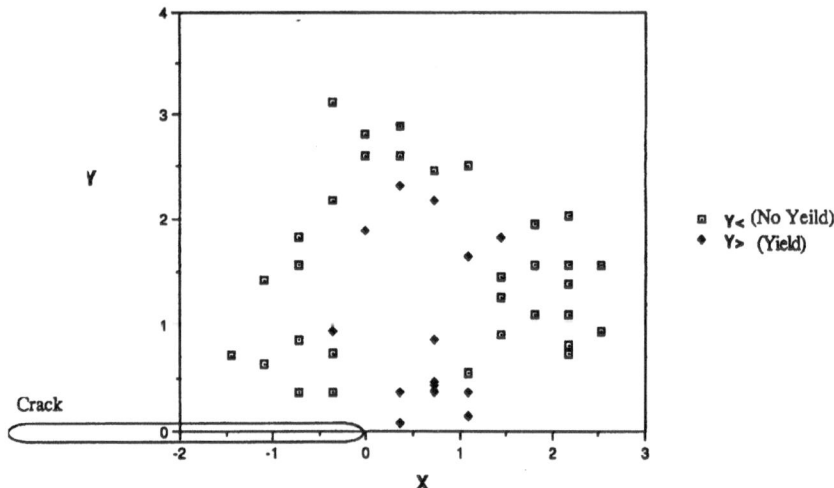

Figure 12. The interfacial yeild zone at a remote to CP Ti yeild stress ratio of 0.83.

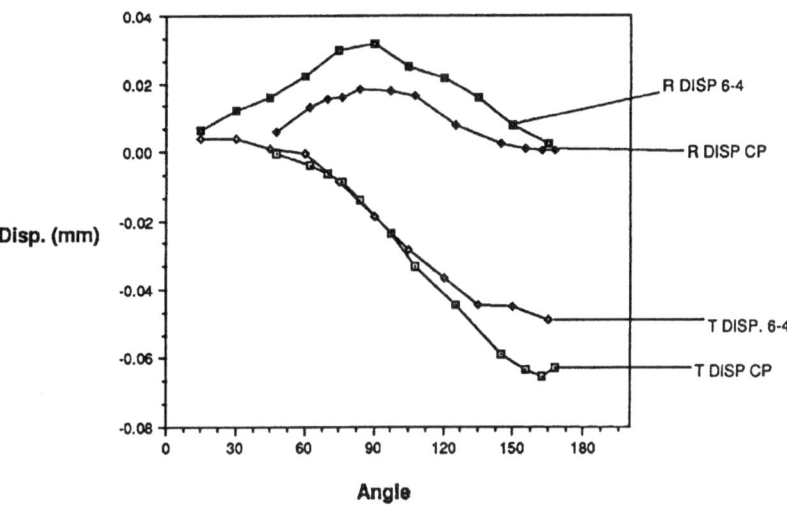

Figure 13. The radial and theta displacement fields for the diffusion bond line crack at a remote stress to CP Ti yield stress ratio of 0.83. The radius, R, to crack length, a, ratio is 0.09.

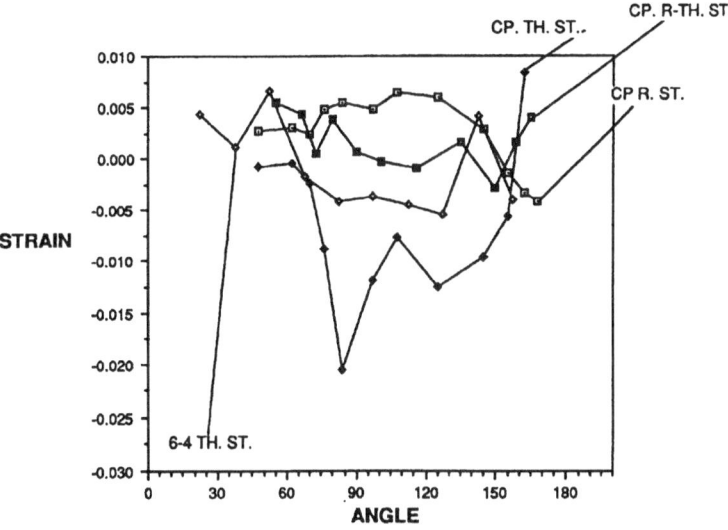

Figure 14. The R, theta and r-theta strain for a remote stress to CP Ti yield stress ratio of 0.83 and an R/a ratio of 0.09.

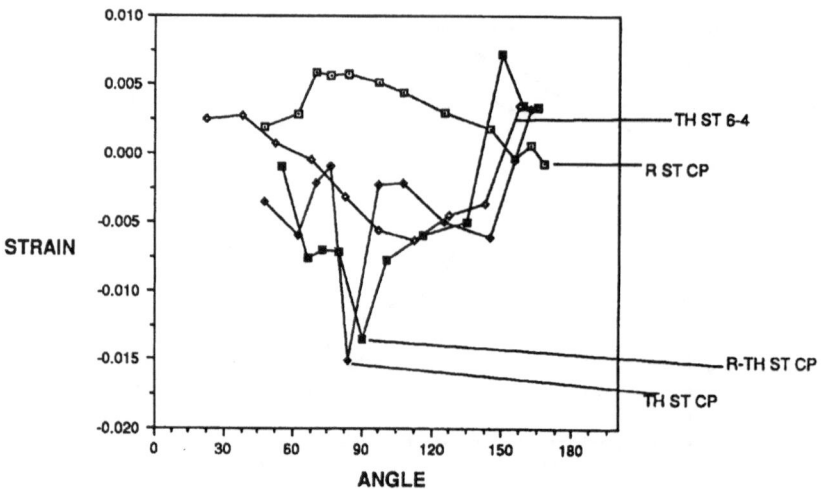

Figure 15. The R, theta and R-theta strain fields for a remote stress to CP Ti yiled stress ratio of 0.83 and an R/a ratio of 0.21.

Figures 16 and 17 detail Log(U) verses radius from the crack tip for the 6Al-4V and CP Ti respectively. Assuming a separable form of the displacement field as:

$$U = r^{\lambda_u^1} f(\theta)$$

(2)

then:

(3)

$$\log(U) = \lambda_u^1 \log(r) + \log(f(\theta))$$

when plotted as the log(U) verse Log(r) the slope linear line should yield the displacement first eigenvalue. For HRR fields then first eigenvalue is of the form:

$$\frac{1}{n+1}$$

where: n is the Ramberg - Osgood strain hardening coefficient,

for the CP Ti, $11 < n < 14$ which should theoretically yield a eigenvalue between 0.063-0.083. For elastic materials the slope is the classical 1/2. Figure 16 details the elastic 6Al-4V mate-

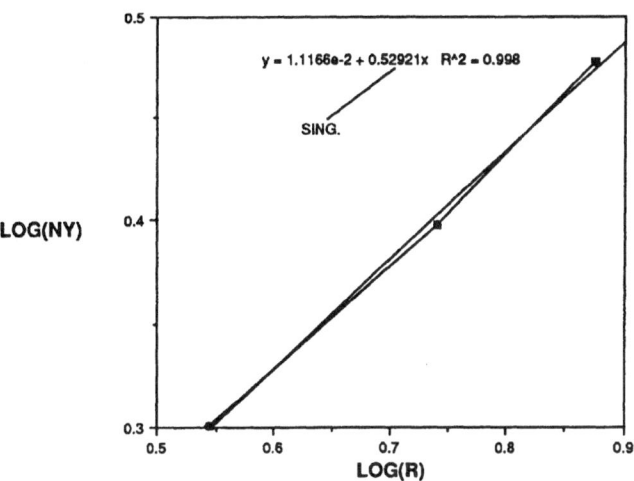

Figure 16. The **singularity** order for the 6Al-4V Ti side of the through crack.

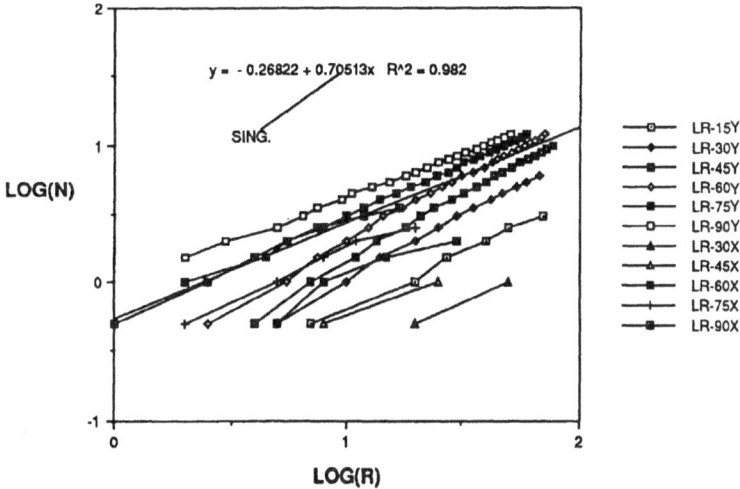

Figure 17. The singularity order for the CP Ti side of the through crack.

rial showing that indeed the elastic field singularity order prevails as the slope is 0.54. Figure 17 shows that the average slope of the yielding CP is 0.7, grossly different form the HRR values. Two possible explanations are currently offered for the variance. First the center cracked geometry is one of minimal tri or biaxial constraint. Even though the plastic zone has not reached the free edge of the material, the zone growth is not constrained with increasing external load. The result of this low constraint geometry is that the asymptotic field never forms in an HRR fashion. Second, the initial notch has a finite root radius. McMeeking[12] has noted that a finite root radius will result in HRR fields only very localized to the crack tip perhaps outside the limits of detection by moire interferometry. Current work is being conducted with a bending geometry to force higher triaxial constraint as well as electro discharge machined notches for a smaller root radius.

CONCLUSIONS

Localized displacement fields have been presented for a number of fundamental and practice weld geometries and material systems. Agreement with computational trends is shown for mixity effects local to the crack tip. Methodologies have been outlined for converting the full field data to contours of strain. Strain fields have been examined for a fundamental generalized plane stress geometry showing shear band instabilities propagating from the diffusion bond line. Near tip asymptotic fields have been examined for the diffusion bond line crack geometry. Conversion of cartesian X-Y strain to radial strain fields has been presented. Qualitative trends of angular variations agreeing with HRR fields in strain has been outlined. However, the singularity order of these field differs from the HRR order. Low constraint and blunt notch effects are attributed to this large difference. Current work is utilizing higher constraint bend geometries. The displacement field for a typical field weldment consisting of 9 - 2 1/4 chrome K geometry was shown revealing the complicated nature of deformation in practical weldments.

ACKNOWLEDGEMENTS

This work was support under the U.S. Dept. of Energy through DOE contract no. DE-AC07-76ID01570, Dr. O.A. Manley program manager. Thanks is given to Dr. S. Graham of Materials Engineering Associates, Mr. R. Lloyd of EG&G Idaho Inc. for helpful comments on the diffusion bond results and Mr. K. Perry for data reduction.

BIBLIOGRAPHY

1.   Blume, J.A. and Shih, C.F. , "The Singular Behavior of a Bimaterial Strip", Proceedings, 1988 ASCE Engineering Mechanics Specialty Conf., R.A. Heller, ed, held at Virginia Polytechnic Inst., p.72, May 23-25, 1988.

2. Chavez, S.A., Determination of Mechanical Properties in the heat Affected Zone of a Dissimilar Metal, Ph.D. Thesis, Univ. Idaho, March 1989.

3. Matic, P. and Jolles, M., The Influence of Weld metal Properties, Weld Ge ometry and Applied Load on Weld System Performance, Naval Research Laboratory Memorandum Report 5987, 1988.

4. Prinaris, A.A. and Saouma, V.E., Elasto Plastic Fracture of Welded Plates, Bureau of Mine Structural Research Series Report 88-90, March 1988.

5. Shih, C.F. and Asaro, R.J., "Elastic Plastic Analysis of Cracks on Bimaterial Inter faces- Part I Small Scale Yielding", ASME J. Applied Mech., vol 55, 1989, pp. 299-319.

6. Zywicz, E. and Parks, D.M., "Elastic Yield Zone Around and Interfacial Crack Tip", ASME J. Applied Mech., vol. 56, pp. 577-584, 1989.

6. Chaing, F.P. , Lu, H., and Yan, X.T., "Photoelastic Analysis of a Crack at a Bimaterial Interface", Proceedings Int. Conf. on Frac. - 7, A.Salama, ed., vol. 4, pp. 3063-3072, 1988.

8. Liechti, K.M. and Chai, Y.S., "Interface Crack Initiation Under Biaxial Loading", Pro ceedings, 1989 Soc. for Exp. Mech. Spring Meeting, pp. 954-961, 1989.

9. Post, D., "Moire Interferometry", Ch.7, Handbook on Experimental Mechanics, A. Kobayashi, ed., McGraw Hill, 1988.

10. Chavez, S.A., Deason, V.A., and Epstein, J.S., "Use of Moire Interferometry in Weldments", Proceedings, ASM Int. Conf. on trends in Welding Research, pp. 533-537, 1986.

11. Epstein, J.S., Graham, S.A. and Reuter, W.G., "Asymptotic Crack Tip Fields in Diffu sion Bonded 6Al-4V/Commercially Pure Titanium Subjected to Far Field Tension", Proceedings, 1990 Southeastern Conf. on Theoretical and Applied Mechanics, (SEC TAM), in press, S. Hanagud, ed. Atlanta, GA, USA, 1990.

12. McMeeking, R.M., "Finite Deformation Analysis of Crack Tip opening in Elastic Plastic Materials and Implications for Fracture", J. Mech. Phys. Sol., vol. 25, pp. 357-381, 1977.

STRAIN AND SHAPE MEASUREMENTS BY SCANNING MOIRE METHOD AND FOURIER TRANSFORM MOIRE METHOD

YOSHIHARU MORIMOTO

Department of Mechanical Engineering,
Faculty of Engineering Science,
Osaka University

Toyonaka, Osaka 560, JAPAN

ABSTRACT

Image processing is a powerful technique for analyzing the fringe patterns obtained by whole-field methods such as geometric moire, holographic interferometry, photoelasticity, etc. In addition to eliminating tedious and time consuming calculations, image processing creates the possibility of new methods of analysis. In this paper, two new moire methods using image processing are presented. One is the scanning moire method in which the scanning lines of a TV camera or the sampling points of an image processor interfere with the model grating lines to create a moire pattern. The other is the Fourier transform moire method which uses the shifted first harmonic of the Fourier spectra of a model grating to yield a sinusoidal brightness distribution corresponding to fractional fringe orders. These methods produce clear moire patterns and permit automated high speed analysis. The outline of these methods and some applications are given.

INTRODUCTION

The conventional analysis of a fringe pattern obtained by geometric moire, moire interferometry, holographic interferometry, etc. requires considerable time for developing photographic films and measuring the coordinates of fringe points. Furthermore, although the mismatch moire method is useful for wide-range strain measurement, it is troublesome to change the master grating pitch and to adjust the position of the master gratings. By introducing image processing [1], the analysis of fringe patterns becomes easier. However, the inconvinience of developing the photographic films and of the changing the pitch for the mismatch method remains.

In this paper, two new moire methods are developed in order to resolve these problems and to analyze the fringe pattern faster and more accurately. One is the scanning moire method [2-4] in which a moire fringe pattern appears as the interference between model grating lines and scanning lines of a TV camera. By thinning-out the scanning lines by image processing, a wide range of strain levels can be analyzed. The other is the Fourier transform moire (and grid) method (abbreviated to FTMM or FTMGM) [5-8]. Since the phase of the grating lines or the fringe lines are used in this method and it is obtained for all pixel points, an accurate displacement distribution is obtained.

The scanning moire method and the FTMGM are explained as the interference between the pixels of an image processor and the model grating lines. Both methods are useful if the deformed model grating or the nearly equally spaced lines and the fringes of moire interferometry or holographic interferometry including a carrier pattern can be recorded by a TV camera. In this paper, these images are analyzed with a personal computer, and the outline of the methods and some applications are shown.

SCANNING MOIRE METHOD

In the conventional geometric moire method, a moire pattern appears as the interference between a model grating and a master grating. However, a moire pattern also appears when the model grating lines are sampled by the scanning lines of a TV camera or by the pixels of an image processor. If the sampling points of the image processor are regarded as the master grating with X and Y-directional crossed lines, the geometric relationship between the model grating lines, the sampling points and fringe lines is obtained in the same way as in the conventional geometric moire. The phenomenon of moire fringe appearance by the scanning moire method can be explained by using sampling theory and the Fourier transform of the image [3], or by using the geometric relationship [4].

The phenomenon in which moire fringes appear by a thinning-out process in which the sampling points of an image processor are thinned-out is explained as follows. Figure 1 illustrates the schematic explanation of the appearance of moire fringes by the scanning moire method. Figure 1(a) shows the positions of the sampling points of an image processor as black points. Figure 1(b) shows the undeformed grating lines drawn or projected on a specimen. Before deformation, the TV camera is adjusted so that the grating lines are perpendicular to the horizontal scanning lines. Each line of the model grating coincides with each corresponding horizontal sampling point of the image processor. The model grating is sampled at the points of Figure 1(a) by the image processor. If a sampling point is on a black line of the model grating, a black point image is obtained. If it is on a white line, a white point image is obtained. Figure 1(c) shows the resultant image obtained by these processes. In this case, all point images are black. It does not show a moire pattern.

In a case where the specimen has deformation, and the strain of the specimen is small as shown in Figure 1(d), the resultant image obtained by sampling is shown in Figure 1(e). The pitch of the model grating is nearly equal to the pitch of the sampling points so that the white point images and the black point images appear in groups which produce a moire fringe pattern.

In a case of large deformation as shown in Figure 1(f), the pitch of the model grating is about 2.6 times larger than that of the scanning lines. The sampled images of Figure 1(f) are shown in Figure 1(g), in which a moire pattern cannot be recognized. However, if the point images from alternate horizontal sampling points are thinned-out, as shown in Figure 1(h), the images are separated into white and black groups showing moire fringes. Conversely, Figure 1(i) shows the image which consists of the line images thinned-out in Figure 1(h), and different with Figure 1(h) by π in phase in terms of both the thinning-out and the moire fringes. Moreover, if every horizontal third sampling point is picked up from Figure 1(g), the pattern shown in Figure 1(j) is obtained and is a different moire pattern than Figure 1(h). In Figure 1(j), the first sampling point of the continuous three points of Figure 1(g) is selected. However, if the second sampling point of the three points is selected, the image in Figure 1(k) is obtained. Figure 1(l) shows the image obtained by selecting the third point from the three

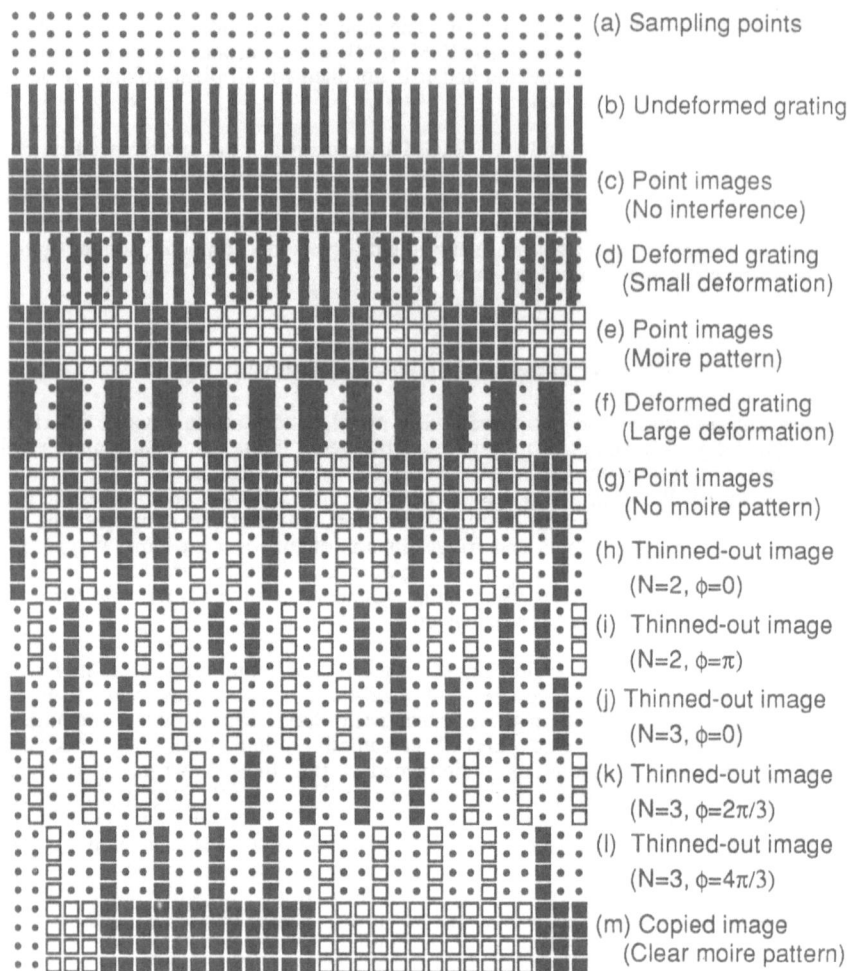

(a) Sampling points

(b) Undeformed grating

(c) Point images
(No interference)

(d) Deformed grating
(Small deformation)

(e) Point images
(Moire pattern)

(f) Deformed grating
(Large deformation)

(g) Point images
(No moire pattern)

(h) Thinned-out image
(N=2, φ=0)

(i) Thinned-out image
(N=2, φ=π)

(j) Thinned-out image
(N=3, φ=0)

(k) Thinned-out image
(N=3, φ=2π/3)

(l) Thinned-out image
(N=3, φ=4π/3)

(m) Copied image
(Clear moire pattern)

Figure 1.   Appearance of moire fringes by scanning moire method: The images sampled by
sampling points show moire patterns. Thinned-out images show different patterns
according to the thinning-out index N and the phase φ

points. The phase of the moire fringes is according to the phase of the thinning-out. The
direction that increases the fringe order can be determined by increasing the phase of the
thinning-out. The direction of the fringe movement when the phase of the thinning-out is
increased indicates the direction increasing the fringe order. If the fringes in Figure 1(j) are
regarded as having the continuous integer fringe orders, the fringes in Figure 1(k) have the
order of the integer fringe order + 1/3. In this case, the fringes move to left and the fringe
order increases toward the negative direction. Therefore if the phase of the thinning-out is
changed, fractional fringe orders are obtained.

Finally, if all of the sampling point images which were thinned-out in Figure 1(l) are
replaced by the proceeding point images which are picked up in Figure 1(l), the moire fringes
becomes clearer as shown in Figure 1(m). In this process, however, the center positions of the
resultant fringes are shifted by a half of N pixels. These thinning-out processes corresponds

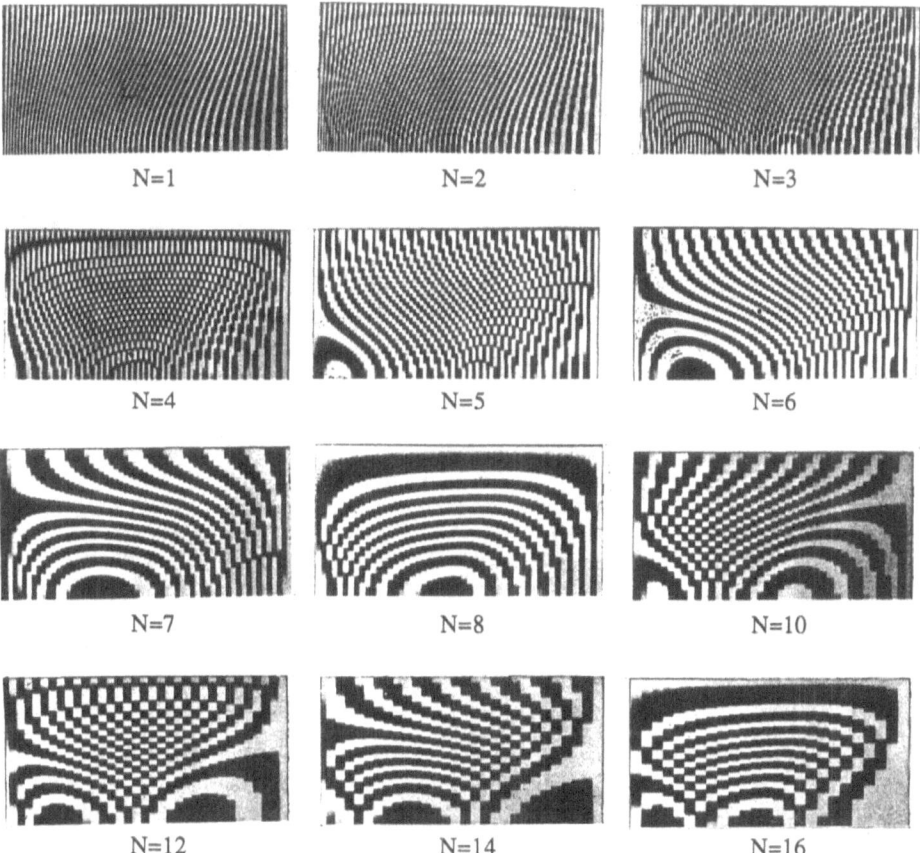

Figure 2. Moire patterns according to thinning-out index N

to a mismatch in the conventional geometric moire method. Although Figure 2 shows only two brightness values of white and black, a gray halftone image is usually obtained. The contrast of the moire fringes depends on the size of the sampling point and the brightness distribution of the model grating.

When the pitch of the sampling points of an image processor is considerably smaller than that of the model grating lines, the image shows only the grating lines and no clearly discernable moire pattern. In such a case, if the sampling points are periodically picked up, the image shows a moire pattern. By changing the pitch of picking up (i.e. the thinning-out index N), different moire patterns can be obtained. This corresponds to the mismatch moire method. It is easy to change the pitch using image processing.

Figure 2 shows a simulated example. Figure 2(a) is the original image of a deformed grating. The grating lines before deformation are parallel and equally spaced. The pitch of the grating lines is 4 times of that of the sampling points. This grating yields deformation as shown in Figure 2(a). The maximum Euler strain is 50%, and the minimum Euler strain is -100 %. The average strain is 0.

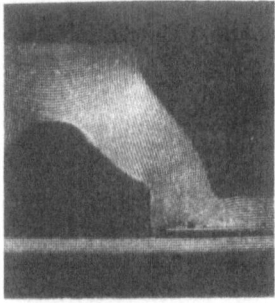
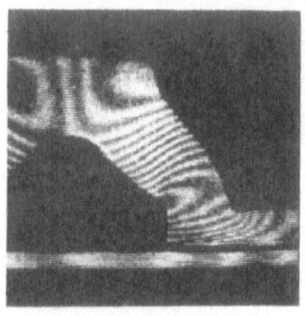

(a)Deformed model grating   (b) Moire pattern thinned-out   (c) Moire pattern thinned-out
                                      in X-direction   in Y-direction

Figure 3. Moire pattern of rubber fender

Every Nth pixel is picked up and copied on the following N-1 thinned-out pixels. N is called as the thinning-out index. If the pixels are thinned-out by different N values, different moire fringe patterns appear as shown in Figure 2. The strains are calculated from the fringe spacing. Where the fringe spacing is large, the strain data density is low. If only one image obtained from one N value is used in analysis, the distribution of the strain data is not uniform. If N is changed, the distribution changes. By combining the strain data obtained from many values of N, almost uniform and high density distribution of the strain data is obtained. The master grating pitch can be easily changed by thinning-out the sampling points. The moire patterns obtained by different thinning-out indeces, N, provide abundant information for the analysis of strain distributions which are calculated from the discrete data by spline functions.

In the case of a two-dimensional analysis using a cross grating with X and Y-directional lines on a specimen, the master grating direction can be easily changed by changing the direction of thinning-out. If the image is thinned-out in the X-direction, the X-directional moire fringe pattern is obtained and similarly for the Y-direction.

Figure 3 shows an example for a rubber fender of a quay to protect a ship from shock. A cross grating is printed on its end surface. Both the X and Y-directional pitches of the grating are 1 mm/line. The right half of the rubber fender is recorded by a TV camera and digitized by an image processor. The image size is a square of 480x480 pixels. The TV camera is set so that the straight lines of both gratings are parallel to the X and Y-axes, respectively, and each grating pitch is with 4 pixels in length before deformation. In other words, the pitch of this grating is 4 times wider than that of the sampling. A vertical compressive force is applied through rigid flat steel plates from the top and bottom of the specimen. The fender is compressed by Y-directional load of 13.6 kN. Figure 3(a) is the deformed grating. Figure 3(b) is the thinned-out image with a thinning-out index N=4 in X-direction. It shows the X-directional equal displacement contour lines. Figure 3(c) is the thinned-out image with N=4 in Y-direction. It shows the Y-directional equal displacement lines.

The scanning moire method makes it possible to synthesize an image processing system showing a moving moire pattern in real time because the calculations are simple and therefore fast. A real time displacement and strain analyzer has been developed using this method [9].

FOURIER TRANSFORM MOIRE METHOD

When two waves of sounds are superposed, interference or beating occurs. The moire pattern created by two of gratings corresponds to beating of the two sound waves and the beat frequency is the difference between the frequencies of the two sound waves. Correspondingly, the spatial frequency of moire is the difference between the spatial frequencies of the two

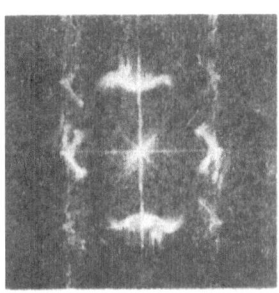

(a) Fourier spectra of Figure 3(a)

(b) Extracted and shifted X-directional first harmonic of figure 4(a)

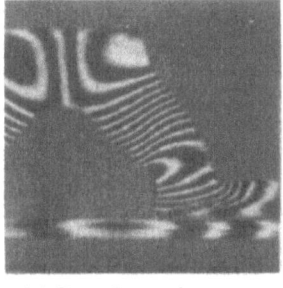

(c) Complex moire pattern (120 pixels shift, real part)

(d) Complex moire pattern (120 pixels shift, imaginary part)

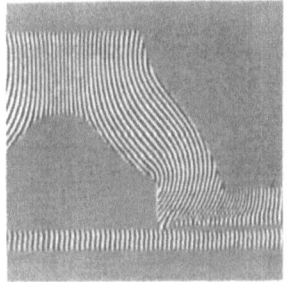

(e) Complex moire pattern (60 pixels shift, real part)

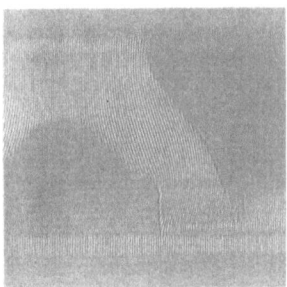

(f) Complex moire pattern (0 pixels shift, real part)

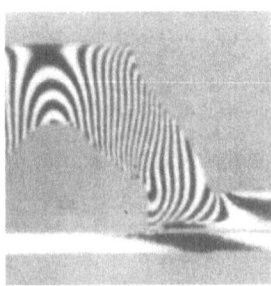

(g) Y-directional complex moire pattern (120 pixels shift, real part)

Figure 4. Analysis of rubber fender by FTMGM

gratings. From this idea, we have proposed the Fourier transform moire method (FTMM) and the Fourier transform moire and grid method (FTMGM) [5-8]. In this method, the frequency spectra of the deformed model grating is calculated using the Fourier transform. The constant frequency of the master grating is subtracted and by performing the inverse Fourier transform of the result, a complex moire pattern is obtained. By using only the first harmonic from the spectra, a very clear moire pattern with a sinusoidal brightness distribution is obtained. The phase of the brightness function of the fringe corresponds to the displacement of the grating. By analyzing the phase, the displacement and strain distributions, or the shape of an object can be determined. The theoretical treatment is detailed in References 5-8. An example of this method is now given using the image of Figure 3(a).

Displacement and Strain Measurement

Computation of discrete Fourier transform (DFT): The Fourier power spectra of the deformed grating are shown in Figure 4(a). In this figure, the horizontal and vertical directions show the X-and Y-directional frequencies Ω_X and Ω_Y, respectively. The origin of the frequency is centered. The brightness intensity of this figure corresponds to the power intensity of the frequency.

Extraction and shifting of first harmonic: In this case, the image size is 480x480 pixels and the pitch is 4 pixels so that the first harmonic frequency is 120. The first harmonic frequency of the X-directional grating before deformation appears at the point of 120 pixels in Ω_X direction and 0 pixels in Ω_Y direction from the frequency origin. The first Ω_X directional harmonic frequency of the deformed grating exists about the point (120,0). The first harmonic is extracted using a circular shape Hamming window filter whose radius is 90 pixels and is centered at the point (120,0). It is then shifted by 120 pixels (toward the origin, i.e. nonmismatch moire method), 60 pixels (i.e. mismatch moire method) or 0 pixels (nonshift, i.e. grid method). Figure 4(b) shows the first harmonic after shifting by 120 pixels toward the origin.

Inverse discrete Fourier transform (IDFT): From the inverse discrete Fourier transform of the shifted first harmonic, complex moire patterns appear as shown in Figures 4(c), (d), (e) and (f). Figure 4(c) shows the real part and (d) shows the imaginary part corresponding to 120 pixels shifting. The difference of the moire fringe phase between Figures 4(c) and (d) is $\pi/2$. These fringes show the X-directional equal displacement lines. Figure 4(e) and (f) show the real parts of the complex moire patterns corresponding to shifting 60 and 0 pixels, respectively. Figure 4(e) shows the mismatched moire fringe pattern. The number of fringe lines is larger than that of Figure 4(c). Figure 4(f) shows only the original grating lines for X-directional

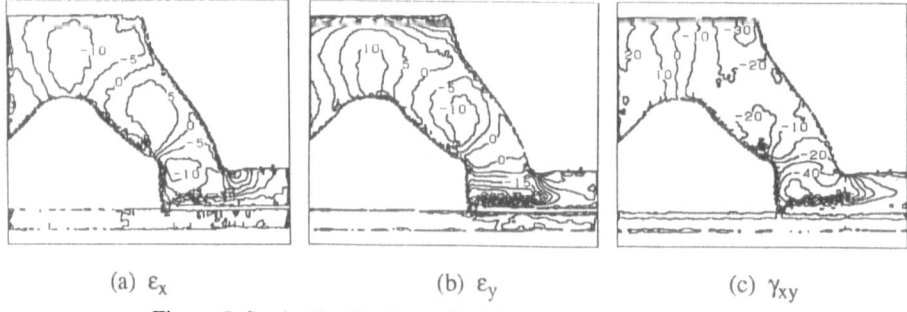

(a) ε_x (b) ε_y (c) γ_{xy}

Figure 5. Strain distributions of rubber fender obtained by FTMGM

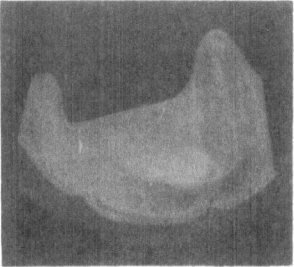

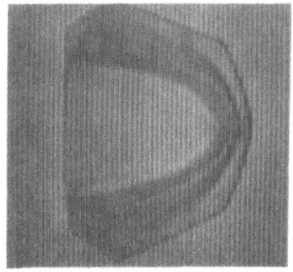

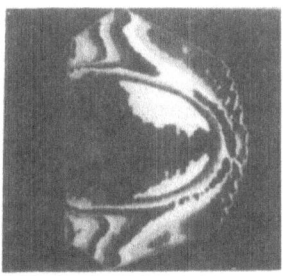

(a) Artificial tooth bed mold   (b) Grating projected on   (c) Contour lines
tooth bed mold

Figure 6. Shape analysis by FTMGM

analysis. However, the brightness distribution of the grating is changed from rectangular to sinusoidal, and the grating becomes clearer by eliminating dot noise, shading, and Y-directional grating lines. The displacement is also analyzed from the phase of the sinusoidal grating lines. This corresponds to a grid method.

Figure 4(g) shows the case of Y-directional analysis. The first harmonic is extracted within a circle whose radius is 90 pixels and the center is at the point (0,120) of Figure 4(a), by using the Hamming window filter, and the circle is shifted by 120 pixels toward the origin along the Ω_Y axis. Computation of the inverse Fourier transform of the shifted Y-directional first harmonic gives the Y-directional complex moire pattern. It corresponds to the Y-directional equal displacement lines.

Computation of phases of complex moire pattern: The phase of the brightness of the complex moire pattern corresponds to the displacement of the object. The phase is obtained by calculating the ratio of the brightness of the imaginary part to that of the real part at each point.

Determination of strain distribution: Strain is obtained by differentiating the displacement. The strain distributions are shown in Figure 5.

Shape Measurement

By applying this method to stereoscopic measurement method using a projected grating, the shape of an object can be measured. The grating projected on the object is recorded on an image processor. The image of the grating lines is deformed according to the shape of the object. By using the FTMGM, the displacements of the grating lines can be analyzed. These displacement correspond to the height of the object surface. Figure 6 shows an example which determines the shape of the mold of an artificial tooth bed.

Figure 6(a) is a mold of an artificial tooth bed. A one-dimensional grating is projected on the mold. The deformed grating image recorded by a TV camera is shown in Figure 6(b). This image is analyzed by the FTMGM to obtain the displacements of the deformed grating lines. The height contour lines of the object obtained from the displacements is shown in Figure 6(c).

CONCLUSIONS

Two methods for measuring displacement and strain distributions or surface topography using image processing have been presented. These are the scanning moire method and the Fourier transform moire and grid method (FTMGM).

The merits of the scanning moire method compared with the geometric moire method are as follows:

(1) It is not necessary to prepare a master grating on a specimen or to develop a photographic film and print. A clear moire pattern without original grating lines is obtained.

(2) The mismatch method is easy to perform by changing the pitch ratio of the master grating to the model grating using a zoom lens.

(3) The mismatch method is also easily performed by systematically picking up the specific scanning lines from the image instead of using a zoom lens.

(4) By changing the phase of the thinning-out, the direction increasing the fringe order is easily checked.

(5) It is easy to separate X and Y-directional fringes in the two-dimensional scanning moire method using a cross grating by changing the direction of the thinning-out.

(6) The procedure to obtain moire patterns is simple and makes real-time display possible.

The Fourier transform moire method has the following advantages:

(1) Phase information at all points gives accurate results and make the analysis simple.

(2) It is easy to separate the X and Y-directional fringes by selecting the X or Y-directional first harmonic.

(3) Clear moire patterns are obtained because this process eliminates high and very low frequency noise.

(4) It is easy to achieve mismatch moire patterns by varying the amount of shift.

(5) Automated and high speed analysis can be performed.

(6) This method appears to be useful to analyze almost equally spaced patterns such as carrier patterns obtained by holographic interferometry or moire interferometry to measure small displacement.

ACKNOWLEDGMENTS

The authors appreciate the advice of Professor Hayashi, Professor Seguchi and the assistance of the students at Osaka University in the experimental work and programming. This work was partially supported by the grant-in-aid for scientific research from the Ministry of Education of Japan. The image processing was partially performed at the Image Processing Laboratory of the Faculty of Engineering Science, Osaka University.

675

REFERENCES

1. Morimoto, Y., Image Processing Aided Analyses of Stress, Strain, Deformation and Shape, Trans. of JSME (in Japanese), 1989, **55**(511), pp.365-72.

2. Morimoto, Y. and Hayashi, T., Deformation Measurement during Powder Compaction by a Scanning Moire Method, Exp. Mech., 1984, **24**(2), pp.112-6.

3. Morimoto, Y., Hayashi, T. and Yamaguchi, N., Strain Measurement by Scanning-moire Method, Trans. of JSME (in Japanese), 1984, **50**(451), pp.489-94: Bull. of JSME (Translation in English), 1984, **27**(233), pp.2347-52.

4. Morimoto, Y. and Hayashi, T., Scanning moire method, Proc. of the Int. Sym. on Photoelasticity, 1986,Tokyo, Springer-Verlag, pp.47-52.

5. Morimoto, Y., Seguchi, Y. and Higashi, T., Moire Analysis of Strain by Fourier Transform, Trans. of JSME, 1988, **54**(504A), pp.1546-52.;JSME Int. J., Ser I (English translation), 1989, **32**(4), pp.540-6.

6. Morimoto, Y., Seguchi, Y. and Higashi, T., Application of Moire Analysis of Strain Using Fourier Transform, Opt. Eng., 1988, **27**(8), pp.650-6.

7. Morimoto, Y., Seguchi, Y. and Higashi, T., Strain Analysis by Moire Method and Grid Method Using Fourier Transform, Compu. Mech. 1990, (to be published)

8. Morimoto, Y., Seguchi, Y. and Higashi, T., Two-dimensional Moire Method and Grid Method Using Fourier Transform, Exp. Mech., 1989,**29**(4), pp.399-404.

9. Morimoto, Y., Seguchi Y. and Suese, N., Real-time Analyzer of Displacement and Strain Distributions Using Moire Method, Proc. of 1989 SEM Fall Conf. 1989, pp.17-20.

THE MEASUREMENT OF PLASTIC-ELASTIC STRAINS AT WELD TOES USING MOIRÉ INTERFEROMETRY

KIM ELLIOTT
Department of Civil Engineering, University of Nottingham,
Nottingham, NG7 2RD, United Kingdom

ABSTRACT

Using moiré interferometry, elastic, plastic-elastic and residual plastic strains have been measured at, and near to weld toes in real steel weldments. The shapes and loadings of the models are as in tubular joints used in offshore structures. Plastic-elastic strain concentration factors of up to 17 were measured at weld toes in a joint where the elastic value was 4. Strain gradients upto $10000\mu\varepsilon$/mm were measured near to the edges of Luders bands. Plastic strain reversals and strains in Luders bands were measured. Contours of plasticity at different load magnitudes are given. Moiré interferometry is an excellent tool for studying inhomogeneous materials.

THE RELEVANCE OF PLASTIC-ELASTIC STRAINS IN FATIGUE DESIGN

The development of large plastic strains at, and near to, surface irregularities (eg welds, cracks) or geometric discontinuities (eg fillets) precedes structural failures in many engineering components. The precise measurement of the strains enables the effects of geometry, material properties and loading conditions to be assessed with respect to an appropriate failure criteria. Although the most important values are the magnitude and position of the maximum plastic strain, post-yield behaviour can often be predicted if the distributions of plastic-elastic and residual-plastic strains in the surface and through the thickness of a component are known. The differences between an elastic response, which is readily measurable using finite elements or frozen-stress photoelasticity, and plastic behaviour is not obvious. Material inhomogeniety, non-isotropy and residual welding stresses have little effect on elastic values. Because these effects cannot be predicted from elastic conditions, experimental investigations are required.

Considerable investments have been made into research aspects associated with the fatigue behaviour of tubular steel jacket structures manufactured from circular hollow sections. See Fig 1. Despite their enormous size, these structures are prone to fatigue failures on a microscopic scale - at the toes of the fillet welds joining the various tubes. It is well established that the fatigue performance of fillet welded joints can be related, by the use of an appropriate S-N design curve, to the maximum stress (or strain) range $\Delta\sigma$ (or $\Delta\varepsilon$) endured by the joint. The assumption is that $\Delta\sigma$ is the product of the elastic stress concentration factor (SCF) at the fatigue crack initiation site and the nominal stress range $\Delta\sigma_{nom}$ in one of the loaded members, usually a brace tube. Alternatively, strain range $\Delta\varepsilon$ may be used. The assumption is that this is the product of the elastic strain concentration factor (SNCF) and $\Delta\varepsilon_{nom}$.

However, it has been observed (1, 2) that during the first load cycle in fatigue tests on fillet welded tubular joints, local yielding has occurred at the toe of some of the welds. Strains exceeding yield have been measured by small gauges placed as near as possible to the toe, typically 1.5 mm from the toe. It is unlikely that actual plastic-elastic strains have ever been measured at *real* weld toes. Attempts to model the plastic-elastic behaviour of 3-d joints using finite element methods etc. must consider the mechanical properties of the different materials near the toe of the weld. These are shown in Fig 1 as weld metal (WM), heat affected zones (HAZ) and parent plate (PP). The strains in these different material zones must be measured for typical tubular joint loading. Both elastic and plastic-elastic strains will be present and these inevitably lead to residual plastic strains. The relevance of these measurements to fatigue work is in the determination of the true strain amplitude $\Delta\varepsilon'$, ie plastic-elastic strains minus residual-plastic strains. The difference between $\Delta\varepsilon'$ and $\Delta\varepsilon$ (that which is obtained from elastic analysis) must be established.

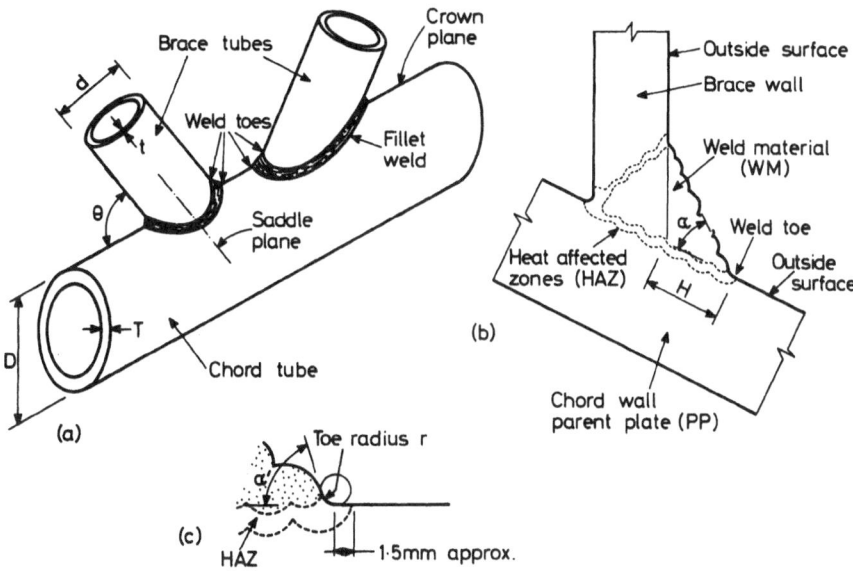

Figure 1   Tubular joint geometry and weld toe microgeometry

In this present work, plastic-elastic strains have been measured in the edges of flat, 2-d real steel weldments using moiré interferometry. See Plate 1. The 2-d models represent the saddle position (see Fig 1) in a 90° brace-to-chord connection in a K-type tubular joint. The models were loaded in their own plane so that the elastic stress distributions in their edges were as near as possible to those obtained in the same position in a 3-d frozen stress photoelastic K-type tubular joint analysed by Fessler and Little (3). This technique of using the planes of symmetry in 3-d models for use in 2-d plane stress analysis is well established and reported (4, 5).

DESIGN OF STEEL MODELS

The shapes of the models are given in Fig 2. The models were cut 4 mm in thickness from a 250 mm long weldment manufactured by British Steel Corporation, Swinden Laboratories, Rotherham using grade 50 D offshore quality plate and 4 mm diameter grade E51 welding electrodes. Mechanical and chemical properties for all materials and a full manufacturing specification is reported elsewhere (4). Two steel plates were joined at 60° to each other to give the correct angle at the saddle position of a 90° brace-to- chord connection with d/D = 0.5.

Plate 1   Moire interferometer and loading rig (on left)

This was the same d/D ratio as used in the parent 3-d K-joint (3). Flat plates were used because the curvature of the tubes in the 3-d model, given by D/T = 25, was small. The thickness of the plate representing the chord wall was 50 mm, a typical value in jacket structures, and that of the brace wall was 25 mm, giving t/T = 0.5. A 60° weld preparation was made to the brace wall plate. The weldment was designed to AWS recommendations (6) with a weld projection onto the outside chord wall of 1.75 t.

Two types of welds were made at different ends of the same weldment. These are:
i)   Uncontrolled profile , Fig 2a, in which the profile received NO particular profile control at the toe.
ii)   Controlled profile, Fig 2b, in which the uncontrolled weld was dressed with a small triangular fillet at the chord wall end of the weld only. This additional fillet was designed in accordance with the recommendations made by Marshall (7) to prevent drastic reductions in fatigue strength caused by sharp acute weld toes in thick-walled sections.

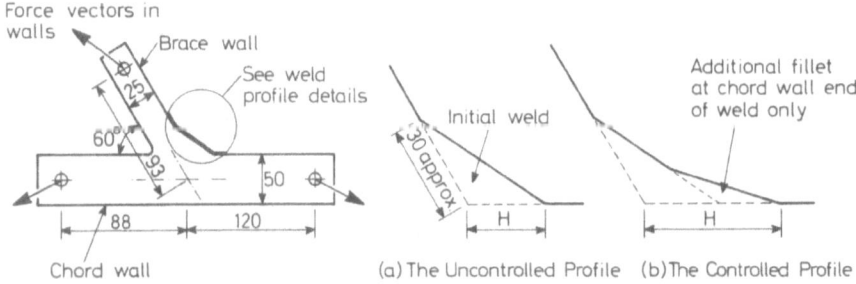

Figure 2   Shapes of 2-d steel models

A 12 mm thick section was cut from a part of the uncontrolled profile weldment and was taken for heat treatment . The specimen was placed in an electrically heated furnace and soaked at 620°C for 12 minutes prior to controlled cooling. Three models were prepared for testing in the following schedule:

Model No	Weld Profile	Heat Treatment
1	Uncontrolled	No
2	Uncontrolled	Yes
3	Controlled	No

The shapes and dimensions of the models are given in Table 1. The upper yield strength and ultimate tensile strength of the WM, HAZ and PP for Model No 1 are given in Table 2. These values were obtained using 3 mm diameter tensile testing coupons cut from material immediately adjacent to the models tested.

EXPERIMENTAL TECHNIQUE

In-plane strains were determined from two separate displacement fields using moiré interferometry. This optical method of whole field strain analysis was developed by Post and is well documented (8). Briefly, moiré interferometry combines the concepts of geometrical moiré, diffraction and optical interferometry. Patterns of interference, known as *fringes* are produced by the relative movement of a *real* grating attached to the steel models and a virtual or *reference* grating that is established by two coherent beams of light. The fringes are contour maps of points of equal in-plane displacements in the surface of the model. There is no shear lag.

The Optical System
The essential features of the two-beam moiré interferometer are shown in Fig 3. A highly reflective phase-type diffraction cross-grating was firmly bonded to one of the faces of the steel

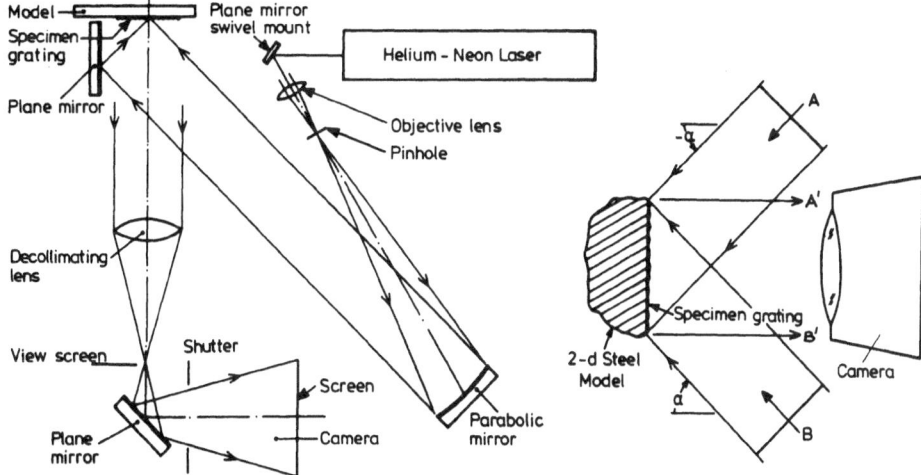

Figure 3   Moiré interferometer optical layout

models. The grating, measuring 50 mm x 35 mm, was attached to the model in the vicinity of the chord weld toe where maximum strains were anticipated from elastic results (4). The frequency of the grating is large; 1200 lines per mm and the sensitivity of measuring displacements using this system is in the order of tenths of a micron per fringe. When the model is loaded, the grating deforms and its frequency changes.

Two beams of coherent light, A and B in Fig 3, illuminate the specimen obliquely. They emerge from the reflective grating, as A'and B', with warped wavefronts. They coexist in space and generate optical interference. The directions of the emerging beams are prescribed by diffraction theory. Global body translations, unwanted in strain analysis, are eliminated using an auxiliary reference grating positioned in the same field of view as the grating on the steel model. The auxiliary grating would remain unstressed, undeformed and coincident with the model during the test. The design and specification of the apparatus, the methods of preparation of the model grating and the numerous optical alignment procedures necessary in this careful work, are fully described elsewhere (4).

Model Loading Frame
The test frame is shown in Plate 1. Its load capacity is 30 kN, sufficient to produce a tensile stress in the brace wall of 300 N/mm^2. The frame was manufactured from high tensile (En 24) plate. It was sufficiently light weight to be man-handled into position and be "balanced" on the edge of the moiré optical table. The frame was suspended by a set of springs so that fine physical alignment, to 0.001 radians, was achieved in three mutually perpendicular directions by turning micrometer barrels. The models were loaded through pairs of matched links and dowel pins. A 30 kN capacity load cell, accurate to 0.1%, was used to measure the loading applied to the brace wall. Other details of the apparatus are reported elsewhere (4).

INTERPRETATION OF THE MOIRÉ FRINGE PATTERN

A two-beam system measures only the displacements perpendicular to the virtual reference grating. This was vertical in this work To measure the orthogonal strains ε_{xx} and ε_{yy} (defined in Fig 4) it is necessary to present the real model grating to the virtual grating in two mutually

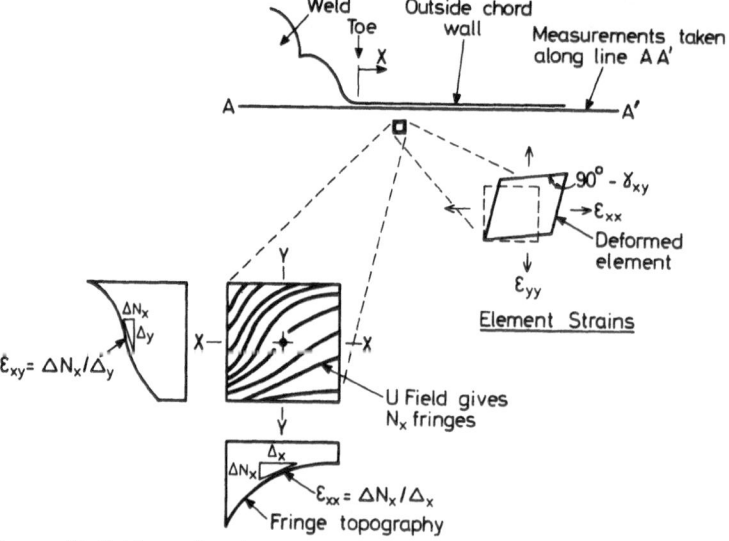

Figure 4   Definitions of strain measurements

perpendicular positions. Using the cartesian notation in Fig 4 the displacement field observed in the x plane (with the chord wall horizontal) is called the *u-field*, and that in the y plane is the *v-field*. Moiré fringes in the u and v displacement fields are thus described N_x and N_y

respectively. Fig 5 shows examples of the N_x and N_y fringe patterns used in the analysis of model No 1. In moiré work, derivatives of displacements are used to calculate strains and therefore fringe order gradients are measured. Because each moiré fringe, of order N_x (or N_y), represents the loci of points of equal in-plane displacement u (or v), it is shown (8) that

$$u = \frac{1}{f} N_x \quad \text{and} \quad v = \frac{1}{f} N_y$$

where f is the frequency of the virtual grating equal to 2400 lines per mm, ie twice the frequency of the model grating as given by the equations of diffraction.

If the incremental displacement Δu between two points Δx apart is depicted in the fringe pattern by a small change in fringe order ΔN_x, then

$$\Delta u = \frac{1}{f} \Delta N_x \quad \text{but} \quad \varepsilon_{xx} = \Delta u / \Delta x$$

$$\text{thus} \quad \varepsilon_{xx} = \frac{1}{f} \frac{\Delta N}{\Delta x} \quad \text{and, in the limit} \quad \varepsilon_{xx} = \frac{1}{f} \frac{\partial N_x}{\partial x}$$

Thus, the fringe gradient, measured in a prescribed direction, is directly proportional to the strain in that same direction. It follows that a shear strain ε_{xy} (see Fig 4) may also be obtained from the fringes in the u-field, as follows

$$\varepsilon_{xy} = \frac{1}{f} \frac{\partial N_x}{\partial y}$$

and, from the v field

$$\varepsilon_{yy} = \frac{1}{f} \frac{\partial N_y}{\partial y} \quad \text{and} \quad \varepsilon_{yx} = \frac{1}{f} \frac{\partial N_y}{\partial x}$$

Thus, the direct (ε_{xx}, ε_{yy}) and shear ($\gamma_{xy} = \varepsilon_{xy} + \varepsilon_{yx}$) strains at any point in the field are fully described.

STRAIN DISTRIBUTIONS NEAR TO CHORD WALL WELD TOES

Elastic (open symbols) and plastic-elastic strain distributions near to the weld toes in models Nos 1 to 3 are given in Figs 6 to 8, respectively. All values are maximum principal strain indices $J_1 = \varepsilon_1 / \varepsilon_{nom}$, where ε_1 is computed from ε_{xx}, ε_{yy} and γ_{xy}. The direction Φ of J_1 is given in Fig 9 for model No 2 only; values of Φ in other models were similar. Measurements were taken in line A-A, defined in Fig 4, with the origin x=o at the weld toe. Some of the strains presented are in the body of the weld, ie x< 0, and are not important in fatigue design. However, they are included to show the rapid decrease in strain away from the weld toe.

Measurements were taken at several load stations, but only the more significant ones are shown here. The load stations used correspond with i) a fully elastic condition, ii) the "observed" first yield, and iii) typical design values for brace wall stress of 200 N/mm^2 in models 1 to 3, and up to 270 N/mm^2 in model No 3. To generalize the results, the nominal strain in the brace wall is divided by the uniaxial yield strain in that wall, ie $\varepsilon_{nom} / \varepsilon_{yield}$.

682

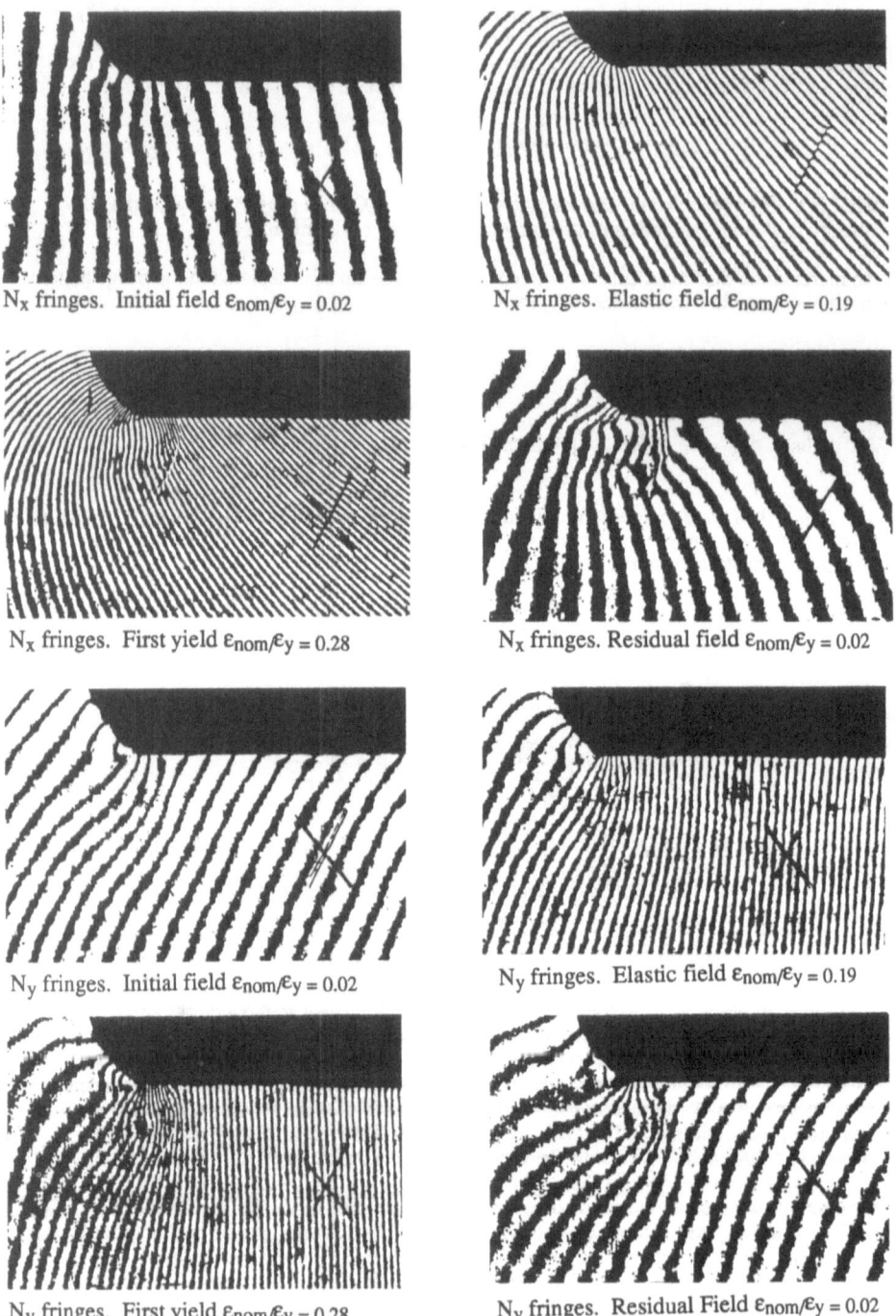

N_x fringes. Initial field $\varepsilon_{nom}/\varepsilon_y = 0.02$

N_x fringes. Elastic field $\varepsilon_{nom}/\varepsilon_y = 0.19$

N_x fringes. First yield $\varepsilon_{nom}/\varepsilon_y = 0.28$

N_x fringes. Residual field $\varepsilon_{nom}/\varepsilon_y = 0.02$

N_y fringes. Initial field $\varepsilon_{nom}/\varepsilon_y = 0.02$

N_y fringes. Elastic field $\varepsilon_{nom}/\varepsilon_y = 0.19$

N_y fringes. First yield $\varepsilon_{nom}/\varepsilon_y = 0.28$

N_y fringes. Residual Field $\varepsilon_{nom}/\varepsilon_y = 0.02$

Fig 5   Moiré fringe patterns near to weld toe in model no 1. Scale x 10 (approx)

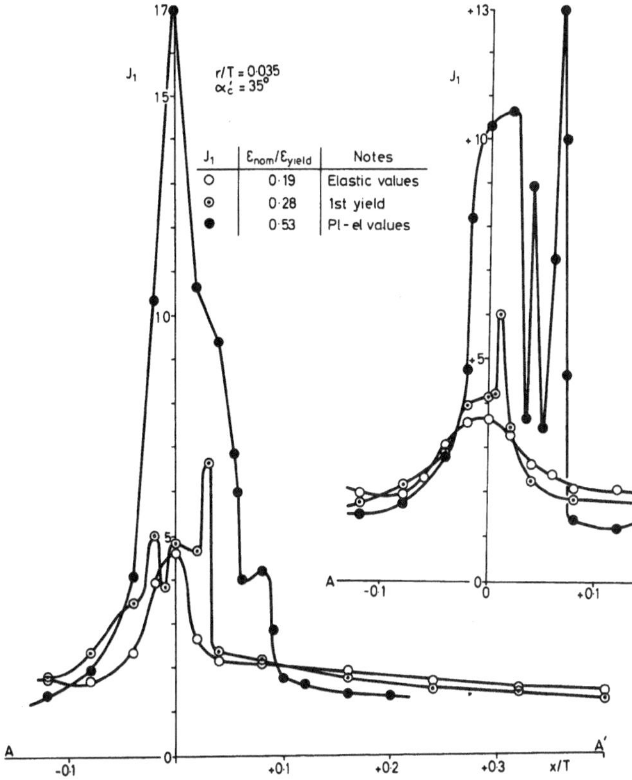

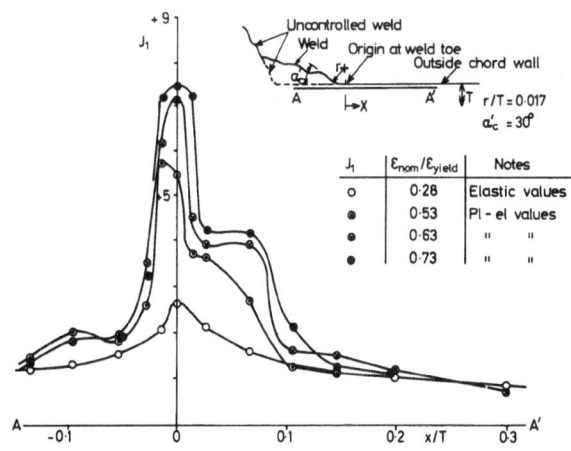

Fig 6   Model no 1

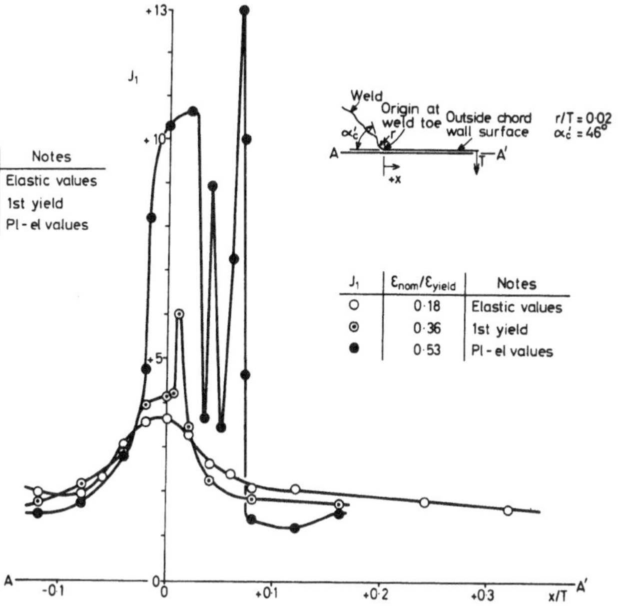

Fig 7   Model no 2

Fig 8   Model no 3

Figures 6 to 8   Elastic and plastic-elastic principal strain indices

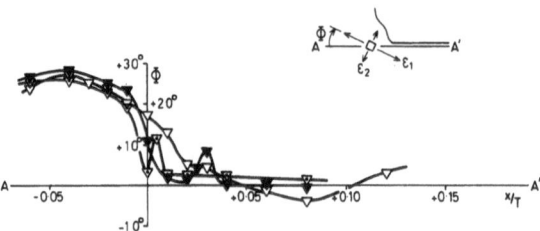

Figure 9   Direction of principal strains in model no 2

Discussion on Strain Distributions

Model No 1 - Uncontrolled Profile, No Heat Treatment. Elastic values may be compared with results obtained using 3-d (3) and 2-d (4) photoelastic techniques. Agreement in the linear region (different weld toe profiles prevent a direct comparison at the toe) is within ±10%. The extrapolated elastic SNCF = 2.22, which gave an elastic SCF = 2.65 is in close agreement with the photoelastic result which gave SCFs between 2.1 and 2.5.

First yield was observed in the parent plate at a position approximately at the edge of the HAZ/PP boundary at a loading $\varepsilon_{nom}/\varepsilon_{yield} = 0.28$. The strain at this position was 3460 $\mu\varepsilon$, ie 1.73 times the uniaxial yield strain in the HAZ material. Local yielding was identified in the u-field as a small slip line (about 2 mm long) in which the fringes were closely packed. The slip line followed the arc of the HAZ/PP boundary, which had been previously identified by etching. At the higher loading, a SNCF = 17 was measured at the toe. The actual strain here was 1.67%. Plastic strains were confined to within a distance of about 4.5 mm (0.09T) from the toe. Beyond this region strain indices were upto 25% less than values measured in the elastic condition. This would result in an under-predictions in SNCFs in joints where extensive localised plasticity was present and undetected.

Model No 2 - Uncontrolled Profile, Heat Treated. As expected, elastic values were similar to those in model No 1 except at the toe where, despite a smaller weld toe radius and steeper weld angle, the peak SNCF was lower. No explanation can be given for this.

First yield was observed very close to the weld toe and in the HAZ. Because of difficulties in locating the exact position of the weld toe, first yield may have occurred anywhere between 0.2 mm and 0.7 mm from the toe; the exact position is probably not important. The load magnitude $\varepsilon_{nom}/\varepsilon_{yield} = 0.36$ represents an increase of 28% in the loading in model No 1 at the same point in the test. The yield strain in the model was 4020 $\mu\varepsilon$, equivalent to an increase of about 115 N/mm^2 in model 1. This probably represents the additional stress limit available in the heat treated condition. Results are given for $\varepsilon_{nom}/\varepsilon_{yield} = 0.53$ for comparison with model No 1. Plastic-elastic strains were measured within 4 mm of the weld toe. In this region, *three* peak values of SNCF = 10.7, 8.9 and 13.0 were measured; the latter being furthest from the toe and nearer to the HAZ/PP boundary. In between these peaks strain indices were only about 30% greater than elastic values. Strain indices outside the plastic zone were, as in model No 1, smaller than elastic values.

Model No 3 - Controlled Profile, No Heat Treatment. Elastic values were smaller than in model No 1 because of a shift in the position of the weld toe (used as origin here) into a lower stress field. The agreement with photoelastic results is within 7%. The *onset* of yielding was not observed in this model. It was not apparent that yielding had occurred until considerable plastic strains were present. The controlled weld profile is obviously beneficial in this respect. Plastic-elastic SNCFs were smaller than in other models, and the rate of increase in strains with loading was less than 50% of former values. The spread of plasticity in the surface of the

Plastic-elastic SNCFs were smaller than in other models, and the rate of increase in strains with loading was less than 50% of former values. The spread of plasticity in the surface of the chord wall (upto 7 mm from the toe) was little influenced by the different material properties in and around the HAZ.

RESIDUAL PLASTIC STRAINS

Residual plastic strain distributions for all models are given in Fig 10. The strains were measured at $\varepsilon_{nom}/\varepsilon_{yield} = 0.02$, the smallest practical value. The corresponding fringe pattern in model No 1 is shown in Fig 5. The residual strains outside the plastic zone were negative.

From this figure and Figs 6 to 8, it is possible to deduce the reductions in strains from the plastic-elastic condition at (say) $\varepsilon_{nom}/\varepsilon_{yield} = 0.53$ to the residual condition at $\varepsilon_{nom}/\varepsilon_{yield} = 0.02$ - a decrease in $\varepsilon_{nom} \cong 950$ $\mu\varepsilon$. This is shown in Fig 11 for model No 1 in which the hatched area represents the reduction due to elastic memory *and* plastic reversals. These two quantities can be separated because the elastic strain reduction is given by the elastic strain index (at any point) multiplied by $\Delta\varepsilon_{nom}$ (in this case = 950$\mu\varepsilon$). This is shown in Fig 12. The strain represented by the hatched area in Fig 12 is due to a plastic reversal.

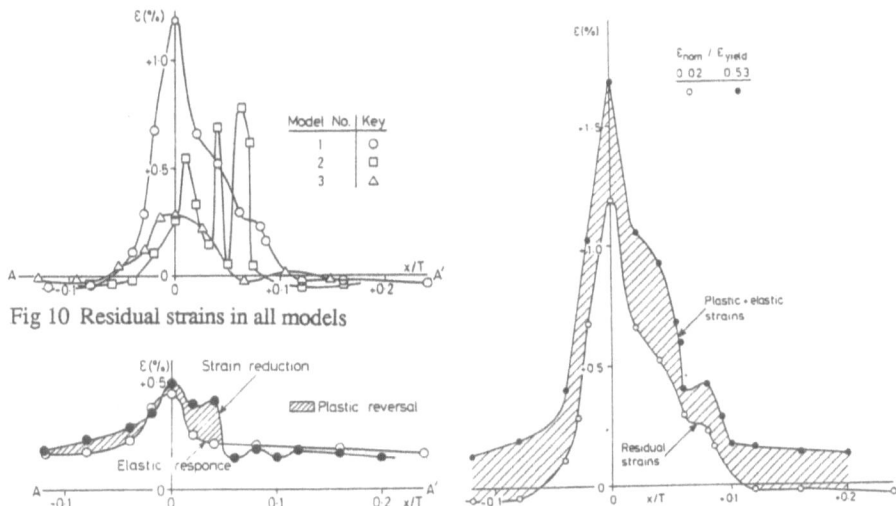

Fig 10 Residual strains in all models

Fig 12 Differences between true strain
reduction and that predicted using
an elastic response

Fig 11 Plastic-elastic and residual-plastic
strain magnitudes in model no 1

Discussion on Residual Plastic Strains

Residual plastic strains upto 2800$\mu\varepsilon$ in magnitude were measured in all models, even though the mean axial strain in the brace wall had not generally exceeded 53% of yield. Residual strains and the extent of plasticity were greater in the uncontrolled than in the controlled welds for the same maximum loading condition. Note that absolute values are appropriate only to the loadings used in this work. The most important result is that the true strain range $\Delta\varepsilon'$, shown hatched in Fig 11, is NOT equal to the product of the elastic SNCF and the nominal load range, ie $\Delta\varepsilon' \neq$ SNCF x $\Delta\varepsilon_{nom}$. The difference is a plastic strain reversal, upto 3900$\mu\varepsilon$ in the more highly strained regions of model No 2. This has undoubtedly been overlooked in fatigue design, where if $\Delta\varepsilon$ at the weld toe had been computed from elastic values alone the error in $\Delta\varepsilon'/\Delta\varepsilon$ would be 1.15, 2.1 and 1.1 in models 1 to 3, respectively.

PLASTIC ZONES

Regions of plastic deformation were traced from the residual moiré fringe patterns in the u-field at a load of $\varepsilon_{nom}/\varepsilon_{yield} = 0.02$. Contours of plastic deformations in models Nos 1 and 2 are given in Fig 13. It is assumed that the extent of plasticity at the high loadings is equal to the extent of residual plasticity. This is confirmed in Fig 11. Figure 13 shows plasticity spreading more rapidly in the parent plate than in the weld. In all models, plasticity was partially arrested in the HAZ. As a result, plastic deformation extended into the body of the weld for less than 2.5 mm (on average) from the toe.

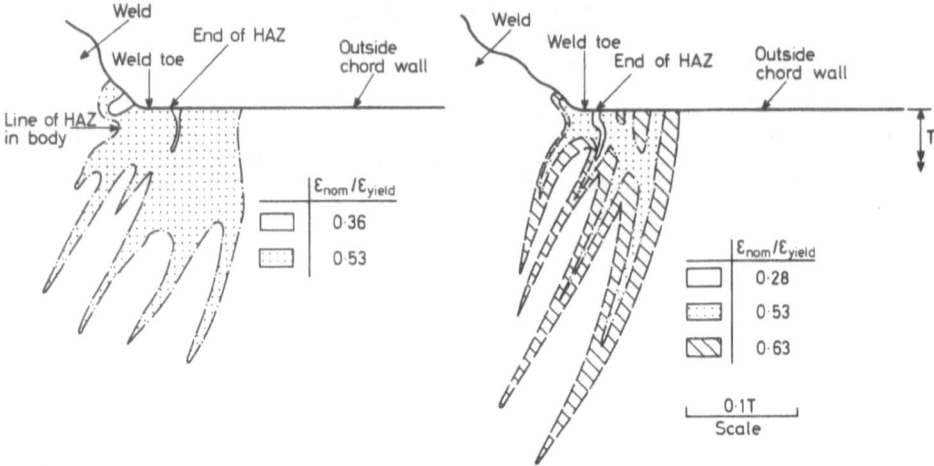

Figure 13   Zones of plasticity at different loadings in model no1 (left) and no 2 (right)

Discussion on Plastic Zones

Informative data on the behaviour of inhomogeneous materials, composites etc, can be obtained from the moiré fringe patterns alone. Sudden changes in the direction of fringes show slip planes and rapid changes in fringe density reveal large strain gradients. Strain gradients of upto 10,000 µɛ/mm were detected over distances of about 0.3 mm. A prominent feature in these models was the presence of *Luders bands*, typically 4 or 5 in number and 0.5 mm to 1.0 mm in width. Strains in the bands were in the order of 15,000 µɛ, whilst in between the bands the material remained essentially elastic. The strains near to the leading tip of the bands were approximately 5000 µɛ and were independent of load magnitude and model geometry. The bands followed roughly the path of the maximum principal stress trajectory measured by the author in some similar 2-d models using finite element methods (4).

ERRORS

In the measurement of large weld toe strains the significant errors were i) accuracy of loading, ii) fringe gradient measurement (clarity in photographs) and iii) edge profile effects (uncertainty of exact location of toe). These accumulated to 12.5%. Errors in the measurement of elastic and residual plastic strains were 4% to 6%. Out-of-plane undulations due to Poisson effects (necking) were measured (using Talysurf apparatus) and were found to produce extraneous strains of less than 5µɛ.

CONCLUSIONS

Using moiré interferometry, elastic, plastic-elastic and residual plastic strains near to weld toes in real steel weldments were measured. Whole field distributions of strain were obtained. The shapes and loadings of the 2-d steel models were as for a saddle connection in a 90° brace tube in a K-type tubular joint. Three models were tested - two with *uncontrolled* weld profiles, and one with an additional fillet at the chord end of the weld, called a *controlled* weld profile. One of the former type was heat treated. The conclusions are:-

1   Elastic results showed good agreement with photoelastic values. Elastic SNCFs at weld toes were 2.65 to 4.60,the large values being measured in the uncontrolled welds.

2   First yield was observed near to the HAZ/parent plate boundary, and NOT at the weld toe where elastic strains were maximum. Strains at first yield were 1.7 to 2.0 times the unixial yield strain for the material.

3   SNCFs equal to 17.0, 13.0 and 7.4 were measured in different models for a mean axial loading in the brace wall of 200 N/mm^2, or 53% of the uniaxial yield stress. This is a typical design brace loading.

4   Large residual strains, up to 2800 $\mu\varepsilon$, were measured. The true strain range, ie the difference between the plastic-elastic and residual strains, was NOT equal to the elastic SNCF multiplied by the nominal brace wall strain range. The difference was 10% in the as-welded models and 110% in the heat treated model. This is an important discovery with implications in fatigue life assessment.

5   Luder's bands were observed and detailed measurements of strains in them are given.

NOTATION

D, d, T, t, θ	tubular joint dimensions defined in Fig 1.
H, r α, α'	weld profile dimensions defined in Fig 1.
f	frequency of virtual reference grating
u, v	components of displacements parallel to x and y axes, respectively (Fig 4)
x	distance from weld toe
A	cross sectional area of brace wall
E	Young's Modulus (assume 200 kN/mm^2)
J	strain index = $\varepsilon/\varepsilon_{nom}$
N	fringe order
P	axial load applied to brace wall
SNCF	strain concentration factor = J_{max}
SCF	stress concentration factor
ε	strain
ε_{nom}	nominal strain in brace wall = P/AE
σ	stress
σ_{nom}	nominal stress in brace wall = P/A
γ	shear strain
Δ	range, increment

ACKNOWLEDGEMENT

This work was supported by the Science and Engineering Research Council and was carried out in the Department of Mechanical Engineering, University of Nottingham. The author wishes to thank Professor H Fessler for guidance, and the photoelastic laboratory technicians for their skilled assistance.

REFERENCES

1   Bouwkamp, J G and Mukhopadhyay A, Effect of tensile overstrain on fatigue life of welded tubular joints, Offshore Technology Conference, Paper OTC 3255, Texas, May 1978.

2   Wylde, J G, Fatigue tests on welded tubular K and KT joints under out-of-plane and axial loading, Final Contract Report 3612/4/84, The Welding Institute, Cambridge, UK, July 1984.

3   Fessler, H and Little W J G, Fillet stresses in tubular joints obtained by photoelastic techniques, Proc. International Symposium on the Integrity of Offshore Structures, Glasgow, April, 1987.

4   Elliott, K S, Stresses at weld toes in tubular joints in offshore structures, PhD thesis, University of Nottingham, 1987.

5   Elliott, K S and Fessler, H, Stresses at weld toes in non-overlapped K type tubular joints, Fatigue in Offshore Structures, IMechE, London, 1985.

6   American Welding Society, Structural welding code AWS D1-1-84, 1984.

7   Marshall, P, Connections for welded tubular structures, The Houdremont Lecture, Proc 2nd Int Conf on Welding of Tubular Structures, Boston, Mass, 1984.

8   Post, D, Moire interferometry, SESA Handbook on Experimental Mechanics, Prentice Hall, 1986.

TABLE 1   Shapes and dimensions of steel models

Model No	t (mm)	T (mm)	Weld profile at Chord End of Weld			
			H/T	r/T	α	α'
1	25.15	49.65	0.44	0.035	37°	35°
2	25.15	49.65	0.44	0.020	37°	46°
3	25.20	49.85	0.77	0.017	18°	30°

TABLE 2   Mechanical properties of steel models

Designations	Position	Yield Stress (N/mm^2)	Yield / UTS
PP	Chord Wall	314	0.67
		324	0.68
PP	Brace Wall	352	0.70
		344	0.69
HAZ	Adjacent chord wall	392	0.74
		404	0.77
HAZ	Adjacent brace wall	475	0.84
		502	0.88
WM	Weld	479	0.82
		457	0.79

AN OPTICAL TECHNIQUE TO MEASURE MICRO DISPLACEMENTS

C.A.SCIAMMARELLA, G. BHAT AND A. ALBERTAZZI, JR.
Illinois Institute of Technology
Department of Mechanical and Aerospace Engineering
Chicago,l IL 60616

ABSTRACT

An optical microscope together with a CCD camera and an imaging system is utilized to measure displacements in the microscopic range. Although the strains may be large, the local displacements are very small. The displacement sensitivity is limited by the   optical technique used to measure displacements and the larger are the magnifications, the more sensitive must be the technique to determine displacements. To increase the displacement sensitivity a new method has been developed. Two examples of application are given, one to the field of composites and the other to the area of fracture mechanics.

INTRODUCTION

The mechanical behavior of materials at the microscopic level has become a topic of increasing interest. This is particularly true if one wants to investigate the variables that affect structure sensitive properties, and how these variables can be manipulated to obtain materials with prescribed properties. Of particular interest are the optical techniques since they are specially suitable for this kind of research. Holographic interferometry, speckle techniques and moire are three useful alternatives for microscopy work.

BASIC PROPERTIES OF THE MICROSCOPE

We are going to look at some of the basic properties that must be considered in the design of a microscope to analyze deformations, leaving aside specific topics such as aberrations which are dealt with in the optical literature.

There are two separate measurements that are involved in measuring displacements and their derivatives, spatial location and displacement. To clarify the whole process one has to recall that the aim of the investigation is to obtain the displacement function on a surface. The function is obtained by sampling it at given intervals. Under ideal conditions, the answer to the question as to how frequent the sampling must be, is given by the Whittaker-Shannon sampling theorem; the so-called Nyquist sampling interval must be equal to,

$$x_{Ny} = \frac{1}{w}$$
(1)

where w is the spectral width of the function. If one considers that the high frequencies of a band limited function vary as $\exp(i\,wx)$, the derivative of the function is proportional to the bandwidth:

$$\frac{de^{i\pi wx}}{dx} = i\pi w e^{i\pi wx}$$
(2)

Hence, the sampling rate must be increased when the function has steep gradients. If one looks at the scaling and the differentiation expressions corresponding to the Fourier transforms,

$$\frac{df\left(\frac{x}{b}\right)}{d\left(\frac{x}{b}\right)} \xrightarrow{F} (i2\pi b\xi)F(b\xi))$$
(3)

one can see that one has to enlarge the image in such a way that the sampling interval comes within the resolution interval of the optical system; otherwise the detail of the displacement surface will be lost. The condition resulting from (3) guarantees that the spatial resolution is sufficient to follow the change experienced by the function, but corresponds to only one of the two variables to be measured. Indeed, the displacement resolution needs to be such that one can measure the displacements in two very close points. If one considers the definition of derivative, calling u the displacement in the x-direction

$$\frac{du(x)}{dx} = \lim_{\Delta x \to 0} \frac{u(x+\Delta x) - u(x)}{\Delta x}$$
(4)

one can see that as x becomes smaller the ability to measure u must be increased in such a way that,

$$\Delta x \Delta u = \text{constant} \qquad (5)$$

Then the closer the points are, the higher the sensitivity to measure the displacements must be. Unfortunately, in the techniques of interest, holography, speckle, moire, the displacement sensitivity is independent of the magnification. Hence, one must provide the means to increase the sensitivity in such a way the displacement sensitivity can match the increased spatial resolution. If this is not done, the added magnification will not help in obtaining the displacement information.

A TECHNIQUE TO MEASURE FRACTIONAL ORDERS

The light intensity in a moire fringe pattern whether it is caused by a real grating (moire) or a virtual grating (holographic moire) [1] can be represented by the following equation,

$$I(x,y,\alpha) = I_0(x,y) + \sum_{n=1}^{n=N} I_n(x,y) \cos[2n\pi f_x x + 2n\pi f_y y + \alpha_n(x,y)] \qquad (6)$$

$I_0(x,y)$ is the background intensity; it may include local random values, but on the average $<I_0(x,y)>$, it is assumed to vary slowly when compared to the fringe intensity variation. The $I_n(x,y)$ are the intensity variation of the different harmonics, again assumed to vary slowly on the average when compared to the fringes. f_x and f_y are the frequencies of the deformed grating in the x and y directions, $\alpha_n(x,y)$ is a phase term that depends on the origin of the adopted coordinate system, but once the origin of coordinates has been selected it can be modified by introducing changes in the optical system used to record the patterns. The changes of phase can be thought of as translations of the reference grating [2], and the phase angle equivalent to a translation can be represented by,

$$\alpha = 2\pi f_{ox} \Delta x \qquad (7)$$

The Fourier transform of the intensity can be computed by two steps that include one of the coordinate axis at a time.

$$\overline{I}_x[(\xi,y)] = \int_{-\infty}^{+\infty} I(x,y)e^{-i2\pi\xi x}dx \tag{8}$$

$$\overline{I}(\xi,\eta) = \int_{-\infty}^{+\infty} I_x(\xi,y)e^{-i2\pi\eta y}dy \tag{9}$$

It is interesting to observe that the original intensity can be obtained from one of the transforms,

$$I(x,y) = \int_{-\infty}^{+\infty} \overline{I}_x(\xi,y)e^{i2\pi\xi x}d\xi \tag{10}$$

Taking into consideration (7), one can introduce in (8) the change of variables

$$2\pi\xi = \alpha \tag{11}$$

and one can obtain

$$\overline{I}_x(\xi)\mid_{\ell m} = \frac{p}{2\pi} \int_{-\pi}^{+\pi} I(x_\ell, y_m, \alpha)e^{-i\alpha}d\alpha \tag{12}$$

where p is the grating pitch (real or virtual) and $f_{ox} = 1/p$

That is, by changing α, one can get the data necessary to compute the Fourier transforms of the intensity.

The following observation can be made; the optical system used to observe the patterns can be considered as the analog equivalent of a parallel filter used to compute Fourier transforms. Furthermore, the values of the transforms are given for the local coordinates of the point under analysis. Since I(x,y) is a real function, the real and the imaginary components of the Fourier transform are given by,

$$Re\overline{I}_x(f_{ox})\mid_{\ell m} = \frac{p}{2\pi} \int_{-\pi}^{+\pi} I(x_\ell, y_m, \alpha)\cos\alpha\, d\alpha \tag{13}$$

and

$$\text{Im } \bar{I}_x(f_{ox})|_{\ell m} = -\frac{p}{2\pi} \int\limits_{-\pi}^{+\pi} I(x_\ell, y_m, \alpha) \sin \alpha \, d\alpha \qquad (14)$$

From (13) and (14)

$$\left| \bar{I}_{\ell m}(f_{ox}) \right| = \sqrt{[\text{Re } \bar{I}_{\ell m}(f_{ox})]^2 + [\text{Im } \bar{I}_{\ell m}(f_{ox})]^2} \qquad (15)$$

and

$$\phi_x|_{\ell m} = \text{arc tg } \frac{\text{Im } \bar{I}_{\ell m}(f_{ox})}{\text{Re } \bar{I}_{\ell m}(f_{ox})} \qquad (16)$$

Once the phases have been obtained,

$$u = \frac{p}{2\pi} \phi_x \qquad (17)$$

and

$$\epsilon_x = \frac{p}{2\pi} \frac{\partial \phi_x}{\partial x} \qquad (18)$$

The component of the shear strain is given by

$$\frac{\partial u}{\partial y} = \frac{p}{2\pi} \frac{\partial \phi_x}{\partial y} \qquad (19)$$

Similar equations can be applied to the y-axis and thus one can obtain the rest of the components of the strain tensor. The described operation can be performed at selected points and thus the displacements can be interpolated. The ability to perform this operation depends on the signal to noise ratio. Consequently, one must have good quality images with high contrast.

The Fourier transform has the effect of providing a filtering action that removes all other harmonics but the one that is observed.

The technique has been applied to two different techniques; holographic moire and to moire. Each of these applications will be discussed separately.

OPTO-ELECTRONIC HOLOGRAPHY

The recording of holographic interference patterns by means of a TV camera, is referred in the literature as electronic speckle pattern interferometry (ESPI). The patterns are recorded in the same way as in conventional lens holography with an almost in line reference beam. The basic equation of holographic interferometry is valid. The difference with conventional holography arises in the process of reconstruction. While in conventional holographic interferometry the reconstruction is done by illuminating the hologram with a reconstruction beam, in opto-electronic holography the reconstruction is done by purely electronic means. The reconstructed holographic interferogram is a two-dimensional image, and of course this is a big difference with optical holography.

Conventional holographic microscopy has been the subject of many applications to different fields. While some of these applications take advantage of the features of holography, a great deal of effort has gone to suppress the presence of speckles and different approaches have been used to obtain good quality microscopic images. In the case of holographic interferometry the displacement information is encoded in the speckles and consequently in the primary recorded images, the speckles must have optimum visibility. This of course implies a deterioration of the quality of the image, unless some kind of averaging techniques are applied that preserve the displacement information and reduce the speckles. In the present approach, the applied solution has been to record two images, one to measure displacements and a separate one to register surface features. The two images have been matched afterwards.

The phase interpolation technique that has been previously described was applied to the measurement of displacements at the microscopic level. Due to the heavily speckled images a particular problem arises in data processing. The displacement information appears as a trend in an otherwise random signal, the randomness being caused by the random jump of phase from speckle to speckle. The trend has been separated from the random jumps by applying a digital low pass filter. After smoothing, the displacement curves have been differentiated by using differentiating filters. The obtained strains have been again smoothed by applying digital low pass filters.

APPLICATION OF THE OPTO-ELECTRONIC MICROSCOPIC HOLOGRAPHY

The experimental setup is shown in Figure 1. A tensile specimen of a composite made with chopped glass fibers in an epoxy resin matrix is used for the purpose of analysis. The region of analysis is the neighborhood of a glass fiber oriented perpendicular to the tension direction (Figure 2); the magnification is X1500. The numerical aperture of the microscope is 0.3, which correspond to a resolution of 1.05 microns. The illumination angle is 70 degrees, which correspond to a displacement sensitivity of 0.337 microns per fringe.

The fringe pattern shifting was achieved by a technique described in [2].

The fringe pattern is recorded by a CCD camera with 512X480 pixels. An initial image is grabbed and stored in the buffer memory. To reduce the electronic noise the image is read 64 times. After the specimen is loaded, the initial and final images are compared by displacing them relative to each other until they are within the correlation radius. The remaining rigid body motion does not affect the results, since it is eliminated during the processing stage. Figure 3 shows the flow diagram of the program used for data processing. Figures 4 and 5 show the displacement and the strain contour lines around the fiber. Figures 6 and 7 show the three dimensional views of the displacement and strain fields. Figures 8 and 9 show the displacements and strains plotted along selected cross sections.

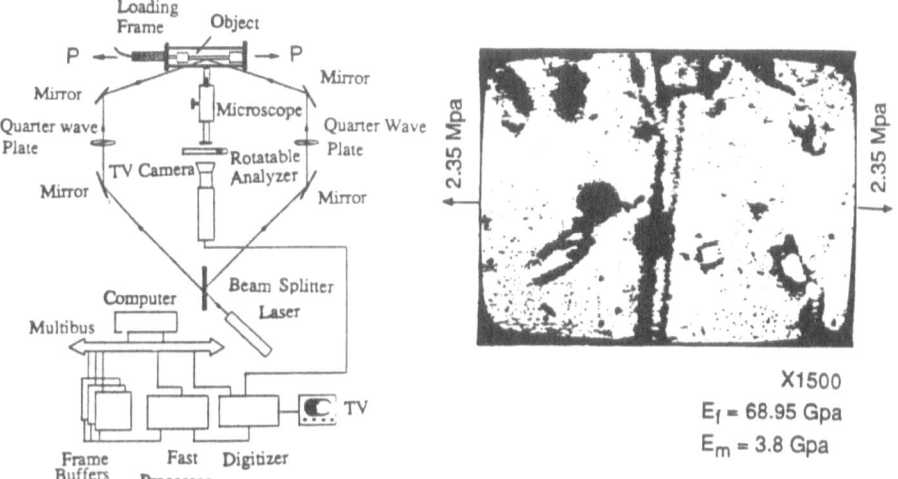

Figure 1. Electro-optical system
for micro-strain measurement.

Figure 2. Glass fiber-epoxy matrix.

X1500
$E_f = 68.95$ Gpa
$E_m = 3.8$ Gpa

COMPUTER ASSISTED MOIRE

In conventional moire technique, the displacement fringes are generated by optically comparing the deformed object grating with a reference grating and the resulting fringe pattern is recorded on film. In computer assisted moire [3], the object grating is recorded by a TV camera. The recorded deformed shape is compared with a grating internally generated by the computer. To increase the accuracy of the technique, the initial undeformed shape of the grating is recorded and is compared to a computer generated reference grating. In this way, one can remove the errors

696

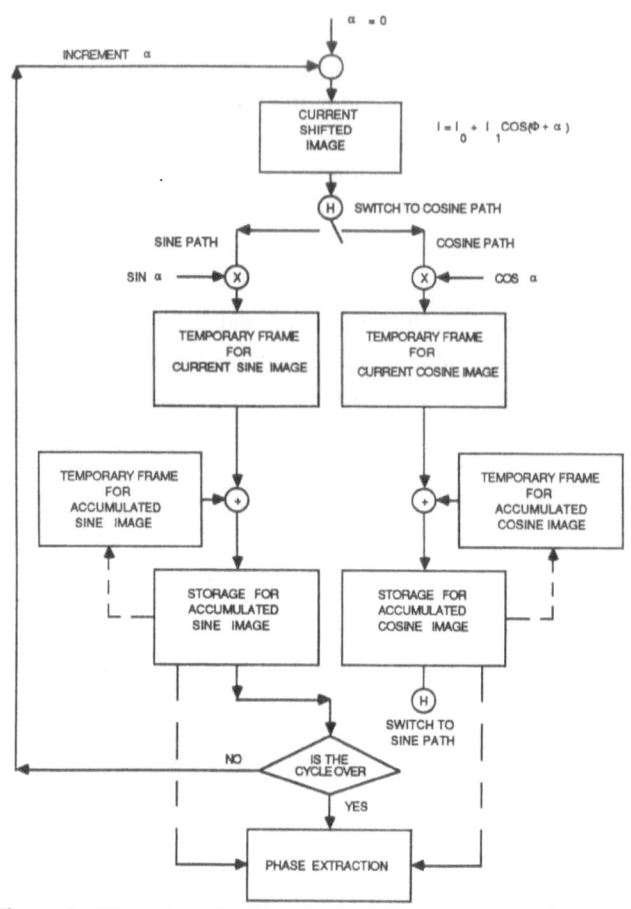

Figure 3. Flow chart for the program to measure displacements.

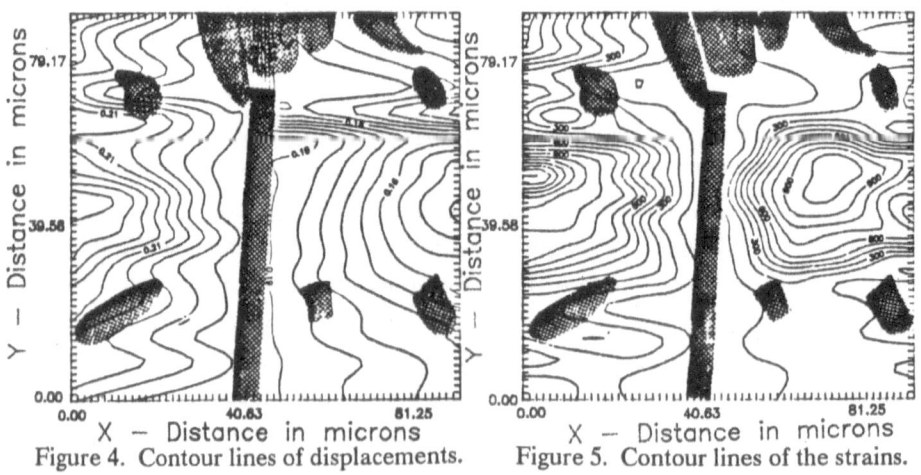

Figure 4. Contour lines of displacements.   Figure 5. Contour lines of the strains.

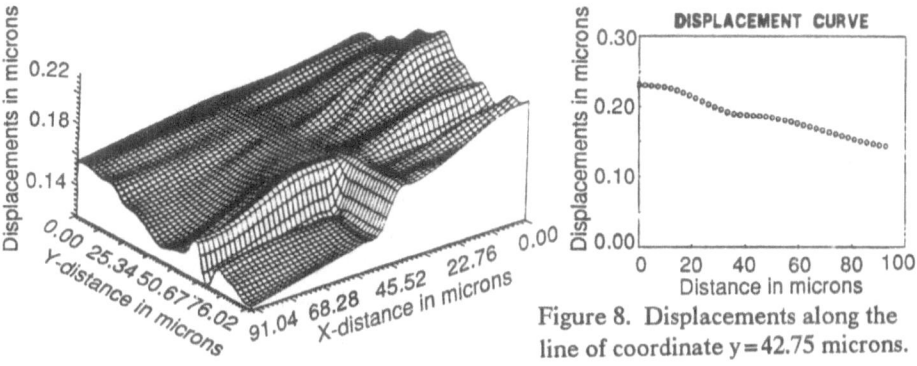

Figure 6. 3-D view of displacements.

Figure 8. Displacements along the line of coordinate y = 42.75 microns.

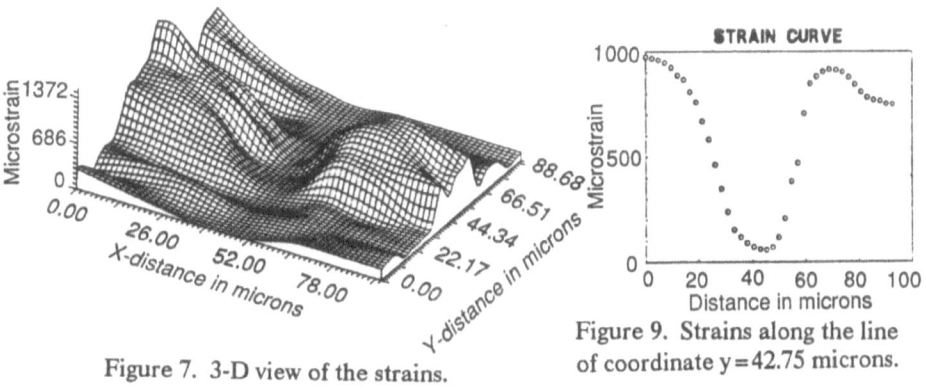

Figure 7. 3-D view of the strains.

Figure 9. Strains along the line of coordinate y = 42.75 microns.

coming from the initial imperfections of the grating and from the aberrations of the imaging optical system. An initial mismatch fringe system is used to take care of the order determination as the phase information is retrieved.

APPLICATION OF THE COMPUTER ASSISTED MOIRE TO MICROSCOPY

Figure 10 shows the optical setup used in the experiment and Figure 11 shows a view of the actual system. The application in this case is the analysis of the displacement at the crack tip of a compact tension specimen. The dimensions of the specimen are shown in Figure 10. The specimen was loaded on a servohydraulic testing machine. The applied load was 60% of the critical load corresponding to the toughness of the material, 601-T451 aluminum.

A 80 line per mm grating was cemented to the surface. A CCD camera with 800x490 pixels was used to record the images. The processing system has 512x512 pixels. The crack tip was observed at different magnifications; the observed areas are 2130x1640 μm, 894x689 μm, 123x95 μm, 74x57 μm. The final fields were obtained by combining information of the different magnifications and by matching boundary data. The numerical aperture of the microscope was 0.3 with a resolution of roughly 1 μm.

Before the moire patterns were generated, the recorded object gratings were filtered using band pass digital filters. Mismatch fringes were generated using internally generated reference gratings. The fringe shifting was produced by shifting the reference gratings. The final displacement fields were obtained by comparing the carrier fringes before and after the loading. The smoothed displacement curves were differentiated by means of differentiating filters and the resulting strain curves were smoothed by low pass filters. Two dimensional smoothing was performed while plotting the corresponding displacement and strain fields. Rigid body rotations originated by the deformation of the specimen were corrected. The actually observed displacements are small enough in most of the image that it was not necessary to introduce corrections due to changes in the geometry. Figure 12 shows the moire pattern corresponding to the v-displacement for the 2130x1640 μm image, with the fringes corresponding to 1/30 of the fringe pitch or .40 microns. Figure 13 shows the u-displacement field. Figure 14 shows the v-displacement field in the neighborhood of the crack tip.

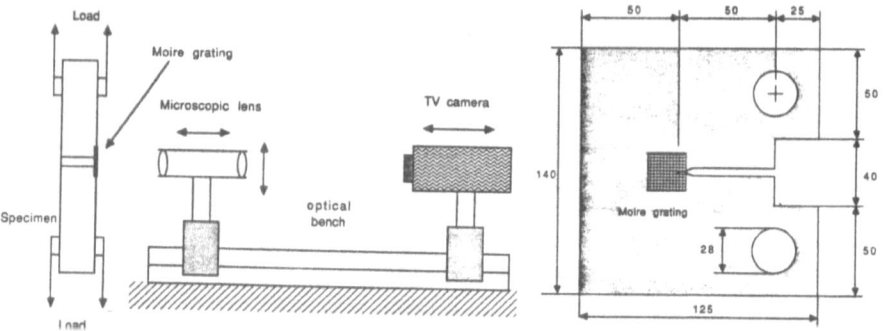

Figure 10. Setup used to measure displacements at the tip of a crack in a compact tension specimen.

699

Figure 11. View of the setup used to measure displacements at a crack tip while the specimen is loaded in a testing machine.

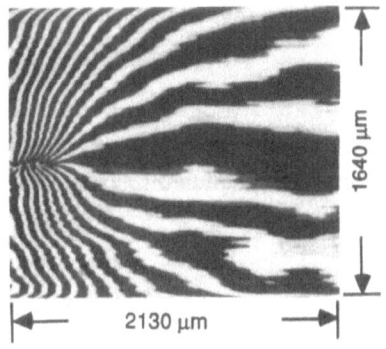

Figure 12. v-displacement field in the neighborhood of the crack tip.

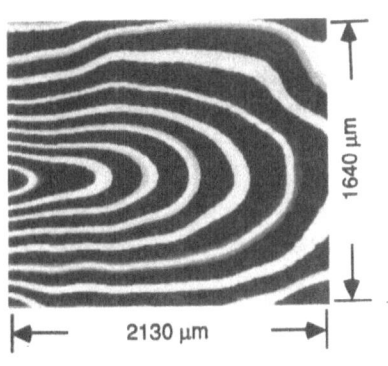

Figure 13. u-displacement field in the neighborhood of the crack tip.

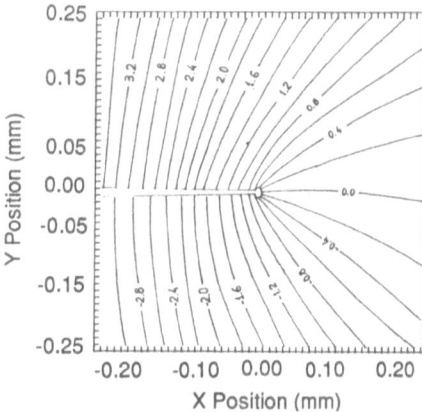

Figure 14. v-isotetic lines in the neighborhood of the crack tip. Crack opening shown

CONCLUSIONS

The combined use of an optical microscope together with a TV camera recording and an image processing system yields an extremely useful tool for the measurement of displacements at the microscopic level. To operate such a system, it is necessary to increase the sensitivity of the techniques used to measure displacements. This has been achieved by a fringe shifting technique that makes it feasible to measure fractional fringe orders. Two technical applications, one to the field of composites and the other to the field of fracture mechanics provide a direct proof of the capabilities of the developed technology.

ACKNOWLEDGMENT

The research work presented in this paper has been sponsored by the National Science Foundation through Grant No. CES-8816096.

REFERENCES

1.   Sciammarella, C.A. and Bhat, G., "A generalization of the theory of fringe patterns containing displacement information," to be presented in the 1990 SEM meeting and to be published in the corresponding proceedings.

2.   Sciammarella, C.A. and Lurowitz, N., "Multiplication and interpolation of moire, fringe orders by purely optical techniques," J. Appl. Mech. 34, Series E(2), June 1967.

3.   Sciammarella, C.A. and Bhat, G., "Computer assisted 3-D moire," to be presented at the CSME Mechanical Engineering Forum, Canada, 1990.

OPTICAL ANALYSIS OF DUCTILE FRACTURE OF METALS

F.P.Chiang & S.Li
Laboratory for Experimental Mechanics Research
State University of New York at Stony Brook
Stony Brook, New York 11794-2300

ABSTRACT

The optical technique capable of recording 3-D deformation is applied to different cracked aluminum alloy specimens under tension. The experimental results are compared with the theories at the corresponding deformation stages. The singularity forms agree well with HRR field for stationary cracks and agree with the solutions proposed by Gao and Hwang when crack grows.

INTRODUCTION

One of the major concerns of fracture mechanics is the deformation field at the crack tip. For power law hardening materials with yielded stationary cracks, the fields are often described by HRR equations [1,2]. In the case of growing crack, where the deformation is considered to be history dependent, the crack tip fields have been investigated by Amazigo and Hutchinson [3], Gao, Hwang and Dai [4,5].

Experimentally, a number of investigators have studied the deformation field of metallic specimens undergoing elastic-plastic fracture [6-9]. In this paper, we try to understand the nature of the prevailing deformation fields at the crack tip as it goes through different deformation stages. We applied a combined moire technique to different aluminum alloys to map the displacement singularity fields at the crack tip. The results are compared with HRR singularity and the singularity forms provided by Gao, Hwang and Dai (referred to as GHD field) respectively.

OPTICAL METHOD

The experimental arrangement of the system is as shown in Fig.1. Two optical methods were employed in this study. One is the in-plane moire method [10] whereby

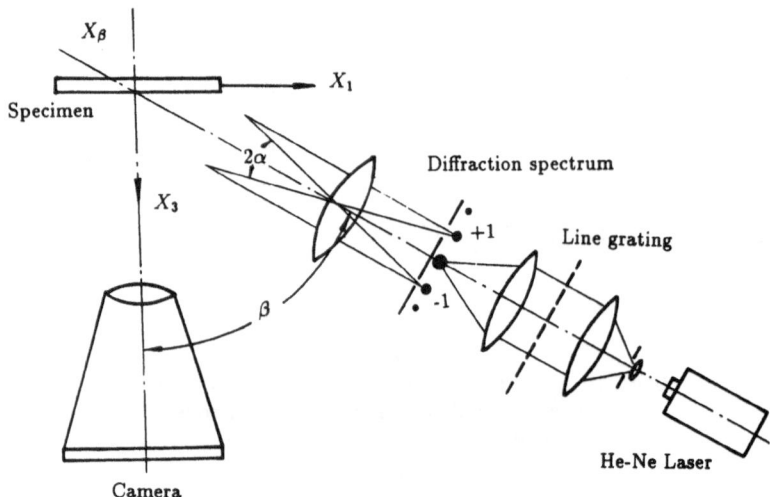

Figure 1. Experimental arrengement

a cross grating is printed on the specimen surface. Under load the deformed grating is ghotographed and superimposed with an undeformed grating. The pair is then inserted at the input plane of an optical Fourier filtering system [11] as shown in Fig.2. At the focal plane of the first transform lens, a cross array of diffraction orders are displayed. By filtering out all but the first horizontal and vertical orders to pass through the second transform lens sequentially for image reconstruction, moire fringes representing displacement contours along x (horizontal) and y (vertical) directions, respectively, are obtained.

These fringes are governed by the following equations:

$$u_1 = N_1 p_1, \qquad u_2 = N_2 p_2$$

where u_1 and u_2 are displacement components along x and y directions, respectively, p_1 and p_2 are the grating pitches (usually $p_1 = p_2 = p$), and N_1 and N_2 are the resulting fringe orders.

The second optical method involves the projection of a virtual grating onto the specimen surface as shown in Fig.1. The He-Ne laser is expanded to impinge upon a line grating. The diffracted waves are collected by a second lens which forms various diffraction orders as shown. All but the two first orders are blocked by a mask. The light coming from the first two orders are collected by a third lens to form two collimated beams with an angle 2α impinging upon the specimen surface. A volume of virtual grating is created within the region of the two intersecting beams. Depending on the orientation of the specimen, a grating of certain pitch is thus formed onto the specimen surface. Provided that the projection angle β is not zero, thickness change of the specimen under load will result in variations of the projected grating pitch. When this "deformed grating" is photographed and optically processed as described before, a moire pattern representing the displacement contours of the thickness change can be obtained and the fringes are governed by the following equation [10]:

$$u_3 = \frac{N_3 \lambda}{4 \tan \alpha \tan \beta} = N_3 p_3$$

where u_3 is the displacement along the z-direction, α and β are the angles as shown in Fig.2. λ is the wavelength of the laser beam used to generate the grating. N_3 and p_3 are the fringe orders and grating pitch, respectively.

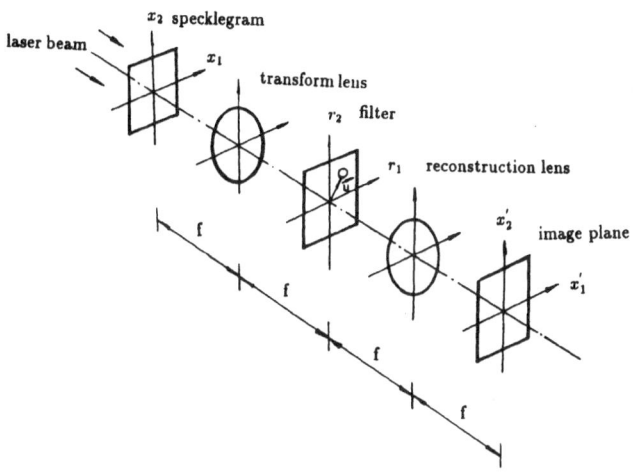

Figure 2. Schematic for optical spatial filtering

THE EXPERIMENT

Several different aluminum alloys were selected for the investigation. The results of Al2024-0 and Al ALM are presented. They are assumed to be power hardening materials and governed by the Ramber-Osgood law:

$$\frac{\epsilon}{\epsilon_0} = \frac{\sigma}{\sigma_0} + \alpha(\frac{\sigma}{\sigma_0})^n$$

where σ_0 and ϵ_0 are yield stress and yield strain, respectively. α is a constant and n the hardening exponent. The above parameters were all calibrated and listed in the table below.

TABLE 1
Material constants for Al 2024-0 and Al ALM

material	σ_0 MPa	ϵ_0	n	α
Al 2024-0	50.0	0.0008	3.0	1.6
Al ALM	80.7	0.0011	7.7	0.115

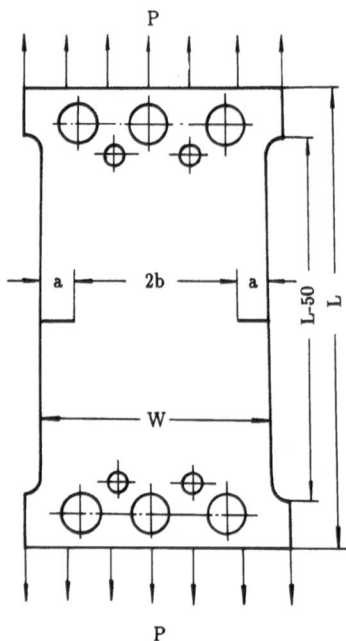

P

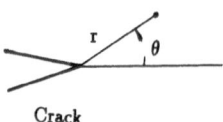

Crack

crack length: a
ligament: b
thickness: t

$\sigma = \frac{P}{2bt}$

$\sigma_\infty = \frac{P}{2(a+b)t}$

TABLE 2
Dimensions of the specimens (mm)

material	a	b	W	L	t
2024-0	11	27.1	76.2	254	3.18
ALM	5.23	34.77	80.0	216	1.40

Figure 3. Specimen configuration

The geometry of the specimens are shown in Fig.3. Double edge cracks were made, however, because of symmetry, only one edge crack was studied.

A cross grating of 20 lines/mm was printed on the specimen surface in the region surrounding the crack tip. Superimposing on top of it is a laser projected grating. The deformed gratings at different stages of loading were recorded on photographic films which were subsequently Optical-Fourier processed as described in the proceeding section. A typical set of fringes representing the displacement contours of u_1, u_2 and u_3 are shown in Fig.4.

THEORETICAL CRACK TIP FIELDS

Some of the known theoretical solutions of the crack tip fields are listed and rearranged for comparison. Since the experiments provide the information on displacement, we compare them directly to theoretical predictions wherever there are sufficient data. Strain values were obtained via numerical differentiation using an cubic spline routine. We compared the strain at the regions where the polar angle $\theta = 0°$.

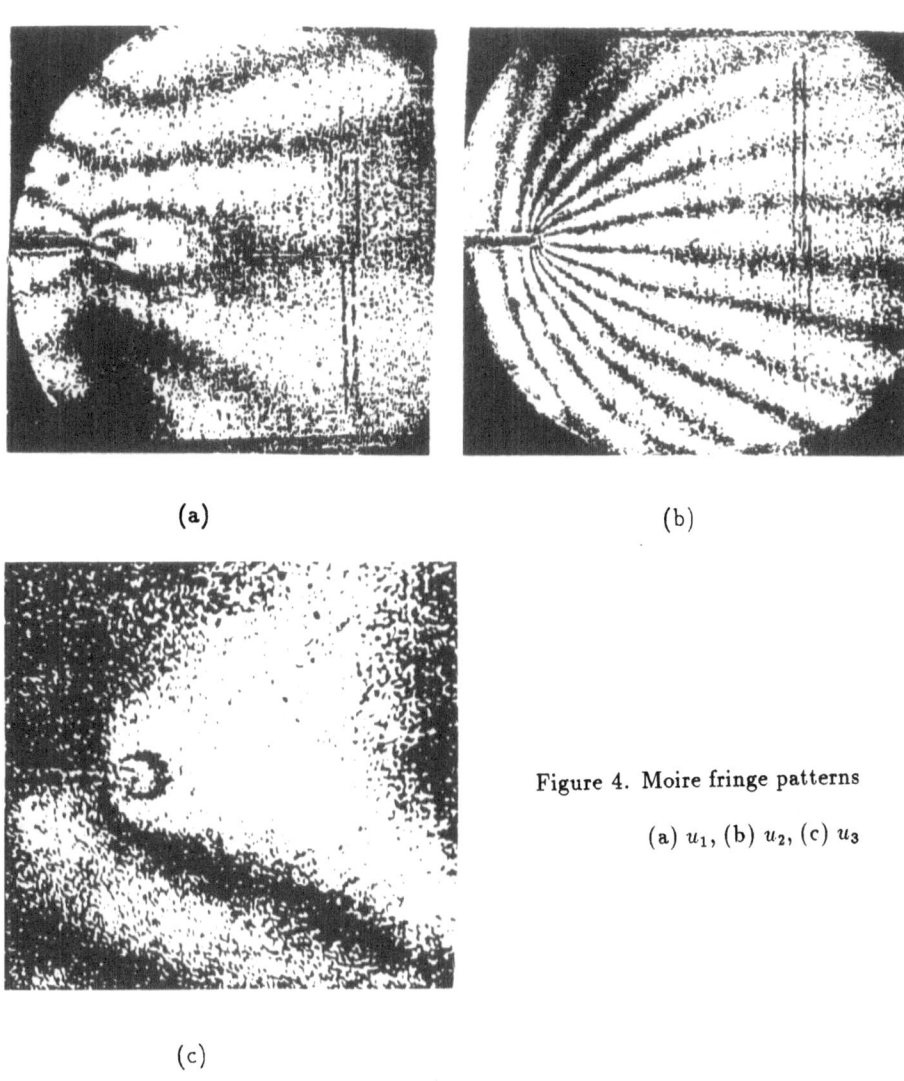

(a)

(b)

(c)

Figure 4. Moire fringe patterns

(a) u_1, (b) u_2, (c) u_3

The HRR equations are assumed to describe the plastic deformation field at the tip of a stationary crack. The governing equations for displacement and strain are

$$u_i = \alpha \epsilon_o r \left[\frac{J}{\alpha \epsilon_o \sigma_o I_n r} \right]^{\frac{n}{n+1}} \tilde{u}_i(\theta, n) \qquad (1)$$

$$\epsilon_{ij} = \alpha \epsilon_o \left[\frac{J}{\alpha \epsilon_o \sigma_o I_n r} \right]^{\frac{n}{n+1}} \tilde{\epsilon}_{ij}(\theta, n) \qquad (2)$$

where J is Rice's path-independent integral, I_n, $\tilde{u}_i$ and $\tilde{\epsilon}_{ij}$ are all constants determined by hardening exponent n and the prevailing state of stress surrounding the crack tip. The latter two parameters are also the function of the polar angle θ.

By taking the logarithm of equations (1) and (2), we have

$$\log u_i = C_1 + \frac{1}{n+1} \log r \tag{3}$$

$$\log \epsilon_{ij} = C_2 - \frac{n}{n+1} \log r \tag{4}$$

where

$$C_1 = \log[(\alpha\epsilon_o)^{\frac{1}{n+1}} (\frac{J}{\sigma_o I_n})^{\frac{n}{n+1}} \tilde{u}_i(\theta, n)] \tag{5}$$

$$C_2 = \log[(\alpha\epsilon_o)^{\frac{n}{n+1}} (\frac{J}{\sigma_o I_n})^{\frac{n}{n+1}} \tilde{\epsilon}_{ij}(\theta, n)] \tag{6}$$

It is seen that if displacement or strain distribution is plotted along various directions in log-log scale, the singularity of the crack tip field is a straight line with the slope $\frac{1}{n+1}$ for displacement and $\frac{n}{n+1}$ for strain. We tested the validity of HRR singularity by searching for these slopes. A sample plot is given in Fig.5.

As for the growing crack, we compare our results with the solutions provided by Gao, Hwang and Dai[4,5]. The GHD equations for displacement and strain are

$$u_i = r(\ln \frac{A}{r})^{\frac{n}{n-1}} \tilde{u}_i \tag{7}$$

$$\epsilon_{ij} = (\ln \frac{A}{r})^{\frac{n}{n-1}} \tilde{\epsilon}_{ij} \tag{8}$$

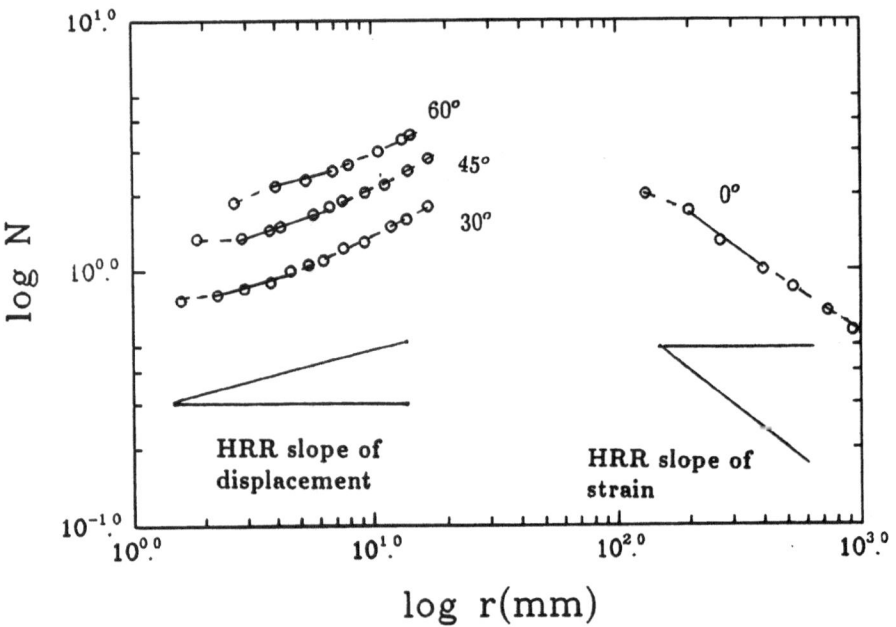

Figure 5. Plots for testing the HRR singularity

where A is a constant, $\bar{u}_i$ and $\bar{\epsilon}_{ij}$ are determined by hardening exponent n, **polar angle** θ and material constant α, E, ν etc. Again if equations (7) and (8) are rearranged and taken logarithm we have:

$$\left(\frac{u_i}{r}\right)^{\frac{n-1}{n}} = (logA - logr)(ln10)^{-1}\bar{u}_i^{\frac{n-1}{n}} \tag{9}$$

$$(\epsilon_{ij})^{\frac{1}{\alpha+1}} = (logA - logr)(ln10)^{-1}\bar{\epsilon}_{ij}^{\frac{n-1}{n}} \tag{10}$$

Thus if we plot $\left(\frac{u_i}{r}\right)^{\frac{n-1}{n}}$ or $\epsilon_{ij}^{\frac{n-1}{n}}$ versus $log\,r$, they are simply straight lines. We search for straight line portion of these plots to see if GHD singularity do indeed exist. A sample of this plot is given in Fig.6.

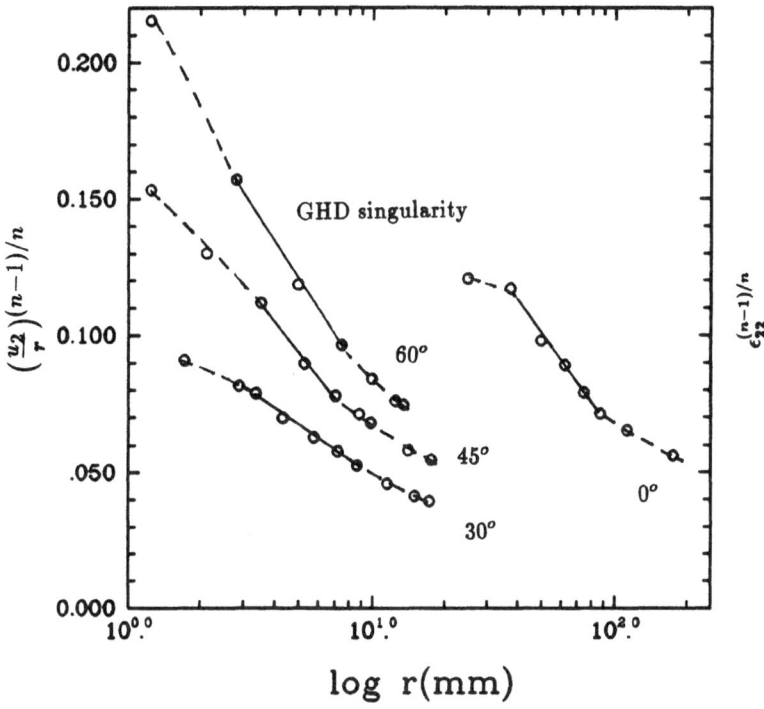

Figure 6. Plots for testing the GHD singularity

RESULT AND DISCUSSION

Respective curves along various directions were plotted and various direction angles we sought the segments of the curves which conformed with the corresponding theoret-

708

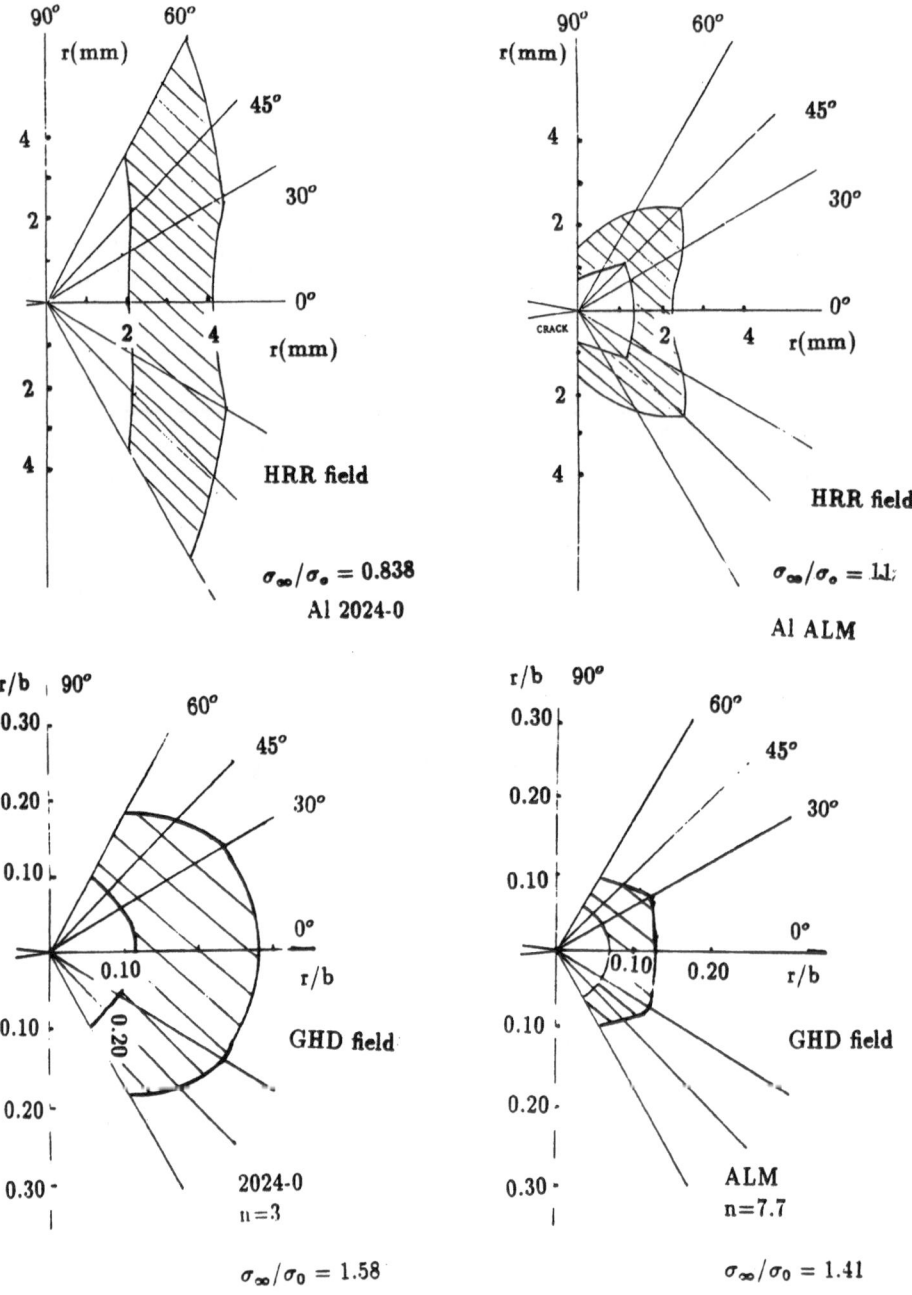

Figure 7. Mapped region where experimental data agree with the HRR or the GHD sigularity

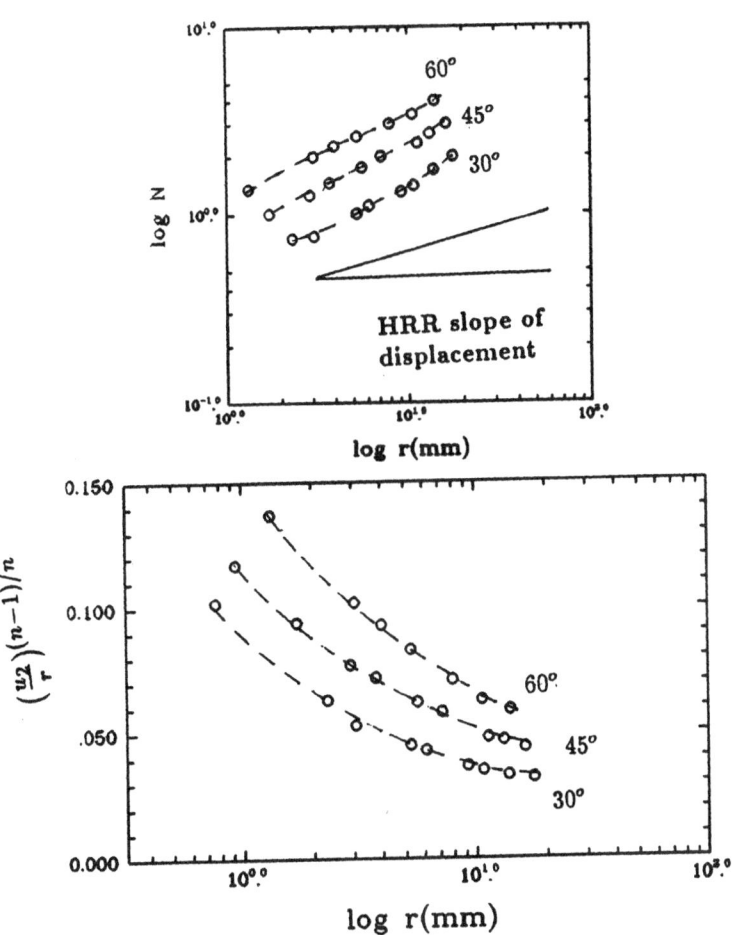

Figure 8. The stage where neither HRR nor GHD singularity can be located

ical singularity as shown in Fig.5 and Fig.6 and mapped the regions as given in Fig.7. The results indicate that Before the crack extends, there is a region at the crack tip wherein the experimental data agree with the HRR singularity. This region becomes smaller if hardening exponent is increased. After the crack has extended, experiment data agree with GHD singularity field. The size of the region is also reduced for less hardening material. However, at the stage slightly before and slightly after the onset of extension, neither a HRR nor a GHD singularity field can be located. Example plots at this stage of crack tip deformation is shown in Fig.8.

We may conclude from the experiments that the HRR field prevails when crack deforms plastically but remains stationary; and the GHD field exists under certain loading after crack extension. Most interesting of all is the fact that there is a transitional stage of deformation for which there is no valid theoretical model.

Acknowlegement

We thank the financial support of this work provided by the Office of Naval Research, Mechanics Division through contract No. N0001482K0566. (Scientific Officer, Dr. Y. Rajapakse)

REFERENCES

[1] Hutchinson, J.W., Singularity behavior at the end of a tensile crack in a hardening material. J. Mech. Phys. Solids, 1968, 16, pp.13-31.

[2] Rice, J.R. and Rosengren, G.F., Plane strain deformations. J. Mech. Phys. Solids, 1968, 16, pp. 337-347.

[3] Amazigo, J.C. and Hutchinson, J.W., Crack-tip fields in steady crack growth with linear strain-hardening. J. Mech. Phys. Solids, 1977, 25, pp. 81-97.

[4] Gao, Y.C. and Hwang, K.C., Elastic-plastic fields in steady crack growth in a strain hardening material. In Proceedings of ICF5, ed. D. Franois, 2, Pergamon press, 1981, pp. 669-682.

[5] Dai, Y. and Hwang, K.C., A numerical investigation of unsteady crack growth in power-hardening material. ICF Int. Symp. on Fracture Mechanics, Beijing, China, 1983, pp. 315-311.

[6] Kobayashi, A.S. and Kang, B.S.F, Stable crack growth in aluminum tensile specimens. Exp. Mech., 1986, 27(3), pp. 523-526.

[7] Rosakis, A.J. and Ravi-Chander, K., On crack-tip stress state: an experimental evaluation of three-dimensional effects. Int. J. Solids Structures, 1986, 22 No 2, pp. 121-134.

[8] Chiang, F.P. and Hareesh, T.V., Three dimensional crack tip deformation: an experimental study and comparison to HRR field. Int. J. Fracture, 1988, 36, pp. 243-257.

[9] Chiang, F.P., Hareesh, T.V., Liu, B.C. and Li, S., Optical analysis of HRR field. Optical Eng., 1988, 27, pp. 625-629.

[10] Chiang, F.P., Moire method of strain analysis. Ch. VI, Manual of Experimental Stress Analysis, 3rd ed., ed. A.S. Kobayashi, Society for Experimental Stress Analysis, Brookfield Center, conn, 1978.

[11] Chiang, F.P., Techniques of optical spatial filtering applied to the processing of moire fringe patterns. Exp. Mech., 1969, 6(11), pp. 523-526.

SHARP V-NOTCHES SUBJECTED TO AXIAL AND SHEAR STRESSES

GIUSEPPE DEMELIO, CARMINE PAPPALETTERE
Dipartimento di Progettazione e Produzione Industriale
University of Bari - Italy

ABSTRACT

In the last few years, stress field criteria, similar to those used in linear elastic fracture mechanics, have been proposed for fatigue design of structures with sharp V-notches under symmetrical (pure tension) or anti-symmetrical (pure shear) conditions. For the case of combined conditions, an analytical study has recently been performed. In this paper particular structures with notches, subjected to combinations of tension and shear stresses are studied, using numerical and photoelastic techniques, in order to confirm also in this case the possibility of applying stress field criteria to fatigue designs. Interesting applications can be proposed for butt and T welded joints.

INTRODUCTION

The singularity existing in angular corners in condition of symmetry (pure tension) was first studied by Williams in 1952 [1] by means of analytical methods. In 1972 Gross and Mendelson reconsidered the problem [2], investigating also the case of anti-symmetrical conditions (pure shear). Recently the degrees of singularity of symmetrical sharp V-notches have been tested using numerical finite and boundary element methods [3,7], and experimental photoelastic and moiré-holographic techniques [4-7].
The case of a singularity under both tension and shear stresses has been studied analytically in Refs. 8 and 9. Ref. 9 reports numerical values of the degrees of V-notch singularities in anti-symmetrical conditions.
All the degrees of singularity versus opening angles of the notch obtained by the above-mentioned Authors are reported in the graph in Fig. 1.
The Authors of Refs. 3,7 have also proposed to use stress field criteria, similar to those of linear elastic fracture mechanics

712

(LEFM), for the fatigue design of structures with sharp V-notches (zero or very small notch bottom radiuses) . The reason for this is that maxima criteria cannot easily be applied due to the very high stress concentrations.

When tension and shear stresses act simultaneously the problem is more complicated. In such a case, as shown in Ref. 9, moving away from the notch bottom, the degree of singularity varies from that of symmetrical conditions to that of anti-symmetrical conditions. On the other hand, this behaviour is very slight, so that it can be considered constant in a limited area.

Therefore, if for similar geometries having the same values of symmetrical and anti-symmetrical conditions, this area is assumed to be constant (i.e. starting and ending at the same distances from the notch bottom) the stress level can give substantial help to the fatigue design of structures with sharp notches.

Field criteria can thus be applied to structures like those shown in Fig. 2, which are subjected to both tension and shear stresses. For particular values of the notch opening angle, they are similar to structures for which fatigue design using field criteria have been empirically proposed, such as cruciform welded joints [10,11] (finite element, photoelastic and strain gauge analyses), lap joints [12] (finite element and photoelastic analyses) and butt welded joints [13] (moiré-holographic analyses).

In this paper the stress distributions near the notch bottom of structures like those shown in Fig. 2 have been determined by

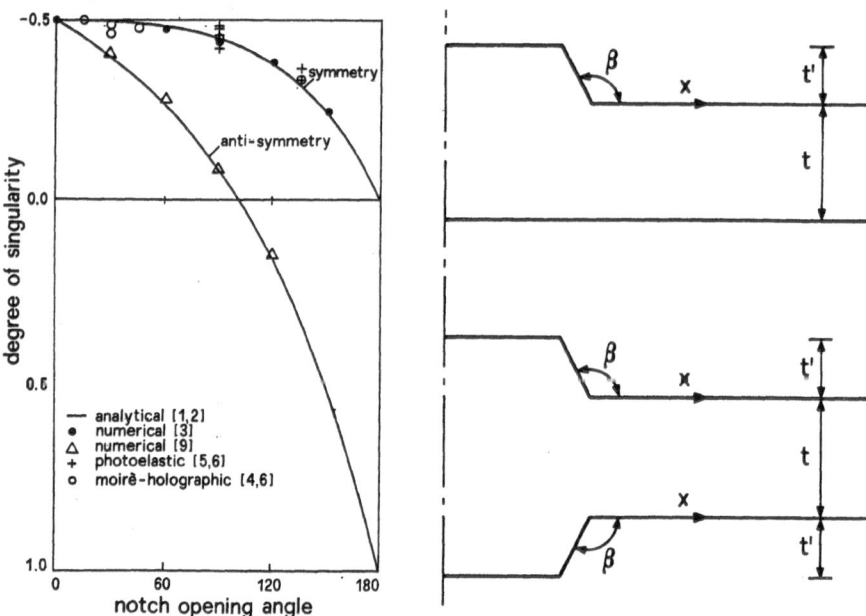

Fig. 1: Degrees of singularity for sharp V-notches.

Fig. 2: Structures studied.

means of boundary element [14] and photoelastic two-dimensional transmission techniques. The results obtained confirm the possibility of using field criteria for the fatigue design of the notched structures studied.

MODELS AND RESULTS

Fig. 2 reports the structures analysed, which will be called single and double joints from now on. Such types of structures were chosen in order to have a notch apex under combined tension and shear stresses, i.e. in condition of both symmetry and anti-symmetry.

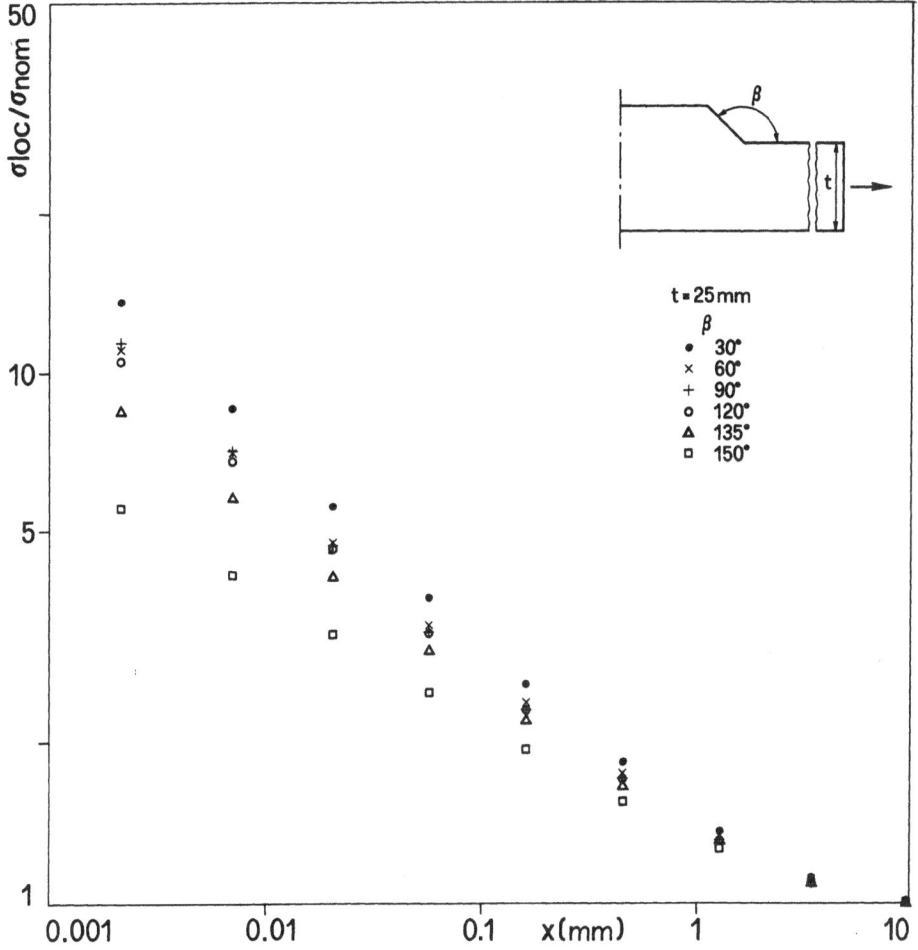

Fig. 3: Numerical results for single joints with t=25mm.

The analysis was performed using the boundary element method [14] and the two-dimensional transmission photoelastic technique. The specimens were made of polycarbonate sheets, annealed as reported in Ref. 15.

The numerical models had a bottom notch radius of zero. The experimental specimens were obtained using a CO_2 laser keeping this radius very small (about 0.2mm), so that – as demonstrated in Ref. 7 – it would not affect the proposed method of fatigue design using stress field criteria.

The numerical analysis results for single joints with a t=25mm thick main plate are reported in Fig. 3, while Fig. 4 shows the results for double joints with main plate having thicknesses of t=25mm and t=50mm respectively.

The photoelastic analysis results for t=25mm and t=50mm thick

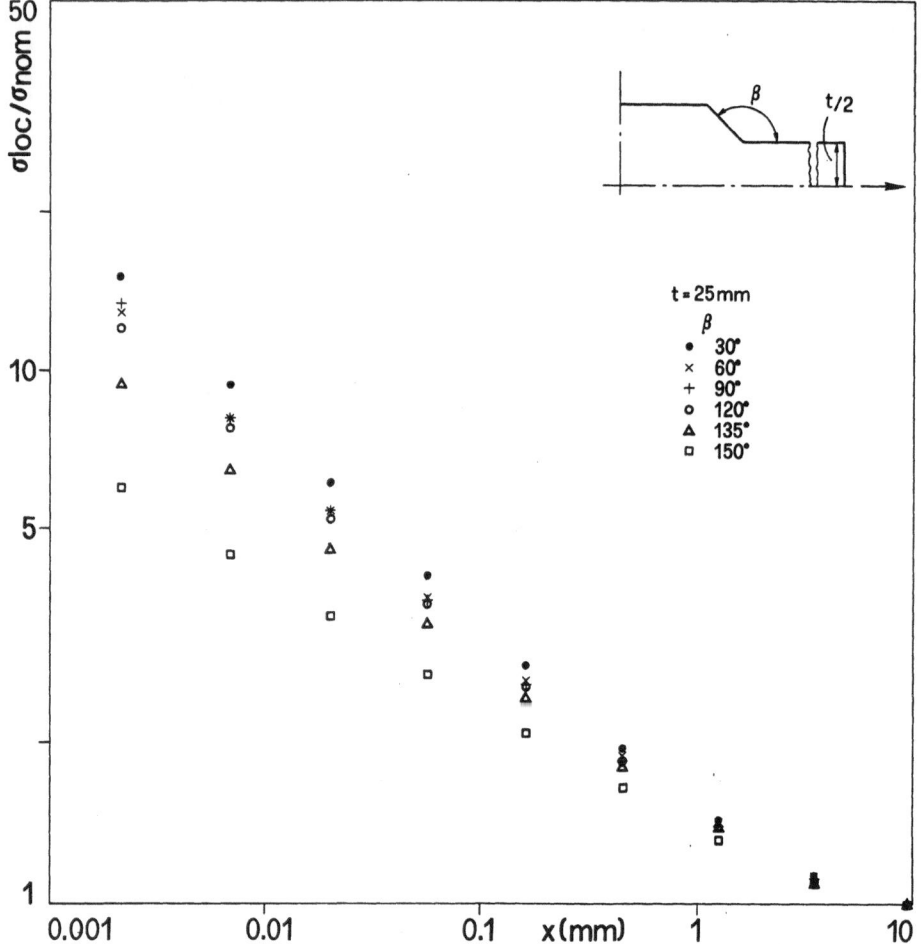

Fig. 4a: Numerical results for double joints with t=25mm.

single joints and t=25mm double joints are reported in Figs. 5
and 6 respectively. Fig. 7 shows the photoelastic results
obtained for single joints having an opening angle of ß=135°, a
thickness of t=50mm and a thickness t' varying from 5mm to 35mm
up to a T-joint.

All the results are reported in double logarithmic graphs in
terms of the ratio between the local stresses close to the
notch bottom and the stress far away from it, versus the distance from it.

In some cases the photoelastic results were translated to be
fitted on the numerical results. This is justified because the
aim of this work was to show the stress distribution near the
tip of the notch and not to find its absolute values. Misalign-
ments of loading rigs originated slight secondary bending which
must not be taken into account.

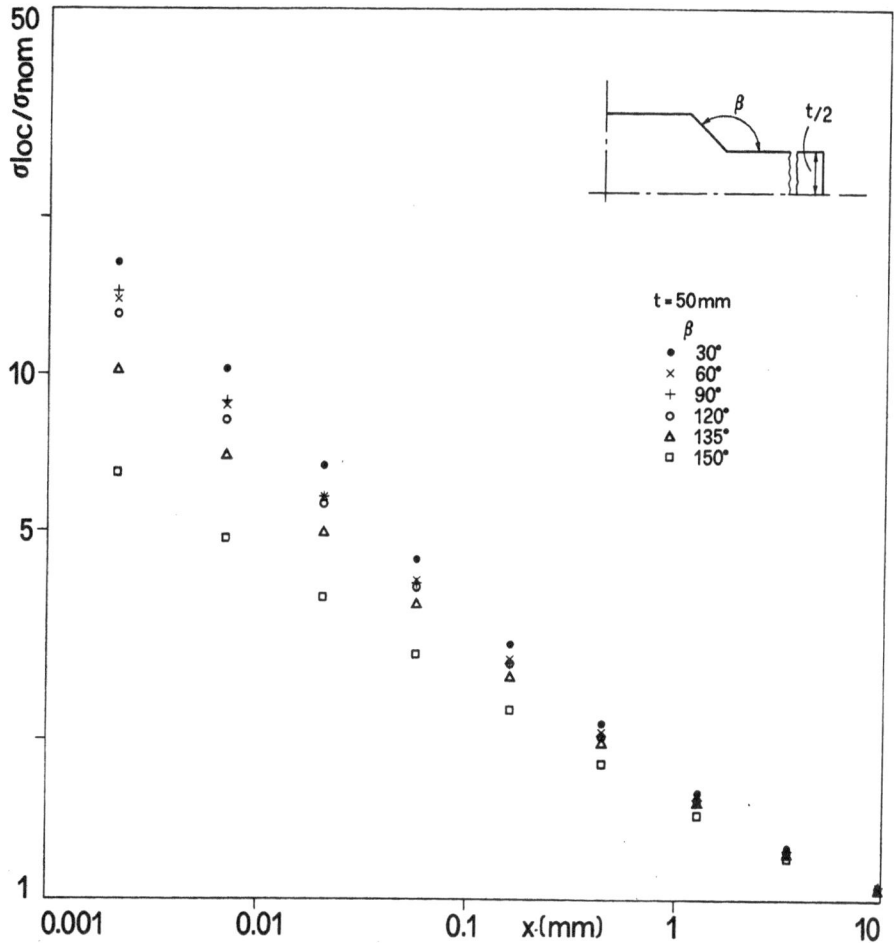

Fig. 4b: Numerical results for double joints with t=50mm.

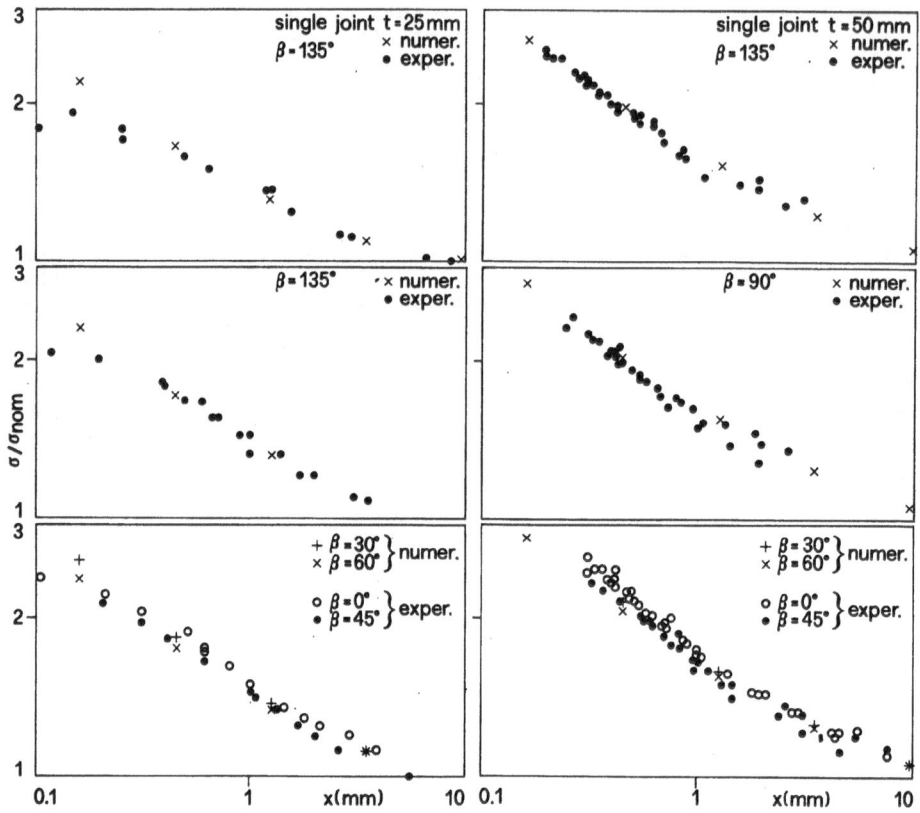

Fig. 5: Photoelastic results for single joints with t=25mm and t=50mm.

CONCLUSIONS

From the results obtained the following conclusions can be drawn:

- photoelastic and numerical results agree well;
- as already shown in the analytical work of Ref. 9, the behaviour of the stress fields obtained confirm that the contribution of the anti-symmetrical field must be taken into account in comparing local stress distributions for the fatigue design of structures using LEFM-like criteria; in fact, in the structures under study, the symmetrical field affected only zones very close to the singularity. These zones are very small and almost surely fall within the plastic area, so it is not correct to take them into account, as other Authors have done [16];
- stress fields for fatigue design can be compared in structures with notches having both the same opening angle and the same combination of symmetrical and anti-symmetrical conditions;

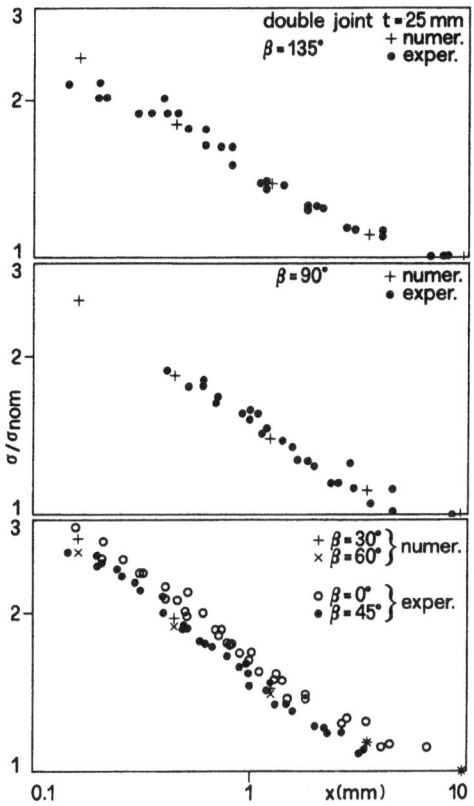

Fig. 6: Photoelastic results for double joints with t=25mm.

- a notch bottom radius, as long as it is small, does not
  influence the possibility of applying stress field criteria,
  as shown in Ref. 7;
- numerical or experimental determination of the stress field
  on the edges of the notch makes it possible to know the whole
  field near the tip of the notch using the formula reported in
  Refs. 8 and 9;
- the structures here examined can also be assumed to be butt
  or T welded joints; therefore field criteria can be extended
  to these types of structures as well;
- the use of strain gauges for determining the level of the
  stress field on the edge of the notch is particularly interesting since their application is both easy and economical;
  their grid length must be constant and their position with
  respect to the singularity must always be the same; besides,
  measurements must always be statistically validated, particularly for welded structures with normal electrodes.

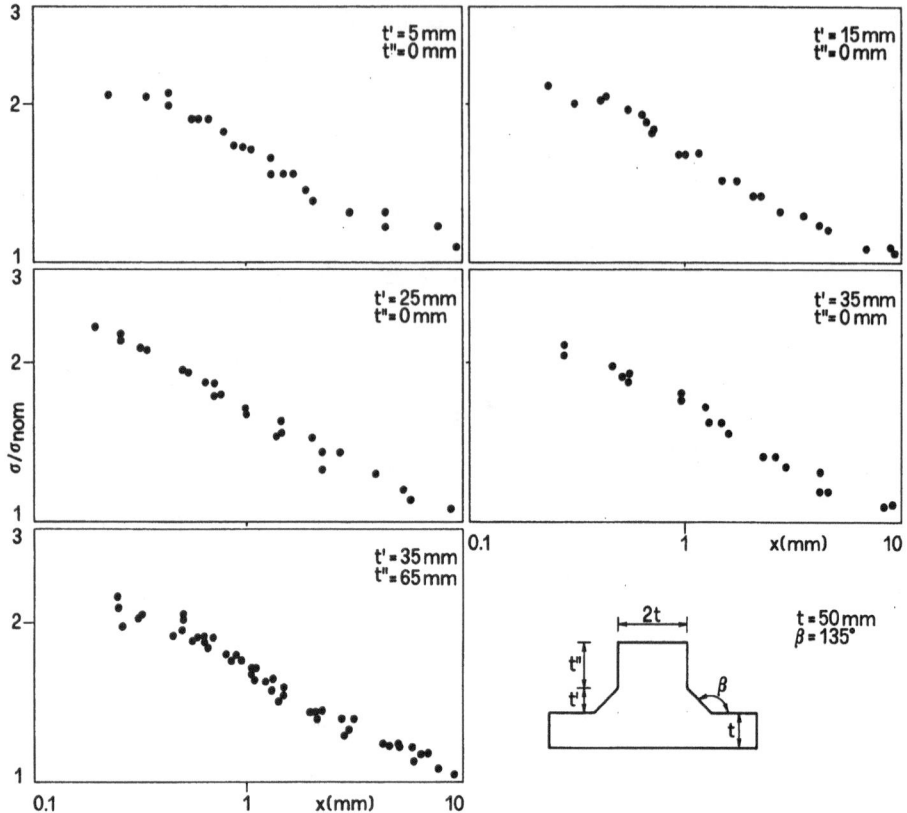

Fig. 7: Photoelastic results for single joints with t= 50mm, ß=135° and t´ varying from 5mm to 35mm up to a T-joint.

REFERENCES

[1] Williams M.L.: Surface Stress Singularities from Various Boundary Conditions in Angular Corner of Plates in Extension. *Journal of Applied Mechanics*, Vol. 74, December 1952, pp. 526-528.

[2] Gross B., Mendelson A.: Plane Elastostatic Analysis of V-Notched Plates. *International Journal of Fracture Mechanics*, Vol. 8, n. 3, September 1972, pp. 267-276.

[3] Atzori B.: Meccanica della frattura o effetto d´intaglio nella progettazione a fatica. XIII Convegno Nazionale A.I.A.S., Bergamo, 23-27 settembre 1985.

[4] Ginesu F., Pappalettere C.: Real-Time Moiré-Holographic Analysis of Plates with Sharp V-Notches in Tension. *Experimental Stress Analysis*, Martinus Nijhoff Publishers, Dordrecht, The Netherlands, 1986, pp. 387-395.

[5] Ginesu F., Pappalettere C: Indagine sperimentale su provini con intagli a V. XIV Convegno Nazionale AIAS, Catania 23-27 settembre 1986, pp. 245-256.

[6] Di Chirico G., Ginesu F., Pappalettere C.: A Contribution to the Analysis of Stress Fields near V-Notches. 1987 SEM Spring Conference on Experimental Mechanics, Houston, Texas, USA, 14-19 June 1987, pp. 446-452.

[7] Atzori B., Demelio G., Pappalettere C.: Stress Fields near V-Notches. 1989 SEM Spring Conference on Experimental Mechanics, Cambridge, MA, USA, 29 May - 1 June 1989, pp. 159-165.

[8] Demelio G., Pappalettere C.: Una nuova formulazione del campo di singolarità originato da intagli a V acuti in lastre piane. To be published.

[9] Atzori B., Demelio G., Pappalettere C.: Effetto di solle-citazioni composte sullo stato di tensione in prossimità di singolarità. Conference on "Organi delle Macchine". 7 giugno 1990, Milano.

[10] Atzori B., Blasi G., Pappalettere C.: Evaluation of Fa-tigue Strength of Welded Structures by Local Strain Meas-urements. Experimental Mechanics, Vol. 25, n. 2, June 1985, pp. 129-139.

[11] Pappalettere C.: Stress Concentration in Cruciform Joints Affected by Secondary Bending. International Conference on the Effects of Fabrication Related Stresses on Product Manufacture and Performance, Cambridge, England, 23-25 September 1985, pp. 399-411.

[12] Atzori B., Pappalettere C.: Application of Numerical and Experimental Techniques for Lap Joint Fatigue Design. International Conference on fatigue of Engineering Materi-als and Structures, Sheffield, England, 15-19 September 1986, pp. 1-6.

[13] Romita E. Trentadue B.: Strain Distribution near Butt Welds. 1990 SEM Spring Conference on Experimental Mechan-ics. Albuquerque, New Mexico, USA, 3-6 June 1990.

[14] Demelio G.: Il metodo degli elementi di contorno ed i microcomputers nell'analisi delle sollecitazioni. XIV Convegno Nazionale AIAS, Catania, 23-27 settembre 1986, pp. 529-537.

[15] Pappalettere C.: Annealing Polycarbonate Sheets. Strain, Vol. 20, n. 4, November 1985.

[16] Carpenter W.C.: A Collocation Procedure for Determining Fracture Mechanics Parameters at a Corner. International Journal of Fracture Mechanics, Vol. 24, 1984, pp. 255-266.

INDEX OF CONTRIBUTORS